International
REVIEW OF
Neurobiology
Volume 41

The Cerebellum
AND
Cognition

International Review of Neurobiology

Volume 41

SERIES EDITORS

RONALD J. BRADLEY
Department of Psychiatry, School of Medicine
Louisiana State University Medical Center
Shreveport, Louisiana, USA

R. ADRON HARRIS
Department of Pharmacology, University of Colorado
Health Sciences Center, Denver, Colorado, USA

PETER JENNER
Biomedical Sciences Division, King's College, London, UK

EDITORIAL BOARD

PHILIPPE ASCHER
ROSS J. BALDESSARINI
TAMAS BARTFAI
COLIN BLAKEMORE
FLOYD E. BLOOM
DAVID A. BROWN
MATTHEW J. DURING
KJELL FUXE
PAUL GREENGARD
SUSAN D. IVERSEN

KINYA KURIYAMA
BRUCE S. MCEWEN
HERBERT Y. MELTZER
NOBORU MIZUNO
SALVADOR MONCADA
TREVOR W. ROBBINS
SOLOMON H. SNYDER
STEPHEN G. WAXMAN
CHIEN-PING WU
RICHARD J. WYATT

The Cerebellum
AND
Cognition

EDITED BY

JEREMY D. SCHMAHMANN
Department of Neurology
Massachusetts General Hospital and Harvard Medical School
Boston, Massachusetts, USA

ACADEMIC PRESS

San Diego   London   Boston   New York   Sydney   Tokyo   Toronto

Front cover photograph: Lateral cerebellar activation across verb generation tasks. (For more details, see Chapter 10, Figure 3.)

This book is printed on acid-free paper. ∞

Copyright © 1997 by ACADEMIC PRESS

All Rights Reserved.
No part of this publication may be reproduced or transmitted in any form or by any means, electronic or mechanical, including photocopy, recording, or any information storage and retrieval system, without permission in writing from the Publisher.
The appearance of the code at the bottom of the first page of a chapter in this book indicates the Publisher's consent that copies of the chapter may be made for personal or internal use of specific clients. This consent is given on the condition, however, that the copier pay the stated per copy fee through the Copyright Clearance Center, Inc. (222 Rosewood Drive, Danvers, Massachusetts 01923), for copying beyond that permitted by Sections 107 or 108 of the U.S. Copyright Law. This consent does not extend to other kinds of copying, such as copying for general distribution, for advertising or promotional purposes, for creating new collective works, or for resale. Copy fees for pre-1997 chapters are as shown on the title pages, if no fee code appears on the title page, the copy fee is the same as for current chapters.
0074-7742/97 $25.00

Academic Press
a division of Harcourt Brace & Company
525 B Street, Suite 1900, San Diego, California 92101-4495, USA
http://www.apnet.com

Academic Press Limited
24-28 Oval Road, London NW1 7DX, UK
http://www.hbuk.co.uk/ap/

International Standard Book Number: 0-12-366841-7 (case)

International Standard Book Number: 0-12-625660-8 (paper)

Transferred to digital printing 2006
97 98 99 00 01 02 BB 9 8 7 6 5 4 3 2 1

To my mother, Bella,
to the memory of my late father, Oscar,
and to my family.

BRIEF CONTENTS

SECTION I HISTORICAL OVERVIEW
Rediscovery of an Early Concept   4

SECTION II ANATOMIC SUBSTRATES
The Cerebrocerebellar System   31
Cerebellar Output Channels   61
Cerebellar–Hypothalamic Axis   84

SECTION III PHYSIOLOGICAL OBSERVATIONS
Amelioration of Aggression   111
Autonomic and Vasomotor Regulation   122
Associative Learning   152
Visuospatial Abilities   191
Spatial Event Processing   217

SECTION IV FUNCTIONAL NEUROIMAGING STUDIES
Linguistic Processing   233
Sensory and Cognitive Functions   255
Skill Learning   273

SECTION V CLINICAL AND NEUROPSYCHOLOGICAL OBSERVATIONS
Executive Function and Motor Skill Learning   297
Verbal Fluency and Agrammatism   325
Classical Conditioning   342
Early Infantile Autism   367
Olivopontocerebellar Atrophy and Friedreich's Ataxia   388
Posterior Fossa Syndrome   412
Cerebellar Cognitive Affective Syndrome   433

Inherited Cerebellar Diseases   441
Neuropsychological Abnormalities in Cerebellar Syndromes   456

SECTION VI THEORETICAL CONSIDERATIONS
Cerebellar Microcomplexes   475
Control of Sensory Data Acquisition   490
Neural Representations of Moving Systems   516
How Fibers Subserve Computing Capabilities   536
Cerebellar Timing Systems   556
Attention Coordination and Anticipatory Control   575
Context–Response Linkage   600
Duality of Cerebellar Motor and Cognitive Functions   613

SECTION VII FUTURE DIRECTIONS
Therapeutic and Research Implications   637

CONTENTS

BRIEF CONTENTS vii
CONTRIBUTORS xix
FOREWORD xxiii
PREFACE xxvii
ACKNOWLEDGMENTS xxxi

SECTION I
HISTORICAL OVERVIEW

Rediscovery of an Early Concept
JEREMY D. SCHMAHMANN

I.	Introduction	4
II.	First Indications: Autonomic, Reticular, and Hypothalamic Modulation . . .	5
III.	Sensory and Associative Connections: Early Evidence	8
IV.	The Cerebellum Influences Emotions	11
V.	The Reexploration of Early Findings	13
VI.	Basal Ganglia and Cognition: An Analogy	14
VII.	The Modern Era: New Ideas and Methods	14
VIII.	Old Concepts Revisited	17
IX.	Conclusions	18
	References	18

SECTION II
ANATOMIC SUBSTRATES

The Cerebrocerebellar System
JEREMY D. SCHMAHMANN AND DEEPAK N. PANDYA

I.	Introduction	31
II.	The Feedforward Limb of the Cerebrocerebellar System	33

III.	The Feeback Limb of the Cerebrocerebellar System................	49
IV.	Climbing Fibers and Cognition: Is There an Anatomic Substrate?......	52
V.	Conclusions..	53
	References...	55

Cerebellar Output Channels
Frank A. Middleton and Peter L. Strick

I.	Introduction...	61
II.	Anatomical Studies...	63
III.	Physiological Studies...	71
IV.	Synthesis..	77
	References...	79

Cerebellar–Hypothalamic Axis: Basic Circuits and Clinical Observations
Duane E. Haines, Espen Dietrichs, Gregory A. Mihailoff, and E. Frank McDonald

I.	Introduction...	84
II.	Hypothalamocerebellar Projections and Related Neurotransmitters.....	85
III.	Cerebellar Projections to the Hypothalamus	92
IV.	Collaterals of Hypothalamocerebellar Fibers	94
V.	Indirect Hypothalamocerebellar Connections Mediated by the Basilar Pontine Nuclei and Lateral Reticular Nucleus.................	94
VI.	Clinical Evidence of Cerebellar Influence on Visceral Function........	96
VII.	Conclusions..	100
	References...	103

SECTION III
PHYSIOLOGICAL OBSERVATIONS

Amelioration of Aggression: Response to Selective Cerebellar Lesions in the Rhesus Monkey
Aaron J. Berman

I.	Introduction...	111
II.	Effect of Cerebellar Lesions on Emotional Behavior in the Rhesus Monkey..	112
	References...	117

Autonomic and Vasomotor Regulation

DONALD J. REIS AND EUGENE V. GOLANOV

I.	Introduction	122
II.	Regulation of Cerebral Blood Flow and Metabolism by the Fastigial Nucleus	122
III.	Neuroprotection Evoked from the Fastigial Nucleus	133
IV.	Conclusions	139
	References	142

Associative Learning

RICHARD F. THOMPSON, SHAOWEN BAO, LU CHEN, BENJAMIN D. CIPRIANO, JEFFREY S. GRETHE, JEANSOK J. KIM, JUDITH K. THOMPSON, JO ANNE TRACY, MARTHA S. WENINGER, AND DAVID J. KRUPA

I.	Introduction	152
II.	The Conditioned Response Pathway	152
III.	The Unconditioned Stimulus Pathway	154
IV.	The Conditioned Stimulus	154
V.	The Cerebellum and the Reflex Eyeblink Response	156
VI.	Purkinje Neuron Activity	156
VII.	Cerebellar Cortical Lesions	157
VIII.	Decerebration	158
IX.	Locus of the Long-Term Memory Trace	159
X.	Issues of Performance	171
XI.	Putative Noncerebellar Sites of Memory Storage	172
XII.	Neural Substrate of the Error-Correcting Algorithm in Classical Conditioning	173
XIII.	Supervised Learning and the Cerebellum	175
XIV.	Putative Mechanisms	176
XV.	Cerebellar Involvement in Other Forms of Memory	179
	References	181

Visuospatial Abilities

ROBERT LALONDE

I.	Introduction	191
II.	Evaluations of Spatial Learning	192
III.	Conclusions	208
	References	208

Spatial Event Processing

Marco Molinari, Laura Petrosini, and Liliana G. Grammaldo

I.	Morris Water Maze (MWM) for Spatial Function Studies...............	217
II.	Searching Strategies	218
III.	Neural Structure for Spatial Event Processing	219
IV.	MWM Performance of Hemicerebellectomized Rats	220
V.	Acquisition vs Retention...	225
VI.	Cerebellar Contribution to Spatial Event Processing..................	225
VII.	Cerebellum and Spatial Procedure Development	226
	References	227

SECTION IV
FUNCTIONAL NEUROIMAGING STUDIES

Linguistic Processing

Julie A. Fiez and Marcus E. Raichle

I.	Introduction..	233
II.	Articulatory Processes...	234
III.	Selection and Production of Verbal Responses	240
IV.	Verbal Learning	245
V.	Future Directions	249
	References	250

Sensory and Cognitive Functions

Lawrence M. Parsons and Peter T. Fox

I.	Introduction..	255
II.	Early Studies of Cerebellar Function in Cognition	257
III.	Dissociating Perceptual/Cognitive and Somatomotor Functions within Cerebellar Regions	258
IV.	Double Dissociation of Cerebellar Function and Motor Processing	260
V.	Implication for Hypothesis about Cerebellar Function................	265
	References	267

Skill Learning

Julien Doyon

I.	Introduction..	273
II.	Conceptual Framework	274

III.	Motor and Visuomotor Skill Learning and the Cerebellum	276
IV.	Discussion	286
	References	292

SECTION V
CLINICAL AND NEUROPSYCHOLOGICAL OBSERVATIONS

Executive Function and Motor Skill Learning
MARK HALLETT AND JORDON GRAFMAN

I.	Introduction	297
II.	Motor Learning	298
III.	Executive Function	308
IV.	Conclusions	318
	References	319

Verbal Fluency and Agrammatism
MARCO MOLINARI, MARIA G. LEGGIO, AND MARIA C. SILVERI

I.	Historical Background	325
II.	Dysarthria	326
III.	Verbal Fluency	327
IV.	Fluency Strategies	330
V.	Agrammatism	332
VI.	Dysgraphia	335
VII.	Hypothesis for a Cerebellar Role in Language	336
	References	336

Classical Conditioning
DIANA S. WOODRUFF-PAK

I.	Introduction	342
II.	Eyeblink Classical Conditioning in Patients with Cerebellar Lesions	344
III.	Normal Aging and Eyeblink Conditioning: Cerebellar Purkinje Cell Loss	350
IV.	Position Emission Tomography Detection of Cerebellar Involvement in Eyeblink Conditioning	353
V.	Eyeblink Conditioning and Other Neuropsychological Tasks	354

VI.	Eyeblink Classical Conditioning in Noncerebellar Lesions, Neurodegenerative Disease, and Other Syndromes	356
VII.	Summary and Conclusions	361
	References	362

Early Infantile Autism

MARGARET L. BAUMAN, PAULINE A. FILIPEK, AND THOMAS L. KEMPER

I.	Introduction	367
II.	Neuroimaging	368
III.	Microscopic Observations in the Cerebellum and Related Olive	371
IV.	Implications of Cerebellar Abnormalities in Autism	372
V.	Conclusion	382
	References	383

Olivopontocerebellar Atrophy and Friedreich's Ataxia: Neuropsychological Consequences of Bilateral versus Unilateral Cerebellar Lesions

THÉRÈSE BOTEZ-MARQUARD AND MIHAI I. BOTEZ

I.	Introduction	388
II.	Neuropsychological and Neurobehavioral Studies of Patients with Bilateral Cerebellar Damage	389
III.	Reaction Time and Movement Time Assessment in Patients with Bilateral Cerebellar Damage	394
IV.	Single Photon Emission Computed Tomography Studies and Neuropsychology of the Cerebellum	400
V.	Neuropsychological Findings in Patients with Unilateral Cerebellar Damage	402
VI.	Negative Findings	405
VII.	Conclusion and Summary	405
	References	407

Posterior Fossa Syndrome

IAN F. POLLACK

I.	Introduction	412
II.	Patient Population	413
III.	Discussion	422
	References	429

Cerebellar Cognitive Affective Syndrome

JEREMY D. SCHMAHMANN AND JANET C. SHERMAN

I.	Introduction.	433
II.	Patient Selection and Methods of Study	434
III.	Subjects.	435
IV.	Results.	435
V.	Discussion.	437
	References	440

Inherited Cerebellar Diseases

CLAUS W. WALLESCH AND CLAUDIUS BARTELS

I.	Introduction.	441
II.	Animal Models.	442
III.	Human Pathology	443
IV.	Discussion.	449
V.	Conclusion.	450
	References	451

Neuropsychological Abnormalities in Cerebellar Syndromes— Fact or Fiction?

IRENE DAUM AND HERMANN ACKERMANN

I.	Introduction.	456
II.	Clinical Observations after Cerebellar Damage	457
III.	Cerebellar Involvement in Motor Learning	459
IV.	Cerebellar Involvement in Temporal Processing	461
V.	Cerebellar Involvement in Higher Cognitive Functions	462
VI.	Methodological Considerations and Possible Directions.	465
	References	468

SECTION VI
THEORETICAL CONSIDERATIONS

Cerebellar Microcomplexes

MASAO ITO

I.	Introduction.	475
II.	Evolutionary View	476

III.	Roles in Spinal Cord and Brain Stem Functions	477
IV.	Roles in Voluntary Movements	488
V.	Possible Roles in Thought	481
VI.	Cerebellar Microcomplexes	483
VII.	Comments	485
	References	486

Control of Sensory Data Acquisition

JAMES M. BOWER

I.	Introduction	490
II.	Whiskers of the Rat as Viewed by the Cerebellum	490
III.	What Is the Cerebellum Controlling?	498
IV.	Implications for Cerebellar Function as a Whole	501
V.	Conclusion	507
	References	508

Neural Representations of Moving Systems

MICHAEL PAULIN

I.	Introduction	516
II.	What Is State Estimation?	516
III.	Organization of the Cerebellum and the Motor System	520
IV.	Oculomotor Vermis	521
V.	The Vestibulo-Ocular Reflex	524
VI.	Cerebellum and Motor Learning	525
VII.	Bower's Hypothesis	526
VIII.	Cerebellum and Cognition	528
IX.	Implementation	528
X.	Conclusions	531
	References	532

How Fibers Subserve Computing Capabilities: Similarities between Brains and Machines

HENRIETTA C. LEINER AND ALAN L. LEINER

I.	Introduction: "Hardware" and "Software" Capabilities	536
II.	"Hardware" in the Human Cerebellum	539
III.	"Software" Capabilities Inherent in Cerebro-Cerebellar Connections	542

IV.	Conclusions: Combined "Hardware" and "Software" Capabilities......	548
	References	551

Cerebellar Timing Systems
Richard Ivry

I.	A Modular Approach to Coordination.............................	556
II.	Cerebellar Contribution to Movement Timing......................	558
III.	Perceptual Deficits in the Representation of Temporal Information	561
IV.	Timing Requirements in Sensorimotor Learning	563
V.	Characterizing the Cerebellar Timing System	565
VI.	Interpreting Cerebellar Activation in Neuroimaging Studies: A Challenge for the Timing Hypothesis?..............................	568
VII.	Conclusions	570
	References	571

Attention Coordination and Anticipatory Control
Natacha A. Akshoomoff, Eric Courchesne, and Jeanne Townsend

I.	Attention and the Cerebellum	575
II.	Shifting Attention and the Cerebellum	578
III.	The Cerebellum and Attention Orienting...........................	584
IV.	Conclusions	592
	References	594

Context–Response Linkage
W. Thomas Thach

I.	Motor Learning and the Cerebellum.............................	600
II.	Cognitive Functions of the Cerebellum............................	604
III.	Conclusions	608
	References	609

Duality of Cerebellar Motor and Cognitive Functions
James R. Bloedel and Vlastislav Bracha

I.	Cerebellum and Cognition: A Historical Perspective..................	613
II.	Motor Function and Cognition: A Dichotomy Worth Saving?	617

III.	Task Dependency and Context Dependency: Determinants of Cerebellar Involvement in Regulating Behaviors	620
IV.	Cerebellar Functions: Implications from Distributed Circuits	624
V.	Conclusions	629
	References	630

SECTION VII
FUTURE DIRECTIONS

Therapeutic and Research Implications
JEREMY D. SCHMAHMANN

I.	Introduction.	637
II.	Therapeutic Implications.	638
III.	Research Implications.	641
IV.	Conclusions	643
	References	644

INDEX 649

CONTENTS OF RECENT VOLUMES 661

CONTRIBUTORS

Numbers in parentheses indicate the pages on which the authors' contributions begin.

Hermann Ackermann (455), Department of Neurology, University of Tübingen, D-72074 Tübingen, Germany

Natacha A. Akshoomoff (575), Department of Psychology, Georgia State University, University Plaza, Atlanta, Georgia 30303

Shaowen Bao (151), Neuroscience Program, University of Southern California, Los Angeles, California 90089

Claudius Bartels (441), Department of Neurology, Otto-von-Guericke University, Magdeburg, D-39120, Germany

Margaret L. Bauman (367), Massachusetts General Hospital, Boston, Massachusetts 02114

Aaron J. Berman (111), Mount Sinai Services at Elmhurst Hospital Center, Elmhurst, New York 11373

James R. Bloedel (613), Division of Neurobiology, Barrow Neurological Institute, Phoenix, Arizona 85013

Mihai I. Botez (387), Neurology Service and Neurobiology Laboratory, Department of Medicine, Hôtel-Dieu Hospital and University of Montreal, Montréal, Quebec H2W1T8, Canada

Thérèse Botez-Marquard (387), Neurology Service and Neurobiology Laboratory, Department of Medicine, Hôtel-Dieu Hospital and University of Montreal, Montréal, Quebec H2W1T8, Canada

James M. Bower (489), California Institute of Technology, Division of Biology, Pasadena, California 91125

Vlastislav Bracha (613), Division of Neurobiology, Barrow Neurological Institute, Phoenix, Arizona 85013

Lu Chen (151), Neuroscience Program, University of Southern California, Los Angeles, California 90089

Benjamin D. Cipriano (151), Neuroscience Program, University of Southern California, Los Angeles, California 90089

Eric Courchesne (575), Department of Neuroscience, University of California, San Diego, La Jolla, California 92093

Irene Daum (455), Institute of Medical Psychology and Department of Neurology, University of Tübingen, D-72074 Tübingen, Germany

Espen Dietrichs (83), Department of Neurology, Ulleval Hospital, University of Oslo, N-0407 Oslo, Norway

Julien Doyon (273), Department of Psychology and Rehabilitation Research Group, Laval University, Québec City, Quebec G1J 1Z4, Canada

Julie A. Fiez (233), Department of Psychology, University of Pittsburgh, and the Center for the Neural Basis of Cognition, University of Pittsburgh and Carnegie Mellon University, Pittsburgh, Pennsylvania 15260

Pauline A. Filipek (367), Departments of Neurology and Pediatrics, UCI Medical Center, University of California, Irvine, California 92668

Peter T. Fox (255), Research Imaging Center, University of Texas Health Science Center, San Antonio, Texas 78284

Eugene V. Golanov (121), Department of Neurology and Neuroscience, Cornell University Medical College, New York, New York 10021

Jordan Grafman (297), Cognitive Neuroscience Section, National Institute of Neurological Disorders and Stroke, National Institutes of Health, Bethesda, Maryland 20892

Liliana G. Grammaldo (217), Institute of Neurology, Catholic University, 00168 Rome, Italy

Jeffrey S. Grethe (151), Neuroscience Program, University of Southern California, Los Angeles, California 90089

Duane E. Haines (83), Department of Anatomy, University of Mississippi Medical Center, Jackson, Mississippi 39216

Mark Hallett (297), Human Motor Control Section, Medical Neurology Branch, National Institute of Neurological Disorders and Stroke, National Institutes of Health, Bethesda, Maryland 20892

Masao Ito (475), Frontier Research Program, The Institute of Physical and Chemical Research, Wako, Saitama 351-01, Japan

Richard Ivry (555), Department of Psychology, University of California, Berkeley, California 94720

Thomas L. Kemper (367), Department of Neurology, and Departments of Anatomy and Neurobiology, and Neuropathology, Boston University School of Medicine, Boston, Massachusetts 02118

Jeansok J. Kim (151), Department of Psychology, Yale University, New Haven, Connecticut

David J. Krupa (151), Neuroscience Program, USC, Los Angeles, California 90089

Robert Lalonde (191), Neurology Service, Unit of Behavioral Neurology, Neurobiology, and Neuropsychology, Hôtel-Dieu Hospital Research Center, Montréal, Quebec, H2W1T8, Canada

Maria G. Leggio (325), Department of Neurology, University of Rome "La Sapienza," 00185 Rome, Italy

Alan L. Leiner (535), Channing House, Palo Alto, California 94301

Henrietta C. Leiner (535), Channing House, Palo Alto, California 94301

E. Frank McDonald (83), Department of Neurology, University of Mississippi Medical Center, Jackson, Mississippi 39216

Frank A. Middleton (61), Department of Physiology, State University of New York, Health Science Center, Syracuse, New York 13210

Gregory A. Mihailoff (83), Department of Anatomy, University of Mississippi Medical Center, Jackson, Mississippi 39216

Marco Molinari (217, 325), Institute of Neurology, Catholic University, 00168 Rome, Italy

Deepak N. Pandya (31), Department of Anatomy and Neurobiology, Boston University School of Medicine, Boston, Massachusetts 02118

Lawrence M. Parsons (255), Research Imaging Center, University of Texas Health Sciences Center, San Antonio, Texas 78284

Michael Paulin (515), Department of Zoology and Center for Neuroscience, University of Otago, Dunedin, New Zealand

Laura Petrosini (217), Department of Psychology, University of Rome "La Sapienza," 00185 Rome, Italy

Ian F. Pollack (411), Department of Neurosurgery and University of Pittsburgh Cancer Institute Brain Tumor Center, University of Pittsburgh School of Medicine and Children's Hospital of Pittsburgh, Pittsburgh, Pennsylvania 15213

Marcus E. Raichle (233), Departments of Neurology and Neurological Surgery, Anatomy, and Neurobiology and Radiology, Washington University School of Medicine, St. Louis, Missouri 63103

Donald J. Reis (121), Department of Neurology and Neuroscience, Cornell University Medical College, New York, New York 10021

Jeremy D. Schmahmann (3, 31, 433, 637), Department of Neurology, Massachusetts General Hospital and Harvard Medical School, Boston, Massachusetts 02114

Janet Sherman (433), Department of Neuropsychology, Massachusetts General Hospital and Harvard Medical School, Boston, Massachusetts 02114

Maria Silveri (325), Institute of Neurology, Catholic University of Rome, 00168 Rome, Italy

Peter L. Strick (61), Research Service, VA Medical Center and Departments of Neurosurgery and Physiology, State University of New York Health Science Center at Syracuse, Syracuse, New York 13210

W. Thomas Thach (599), Department of Anatomy and Neurobiology, Washington University School of Medicine, St. Louis, Missouri 63110

Judith K. Thompson (151), Program in Neural Science, Department of Psychology, Indiana University, Bloomington, Indiana 47401

Richard F. Thompson (151), Neuroscience Program, University of Southern California, Los Angeles, California 90089

Jeanne Townsend (575), Department of Neurosciences, University of California, San Diego, La Jolla, California 92093

Jo Anne Tracy (151), Department of Neurobiology, Duke University Medical Center, Durham, North Carolina 27706

Claus W. Wallesch (441), Department of Neurology, Otto-von-Guericke University, Magdeburg, D-39120, Germany

Martha S. Weninger (151), Neuroscience Program, University of Southern California, Los Angeles, California 90089

Diana Woodruff-Pak (341), Department of Psychology, Temple University, Philadelphia, Pennsylvania 19122; and Laboratory of Cognitive Neuroscience, Philadelphia Geriatric Center, Philadelphia, Pennsylvania 19141

FOREWORD

It is with regret that my retirement and, hence, ready access to necessary resources caused me to decline Dr. Schmahmann's invitation to contribute a chapter to this volume. During the latter years of my professional career, as a result of findings in the Tulane laboratory, I was increasingly intrigued with the role of the cerebellum in behavior. So it was gratifying when I was given the option of writing a short foreword.

Reviewing the contents of this book precipitated the recall of personal incidents regarding the functions of the cerebellum. During my neurology residency at The Neurological Institute from 1940 to 1942, it was conventional wisdom that the cerebellum was implicated in motor function, specifically, muscle tone. At that time, a resident colleague, Averill Stowell, decided to interrupt his training in neurosurgery and gain some experience in neurophysiological research. He chose to go to Baltimore and work under Raymond Snider. During his research fellowship, Stowell, on a return visit to New York, told us he had participated in a project in which sensory-evoked potentials were being recorded in the cerebellum. We were astonished—and skeptical. It was many years later that my attention was drawn again to Snider's work. His studies, as well as our own at Tulane, indicated that the cerebellum and the hippocampus were related and possibly had a role in epileptic seizures. It was while Snider was demonstrating the effects of cerebellar ablation on behavior in the cat that I had the pleasure of a brief collaboration with him. Snider's findings implicating the cerebellum in seizures were the impetus for Irving Cooper's use of cerebellar stimulation in the treatment of epilepsy.

Although the early reports of autonomic responses to cerebellar stimulation were of passing interest to me, as were Robert Dow's early reports concerning cerebellar function, which long preceded his extensive review of nonmotor functions of the cerebellum, it was not until the early 1960s that a series of events, associated with reports from Harry Harlow's laboratory of severe behavioral pathology in rhesus monkeys raised in sensory isolation, focused my attention on a possible role for the cerebellum in behavior. James Prescott, who was with the NIH at the time, postulated that somatosensory input, particularly movement stimulation, was important in the development of appropriate emotional behavior, whereas its absence could result in violent-aggressive behavior. Supporting Prescott's contention were the studies of William Mason, who was working with Harlow, which suggested that the absence of somatosensory stimulation, particularly movement, was the ingredient responsible for the disturbed behavior of sensory-isolated monkeys.

In the Tulane laboratory, we had accumulated extensive data correlating aberrant brain activity, principally in subcortical forebrain structures, with pathological behavior and seizures in patients as well as in monkeys. Mason's findings directed our attention to a possible role for the cerebellum and somatosensory systems, particularly the vestibular proprioceptive system, as possibly being implicated in abnormal emotional behavior. Prescott was able to arrange for us to obtain some of Harlow's monkeys that had severe behavioral pathology. In addition to implanting electrodes into the forebrain sites of these monkeys where, in our studies, activity had been shown to correlate with behavior, we also implanted deep electrodes for long-term study into deep cerebellar nuclei, over the cerebellar surface, and into the somatosensory thalamus. Monkeys with normal behavior were similarly prepared with electrodes and served as controls. The pathological recordings from the cerebellum and somatosensory thalamus of the Harlow monkeys often synchronized with aberrant recordings in the forebrain structures, where we had demonstrated in animals and patients that activity correlated with behavior, whereas recordings from control monkeys were normal.

These initial findings in isolation-raised monkeys were the basis for our extensive anatomical and physiological studies in the cat and monkey, in which we demonstrated numerous monosynaptic connections between the cerebellum and the forebrain structures where activity had been correlated with emotion, as well as physiological experiments that indicated a functional relationship between the cerebellum and numerous forebrain sites. Our aim was to demonstrate a functional neural system for emotion, sensory perception, and possibly memory. We contended that this neural network could form a bridge for relating brain function to mental activity. In time we extended these studies, as part of our treatment program using deep electrode techniques, to a few intractably ill, psychotic patients. In these patients, electrodes were implanted into the deep midline cerebellum and into several forebrain sites where brain activity had previously been shown to correlate with emotion. As in the monkey, a functional relationship was demonstrated between the cerebellum and the subcortical forebrain sites, and with the cingulate gyrus.

The functional relationship between the paleocerebellum and the forebrain structures, in which cerebellar stimulation inhibited unit activity in the hippocampus, became the basis for studies in which we used continuous stimulation in the treatment of a few patients whose violent behavior was uncontrollable. In using the continuous cerebellar stimulation, we followed the procedure introduced by Cooper, and also used by Stanley Larson and Anthony Sances, involving stimulation of the lateral cerebellar surface to control seizures in epileptic patients. Our stimulus, for the purpose of behavior control, was to the vermis, which had been shown in our animal studies to play a more influential role in affecting forebrain structures. Our results suggested that the cerebellum plays a modulating role in affecting emotional behavior, somewhat analogous to the manner in which it functions to modulate motor activity.

This volume updates the rapidly accumulating data on functions of the cerebellum other than control of motor activity. It unquestionably will serve as a stimulus for additional productive studies.

Robert G. Heath

PREFACE

The traditional view of the cerebellum is that it coordinates gait and voluntary movement, is responsible for balance and posture, and is important in speech and control of gaze. This understanding of the cerebellum is rooted in medical and neuroscience textbooks and curricula. It has been the central focus of clinical interest in the cerebellum and the driving hypothesis behind experimental investigations of its function.

Against this background there has been a consistent, but minority, view that the role of the cerebellum is not limited to motor control. Despite convincing demonstrations over the years that the cerebellum participates in a number of nonmotor activities, this view has not gained wide acceptance. The motor deficits from cerebellar lesions are so dramatic in their fully expressed form, and the intellectual faculties seemingly so spared (we have been taught), that it has seemed unnecessary to look for properties of the cerebellum other than those related to movement.

It now appears that both these tenets may be inaccurate. Not all cerebellar lesions produce deficits of motor performance, and patients with lesions confined to the cerebellum fail some cognitive tests when studied appropriately. Now, clinical neuropsychology, neuroimaging techniques, and other recent advances in neuroscience have made it possible to explore both old and new questions about the province of the cerebellum outside of the motor domain.

This volume presents the current facts, theories, and controversies concerning the cerebellum. The authors of this book have collectively defined this debate and their selection reflects the evolving understanding of this field. In recognizing that many investigators are currently pursuing this line of research, contributing authors have endeavored to discuss the relevant findings from other laboratories as well as their own observations in order to provide a comprehensive account.

The term "cognition" usually refers to thought processes such as executive function, learning, memory, visual analysis, and language. In this volume, these are dealt with in some detail. In addition, emotion, personality, and behavior are discussed in both experimental and clinical contexts, and autonomic and vascular regulation are considered as well. Thus, whereas the major focus of this volume is related to cognition and affect, other nonmotor areas that seem to be subserved by the cerebellum are also addressed.

Section I traces the origins and evolution of ideas about the nonmotor contributions of the cerebellum. The history of this field from its inception to the present time is presented, and attention is drawn to the hypotheses and clinical and experimental observations of the early investigators.

Connectional neuroanatomy has provided a wealth of information concerning the anatomic substrates that may support the cerebellar contribution to cognition. Understanding these pathways is critical to the rationale presented in the subsequent sections and to the consideration of their validity. Section II is therefore devoted to a discussion of these anatomic systems.

Experimental observations in animal models of cerebellar disease and the modulation of autonomic, affective, and other complex behavioral paradigms by manipulation of the cerebellum stand in contrast to narrow notions of motor control as the principal concern or consequence of cerebellar activity. The findings from this work are considered in Section III.

Functional neuroimaging has provided much of the impetus for the exploration of the hypotheses and concepts derived from systems neuroanatomy, theoretical modeling, and experimental observations. The demonstration of cerebellar activation by nonmotor tasks, first noted incidentally and then studied as a specific entity in its own right, has essentially validated the new questions. Why is the cerebellum being activated, where, and under what conditions? The first decade of research utilizing this methodology has revealed insights and posed new questions about the cerebellum, which are discussed in Section IV.

Anecdotal reports of cerebellar abnormalities being found in association with changes in mental and behavioral states have largely given way to systematic analyses utilizing contemporary techniques to investigate behavioral syndromes in patients with cerebellar lesions. The chapters in Section V that describe the cognitive and affective changes seen in association with cerebellar disorders represent a new area of research in clinical neuroscience.

How does one account for these observations? What are the implications for the role of the cerebellum in motor control? What is the specific contribution of the cerebellum to nonmotor processing? Theoretical notions about the cerebellum discussed in Section VI help to conceptualize its role, but they also challenge many established ideas about the cerebellum. Nonmotor activity may be regulated by the cerebellum, and indeed, voluntary movement may be only one of a great many nervous system functions that benefit from a cerebellar influence.

The implications that the recent research findings may have for therapeutic strategies and for possible avenues for future investigation are discussed in the concluding section.

This volume attempts to define this exciting new area of cognitive neuroscience focused on the cerebellum and to present a cohesive statement of the current status of the field. Renewed interest in the relationship between the cerebellum and cognitive and affective processing has caught the imagination of both the neuroscience community and the interested public. This field, once alluded to as an afterthought and then as an emerging concept, appears now to have come of age.

Jeremy D. Schmahmann

ACKNOWLEDGMENTS

The successful completion of this volume was dependent upon the committed efforts of all the contributors, and I am grateful to them. It has also been a pleasure working with Graham Lees and the staff at Academic Press. I thank my teachers and mentors, Thomas D. Sabin and Deepak N. Pandya, and my colleagues in the Neurology Department of the Massachusetts General Hospital. Janet Sherman and David Caplan have been particularly helpful in our frequent discussions about the cerebellum and its role in cognition, and Marygrace Neal has been invaluable in helping me coordinate this project. Thanks also to the students and laboratory assistants with whom I have worked on clinical and neuroanatomy projects over the years, and also to my patients without whose willing cooperation clinical research would be impossible.

SECTION I
HISTORICAL OVERVIEW

REDISCOVERY OF AN EARLY CONCEPT

Jeremy D. Schmahmann

Department of Neurology, Massachusetts General Hospital and Harvard Medical School, Boston, Massachusetts 02114

I. Introduction
   A. The Cerebellum and Volitional Movement
   B. Clinical Reports from a Different Perspective
   C. Physiological Studies of Cerebellar Participation in Nonmotor Functions
II. First Indications: Autonomic, Reticular, and Hypothalamic Modulation
   A. Cerebellar Influences on Autonomic Phenomena
   B. Sham Rage and Cerebellar Modulation
   C. Reticular Connections and Physiology
III. Sensory and Associative Connections: Early Evidence
   A. The Corticopontine Pathway: First Awareness of Its Role in Nonmotor Functions
   B. The Neodentate Nucleus
   C. Peripheral Sensory Afferents to the Cerebellum
   D. Sensory, Visual, and Auditory Connections with the Cerebellum
IV. The Cerebellum Influences Emotions
   A. Sensorimotor Integration: A Developmental Hypothesis Related to the Cerebellum
   B. Cerebellum and Emotional Modulation
V. The Reexploration of Early Findings
VI. Basal Ganglia and Cognition: An Analogy
VII. The Modern Era: New Ideas and Methods
   A. Clinical Observations in the Era of Anatomic Neuroimaging
   B. Experimental Investigations of the Cerebellar Role Specifically in Nonmotor and Cognitive Processes
   C. Contemporary Investigations
VIII. Old Concepts Revisited
IX. Conclusions
   References

The study of the cerebellum has been dominated by interest in its role in movement and motor control. From the earliest days of the neuroscientific era, however, clinical reports and physiological and behavioral investigations have suggested that overt motor dysfunction is but one manifestation of cerebellar disease. The nature of cerebellar involvement in autonomic, sensory, and cognitive functions has been investigated for many years, and possible mechanisms that could subserve this relationship have been specifically addressed. This work has not been incorporated into the mainstream of neuroscience or clinical neurological thinking.

This chapter traces the history of these early investigations that demonstrated the need to revise the notion that cerebellar function is confined to the motor realm. The collaboration across disciplines and the advances in the methods and concepts of contemporary neuroscience have facilitated the maturation of this field of inquiry. The "new" story of the cerebellum and cognition, in fact, represents the evolution of a century-old revolutionary concept.

I. Introduction

A. The Cerebellum and Volitional Movement

The study of the cerebellum has been dominated by interest in its role in movement and motor control. Luigi Rolando (1809) first demonstrated that ablation of the cerebellum results in disturbances of posture and voluntary movement, and Marie-Jean-Pierre Fluorens (1824) showed that the cerebellum is responsible for the coordination of voluntary movement and gait. Luigi Luciani (1891), David Ferrier and William Aldren Turner (1893), and J. S. Rissien Russell (1894) established the nature and extent of motor deficits produced by cerebellar lesions in different locations in monkey. The clinical reports of Sanger Brown (1892), Pierre Marie (1893), Joseph Francois Felix Babinski (1899, 1902), and Gordon Holmes (1907) concerning cerebellar degeneration in patients established the critical role of cerebellum in the coordination of extremity movement, gait, posture, equilibrium, and speech. Holmes (1939) later analyzed the motor and speech deficits resulting from focal cerebellar injury, and much of his terminology and skills of neurologic evaluation remain in contemporary use. These clinical notions have been refined through the years by a number of clinical–pathological studies (including, e.g., Amici *et al.*, 1976; Lechtenberg and Gilman, 1978; Amarenco and Hauw, 1990a,b).

The history of the study of cerebellar physiology until the mid-1950s is extensively documented in the monograph of Dow and Moruzzi (1958) which also recognized the previous historical synopses of Luciani (1891) and André-Thomas (1897). Readers are referred to Dow and Moruzzi (1958) for a comprehensive review of the early studies in this field. Subsequent studies have advanced the understanding of cerebellar function in the realm of motor control (Chambers and Sprague, 1955a,b; Eccles *et al.*, 1967; Ito, 1984; Evarts and Thach, 1969; Houk and Wise, 1995; Stein and Glickstein, 1992) and motor learning (Marr, 1969; Albus, 1971; Ito, 1982).

B. CLINICAL REPORTS FROM A DIFFERENT PERSPECTIVE

An entirely separate, long-standing view of cerebellar function has been overshadowed by its role in the motor system. From the earliest days of clinical case reporting, instances of mental and intellectual dysfunction were described in the setting of cerebellar pathology (see Table I). Investigators at that time lacked the necessary clinical and pathological tools to provide a clear understanding of their patients' lesions and psychiatric/cognitive disturbances. Consequently, their anecdotal reports have been essentially ignored. The result has been that the possibility of a causal relationship between cerebellar dysfunction and cognitive/psychiatric pathology has either been summarily dismissed or not considered for lack of awareness of the question having been posed.

C. PHYSIOLOGICAL STUDIES OF CEREBELLAR PARTICIPATION IN NONMOTOR FUNCTIONS

In addition to this clinical background, a substantial body of experimental evidence dating back to the early part of the 19th century indicates that the cerebellum is involved in a number of nonmotor functions. The motor bias with respect to the study of cerebellum has been so overwhelming that this work, albeit conducted by eminent neurophysiologists, has also been omitted from mainstream thinking about the functions of the cerebellum.

The approach adopted in this chapter is mostly chronological in order to convey the sense of how these experimental findings and concepts have evolved through the years.

II. First Indications: Autonomic, Reticular, and Hypothalamic Modulation

A. CEREBELLAR INFLUENCES ON AUTONOMIC PHENOMENA

A close relationship between the cerebellum and the autonomic nervous system ("vegetative phenomena") was established early on. Changes in pupil diameter were produced in the monkey by stimulation of the fastigial and interpositus nuclei (Sachs and Fincher, 1927) and in cat by stimulation of the cerebellar anterior lobe (Moruzzi, 1941) and cerebellar white matter adjacent to the nuclei (Hare *et al.*, 1937; Chambers, 1947; Emerson *et al.*,

TABLE I
SELECTED CLINICAL REPORTS FROM THE EARLY 19th CENTURY UNTIL THE ERA OF ANATOMIC NEUROIMAGING[a]

Investigator	Cerebellar lesion	Behavioral manifestation
Combettes (1831)	Agenesis	Delayed development, aberrant behavior
Andral (1848)	Agenesis, left hemisphere	"Imbecile, weakness of character"
Vulpian (1866)	Atrophy	Aberrant behavior
Otto (1873)	Agenesis	Low intelligence, aberrant/deviant behavior
Ferrier (1876)	Agenesis	Feeble minded
Doursout (1891)	Atrophy	"Idiocy, irritability, brutality"
Fusari (1892)	Near complete agenesis	Mental retardation ("grave imbecility")
Neff (1894)	Atrophy	Mental deficiency
Bond (1895)	Atrophy	"Foolishness"
Londe (1895)	Spastic ataxia/?olivopontcerebellar atrophy	Mental difficulties
Classen (1898)	Atrophy	Mental deficiency
Whyte (1898)	Friedreich's ataxia	Mental impairment
Anton (1903); Anton and Zingerle (1914)	Agenesis	Delayed development, mental retardation
Batten (1905)	Agenesis	Mental retardation
Vogt and Astwazaturow (1912)	Hypoplasia of hemispheres	Mental retardation
Beyerman (1917)	Agenesis/"congenital atrophy"	Mental retardation
Schob (1921)	Agenesis/"congenital atrophy"	Mental retardation
Curschmann (1922)	Hereditary cerebellar ataxia	Mental impairment
Koster (1926)	Hypoplasia, cerebellar hemispheres	Mental retardation
Walter and Roese (1926)	Hereditary ataxia	Mental impairment
Santha (1930)	Agenesis/"congenital atrophy"	Mental retardation
Scherer (1933)	Agenesis/"congenital atrophy"	Mental retardation
Akelaitis (1938)	Cortical atrophy	Dementia (late stages)

(*continued*)

TABLE I (Continued)

Investigator	Cerebellar lesion	Behavioral manifestation
Rubinstein and Freeman (1940)	Near complete agenesis	Mild mental retardation, poor recent memory, delusions
Knoepfel and Macken (1947)	Degeneration	Psychosis
Jervis (1950)	Agenesis/"congenital atrophy"	Mental retardation
Schut (1950)	Olivopontocerebellar atrophy	Intellectual difficulty (late stages)
Mutrux et al. (1953)	Agenesis/"congenital atrophy"	Mental retardation
Gillespie (1965)	Degeneration with aniridia	Mental retardation ("oligophrenia")
Carpenter and Schumacher (1966)	Familial infantile cerebellar atrophy	Mental retardation
Aguilar et al. (1968)	Ataxia-telangiectasia	Mental deficiency (late stages)
Joubert et al. (1969)	Familial agenesis of the vermis	Mental retardation
Keddie (1969)	Cortical atrophy	Paranoid psychosis
Hoffman et al. (1971)	Hereditary late onset cerebellar degeneration	Impaired intellect (late stages)
Landis et al. (1974)	Olivopontocerebellar atrophy	Mild cognitive impairment

^a These reports described an association between cerebellar lesions and mental and intellectual dysfunction. Details concerning the pathology of the cerebellum and other brain structures were seldom comprehensive, and clinical descriptions of the cognitive and psychiatric presentations were often limited.

1956). An inhibitory effect on vasomotor tone resulting in a decrease in blood pressure was demonstrated in the decerebrate cat by stimulation of the vermis of the anterior lobe (Moruzzi, 1938, 1940; Wiggers, 1942). Vasomotor effects were subsequently reproduced by a number of investigators (Rasheed et al., 1970; Reis et al., 1973; Berntson et al., 1973; Ball et al., 1974; Martner, 1975).

B. SHAM RAGE AND CEREBELLAR MODULATION

The phenomenon of sham rage in the acute thalamic cat was described by Bard in 1928. This behavior resembles that of an infuriated animal and is thought to be mediated by the hypothalamus; hence "autonomic hypothalamic outbursts" (Zanchetti and Zoccolini, 1954). It occurs episodically and is characterized by autonomic activity (increased blood pressure, pupil dilatation, retraction of the upper eyelid) and struggling movements of the body with lashing of the tail. Moruzzi (1947) determined that when cerebellar cortical stimulation was timed to occur during an outburst of spontaneously occurring sham rage, both autonomic and somatic components were inhibited, but they reappeared more strongly during the rebound phase after cessation of the cerebellar stimulation. Furthermore, cerebellar stimulation during a period of quiescence provoked an outburst of sham rage. Zanchetti and Zoccolini (1954) later elicited typical sham rage with stimulating electrodes placed in the rostral and central parts of the fastigial nucleus.

C. RETICULAR CONNECTIONS AND PHYSIOLOGY

Stimulation of the fastigial nucleus by Moruzzi and Magoun (1949) and of the cortex of the cerebellar anterior lobe (Mollica et al., 1953) produced generalized arousal of the electroencephalogram, and Snider et al. (1949) demonstrated an inhibitory influence of the cerebellum on the inhibitory part of the brain stem reticular formation. Fastigial nucleus ablation produced a state of constant hyperactivity both in monkey (Carpenter, 1959) and in cat (Sprague and Chambers, 1959), and the role of the fastigial nucleus in regulation of the sleep–wake cycle was shown by Manzoni et al. (1968) and verified by subsequent studies (Steriade et al., 1971; Cunchillos and De Andrés, 1982). Later anatomic investigations confirmed physiological data showing connections between the cerebellum and reticular, hypothalamic, and limbic structures that had been postulated to be the substrates for cerebellar modulation of arousal, attention, sleep, and sham rage.

III. Sensory and Associative Connections: Early Evidence

A. THE CORTICOPONTINE PATHWAY: FIRST AWARENESS OF ITS ROLE IN NONMOTOR FUNCTIONS

An early discussion of the relationship between the cerebellum and human behavior beyond motor control was presented by Andrew Arthur Abbie in 1934. His study of the cerebrocerebellar system focused on the anatomy of the corticopontine pathway. He observed degeneration in the cerebral peduncle and basis pontis of the human brain following large lesions involving the parietal, temporal, and occipital lobes. Abbie did not know the precise origins or terminations of this pathway, but he was intrigued by the existence of this tract connecting nonmotor areas of the cerebral hemisphere with the pons. He suggested that this pathway "weav(es) all sensory impulses into a homogeneous fabric and translat(es) the resultant in muscular response which is accurately coordinated and acutely adapted to the requirements of the situation as a whole. To it man owes the possibility of his highest powers as expressed in his work, in sport, and in art" (Abbie, 1934). It seems unlikely that Abbie had considered a direct role for the cerebellum in cognitive tasks, but he drew attention to at least the possible contribution of the cerebellum to motor performance that incorporates or reflects a creative purpose.

B. The Neodentate Nucleus

Robert S. Dow (1942, 1974) also relied on anatomic principles to suggest a cerebellar incorporation into nonmotor circuitry. He determined that the dentate nucleus of the cerebellum could be divided into two components on the basis of differential staining properties and microscopic anatomy. The lateral part of the dentate (the "neodentate") is phylogenetically and more recently developed than the medial part (Dow, 1942). The coincidence in evolution of the appearance of the neodentate and the expanded lateral cerebellar hemisphere, with the expansion of the frontal and temporal association areas, later led Dow (1974, 1988) to postulate that these cerebral and cerebellar regions were anatomically interconnected and therefore functionally relevant.

C. Peripheral Sensory Afferents to the Cerebellum

It had been known from the work of Sherrington (1906) that the cerebellum receives afferents from the proprioceptive system. In 1939, Dow demonstrated that stimulation of the sciatic and saphenous nerves in the cat resulted in cerebellar sensory potentials, and in the rat he showed (Dow and Anderson, 1942) that both proprioceptive and cutaneous stimulation also resulted in cerebellar action potentials. Dow and Moruzzi (1958, p. 368) later reached the conclusion that "the arrival of splanchnic or vagal volleys and of auditory or visual impulses to the cerebellar cortex, and the possible modification of the reflex activity of the vasomotor centers by cerebellifugal volleys, justify the hypothesis that a hitherto unknown control may be exerted by the cerebellum in the sensory sphere and on autonomic functions." Further, they were of the opinion that important facts may have been missed in ablation experiments and routine clinical observations because the observers were not looking for these unexpected functional relationships and therefore no adequate tests were developed to bring them out. Other aspects of cerebellar function might be unveiled, Dow and Moruzzi believed, if the effects of experimental or clinical lesions were analyzed by means of "adequate physiological and psychological tests" (Dow and Moruzzi, 1958, pp. 373 and 374).

D. Sensory, Visual, and Auditory Connections with the Cerebellum

Ray S. Snider and Clinton N. Woolsey contributed substantially to the evolving concepts of cerebellar function. They demonstrated in cat (Snider and Stowell, 1942; Hampson *et al.*, 1946, 1952) and monkey (Snider and Stowell, 1944) that there are topographically organized cerebellar tactile receiving areas responsive to both proprioceptive input and cutaneous stimulation. Snider and Eldred (1948) and Snider and Stowell (1942) also demonstrated visual and auditory projections to the cerebellar vermis, and that visual projections are conveyed via the tectum (Snider, 1945). Anatomical studies (Sunderland, 1940; Brodal and Jansen, 1946) of the feedforward loop of the cerebrocerebellar system, and electrophysiological experiments (Henneman *et al.*, 1948) of the feedback limb, were also influential in shaping Snider's (1950) conclusions that there are dual projections to the cerebellum: one from end organs and one from related sensory and motor areas of the cerebral hemispheres. Further, in view of these findings "a much broader interpretation of cerebellar function must be used than has hitherto been the case in past." He saw the cerebellum as "the great modulator of neurologic function" and predicted for it a role not only in the field of neurology, but also in psychiatry. Later work on connections linking the cerebellum with the locus ceruleus and limbic structures, hippocampus, septum, and amygdala (Maiti and Snider, 1975; Mitra and Snider, 1975; Snider, 1975; Snider and Maiti, 1975, 1976), and on the fracturing of somatosensory representation in the cerebellar cortex (Welker, 1987) further supported his contention that notions of cerebellar function needed to be revised.

Snider was aware that not all lesions of the cerebellum produce ataxia. He noted [in his remarks in the paper of Hennemann *et al.* (1952, p. 332)] that "one can remove considerable masses of cerebellar tissue without producing any apparent deficits." This conclusion was also apparent to Dow (1974) who commented that it was particularly true for the lateral cerebellar cortex and the dentate nucleus. He posed the question that "if lesions or cooling of the dentate nucleus alone are not productive of the classical signs of cerebellar ataxia, what methods can one employ to unravel the functions of this part of the cerebellum so large in man and so selectively related to the association areas of the cerebral cortex?" (Dow, 1974, p. 115).

IV. The Cerebellum Influences Emotions

A. SENSORIMOTOR INTEGRATION: A DEVELOPMENTAL HYPOTHESIS RELATED TO THE CEREBELLUM

The 1970s saw increased interest in nonmotor functions of the cerebellum, and a number of studies addressed different aspects of this issue. The hypotheses of James W. Prescott (1971) were influential in directing some of these experiments (R. G. Heath and A. J. Berman, personal communication, 1996). Prescott's views were similar in principle to those of earlier developmental psychologists, most notably Jean Piaget (Gruber and Voneche, 1977), that movement is intricately bound with sensation and with intellectual and emotional growth. Prescott reached the conclusion that the cerebellum participates in emotional development and that it is a "master integrating and regulatory system for sensory-emotional and motor processes." His rationale was as follows. He asserted that maternal–social deprivation of neonatal animals ("the Harlow monkeys," Harlow and McKinney, 1971) is fundamentally a form of somatosensory input deficit. Prescott's theory was that the physiological effect of this deprivation would be a reduction in the number of cerebellar neurons, and those neurons that survived would operate under a condition of "denervation supersensitivity." Consequently, he reasoned, the psychopathological characteristics that these animals manifest (rhythmic rocking, head banging) reflect the effects of sensory deafferentation at a critical period. Minimal sensory stimuli thus provoke the hyperreactive cerebellar neurons to generate unusual movement patterns. An interesting development in this line of reasoning is the similarity between the behavior of the Harlow monkeys and the rocking and head banging behavior of patients with autism (Rapin, 1994) who were later shown to have pathological findings in cerebellar and limbic structures (Bauman and Kemper, 1985).

B. CEREBELLUM AND EMOTIONAL MODULATION

The thesis of Prescott, extraordinary as it may be, received some experimental support from Heath's (1972) recording of abnormal electrical potentials in both the dentate nucleus and the anterior septum of these Harlow monkeys. Furthermore, Prescott's hypotheses prompted further studies of the relationship between the cerebellum and emotional states. Heath and colleagues sought to determine whether there is a clinically relevant relationship between the cerebellum and psychopathology, and what anatomic substrates exist that could support this idea. He used electrophysiologic recordings and demonstrated fastigial nucleus connections with the septum (Heath, 1973) as well as with the hippocampus and amygdala (Heath and Harper, 1974; Heath et al., 1978), in accordance with the earlier observations of Snider and colleagues referenced earlier and with those of Anand et al. (1959) who showed connections between cerebellum and limbic structures. Reciprocal cerebellar connections with the hypothalamus and mammillary bodies were to be convincingly shown anatomically in later studies. Heath then turned his attention to patients in order to address this question of a link between the cerebellum and emotion. Recording from electrodes implanted in the fastigial nucleus of an emotionally disturbed patient, Heath et al. (1974) observed that increased neuronal discharges correlated with the patient's experience of fear and anger. A cerebellar influence on human emotional experience had also been described previously by Nashold and Slaughter (1969). When these authors stimulated the dentate nucleus and superior cerebellar peduncle, their patient reported a subjective experience of unpleasant sensations and of feeling scared. In 1977, Heath produced amelioration of aggression in 10 out of 11 patients with severe emotional dyscontrol by chronically stimulating the cerebellar vermis through subdurally implanted electrodes. At follow-up 6 to 16 months later, 10 out of the 11 patients were reported to be markedly improved. He ascribed these behavioral effects from cerebellar lesions and stimulation studies to connections with the limbic system. The conclusion that Heath (1977) reached from his clinical and experimental observations was that the cerebellum functions in the nervous system as an emotional pacemaker, necessary for the modulation of normal behavior.

Prescott's hypotheses also prompted the study in monkeys, reported by Berman et al. (1978), of the question of a link between the cerebellum and aggression. The conclusion, derived from their lesion-behavior study, that the vermis and archicerebellum are concerned with aggression provided confirmation of similar but less detailed observations published by Peters and Monjan (1971).

The technique of chronically implanted subdural electrodes delivering cerebellar cortical stimulation in humans was pioneered by Cooper and colleagues (1974, 1978). Dow had previously shown (Dow et al., 1962; Reimer et al., 1967) that cerebellar cortical stimulation had an ameliorating influence on experimental epilepsy in cats. Cooper's studies demonstrated that cerebellar cortical stimulation achieved seizure control in his patients, but it also had the unexpected side effect of improving aggression, anxiety, and depression (Riklan *et al.*, 1974). The understanding of the type, cause, and locus in the cerebral hemispheres of the seizure disorder was not described by these investigators, and the cumulative damage caused to Purkinje cells by this treatment modality rendered it obsolete. The hypothesis, however, that the cerebellum modulates seizure control and behavior was clearly stated and specifically addressed, if not conclusively so.

V. The Reexploration of Early Findings

At this time it became apparent to some that a major avenue of investigation into cerebellar function had been overlooked, and discussion papers attempting to draw attention to this issue began to appear. Martner (1975) provided detailed descriptions of the autonomic effects of fastigial nucleus stimulation in cats including changes in blood pressure, heart rate, and gastric motility, as well as oral behaviors such as grooming and chewing. Evaluations in the awake behaving cat model provided further verification of earlier findings. Fastigial nucleus stimulation reproduced the autonomic reactions recorded earlier in anesthetized or decerebrate preparations. In addition, fastigial stimulation produced complex oral behaviors such as eating and gnawing, as well as intense grooming behaviors (Berntson *et al.*, 1973; Ball *et al.*, 1974; Watson, 1978). Higher intensities of stimulation produced aggressive behavior manifesting as predatory attack that was target specific and purposeful, as well as sham rage (Rasheed *et al.*, 1970; Reis *et al.*, 1973). These responses were specific to the fastigial nucleus and were not elicited from the dentate.

Watson (1978) provided a most comprehensive and detailed review of the available literature dealing specifically with the subject of nonmotor functions of the cerebellum. He concluded that a large amount of evidence underscored the insufficiency of strictly postural and motor control concepts. Further, he suggested that "the cerebellum may contribute to sensory processing, learning, performance, emotion, motivation, and reward." Watson was hopeful that the relatively uncharted territory of the question of the cerebellum and behavior would be explored because it promised the discovery of important new data. Frick (1982) later attempted to understand specifically the effect of cerebellum on mood and psychosis. He did so by conceptualizing the archicerebellum (or "vestibulocerebellum") in terms of its relationship to, in psychoanalytic terms, the "ego." This thesis was supported largely by his reference to many of the anatomic/physiologic observations discussed by Watson (1978) and alluded to in this chapter.

VI. Basal Ganglia and Cognition: An Analogy

Around this time, i.e., the late 1970s and early 1980s, accepted notions of the basal ganglia were evolving in a manner that would ultimately influence later investigators interested in the cerebellum. The basal ganglia had been considered as quintessential motor structures, and elegant treatises on this topic (Denny-Brown, 1966) explored the known anatomic and physiologic relationships in reference to extrapyramidal movement disorders such as Parkinson's disease. Case reports describing cognitive and emotional impairments with lesions of the basal ganglia challenged accepted notions. Disturbances of language, attention, motivation, and visuospatial function were shown to be produced by caudate nucleus infarcts (Damasio et al., 1980; Pardal et al., 1985; Caplan et al., 1990), and depression and intellectual decline began to be documented in patients with Parkinson's disease (Freedman, 1990). These clinical findings were matched by connectional studies showing that basal ganglia structures were linked in a precise topographically organized fashion with the associative cortices in the cerebral hemispheres (Goldman and Nauta, 1977; Yeterian and Van Hoesen, 1978). Further, there appeared to be multiple loops of communication between certain basal ganglionic regions and motor, premotor, and prefrontal cortices (Alexander et al., 1986). These connections, supported by physiologic studies, provided a plausible anatomic substrate that could support the kinds of behaviors described in the clinical setting and in experimental animals.

For some investigators, including this author, the shift in thinking about the basal ganglia was a pivotal occurrence. If basal ganglia, previously thought of as so fundamentally motor, are indeed relevant for cognition and emotion, what about the cerebellum? Could the cerebellum be similar to the basal ganglia in the sense that it may be involved in cognition as well as movement, but at the same time be different because of its unique structure and physiology?

VII. The Modern Era: New Ideas and Methods

A. Clinical Observations in the Era of Anatomic Neuroimaging

Clinical and experimental reports linking the cerebellum with nonmotor function began to appear in the literature of the late 1970s and early

1980s. Heath and colleagues (1979) extended their earlier observations by studying the pathoanatomical correlations of psychiatric disease. They used the relatively new modality of computerized tomographic brain imaging in 85 schizophrenic patients, specifically focusing on the cerebellum. Abnormalities in the cerebellum were detected in 40% of this series. Abnormal radiologic features were noted particularly in the cerebellar vermis and included atrophy in some cases and mass lesions in others. The cerebral hemispheres in these patients appeared radiographically normal. The presence of cerebellar abnormalities, particularly in the vermis, of patients with schizophrenia was suggested by others as well (Weinberger et al., 1980; Moriguchi, 1981; Lippmann et al., 1981; Joseph et al., 1985). Comprehensive discussions of the possible pathogenic role of the cerebellum in schizophrenia may be found in Snider (1982) and Taylor (1991). Clinical case studies at this time reported an association between cerebellar abnormality and intellectual or emotional dysfunction. Bolthauser and Isler (1977) confirmed the earlier report of Joubert et al. (1969) describing mental retardation in children with dysplasia of the cerebellar vermis, and Cutting (1976) observed mania in a child with cerebellar degeneration. Kutty and Prendes (1981) reported psychotic behavior in adults with cerebellar degeneration. Hamilton et al. (1983) reported psychotic behavior and cognitive deficits in patients who were found at autopsy to have cerebellar degeneration, infarct, or tumor. Cognitive deficits were described in patients with Friedreich's ataxia (Fehrenbach et al., 1984), but this has not been consistently reported.

B. EXPERIMENTAL INVESTIGATIONS OF THE CEREBELLAR ROLE SPECIFICALLY IN NONMOTOR AND COGNITIVE PROCESSES

A number of experimental observations in the 1980s in many disciplines within the neurosciences helped anchor this field and facilitated its further development. Classically conditioned learning was shown to be dependent on the cerebellum in extensive studies of the rabbit nictitating membrane response (Thompson, 1988) and of the acoustic startle response of the rat (Leaton and Supple, 1986), and visual spatial navigational skills were impaired in the mutant mouse model (Lalonde and Botez, 1986). Anatomic substrates for the cerebellar modulation of cognitive processing began to be demonstrated by the finding of organized projections from associative and paralimbic cerebral areas to the feedforward limb of the cerebrocerebellar system (Schmahmann and Pandya, 1987, 1989a,b; reviewed in Schmahmann, 1991, 1996), and Dow's (1974) earlier hypothesis concerning a possible relationship between the neodentate nucleus and the prefrontal cortex was revitalized and extended in his collaboration with Leiner *et al.* (1986, 1989). A neuropathologic study (Bauman and Kemper, 1985) and neuroimaging (Courchesne *et al.*, 1988) of patients with early infantile autism revealed abnormalities in the cerebellum, and these anatomic correlations and their clinical relevance remain the subject of ongoing study.

C. CONTEMPORARY INVESTIGATIONS

A central development in this field has been the emergence of functional neuroimaging modalities as powerful investigative tools. These new technologies produced interesting but baffling results that needed to be explained in "novel" terms. The functional activation within the cerebellum by tasks of linguistic processing (Petersen *et al.*, 1988) and mental imagery of imagined movement (Ryding *et al.*, 1993) challenged the widely held assumption that cerebellar activation should reflect only perceptible movement.

The issue of a cerebellar role beyond motor control has sparked the interest of the contemporary neuroscience community (Cole, 1994; Dow, 1995; Baringa, 1996). This past decade has been witness to a plethora of studies across disciplines designed to address both new and old questions about the functions of the cerebellum. Anatomical studies show that the cerebellum is linked to associative and paralimbic cerebral areas, including the prefrontal cortex and the hypothalamus, by both feedforward and feedback loops (Haines and Dietrichs, 1984; Schmahmann, 1991, 1996; Middleton and Strick, 1994; Schmahmann and Pandya, 1995, 1997). Physiological and behavioral observations have explored the cerebellar influence on autonomic functions (Golanov and Reis, 1995), conditioned learning (Thompson and Krupa, 1994), and visual–spatial skills (Lalonde and Botez, 1986; Petrosini *et al.*, 1996). Functional neuroimaging reveals cerebellar activation in a multiplicity of tasks, including linguistic processing (Petersen *et al.*, 1988; Klein *et al.*, 1995), mental imagery (Ryding *et al.*, 1993; Mellet *et al.*, 1996; Parsons *et al.*, 1995) cognitive flexibility (Kim *et al.*, 1994), sensory discrimination (Gao *et al.*, 1996), classical conditioning (Logan and Grafton, 1995), motor learning (Seitz and Roland, 1992; Jenkins *et al.*, 1994; Rauch *et al.*, 1995), verbal memory (Grasby *et al.*, 1993; Andreasen *et al.*, 1995), working memory (Klingberg *et al.*, 1995), attention (Allen *et al.*, 1997), and emotional states (Reiman *et al.*, 1989; Bench *et al.*, 1992; Dolan *et al.*, 1992; George *et al.*, 1995; Mayberg *et al.*, 1995). Recent clinical evaluations are distinct from their predecessors in that they have used neuropsychological parameters that were applied with varying degrees of sophistication, computerized tomographic scans, and magnetic resonance imaging to understand the nature of the interaction between the cerebellum and intellectual/behavioral function. These studies reveal that patients with cerebellar lesions exhibit selected patterns of cognitive and emotional dysfunction, including deficits in planning and executive functions (Kish et al., 1988, 1994; Botez et al., 1989; Bracke-Tolkmitt et al., 1989; Grafman et al., 1992; Appollonio et al., 1993), motor learning (Sanes et al., 1990; Molinari et al., 1995), visual–spatial ability (Botez et al., 1985, 1989; Wallesch and Horn, 1990), linguistic processing (Fiez et al., 1992; Silveri et al., 1994; van Dongen et al., 1994), attention shifting (Akshoomoff and Courchesne, 1994), and emotion (Bauman and Kemper, 1985; Courchesne et al., 1988; Pollack, 1995). The complete characterization of what has been termed the cerebellar cognitive–affective syndrome (see J. D. Schmahmann and J. C. Sherman, this volume) continues to receive attention and is in the process of being further defined. Theories of how the cerebellum may modulate nonmotor functions have now also begun to be tested experimentally (Gao et al., 1996; Helmuth and Ivry, 1996; Allen et al., 1997).

VIII. Old Concepts Revisited

Perhaps it should not be surprising that the cerebellum may contribute to sensory, affective, autonomic, and cognitive functions as well as to motor control. John Hughlings Jackson (1887) wrote of the continuum from movement to thought. He agreed with Ferrier's (1876) statement that "mental operations, in the last analysis, must be merely the subjective side of sensory and motor substrata." In Jackson's (1887) view, movement was the externally visible manifestation of internal neuronal activity. Thought was as much a product of that neuronal activity, but the overt manifestations, he stated, were not readily detected by the observer. Thus movement of a limb, and movement of an idea, occupy different positions on the same scale. "Before I put out my arm voluntarily I must have a 'dream' of the hand as being already put out. So too, before I can think of now putting it out I must have a like 'dream,' for the difference betwixt thinking of now doing and now actually doing is, like the difference betwixt internal speech and external speech, only one of degree; in one there is slight discharge of a certain series of nervous arrangements, in the other strong discharge of that series " (Jackson, 1879–1980). Piaget (Gruber and Voneche, 1977) viewed movement as being intricately bound with sensation and with intellectual and emotional growth. Sensorimotor, cognitive, and affective systems all incorporate cerebellar input, and the evolving understanding that these functions are likely to be influenced by the cerebellum is harmonious with these concepts.

IX. Conclusions

The notion that the cerebellum is essential to nonmotor functions did not, therefore, arise anew and without precedent in the last few years. From the earliest days of the neuroscience era that had its origins in the last century, clinical reports have suggested this association, and investigators have been aware of the possibility that overt motor dysfunction is but one manifestation of cerebellar disease. This chapter has aimed at presenting a summary of the ideas and experimental observations that have influenced our growing understanding of the cerebellar contributions to nervous system function. The collaboration across disciplines and the advances in the methods and concepts of contemporary neuroscience have facilitated the maturation of this field of inquiry. Thus it would appear that the "new" story of the cerebellum and cognition in fact represents the evolution of a century-old revolutionary concept.

Acknowledgments

The author is grateful to Drs. Eric Courchesne, Robert G. Heath, Richard F. Thompson, and Henrietta and Alan Leiner for their helpful comments regarding this manuscript.

References

Abbie, A. A. (1934). The projection of the forebrain on the pons and cerebellum. *Proc. Roy. Soc. Lond. Ser. B* **115**, 504–522.
Aguilar, M. J., Kamoshita, S., Landinbg, B. H., and Boder, E. (1968). Pathological observations in ataxia-telangiectasia: A report on five cases. *J. Neuropathol. Exp. Neurol.* **27**, 659–676.
Akelaitis, A. J. (1938). Hereditary form of primary parenchymatous atrophy of the cerebellar cortex associated with mental deterioration. *Am. J. Psychiat.* **94**, 1115–1140.
Akshoomoff, N. A., and Courchesne, E. (1994). Intramodality shifting attention in children with damage to the cerebellum. *J. Cogn. Neurosci.* **6**, 388–399.
Albus, J. S. (1971). A theory of cerebellar function. *Math. Biosc.* **10**, 25–61.
Alexander, G. E., De Long, M. R., and Strick, P. L. (1986). Parallel organization of functionally segregated circuits linking basal ganglia and cortex. *Annu. Rev. Neurosci.* **9**, 357–381.

Allen, G., Buxton, R. B., Wong, E. C., and Courchesne, E. (1997). Attention activates the cerebellum independently of motor involvement. *Science*, in press.

Amarenco, P., and Hauw, J. J. (1990a). Cerebellar infarction in the territory of the superior cerebellar artery: Clinicopathological study of 33 cases. *Neurology* **40**, 1383–1390.

Amarenco, P., and Hauw, J. J. (1990b). Cerebellar infarction in the territory of the anterior and inferior cerebellar artery: A clinicopathological study of 20 cases. *Brain* **113**, 139–155.

Amici, R., Avanzini, G., and Pacini, L. (1976). Cerebellar Tumors. "Clinical Analysis and Physiopathological Correlations," Vol. 4. Karger, New York.

Anand, B. K., Malhotra, C. L., Singh, B., and Dua, S. (1959). Cerebellar projections to the limbic system. *J. Neurophysiol.* **22**, 451–458.

Andral, G. (1848). "Clinique Médicale," 4th Ed., Vol 5. Fortin, Masson et Cie, Paris.

Andreasen, N. C., O'Leary, D. S., Arndt, S., Cizadlo, T., Hurtig, R., Rezai, K., Watkins, G. L., Ponto, L., and Hichwa, R. (1995). Short term and long term verbal memory: A positron emission tomography study. *Proc. Natl. Acad. Sci. USA* **92**, 5111–5115.

Andre-Thomas (1897). Le cervelet: Etude anatomique, clinique et physiologique. G. Steinheil, Paris.

Anton, G. (1903). Ueber einen Fall von beiderseitigem Kleinhirnmangel mit kompensatorischer Vergrösserung anderer Systeme. *Wien klin Wchnschr* **16**, 1349–1354.

Anton, G., and Zingerle, H. (1914). Genaue Beschreibung eines Falles von beiderseitigem Kleinhirnmangel. *Arch. Psychiat. Nervenkr.* **54**, 8–75.

Appollonio, I. M., Grafman, J., Schwartz, V., Massaquoi, S., and Hallett, M. (1993). Memory in patients with cerebellar degeneration. *Neurology* **43**, 1536–1544.

Babinski, J. F. F. (1899). De l'asynergie cerebelleuse. *Rev. Neurol.* **7**, 806–816.

Babinski, J. F. F. (1902). Sur le role du cervelet dans les actes volitionnels necessitant une succession rapide de mouvements (diadococinese). *Rev. Neurol.* **10**, 1013–1015.

Ball, G., Micco, D., Jr., and Berntson, G. (1974). Cerebellar stimulation in the rat: Complex stimulation bound oral behaviors and self-stimulation. *Physiol. Behav.* **13**, 123–127.

Bard, P. (1928). A diencephalic mechanism for the expression of rage with special reference to the sympathetic nervous system. *Am. J. Physiol.* **84**, 490–515.

Barinaga, M. (1996). The cerebellum: Movement coordinator or much more? *Science* **272**, 482–483.

Batten, F. E. (1905). Ataxia in childhood. *Brain* **28**, 484–505.

Bauman, M., and Kemper, T. L. (1985). Histoanatomic observations of the brain in early infantile autism. *Neurology* **35**, 866–874.

Bench, C. J., Friston, K. J., Brown, R. G., Scott, L. C., Frackowiak, R. S. J., and Dolan, R. J. (1992). The anatomy of melancholia: Focal abnormalities of cerebral blood flow in major depression. *Psychol. Med.* **22**, 607–615.

Berman, A. J., Berman, D., and Prescott, J. W. (1978). The effects of cerebellar lesions on emotional behavior in the rhesus monkey. *In* "The Cerebellum, Epilepsy and Behavior" (I. S. Cooper, M. Riklan, and R. S. Snider, eds.), pp. 277–284. Plenum Press, New York.

Berntson, G., Potolicchi, S., Jr., and Miller, N. (1973). Evidence for higher functions of the cerebellum: Eating and grooming elicited by cerebellar stimulation in cats. *Proc. Natl. Acad. Sci. USA* **70**, 2497–2499.

Beyerman, W. (1917). Ueber angeborene Kleinhirnstörungen. *Arch. Psychiat. Nervenkr.* **57**, 610–658.

Bolthauser, E., and Isler, W. (1977). Joubert syndrome: Episodic hyperpnea, abnormal eyemovements, retardation and ataxia, associated with dysplasia of the cerebellar vermis. *Neuropaediatrie* **8**, 57–66.

Bond, C. H. (1895). Atrophy and sclerosis of the cerebellum. *J. Mental. Sci.* **41**, 409–420.

Botez, M. I., Botez, T., Elie, R., and Attig, E. (1989). Role of the cerebellum in complex human behavior. *Ital. J. Neurol. Sci.* **10,** 291–300.

Botez, M. I., Gravel, J., Attig, E., and Vezina, J.-L. (1985). Reversible chronic cerebellar ataxia after phenytoin intoxication: Possible role of cerebellum in cognitive thought. *Neurology* **35,** 1152–1157.

Bracke-Tolkmitt, R., Linden, A., Canavan, A. G. M., Rockstroh, B., Scholz, E., Wessel, K., and Diener, H.-C. (1989). The cerebellum contributes to mental skills. *Behav. Neurosci.* **103,** 442–446.

Brodal, A., and Jansen, J. (1946). The ponto-cerebellar projection in the rabbit and cat. *J. Comp. Neurol.* **84,** 31–118.

Brown, S. (1892). On hereditary ataxia, with a series of twenty-one cases. *Brain* **15,** 250–282.

Caplan, L. R., Schmahmann, J. D., Kase, C. S., Feldmann, E., Baquis, G., Greenberg, J. P., Gorelick, P. B., Helgason, C., and Hier, D. B. (1990). Caudate infarcts. *Arch. Neurol.* **47,** 133–143.

Carpenter, M. B. (1959). Lesions of the fastigial nuclei in the rhesus monkey. *Am. J. Anat.* **104,** 1–33.

Carpenter, S., and Schumacher, G. A. (1966). Familial infantile cerebellar atrophy associated with retinal degeneration. *Arch. Neurol.* **14,** 82–94.

Chambers, W. W. (1947). Electrical stimulation of the interior of the cerebellum in the cat. *Am. J. Anat.* **80,** 55–93.

Chambers, W. W., and Sprague, J. M. (1955a). Functional localization in the cerebellum. I. Organization in longitudinal corticonuclear zones and their contribution to the control of posture, both extrapyramidal and pyramidal. *J. Comp. Neurol.* **103,** 105–129.

Chambers, W. W., and Sprague, J. M. (1955b). Functional localization in the cerebellum. II. Somatotopic organization in cortex and nuclei. *Arch. Neurol. Psychiat.* **74,** 653–680.

Classen, K. (1898). Ueber familiare kleinhirnataxie. *Centralbl. f. innere Med.* **19,** 1209–1217.

Cole, M. (1994). The foreign policy of the cerebellum. *Neurology* **44,** 2001–2005.

Combettes (1831). Absence complète du cervelet, des pédoncules postérieurs et de la protubérance cérébrale chez une jeune fille morte dans sa onzième année. *Bull. Soc. Anat. Paris* **5,** 148–157.

Cooper, I. S., Amin, L., Gilman, S., and Waltz, J. M. (1974). The effect of chronic stimulation of cerebellar cortex on epilepsy in man. *In* "The Cerebellum, Epilepsy and Behavior" (I. S. Cooper, M. Riklan, and R. S. Snider, eds.), pp. 119–172. Plenum Press, New York.

Cooper, I. S., Riklan, M., Amin, I., and Cullinan, T. (1978). A long term follow-up study of cerebellar stimulation for the control of epilepsy. *In* "Cerebellar Stimulation in Man" (I. S. Cooper, ed.), pp. 19–38. Raven Press, New York.

Courchesne, E., Yeung-Courchesne, R., Press, G. A., Hesselink, J. R., and Jernigan, T. L. (1988). Hypoplasia of cerebellar vermal lobules VI and VII in autism. *N. Engl. J. Med.* **318,** 1349–1354.

Cunchillos, J. D., and De Andres, I. (1982). Particition of the cerebellum in the regulation of the sleep-wakefulness cycle: Results in cerebellectomized cats. *Electroencephal. Clin. Neurophysiol.* **53,** 549–558.

Curschmann, H. (1922). Zur Kenntnis der hereditären cerebellaren Ataxie. *Dtsch. Z. Nervenheilk.* **75,** 224–229.

Cutting, J. C. (1976). Chronic mania in childhood: Case report of a possible association with a radiological picture of cerebellar disease. *Psychol. Med.* **6,** 635–642.

Damasio, A. R., Damasio, H., and Chui, H. C. (1980). Neglect following damage to frontal lobe or basal ganglia. *Neuropsychologia* **18,** 123–132.

Denny-Brown, D. (1966). "The Cerebral Control of Movement." Liverpool University Press, Liverpool.

Dolan, R. J., Bench, C. J., Scott, R. G., Friston, K. J., and Frackowiak, R. S. J. (1992). Regional cerebral blood flow abnormalities in depressed patients with cognitive impairment. *J. Neurol. Neurosurg. Psychiat.* **55,** 768–773.

Doursout (1891). Note sur quelques cas d'atrophie et d'hypertrophie du cervelet: Annales médicopsychologiques. *J. Aliénat. Mental Méd. Aliénés* **13,** 345–362.

Dow, R. S. (1939). Cerebellar action potentials in response to stimulation of various afferent connections. *J. Neurophysiol.* **2,** 543–555.

Dow, R. S. (1942). The evolution and anatomy of the cerebellum. *Biol. Rev. Cambrid. Phil. Soc.* **17,** 179–220.

Dow, R. S. (1974). Some novel concepts of cerebellar physiology. *Mt. Sinai J. Med. (NY)* **41,** 103–119.

Dow, R. S. (1988). Contribution of electrophysiological studies to cerebellar physiology. *J. Clin. Neurosphysiol.* **5,** 307–323.

Dow, R.S. (1995). Cerebellar cognition. *Neurology* **45,** 1785–1786.

Dow, R. S., and Anderson, R. (1942). Cerebellar action potentials in response to stimulation of proprioceptors and exteroceptors in the rat. *J. Neurophysiol.* **5,** 363–372.

Dow, R. S., Fernandez-Guardiola, A., and Manni, E. (1962). The influence of the cerebellum on experimental epilepsy. *EEG Clin. Neurophysiol.* **14,** 383–398.

Dow, R. S., and Moruzzi, G. (1958). "The Physiology and Pathology of the Cerebellum." University of Minnesota Press, Minneapolis, MN.

Eccles, J. C., Ito, M., and Szentagothai, J. (1967): "The Cerebellum as a Neuronal Machine." Springer-Verlag, New York/Heidelberg.

Emerson, J. D., Bruhn, J. M., Emerson, G. M., and Foley, J. O. (1956). Autonomic responses to cerebellar stimulation in the suprathalamic decerebrate (decorticate) cat. *Fed. Proc.* **15,** 58.

Evarts, E. V., and Thach, W. T. (1969). Motor mechanisms of the CNS: Cerebrocerebellar interrelation. *Annu. Rev. Physiol.* **31,** 451–498.

Fehrenbach, R. A., Wallesch, C.-W., and Claus, D. (1984). Neuropsychological findings in Friederich's ataxia. *Arch. Neurol.* **41,** 306–308.

Ferrier, D. (1876). "The Functions of the Brain." GP Putnam's Sons, New York.

Ferrier, D., and Turner, W. A. (1893). A record of experiments illustrative of the symptomatology and degenerations following lesions of the cerebellum and its peduncles and related structures in monkeys. *Phil. Tran. Roy. Soc. B* **185,** 719–778.

Fiez, J. A., Petersen, S. E., Cheney, M. K., and Raichle, M. E. (1992). Impaired non-motor learning and error detection associated with cerebellar damage. *Brain* **115,** 155–178.

Flourens, P. (1824). Recherches experimentales sur les Proprietes et les Fonctions du Systeme Nerveux dons les Animaux Vertebres, Ed. 1, Paris, Crevot.

Freedman, M. (1990). Parkinson's disease. *In* "Subcortical Dementia" J. L. Cummings, ed.), pp. 108–122. Oxford Univ. Press, New York.

Frick, R. B. (1982). The ego and the vestibulocerebellar system. *Psychoanal. Q.* **51,** 93–122.

Fusari, R. (1892). Caso di mancanza quasi totale del cervelletto. *Mem. R. Accad. Sci. Institut. Bologna* **2,** 643–658.

Gao, J.-H., Parsons, L. M., Bower, J. M., Xiong, J., Li, J., and Fox, P. T. (1996). Cerebellum implicated in sensory acquisition and discrimination rather than motor control. *Science* **272,** 545–547.

George, M. S., Ketter, T. A., Parekh, P. I., Horwitz, B., Herscovitch, P., and Post, R. M. (1995). Brain activity during transient sadness and happiness in healthy women. *Am. J. Psychiat.* **152,** 341–351.

Gillespie, F. D. (1965). Aniridia, cerebellar ataxia and oligophrenia in siblings. *Arch. Opthal.* **73,** 338–341.

Golanov, E. V., and Reis, D. J. (1995). Vasodilation evoked from medulla and cerebellum is coupled to bursts of cortical EEG activity in rats. *Am. J. Physiol.* **268**, R454–467.

Goldman, P. S., and Nauta, W. J. H. (1977). An intricately patterned pre-frontocaudate projection in the rhesus monkey. *J. Comp. Neurol.* **171**, 369–386.

Grafman, J., Litvan, I., Massaquoi, S., Stewart, M., Sirigu, A., and Hallett, M. (1992). Cognitive planning deficit in patients with cerebellar atrophy. *Neurology* **42**, 1493–1496.

Grasby, P. M., Frith, C. D., Friston, K. J., Bench, C. J., Frackowiak, R. S. J., and Dolan, R. J. (1993). Functional mapping of brain areas implicated in auditory-verbal memory function. *Brain* **116**, 1–20.

Gruber, H. E., and Voneche, J. J. (1977). "The Essential Piaget." Basic Books, New York.

Haines, D. E., and Dietrichs, E. (1984). An HRP study of hypothalamo-cerebellar and cerebello-hypothalamic connections in squirrel monkey (*Saimiri sciureus*). *J. Comp. Neurol.* **229**, 559–575.

Hamilton, N. G., Frick, R. B., Takahashi, T., and Hopping, M. W. (1983). Psychiatric symptoms and cerebellar pathology. *Am. J. Psychiat.* **140**, 1322–1326.

Hampson, J. L., Harrison, C. R., and Woolsey, C. N. (1946). Somatotopic localization in the cerebellum. *Fed. Proc.* **5**, 41.

Hampson, J. L., Harrison, C. R., and Woolsey, C. N. (1952). Cerebro-cerebellar projections and somatotopic localization of motor function in the cerebellum. *Res. Publ. Assn. Nerv. Ment. Dis.* **30**, 299–316.

Hare, E. K., Magoun, H. W., and Ranson, S. W. (1937). Localization within the cerebellum of reactions to faradic cerebellar stimulation. *J. Comp. Neurol.* **67**, 145–182.

Harlow, H. F., and McKinney, W. T. (1971). Nonhuman primates and psychoses. *J. Autism. Child. Schizophr.* **1**, 368–375.

Heath, R. G. (1972). Electroencephalographic studies in isolation-raised monkeys with behavioral impairment. *Dis. Nerv. Sys.* **33**, 157–163.

Heath, R. G. (1973). Fastigial nucleus connections to the septal region in monkey and cat: A demonstration with evoked potentials of a bilateral pathway. *Biol. Psychiat.* **6**, 193–196.

Heath, R. G. (1977). Modulation of emotion with a brain pacemaker: Treatment for intractable psychiatric illness. *J. Nerv. Ment. Dis.* **165**, 300–317.

Heath, R. G., Cox, A. W., and Lustick, L. S. (1974). Brain activity during emotional states. *Am. J. Psychiat.* **131**, 858–862.

Heath, R. G., Dempesy, C. W., Fontana, C. J., and Myers, W. A. (1978). Cerebellar stimulation: Effects on septal region, hippocampus, and amygdala of cats and rats. *Biol. Psychiat.* **13**, 501–529.

Heath, R. G., Franklin, D. E., and Shraberg, D. (1979). Gross pathology of the cerebellum in patients diagnosed and treated as functional psychiatric disorders. *J. Nerv. Ment. Dis.* **167**, 585–592.

Heath, R. G., and Harper, J. W. (1974). Ascending projections of the cerebellar fastigial nucleus to the hippocampus amygdala and other temporal lobe sites: Evoked potential and histological studies in monkeys and cats. *Exp. Neurol.* **45**, 2682–2687.

Helmuth, L., and Ivry, R. (1966). When two hands are better than one: Reduced timing variability during bimanual movements. *J. Exp. Psychol. Hum. Percept. Perform.* **22**, 278–293.

Henneman, E., Cooke, P., and Snider, R. S. (1948). Cerebellar projections to the cerebral cortex in the cat and monkey. *Am. J. Physiol.* **155**, 443.

Henneman, E., Cooke, P. M., and Snider, R. S. (1952). Cerebellar projections to the cerebral cortex. *In* "Patterns of Organization in the Central Nervous System" (P. Bard, ed.) *Res. Publ. Assn. Nerv. Ment. Dis.* **30**, 317–333.

Hoffman, P. M., Stuart, W. H., Earle, K. E., and Brody, J. A. (1971). Hereditary late-onset cerebellar degeneration. *Neurology* **21**, 771–777.

Holmes, G. (1907). A form of familial degeneration of the cerebellum. *Brain* **30**, 466–488.
Holmes, G. (1939). The cerebellum of man (Hughlings Jackson memorial lecture). *Brain* **62**, 1–30.
Houk, J. C., and Wise, S. P. (1995). Distributed modular architectures linking basal ganglia, cerebellum, and cerebral-cortex: Their role in planning and controlling action. *Cerebr. Cortex* **5**, 95–110.
Ito, M. (1982). Questions in modeling the cerebellum. *J. Theor. Biol.* **99**, 81–86.
Ito, M. (1984). "The Cerebellum and Neural Control." Raven Press, New York.
Jackson, J. H. (1887). Remarks on evolution and dissolution of the nervous system. *In* "Selected Writings of John Hughlings Jackson (1958)" (J. Taylor, ed.), pp. 76–91. Basic Books, New York.
Jackson, J. H. (1879–1880). On affections of speech from disease of the brain. *In* "Selected Writings of John Hughlings Jackson (1958)" (J. Taylor, ed.), pp. 184–204. Basic Books, New York.
Jenkins, I. H., Brooks, D. J., Nixon, P. D., Frackowiak, R. S. J., and Passingham, R. E. (1994). Motor sequence learning: A study with positron emission tomography. *J. Neurosci.* **14**, 3775–3790.
Jervis, G. A. (1950). Early familial cerebellar degeneration (report of three cases in one family). *J. Nerv. Ment. Dis.* **111**, 398–407.
Joseph, A. B., Anderson, W. H., and O'Leary, D. H. (1985). Brainstem and vermis atrophy in catatonia. *Am. J. Psychiat.* **142**, 352–354.
Joubert, M., Eisenring, J. J., Robb, J. P., and Andermann, F. (1969). Familial agenesis of the cerebellar vermis: A syndrome of episodic hyperpnea, abnormal eye movements, ataxia, and retardation. *Neurology* **19**, 813–825.
Keddie, K. M. G. (1969). Hereditary ataxia, presumed to be of the Menzel type, complicated by paranoid psychosis, in a mother and two sons. *J. Neurol. Neurosurg. Psychiat.* **32**, 82–87.
Kim, S. G., Ugurbil, K., and Strick, P. L. (1994). Activation of a cerebellar output nucleus during cognitive processing. *Science* **265**, 949–951.
Kish, S. J., El-Awar, M., Schut, L., Leach, L., Oscar-Berman, M., and Freedman, M. (1988). Cognitive deficits in olivopontocerebellar atrophy: Implications for the cholinergic hypothesis of Alzheimer's dementia. *Ann. Neurol.* **24**, 200–206.
Kish, S. J, El-Awar, M., Stuss, D., Nobrega, J., Currier, R., Aita, J. F., Schut, L., Zoghbi, H. Y., and Freedman, M. (1994). Neuropsychological test performance in patients with dominantly inherited spinocerebellar ataxia: Relationship to ataxia severity. *Neurology* **44**, 1738–1746.
Klein, D., Milner, B., Zatorre, R. J., Meyer, E., and Evans, A. C. (1995). The neural substrates underlying word generation: A bilingual functional-imaging study. *Proc. Natl. Acad. Sci. USA* **92**, 2899–2903.
Klingberg, T., Roland, P. E., and Kawashima, R. (1995). The neural correlates of the central executive function during working memory: A PET study. *Hum. Brain Mapp. Suppl.* **1**, 414.
Knoepfel, H. K., and Macken, J. (1947). Le syndrome psycho-organique dans les hérédoataxies. *J. Belge. Neurol. Psychiat.* **47**, 314–323.
Koster, S. (1926). Two cases of hypoplasia ponto-cerebellaris. *Acta Psychiat. Neurol.* **1**, 47–83.
Kutty, I. N., and Prendes, J. L. (1981). Single case study: Psychosis and cerebellar degeneration. *J. Nerv. Ment. Dis.* **169**, 390–391.
Lalonde, R., and Botez, M. I. (1986). Navigational deficits in weaver mutant mice. *Brain Res.* **398**, 175–177.
Landis, D. M. D., Rosenberg, R. N., Landis, S. C., Schut, L., and Nyhan, W. L. (1974). Olivopontocerebellar degeneration. *Arch. Neurol.* **31**, 295–307.

Leaton, R. N., and Supple, W. F. (1986). Cerebellar vermis: Essential for long-term habituation of the acoustic startle response. *Science* **232**, 513–515.

Lechtenberg, R., and Gilman, S. (1978). Speech disorders in cerebellar disease. *Ann. Neurol.* **3**, 285–290.

Leiner, H. C., Leiner, A. L., and Dow, R. S. (1986). Does the cerebellum contribute to mental skills? *Behav. Neurosci.* **100**, 443–454.

Leiner, H. C., Leiner, A. L., and Dow, R. S. (1989). Reappraising the cerebellum: What does the hindbrain contribute to the forebrain? *Behav. Neurosci.* **103**, 998–1008.

Lippmann, S., Manshadi, M., Baldwin, H., Drasin, G., Rice, J., and Alrajeh, S. (1981). Cerebellar vermis dimensions on computerized tomographic scans of schizophrenic and bipolar tients. *Am. J. Psychiat.* **139**, 667–668.

Logan, C. G., and Grafton, S. T. (1995). Functional anatomy of human eyeblink conditioning determined with regional cerebral glucose metabolism and positron emission tomography. *Proc. Natl. Acad. Sci. USA* **92**, 7500–7504.

Londe, P. (1895). Maladies familiales du système nerveux. De l'hérédo-ataxie cérébelluese. Thèse, No 158. Bataille et Cie, Paris.

Luciani, L. (1891). Il cervelletto: Nuovi studi di fisiologia normale e pathologica. Firenze: Le Monnier. German translation, Leipzig: E. Besold, 1893.

Maiti, A., and Snider, R. S. (1975). Cerebellar control of basal forebrain seizures: Amygdala and hippocampus. *Epilepsia* **16**, 521–533.

Manzoni, T., Sapienza, S., and Urbano, A. (1968). EEG and behavioral sleeplike effects induced by the fastigial nucleus in unrestrained unanesthetized cats. *Arch. Ital. Biol.* **106**, 61–72.

Marie, P. (1893). Sur l'hérédo-ataxie cérébelleuse. *Sem. Méd.* **13**, 444–447.

Marr, D. (1969). A theory of cerebellar cortex. *J. Physiol.* **202**, 437–470.

Martner, J. (1975). Cerebellar influences on autonomic mechanisms. *Acta Physiol. Scand.* (*Suppl.*) **425**, 1–42.

Mayberg, H. S., Liotti, M., Jerabek, P. A., Martin, C. C., and Fox, P. T. (1995). Induced sadness: a PET model of depression. *Hum. Brain Mapp. Suppl.* **1**, 396.

Mellet, E., Tzourio, N., Crivello, F., Joliot, M., Denis, M., and Mazoyer, B. (1996). Functional anatomy of spatial mental imagery generated from verbal instructions. *J. Neurosci.* **16**, 6504–6512.

Middleton, F. A., and Strick, P. L. (1994). Anatomical evidence for cerebellar and basal ganglia involvement in higher cognitive function. *Science* **266**, 458–451.

Mitra, J., and Snider, R. S. (1975). Effects of hippocampal after-discharges on Purkinje cell activity. *Epilepsia* **16**, 235–243.

Molinari, M., Solida, A., Leggio, M. G., Petrosini, L., Ciorra, R., and Gainotti, G. (1995). Procedural learning in patients with focal cerebellar lesions. *Soc. Neurosci. Abstr.* **21**, 270.

Mollica, A., Moruzzi, G., and Naquet, R. (1953). Decharges reticulaires induites par la polarisation du cervelet: Leurs rapports avec le tonus postural et la reaction d'eveil. *EEG Clin. Neurophysiol.* **5**, 571–584.

Moriguchi, I. (1981). A study of schizophrenic brains by computerized tomography scans. *Folia Psychiat. Neurol. Jpn.* **35**, 55–72.

Moruzzi, G. (1938). Action inhibitrice du paleocervelet sur les reflexes circulatoires et respiratories d'origine sino-carotidienne. *Comp. Rend. Soc. Biol.* **128**, 533–539.

Moruzzi, G. (1940). Paleocerebellar inhibition of vasomotor and respiratory carotid sinus reflexes. *J. Neurophysiol.* **3**, 20–32.

Moruzzi, G. (1941). Sui rapporti fra cervelletto e corteccia cerebrale. I. Azione d'impulsi cerebellari sulle attività corticali motrici dell'animale in narcosi cloralocsica. *Arch. Fisiol.* **41**, 87–139.

Moruzzi, G. (1947). Sham rage and localized autonomic responses elicited by cerebellar stimulation in the acute thalamic cat. *In* "Proc. XVII Internat. Congress Physiol. Oxford," pp. 114–115.
Moruzzi, G., and Magoun, H. W. (1949). Brainstem reticular formation and activation of the EEG. *EEG Clin. Neurophysiol.* **1**, 455–473.
Mutrux, S., Martin, F., and Chesni, Y. (1953). Ataxie congénitale et familiale associée è des malformations somatiques, è des troubles oculaires et è des troubles mentaux. *J. Génét. Hum.* **2**, 103–116.
Nashold, B. S., and Slaughter, D. G. (1969). Effects of stimulating or destroying the deep cerebellar regions in man. *J. Neurosurg.* **31**, 172–186.
Neff, I. H. (1894). A report of thirteen cases of ataxia in adults with hereditary history. *Am. J. Insan.* **51**, 365–373.
Otto, A. (1873). Ein Fall von Verkümmerung des Kleinhirns. *Arch. Psychiat. Nervenkr.* **4**, 730–746.
Pardal, M. M., Micheli, F., Asconape, J., and Paradiso, G. (1985). Neurobehavioral symptoms in caudate hemorrhage: Two cases. *Neurology* **35**, 1806–1807.
Parsons, L. M., Fox, P. T., Downs, J. H., Glass, T., Hirsch, T. B., Martin, C. C., Jerabek, P. A., and Lancaster, J. L. (1995). Use of implicit motor imagery for visual shape discrimination as revealed by PET. *Nature* **375**, 54.
Peters, M., and Monjan, A. A. (1971). Behavior after cerebellar lesions in cats and monkeys. *Physiol. Behav.* **6**, 205–206.
Petersen, S. E., Fox, P. T., Posner, M. I., Mintum, M. A., and Raichle, M. E. (1988). Positron emission tomographic studies of the cortical anatomy of single-word processing. *Nature* **331**, 585–589.
Petrosini, L., Molinari, M., and Dell'Anna, M. E. (1996). Cerebellar contribution to spatial event processing: Morris water maze and T-maze. *Eur. J. Neurosci.* **9**, 1896–1996.
Pollack, I. F. (1995). Mutism and pseudobulbar symptoms after resection of posterior fossa tumors in children: Incidence and pathophysiology. *Neurosurgery* **37**, 885–893.
Prescott, J. W. (1971). Early somatosensory deprivation as ontogenic process in the abnormal development of the brain and behavior. *In* "Medical Primatology 1970" (E. I. Goldsmith and J. Moor-Jankowski, eds.). Karger, Basel, Switzerland.
Rapin, I. (1994). Introduction and overview. *In* "The Neurobiology of Autism" (M. L. Bauman and T. L. Kemper, eds.), pp. 1–17. Johns Hopkins Univ. Press, Baltimore.
Rasheed, B. M. A., Manchanda, S. K., and Anand, B. K. (1970). Effects of the stimulation of paleocerebellum on certain vegetative functions in the cat. *Brain Res.* **20**, 293–308.
Rauch, S. L., Savage, C. R., Alpert, N. M., Brown, H. D., Curran, T., Kendrick, A., Fischman, A. J., and Kosslyn, S. M. (1995). Functional neuroanatomy of implicit sequence learning studied with PET. *Hum. Brain Mapp. Suppl.* **1**, 409.
Reiman, E. M., Raichle, M. E., Robins, E., Mintun, M. A., Fusselman, M. J., Fox, P. T., Price, J. L., and Hackman, K. A. (1989). Neuroanatomical correlates of a lactate-induced anxiety attack. *Arch. Gen. Psychiat.* **46**, 493–500.
Reimer, G., Grimm, R. J., and Dow, R. S. (1967). Effects of cerebellar stimulation in cobalt induced epilepsy in the cat. *EEG Clin. Neurophysiol.* **23**, 456–462.
Reis, D. J., Doba, N., and Nathan, M. A. (1973). Predatory attack, grooming and consummatory behaviors evoked by electrical stimulation of cat cerebellar nuclei. *Science* **182**, 845–847.
Riklan, M., Marisak, I., and Cooper, I. S. (1974). Psychological studies of chronic cerebellar stimulation in man. *In* "The Cerebellum, Epilepsy and Behavior" (I. S. Cooper, M. Riklan, and R. S. Snider, eds.). pp. 285–342. Plenum Press, New York.
Rolando, L. (1809). Saggio sopra la vera struttura del cerbello dell'uome e degli animali e sopra le funzoini del sistema nervosa. Sassari: Stampeia da S.S.R.M. Privilegiata. (Quoted in Dow and Moruzzi, 1958.)

Rubinstein, H. S., and Freeman, W. (1940). Cerebellar agenesis. *J. Nerv. Ment. Dis.* **92**, 489–502.
Russell, J. S. R. (1894). Experimental researches into the functions of the cerebellum. *Phil. Trans. R. Soc. Lond. B* **185**, 819–861.
Ryding, E., Decety, J., Sjohom, H., Sternberg, G., and Ingvar, D. H. (1993). Motor imagery activates the cerebellum regionally: A SPECT rCBF study with tc-99m HMPAO. *Cogn. Brain Res.* **1**, 94–99.
Sachs, E., and Fincher, E. F. (1927). Anatomical and physiological observations on lesions in the cerebellar nuclei in *Macacus rhesus*. *Brain* **50**, 350–356.
Sanes, J. N., Dimitrov, B., and Hallett, M. (1990). Motor learning in patients with cerebellar dysfunction. *Brain* **113**, 103–120.
Santha, K. (1930). Ueber das Verhalten des Kleinhirns in einem Falle von endogen-afamiliärer Idiotie. (Zur Differentialdiagnose der Marieschen und der sonstigen endogenen Kleinhirnerkrankungen nebst Beitrag zur Lehre der Diplomyelie.) *Z. Neurol. Psychiat.* **123**, 717–793.
Scherer, H. J. (1933). Beiträge zur pathologischen anatomie des kleinhirns: Genuine kleinhirnatrophien. *Z. Neurol. Psychiat.* **145**, 335–405.
Schmahmann, J. D. (1991). An emerging concept: The cerebellar contribution to higher function. *Arch. Neurol.* **48**, 1178–1187.
Schmahmann, J. D. (1996). From movement to thought: Anatomic substrates of the cerebellar contribution to cognitive processing. *Hum. Brain Mapp.* **4**, 174–198.
Schmahmann, J. D., and Pandya, D. N. (1987). Posterior parietal projections to the basis pontis in the rhesus monkey: Possible anatomical substrate for the cerebellar modulation of complex behavior? *Neurology* **37**(Suppl. 1), 291.
Schmahmann, J. D., and Pandya, D. N. (1989a). Anatomical investigation of projections to the basis pontis from posterior parietal association cortices in rhesus monkey. *J. Comp. Neurol.* **289**, 53–73.
Schmahmann, J. D., and Pandya, D. N. (1989b). Projections to the basis pontis from superior temporal sulcus (STS) in the rhesus monkey. *Soc. Neurosci. Abstr.* **15**(1), 73.
Schmahmann, J. D., and Pandya, D. N. (1995). Prefrontal cortex projections to the basilar pons: Implications for the cerebellar contribution to higher function. *Neurosci. Lett.* **199**, 175–178.
Schmahmann, J. D., and Pandya, D. N. (1997). Anatomic organization of the basilar pontine projections from prefrontal cortices in rhesus monkey. *J. Neurosci.* **17**, 438–458.
Schob, F. (1921). Weitere Beiträge zur Kenntnis der Friedreich-ähnlichen Krankheitsbilder. *Z. Neurol. Psychiat.* **73**, 188–238.
Schut, J. W. (1950). Hereditary ataxia. *Arch. Neurol. Psychiat.* **63**, 535–568.
Seitz, R. J., and Roland, P. E. (1992). Learning of sequential finger movements in man: A combined kinematic and positive emission tomography (PET) study. *Eur. J. Neurosci.* **4**, 154–165.
Sherrington, C. (1906). "The Integrative Action of the Nervous System." Yale Univ. Press, New Haven, CT.
Silveri, M. C., Leggio, M. G., and Molinari, M. (1994). The cerebellum contributes to linguistic production: A case of agrammatic speech following a right cerebellar lesion. *Neurology* **44**, 2047–2050.
Snider, R. S. (1945). A tectocerebellar pathway. *Anat. Rec.* **91**, 299.
Snider, R. S. (1950). Recent contributions to the anatomy and physiology of the cerebellum. *Arch. Neurol. Psych.* **64**, 196–219.
Snider, R. S. (1975). A cerebellar-ceruleus pathway. *Brain Res.* **88**, 59–63.
Snider, R. S., and Eldred, E. (1948). Cerebral projections to the tactile, auditory, and visual areas of the cerebellum. *Anat. Rec.* **100**, 82.

Snider, R. S., and Maiti, A. (1975). Septal afterdischarges and their modification by the cerebellum. *Exp. Neurol.* **49,** 529–539.

Snider, R. S., and Maiti, A. (1976). Cerebellar contributions to the Papez circuit. *J. Neurosci. Res.* **2,** 133–146.

Snider, R. S., McCulloch, W. S., and Magoun, H. W. (1949). A cerebello-bulbo-reticular pathway for suppression. *J. Neurophysiol.* **12,** 325–334.

Snider, R. S., and Stowell, A. (1942). Evidence of tactile sensibility in the cerebellum of the cat. *Fed. Proc.* **1,** 82.

Snider, R. S., and Stowell, A. (1944). Electro-anatomical studies on a tactile system in the cerebellum of monkey (*Macaca mulatta*). *Anat. Rec.* **88,** 457.

Snider, S. R. (1982). Cerebellar pathology in schizophrenia: Cause or consequence? *Neurosci. Behav. Rev.* **6,** 47–53.

Sprague, J. M., and Chambers, W. W. (1959). An analysis of cerebellar functions in the cat as revealed by its partial and complete destruction and its interaction with cerebral cortex. *Arch. Ital. Biol.* **97,** 68–88.

Stein, J. F., and Glickstein, M. (1992). Role of the cerebellum in visual guidance of movement. *Physiol. Rev.* **72,** 967–1017.

Steriade, M., Apostol, V., and Oakson, G. (1971). Control of unitary activities in cerebellothalamic pathway during wakefulness and synchronized sleep. *J. Neurophysiol.* **34,** 389–413.

Sunderland, S. (1940). The projection of the cerebral cortex on the pons and the cerebellum in the macaque monkey. *J. Anat.* **74,** 201–226.

Taylor, M. A. (1991). The role of the cerebellum in the pathogenesis of schizophrenia. *Neuropsychiat. Neuropsychol. Behav. Neurol.* **4,** 251–280.

Thompson, R. F. (1988). The neural basis of basic associative learning of discrete behavioral responses. *Trends Neurosci.* **11,** 152–155.

Thompson, R. F., and Krupa, D. J. (1994). Organization of memory traces in the mammalian brain. *Annu. Rev. Neurosci.* **17,** 519–549.

van Dongen, H. R., Catsman-Berrevoets, C. E., and van Mourik, M. (1994). The syndrome of 'cerebellar' mutism and subsequent dysarthria. *Neurology* **44,** 2040–2046.

Vogt, H., and Astwazaturow, M. (1912). Ueber angeborene Kleinhirnerkrankungen mit Beiträgen zur Entwickelungsgeschichte des Kleinhirns. *Arch. Psychiat. Nervenkr.* **49,** 75–203.

Vulpian, A. (1866). "Lecons sur la physiologie général et comparée du système nerveux faites au muséum d'histoire naturelle." Baillière, Paris.

Wallesch C.-W., and Horn, A. (1990). Long-term effects of cerebellar pathology on cognitive functions. *Brain Cogn.* **14,** 19–25.

Walter, F. K., and Roese, H. F. (1926). Ein Beitrag zur Kenntnis der "hereditären ataxie (Friedreich-Nonne-Marie)." *Dtsch. Z. Nervenheilk.* **92,** 8–27.

Watson, P. J. (1978). Nonmotor functions of the cerebellum. *Psychol. Bull.* **85,** 944–967.

Weinberger, D. R., Kleinman, J. E., Luchins, D. J., Bigelow, L., and Wyatt, R. (1980). Cerebellar pathology in schizophrenia: A controlled postmortem study. *Am. J. Psychiat.* **137,** 359–361.

Welker, W. (1987). Spatial organization of somatosensory projections to granule cell cerebellar cortex: Functional and connectional implications of fractured somatotopy (summary of Wisconsin studies). *In* "New Concepts in Cerebellar Neurobiology" (J. S. King, ed.), pp. 239–280. Liss, New York.

Whyte, J. M. (1898). Four cases of Friederich's ataxia, with a critical digest of recent literature on the subject. *Brain* **21,** 72–136.

Wiggers, K. (1942). De invloed van het cerebellum op de vegetatieve functies. J. H. Kok N. V., Kampden.

Yeterian, E. H., and Van Hoesen, G. W. (1978). Cortico-striate projections in the rhesus monkey: The organization of certain cortico-caudate connections. *Brain Res.* **139,** 43–63.

Zanchetti, A., and Zoccolini, A. (1954). Autonomic hypothalamic outbursts elicited by cerebellar stimulation. *J. Neurophysiol.* **17,** 475–483.

SECTION II
ANATOMIC SUBSTRATES

THE CEREBROCEREBELLAR SYSTEM

Jeremy D. Schmahmann* and Deepak N. Pandya†

*Department of Neurology, Massachusetts General Hospital and Harvard Medical School, Boston, Massachusetts 02114; and †Department of Anatomy and Neurobiology, Boston University School of Medicine, Boston, Massachusetts 02118

I. Introduction
II. The Feedforward Limb of the Cerebrocerebellar System
   A. Corticopontine Projections
   B. Pontocerebellar Projections
III. The Feedback Limb of the Cerebrocerebellar System
IV. Climbing Fibers and Cognition: Is There an Anatomic Substrate?
V. Conclusions
   References

If there is a cerebellar contribution to nonmotor function, particularly to cognitive abilities and affective states, then there must be corresponding anatomic substrates that support this. The cerebellum is strongly interconnected with the cerebral hemispheres in both feedforward (cerebral hemispheres to cerebellum) and feedback directions. This relationship has long been recognized, particularly with respect to the motor and sensory cortices. Investigations performed over the last decade, however, have demonstrated for the first time the organization and strength of the connections that link the cerebellum with areas of the cerebral cortex known to be concerned with higher order behavior rather than with motor control. The feedforward projections from these higher order areas, namely the associative and paralimbic cortices, seem to be matched, at least in the limited but definite demonstrations to date, by cerebellar projections back to these same areas. These observations are important because they are congruent with the notion that cognitive functions are distributed among multiple cortical and subcortical nodes, each of which functions in concert but in a unique manner to produce an ultimate behavior pattern. This chapter describes the neural circuitry postulated to subserve the cerebellar contribution to nonmotor processing, particularly cognitive and affective modulation, and discusses the theoretical implications of these anatomic findings.

I. Introduction

If there is a cerebellar contribution to nonmotor function, particularly to cognitive abilities and affective states, then there must be corresponding anatomic substrates that support this. The cerebrocerebellar circuit consists of a feedforward, or afferent limb, and a feedback, or efferent limb. The feedforward limb is composed of the corticopontine and pontocerebellar mossy fiber projections; the feedback loop is the cerebellothalamic and thalamocortical pathways (Fig. 1). Our conceptual approach (Schmahmann, 1991) holds that the cerebellum modifies behaviorally relevant information that it has received from the cerebral cortex via the corticopontine pathway and then redistributes this now "cerebellar-processed" information back to the cerebral hemispheres. For this reason, both limbs (feedforward, and feedback) of this cerebrocerebellar circuit are essential to the discussion of the cerebellar contribution to nonmotor processing.

A second feedforward system links the cerebral cortex with the red nucleus, from where the central tegmental tract leads to the inferior olivary

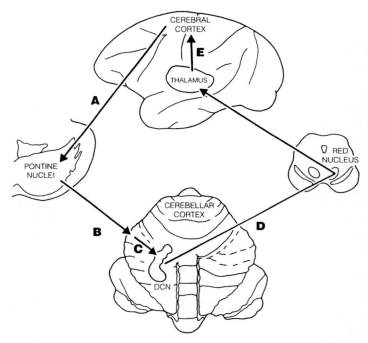

FIG. 1. Diagram of the cerebrocerebellar circuit. Feedforward limb: The corticopontine pathway (**A**) carries associative, paralimbic, sensory, and motor information from the cerebral cortex to the neurons in the ventral pons. The axons of these pontine neurons reach the cerebellar cortex via the pontocerebellar pathway (**B**). Feedback limb: The cerebellar cortex is connected with the deep cerebellar nuclei (DCN, **C**), which project via the red nucleus to the thalamus (the cerebello-thalamic projection, **D**). The thalamic projection back to cerebral cortex (**E**) completes the feedback circuit. From Schmahmann (1994), with permission.

nucleus and then through the climbing fiber system to the cerebellar cortex. This second afferent arc has more restricted relevance to the discussion of the relationship between the cerebellum and cognition. Input from serotonin, norepinephrine and dopamine containing brain stem structures constitutes another substantial source of cerebellar afferents.

This chapter describes the neural circuitry postulated to subserve the cerebellar contribution to nonmotor processing, particularly cognitive and affective modulation. The information presented here is derived from experiments in the nonhuman primate, and whereas there is ample precedent to extrapolate this to an understanding of similar systems in humans, the inherent limitations of this approach are readily acknowledged (cf. H. G. Leiner and A. L. Leiner, this volume).

II. The Feedforward Limb of the Cerebrocerebellar System

A. CORTICOPONTINE PROJECTIONS

The cerebral cortex projection to the basilar pons (the corticopontine pathway) is the obligatory first stage in the feedforward limb of the cerebrocerebellar loop. The corticopontine pathway originates in neurons in layer Vb of the cerebral cortex (Glickstein et al., 1985), the axons of which enter the internal capsule, descend into the cerebral peduncle, and terminate around neurons that occupy the ventral half of the pons. Based in part on its cellular architecture, the basilar pons of the monkey has been parcellated into different nuclear groups (Nyby and Jansen, 1951; Schmahmann and Pandya, 1989, 1991). This subdivision on anatomic grounds appears to be reflected to some extent also in the pontine connections of the different cerebral cortical fields. Following the injection of anterograde tracer (radiolabeled amino acids) in the cerebral hemisphere of a monkey, the tracer passes via the corticopontine projection into the pons where it is distributed in discrete clusters (Figs. 2 and 3). The terminations in the pons derived from different cerebral regions appear to interdigitate with each other, but in general they do not seem to overlap. Motor, premotor, and supplementary motor regions, as well as primary somatosensory cortices, have been shown to send their efferents to the cerebellum via this route (Nyby and Jansen, 1951; Brodal, 1978, 1981; Glickstein et al., 1985; Shook et al., 1990; Schmahmann and Pandya, 1995a); however, the origins of the corticopontine pathway are not limited to these sensorimotor cortices.

The studies described here address the question whether the cerebellum receives the type of information from the cerebral cortex that could facili-

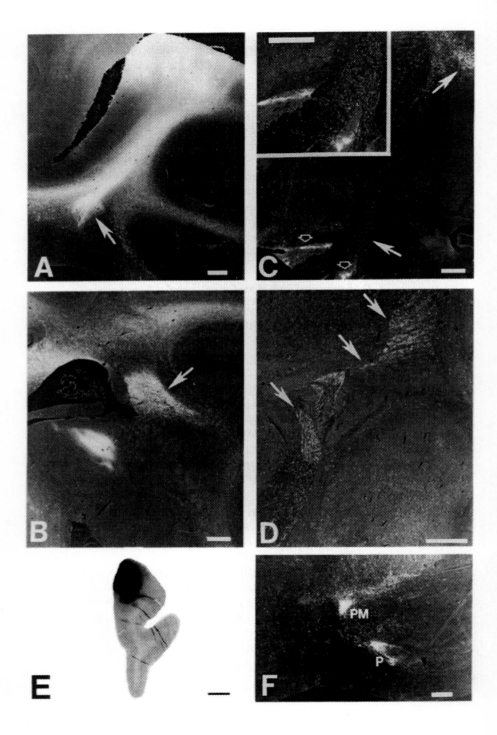

tate a cerebellar modulation of cognitive ability and affective states. Consequently, the cerebral cortical regions studied with respect to their pontine projections were association areas and paralimbic cortices. Association areas in the parietal, temporal, and frontal lobes are responsible for a number of highly complex cognitive operations and, when lesioned in humans and animals, result in clinical syndromes that are now part of classic neurological teaching (Mesulam, 1985; Yeterian and Pandya, 1985). Similarly, the paralimbic areas of the parahippocampal gyrus and cingulate cortices are concerned with motivation and drive, and are thought to play a role in emotionally relevant memory (Nadel, 1991) and in conditions such as obsessive compulsive disorder (Rauch *et al.*, 1994; Spangler *et al.*, 1996). The demonstration of an anatomic link between these higher order cerebral areas and the cerebellum would provide support for the contention that the cerebellum is a contributing node in the distributed neural circuitry subserving cognition.

1. *Parietopontine Connections*

Association areas of the posterior parietal cortices are critical for directed attention, visuospatial analysis, and vigilance in the contralateral hemispace. Clinical observations have demonstrated that posterior parietal lesions are associated with disturbances of complex visuospatial integration, trimodal neglect of the contralateral body and extrapersonal space, alien hand syndrome, impaired language, apraxia, and agnosia. The multimodal and motivationally relevant properties of the posterior parietal neurons have also been shown in behavioral and electrophysiological studies (Denny-Brown and Chambers, 1958; Geschwind 1965a,b; Mountcastle *et al.*, 1977; Hyvarinen, 1982; Pandya *et al.*, 1988).

Both the superior parietal lobule (SPL) and the inferior parietal lobule (IPL) are thought to be involved in the sequential processing of somatotopically organized information received from the adjacent primary somatosensory cortices. The upper bank of the intraparietal sulcus is part of the somatic sensory association area, whereas the lower bank is concerned with the somatosensory modality as well as with visual and vestibular information.

FIG. 2. Dark-field photomicrographs **A** through **D** show the trajectory to the pons of radiolabeled fibers (indicated by arrows) derived from an injection of tracer in areas PG and Opt in the caudal part of the inferior parietal lobule of a rhesus monkey. The terminations (open arrowheads) are seen in the pons (in **C,** and shown at higher magnification in the inset). Bar, 1 mm. Light-field photomicrograph **E** shows a coronal section through an isotope injection in the prefrontal cortex in area 9/46d (bar, 20 mm), and the terminations in the paramedian nucleus in level II of the pons are seen in the dark-field photomicrograph in **F** (bar, 2 mm). (Modified from Schmahmann and Pandya 1992, 1997; with permission.)

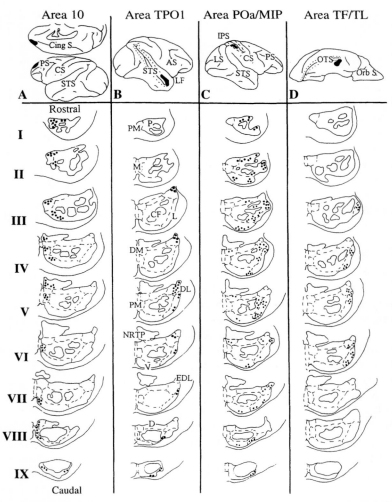

FIG. 3. Diagram of the projections to the basis pontis from selected regions within cerebral association areas. Each cerebral area is connected with a unique and distributed subset of pontine neurons. The projections appear to be arranged in an interdigitating, but not overlapping manner. Radiolabeled amino acids (shaded black area) were placed in the medial and lateral parts of the rostral prefrontal cortex (area 10) (**A**); in the cortex buried within the rostral upper bank of the superior temporal sulcus (area TPO$_1$) (**B**); in cortex buried within the lower bank of the intraparietal sulcus (area POa, or MIP) (**C**); and in the parahippocampal gyrus (areas TF/TL) (**D**). Terminations of the anterogradely transported label are represented by the black dots in the ipsilateral half of the basis pontis. The pons is depicted from rostral level I to caudal level IX, according to Nyby and Jansen (1951) as modified by Schmahmann and Pandya (1988, 1989). Abbreviations for the cerebral cortex: AS, arcuate sulcus; Cing S, cingulate sulcus; CS, central sulcus; IPS, intraparietal sulcus; LF, lateral (Sylvian) fissure; LS,

The SPL is principally concerned with intramodality associative functions such as multiple joint position sense, touch, and proprioception. The most caudal region of the SPL (area PGm) has been shown to be a site of convergence of somatosensory, kinesthetic, visual, and auditory information and it has connections with the prefrontal cortex and the cingulate gyrus. Areas PG and Opt in the caudal parts of the IPL have reciprocal connections with paralimbic cortices (parahippocampal gyrus, presubiculum, perirhinal cortex, and cingulate gyrus) and multimodal zones in the temporal and frontal lobes and are strongly implicated in the neglect syndrome. Unlike the rostral part of the superior and inferior parietal lobules, which are connected with modality-specific thalamic nuclei, areas PGm and PG/Opt have thalamic connections predominantly with associative thalamic nuclei. Area Opt alone has connections with the lateral dorsal nucleus, as well as with the anterior nucleus of thalamus, which are considered part of the limbic system circuitry. These caudally located multimodal zones in PG/Opt and, to a lesser degree, in PGm subserve highly complex, nonmodality-specific functions that are invested with emotional and motivational significance.

Earlier studies (Nyby and Jansen, 1951; Brodal, 1978) minimized the extent of parietal association cortex projections to the pons. It appears from more recent studies, however, that there are consistent pontine projections from these regions (Glickstein *et al.,* 1985; May and Anderson, 1986; Schmahmann and Pandya, 1989). Our investigation demonstrated strong pontine connections with the posterior parietal association cortices, with differential projections to the pontine nuclei arising from superior versus inferior and rostral versus caudal parietal regions (Schmahmann and Pandya, 1989). The projections are directed most heavily toward the peripeduncular and lateral nuclei. Lesser but nevertheless substantial projections are found in the intrapeduncular, ventral, dorsolateral, extreme dorsolateral, and dorsal nuclei. Dorsomedial, paramedian, and reticular (nucleus reticularis tegmenti pontis, NRTP) nuclei receive only minor projections (Figs. 3C and 4). SPL projections are relatively widespread with respect to the more focused IPL projections. IPL projections are, in general, situated more laterally and at more rostral levels of the pontine nuclei than are those of the SPL. The sulcal cortex of the SPL favors the dorsolateral,

---

lunate sulcus; Orb S, orbital sulcus; OTS, occipitotemporal sulcus; PS, principalis sulcus; STS, superior temporal sulcus. Abbreviations for the pontine nuclei: CF, corticofugal fibers; D, dorsal; DL, dorsolateral; DM, dorsomedial; EDL, extreme dorsolateral; L, lateral; M, median; NRTP, nucleus reticularis tegmenti pontis; PM, paramedian; P, peduncular; V, ventral. (From Schmahmann, 1996; with permission.)

extreme dorsolateral, and ventral nuclei compared to the light projections to these nuclei from the convexity of the SPL. The sulcal cortex of the IPL differs from the gyral cortex in favoring the ventral and extreme dorsolateral nuclei. The intrapeduncular nucleus receives projections only from rostral IPL, whereas the lateral nucleus receives projections preferentially from caudal IPL. The pontine projections from area PGm in the medial SPL are unique in the parietal lobe in that they include the paramedian nucleus. There is, therefore, convincing evidence of pontine projections arising from the rostral parietal lobe as well as from the multimodal regions located caudally in both the superior and the inferior parietal lobules.

2. *Temporopontine Connections*

The role of the temporal lobe with respect to language, memory, and complex behaviors has been well established, and confusional states, highly structured visual hallucinations, and the Klüver–Bucy syndrome consequent upon lesions in this area are also recognized (Mesulam, 1985). In the monkey, anatomical and physiological observations have shown that the temporal lobe contains unimodal and multimodal areas which appear to contribute differentially to the organization of behavior (Desimone and Ungerleider, 1989). The cortex in the upper bank of the superior temporal sulcus (STS) has been shown to be concerned with multiple sensory modalities: vision, somatic sensation, and audition (Gattass and Gross, 1981; Desimone and Ungerleider, 1986; Bayliss *et al.*, 1987). It has connections with association areas of the frontal (Seltzer and Pandya, 1989) and parietal

FIG. 4. Composite color-coded summary diagram illustrating the distribution within the basilar pons of the rhesus monkey of projections derived from associative cortices in the prefrontal (purple), posterior parietal (blue), temporal (red), and parastriate and parahippocampal regions (orange), and from motor, premotor, and supplementary motor areas (green). The medial (**A**), lateral (**B**), and ventral (**C**) surfaces of the cerebral hemisphere are at the upper left. The plane of section through the basilar pons is at the lower left, and the rostrocaudal levels of the pons I through IX are on the right. Cerebral areas that have been demonstrated to project to the pons by other investigators using either anterograde or retrograde tracers are depicted in white; those areas studied with both anterograde and retrograde studies and found to have no pontine projections are shown on the hemispheres in yellow; and those with no pontine projections according to retrograde studies by other investigators are shaded in gray. The dashed lines in the hemisphere diagrams represent the sulcal cortices. In the pons diagrams the dashed lines represent the pontine nuclei, and the solid lines depict the traversing corticofugal fibers. The associative corticopontine projections are substantial. There is a complex mosaic of terminations in the pons, and each cerebral cortical region has preferential sites of pontine terminations. There is considerable interdigitation of the terminations from some of the different cortical sites, but almost no overlap. (From Schmahmann, 1996; with permission.)

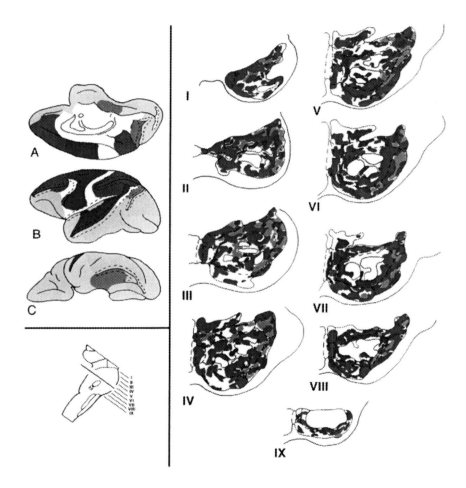

cortices (Seltzer and Pandya, 1984), as well as with limbic-related structures at the medial and inferior frontal convexity and parahippocampal and cingulate gyri (Pandya et al., 1981, 1988; Barnes and Pandya 1992), and contains neurons which respond preferentially to more than one modality and to complex stimuli such as faces (Perrett et al., 1982). In contrast, the superior temporal gyrus appears to be an association area confined to the auditory realm (Merzenich and Brugge,1973; Galaburda and Pandya, 1983; Colombo et al., 1990), and the depth of the STS is an important association area for the somatosensory modality (Seltzer and Pandya, 1984).

Projections from the temporal lobe to the basis pontis are derived predominantly from the upper bank of the STS, with a lesser contribution from the superior temporal gyrus (Schmahmann and Pandya, 1991). The extreme dorsolateral, dorsolateral, and lateral pontine nuclei are the major recipients of efferents from each of the subdivisions of the upper bank of the STS, with the most caudal part also projecting to the peripeduncular, ventral, and rostral intrapeduncular nuclei (Figs. 3B and 4). Pontine projections from the superior temporal gyrus and the medial portion of the supratemporal plane, including the second auditory area, AII, are less intense than from the upper bank of the STS and are directed mainly to the peripeduncular, dorsolateral, and lateral pontine nuclei. The most rostral STG spares the peduncular and dorsolateral nuclei and projects instead to the extreme dorsolateral nucleus. These different patterns of projection to the pons, particularly to the extreme dorsolateral nucleus, from the multimodal regions of the upper bank of the STS on the one hand and from the auditory association areas of the STG on the other hand serve to set these cortical regions apart.

The inferotemporal region and the lower bank of the STS, which are strongly interconnected, contain neurons that are functionally unimodal within the visual system, subserve mainly central vision, and seem to be involved in object recognition (Mishkin et al., 1983; Van Essen, 1985; Desimone and Ungerleider, 1986; Pandya et al., 1988). The lower bank of the STS is quite different from the upper bank and the supratemporal plane because, with the exception of the caudally located area MT (Ungerleider et al., 1984), it contributes no projections to the basis pontis (Glickstein et al., 1985; Schmahmann and Pandya, 1991). These observations appear to correlate with those in the monkey visual corticopontine system in which the pontine connections of the visual cortex are derived either predominantly or exclusively from areas devoted to the peripheral visual field rather than from areas subserving central or focal representation (Nyby and Jansen, 1951; Kuypers and Lawrence, 1967; Brodal, 1978; Glickstein et al., 1985, 1980).

3. Parastriate, Occipitotemporal, and Parahippocampal Projections to Pons

The dichotomy in the pontine connectivity between visual motion (where) versus visual feature discrimination (what) systems (Ungerleider and Mishkin, 1982) in the superior temporal region and the STS is also observed in the parastriate/occipitotemporal pontine system. The dorsal visual stream includes the dorsal part of the prelunate gyrus, the caudal part of the lower bank of the STS (area MT), the polymodal convergence zones in the lower bank of the intraparietal sulcus, and the paralimbic-associated polymodal parts of the caudal inferior parietal lobule. The ventral visual stream is directed inferiorly from the parastriate zone into the ventral prelunate gyrus, the rostral lower bank of the STS, and the middle and inferior temporal gyri. Whereas the dorsal stream is concerned with the spatial features of objects or events in the periphery of the visual field, the ventral stream is more involved with the identification of objects and their characteristics, such as form, color, and orientation, and with stimuli occurring in the central part of the visual field (Mishkin *et al.*, 1983; Van Essen, 1985; Desimone and Ungerleider, 1989).

Pontine projections are derived from the dorsal stream regions and are distributed in the lateral, peripeduncular, and dorsolateral pontine nuclei. The medial prelunate region projects to the rostral half of the pons. The dorsal prelunate gyrus projects throughout the rostrocaudal extent of the pons, including projections to the extreme dorsolateral nucleus, and, to a lesser extent, to the nucleus reticularis tegmenti pontis (Fig. 3). In contrast, the ventral trend (ventral prelunate gyrus, inferotemporal region, and inferior temporal gyrus as well as the lower bank of the STS) has no pontine connections (Schmahmann and Pandya, 1993). These findings extend to the occipitotemporal regions the dichotomy in the pattern of pontine projections arising from the occipital lobe, the STS, and the superior temporal region (Brodal, 1978; Glickstein *et al.*, 1985; Schmahmann and Pandya, 1991). That is to say, visual cortical areas (primary or associative) that are concerned with the peripheral visual field, visual spatial parameters, and visual motion project to the pons, whereas regions concerned with the central visual field and visual object identification do not.

This peripheral field/spatially oriented peristriate projection to the pons also applies to the pontine projections from the posterior parahippocampal regions. Whereas the inferior temporal gyrus in the ventral visual stream does not project to the pons, the medially adjacent mid and caudal parts of the parahippocampal region have pontine connections with the lateral, dorsolateral, and peripeduncular pontine nuclei (Figs. 3D and 4), similar to those seen from the dorsal prelunate gyrus. The posterior parahippocampal gyrus is responsive to events in the periphery rather than the central visual field and is concerned with the spatial aspects of memory (Nadel, 1991). The pontine afferents from the posterior parahippocampal gyrus may therefore facilitate a cerebellar contribution to visual spatial memory, particularly when invested with motivational valence.

4. *Prefrontopontine Connections*

The prefrontal cortex (PFC) has repeatedly been shown in both humans and nonhuman primates to be an essential component of the normal integration of higher order behavior. This includes such functions as planning, foresight, judgment, attention, language, and working memory. On the basis of behavioral studies, different functional attributes have been ascribed to orbital, medial, peri-principalis, and periarcuate prefrontal regions. The dorsolateral and medial convexities are important for kinesthetic, motivational, and spatially related functions, including spatial memory, whereas inferior prefrontal and orbital areas are more related to autonomic and emotional response inhibition, stimulus significance, and object recognition and memory (Milner, 1964; Luria, 1966; Fuster, 1980; Stuss and Benson, 1986). This multiplicity of functional processes is matched by a connectional heterogeneity such that each of the prefrontal subdivisions has a different set of connections with cortical as well as with subcortical structures (Barbas and Mesulam, 1985; Barbas *et al.*, 1991; Barbas and Pandya, 1991; Pandya and Yeterian, 1991; Cavada and Goldman-Rakic, 1989; Boussaoud *et al.*, 1991; Eblen and Graybiel, 1995). Given the importance of the prefrontal regions for complex cognitive operations and hypotheses concerning the cerebellar contribution to cognition, it is critical to know whether these PFC regions participate in the anatomic circuitry linking the cerebral cortex with the cerebellum.

Despite the interest in projections from premotor and prefrontal areas to pons in the nonhuman primate (DeVito and Smith, 1964; Künzle and Akert, 1977; Wiesendanger *et al.*, 1979; Leichnetz *et al.*, 1984; Shook *et al.*, 1990; Stanton *et al.*, 1988), there were a number of persistent questions with respect to the degree to which the PFC participates in the feedforward loop of the cerebrocerebellar system. Schmahmann and Pandya (1995b, 1997) demonstrated that connectional heterogeneity of the prefrontal cortex also exists in the corticopontine pathway. The dorsal lateral convexity and the medial prefrontal cortex provide the majority of the pontine efferents. A smaller projection arises from the ventral lateral convexity. The overall pattern of terminations of the prefrontopontine fibers resembles that from other parts of the cerebral cortex in that labeled terminals are distributed in discrete patches in the gray matter of the basilar pons (Figs. 2E and 2F). The projections are most prominent and occupy the rostrocaudal extent of the pons when derived from dorsal area 46 (9/46d), areas 8Ad and 9 at the dorsolateral convexity, and area 10 at the dorsolateral and medial convexities (Fig. 3A). Somewhat smaller pontine projections in the rostral one-third to one-half of the pons are observed from areas 8B and 9 at the medial surface of the hemisphere, ventral area 46 (9/46v) adjacent to the caudal half of the principal sulcus, and from area 32. Restricted pontine projections arise from area 45B in the rostral bank of the inferior limb of the arcuate sulcus (terminology of Petrides and Pandya, 1994).

Projections in the pons favor the paramedian and medial parts of the peripeduncular pontine nuclei. Some terminations are also present in the medial aspect of the ventral nucleus, the dorsomedial nucleus, and the nucleus reticularis tegmenti pontis (Fig. 4). The median nucleus receives a projection from area 9 medially. Topographic organization within this general framework is discernible. More medial prefrontal areas send projections to the most medial pontine regions, whereas pontine terminations tend to shift away from the midline following lateral prefrontal injections. Furthermore, each cortical area appears to have a unique complement of pontine nuclei with which it is connected (Fig. 5).

These anatomic data therefore reveal an organized and consistent projection from the prefrontal cortices of the rhesus monkey into the feedforward limb of the cerebrocerebellar circuit. These projections are derived from areas concerned with attention as well as with conjugate eye movements (area 8A), the spatial attributes of memory and working memory (area 46), planning, foresight, and judgment (area 10), motivational behavior and decision-making capabilities (areas 9 and 32), and from areas considered to be homologous to the language area in human (area 45B) (Brodmann, 1909; Astruc, 1971; Künzle and Akert, 1977; Glickstein *et al.*, 1985; Stanton *et al.*, 1988; Goldman-Rakic and Friedman, 1991; Pandya and Yeterian, 1991; Petrides and Pandya, 1994; Petrides, 1995; Schmahmann and Pandya, 1995b, 1997).

Not all regions of the PFC project to the pons. There is a lack of pontine input from the ventrolateral and orbitofrontal cortices, including areas 11, 47/12, 14, and 46v below the rostral part of the principal sulcus (Fig. 5) (Nyby and Jansen, 1951; Brodal, 1978; Glickstein *et al.*, 1985; Schmahmann and Pandya, 1995b, 1997). These areas resemble the inferotemporal region and the ventral bank of the superior temporal sulcus with which they are interconnected in that they are concerned with object memory, feature discrimination, and certain aspects of motivation (Desimone and Ungerleider, 1989; Barbas and Pandya, 1991) and they also have no pontine connections. The dichotomy in the spatial (where) versus object (what) apparent in the corticopontine projections from other association areas thus appears to be conserved for the prefrontal areas as well.

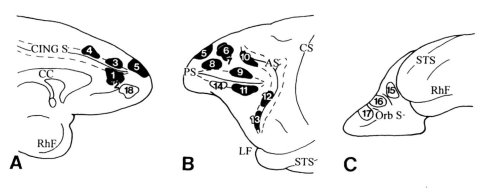

Case Number	Area	Rostro-caudal level	Median	Paramedian	Peri-peduncular	Dorso-medial	Intra-peduncular	Ventral	Lateral	NRTP	Dorsal	Dorso-lateral	Extreme Dorso-lateral
1,2	32	I - III	-	++	+	-	-	-	-	-	-	-	-
3	9(med)	I - VII	++	+	+	+	-	-	-	++	-	-	-
4	8B	I-IV	-	++	+	-	-	-	-	+	-	-	-
5,8	10	I - IX	-	+++	++	-	-	+	+	++	-	-	-
6,7,8	9(lat)	I - IX	-	+++	+	++	-	+	-	-	-	-	-
9	9/46d	I - VIII	-	++	++	++	+	+	-	++	-	-	-
10	8Ad	I - IX	-	+++	++	-	-	+	+	+	-	-	-
11	9/46v	I - V	-	+	+	-	-	-	-	-	-	-	-
12,13	45B	I - III	-	+	+	-	-	-	-	-	-	-	-
14	46(vent)	-	-	-	-	-	-	-	-	-	-	-	-
15	47/12	-	-	-	-	-	-	-	-	-	-	-	-
16	47/12	-	-	-	-	-	-	-	-	-	-	-	-
17	11	-	-	-	-	-	-	-	-	-	-	-	-
18	14	-	-	-	-	-	-	-	-	-	-	-	-

FIG. 5. Diagrams and table illustrating the medial (**A**), lateral (**B**), and orbital (**C**) surfaces of the frontal lobe of a rhesus monkey to show sites of injection of the isotope-labeled amino acid tracer in 18 animals, and the resulting distribution pattern of terminations within the nuclei of the ipsilateral basilar pons. Numbers in the injection sites correspond to the individual cases. Injections that resulted in terminations in the basilar pons are shaded in black. Those that did not result in label in the pons are unshaded. Terminations were present in different rostrocaudal levels of the pons (I to IX), as well as in characteristic sets of pontine nuclei. The strength of projection in each pontine nucleus is graded absent ($-$), mild ($+$), moderate ($++$), or strong ($+++$). The injections in cases 1 and 2 were placed in area 32; case 3 in area 9 medially; case 4 in area 8B medially; case 5 in area 10 at both the medial and dorsolateral convexities; cases 6 and 7 in area 9 at the lateral convexity; case 8 in the rostral part of area 9 (lateral) and dorsal area 10; case 9 in area 9/46d; case 10 in area 8Ad; case 11 in area 9/46v; and cases 12 and 13, respectively, in the dorsal and ventral parts of area 45B. In case 14 the isotope was injected in area 46 below the principal sulcus (46v); case 15 in area 12; cases 16 and 17 in area 11; and case 18 in area 14. Cortex within the walls of the cingulate sulcus, principal sulcus, and arcuate sulcus is represented by dotted lines. 9 (med), area 9 at the medial convexity; 9 (lat), area 9 at the dorsolateral convexity; 46 (vent), area 46 below the principal sulcus. (From Schmahmann and Pandya, 1997; with permission.)

5. Paralimbic and Autonomic Connections with Pons

The lack of pontine input from the ventrolateral prefrontal and orbitofrontal cortices may at first be seen as problematic for the hypothesis that the cerebellum contributes to the modulation of affect and emotional states (Snider, 1950; Heath, 1977; Berman *et al.*, 1978; Watson, 1978; Schmahmann, 1991, 1996), J. D. Schmahmann and J. C. Sherman, this volume). There is, however, a pontine projection from rostral cingulate area 32 as described earlier. In addition, Vilensky and Van Hoesen (1981) speculated on the potential influence of the limbic system on the cerebellum by virtue of their demonstration of pontine projections from the cingulate gyrus. The rostral cingulate (areas 35, 25, and 24) projects to the dorsomedial, medial, and ventromedial pons, and the caudal cingulate gyrus (areas 23 and the retrosplenial cortex) projects to the lateral and ventrolateral pons. This cingulopontine projection has been observed in cats as well (Aas and Brodal, 1988).

A direct and reciprocal hypothalomo-cerebellar projection has been identified in the monkey (Haines and Dietrichs, 1984; D. E. Haines *et al.*, this volume). The ansiform and paramedian lobules and the paraflocculus are connected with the lateral and posterior hypothalamic areas; the anterior lobe is connected with these as well as the ventromedial, dorsomedial, and dorsal hypothalamic nuclei; and deep cerebellar nuclei project to the contralateral posterior and lateral hypothalamic nuclei. Hypothalamic inputs to pons in the cat are derived from the posterior and dorsal hypothalamic regions which project medially and dorsomedially within the caudal third of the pontine nuclei, and a more sparse projection is also seen laterally (Aas and Brodal, 1988).

The medial mammillary bodies implicated in the amnestic syndrome project not only to the pons but also directly to the cerebellum. In cat, the medial mammillary nucleus projects ventromedially at all rostrocaudal levels of the pontine nuclei (Aas and Brodal, 1988). In monkey, lateral mammillary and supramammillary nuclei project to cerebellar ansiform and paramedian lobules, paraflocculus, and anterior lobe, whereas the medial mammillary nucleus receives projections from all contralateral cerebellar nuclei (Haines and Dietrichs, 1984). In the monkey, Schmahmann and Pandya (1993) also demonstrated projections to the pons from the posterior parahippocampal regions, which have been implicated particularly in the spatial aspects of memory. These anatomical data are complemented by physiological observations suggesting the existence of pathways linking several components of the limbic system (septal nuclei, hippocampus, and amygdala) with the cerebellum (Anand *et al.*, 1959; Harper and Heath, 1973; Snider and Maiti, 1976) and by the reciprocal connections

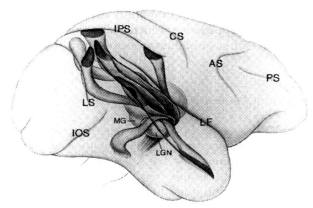

FIG. 6. Artist's representation of the trajectory and organization of fibers from the parietal, temporal, and occipital association areas as they course toward the critical point in the white matter above and medial to the midportion of the lateral geniculate nucleus before descending into the cerebral peduncle. (From Schmahmann and Pandya, 1992; with permission.)

between the cerebellum and the brain stem catecholaminergic and serotoninergic nuclei that have widespread projections to the cerebral cortex, including the higher order areas (Dempsey et al., 1983; Marcinkiewicz et al., 1989).

It is therefore apparent that the pontocerebellar system indeed receives a sizable input from limbic-related cortices. These findings may help explain the autonomic phenomena produced in animals by cerebellar stimulation (Martner, 1975) and also provide a plausible anatomic substrate for a cerebellar role in the modulation of affect (Heath, 1977; Berman et al., 1978; Schmahmann, 1996; J. D. Schmahmann and J. C. Sherman, this volume).

6. *Course of the Fiber Pathways to Pons*

The fibers destined for the basis pontis travel from their sites of origin through the white matter of the cerebral hemispheres toward the cerebral peduncle and assume a characteristic trajectory (Schmahmann and Pandya, 1992). The postrolandic parasensory association fibers travel toward the cerebral peduncle with a topographically organized arrangement and converge in the white matter of the posterior limb of the internal capsule above and medial to the midpoint of the lateral geniculate nucleus (LGN) (Fig. 6). The prefrontopontine fibers are also well organized in the prefrontal white matter and the anterior limb of the internal capsule as they pass to the cerebral peduncle before terminating in the basilar pons (Schmahmann and Pandya, 1994, 1997). Taken together with the observations concerning termination patterns of these associative corticopontine projections, it would appear that the corticopontine system consists of segregated and partially overlapping pathways, which are to some extent distinguishable anatomically at each stage of their trajectory from origin to destination.

B. PONTOCEREBELLAR PROJECTIONS

Only limited information is available regarding the pontocerebellar projection in the nonhuman primate. The anatomical evaluation of the cerebrocerebellar communication has been more completely studied in cat than in monkey, and it is apparent that species differences limit the validity of these findings with respect to the organization of these pathways in human. Precisely where in the cerebellum the associative and paralimbic inputs that project to the pons are distributed still needs to be elucidated. Certain principles of organization have nevertheless been established. For example, it is known from physiological studies that the parietal and prefrontal cortices are functionally related to the neocerebellar hemispheres, and auditory and visual inputs are received in vermal lobules VI and VII (Allen and Tsukuhara, 1974). Anatomic and physiologic studies in the monkey indicate that the dorsal paraflocculus, the uvula, and the vermal visual area [vermal lobule VII of Larsell (1970)] receive information from visually responsive neurons in the dorsolateral pontine region and the nucleus reticularis tegmenti pontis (Brodal, 1979, 1980; Stein and Glickstein, 1992; Glickstein et al., 1995).

In his horseradish peroxidase (HRP) retrograde labeling study of the pontocerebellar projection in the monkey, Brodal (1979) determined that the anterior lobe (mainly lobule V of Larsell) receives input from medial parts of the caudal pons; the vermal visual area (lobules VII–VIIIA) from two cell groups located in the dorsomedial and dorsolateral pons; vermal lobule VIIIB from the intrapeduncular nucleus; crus I of the ansiform lobule from medial parts of the rostral pons; and crus II from the medial, ventral, and lateral pons. The hemispheres have relatively greater pontine input than the rostral vermis. Based on prior corticopontine studies (Brodal, 1978), Brodal concluded that the anterior lobe and lobulus simplex (Larsell lobes I–VI) receive afferents from the motor and premotor cortices and, to a small extent, from the parietal lobe. The premotor and prefrontal cerebral regions are linked with crus I of the cerebellar ansiform lobule (Larsell VII–VIII), the motor cortex is linked with crus II, and, in agreement with earlier physiological work of Allen and Tsukuhara (1974) and Sasaki et al., (1975), the somatosensory and parietal association areas are linked with the paramedian lobule (Fig. 7).

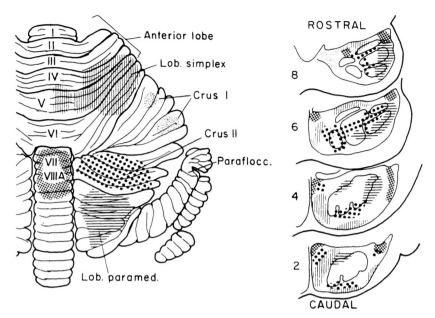

FIG. 7. Summary diagram of pontocerebellar projections in the monkey (reprinted from Brodal, 1979; with permission). In that study, tracer was injected into the cerebellar cortex and the distribution of labeled neurons in the pons was noted. Each cerebellar region appears to receive input from distinct and only partially overlapping sets of neurons in the pons.

From the clustering of labeled cells in the pontocerebellar projection, Brodal (1979) concluded that there was a high degree of order, with each cerebellar subdivision receiving input at least partly from its own pontine territory. One small part of the cerebellum receives input from several discrete pontine cell groups situated far apart, a finding that has been observed in our preliminary investigations of pontocerebellar projections (Fig. 8). In addition, based on the divergent and convergent patterns of corticopontine and pontocerebellar projections, Brodal (1979) concluded that information from one small part of the cerebral cortex is distributed to numerous discrete sites in the cerebellar cortex, where it is combined with other specific kinds of information.

These general organizational principles notwithstanding, detailed understanding of the pontocerebellar system is still not available. Much remains to be elucidated regarding the details of the pontine afferents to defined regions of the cerebellum and with respect to the cerebral and cerebellar connections of individual basilar pontine regions. There is essentially no information available, for example, concerning the transfer of

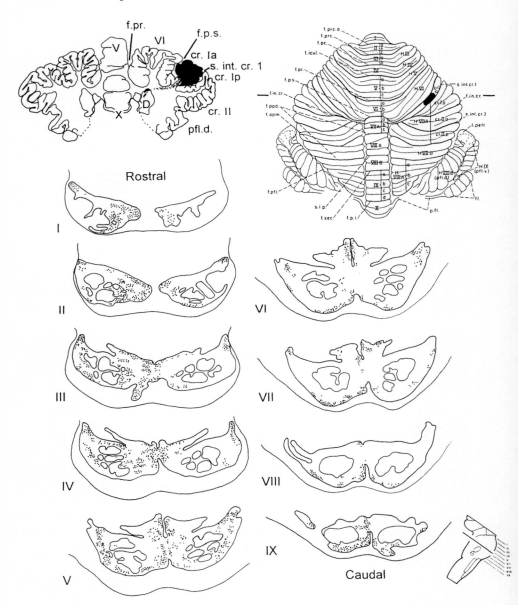

FIG. 8. The distribution of labeled neurons (black dots) in the basilar pons following injection of tracer (WGA-HRP, black shading) into crus I anterior of hemisphere lobule VIIA of a rhesus monkey cerebellum (from Schmahmann, 1996; with permission). The transverse plane (top left) and flattened map of the cerebellum (top right) show the injection site. Transverse levels I through IX of the pons (plane of section at lower right) show labeled

associative information from the pons to the cerebellum. Higher order information is distributed in complex but specific patterns throughout the basilar pons, but the manner in which this information is conveyed to the cerebellum and the corresponding organization within the cerebellum have not yet been studied. Furthermore, the fractured somatotopy that has been discerned in the sensory afferents to the cerebellum (Kassel et al., 1984; Bower and Kassel, 1990) may apply to the associative system as well, but this possibility has not been evaluated.

III. The Feedback Limb of the Cerebrocerebellar System

The feedback loop of the cerebrocerebellar system is composed of the cerebellar corticonuclear projection, efferents from deep cerebellar nuclei *en passant* through the red nucleus to the thalamus, and the thalamocortical relay (Fig. 1). The intricacies of the cerebellar cortex itself are beyond the scope of this discussion except to state that elegant models of cerebellar function (Marr, 1969; Albus, 1971; Ito, 1982) have been based on the structural consistency of the cortex and its physiology (Eccles et al., 1967; Thach, 1968; Palay and Chan-Palay, 1974). Neurotransmitter/modulator/peptide differences in neuronal subtypes of cerebellar cortex are increasingly being identified (Oertel, 1993) and a mediolateral zonal pattern of organization of the cortex has been defined (Voogd, 1967; Oscarsson, 1979; Dore et al., 1990) that correlates with connectional specificity in the olivary projections to the cerebellum (Voogd and Bigaré, 1980). These chemical–morphological variations provide some hope that the otherwise homogeneous-appearing cortex can be subdivided by methods other than gross anatomic descriptions and topographically organized connectional relationships.

The corticonuclear projection consists of axons of the cerebellar Purkinje cells, the only neuron responsible for efferents from the cerebellar cortex, that traverse the cerebellar white matter and terminate in the deep cerebellar nuclei. The topographic arrangement of the corticonuclear pro-

---

neurons bilaterally, but with a contralateral predominance. Neurons are distributed in multiple, but distinct pontine regions following the injection in this single folium. This arrangement seems to allow for many cerebral areas to communicate with a single folium. Incidentally noted is the anterograde transport of label from the injection to the dentate nucleus. cr. Ia, crus I anterior; cr. Ip, crus I posterior; cr. II, crus II; D, dentate nucleus; f.pr., primary fissure; f.p.s., superior posterior fissure; s.int.cr. I, internal sulcus of crus I. Roman numerals V, VI, and X refer to the cerebellar lobules according to Larsell (1970).

jection is such that the midline cortex projects to medial nuclear regions (fastigial nucleus), the lateral hemisphere projects to the dentate, and the intervening cortex corresponds with the interposed nuclei in a predictable mediolateral pattern. The flocculonodular lobe additionally has direct connections with the vestibular nuclei, and the anterior interpositus with the red nucleus (Jansen and Brodal, 1940; Brodal, 1981; Haines et al., 1982). Ito (1982) utilized the repeating sequence of cortical organization and the predictable corticonuclear arrangement to postulate the concept of a corticonuclear microcomplex acting as the essential functional unit of the cerebellum.

Dow (1942, 1974) drew attention to the differential organization of the dentate nucleus in humans and anthropoid apes as compared to that of lower primates and subprimate species. Referencing earlier work in the field, he noted that the dentate nucleus "in man and anthropoid apes consists of two parts, a dorsomedial microgyric, magnocellular older part, which is homologous to the dentate nucleus of lower forms, and a very much expanded new part which comprises the bulk of the dentate nucleus in man and higher apes, the ventro-lateral macrogyric parvi-cellular part." Dow expanded further on how these two parts of the dentate differ with respect to a number of morphologic and embryologic properties and then postulated, marshaling some early physiology and degeneration studies in humans, that the newer part of the dentate (the "neodentate") expanded in concert with, and was connected to, the frontal, temporal, and parietal association areas of higher primates and humans.

At the time that Dow formulated these hypotheses, it was understood that cerebellar-thalamic projections arose exclusively from the dentate nucleus and were conveyed through the ventrolateral thalamic nucleus to the motor cortex (Henneman et al., 1952). Subsequent studies employing newer anatomic techniques demonstrated that the dentate may be assisted in this role by thalamic efferents also from the fastigial and the interpositus nuclei (Batton et al., 1977; Stanton, 1980; Brodal, 1981). A more detailed understanding is needed regarding the precise topographical relationships between each cerebellar nucleus and its corresponding complement of thalamic terminations. Certain principles of organization of the cerebello-thalamic projection have been defined, however. There appear to be differential anterior versus posterior dentate nucleus projections to thalamus, and each cerebellar nuclear region projects to a few (between 3 and 7) rostrocaudally oriented rod-like aggregates situated within a dorsoventral curved lamella in the thalamus (Thach and Jones, 1979).

The classic cerebellar recipient motor thalamic nuclei [the pars oralis of the ventral posterolateral nucleus, VPLo; the caudal and pars postrema aspects of the ventrolateral nucleus, VLc and VLps; and nucleus X, in the terminology of Olszewski (1952)] are not alone in receiving input from the cerebellum. Nonmotor thalamic nuclei have a considerable cerebellar input as well. These include the intralaminar nuclei, particularly centralis lateralis (CL), as well as the paracentralis (Pcn) and centromedian–parafascicular (CM-Pf) complex, and the medial dorsal nucleus (Strick, 1976; Batton *et al.*, 1977; Thach and Jones, 1979; Stanton, 1980; Kalil, 1981; Wiesendanger and Wiesendanger, 1985; Ilinsky and Kultas-Ilinsky, 1987; Orioli and Strick, 1989). The CL nucleus, like other intralaminar nuclei, has widespread cortical connections, including the posterior parietal cortex, the multimodal regions of the upper bank of the superior temporal sulcus, the prefrontal cortex, the cingulate gyrus, and the primary motor cortex (Kievit and Kuypers, 1977; Yeterian and Pandya, 1985, 1989; Vogt and Pandya, 1987; Schmahmann and Pandya, 1990; Siwek and Pandya, 1991); and Pcn nucleus projections include the parahippocampal gyrus (G. Blatt, D. L. Rosene, and D. N. Pandya, personal communication) (Fig. 9).

The medial dorsal (MD) thalamic nucleus, which is the major site of thalamic connections with the frontal lobe, also receives cerebellar input.

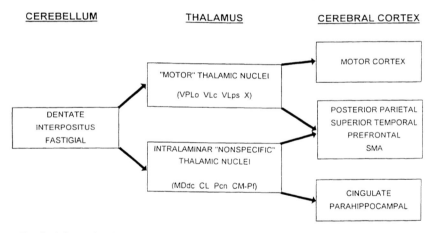

FIG. 9. Schematic of the feedback limb of the cerebrocerebellar system. Deep cerebellar nuclei project to traditionally "motor" thalamic nuclei (VPLo, VLc, VLps, and X), as well as to "nonspecific" intralaminar (CL, Pcn, CM-Pf) and medial dorsal (MDdc) nuclei. Motor nuclei, in turn, project to motor and premotor cortex, but also to the supplementary motor area and prefrontal, posterior parietal, and multimodal temporal regions. Intralaminar nuclei (including MDdc) have widespread projections, including to association and limbic cortices (reprinted from Schmahmann, 1994; with permission). CL, centralis lateralis; CM, centromedian; MDdc, medial dorsal nucleus, pars densocellularis; Pcn, paracentralis; Pf, parafascicularis; SMA, supplementary motor area; VLc, ventral lateral, pars caudalis; VLo, ventral lateral, pars oralis; VLps, ventral lateral, pars postrema; X, nucleus X (thalamic terminology of Olszewski, 1952).

It receives projections from the cerebellum mainly in its paralaminar parts, i.e., in the laterally situated pars multiformis (MDmf), and more caudally in the pars densocellularis (MDdc) (Stanton, 1980; Ilinsky and Kultas-Ilinsky, 1987). These, in turn, have reciprocal connections with area 8, area 46 at both banks of the principal sulcus, and area 9 in the frontal lobe (Giguere and Goldman-Rakic, 1988; Barbas et al., 1991; Siwek and Pandya, 1991), but also with the cingulate gyrus, posterior parietal cortex, and multimodal parts of the superior temporal sulcus (Yeterian and Pandya, 1985, 1989; Vogt and Pandya, 1987; Schmahmann and Pandya, 1990).

The traditionally motor thalamic nuclei have projections to regions of the cerebral cortex outside the primary and supplementary motor areas, including the prefrontal periarcuate cortex (Kievit and Kuypers, 1977; Künzle and Akert, 1977; Stanton et al., 1988). This was confirmed more directly in a transsynaptic retrograde tracer study of area 46 of the prefrontal lobe (cerebral cortex injected with tracer, label followed back to neurons in thalamus and further back to cerebellar dentate neurons) (Middleton and Strick, 1994). The upper bank of the superior temporal sulcus and the superior temporal region have reciprocal connections with the intralaminar nuclei (CL, Pcn, MDdc) (Yeterian and Pandya 1989, 1991) and receive projections from the ventrolateral nucleus (Yeterian and Pandya, 1989). The posterior parietal cortex receives projections from the ventrolateral nucleus (VLps more than VLc), as well as from nucleus X and VPLo (Schmahmann and Pandya, 1990) (Fig. 9).

Using direct transneuronal techniques, it remains to be shown how much of the cerebellar input to thalamus is conveyed to these associative cortices. Use of the novel viral tracer techniques (see F. A. Middleton and P. L. Strick, this volume) should prove valuable in refining current notions of the feedback limb of the cerebro-cerebellar system. Nevertheless, it would appear from available anatomic evidence that the cerebellar recipient "motor" thalamic nuclei project not only to the motor cortices, but also to the associative areas in the posterior parietal, superior temporal, and prefrontal cortice. Furthermore, the intralaminar nuclei, which are themselves a recipient of cerebellar efferents, project widely throughout the cerebral cortex, including the motor, associative, and paralimbic regions.

IV. Climbing Fibers and Cognition: Is There an Anatomic Substrate?

A central feature of the Marr (1969)–Albus (1971) theory of motor learning is the interaction between mossy fiber and climbing fiber systems. It has been suggested that learning is an important mechanism whereby the cerebellum also modulates nonmotor behavior. Mossy fibers to the cerebellum arise largely from neurons in the basilar pons. The inferior olive is the sole source of the climbing fiber input to the cerebellum. The cerebral afferents of the pontine (mossy fiber) and olivary (climbing fiber) systems are markedly different. The pontine system, as described earlier, is derived in large part from the cerebral hemispheres, including the association areas. In the nonhuman primate, the inferior olive receives much of its descending input from the parvicellular red nucleus. Afferents of the parvicellular red nucleus are derived most heavily from motor, premotor, and supplementary motor cortices and to some extent from the postcentral gyrus and area 5 in the superior parietal lobule. They are not derived to any convincing degree (at least in studies to date) from the associative or paralimbic cortices (Kuypers and Lawrence 1967; Saint-Cyr and Courville 1980; Humphrey *et al.*, 1984; Kennedy *et al.*, 1986). Archambault (1914) reported rubral connections with the infratemporal cortices in humans. This improbable pathway has not been confirmed, however, and cannot reliably be used at this time. The zona incerta, which projects to the inferior olive (Saint-Cyr and Courville, 1980; Cintas *et al.*, 1980), has been reported to receive projections from prefrontal cortices (Kuypers and Lawrence 1967; Shammah-Lagnado *et al.*, 1985), suggesting that there may be some indirect prefrontal input to the olivary system.

Shah, Schmahmann, and Pandya (unpublished observations) investigated this question of the nature of the cerebral afferents to the red nucleus and the zona incerta. A preliminary review of previously performed anterograde tracer experiments in the monkey reveals that there are prominent and topographically organized projections from the precentral motor cortex to the parvicellular and magnocellular divisions of the red nucleus. Additionally, there is substantial input from the supplementary motor area to the parvicellular division. In contrast, no projections to the red nucleus (parvicellular or magnocellular divisions) were seen to arise from associative or paralimbic cerebral cortices. However, significant projections to the zona incerta were observed from the cingulate cortex, as well as from the posterior parietal, prefrontal, and parastriate association areas. The rostral cingulate cortex, area 24, showed projections throughout the rostral to caudal extent of the zona incerta. More limited but definite projections to the zona incerta were noted from areas PF and PG in the inferior parietal lobule, as well as from area PGm at the medial convexity of the superior parietal lobule. Zona incerta projections were observed to arise from the medial prestriate cortex bordering area PGm and area PO (subdivision of area 19). Additionally, prefrontal cortex projections to the zona incerta arose from area 9/46d as well as from area 9 at the medial convexity.

The possibility of interaction between mossy fiber and climbing fiber systems in learning nonmotor tasks is therefore maintained by virtue of the associative projections to the zona incerta, which in turn projects to the inferior olivary nucleus. The Marr–Albus hypothesis has been substantiated for motor learning, and this theory could conceivably be adapted for nonmotor learning and other higher order functions as well.

V. Conclusions

The new understanding of the anatomy of the cerebrocerebellar system presented in this chapter is consistent with the hypothesis that the cerebellum is incorporated into the neural circuitry subserving cognitive and affective operations. The anatomic circuitry that links the associative and paralimbic cerebral cortices with the cerebellum appears to be directed in both a feedforward and a feedback manner. Those cerebral areas that commit efferents to the cerebellum via the corticopontine circuit receive input back from the cerebellum by way of the thalamus. This system is composed of multiple loops or channels that are highly organized and topographically precise, despite their complexity. These anatomic avenues facilitate cerebrocerebellar communication concerning multiple specific kinds of highly processed, multimodal information. These channels of communication in the cerebrocerebellar system are reminiscent of the multiple parallel but partially overlapping circuits described between the frontal lobe and the basal ganglia (Alexander and Crutcher, 1990; Goldman-Rakic, 1988). Both of these major circuits (cerebral–cerebellar and cerebral–basal ganglia) appear to be discretely organized into anatomical subsystems. In addition, they both (as postulated here for the cerebellum) contribute to, and are integral components of, differentially organized functional subsystems within the framework of distributed neural circuits. Dow and Moruzzi (1958, p. 371) believed that "the same anatomical tools may be utilized, alone or in combination, for quite different purposes" in the cerebellar cortex. Based on the findings presented in this chapter, the proposed net effect of these multiple streams of diverse information reaching into and being sent back from the cerebellum is that the cerebellum integrates multiple internal representations with external stimuli and self-generated responses. The cerebellar contribution to these different subsystems permits the ultimate production of harmonious sensorimotor, cognitive, and affective/autonomic behaviors.

The mechanism of the cerebellar contribution to behavior remains to be determined. In agreement with the earliest investigators of the cerebellar contribution to voluntary movement (Fluorens, 1824), the role of the cerebellum in cognition, affect, and autonomic function is viewed to be one of modulation rather than generation, i.e., the cerebellum serves as an oscillation dampener, maintaining function steadily around a homeostatic baseline and smoothing out performance (Schmahmann, 1996). It is possible that the cerebellum serves to correlate motor acts with mood states and unconscious motivation, thus facilitating nonverbal communication. It may also transpire that in the same way as the cerebellum regulates the rate, force, rhythm, and accuracy of movements, so may it regulate the speed, capacity, consistency, and appropriateness of mental or cognitive processes. Dysmetria of movement is then matched by an unpredictability and illogic to social and societal interaction. The overshoot and inability in the motor system to check parameters of movement may thus be equated in the cognitive/affective realm with "dysmetria of (or ataxic) thought," a mismatch between reality and perceived reality, and erratic attempts to correct errors of thought or behavior.

References

Aas, J. E., and Brodal, P. (1988). Demonstration of topographically organized projections from the hypothalamus to the pontine nuclei: An experimental study in the cat. *J. Comp. Neurol.* **268**, 313–328.

Albus, J. S. (1971). A theory of cerebellar function. *Math. Biosc.* **10**, 25–61.

Alexander, G. E., and Crutcher, M. D. (1990). Substrates of parallel processing. *Trends Neurosci.* **13**, 266–271.

Allen, G. I., and Tsukuhara, N. (1974). Cerebrocerebellar communication systems. *Physiol. Rev.* **54**, 957–1008.

Anand, B. K., Malhotra, C. L., Singh, B., and Dua, S. (1959). Cerebellar projections to the limbic system. *J. Neurophysiol.* **22**, 451–458.

Archambault, L. (1914–1915). Les connexiones corticales du noyau rouge. *Nouv. Iconograph. Salpetr.* **27**, 187–225.

Astruc, J. (1971). Corticofugal connections of area 8 (frontal eye lid) in macaca mulatta. *Brain Res.* **33**, 241–256.

Barbas, H., Haswell Henion, T. H., and Cermon, C. R. (1991). Diverse thalamic projections to the prefrontal cortex in the rhesus monkey. *J. Comp. Neurol.* **313**, 65–94.

Barbas, H., and Mesulam, M.-M. (1985). Cortical afferent input to the principalis region of the rhesus monkey. *Neuroscience* **15**, 619–637.

Barbas, H, and Pandya, D. N. (1991). Architecture and connections of the prefrontal cortex in rhesus monkey. *In* "Frontal Lobe Function and Dysfunction" (H. S. Levin, H. M. Eisenberg, and A. L. Benton, eds.), pp. 35–58. Oxford University Press, New York.

Barnes, C. L., and Pandya, D. N. (1992). Efferent cortical connections of multimodal cortex of the superior temporal sulcus in the rhesus monkey. *J. Comp. Neurol.* **318**, 222–244.

Batton, R. R., III, Jayaraman, A., Ruggiero, D., and Carpenter, M. B. (1977). Fastigial efferent projections in the monkey: An autoradiographic study. *J. Comp. Neurol.* **174**, 281–306.

Bayliss, G. C., Rolls, E. T., and Leonard, C. M. (1987). Functional subdivisions of the temporal lobe neocortex. *J. Neurosci.* **7**, 330–342.

Berman, A. J., Berman, D., and Prescott, J. W. (1978). The effect of cerebellar lesions on emotional behavior in the rhesus monkey. In "The Cerebellum, Epilepsy and Behavior" (I. S. Cooper, M. Riklan, and R. S. Snider, eds.), pp. 277–284. Plenum, New York.

Boussaoud, D., Desimone, R., and Ungerleider, L. G. (1991), Visual topography of area TEO in the macaque. *J. Comp. Neurol.* **306,** 554–575.

Bower, J. M., and Kassel, J. (1990). Variability in tactile projection patterns to the cerebellar folia crus IIA of the Norway rat. *J. Comp. Neurol.* **302,** 768–778.

Brodal, A. (1981). "Neurological Anatomy in Relation to Clinical Medicine." Oxford University Press, New York.

Brodal, P. (1978). The corticopontine projection in the rhesus monkey: Origin and principles of organization. *Brain* **101,** 251–283.

Brodal, P. (1979). The pontocerebellar projection in the rhesus monkey: An experimental study with retrograde axonal transport of horseradish peroxidase. *Neuroscience* **4,** 193–208.

Brodal, P. (1980). The projection from the nucleus reticularis tegmenti pontis to the cerebellum in the rhesus monkey. *Exp. Brain Res.* **38,** 29–36.

Brodmann, K. (1909). "Vergleichende Lokalisationslehre der Grosshirnrinde in inhren Prinzipien dargestellt auf Grund des Zellenbaues." Leipzig: J A Barth, 1909, xii.

Cavada, C., and Goldman-Rakic, P. S. (1989). Posterior parietal cortex in rhesus monkey. II. Evidence for segregated corticocortical networks linking sensory and limbic areas with the frontal lobe. *J. Comp. Neurol.* **287,** 422–445.

Cintas, H. M., Rutherford, J. G., and Gwyn, D. G. (1980). Some midbrain and diencephalic projections to the inferior olive in the rat. In "The Inferior Olivary Nucleus: Anatomy and Physiology"(J. Courville, C. de Montigny, and Y. Lamarre, eds.), pp. 73–96. Raven Press, New York.

Colombo, M., D'Amato, M. R., Rodman, H. R., and Gross, C. G. (1990). Auditory association cortex lesions impair auditory short term memory in monkeys. *Science* **247,** 336–338.

Dempsey, C. W., Tootle, D. M., Fontana, C. J., Fitzjarrell, A. T., Garey, R. E., and Heath, R. G. (1983). Stimulation of the paleocerebellar cortex of the cat: Increased rate of synthesis and release of catecholamines at limbic sites. *Biol. Psychiat.* **18,** 127–132.

Denny-Brown, D., and Chambers, R. A. (1958). The parietal lobe and behavior. *Res. Publ. Nerv. Ment. Dis.* **36,** 35–117.

Desimone, R., and Ungerleider, L. G. (1986). Multiple visual areas in the caudal superior temporal sulcus of the macaque. *J. Comp. Neurol.* **248,** 164–189.

Desimone, R., and Ungerleider, L. G. (1989). Neural mechanisms of visual processing in monkeys. In "Handbook of Neuropsychology" (F. Boller and J. Grafman, eds.), Vol. 2, pp. 267–299. Elsevier, New York.

DeVito, J. L., and Smith, O. A. (1964). Subcortical projections of the prefrontal lobe of the monkey. *J. Comp. Neurol.* **123,** 413–424.

Dore, L., Jacobson, C. D., and Hawkes, R. (1990). Organization and postnatal development of Zebrin II antigenic compartmentation in the cerebellar vermis of the grey opossum, *Monodelphis domestica. J. Comp. Neurol.* **291,** 431–449.

Dow, R. S. (1942). The evolution and anatomy of the cerebellum. *Biol. Rev.* **17,** 179–220.

Dow, R. S. (1974). Some novel concepts of cerebellar physiology. *Mt. Sinai J. Med.* **41,** 103–119.

Dow, R. S., and Moruzzi, G. (1958). "The Physiology and Pathology of the Cerebellum." University of Minnesota Press, Minneapolis, MN.

Eblen, F., and Graybiel, A. M. (1995). Highly restricted origin of prefrontal cortical inputs to striosomes in the macaque monkey. *J. Neurosci.* **15,** 5999–6013.

Eccles, J. C., Ito, M., and Szentagothai, J. (1967). "The Cerebellum as a Neuronal Machine." Springer-Verlag, New York/Heidelberg.

Fuster, J. M. (1980). "The Prefrontal Cortex: Anatomy, Physiology and Neurophysiology of the Frontal Lobe." Raven Press, New York.

Galaburda, A. M., and Pandya, D. N. (1983). The intrinsic architectonic and connectional organization of the superior temporal region of the rhesus monkey. *J. Comp. Neurol.* **221,** 169–184.

Gattass, R., and Gross, C. G. (1981). Visual topography of striate projection zone (MT) in posterior superior temporal sulcus of the macaque. *J. Comp. Neurol.* **46,** 621–638.

Geschwind, N. (1965a). Disconnexion syndromes in animals and man. Part I. *Brain* **88,** 237–294.
Geschwind, N. (1965b). Disconnexion syndromes in animals and man. Part II. *Brain* **88,** 585–644.
Giguere, M., and Goldman-Rakic, P. S. (1988). Mediodorsal nucleus: Areal, laminar, and tangential distribution of afferents and efferents in the frontal lobe of rhesus monkeys. *J. Comp. Neurol.* **277,** 195–213.
Glickstein, M., Cohen, J. L., Dixon, B., Gibson, A., Hollins, M., LaBossiere, E., and Robinson, F. (1980). Corticopontine visual projections in macaque monkeys. *J. Comp. Neurol.* **190,** 209–229.
Glickstein, M., May, J. G., and Mercier, B. E. (1985). Corticopontine projection in the macaque: The distribution of labelled cortical cells after large injections of horseradish peroxidase in the pontine nuclei. *J. Comp. Neurol.* **235,** 343–359.
Goldman-Rakic, P. S. (1988). Topography of cognition: Parallel distributed networks in primate association cortex. *Annu. Rev. Neurosci.* **11,** 137–156.
Goldman-Rakic, P. S., and Friedman, H. R. (1991). The circuitry of working memory revealed by anatomy and metabolic imaging. In "Frontal Lobe Function and Dysfunction" (H. S. Levin, H. M. Eisenberg, and A. L. Benton, eds.), pp. 72–91. Oxford University Press, New York.
Haines, D. E., and Dietrichs, E. (1984). An HRP study of hypothalamo-cerebellar and cerebello-hypothalamic connections in squirrel monkey (*Saimiri sciureus*). *J. Comp. Neurol.* **229,** 559–575.
Haines, D. E., Patrick, G. W., and Satrulee, P. (1982). Organization of cerebellar corticonuclear fiber systems. In "The Cerebellum: New Vistas" (S. L. Palay and V. Chan-Palay, eds.), pp. 320–371. Springer-Verlag, Berlin.
Harper, J. W., and Heath, R. G. (1973). Anatomic connections of the fastigial nucleus to the rostral forebrain in the cat. *Exp. Neurol.* **39,** 285–292.
Heath, R. G. (1977). Modulation of emotion with a brain pacemaker. *J. Nerv. Ment. Dis.* **165,** 300–317.
Henneman, E., Cooke, P. M., and Snider, R. S. (1952). Cerebellar projections to the cerebral cortex. In "Patterns of Organization in the Central Nervous System" (P. Bard, ed.). *Res. Publ. Ass. Nerv. Ment. Dis.* **30,** 317–333.
Humphrey, D. R., Gold, R., and Reed, D. J. (1984). Sizes, laminar and topographic origins of cortical projections to the major divisions of the red nucleus in the monkey. *J. Comp. Neurol.* **225,** 75–94.
Hyvarinen, J. (1982). Posterior parietal lobe of the primate brain. *Physiol. Rev.* **62,** 1060–1129.
Ilinsky, I. A., and Kultas-Ilinsky, K. (1987). Sagittal cytoarchitectonic maps of *Macaca mulatta* thalamus with a reviewed nomenclature of the motor-related nuclei validated by observations on their connectivity. *J. Comp. Neurol.* **262,** 331–364.
Ito, M. (1982). Questions in modeling the cerebellum. *J. Theor. Biol.* **99,** 81–86.
Jansen, A., and Brodal, A. (1940). Experimental studies on the intrinsic fibers of the cerebellum. II. The cortico-nuclear projection. *J. Comp. Neurol.* **73,** 267–321.
Jones, E. G., and Powell, T. P. S. (1970). An anatomical study of converging sensory pathways within the cerebral cortex of the monkey. *Brain* **93,** 793–820.
Kalil, K. (1981). Projections of the cerebellar and dorsal column nuclei upon the thalamus of the rhesus monkey. *J. Comp. Neurol.* **195,** 25–50.
Kassel, J., Shambes, G. M., and Welker, W. (1984). Fractured cutaneous projections to the granule cell layer of the posterior cerebellar hemisphere of the domestic cat. *J. Comp. Neurol.* **225,** 458–468.
Kennedy, P. R., Gibson, A. R., and Houk, J. C. (1986). Functional and anatomic differentiation between parvicellular and magnocellular regions of red nucleus in the monkey. *Brain Res.* **364,** 124–136.
Kievet, J., and Kuypers, H. G. J. M. (1977). Organization of the thalamocortical connections to the frontal lobe in the rhesus monkey. *Exp. Brain Res.* **29,** 299–322.
Künzle, H., and Akert, K. (1977). Efferent connections of cortical area 8 (frontal eye field) in *Macaca fascicularis*: A reinvestigation using the autoradiographic technique. *J. Comp. Neurol.* **173,** 147–164.

Kuypers, H. G. J. M., and Lawrence, D. G. (1967). Cortical projections to the red nucleus and the brainstem in the rhesus monkey. *Brain Res.* **4,** 151–188.
Larsell, O. (1970). "The Comparative Anatomy and Histology of the Cerebellum from Monotremes through Apes" (J. Jansen, ed.). University of Minnesota, Minneapolis.
Leichnetz, G. R., Smith, D. J., and Spencer, R. F. (1984). Cortical projections to the paramedian and basilar pons in the monkey. *J. Comp. Neurol.* **228,** 388–408.
Luria, AR. (1966). "Higher Cortical Functions in Man." Basic Books, New York.
Marcinkiewicz, M., Morocos, R., and Chretien, M. (1989). CNS connections with the median raphe nucleus: Retrograde tracing with WGA-apoHRP-gold complex in the rat. *J. Comp. Neurol.* **289,** 11–35.
Marr, D. (1969). A theory of cerebellar cortex. *J. Physiol.* **202,** 437–470.
Martner, J. (1975). Cerebellar influences on autonomic mechanisms. *Acta Physiol. Scand. (Suppl.)* **425,** 1–42.
May, J. G., and Anderson, R. A. (1986). Different patterns of corticopontine projections from separate cortical fields within the inferior parietal lobule and dorsal prelunate gyrus of the macaque. *Exp. Brain Res.* **63,** 265–278.
Merzenich, M. M., and Brugge, J. F. (1973). Representation of the cochlear partition on the superior temporal plane of the macaque monkey. *Brain Res.* **50,** 275–296.
Mesulam, M. M. (1985). "Principles of Behavioral Neurology." Davis, Philadelphia.
Middleton, F. A., and Strick, P. L. (1994). Anatomical evidence for cerebellar and basal ganglia involvement in higher cognitive function. *Science* **266,** 458–451.
Milner, B. (1964). Some effects of frontal lobectomy in man. *In* "The Frontal Granular Cortex and Behavior" (J. M. Warren and K. Akert, eds.), pp. 313–334. McGraw Hill, New York.
Mishkin, M., Ungerleider, L. G., and Macko, K. A. (1983). Object vision and spatial vision: Two cortical pathways. *Trends Neurosci.* **6,** 414–417.
Mountcastle, V. B., Talbot, W. H., and Yin, T. C. T. (1977). Parietal lobe mechanisms for directed visual attention. *J. Neurophysiol.* **40,** 362–389.
Nadel, L. (1991). The hippocampus and space revisited. *Hippocampus* **1,** 221–229.
Nyby, O., and Jansen, J. (1951). An experimental investigation of the corticopontine projections in *Macaca mulatta*. Skrifter utgitt av det Norske Vedenskapsakademie: Oslo: 1. Mat. Naturv. Klasse **3,** 1–47.
Oertel, W. H. (1993). Neurotransmitters in the cerebellum: Scientific aspects and clinical relevance. *In* "Inherited Ataxias" (A. E. Harding and T. Duefel, eds.). *Adv. Neurol.* **61,** 33–75.
Olszewski, J. (1952). "The Thalamus of the *Macaca mulatta*." Karger, Basel.
Orioli, P. J., and Strick, P. L. (1989). Cerebellar connections with the motor cortex and the arcuate premotor area: An analysis employing retrograde transneuronal transport of WGA-HRP. *J. Comp. Neurol.* **288,** 621–626.
Oscarsson, O. (1979). Functional units of the cerebellum: Sagittal zones and microzones. *Trends Neurosci.* **2,** 143–145.
Palay, S. L., and Chan-Palay, V. (1974). "Cerebellar Cortex." Springer-Verlag, New York.
Pandya, D. N., and Kuypers, H. G. J. M. (1969). Cortico-cortical connections in the rhesus monkey. *Brain Res.* **13,** 13–16.
Pandya, D. N., Van Hoesen, G. W., and Mesulam, M.-M. (1981). Efferent connections of the cingulate gyrus in the rhesus monkey. *Exp. Brain Res.* **42,** 319–330.
Pandya, D. N., Seltzer, B., and Barbas, H. (1988). Input-output organization of the primate cerebral cortex. *In* "Comparative Primate Biology" (H. D. Steklis and J. Erwin, eds.), Vol. 4, pp. 39–80. A. R. Liss, New York.
Pandya, D. N., and Yeterian, E. G. (1991). Prefrontal cortex in relation to other cortical areas in rhesus monkey: Architecture and connections. *Prog. Brain Res.* **85,** 3–94.
Perrett, D., Rolls, E. T., and Caan, W. (1982). Visual neurons responsive to faces in the monkey temporal cortex. *Exp. Brain Res.* **47,** 329–342.
Petrides, M. (1995). Impairments of nonspatial, self-ordered, and externally ordered working memory tasks after lesions of the mid-dorsal part of the lateral frontal cortex in the monkey. *J. Neurosci.* **15,** 359–375.

Petrides, M., and Pandya, D. N. (1994). Comparative architectonic analysis of the human and the macaque frontal cortex. *In* "Handbook of Neuropsychology" (R. Boller and J. Grafman, eds., Vol. 9, pp. 17–57. Elsevier, New York.
Rauch, S. L., Jenike, M. A., Alpert, N. M., Baer, L., Breiter, H. C., Savage, C. R., and Fischman, A. J. (1994). Regional cerebral blood flow measured during symptom provocation in obsessive-compulsive disorder using oxygen 15-labeled carbon dioxide and positron tomography. *Arch. Gen. Psychiat.* **51,** 62–70.
Saint-Cyr, J. A., and Courville, J. (1980). Projections from the motor cortex, midbrain, and vestibular nuclei to the inferior olive in the cat: Anatomical and functional correlates. *In* "The Inferior Olivary Nucleus: Anatomy and Physiology" (J. Courville, C. DeMontigny, and Y. Lamarre, eds.), pp. 97–124. Raven Press, New York.
Sasaki, K., Oka, H., Matsuda, Y., Shimono, T., and Mizuno, N. (1975). Electrophysiological studies of the projections from the parietal association area to the cerebellar cortex. *Exp. Brain Res.* **23,** 91–102.
Schmahmann, J. D. (1991). An emerging concept: The cerebellar contribution to higher function. *Arch. Neurol.* **48,** 1178–1187.
Schmahmann, J. D. (1994). The cerebellum in autism: Clinical and anatomic perspectives. *In* "The Neurobiology of Autism" (M. L. Bauman and T. L. Kemper, eds.), pp. 195–226. Johns Hopkins University Press, Baltimore.
Schmahmann, J. D. (1996). From movement to thought: Anatomic substrates of the cerebellar contribution to cognitive processing. *Hum. Brain Mapp.* **4,** 174–198.
Schmahmann, J. D., and Pandya, D. N. (1989). Anatomical investigation of projections to the basis pontis from posterior parietal association cortices in rhesus monkey. *J. Comp. Neurol.* **289,** 53–73.
Schmahmann, J. D., and Pandya, D. N. (1990). Anatomical investigation of projections from thalamus to the posterior parietal association cortices in rhesus monkey. *J. Comp. Neurol.* **295,** 299–326.
Schmahmann, J. D., and Pandya, D. N. (1991). Projections to the basis pontis from the superior temporal sulcus and superior temporal region in the rhesus monkey. *J. Comp. Neurol.* **308,** 224–248.
Schmahmann, J. D., and Pandya, D. N. (1992). Fiber pathways to the pons from parasensory association cortices in rhesus monkey. *J. Comp. Neurol.* **326,** 159–179.
Schmahmann, J. D., and Pandya, D. N. (1993). Prelunate, occipitotemporal, and parahippocampal projections to the basis pontis in rhesus monkey. *J. Comp. Neurol.* **337,** 94–112.
Schmahmann, J. D., and Pandya, D. N. (1994). Trajectories of the prefrontal, premotor and precentral corticopontine fiber systems in the rhesus monkey. *Soc. Neurosci. Abstr.* **20,** 985.
Schmahmann, J. D., and Pandya, D. N. (1995a). The organization of the motor corticopontine projection in monkey. *Soc. Neurosci. Abstr.* **21,** 410.
Schmahmann, J. D., and Pandya, D. N. (1995b). Prefrontal cortex projections to the basilar pons: Implications for the cerebellar contribution to higher function. *Neurosci. Lett.* **199,** 175–178.
Schmahmann, J. D., and Pandya, D. N. (1997). Anatomic organization and functional implications of the basilar pontine projections from prefrontal cortices in rhesus monkey. *J. Neurosci.* **17,** 438–458.
Seltzer, B., and Pandya, D. N. (1984). Further observations on parieto-temporal connections in the rhesus monkey. *Exp. Brain Res.* **55,** 301–312.
Seltzer, B., and Pandya, D. N. (1989). Frontal lobe connections of the superior temporal sulcus in the rhesus monkey. *J. Comp. Neurol.* **281,** 97–113.
Shammah-Lagnado, S. J., Negrao, N., and Ricardo, J. A. (1985). Afferent connections of the zona incerta: A horseradish peroxidase study in the rat. *Neuroscience* **15,** 109–134.
Shook, L., Schlag-Rey, M., and Schlag, J. (1990). Primate supplementary eye field. I. Comparative aspects of mesencephalic and pontine connections. *J. Comp. Neurol.* **301,** 618–642.
Siwek, D. F., and Pandya, D. N. (1991). Prefrontal projections to the mediodorsal nucleus of the thalamus in the rhesus monkey. *J. Comp. Neurol.* **312,** 509–524.
Snider, R. S., and Maiti, A. (1976). Cerebellar contribution to the Papez circuit. *J. Neurosci. Res.* **2,** 133–146.

Spangler, W. J., Cosgrove, G. R., Ballantine, H. T. Jr., Casem, E. H., Rauch, S. L., Nierenberg, A., and Price, B. H. (1996). Magnetic resonance image-guided stereotactic cingulotomy for intractable psychiatric disease. *Neurosurgery* **38**, 1071–1076.

Stanton, G. B. (1980). Topographical organization of ascending cerebellar projections from the dentate and interposed nuclei in *Macaca mulatta:* An anterograde degeneration study. *J. Comp. Neurol.* **190**, 699–731.

Stanton, G. B., Goldberg, M. E., and Bruce, C. J. (1988). Frontal eye field efferents in the macaque monkey. II. Topography of terminal fields in midbrain and pons. *J. Comp. Neurol.* **271**, 493–506.

Stein, J. R., and Glickstein, M. (1992). Role of the cerebellum in visual guidance of movement. *Physiol. Rev.* **72**, 967–1017.

Strick, P. L. (1976). Anatomical analysis of ventrolateral thalamic input to primate motor cortex. *J. Neurophysiol.* **39**, 1020–1031.

Stuss, D. T., and Benson, D. F. (1986). "The Frontal Lobes." Raven Press, New York.

Thach, W. T. (1968). Discharge of Purkinje and cerebellar nuclear neurons during rapidly alternating arm movements in the monkey. *J. Neurophysiol.* **31**, 785–797.

Thach, W. T., and Jones, E. G. (1979). The cerebellar dentatothalamic connection: Terminal field, lamellae, rods and somatotopy. *Brain Res.* **169**, 168–172.

Ungerleider, L. G., and Mishkin, M. (1982). Two cortical visual systems. *In* "Analysis of Visual Behavior" (D. J. Ingle, M. A. Goodale, and R. J. W. Mansfield, eds.), pp. 549–586. MIT Press, Cambridge, MA.

Ungerleider, L. G., Desmone, R., Galkin, T. W., and Mishkin, M. (1984). Subcortical projections of area MT in the macaque. *J. Comp. Neurol.* **233**, 368–386.

Van Essen, D. C. (1985). Functional organization of primate visual cortex. *In* "Cerebral Cortex" (A. Peters and E. G. Jones, eds.), Vol. 3, pp. 259–329. Plenum, New York.

Vilensky, J. A., and Van Hoesen, G. W. (1981). Corticopontine projections from the cingulate cortex in the rhesus monkey. *Brain Res.* **205**, 391–395.

Vogt, B. A., and Pandya, D. N. (1987). Cingulate cortex of the rhesus monkey. II. Cortical afferents. *J. Comp. Neurol.* **262**, 271–289.

Voogd, J. (1967). Comparative aspects of the structure and fiber connexions of the mammalian cerebellum. *In* "Progress in Brain Research" (C. A. Fox and R. S. Snider, eds.), Vol 25, pp. 94–135. Elsevier, Amsterdam.

Voogd, J., and Bigaré, F. (1980). Topographical distribution of olivary and corticonuclear fibers in the cerebellum: A review. *In* "The Inferior Olivary Nucleus" (E. Courville, C. de Montigny, and Y. Lamarre, eds.), pp. 297–324. Raven Press, New York.

Watson, P. J. (1978). Nonmotor functions of the cerebellum. *Psychol. Bull.* **85**, 944–967.

Wiesendanger, R., Wiesendanger, M., and Ruegg, D. G. (1979). An anatomical investigation of the corticopontine projection in the primate (*Macaca fascicularis* and *Saimiri sciureus*). II. The projection from the frontal and parietal association areas. *Neuroscience* **4**, 747.

Wiesendanger, R., and Wiesendanger, M. (1985). The thalamic connections with medial area 6 (supplementary motor cortex) in the monkey (*Macaca fascicularis*). *Exp. Brain Res.* **59**, 91–104.

Yeterian, E. H., and Pandya, D. N. (1985). Corticothalamic connections of the posterior parietal cortex in the rhesus monkey. *J. Comp. Neurol.* **237**, 408–426.

Yeterian, E. H., and Pandya, D. N. (1989). Thalamic connections of the cortex of the superior temporal sulcus in the rhesus monkey. *J. Comp. Neurol.* **282**, 80–97.

Yeterian, E. H., and Pandya, D. N. (1991). Corticothalamic connections of the superior temporal sulcus in rhesus monkeys. *Exp. Brain Res.* **83**, 268–284.

CEREBELLAR OUTPUT CHANNELS

Frank A. Middleton* and Peter L. Strick*,†

†Veterans Administration Medical Center, and Departments of †Neurosurgery and
*,† Physiology, State University of New York Health Science Center at Syracuse, Syracuse,
New York 13210

I. Introduction
II. Anatomical Studies
   A. Cerebellar Output to Skeletomotor and Oculomotor Areas of Cerebral Cortex
   B. Cerebellar Output to Prefrontal Cortex
III. Physiological Studies
   A. Neuron Recording in Awake Trained Primates
   B. Functional Magnetic Resonance Imaging of the Dentate in Human Subjects
IV. Synthesis
   References

The cerebellum has long been regarded as involved in the control of movement, in part through its connections with the cerebral cortex. These connections were thought to combine inputs from widespread regions of the cerebral cortex and "funnel" them into the motor system at the level of the primary motor cortex. Retrograde transneuronal transport of herpes simplex virus type 1 has recently been used to identify areas of the cerebral cortex that are "directly" influenced by the output of the cerebellum. Results suggest that cerebellar output projects via the thalamus to multiple cortical areas, including premotor and prefrontal cortex, as well as the primary motor cortex. In addition, the projections to different cortical areas appear to originate from distinct regions of the deep cerebellar nuclei. These observations have led to the proposal that cerebellar output is composed of a number of separate "output channels." Evidence from functional imaging studies in humans and single neuron recording studies in monkeys suggests that individual output channels are concerned with different aspects of motor or cognitive behavior.

I. Introduction

It is well established that inputs to the cerebellum arise from multiple areas of the cerebral cortex, including portions of the frontal, parietal, and temporal lobes (e.g., Brodal, 1978; Vilensky and Van Hoesen, 1981; Leichnetz et al., 1984; Glickstein et al., 1985; Schmahmann and Pandya, 1991, 1993, 1995). However, the output of the cerebellum from the deep nuclei was thought to terminate in a single region of the ventrolateral thalamus (e.g., Kemp and Powell, 1971; Asanuma et al., 1983). This thalamic region was believed to project exclusively upon a single cortical area, the primary motor cortex (M1). Thus, according to this view, the function of cerebellar loops with the cerebral cortex was to collect information from widespread areas of cerebral cortex and "funnel" this information into the motor system for use in initiating movement and defining movement parameters (e.g., Evarts and Thach, 1969; Kemp and Powell, 1971; Allen and Tsukahara, 1974; Brooks and Thach, 1981; Asanuma et al., 1983; Ito, 1984).

A number of observations have led some investigators to challenge this point of view. For example, Leiner et al. (1987, 1989, 1991, 1993, and this volume) have suggested that cerebellar output is directed to prefrontal, as well as motor, areas of the cerebral cortex. They noted that, in humans and apes, the dentate nucleus of the cerebellum appears to have increased in size in parallel with the frontal lobe. They argued that this enlargement of the dentate has enabled it to expand its influence beyond the primary motor cortex. In support of their proposal, it is now apparent that cerebellar projections to the thalamus are not limited to a single region of the ventrolateral thalamus, but target other thalamic nuclei as well (e.g., Percheron, 1977; Stanton, 1980; Kalil, 1981; Yamamoto et al., 1992). Some of these thalamic nuclei project to cortical areas other than the primary motor cortex (e.g., Kievit and Kuypers, 1977; Miyata and Sasaki, 1983; Schell and Strick, 1984; Wiesendanger and Wiesendanger, 1985a; Goldman-Rakic and Porrino, 1985; Matelli et al., 1989; Schmahmann and Pandya, 1990; Barbas et al., 1991; Yamamoto et al., 1992; Rouiller, et al., 1994; Lynch et al., 1994; Middleton and Strick, 1994).

Clearly, one of the major unresolved issues of cerebro-cerebellar circuitry is defining the cortical areas that are the targets of cerebellar output. If the cerebellum is to influence cognition or perception, as well as motor control, it must do so through projections from the deep nuclei to thalamocortical circuits concerned with these aspects of behavior. Thus, it is not sufficient to show that the cerebellum receives input from diverse cortical areas. Others have argued that such input functions simply to guide movement (e.g., Evarts and Thach, 1969; Kemp and Powell, 1971; Allen and Tsukahara, 1974; Brooks and Thach, 1981; Asanuma et al., 1983). Instead, the anatomical argument for a cerebellar influence on activities other than the control of movement parameters must be based on the demonstration that cerebellar output targets diverse areas of cerebral cortex such as premotor, prefrontal, and posterior parietal cortex (see also Sasaki *et al.*, 1976, 1979).

In the past, a number of technical limitations have made it difficult to define cerebello-thalamocortical circuits. For example, most studies which examined the pattern of cerebellar terminations in the thalamus did not determine the cortical targets of these thalamic nuclei (however, see Hendry, *et al.*, 1979; Yamamoto *et al.*, 1992; Rouiller *et al.*, 1994). In addition, the lack of standard criteria for defining thalamic borders and a confusing thalamic nomenclature have made comparison of the results from different studies difficult. Consequently, with few exceptions, it has not been possible to determine the full extent of the cortex "directly" influenced by cerebellar output.

We have developed a tracing technique, transneuronal transport of herpes simplex virus type 1 (HSV1), which overcomes many of these problems (see Zemanick *et al.*, 1991; Strick and Card, 1992). This chapter reviews the rationale for using HSV1 as a transneuronal tracer and then presents some recent findings on the organization of cerebellar projections to the frontal lobe. These results indicate that cerebellar output targets not only the primary motor cortex, but also several areas of premotor, oculomotor, and prefrontal cortex. In addition, the projections to these cortical areas appear to originate from distinct regions of the cerebellar nuclei. Thus, we propose that the output from the cerebellum, and specifically that from the dentate, contains multiple "output channels," each of which projects to a distinct cortical area (Strick *et al.*, 1993; Middleton and Strick, 1994, 1996). Section III presents some of our physiological observations that indicate individual output channels are concerned with different aspects of motor or cognitive behavior (Mushiake and Strick, 1993, 1995; Strick *et al.*, 1993; Kim *et al.*, 1994).

II. Anatomical Studies

Transneuronal transport of HSV1 provides a novel method for labeling a chain of synaptically linked neurons (for references and review see Zemanick *et al.*, 1991; Strick and Card, 1992). In fact, this technique is capable of identifying circuits at least three neurons in length (Hoover and Strick, 1993b). In our anatomical studies, we have employed two different strains of HSV1, the H129 and McIntrye-B strains. To test the transport characteristics of these strains, we made localized injections into the arm area of the primary motor cortex of cebus monkey (Figs. 1 and 2). We found that the

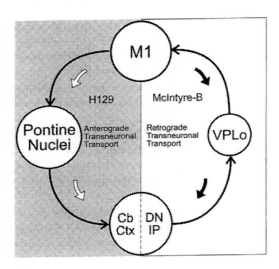

FIG. 1. Patterns of HSV1 transneuronal transport in cerebellar circuits. Different strains of HSV1 are transported transneuronally in different directions. The H129 strain is transported transneuronally in the anterograde direction. After injections of this strain into the primary motor cortex (M1), virus moves from the injection site to label second-order neurons in the pontine nuclei, and then third-order neurons in the cerebellar cortex (Cb Ctx) and the dentate (DN) and interpositus (IP) nuclei. In contrast, the McIntyre-B strain is transported transneuronally in the retrograde direction. After injections of the McIntyre-B strain into M1, virus moves from the injection site to label first-order neurons in the ventrolateral thalamus (VPLo), and then second-order neurons in DN and IP.

H129 and McIntyre-B strains are transported transneuronally in different directions (Zemanick et al., 1991) (Fig. 1).

The H129 strain is transported transneuronally in the anterograde direction. Three days after injections of this virus into the arm area of M1, virus was transported from "first-order" neurons in the injection site to "second-order" neurons in regions of the pontine nuclei known to receive input from the arm area of M1 (Fig. 3A) (e.g., Brodal, 1978; Glickstein et al., 1985). Five days after these injections, multiple patches of "third-order" neurons, labeled by anterograde transneuronal transport, were found in the cerebellar cortex. These patches were located in the granular layer and contained two types of labeled neurons: granule and Golgi cells (Fig. 3B). Both cell types are known to be contacted by mossy fiber afferents that project to the cerebellar cortex from the pontine nuclei (e.g., Allen and Tsukahara, 1974; Brooks and Thach, 1981; Ito, 1984). The majority of the labeled patches were located in vermal and hemispheric lobules V and VI, in and adjacent to the primary fissure (Figs. 3C and 4). Separate labeled patches were found posteriorly in the paramedian lobule (VIIIA) and laterally in lobule VIIB (Figs. 3C and 4).

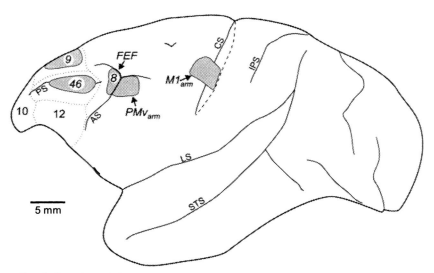

FIG. 2. Location of virus injection sites in the cerebral cortex. Lateral view of a cebus monkey brain. The shaded areas indicate the spread of HSV1 from injections into each cortical area. The numbers 8, 9, 10, 12, and 46 refer to cytoarchitectonic areas of the frontal lobe according to Walker (1940). The dotted lines define the borders between areas. AS, arcuate sulcus; CS, central sulcus; FEF, frontal eye field; IPS, intraparietal sulcus; LS, lateral sulcus; M1$_{arm}$, arm region of the primary motor cortex; PMv$_{arm}$, arm region of the ventral premotor area; PS, principal sulcus; STS, superior temporal sulcus.

Some labeled neurons were also found in portions of the dentate and interpositus nuclei. Evidence shows that the pontine nuclei project directly to the deep cerebellar nuclei in cats (Shinoda *et al.*, 1987); a similar pathway is thought to exist in primates. Thus, we found labeled neurons at third-order sites where one might expect to see them based on the results of prior studies using conventional tracing methods. In summary, the regions of cerebellar cortex and deep nuclei containing labeled neurons correlated well with the sites where evoked potentials have been recorded after stimulation of the arm area of the primary motor cortex (e.g., Sasaki *et al.*, 1977). These results suggest that it will be possible to define how many areas of the cerebral cortex map onto the cerebellar cortex and deep nuclei using the H129 strain as an anterograde transneuronal tracer.

In contrast to the H129 strain, the McIntyre-B strain of HSV1 is transported transneuronally in the retrograde direction (Zemanick *et al.*, 1991; Hoover and Strick, 1993a,b; Middleton and Strick, 1994) (Fig. 1). Three days after injections of this virus into the arm area of M1 (Fig. 2), many labeled neurons were found in subdivisions of the ventrolateral thalamus that are known to innervate M1 (Fig. 5A), such as the nucleus ventralis

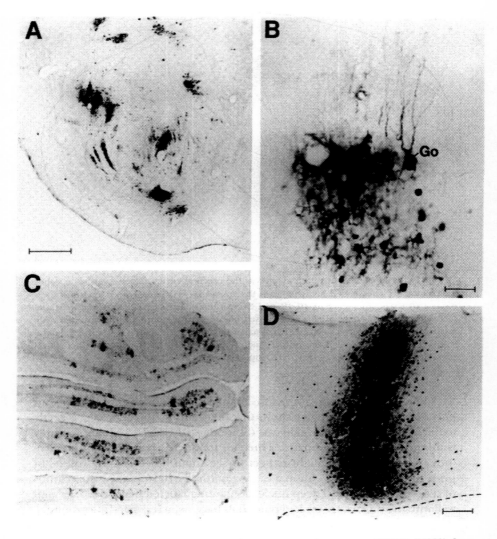

FIG. 3. Neurons labeled by anterograde transneuronal transport of HSV1 (H129) from M1. (A) Labeled neurons in the pontine nuclei (scale bar, 1 mm). (B) A patch of labeled granule cells and a Golgi cell (Go) in cerebellar cortex (scale bar, 30 μm). (C) Multiple patches of labeled neurons in the granular layer of cerebellar cortex (scale in A). (D) A "column" of labeled neurons in cerebral cortex buried within the dorsal bank of the cingulate sulcus. The dashed line indicates the border between white and gray matter (scale bar, 300 μm). (From Zemanick et al., 1991.)

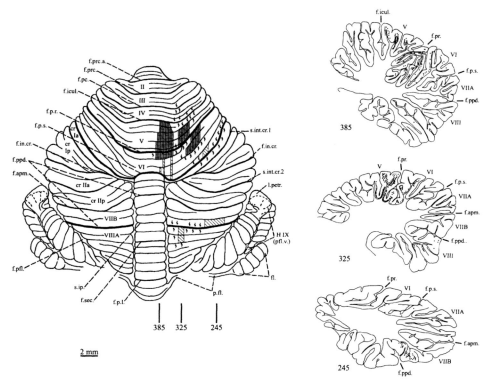

FIG. 4. Distribution of labeled neurons in cerebellar cortex after anterograde transneuronal transport of HSV-1 (H129) from the "arm area" of the primary motor cortex. (Left) Surface reconstruction of the distribution of labeled neurons found on the surface of the cerebellar cortex. The relative density and location of cells are indicated by the density and location of the cross-hatching. Small arrows indicate the location of labeled regions buried within fissures. The lines at the bottom of the figure indicate the location of the three sagittal sections shown on the right. The flattened view of cerebellar cortex and abbreviations used are adapted from Larsell (1970). (Right) Plots of labeled neurons in three sagittal sections taken from vermal (385), intermediate (325), and lateral (245) regions of cerebellar cortex. The small dots indicate the relative density and distribution of labeled neurons.

posterior lateralis pars oralis (VPLo) and nucleus ventralis lateralis pars oralis (VLo) of Olszewski (1952) (for references and review, see Holsapple *et al.*, 1991). Five days after these injections, virus was transported transneuronally in the retrograde direction from first-order neurons in the ventrolateral thalamus to second-order neurons in output nuclei of the cerebellum (i.e., the dentate and interpositus) and the basal ganglia (e.g., the internal segment of the globus pallidus) (Figs. 5B–5D).

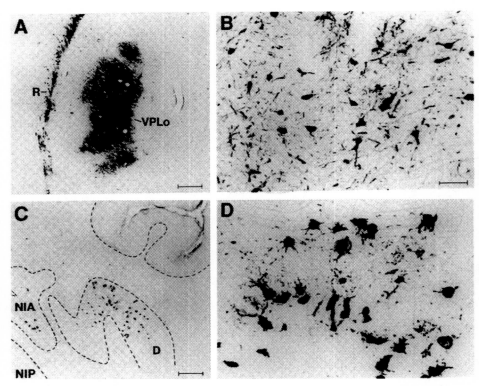

FIG. 5. Neurons labeled by retrograde transneuronal transport of HSV1 (McIntyre-B) from M1. (**A**) Ventrolateral thalamus (VPLo) and the reticular nucleus (R) (scale bar, 1 mm). (**B**) Internal segment of the globus pallidus (scale bar, 100 μm). (**C**) Deep cerebellar nuclei. Dashed lines outline the dentate (D), anterior interpositus (NIA), and posterior interpositus (NIP) nuclei (lower left) and a portion of cerebellar cortex (upper right) (scale bar, 500 μm). No labeled neurons were found in the posterior interpositus (NIP). (**D**) Dentate nucleus (scale in B). (From Zemanick *et al.*, 1991.)

Based on these results, we have used the labeling of second-order neurons by retrograde transneuronal transport of the McIntyre-B strain to map the origin of cerebellar (and basal ganglia) input to different cortical areas (Zemanick *et al.*, 1991; Hoover and Strick, 1993a; Strick *et al.*, 1993; Lynch *et al.*, 1994; Middleton and Strick, 1994). The following sections describe some of our observations on cerebellar projections to the arm representations of the primary motor cortex (area 4) and ventral premotor area (PMv) (area 6), the frontal eye field (FEF) (area 8), and two regions in the prefrontal cortex (areas 9 and 46) (Fig. 1) (Hoover and Strick, 1993a; Strick *et al.*, 1993; Lynch *et al.*, 1994; Middleton and Strick, 1994, 1996).

Virus injections into the arm areas of M1 and PMv, and into the FEF, were made after each area was physiologically mapped using intracortical stimulation.

A. CEREBELLAR OUTPUT TO SKELETOMOTOR AND OCULOMOTOR AREAS OF CEREBRAL CORTEX

Specific portions of the dentate and interpositus contained labeled neurons following injections of McIntrye-B into the arm area of M1. Labeled neurons in the interpositus were located largely in caudal portions of the anterior division of this nucleus. Labeled neurons in the dentate were restricted to dorsal portions of the nucleus at mid rostrocaudal levels (Figs. 5C and 5D and Fig. 6, "M1$_{arm}$"). These regions of the dentate and interpositus are comparable to the sites where neuron activity related to arm movements has been recorded (e.g., Thach, 1978; Wetts et al., 1985; van Kan et al., 1993). Thus, the arm area of the primary motor cortex appears to be influenced by localized "arm areas" in the dentate and interpositus.

Injections of the McIntyre-B strain into the arm representation of the PMv (Fig. 2) labeled many neurons in the dentate, at mid rostrocaudal levels of the nucleus (Figs. 5 and 6) (Strick et al., 1993). These neurons were located ventral to the region of the dentate that contained labeled neurons after injections into the arm area of M1 (compare Fig. 6, "M1$_{arm}$"

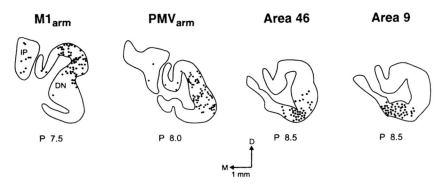

FIG. 6. Origin of cerebellar projections to M1, PMv, area 46, and area 9. Representative coronal sections through the dentate and/or interpositus nuclei of animals that received injections of the McIntyre-B strain of HSV1 into the arm representations of M1 or PMv, or into area 46 or area 9 in the prefrontal cortex. Solid dots indicate the positions of neurons labeled by the retrograde transneuronal transport of virus. Labeled neurons are charted from two adjacent sections whose approximate anterior–posterior location is indicated at the bottom of each section. (Adapted from Zemanick et al., 1991; Strick et al., 1993; Middleton and Strick, 1994, 1997.)

and "PMv$_{arm}$"). Thus, the arm areas in M1 and PMv receive input from different portions of the dentate.

HSV1 injections into the FEF (Fig. 2) labeled neurons in the most caudal third of the dentate nucleus (see Lynch et al., 1994). Prior studies have shown that this region of the dentate contains neurons that display changes in activity correlated with saccadic eye movements (e.g., van Kan et al., 1993). This caudal portion of the dentate is strikingly different from the dentate regions that contained labeled neurons after virus injections into M1 and PMv (Fig. 6). The FEF is known to be an important component of the cortical system that controls voluntary eye movements in primates (for references and review, see Bruce and Goldberg, 1985; Bruce et al., 1985). Thus, dentate projections to skeletomotor and oculomotor areas of cerebral cortex originate from separate regions of the nucleus.

B. Cerebellar Output to Prefrontal Cortex

Our initial studies on input to prefrontal cortex have focused on dorsolateral regions that are included in Walker's (1940) areas 9 and 46 (Fig. 2). These areas of prefrontal cortex have been reported to be involved in "working memory" and in the guidance of behavior based on transiently stored information rather than immediate external cues (e.g., Fuster, 1989; Funahashi et al., 1989, 1993; Goldman-Rakic, 1990; Petrides, 1995). Areas 9 and 46 have both been shown to project to regions of the pontine nuclei (Leichnetz et al., 1984; Glickstein et al., 1985; Schmahmann and Pandya, 1995, 1997; Brodal, 1978) and to receive input from subdivisions of the ventrolateral thalamus (Kievit and Kuypers, 1977; Goldman-Rakic and Porrino, 1985; Barbas et al., 1991; Yamamoto et al., 1992; Middleton and Strick, 1994).

Virus injections into prefrontal cortex labeled many neurons in the dentate nucleus. These neurons were confined to the most ventral portions of the dentate and were concentrated rostrocaudally in the middle third of the nucleus (Fig. 6, "Area 46" and "Area 9"). Within this region of the dentate, neurons labeled after area 9 injections were found largely medial to neurons labeled after area 46 injections. These regions of the dentate clearly differ from the more dorsal regions of the nucleus that were labeled by virus injections in M1 or the PMv (Fig. 6) and the more caudal region of the dentate labeled by injections in the FEF.

Two main conclusions arise from these results. First, the output of the cerebellum can influence skeletomotor, oculomotor, and prefrontal regions of the cerebral cortex. Second, each of these different cortical regions receives input from a different region of the dentate. As a consequence, the dentate nucleus appears to contain a number of distinct "out-

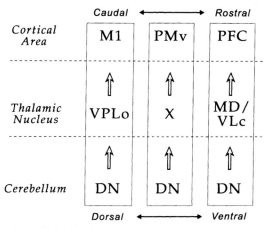

FIG. 7. Output channels in the dentate. Regions of the motor cortex (M1), premotor cortex (PMv), and prefrontal cortex (PFC) are each the target, via the thalamus, of projections from distinct regions of the dentate. The topographic trends in the localization of output channels related to these cortical areas are indicated at the top and bottom of the diagram.

put channels," which project via the thalamus to specific areas of the cerebral cortex (Fig. 7).

III. Physiological Studies

A. NEURON RECORDING IN AWAKE TRAINED PRIMATES

The anatomical findings just described raise an important question. What is the nature of the information conveyed to the cerebral cortex by individual output channels? For example, do output channels that project to motor areas of cortex send signals related to the control of movement, while output channels directed to prefrontal cortex send signals related to some aspect of mnemonic behavior? To begin to address this issue we recorded the activity of single neurons in the dentate nucleus of awake monkeys trained to perform sequential pointing movements under two different task conditions (Mushiake and Strick, 1993, 1995). Briefly, in both conditions, the monkey faced a panel with five touch pads that were numbered 1 to 5 (left to right) (Fig. 8). A small red light-emitting diode (LED) was located over each touch pad. The monkey began a trial by placing his right hand on a hold key for a variable "Hold" period. In the *remembered sequence task* (REM task)(Fig. 8, left), LEDs over three touch

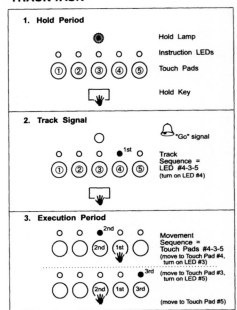

FIG. 8. REM and TRACK tasks. Monkeys faced a panel with five touch pads numbered 1 to 5. A small red LED was located over each touch pad (Instruction LEDs). The monkey began a trial by placing its right hand on a hold key in front of him for a "hold" period of 1.5–2.5 sec. Correct holding was signaled by a green LED (Hold Lamp) over the middle touch pad–LED combination. REM task: Instruction LEDs over three of the touch pads were illuminated in a sequence as an instruction to the monkey. At the end of a variable "Instruction" period of 1.5–2.5 sec, an auditory "Go" signal told the monkey to release the hold key and to press the three touch pads in the same order that the LEDs were illuminated. TRACK task: An instruction LED over a single touch pad was illuminated after a "Hold" period of 2.5–3.5 sec. The auditory "Go" signal was turned on at the same time. Following the onset of this signal, the monkey was required to release the hold key and press the indicated touch pad. As soon as the monkey contacted the first touch pad, an LED over a second touch pad was illuminated and the monkey was required to move to this touch pad. Then, when the monkey contacted the second touch pad, an LED over a third touch pad was illuminated and the monkey was required to move to this touch pad. (Adapted from Mushiake and Strick, 1993.)

pads were illuminated in a pseudorandom sequence as an instruction to the monkey. At the end of a variable "Instruction" period, an auditory "Go" signal told the monkey to release the hold key and press the three touch pads in the same order that they were illuminated. Thus, the sequence of movements that the monkey performed during each trial of the REM task was initially stored in "working memory" and then internally guided.

In the *tracking task* (TRACK task) (Fig. 8, right), an LED over a single touch pad was illuminated after the "Hold" period, and an auditory "Go" signal was turned on at the same time. Following this signal, the monkey was required to release the hold key and press the indicated touch pad. As soon as the monkey contacted the first touch pad, an LED over a second touch pad was illuminated. The monkey was required to quickly move to this second touch pad. When the monkey contacted the second touch pad, an LED over a third touch pad was illuminated and the monkey was required to move to this touch pad. Thus, the sequence of movements that the monkey performed during each trial of the TRACK task was externally cued.

We recorded 172 neurons that were task related during the reaction time (RT) period in the dentate nucleus of two trained monkeys. Most task-related neurons were located in the middle third of the dentate, rostrocaudally. Approximately 60% of the task-related neurons (102/172) were classified as *task independent*. These neurons displayed movement-related activity during the RT period of both REM and TRACK tasks. Most task-independent neurons were located dorsally in the dentate. This region is likely to be within the output channel that projects to M1 (see Figs. 6 and 7). Based on their firing patterns, it is likely that the neurons in this output channel are involved in defining the parameters of movements, independent of whether the movements are internally guided or externally cued.

However, approximately 40% of the task-related neurons in the dentate (70/172) were considered *task dependent* because their activity patterns differed substantially during TRACK and REM tasks. More than 75% of the task-dependent neurons (54/70) were termed *TRACK* neurons because they either displayed activity changes during the RT period only for the TRACK task or their changes in activity were more pronounced (>±50%) for the TRACK task than for the REM task. An example of a TRACK neuron is shown in Fig. 9. The rasters and averages are aligned on the hold key release. They illustrate the activity of this neuron during trials that began with a movement to touch pad 4 (4-5-3, 4-3-5, 4-3-1, 4-2-1). The individual trials in the rasters have been sorted according to the length of the RT period. This neuron displayed little or no modulation in its activity during the RT period of the REM task (Fig. 9, left). In contrast, the same neuron showed a clear increase in activity during the RT period of the TRACK task (Fig. 9, right).

Many of the TRACK neurons were located ventral and lateral to dentate neurons that were task related, but *task independent*. This localization suggests that TRACK neurons are within the output channel that innervates the PMv (see Figs. 6 and 7). Thus, the neurons in this output channel appear

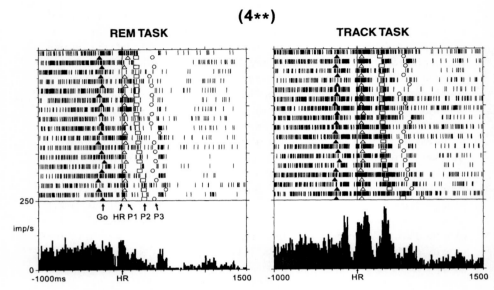

FIG. 9. Responses of a TRACK neuron in the dentate. The rasters and averages are aligned on the hold key release (HR). They illustrate the response of this neuron during trials that began with a movement to touch pad 4 (4**, i.e., 4-5-3, 4-3-5, 4-3-1, 4-2-1). The individual trials in the rasters have been sorted according to the length of the interval between the onset of the "Go" signal and HR (i.e., the RT period). Symbols indicate the onset of different behavioral events: filled triangle, "Go" signal; open triangle, press of the first touch pad in the sequence (P1); open square, press of the second touch pad (P2); open circle, press of the third touch pad (P3). Note that this neuron displayed a phasic increase in activity in the RT period only during the TRACK task. (Adapted from Mushiake and Strick, 1993.)

to be preferentially involved in the generation and control of sequential movements that are visually guided.

Approximately 16% (27/172) of the task-related neurons were *"instruction related"* (I-related), i.e., they displayed changes in activity during the instructed delay period (Fig. 10). Some of these I-related neurons displayed transient changes in activity immediately after the presentation of visual cues (Fig. 10, "Cue" neuron). Other I-related neurons displayed changes in activity only during the delay period following the illumination of the three instruction LEDs (Fig. 10, "Delay" neuron). Approximately one quarter of the I-related neurons (7/27) displayed "Delay" activity that depended on the sequence the animal was preparing to perform. Still other I-related neurons displayed two phases of activity during the instructed delay period (Fig. 10, "Cue + Delay" neuron). I-related neurons tended to be located in ventral regions of dentate. This site appears to be within

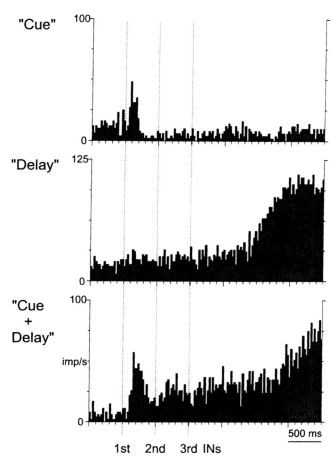

FIG. 10. Responses of I-related neurons in the dentate. The rasters and averages illustrate the activity of three different types of I-related neurons during the instructed delay period (first, second, and third INs). The trials are aligned on the presentation of the third instruction. The bin width for the averages is 20 msec. The trials illustrated all began with the illumination of LED #4. Note that some neurons displayed I-related activity after the presentation of a visual cue ("Cue" neuron, top), during the delay period following cue presentation ("Delay" neuron, middle), or during both the cue and delay periods ("Cue + Delay" neuron, bottom) (H. Mushiake and P. L. Strick, manuscript in preparation).

output channels that innervate prefrontal areas involved in working memory (areas 46 and 9, see Fig. 6). I-related neurons may also be within channels directed toward premotor areas concerned with motor preparation (e.g., the presupplementary motor area, see Wiesendanger and Wiesendanger, 1985b). In any event, the activity patterns of I-related neurons indicate that a portion of dentate output is concerned with higher-order motor and/or cognitive functions. Furthermore, these observations suggest that each output channel sends a unique signal to the cortical area it innervates.

B. Functional Magnetic Resonance Imaging of the Dentate in Human Subjects

The anatomical studies presented provide clear evidence that the dentate innervates regions of prefrontal cortex. This result, along with the physiological observations just described, raised the possibility that a portion of the output from the primate dentate is involved in some aspect of cognitive function. To test whether the human dentate participates in cognitive function, we used functional magnetic resonance imaging to study activation in the dentate while subjects attempted to solve a "pegboard" task (Kim *et al.*, 1994).

Seven healthy volunteers participated in these experiments. During imaging, subjects were asked to use their dominant limb to perform two different tasks. For the first task, termed the "Visually Guided Task," a small pegboard with nine holes was securely positioned over each subject's chest. The board contained four red pegs in the holes at its right end. The task was to move each peg, one hole at a time, to the holes at the opposite end of the board. The second task, termed the "Insanity Task," used the same pegboard as the visually guided task. However, in this case, four red pegs were placed in holes at the right end of the board *and* four blue pegs were placed in holes at the left end. Subjects were instructed to move the four pegs of each color from one end to the other using three rules: (1) move one peg at a time; (2) move to an adjacent open space or jump an adjacent peg (of a different color); and (3) move forwards, never backwards. No subject solved the insanity task during the period of scanning.

The major result of this study was that all seven of the subjects displayed a large bilateral activation in the dentate during attempts to solve the insanity task (Fig. 11B). Furthermore, in every subject, the extent of this activation was three to four times larger than that found during the visually guided task (Fig. 11A). In addition, the ventral portions of the dentate activated by the insanity task appeared to differ in their location from the portions of this nucleus activated during the visually guided task (Fig. 11). These results suggest two important conclusions. First, the cognitive demands associated with attempts to solve the insanity task lead to dentate activation. Second, the ventral regions of the dentate involved in cognitive processing are distinct from the dentate regions involved in the control of

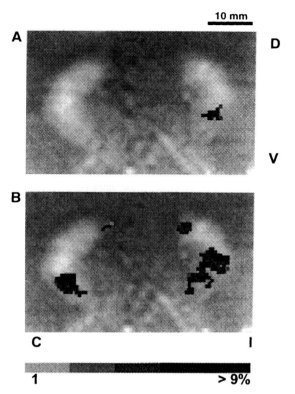

FIG. 11. Activation of dentate nucleus during cognitive processing. Maps of functional activation in the dentate for one subject during the visually guided task (**A**) and during the insanity task (**B**). Dentate nuclei are white crescent-shaped regions with low background signal intensity. Only those activation sites located within the dentate nuclei are shown. C, dentate contralateral to the moving limb; I, dentate ipsilateral to the moving limb; D, dorsal; V, ventral. (Adapted from Kim et al., 1994.)

eye and limb movements, and are potentially within an output channel that innervates prefrontal cortex.

IV. Synthesis

The anatomical and physiological results just described represent a significant departure from prior theories about the functional organization of cerebellar loops with the cerebral cortex. The classical view of these loops is that they provide a means for linking widespread regions of the cerebral cortex, such as prefrontal and posterior parietal cortex, with motor output at the level of the primary motor cortex (e.g., Kemp and Powell, 1971). Our results support an alternative view. Cerebellar output gains access to multiple cortical areas. These areas include regions involved in cognitive function, as well as those involved in the control of movement. In addition, we have proposed that the clustering of output neurons that project to an individual cortical area creates distinct output channels in the dentate. The output channels related to motor areas appear to be separate from those associated with the prefrontal cortex (Figs. 6 and 7).

Our physiological results suggest that individual output channels are concerned with different aspects of behavior. The output channel that projects to M1 may be involved in the control of specific movement parameters. The output channel that innervates the PMv may be concerned with higher order aspects of motor behavior such as the generation of sequential movements based on external cues. Output channels that influence the prefrontal cortex may be involved in cognitive aspects of behavior such as working memory.

Given the topographic organization of individual output channels and our physiological results, it is not hard to imagine that dysfunction of different regions within the dentate would lead to distinct changes in behavior. For example, abundant evidence shows that lesions of sensorimotor regions of the cerebellum produce alterations in motor behavior. In contrast, there are also reports that lesions of other regions of the cerebellum result in cognitive dysfunction (e.g., Botez *et al.*, 1989; Leiner *et al.*, 1991, 1993; Schmahmann, 1991; Ivry and Keele, 1989; Fiez *et al.*, 1992; Petersen and Fiez, 1993; Canavan *et al.*, 1994).

At this point, the full extent of cerebellar influence on the cerebral cortex remains to be determined. One of the immediate goals of our present research is to resolve this question. The outcome of these studies should have an important impact on concepts regarding cerebellar contributions to behavior and provide additional insights into the consequences of cerebellar dysfunction.

Acknowledgments

This work was supported by funds from an Established Investigator Award from the National Alliance for Research on Schizophrenia and Depression (P.L.S.), the Veterans Administration Medical Research Service (P.L.S.), U.S. Public Health Service Grant NS24328 (P.L.S.), and Predoctoral Fellowship MH11262 from the National Institute of Mental Health (F.A.M.).

References

Allen, G. I., and Tsukahara, N. (1974). Cerebrocerebellar communication systems. *Physiol. Rev.* **54,** 957–1006.
Asanuma, C., Thach, W. T., and Jones, E. G. (1983). Distribution of cerebellar terminations in the ventral lateral thalamic region of the monkey. *Brain Res. Rev.* **5,** 237–265.
Barbas, H., Haswell Henion, T. H., and Dermon, C. R. (1991). Diverse thalamic projections to the prefrontal cortex in the rhesus monkey. *J. Comp. Neurol.* **313,** 65–94.
Botez, M. I., Botez, T., Elie, R., and Attig, E. (1989). Role of the cerebellum in complex human behavior. *Ital. J. Neurol. Sci.* **10,** 291–300.
Brodal, P. (1978). The corticopontine projection in the rhesus monkey: Origin and principles of organization. *Brain* **101,** 251–283.
Brooks, V. B., and Thach, W. T. (1981). Cerebellar control of posture and movement. *In* "Handbook of Physiology" (V. B. Brooks, ed.), Vol. II, pp. 877–946. American Physiological Society, Bethesda, MD.
Bruce, C. J., and Goldberg, M. E. (1985). Primate frontal eye fields. I. Single neurons discharging before saccades. *J. Neurophysiol.* **53,** 603–635.
Bruce, C. J., Goldberg, M. E, Bushnell, M. C., and Stanton, G. B. (1985). Primate frontal eye fields. II. Physiological and anatomical correlates of electrically evoked eye movements. *J. Neurophysiol.* **54,** 714–734.
Canavan, A. G. M., Sprengelmeyer, R., Deiner, H. C., and Homberg, V. (1994). Conditional associative learning is impaired in cerebellar disease in humans. *Behav. Neurosci.* **108,** 475–485.
Evarts, E. V., and Thach, W. T. (1969). Motor mechanisms of the CNS: Cerebrocerebellar interrelations. *Annu. Rev. Physiol.* **31,** 451–498.
Fiez, J. A, Petersen, S. E., Cheney, M. K., and Raichle, M. E. (1992). Impaired non-motor learning and error detection associated with cerebellar damage. *Brain* **115,** 155–178.
Funahashi, S., Bruce, C. J., and Goldman-Rakic, P. S. (1989). Mnemonic coding of visual space in the monkeys dorsolateral prefrontal cortex. *J. Neurophysiol.* **61,** 331–349.
Funahashi, S., Bruce, C. J., and Goldman-Rakic, P. S. (1993). Dorsolateral prefrontal lesions and oculomotor delayed-response: Evidence for mnemonic "scotomas." *J. Neurosci.* **13,** 1479–1497.
Fuster, J. M. (1989). "The Prefrontal Cortex," 2nd Ed. Raven Press, New York.
Glickstein, M., May, J. G., and Mercier, B. E. (1985). Corticopontine projection in the macaque: The distribution of labelled cortical cells after large injections of horseradish peroxidase in the pontine nuclei. *J. Comp. Neurol.* **235,** 343–359.
Goldman-Rakic, P. S. (1990). Cellular and circuit basis of working memory in prefrontal cortex of non-human primates. *Prog. Brain Res.* **85,** 325–336.
Goldman-Rakic, P. S., and Porrino, L. J. (1985). The primate mediodorsal (MD) nucleus and its projection to the frontal lobe. *J. Comp. Neurol.* **242,** 535–560.
Hendry, S. H. C., Jones, E. G., and Graham, J. (1979). Thalamic relay nuclei for cerebellar and certain related fiber systems in the cat. *J. Comp. Neurol.* **185,** 679–714.
Holsapple, J. W., Preston, J. B., and Strick, P. L. (1991). The origin of thalamic inputs to the "hand" representation in the primary motor cortex. *J. Neurosci.* **11,** 2644–2654.
Hoover, J. E., and Strick, P. L. (1993a). Multiple output channels in the basal ganglia. *Science* **259,** 819–821.
Hoover, J. E., and Strick, P. L. (1993b). Retrograde transneuronal transport of HSV-1 from primary motor cortex to cerebellar deep nuclei and purkinje cells. *Soc. Neurosci. Abstr.* **19,** 1590.

Ito, M. (1984). "The Cerebellum and Neural Control." Raven Press, New York.
Ivry, R. B., and Keele, S. W. (1989). Timing functions of the cerebellum. *J. Cog. Neurosci.* **1**, 136–152.
Kalil, K. (1981). Projections of the cerebellar and dorsal column nuclei upon the thalamus of the rhesus monkey. *J. Comp. Neurol.* **195**, 25–50.
Kemp, J. M., and Powell, T. P. S. (1971). The connexions of the striatum and globus pallidus: Synthesis and speculation. *Philos. Trans. R. Soc. Lond. Ser. B* **262**, 441–457.
Kievit, J., and Kuypers, H. G. J. M. (1977). Organization of thalamo-cortical connexions to the frontal lobe in the rhesus monkey. *Exp. Brain Res.* **29**, 299–322.
Kim, S.-G., Ugurbil, K., and Strick, P. L. (1994). Activation of a cerebellar output nucleus during cognitive processing. *Science* **265**, 949–951.
Leichnetz, G. R., Smith, D. J., and Spencer, R. F. (1984). Cortical projections of the paramedian tegmental and basilar pons in the monkey. *J. Comp. Neurol.* **228**, 388–408.
Leiner, H. C., Leiner, A. L., and Dow, R. S. (1987). Cerebro-cerebellar learning loops in apes and humans. *Ital. J. Neurol. Sci.* **8**, 425–436.
Leiner, H. C., Leiner, A. L., and Dow, R. S. (1989). Reappraising the cerebellum: What does the hindbrain contribute to the forebrain? *Behav. Neurosci.* **103**, 998–1008.
Leiner, H. C., Leiner, A. L., and Dow, R. S. (1991). The human cerebro-cerebellar system: Its computing, cognitive, and language skills. *Behav. Brain Res.* **44**, 113–128.
Leiner, H. C., Leiner, A. L., and Dow, R. S. (1993). Cognitive and language functions of the human cerebellum. *Trends Neurosci.* **16**, 444–447.
Lynch, J. C., Hoover, J. E., and Strick, P. L. (1994). Input to the primate frontal eye field from the substantia nigra, superior colliculus, and dentate nucleus demonstrated by transneuronal transport. *Exp. Brain Res.* **100**, 181–186.
Matelli, M., Luppino, G., Fogassi, L., and Rizzolatti, G. (1989). Thalamic input to inferior area 6 and area 4 in the macaque monkey. *J. Comp. Neurol.* **280**, 468–488.
Middleton, F. A., and Strick, P. L. (1994). Anatomical evidence for cerebellar and basal ganglia involvement in higher cognitive function. *Science* **266**, 458–461.
Middleton, F. A., and Strick, P. L. (1997). Dentate ouput channels: Motor and cognitive components. *Prog. Brain Res.*, **114**, 555–568.
Miyata, M., and Sasaki, K. (1983). HRP studies on thalamocortical neurons related to the cerebellocerebral projection in the monkey. *Brain Res.* **274**, 213–224.
Mushiake, H., and Strick, P. L. (1993). Preferential activity of dentate neurons during limb movements guided by vision. *J. Neurophysiol.* **70**, 2660–2664.
Mushiake, H., and Strick, P. L. (1995). Cerebellar and pallidal activity during instructed delay periods. *Soc. Neurosci. Abstr.* **21**, 411.
Olszewski, J. (1952). "The thalamus of *Macaca mulatta:* An Atlas for Use with the Stereotaxic Instrument." Karger, Basel.
Percheron, G. (1977). The thalamic territory of cerebellar afferents and the lateral region of the thalamus of the macaque in stereotaxic ventricular coordinates. *J. Hirnforsch.* **18**, 375–400.
Petersen, S. E., and Fiez, J. A. (1993). The processing of single words studied with positron emission tomography. *Annu. Rev. Neurosci.* **16**, 509–530.
Petrides, M. (1995). Impairments on nonspatial self-ordered and externally-ordered working memory tasks after lesions of the mid-dorsal part of the lateral frontal cortex in the monkey. *J. Neurosci.* **15**, 359–375.
Rouiller, E. M., Liang, F., Babalian, A., Moret, V., and Wiesendanger, M. (1994). Cerebellothalamocortical and pallidothalamocortical projections to the primary and supplementary motor cortical areas: A multiple tracing study in macaque monkeys. *J. Comp. Neurol.* **345**, 185–231.

Sasaki, K., Kawaguchi, S., Oka, H., Sakai, M., and Mizuno, N. (1976). Electrophysiological studies on the cerebello-cerebral projections in monkeys. *Exp. Brain Res.* **24,** 495–507.

Sasaki, K., Oka, H., Kawaguchi, S., Jinnai, K., and Yasuda, T. (1977). Mossy fibre and climbing fibre responses produced in the cerebellar cortex by stimulation of the cerebral cortex in monkeys. *Exp. Brain Res.* **29,** 419–428.

Sasaki, K., Jinnai, K., Gemba, H., Hashimoto, S., and Mizuno, N. (1979). Projection of the cerebellar dentate nucleus onto the frontal association cortex in monkeys. *Exp. Brain Res.* **37,** 193–198.

Schell, G. R., and Strick, P. L. (1984). The origin of thalamic inputs to the arcuate premotor and supplementary motor areas. *J. Neurosci.* **4,** 539–560.

Schmahmann, J. D. (1991). An emerging concept: The cerebellar contribution to higher function. *Arch. Neurol.* **48,** 1178–1187.

Schmahmann, J. D., and Pandya, D. N. (1990). Anatomical investigation of projections from the thalamus to the posterior parietal association cortices in the rhesus monkey. *J. Comp. Neurol.* **295,** 299–326.

Schmahmann, J. D., and Pandya, D. N. (1991). Projections to the basis pontis from the superior temporal sulcus and superior temporal region in the rhesus monkey. *J. Comp. Neurol.* **308,** 224–248.

Schmahmann, J. D., and Pandya, D. N. (1993). Prelunate, occipitotemporal, and parahippocampal projections to the basis pontis in rhesus monkey. *J. Comp. Neurol.* **337,** 94–112.

Schmahmann, J. D., and Pandya, D. N. (1995). Prefrontal cortex projections to the basilar pons in rhesus monkey: Implications for the cerebellar contributions to higher function. *Neurosci. Lett.* **199,** 175–178.

Schmahmann, J. D., and Pandya, D. N. (1997). Anatomic organization of the basilar pontine projections from prefrontal cortices in rhesus monkey. *J. Neurosci.* **17,** 438–458.

Shinoda, Y., Sugiuchi, Y., and Futami, T. (1987). Excitatory inputs to cerebellar dentate nucleus neurons from the cerebral cortex in the cat. *Exp. Brain Res.* **67,** 299–315.

Stanton, G. (1980). Topographical organization of ascending cerebellar projections from the dentate and interposed nuclei in *Macaca mulatta:* An anterograde degeneration study. *J. Comp. Neurol.* **190,** 699–731.

Strick, P. L., and Card, J. P. (1992). Transneuronal mapping of neural circuits with alpha herpesviruses. In "Experimental Neuroanatomy: A Practical Approach" (J. P. Bolam, ed.), pp. 81–101. Oxford University Press, Oxford.

Strick, P. L., Hoover, J. E., and Mushiake, H. (1993). Evidence for "output channels" in the basal ganglia and cerebellum. In "Role of the Cerebellum and Basal Ganglia in Voluntary Movement" (N. Mano, I. Hamada, and M. R. DeLong, eds.), pp. 171–180. Elsevier, Amsterdam.

Thach, W. T. (1978). Correlation of neural discharge with pattern and force of muscular activity, joint position, and direction of intended next movement in motor cortex and cerebellum. *J. Neurophysiol.* **41,** 654–676.

van Kan, P. L. E., Houk, J. C., and Gibson, A. R. (1993). Output organization of intermediate cerebellum of the monkey. *J. Neurophysiol.* **69,** 57–73.

Vilensky, J. A., and Van Hoesen, G. W. (1981). Corticopontine projections from the cingulate cortex in the rhesus monkey. *Brain Res.* **205,** 391–395.

Walker, A. E. (1940). A cytoarchitectural study of the prefrontal area of the macaque monkey. *J. Comp. Neurol.* **73,** 59–86.

Wetts, R., Kalaska, J. F., and Smith, A. M. (1985). Cerebellar nuclear activity during contraction and reciprocal inhibition of forearm muscles. *J. Neurophysiol.* **54,** 231–244.

Wiesendanger, R., and Wiesendanger, M. (1985a). The thalamic connections with medial area 6 (supplementary motor cortex) in the monkey (macaca fascicularis). *Exp. Brain Res.* **59,** 91–104.

Wiesendanger, R., and Wiesendanger, M. (1985b). Cerebello-cortical linkage in the monkey as revealed by transcellular labeling with the lectin wheat germ agglutinin conjugated to the marker horseradish peroxidase. *Exp. Brain Res.* **59,** 105–117.

Yamamoto, T., Yoshida, K., Yoshikawa, Y., Kishimoto, Y., and Oka, H. (1992). The medial dorsal nucleus is one of the thalamic relays of the cerebellocerebral response to the frontal association cortex in the monkey: Horseradish peroxidase and fluorescent dye double staining study. *Brain Res.* **579,** 315–320.

Zemanick, M. C., Strick, P. L., and Dix, R. D. (1991). Transneuronal transport of herpes simplex virus type 1 in the primate motor system: Transport direction is strain dependent. *Proc. Natl. Acad. Sci. USA* **88,** 8048–8051.

THE CEREBELLAR–HYPOTHALAMIC AXIS: BASIC CIRCUITS AND CLINICAL OBSERVATIONS

Duane E. Haines,* Espen Dietrichs,† Gregory A. Mihailoff,* and E. Frank McDonald‡

Departments of *Anatomy and ‡Neurology, The University of Mississippi Medical Center, Jackson, Mississippi 39216; and †Department of Neurology, Ullevål Hospital, University of Oslo, Norway

I. Introduction
II. Hypothalamocerebellar Projections and Related Neurotransmitters
   A. Hypothalamocerebellar Cortical Projections
   B. Hypothalamocerebellar Nuclear Projections
   C. Neurotransmitters of the Hypothalamocerebellar Projection
III. Cerebellar Projections to the Hypothalamus
IV. Collaterals of Hypothalamocerebellar Fibers
V. Indirect Hypothalamocerebellar Connections Mediated by the Basilar Pontine Nuclei and Lateral Reticular Nucleus
VI. Clinical Evidence of Cerebellar Influence on Visceral Function
   A. Patient 1
   B. Patient 2
VII. Conclusions
   References

Experimental studies on a variety of mammals, including primates, have revealed direct and reciprocal connections between the hypothalamus and the cerebellum. Although widespread areas of the hypothalamus project to cerebellum, axons arise primarily from cells in the lateral, posterior, and dorsal hypothalamic areas; the supramammillary, tuberomammillary, and lateral mammillary nuclei; the dorsomedial and ventromedial nuclei; and the periventricular zone. Available evidence suggests that hypothalamocerebellar cortical fibers may terminate in relation to neurons in all layers of the cerebellar cortex. Cerebellohypothalamic axons arise from neurons of all four cerebellar nuclei, pass through the superior cerebellar peduncle, cross in its decussation, and enter the hypothalamus. Some axons recross the midline in caudal areas of the hypothalamus. These fibers terminate primarily in lateral, posterior, and dorsal hypothalamic areas and in the dorsomedial and paraventricular nuclei. Evidence of a cerebellar influence on the visceromotor system is presented in two patients with vascular lesions: one with a small defect in the medial cerebellar nucleus and the other with a larger area of damage involving primarily the globose and emboliform nuclei. Both patients exhibited an abnormal visceromotor response. The second, especially, showed abnormal visceromotor activity concurrent with tremor induced by voluntary movement. These experimental and clinical data suggest that the cerebellum is actively involved in the regulation of visceromotor functions.

I. Introduction

A wide range of visceral responses have been reported following experimental stimulation of cerebellar structures or following lesions of the cerebellar cortex and/or nuclei (for reviews see Dow and Moruzzi, 1958; Haines et al., 1984; Dietrichs et al., 1994a). In general, these observations implied a cerebellar influence on visceral functions via pathways that were largely unknown. Although some early investigators offered anatomic or physiologic evidence of tentative connections between the cerebellum and hypothalamus (e.g., Moruzzi, 1940, 1950; Chambers, 1947; Whiteside and Snider, 1953; Chambers and Sprague, 1955a,b; Ban and Inoue, 1957; Raymond, 1958; Martner, 1975), definitive proof of such a projection was first offered by Dietrichs (1984). Following injections of wheat germ agglutinin–horseradish peroxidase (WGA-HRP) into the cerebellar cortex, Dietrichs (1984) reported retrograde labeling of cells in the lateral, posterior, and dorsal hypothalamic areas; in the supraoptic, periventricular, lateral mammillary, tuberomammillary, and ventromedial nuclei; and in the tuber cinereum. This projection, as initially described in cat, is bilateral but with an ipsilateral preponderance.

A subsequent series of studies carried out on a variety of mammals [rat, cat, tree shrew (*Tupaia glis*), squirrel monkey (*Saimiri sciureus*), greater bushbaby (*Galago crassicaudatus*), and rhesus monkey (*Macaca fascicularis*)] by Dietrichs and Haines (see Section II) confirmed and significantly expanded these initial observations. In addition, other investigators using neurophysiologic, tract tracing, and neurochemical methods have described further the nature of direct connections between cerebellar and hypothalamic structures.

Although the distribution of terminal label may vary slightly among species following the placement of tracers in cerebellar or hypothalamic structures, there is no convincing evidence of species-specific patterns in these connections. To the contrary, even though the projections appear to be stronger as one ascends the phylogenetic scale, the general organization is strikingly similar among species. Furthermore, there is evidence of hypothalamocerebellar connections in nonmammalian vertebrates (Bangma and ten Donkelaar, 1982; Künzle, 1983), suggesting that these may indeed be phylogenetically old pathways. Recognizing the general nature of these connections, the ensuing discussion mainly emphasizes patterns that are largely common to all mammalian species studied to date.

II. Hypothalamocerebellar Projections and Related Neurotransmitters

Direct projections from the various hypothalamic nuclei/areas to the cerebellum have been reported in studies on mammals that used HRP or WGA-HRP as retrograde and anterograde tracers, *Phaseolus vulgaris* leucoagglutinin (PHA-L) or [^{3}H]leucine as an anterograde tracer, and neurophysiologic or fluorescent-labeling methods (e.g., Bratus and Yoltukhovsky, 1986; Dietrichs, 1984; Dietrichs and Haines, 1984, 1985a,b, 1989; Dietrichs *et al.*, 1985b, 1992, 1994b; Haines and Dietrichs, 1984, 1987; Haines *et al.*, 1984, 1985, 1986, 1990; Wang *et al.*, 1994; Ter Horst and Luiten, 1986). Collectively these reports have described (a) hypothalamic projections to the cerebellar cortex and (b) hypothalamic projections to the cerebellar nuclei.

A. HYPOTHALAMOCEREBELLAR CORTICAL PROJECTIONS

Following injections of retrograde tracers into the cerebellar cortex, labeled cells are found bilaterally in the hypothalamus, but with an ipsilateral preponderance (Figs. 1 and 2). Studies reporting ipsilateral/contralateral cell counts (Dietrichs and Haines, 1984; Haines and Dietrichs, 1984) indicate that this ratio may vary from less than 1.4:1 to more than 3.2:1. Dedicated quantitative investigations have not been conducted, and it must be acknowledged that laterality may also be affected by the location of the cortical injection site. For example, the ipsilateral/contralateral ratio of labeled hypothalamic neurons following injections in more caudal and lateral cortex regions (such as the paraflocculus) is about 2 to 3:1 whereas the ratio following injections in more anterior regions of cortex (such as the anterior lobe or ansiform lobule) is about 1 to 2:1.

Hypothalamic neurons that project directly to the cerebellar cortex are found primarily in the lateral (LHAr), dorsal (DHAr), and posterior (PHAr) hypothalamic areas; the lateral mammillary (LMNu), supramammillary (SMNu), and tuberomammillary (TMNu) nuclei; the dorsomedial (DMNu) and ventromedial (VMNu) nuclei; and in cells of the periventricular zone/ nucleus (PVZo) (Figs. 1 and 2). Labeled neurons are less frequently, or less consistently, found within the medial mammillary nucleus (MMNu; although they were found on the periphery of this cell group) and in the

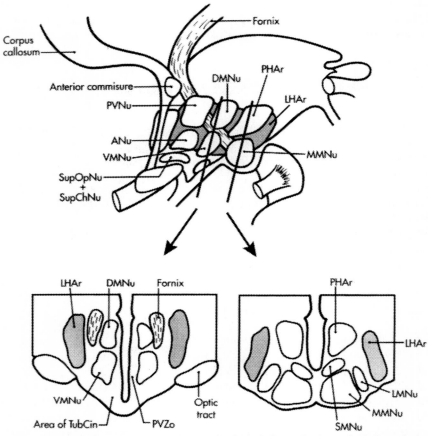

FIG. 1. Diagrammatic representation of the hypothalamus in the sagittal plane (upper) and two cross sections (lower) corresponding to the approximate positions of the heavy lines in the sagittal drawing. The main nuclei/areas of the hypothalamus that project to the cerebellar cortex and/or nuclei and those that receive input from the cerebellar nuclei are shown. Abbreviations for Figs. 1–3: ANu, anterior hypothalamic nucleus; DHAr, dorsal hypothalamic area; DMNu, dorsomedial hypothalamic nucleus; DNu, dentate (lateral cerebellar) nucleus; ENu, emboliform (anterior interposed) nucleus; FNu, fastigial (medial cerebellar) nucleus; GNu, globose nucleus; LHAr, lateral hypothalamic area; LMNu, lateral mammillary nucleus; MMNu, medial mammillary nucleus; PHAr, posterior hypothalamic area; PVNu, paraventricular nucleus; PVZo, periventricular zone; SMNu, supramammillary nucleus; SupChNu, suprachiasmatic nucleus; SupOpNu, supraoptic nucleus; TMNu, tuberomammillary nucleus; TubCin, tuber cinereum; and VMNu, ventromedial nucleus.

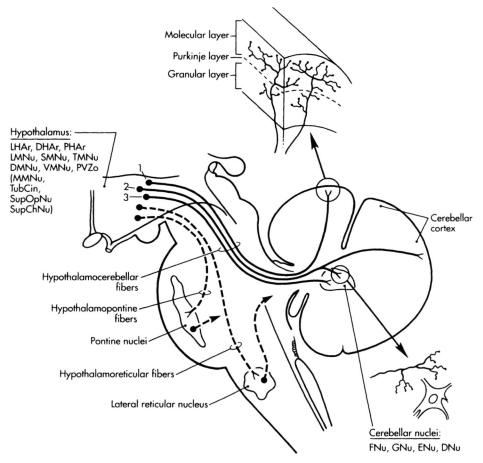

FIG. 2. Diagrammatic representation of efferent projections of the hypothalamus to the cerebellum and to the pontine nuclei and the lateral reticular nucleus. Those cell groups listed under hypothalamus and out of parentheses are the prime source of hypothalamocerebellar fibers; those listed in parentheses give rise to fewer projections. Cell numbers 1, 2, and 3 are indicative of (1) hypothalamic cells that project only to the cortex, (2) hypothalamic cells that project to the cortex and send collaterals into the cerebellar nuclei, and (3) hypothalamic cells that project only to the cerebellar nuclei. Hypothalamospinal fibers are not shown.

tuber cinereum and supraoptic or suprachiasmatic nuclei. These labeled cells have somata that range in size from 15 to 30 μm and are variously described as being round, oval, triangular, or fusiform shaped (Dietrichs,

1984; Haines and Dietrichs, 1984; Dietrichs and Haines, 1984; Haines et al., 1985; see also Haines et al., 1990).

There is not a clear-cut topography in this projection. Relatively small or restricted injection sites, such as those in the paraflocculus (Dietrichs and Haines, 1984, Haines and Dietrichs, 1984) or caudal vermis (Haines et al., 1985), result in a widespread distribution of labeled cells in several hypothalamic nuclei/areas. Likewise, small injections in the hypothalamus (Dietrichs and Haines, 1985a), especially those using iontophoretic methods (Haines and Dietrichs, 1984), result in anterograde labeling in comparatively widespread areas of the cerebellar cortex.

The placement of anterograde tracers in the hypothalamus (WGA-HRP, PHA-L, [^{3}H]leucine) results in the labeling of axons to the cerebellar cortex, further substantiating a direct hypothalamocerebellar cortical pathway (Fig. 2). Following pressure or iontophoretic injections in the hypothalamus (Dietrichs and Haines, 1985b; Haines et al., 1986; Ter Horst and Luiten, 1986; see also reviews by Haines and Dietrichs, 1987; Dietrichs and Haines, 1989), anterogradely labeled axons course caudally and pass through the periventricular gray and then arch dorsally to enter the base of the cerebellum. Labeled axons traverse the subcortical white matter and branch within all layers (granular, molecular, Purkinje cell) of the cerebellar cortex (Fig. 2). Although the ultrastructural characteristics of hypothalamocerebellar cortical terminals have not been described, the branching pattern of these fibers in the cortex (see Haines and Dietrichs, 1987; Dietrichs and Haines, 1989) and the response patterns of Purkinje and non-Purkinje cells to LHAr stimulation (Wang et al., 1994) indicate that synaptic contacts are made with Purkinje cells and other neurons of the molecular and granular layers. Hypothalamocerebellar cortical fibers share many morphological similarities in the cerebellar cortex (distribution, orientation, branching patterns, potential synaptic sites) with noradrenergic, histaminergic, serotoninergic, and some corticotrophin-releasing factor immunoreactive axons (e.g., Hökfelt and Fuxe, 1969; Hökfelt et al., 1984; Mugnaini and Dahl, 1975; Chan-Palay, 1975; Moore and Card, 1984; Bishop et al., 1985a,b; Cummings, 1989; Panula et al., 1988, 1989a,b, 1993). On the basis of these structural and potential functional similarities, especially recognizing the location of these fibers in the cerebellar cortex, it has been suggested that *all* nonmossy and nonclimbing fibers may be classified as *multilayered fibers* (Dietrichs and Haines, 1985a, 1989; Haines et al., 1986; Dietrichs et al., 1994a).

Diffuse labeling of cerebellar cortical fibers is seen following injections that include several hypothalamic nuclei/areas (e.g., Dietrichs and Haines, 1985a; Haines et al., 1986), although more labeled axons appear to enter the vermis and flocculonodular lobe. In contrast, iontophoretic placement of PHA-L in a single hypothalamic nucleus (the DMNu of rat) resulted in labeled fibers, and presumed terminal boutons, in granular and molecular layers of only the ansiform lobule (Ter Horst and Luiten, 1986). This observation suggests that *individual* hypothalamic nuclei may have preferred target areas in the cerebellar cortex. The observation of Wang *et al.* (1994) that only 1 of 27 cerebellar cortical cells responding to LHAr stimulation *also* responded to VMNu stimulation (lack of convergence) supports this view.

The observation of a direct anatomical projection from hypothalamic nuclei to the cerebellar cortex has been corroborated by Bratus and Yoltukhovsky (1986), Supple (1993), and Wang *et al.* (1994) using neurophysiologic techniques. Stimulation of the VMNu or of the LHAr resulted in short latency (monosynaptic) as well as long latency (polysynaptic) responses of Purkinje cells, and other cells, in the cerebellar cortex; the former indicating a direct hypothalamocerebellar cortical connection. Wang *et al.* (1994) differentiated the responses of Purkinje cells (complex spikes) from those of non-Purkinje cells (simple spikes) and reported that more non-Purkinje cells responded to hypothalamic stimulation than Purkinje cells. Based on the distance from the hypothalamus to the cerebellar cortex (20 mm) and on the conduction velocity of central nerve fibers (about 1 m/sec), these authors concluded that all responses <20 msec were indicative of a direct hypothalamocerebellar connection; this representing about half of all responses. Bratus and Yoltukhovsky (1986) suggested that the monosynaptic projection from VMNu to the cerebellar cortex is primarily excitatory, while those from LHAr may be excitatory or inhibitory to the cortex. In contrast, Wang *et al.* (1994) reported that the primary monosynaptic response of the cerebellar cortex to both LHAr and VMNu stimulation is inhibition; only 10% of LHAr neurons elicit an excitatory response and about 12–14% of LHAr and VMNu neurons, respectively, elicit mixed responses. Supple (1993) also reported unimodal excitatory, biphasic excitatory/inhibitory, or complex responses of Purkinje cells to hypothalamic stimulation.

B. HYPOTHALAMOCEREBELLAR NUCLEAR PROJECTIONS

Projections from the hypothalamus to the cerebellar nuclei (Fig. 2) have been reported by Dietrichs *et al.* (1985b, 1994b) and Haines *et al.* (1990) following injections or implants of WGA-HRP or of rhodamine-B–isothiocyanate (RITC) in various combinations with Fluoro-Gold (FG) in the cerebellar nuclei. Although an early study revealed retrograde labeling of hypothalamic neurons only after implants in the medial and posterior interposed cerebellar nuclei (none after anterior interposed or lateral cerebellar nuclei implants) (Dietrichs et al., 1985b), subsequent experiments in *Macaca* (Haines et al., 1990) and cat (Dietrichs et al., 1994b) clearly indicated that all cerebellar nuclei receive a direct hypothalamic input. In general, the cerebellar nuclei receive a bilateral projection (with ipsilateral preponderance) primarily from LHAr, PHAr, DHAr, DMNu, SMNu, and the TMNu (Fig. 2). Labeled neurons were also found, albeit in fewer numbers, in the VMNu, LMNu, MMNu, periventricular zone, perifornical area, and in the periphery of the periventricular gray in the caudal hypothalamus (Haines et al., 1990).

The striking similarity of hypothalamic projections to the cerebellar cortex and nuclei suggests that at least some of the fibers may be collateral branches of the other. Dietrichs et al. (1994b) addressed this specific question in a series of experiments using the fluorescent substances RITC and FG. Injections of RITC were made in the vermis, the intermediate cortex, and the lateral cortex of the posterior lobe (in cat) and implants of FG in the corresponding cerebellar nucleus (medial, anterior, and posterior interposed, lateral) on the same side. There was no evidence of overlap between the cerebellar cortical and the nuclear injection sites. Single- and double-labeled cells were found bilaterally, but with a clear ipsilateral preponderance, primarily in the LHAr, but also in PHAr, DHAr, TMNu, and in cells of the periventricular area. About half of all labeled cells contained RITC + FG. These data indicate that three types of hypothalamocerebellar cells may exist (Fig. 2): (1) neurons that project directly to *only* the cerebellar cortex, (2) neurons that project *only* to the cerebellar nuclei, and (3) individual neurons that send axonal branches to the cerebellar cortex *and* to the underlying cerebellar nucleus that is functionally associated with that particular part of cortex.

C. NEUROTRANSMITTERS OF THE HYPOTHALAMOCEREBELLAR PROJECTION

Following injections of D-[^{3}H]aspartate into various regions of the cerebellar cortex in cat, Dietrichs *et al.* (1992) reported negative findings regarding the retrograde labeling of hypothalamic neurons in 7 out of 10 cases. Weak labeling of only four cells in 3 out of 10 cases was not regarded as convincing evidence that aspartate (or glutamate) is a neurotransmitter in the hypothalamocerebellar projection.

However, Panula and co-workers (Airaksinen and Panula, 1988; Airaksinen et al., 1989; Panula et al., 1988, 1989a,b; 1993) and Ericson *et al.* (1987) have offered convincing evidence that histamine is an important neurotransmitter in the hypothalamocerebellar circuit. Three general correlations between the tracing studies and the neurotransmitter experiments merit comment. First, histamine-containing neurons are found in VMNu, DMNu, SMNu, the periventricular zone, and especially the TMNu; all of these hypothalamic areas contain cells known to project to the cerebellar cortex and nuclei. Using a fluorescent label [Fast Blue (FB)] combined with an immunocytochemical method, Ericson *et al.* (1987) demonstrated FB-labeled L-histidine containing neurons in TMNu after cerebellar injections. Second, histamine-positive fibers are especially prominent in the periventricular gray; this is the primary route followed by hypothalamocerebellar fibers. Third, the morphology and distribution of histamine-positive fibers in the cerebellar cortex are markedly similar to that reported in anterograde tracing experiments following injections of WGA-HRP and PHA-L in the hypothalamus. This pattern is one of branching in the granular and molecular layers and some fibers being oriented parallel to the long axes of the folia.

The fact that γ-aminobutyric acid (GABA) and histamine coexist in some hypothalamic cells (Takeda *et al.*, 1984) implicates GABA as a potential neurotransmitter in some hypothalamocerebellar fibers. This may correlate with the inhibitory versus excitatory responses of some Purkinje cells in neurophysiological studies (Wang *et al.*, 1994; Supple, 1993). GABA- and glycine-like immunoreactivity has indeed been found in a small number of hypothalamocerebellar neurons (Dietrichs *et al.*, 1994a). However, retrogradely labeled hypothalamocerebellar cells are also found in areas/nuclei of the hypothalamus that contain neither GABAergic nor histaminergic neurons. Consequently, other, yet to be identified, neuroactive substances are possibly involved in this projection.

One potential candidate for a neurotransmitter in the hypothalamocerebellar system is a hexadecapeptide called "cerebellin." This amino acid was initially isolated from rat cerebellum (Slemmon *et al.*, 1984, 1985). Subsequent studies in rat have shown that cerebellin-like immunoreactivity is greatest in the cerebellum followed next by anterior and posterior portions of the hypothalamus (Burnet *et al.*, 1988). In guinea pig, cerebellin is most concentrated in the cerebellum followed, in order (pmol/g wet tissue), by the stomach, anterior hypothalamus, descending colon, lower esophagus, and posterior hypothalamus. In addition, a cerebellin-like cDNA, which codes for a protein with a high degree of sequence similarity to cerebellin, has been cloned (Wada and Ohtani, 1991). Cerebellin is also found at postsynaptic sites (Mugnaini *et al.*, 1988, Urade *et al.*, 1991), especially in Purkinje cells, and the significant decrease in cerebellin in patients with degenerative cerebellar disease may relate to the loss of Purkinje cells in these individuals (Mizuno *et al.*, 1995). The fact that cerebellin is highly concentrated in hypothalamus and cerebellum, is found in brain synaptosomes, and is released via a calcium-dependent mechanism (Burnet et al., 1988) suggests that it may be a neuroactive substance in the hypothalamocerebellar/cerebellohypothalamic pathway.

III. Cerebellar Projections to the Hypothalamus

Direct (monosynaptic) projections from the cerebellar nuclei to the hypothalamus (Fig. 3) have been reported in studies that used anterograde and retrograde tracing (Haines and Dietrichs, 1984; Dietrichs and Haines, 1984, 1985a,b; Haines et al., 1984, 1985, 1990) and neurophysiologic (Min et al., 1989; Katafuchi and Koizumi, 1990; Pu et al., 1995) techniques.

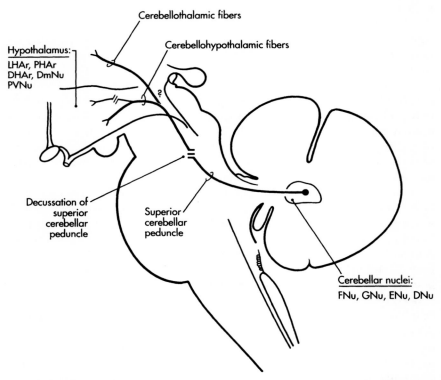

FIG. 3. Diagrammatic representation of cerebellohypothalamic projections. It is unclear (?) whether cerebellothalamic fibers are collaterals of cerebellohypothalamic fibers or vice versa. The fibers which recross in the hypothalamus are sparse in number. Those cell groups listed under hypothalamus are the principal targets of cerebellohypothalamic fibers.

Following injections of WGA-HRP into the cerebellar cortex plus portions of the nuclei, or into only the cerebellar nuclei, two patterns of labeling are seen. First, retrogradely labeled neurons, as described earlier, are found in the various nuclei and areas of the hypothalamus. Second, anterogradely labeled axons pass through the superior cerebellar peduncle, cross in the decussation of the superior cerebellar peduncle, follow the general trajectory of cerebellothalamic fibers, and then separate from the thalamic fasciculus to enter posterior portions of the contralateral hypothalamus (Fig. 3). These labeled axons originate from the injection site in the cerebellar nuclei as there is no evidence that Purkinje cells project to any extracerebellar target other than the vestibular nuclei (Haines et al., 1982; Haines and Dietrichs, 1991). Within the hypothalamus, anterogradely labeled axons distribute within the PHAr, LHAr, and DHAr at more rostral levels and in lateral portions of DMNu (Fig. 3). In all experiments, including those with injection sites restricted to the cerebellar nuclei, retrogradely labeled hypothalamic neurons are sometimes found within the domain of anterogradely labeled cerebellohypothalamic axons. Although synaptic boutons of cerebellar nuclear efferent fibers have not been demonstrated on hypothalamocerebellar cell bodies, the general apposition of anterogradely labeled cerebellar fibers and retrogradely labeled hypothalamic neurons suggests that some reciprocal connections may exist. It has been postulated (Haines et al., 1985, 1990) that the cerebellohypothalamic projection may be a combination of fibers passing directly to the hypothalamus and of fibers that are collaterals of cerebellothalamic axons (Fig. 3).

Cerebellohypothalamic axons cross in the decussation of the superior cerebellar peduncle to terminate contralateral to their origin. However, within the hypothalamus sparse numbers of these fibers recross and appear to terminate in LHAr, PHAr, and DHAr (Dietrichs and Haines, 1985b; Haines et al., 1990) on the side ipsilateral to the cell of origin (Fig. 3).

A direct projection from the fastigial nucleus and the "interpositus nucleus" (anterior or posterior not specified) to the lateral hypothalamic area and paraventricular nucleus has been reported in neurophysiologic studies. Min et al. (1989) and Katafuchi and Koizumi (1990) reported monosynaptic and polysynaptic responses of neurons in the LHAr and paraventricular nucleus, respectively, following intracellular and extracellular stimulation of neurons in the fastigial nucleus. The monosynaptic connections from the fastigial nucleus to the paraventricular nucleus and the LHAr are primarily, or exclusively, inhibitory. In contrast, the response of neurons in the LHAr to fastigial stimulation via the polysynaptic route is primarily excitatory (Min et al., 1989). Following the extracellular stimulation of neurons in the "interpositus nucleus," Pu et al. (1995) reported both monosynaptic and polysynaptic responses of neurons in the LHAr.

The majority of LHAr neurons had short latency inhibitory responses indicative of a direct connection from the interposed nuclei. In addition, most LHAr neurons that responded to stimulation of the "interposed nucleus" were glucose sensitive as seen by their decreased activity during stimulation after intravenous administration of glucose (Pu *et al.*, 1995).

IV. Collaterals of Hypothalamocerebellar Fibers

In addition to projections to the cerebellar cortex and nuclei, the hypothalamus also sends axons to wide areas of the neuraxis, including the basal forebrain, brain stem, and spinal cord (for reviews see Haines *et al.*, 1984; Dietrichs *et al.*, 1994a). A series of experiments, using double-labeling protocols, examined whether hypothalamic cells that projected to cerebellar structures also sent collateral branches to other targets. Injections of FB in the cerebellum, rhodamine (R)-labeled microspheres in the central nucleus of the amygdaloid complex, and nuclear yellow (NY) in the cervical spinal cord (Dietrichs and Haines, 1986) were made. This experimental approach resulted in many single (FB, R, and NY)-labeled neurons in the hypothalamus, but in very few double-labeled cells (FB + R, FB + NY, R + NY). No triple-labeled hypothalamic neurons were seen. Similar experiments with injections of R or diamidino yellow (DY) in the hippocampus and FB in the cerebellum (Dietrichs *et al.*, 1994b) revealed many single (R, FB, or DY)-labeled cells in the hypothalamus but no double-labeled neurons. Comparatively few double-labeled hypothalamic neurons were also seen by Dietrichs and Zheng (1984) following injections of FB and NY in the cerebellum and in cervical and thoracic levels of the spinal cord. These data strongly suggest that hypothalamic neurons which project either to the cerebellar cortex or to the nuclei (or both) are primarily dedicated to that specific connection and are not subservient collaterals of other efferent hypothalamic projections.

V. Indirect Hypothalamocerebellar Connections Mediated by the Basilar Pontine Nuclei and Lateral Reticular Nucleus

Studies employing axonal tracing methods have described hypothalamic projections (Fig. 2) that reach dorsomedial and ventromedial portions of the basilar pontine nuclei (BPN). In cats, such projections originate in the mammillary region and in DHAr, LHAr, and PHAr (for reviews see Aas and Brodal, 1988; Aas, 1989). In the rat, afferents to the medial BPN and portions of the nucleus reticularis tegmenti pontis (NRTP) arise from similar hypothalamic regions and the anterior hypothalamic area (Cruce, 1977; Hosoya and Matsushita, 1981; Mihailoff et al., 1989; Allen and Hopkins, 1990). Comparable hypothalamopontine connections in nonhuman primates have not been reported.

Retrograde tracing studies in cats have shown that BPN neurons receiving hypothalamic input project to cerebellar vermal visual areas (Hoddevik et al., 1977) and to intermediate and lateral portions of the anterior lobe (Brodal and Walberg, 1977). In a study that combined orthograde and retrograde tracing methods in the same animal, pontine afferents from the mammillary region overlapped with BPN neurons that projected to the paraflocculus (Aas and Brodal, 1989). Although similar combined orthograde/retrograde studies have not been undertaken in the rat, it appears that those medial BPN regions receiving hypothalamic input give rise to cerebellar projections to vermal lobules VI and VII (Azizi et al., 1981), VIII (Azizi et al., 1981; Eisenman, 1981), and lobule IX (Eisenman and Noback, 1980; Azizi et al., 1981), as well as hemispheral regions including lobulus simplex, crus I, crus II, and the paramedian lobule (Mihailoff et al., 1981), the paraflocculus (Mihailoff, 1983), and the flocculus (Blanks et al., 1983). Moreover, in the rat, neurons in those medial portions of BPN and NRTP that receive hypothalamic input also give rise to axon collaterals that terminate to some extent in each of the cerebellar nuclei (Mihailoff, 1993).

These *indirect* linkages joining the hypothalamus and cerebellum via the pontine nuclei involve widespread areas of the cerebellar cortex (Fig. 2). In addition, it is noteworthy to recognize that limbic regions of the cerebral cortex, as well as cortical regions involved in autonomic function in cat (Brodal, 1971; Aas and Brodal, 1989) and rat (Wyss and Sripanidkulchai, 1984), are sources of input to those medial BPN regions that receive hypothalamic terminations, thus providing another potential gateway to the cerebellum for autonomic-related information.

Indirect projections from the hypothalamus to the cerebellum may also pass through the lateral reticular nucleus (Fig. 2). Dietrichs et al. (1985a) reported retrogradely labeled cells in the LHAr, AHAr, DHAr, PHAr, in the TMNu, DMNu, and in cells of the periventricular zone after injections of WGA-HRP into the lateral reticular nucleus (LRNu). In addition, the LRNu contained anterogradely labeled fibers that branched and appeared to give rise to terminal boutons, following injections of WGA-HRP into the LHAr. These cases of LHAr injections also demonstrated retrogradely filled cell bodies at the periphery of LRNu at its interface with the surrounding reticular formation.

VI. Clinical Evidence of Cerebellar Influence on Visceral Function

It is well known that lesions of the cerebellum result in a range of somatomotor signs and symptoms. However, the autonomic (visceromotor) correlates of these lesions, some of which may help to localize the area of damage, have not been adequately investigated. Some patients may have visceromotor signs/symptoms that are frequently, if not always, interpreted as being related to increases in intracranial pressure with resultant injury (transient or permanent) to the medulla oblongata or pons.

Two patients (one in Oslo and one in Jackson) with cerebellar lesions and with specific and discrete visceromotor dysfunction concurrent with somatomotor signs and symptoms were examined. Both of these patients had well-localized cerebrovascular lesions (Figs. 4 and 5) that were confirmed by magnetic resonance imaging (MRI). In addition, both patients had no obvious pathology, other than their cerebellar lesion, as seen either during neurological examinations or in their MRI scans. For example, these patients showed no lethargy or alteration of consciousness, general mentation was normal, and the ventricles were of normal size, shape, and general configuration for patients of their ages. Also, details of brain structure, such as general shape and visibility of folia of the cerebellum, or the gyri of the cerebrum, were clearly evident on MRI.

A. Patient 1

This 60-year-old man initially experienced an acute and severe headache accompanied by vertigo and nausea. Two days later, when admitted to an Oslo hospital, he showed bradycardia (heart rate of 40), respiratory alkalosis (pH 7.55), and hyperventilation. A thorough neurological examination further revealed a severe gait ataxia, but no other cerebellar signs. The hyperventilation resolved in about 2 weeks and the ataxia slowly improved, but the bradycardia persisted for 6 months. After this he was lost to long-term follow-up.

Coronal, axial, and sagittal MRI scans of the brain (Fig. 4) revealed a small hematoma in the medial aspect of the cerebellum on the left side. The lesion was located adjacent to the midline, causing a slight bulge of the ventricular wall (Fig. 4B). However, there was no obstruction to cerebrospinal fluid flow and no signs of hydrocephalus. A comparison of the position of this lesion in all three planes indicates that it was centered in the medial (fastigial) cerebellar nucleus. In addition to damaged cells of the fastigial nucleus, it is likely that this lesion destroyed the terminals

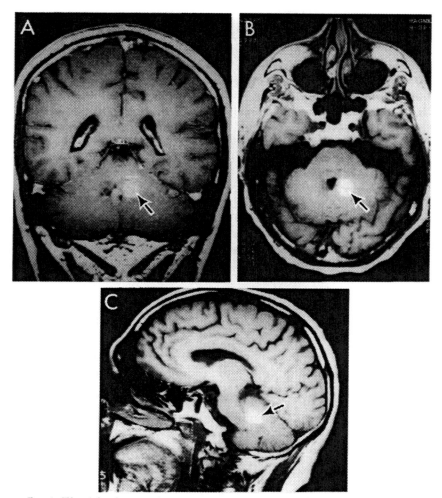

FIG. 4. T1-weighted magnetic resonance images of patient #1 in the coronal (A), axial (B), and sagittal (C) planes which reveal an area of hemorrhage principally involving the left fastigial nucleus.

of afferent fibers in the nucleus and efferent fibers exiting the nucleus via the uncinate fasciculus and juxtarestiform body. Whereas the lateral aspects of this lesion may have impinged on the interposed (emboliform or globose) cerebellar nuclei, the neurological examination suggested that the fastigial nucleus was the primary locus.

B. Patient 2

This 73-year-old man presented with a left-sided intention tremor, more pronounced in the arm than leg, when he was admitted to the University of Mississippi Medical Center in Jackson for dental extractions. Following referral to the neurology service, a detailed examination revealed that he had experienced a prior cerebellar hemorrhage that had been treated conservatively. With the exception of the cerebellar tremor, the neurologic examination revealed no abnormalities and mental function was normal for a patient of this age. The patient (and his wife) indicated that he had simply learned to live with the tremor.

MRI scans of the brain in all three planes (Fig. 5) revealed a lesion in the anterior lobe of the cerebellum on the left involving the cortex and subjacent white matter. As judged from the MRIs, the vascular lesion corresponded to the area supplied by the medial branch of the superior cerebellar artery. The defect appeared to involve parts of the medial (fastigial) cerebellar nucleus, most of the anterior (emboliform) and posterior (globose) interposed cerebellar nuclei, and portions of the superior cerebellar peduncle.

At rest the patient had no tremor, was relaxed, and showed no unusual facial movements or grimaces (Fig. 6). He did not complain of being hot or uncomfortable and his pupils were of normal and equal size and were reactive. The patient had no trouble performing the finger-to-nose movement with his right hand, but was unable to complete this movement with his left hand due to the pronounced intention tremor. While in a standing position the patient was asked to touch his left chest with his right hand and his right chest with his left hand. He completed this movement with the right extremity (Fig. 7A), but was unable to do so with the left extremity because of his tremor (Fig. 7B). During attempted movement on the left the patient had an involuntary facial grimace accompanied by a guttural "pathologic laughter" (Fig. 7B; see Doorenbos *et al.,* 1993). In addition, and *commensurate with* the cerebellar tremor and pathologic laughter, the patient's face flushed and was warm to the touch and both pupils dilated by 1–2 mm.

Not only was the patient's face warm to the touch, as determined by the neurologist, *during the movement,* but the patient also perceived this pathophysiologic event. In another examination, the patient was asked to pat his left knee with his left hand. During this unsuccessful attempt (Fig. 8A; his hand flails around his knee) the patient again had involuntary pathologic laughter, flushing of the face, and his pupils dilated. *Immediately* upon cessation of the tremor (within 2 sec) the patient fans his face with his right hand (Fig. 8B) and pointedly complains of being hot. This same

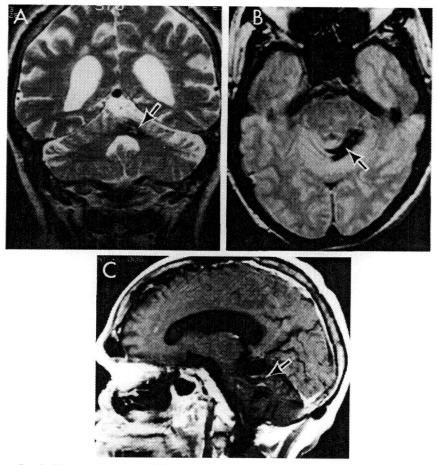

FIG. 5. Magnetic resonance T2W (A) and T1W (B and C) images (MRIs) of the head of patient #2 in coronal (A), axial (B), and sagittal (C) planes. See text for details.

sequence, cessation of tremor–immediate complaint of being hot, was repeated on several other occasions.

Two conclusions can be drawn from these cases. First, *concurrent with* their ataxia or tremor, these patients had obvious abnormal visceromotor manifestations (bradycardia, hyperventilation, flushing of face, pupil dilation). In patient 2 especially, these visceral responses were *not* seen before or after the intention tremor, but were seen *only during* the tremor that was evoked by an attempted movement. These observations support the

Fig. 6. Patient #2 at rest; taken from a videotape. Notice the lack of a facial grimace.

view that these abnormal visceromotor responses are directly related to the lesion in the cerebellum. Second, when considered in light of the known interconnections between the cerebellum and the hypothalamus, these clinical observations suggest that the cerebellum is involved in the regulation of visceromotor activity. Furthermore, these observations also suggest that circuits interconnecting the cerebellum and hypothalamus exist in humans.

VII. Conclusions

It would appear that the cerebellum is involved in functions considerably more diverse than just those related to the somatomotor system. In light

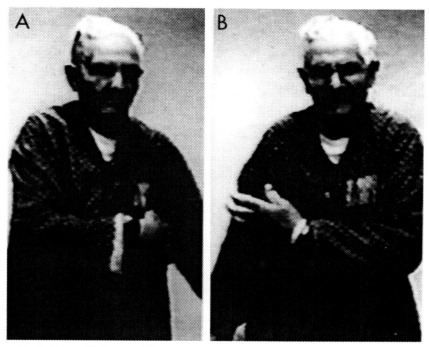

FIG. 7. Patient #2 (from a videotape) attempting a right-hand-to-left chest movement (A) and a left-hand-to-right chest movement (B). The patient is not able to complete the movement on the left side (B) and has an involuntary facial grimace during the attempted movement (compare B with A and with Fig. 5).

of the known visceral responses to cerebellar manipulation, and the now proven direct connections between the cerebellum and the hypothalamus, it has been suggested (Haines and Dietrichs, 1987, 1989; Dietrichs and Haines, 1985b; Dietrichs et al., 1994a) that the cerebellum is proactively involved in regulating the wide range of visceral functions that accompany somatomotor activity. For example, cerebellar nuclear neurons that are activated by collaterals of afferent axons (such as branches of spinocerebellar fibers) may communicate directly with hypothalamic nuclei via cerebellohypothalamic projections; such circuits may be excitatory and/or inhibitory. The hypothalamus, in turn, may directly modulate the activity of cerebellar nuclear or cerebellar cortical neurons via hypothalamocerebellar projections.

Intrinsic to this scheme are potential *feedforward* and *feedback circuits*. Feedforward in the sense that increased somatomotor activity, as indicated by increased proprioceptive input to the cerebellar nuclei, may be relayed directly to the hypothalamus (the primary visceral center in the brain),

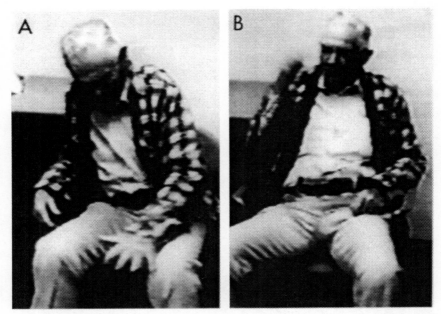

FIG. 8. Patient #2 (from a videotape) attempting unsuccessfully to pat his left knee with his left hand. He is unable to complete this movement (A) and *immediately* upon cessation of the attempted movement (and resulting tremor) he fans his face (B) and specifically complains of being hot.

alerting these cell groups to impending demands on the visceral motor system. Through its many efferent pathways, the hypothalamus elicits immediate visceral motor responses (see Reis *et al.*, 1973; Nakai *et al.*, 1983). Feedback in the sense that the hypothalamus may directly monitor/modulate the activity of cerebellar cortical and nuclear neurons during periods of increased, or fluctuating, demand and may influence the activity of these neurons as needed. In fact, it has been noted that histaminergic projections may modulate the sensitivity level of target cells to other inputs (Pollard and Schwartz, 1987), suggesting that the histaminergic hypothalamocerebellar projection may increase or decrease the responsiveness of cerebellar cortical or nuclear neurons to their other afferent inputs and, consequently, modify the ability of these neurons to influence their target cells.

A growing body of evidence suggesting an active cerebellar involvement in a wide range of nonsomatic and nonmotor activities, including behavioral and cognitive functions, requires a reassessment of how the cerebellum may actually function. Although lesions of the cerebellum (especially those involving cortex plus nuclei) frequently cause somatomotor deficits, there are situations in which cerebellar lesions (particularly those involving only the cortex) result in little or no obvious somatomotor signs or symptoms. In some cases, however, significant parts of the cerebellum have been injured. The main problem seems to be that dysfunctions of the cerebellum, other than somatomotor, have been overlooked because of inattention to their possible existence.

It is hypothesized that the cerebellum maintains within its cortex repetitive circuits that undertake the integration of information arriving, either directly or indirectly, from widely diverse areas of the neuraxis. Individual lobules of the cerebellar cortex receive somatosensory (tactile and proprioceptive) input, information relayed through visceral nuclei, and indirect input from areas of the cerebral cortex concerned with behavior, vision, and auditory function; these are only a few examples. The cerebellar cortex integrates and processes these diverse inputs to elicit the appropriate responses, some of which function in the somatomotor sphere, whereas others may be visceromotor or behavioral. At the very least, a motor response has somatic, visceral, behavioral, and cognitive components. The cerebellum is certainly capable of modulating or influencing *all* of these responses, not just the synergy of skeletal muscles.

Acknowledgments

The authors are indebted to Mr. Michael Schenk for his renderings of Figs. 1 to 3, to Ms. Anu Subramony for library work, to Mr. Bill Armstrong and Mr. Chuck Runyan for photography, and to Ms. Gail Rainer for typing the manuscript. The radiologists who initially read the MRIs are Dr. Søren Bakke (Oslo) and Dr. Joseph Ferguson (Jackson). Dr. Armin F. Haerer and Dr. Eric Undresser provided valuable assistance.

References

Aas, J.-E. (1989). Subcortical projections to the pontine nuclei in the cat. *J. Comp. Neurol.* **282,** 331–354.
Aas, J.-E., and Brodal, P. (1988). Demonstration of topographically organized projections from the hypothalamus to the pontine nuclei: An experimental anatomical study in the cat. *J. Comp. Neurol.* **268,** 313–328.
Aas, J.-E., and Brodal, P. (1989). Demonstration of a mamillo-ponto-cerebellar pathway. *Euro. J. Neurosci.* **1,** 61–74.
Airaksinen, M. S., and Panula, P. (1988). The histaminergic system in the guinea pig central nervous system: An immunocytochemical mapping study using an antiserum against histamine. *J. Comp. Neurol.* **273,** 163–186.

Airaksinen, M. S., Flügge, G., Fuchs, E., and Panula, P. (1989). Histaminergic system in the tree shrew brain. *J. Comp. Neurol.* **286**, 289–310.

Allen, G. V., and Hopkins, D. (1990). Topography and synaptology of mammillary body projections to the mesencephalon and pons in the rat. *J. Comp. Neurol.* **301**, 214–231.

Azizi, S. A., Mihailoff, G. A., Burne, R. A., and Woodward, D. J. (1981). The pontocerebellar system in the rat: An HRP study. I. Posterior vermis. *J. Comp. Neurol.* **197**, 543–558.

Ban, T., and Inoue, K. (1957). Interrelation between anterior lobe of cerebellum and hypothalamus. *Med. J. Osaka. Univ.* **7**, 841–857.

Bangma, G. C., and ten Donkelaar, H. J. (1982). Afferent connections of the cerebellum in various types of reptiles. *J. Comp. Neurol.* **207**, 255–273.

Bishop, G. A., Ho, R. H., and King, J. S. (1985a). Localization of serotonin immunoreactivity in the opossum cerebellum. *J. Comp. Neurol.* **235**, 301–321.

Bishop, G. A., Ho, R. H., and King, J. S. (1985b). An immunohistochemical study of serotonin development in the opossum cerebellum. *Anat. Embryol.* **171**, 325–338.

Blanks, R. H. I., Precht, W., and Torigoe, Y. (1983). Afferent projections to the cerebellar flocculus in the pigmented rat demonstrated by retrograde transport of horseradish peroxidase. *Exp. Brain Res.* **52**, 293–306.

Bratus, N. V., and Yoltukhovsky, M. V. (1986). Electrophysiological analysis of hypothalamic influences on the cerebellar cortex. *Physiol. J.* **32**, 257–263. [Original in Russian]

Brodal, P. (1971). The corticopontine projection in the cat. II. The projection from the orbital gyrus. *J. Comp. Neurol.* **142**, 141–152.

Brodal, P., and Walberg, F. (1977). The pontine projection to the cerebellar anterior lobe: An experimental study in the cat with retrograde transport of horseradish peroxidase. *Exp. Brain Res.* **29**, 233–248.

Burnet, P. W. J., Bretherton-Watt, D., Ghatei, M. A., and Bloom, S. R. (1988). Cerebellin-like peptide: Tissue distribution in rat and guinea-pig and its release from rat cerebellum, hypothalamus and cerebellar synaptosomes *in vitro*. *Neuroscience* **25**, 605–612.

Chambers, W. W. (1947). Electrical stimulation of the interior of the cerebellum in the cat. *Am. J. Anat.* **80**, 55–93.

Chambers, W. W., and Sprague, J. M. (1955a). Functional localization in the cerebellum. I. Organization in longitudinal cortico-nuclear zones and their contribution to the control of posture, both extrapyramidal and pyramidal. *J. Comp. Neurol.* **103**, 105–129.

Chambers, W. W., and Sprague, J. M. (1955b). Functional localization in the cerebellum. II. Somatotopic organization in cortex and nuclei. *Arch. Neurol. Psychiat.* **74**, 653–680.

Chan-Palay, V. (1975). Fine structure of labeled axons in the cerebellar cortex and nuclei of rodents and primates after intraventricular infusions with tritiated serotonin. *Anat. Embryol.* **148**, 235–265.

Cruce, J. A. F. (1977). An autoradiographic study of the descending connections of the mammillary nuclei of the rat. *J. Comp. Neurol.* **176**, 631–644.

Cummings, S. L. (1989). Distribution of corticotropin-releasing factor in the cerebellum and precerebellar nuclei of the cat. *J. Comp. Neurol.* **289**, 657–675.

Dietrichs, E. (1984). Cerebellar autonomic function: Direct hypothalamocerebellar pathway. *Science* **223**, 591–593.

Dietrichs, E., and Haines, D. E. (1984). Demonstration of hypothalamocerebellar and cerebellohypothalamic fibres in a prosimian primate (*Galago crassicaudatus*). *Anat. Embryol.* **170**, 313–318.

Dietrichs, E., and Haines, D. E. (1985a). Do hypothalamocerebellar fibres terminate in all layers of the cerebellar cortex? *Anat. Embryol.* **173**, 279–284.

Dietrichs, E., and Haines, D. E. (1985b). Observations on the cerebello-hypothalamic projection, with comments on non-somatic cerebellar circuits. *Arch. Ital. Biol.* **123**, 133–139.

Dietrichs, E., and Haines, D. E. (1986). Do the same hypothalamic neurons project to both amygdala and cerebellum? *Brain Res.* **364,** 241–248.
Dietrichs, E., and Haines, D. E. (1989). Interconnections between hypothalamus and cerebellum. *Anat. Embryol.* **179,** 207–220.
Dietrichs, E., Haines, D. E., and Qvist, H. (1985a). Indirect hypothalamocerebellar pathway? Demonstration of hypothalamic efferents to the lateral reticular nucleus. *Exp. Brain Res.* **60,** 483–491.
Dietrichs, E., Haines, D. E., Røste, G. K., and Røste, L. S. (1994a). Hypothalamocerebellar and cerebellohypothalamic projections: Circuits for regulating nonsomatic cerebellar activity? *Histol. Histopathol.* **9,** 603–614.
Dietrichs, E., Røste, G. K., Røste, L. S., Qvist, H. L., and Haines, D. E. (1994b). The hypothalamocerebellar projection in the cat: Branching and nuclear termination. *Arch. Ital. Biol.* **132,** 25–38.
Dietrichs, E., Walberg, F., and Haines, D. E. (1985b). Cerebellar nuclear afferents from feline hypothalamus demonstrated by retrograde transport after implantation of crystalline wheat germ agglutinin-horseradish peroxidase complex. *Neurosci. Lett.* **54,** 129–133.
Dietrichs, E., Wiklund, L., and Haines, D. E. (1992). The hypothalamocerebellar projection in the rat: Origin and transmitter. *Arch. Ital. Biol.* **130,** 203–211.
Dietrichs, E., and Zheng, Z.-H. (1984). Are hypothalamo-cerebellar fibres collaterals from the hypothalamo-spinal projection? *Brain Res.* **296,** 225–231.
Dow, R. S., and Moruzzi, G. (1958). "The Physiology and Pathology of the Cerebellum," pp. 50–396. University of Minnesota Press, Minneapolis, MN.
Eisenman, L. M. (1981). Pontocerebellar projections to the pyramis and copula pyramidis in the rat: Evidence for a mediolateral topography. *J. Comp. Neurol.* **199,** 77–86.
Eisenman, L. M., and Noback, C. R. (1980). The pontocerebellar projection in the rat: Differential projections to sublobules of the uvula. *Exp. Brain Res.* **38,** 11–17.
Ericson, H., Watanabe, T., and Köhler, C. (1987). Morphological analysis of the tuberomammillary nucleus in the rat brain: Delineation of subgroups with antibody against L-histidine decarboxylase as a marker. *J. Comp. Neurol.* **263,** 1–24.
Haines, D. E., and Dietrichs, E. (1984). An HRP study of hypothalamocerebellar and cerebellohypothalamic connections in squirrel monkey (*Saimiri sciureus*). *J. Comp. Neurol.* **229,** 559–575.
Haines, D. E., and Dietrichs, E. (1987). On the organization of interconnections between the cerebellum and hypothalamus. *In* "New Concepts in Cerebellar Neurobiology" (J. S. King, ed.), pp. 113–149. A. R. Liss, New York.
Haines, D. E., and Dietrichs, E. (1989). Non-somatic cerebellar circuits: A broader view of cerebellar involvement in locomotion. *J. Motor Behav.* **21,** 518–525.
Haines, D. E., and Dietrichs, E. (1991). Evidence of an x zone in lobule V of the squirrel monkey cerebellum: The distribution of corticonuclear fibers. *Anat. Embryol.* **184,** 255–268.
Haines, D. E., Dietrichs, E., Culberson, J. L., and Sowa, T. E. (1986). The organization of hypothalamocerebellar cortical fibres in the squirrel monkey (*Saimiri sciureus*). *J. Comp. Neurol.* **250,** 377–388.
Haines, D. E., Dietrichs, E., and Sowa, T. E. (1984). Hypothalamocerebellar and cerebellohypothalamic pathways: A review and hypothesis concerning cerebellar circuits which may influence autonomic centers and affective behavior. *Brain Behav. Evol.* **24,** 198–220.
Haines, D. E., May, P. J., and Dietrichs, E. (1990). Neuronal connections between the cerebellar nuclei and hypothalamus in *Macaca fascicularis:* Cerebello-visceral circuits. *J. Comp. Neurol.* **299,** 106–122.
Haines, D. E., Patrick, G. W., and Satrulee, P. (1982). Organization of cerebellar corticonuclear fiber systems. *In* "The Cerebellum: New Vistas" (S. L. Palay and V. Chan-Palay, eds.), pp. 320–371. Springer-Verlag, Berlin.

Haines, D. E., Sowa, T. E., and Dietrichs, E. (1985). Connections between the cerebellum and hypothalamus in tree shrew (*Tupaia glis*). *Brain Res.* **328,** 367–373.

Hoddevik, G. H., Brodal, A., Kawamura, K., and Hashikawa, T. (1977). The pontine projection to the cerebellar vermal visual area studied by means of the retrograde axonal transport of horseradish peroxidase. *Brain Res.* **123,** 209–227.

Hökfelt, T., and Fuxe, K. (1969). Cerebellar monoamine nerve terminals, a new type of afferent fibers to the cortex cerebelli. *Exp. Brain Res.* **9,,** 63–72.

Hökfelt, T., Johansson, O., and Goldstein, M. (1984). Central catecholamine neurons as revealed by immunohistochemistry with special reference to adrenaline neurons. *In* "Handbook of Chemical Neuroanatomy" (A. Bjöorklund and T. Hökfelt, eds.), Vol. 2, pp. 157–276. Amsterdam, Elsevier.

Hosoya, Y., and Matsushita, M. (1981). Brainstem projections from the lateral hypothalamic area in the rat as studied with autoradiography. *Neurosci. Lett.* **24,** 111–116.

Katafuchi, T., and Koizumi, K. (1990). Fastigial inputs to paraventricular neurosecretory neurones studied by extra- and intracellular recordings in rats. *J. Neurophysiol.* **421,** 535–551.

Künzle, H. (1983). Supraspinal cell populations projecting to the cerebellar cortex in the turtle (*Pseudemys scripta elegans*). *Exp. Brain Res.* **49,** 1–12.

Martner, J. (1975). Cerebellar influences on autonomic mechanisms: An experimental study in the cat with special reference to the fastigial nucleus. *Acta Physiol. Scand. Suppl.* **425,** 1–42.

Mihailoff, G. A. (1983). Intra- and interhemispheric collateral branching in the rat pontocerebellar system, a fluorescence double-label study. *Neuroscience.* **10,** 141–160.

Mihailoff, G. A. (1993). Cerebellar nuclear projections from the basilar pontine nuclei and nucleus reticularis tegmenti pontis as demonstrated with PHA-L tracing in the rat. *J. Comp. Neurol.* **330,** 130–146.

Mihailoff, G. A., Burne, R. A., Azizi, S. A., Norell, G., and Woodward, D. J. (1981). The pontocerebellar system in the rat: An HRP study. II. Hemispheral components. *J. Comp. Neurol.* **197,** 559–577.

Mihailoff, G. A., Kosinski, R. J., Azizi, S. A., and Border, B. G. (1989). Survey of non-cortical afferent projections to the basilar pontine nuclei: A retrograde tracing study in the rat. *J. Comp. Neurol.* **282,** 617–643.

Min, B.-I., Oomura, Y., and Katafuchi, T. (1989). Responses of rat lateral hypothalamic neuronal activity to fastigial nucleus stimulation. *J. Neurophysiol.* **61,** 1178–1184.

Mizuno, Y., Takahashi, K., Totsume, K., Ohneda, M., Konno, H., Murakami, O., Satoh, F., Sone, M., Takase, S., Itoyama, Y., and Mouri, T. (1995). Decrease in cerebellin and corticotropin-releasing hormone in the cerebellum of olivopontocerebellar atrophy and Shy-Drager syndrome. *Brain Res.* **686,** 115–118.

Moore, R. Y., and Card, J. P. (1984). Noradrenaline-containing neuron systems. *In* "Handbook of Chemical Neuroanatomy" (A. Björklund and T. Hökfelt, eds.), Vol. 2, pp. 123–156. Elsevier Amsterdam.

Moruzzi, G. (1940). Paleocerebellar inhibition of vasomotor and respiratory carotid sinus reflexes. *J. Neurophysiol.* **3,** 20–31.

Moruzzi, G. (1950). "Problems in Cerebellar Physiology," pp. 3–116. Thomas, Springfield.

Mugnaini, E., and Dahl, A.-L. (1975). Mode of distribution of aminergic fibers in the cerebellar cortex of the chicken. *J. Comp. Neurol.* **162,** 417–432.

Mugnaini, E., Dahl, A.-L., and Morgan, J. I. (1988). Cerebellin is a postsynaptic neuropeptide. *Synapse* **2,** 125–138.

Nakai, M., Iadecola, C., Ruggiero, D. A., Tucker, L. W., and Reis, D. J. (1983). Electrical stimulation of cerebellar fastigial nucleus increases cerebral cortical blood flow without change in local metabolism: Evidence for an intrinsic system in brain for primary vasodilation. *Brain Res.* **260,** 35–49.

Panula, P., Airaksinen, M. S., Pirvola, U., and Kotilainen, E. (1988). Histamine-immunoreactive neurons and nerve fibers in human brain. *Soc. Neurosci Abst.* **14,** 211.

Panula, P., Flügge, G., Pirvola, U., Auvinen, S., and Airaksinen, M. S. (1989a). Histamine-immunoreactive nerve fibers in the mammalian spinal cord. *Brain Res.* **484,** 234–239.

Panula, P., Pirvola, U., Auvinen, S., and Airaksinen, M. S. (1989b). Histamine-immunoreactive nerve fibers and terminals in the rat brain. *Neuroscience* **28,** 585–610.

Panula, P., Takagi, H., Inagaki, N., Yamalodani, A., Tokyama, M., Wada, H., and Kotlainen, E. (1993). Histamine-containing nerve fibers innervate human cerebellum. *Neurosci. Lett.* **160,** 53–56.

Pollard, H., and Schwartz, J.-C. (1987). Histamine neuronal pathways and their functions. *TINS* **10,** 86–89.

Pu, Y.-M., Wang, J.-J., Wang, T., and Yu, Q.-X. (1995). Cerebellar interpositus nucleus modulates neuronal activity of lateral hypothalamic area. *Neuroreport* **6,** 985–988.

Raymond, A. M. (1958). Responses to electrical stimulation of the cerebellum of unanesthetized birds. *J. Comp. Neurol.* **110,** 299–320.

Reis, D. J., Doba, N., and Nathan, M. A. (1973). Predatory attack, grooming, and consummatory behaviors evoked by electrical stimulation of cat cerebellar nuclei. *Science* **182,** 845–847.

Slemmon, J. R., Blacher, R., Danho, W., Hempstead, J. L., and Morgan, J. I. (1984). Isolation and sequencing of the cerebellum specific peptide. *Proc. Natl. Acad. Sci. USA* **81,** 6866–6870.

Slemmon, J. R., Danho, W., Hempstead, J. L., and Morgan, J. L. (1985). Cerebellin: A quantifiable marker for Purkinje cell maturation. *Proc. Natl. Acad. Sci. USA* **82,** 7145–7148.

Supple, W. F. (1993). Hypothalamic modulation of Purkinje cell activity in the anterior cerebellar vermis. *Neuroreport* **4,** 979–982.

Takeda, N., Inagaki, S., Shiosaka, S., Taguchi, Y., Oertel, W., Tohyama, M., Watanabe, T., and Wada, H. (1984). Immunohistochemical evidence for the coexistence of histidine decarboxylase-like and glutamate decarboxylase-like immunoreactivities in nerve cells in the magnocellular nucleus of the posterior hypothalamus of rats. *Proc. Natl. Acad. Sci. USA* **81,** 7647–7650.

Ter Horst, G. J., and Luiten, P. G. M. (1986). The projections of the dorsomedial hypothalamic nucleus in the rat. *Brain Res. Bull.* **16,** 231–248.

Urade, Y., Oberdick, J., Molinar-Rode, R., and Morgan, J. I. (1991). Precerebellin is a cerebellum-specific protein with similarity to the globular domain of complement C1q of B chain. *Proc. Natl. Acad. Sci. USA* **88,** 1069–1073.

Wada, C., and Ohtani, H. (1991). Molecular cloning of rat cerebellin-like protein cDNA which encodes a novel membrane-associated glycoprotein. *Mol. Brain Res.* **9,** 71–77.

Wang, T., Yu, Q.-X., and Wang, J.-J. (1994). Effects of stimulating lateral hypothalamic area and ventromedial nucleus of hypothalamus on cerebellar cortical neuronal activity in the cat. *Chin. J. Physiol. Sci.* **10,** 17–25.

Whiteside, J. A., and Snider, R. S. (1953). Relations of cerebellum to upper brain stem. *J. Neurophysiol.* **16,** 397–413.

Wyss, J. M., and Sripanidkulchai, K. (1984). The topography of the mesencephalic and pontine projections from the cingulate cortex of the rat. *Brain Res.* **293,** 1–15.

SECTION III
PHYSIOLOGICAL OBSERVATIONS

AMELIORATION OF AGGRESSION: RESPONSE TO SELECTIVE CEREBELLAR LESIONS IN THE RHESUS MONKEY[1]

Aaron J. Berman

Emeritus Director, Mt. Sinai Services at Elmhurst Hospital Center,
Elmhurst, New York 11373

I. Introduction
II. The Effect of Cerebellar Lesions on Emotional Behavior in the Rhesus Monkey
References

I. Introduction

I thank Plenum Press for permitting me to reproduce our article (Berman *et al.*, 1978). The study that was reported in that article was performed many years ago and was, in a sense, a product of its historical era. We were fortunate, in the 1960s, to have been given access to animals that had been reared in the same conditions as subjects in the early Harlow studies of maternal deprivation. Today, ethical considerations preclude raising animals in semi-isolation for the many years needed until they become violently aggressive.

In the 1960s, there was already some information suggesting a possible role for cerebellum in the modulation of emotional behavior, but this had received little attention as it was difficult to reconcile with what was then the accepted view of cerebellar function. Aside from our study, I am aware of only one experiment investigating the possibility of a taming effect in monkeys after lesions of the vermis, that reported by Peters and Monjan (1971). Peters (1969) had previously used a similar approach in laboratory cats, but cats tend to be quite tame, and it had therefore been difficult to establish a meaningful baseline of aggression. He did, however, observe a taming effect of cerebellar lesions in four aggressive adult male squirrel monkeys. The animals could be approached and stroked without difficulty after lesions which damaged the cerebellar cortex of the vermis, from the declive to the pyramis. The fighting and aggressive behaviors typical of the

---

[1] Adapted and reprinted from Berman *et al.* (1978), with permission of Plenum Publishing Corp.

animals before surgery were not seen during the 3-month postoperative period.

In our study, we also found a taming effect of cerebellar lesions. However, several differences between our study and that of Peters should be noted. Our animals were rhesus macaques, not squirrel monkeys. Further, our lesions were more extensive than those of Peters. We found a reduction of aggression, however, only in those animals with lesions that included the vermis. It is important also to note that our isolation-reared animals were probably more vicious and aggressive than the animals studied by Peters. They were certainly considerably more aggressive than our feral monkeys. It is likely, then, that the preoperative behaviors of the animals in Peter's and our study were not similar.

Thanks to subsequent research (Schmahmann, 1991, 1996; Leiner *et al.*, 1993) there is now an increasing understanding that the cerebellum, in addition to its role in motor control, is implicated in higher brain functions and affective behavior.

II. The Effect of Cerebellular Lesions on Emotional Behavior in the Rhesus Monkey

Although the relationship between limbic system structures and emotionality is well known, the role of the cerebellum in the control of affective behavior is not usually appreciated (Berman, 1970a,b, 1971). Involvement of the limbic system in the elaboration of emotional behavior has been demonstrated by studies such as those of Kluver and Bucy (1939), Pribram and Bagshaw (1953), and Weiskrantz (1956), in which a taming effect was reported following amygdaloidectomy in the monkey. Reduced emotionality in the monkey has also been reported following cingulectomy (Glees *et al.*, 1950) and posteromedial orbital frontal cortex ablations (Butter *et al.*, 1970).

Cerebellar participation in affective behavior has been indicated through the demonstration of functional connections between cerebellar and limbic structures. Anand *et al.* (1959), working with dogs, showed that stimulation of posterior paleocerebellar structures, including the flocculonodular lobe, evoked potentials in all limbic regions from which recordings were taken. Stimulation of anterior paleocerebellum evoked potentials mainly in orbital cortex, hippocampus, and posterior hypothalamus. Heath (1972a), using evoked potential and mirror focus techniques, demonstrated numerous interconnections between cerebellar nuclei and limbic system structures, including direct back-and-forth connections between the fastigial nuclei and the septal region and hippocampus.

Clark (1939) was the first to show that electrical stimulation of the cerebellum in the unanesthetized cat produced a cringe-like response with a kneading of the claws in a pleasure-like reaction occurring on occasion as a rebound effect. In addition, he noted what appeared to be a hypersensitivity to touch. Chambers (1947) found that stimulation of the midline vermis and fastigial nuclei resulted in hypersensitivity to sound and touch and in attempted escape. Chambers and Sprague (1955) and Sprague and Chambers (1959) reported that cerebellar ablations resulted in persistent pleasure reactions, as demonstrated by constant purring and kneading of the claws, while lesions of the fastigial nucleus resulted in lethargy. Peters (1969) and Peters and Monjan (1971) not only confirmed this work but also reported that squirrel monkeys which showed aggressive cage behavior before surgery became tame and tractable following lesions of the vermis.

One of the best established etiologies of aggression in the monkey is a lack of social interaction during development (Harlow and Harlow, 1962). Monkeys that are raised in individual cages, with visual and auditory contact but no bodily contact with each other, develop a number of bizarre behaviors. They indulge in a great deal of head or body rocking, sucking of body parts, and self-clutching. They are timorous of human approaches and, when placed with their peers, do not indulge in the usual social interactions of young primates. If such infants are maintained in partial isolation, then as adults they manifest violent aggressive behaviors directed at themselves and others.

Mason (1968) demonstrated that much of the abnormal behavior of partially isolated infant monkeys could be prevented by providing them with mobile rather than stationary surrogate mothers. Based on this finding, Prescott (1971) suggested that somatosensory and vestibular stimulation provided by the moving surrogate prevented the development of the abnormal behavior patterns. This interpretation parallels that of many human studies in which it has been suggested that children with autistic behaviors have suffered a lack of proprioceptive, labyrinthine, kinesthetic, and tactile stimulation (Fraiberg and Freedman, 1964; Freedman, 1968; Klein, 1962). It is supported by the finding of Heath (1972b) that isolation-reared adult monkeys show electroencephalographic abnormalities in cerebellar nuclei concerned with these sensory inputs, as well as limbic structures.

Given this evidence linking lack of proprioceptive, tactile, and vestibular input to the occurrence of so-called isolation behaviors in infant monkeys, and considering the abnormal discharge of relevant cerebellar nuclei in adult isolation-reared animals, we investigated the effects, if any, of various cerebellar lesions on the emotional behavior of the monkey. Some attempt was made to evaluate the effects of cerebellar lesions on the social interactions of juvenile rhesus monkeys, but our major attention was focused on the effects of such ablations on indices of aggression in adult male rhesus monkeys.

Six 8-month-old juvenile monkeys from Hazeltine Laboratories, that had been raised in partial social isolation were used to demonstrate the emotional and behavioral disorders consequent to such rearing. These animals were placed in a large observation chamber for a 1-hr period each month in order to observe their social interaction. Three walls and the ceiling of this enclosure were composed of cyclone fencing, to facilitate climbing, while one wall was of clear plastic for ease of monitoring and photographing group behavior. On each occasion, the six animals arranged themselves within the chamber so that little or no body contact occurred. Any body contact that did occur appeared to be accidental and was not followed by any further interchange. In the main, each huddled or sat or rocked as though alone in the cage.

The adult animals used in this study were six highly aggressive ferally reared male rhesus (Bishop, Love, Shiff, Diddle, Throck, and Anesthesia) that had been in our laboratory for several years and had been dropped from other experiments because of their aggressiveness toward laboratory personnel, who were unable to handle them. In addition, two violent, isolation-reared adult male rhesus (Ding and Dong) were contributed by Dr. Gary Mitchell. In studying these mature animals, we used the method of direct confrontation. This entailed placing two monkeys at a a time in the observation cage, while recording their encounter on film for subsequent analysis. This technique resulted in a social response obviously synonymous with the complex of behaviors known as aggression, although several drawbacks to its use were recognized. These included a lack of behavioral control, risk to the animals, and the impossibility of replication. Despite these limitations the method proved most rewarding. In each case, when two of these adult animals were placed in the observation chamber, violent fighting ensued and there was great difficulty in separating the animals. This response occurred even after several pairings of the same animals.

After the initial observations had been made, surgery was performed under endotracheal anesthesia with subjects in the sitting position. A suboccipital approach was used and a dural flap turned, exposing the posterior fossa. At this time mannitol (2 mg/kg of body weight) was administered intravenously, resulting in marked shrinking of the brain, which allowed visualization of the anterior lobe of the cerebellum. It was possible, in fact, with slight gentle retraction on the superior surface of the cerebellum, to visualize the pineal gland and the superior quadrigeminal body. The superficial cerebellar surface was then coagulated, after which approximately 2–3 mm of cortex was removed by suction. Care was taken to avoid direct damage to the deep cerebellar nuclei. Surgery was performed under 6 and 10× magnification with constant monitoring of vital signs. Drawings of the lesions were made before closure (Fig. 1).

Two of the juvenile animals (Wes and Betty) were operated at 11 months of age. In one animal (Wes), a lesion was made similar to that in Ding (see later) which included culmen, lobulus simplex, pyramis, paramedian lobule, uvula, and flocculonodular lobe. In the other animal (Betty), the lesion was similar to that in Dong (see later) and was confined to the neocerebellum. Two months after surgery, when placed in the observation chamber, these two animals began to interact. The initiation of each social contact was made by Wes, while Betty responded to his overtures. No such behavior was elicited from the remaining nonoperated juveniles, who remained withdrawn and nonresponsive.

A variety of lesions were made in the aggressive adult animals in an attempt to delineate cerebellar areas that might selectively alter emotional behaviors. In two cases (Fig. 1A; Ding, isolation reared, and Anesthesia, ferally reared), the superficial cortex of culmen, lobulus simplex, pyramis, paramedian lobule, uvula, and flocculonodular lobe were removed. Two monkeys (Fig. 1C; Dong, isolation reared, and Throck, ferally reared) received neocerebellar lesions; and three ferally reared animals (Fig. 1B; Bishop, Love, and Shiff) had incomplete vermian lesions that included pyramis, uvula, and part of the nodulus. A final ferally reared animal (Fig. 1D; Diddle) was subjected to resection of the cortex of culmen, lobulus simplex, pyramis, uvula, and part of the nodulus.

Postoperatively, the two animals with the lesions which included culmen, lobulus simplex, pyramis, paramedian lobule, uvula, and flocculonodular lobe (Fig. 1A) demonstrated marked behavioral alteration. They became quite docile and showed no signs of aggression in the confrontation situation. They could be handled by the laboratory personnel with no difficulty. The three animals with incomplete vermian lesions (Fig. 1B) continued to be as aggressive as before surgery. The animal in which the lesion was larger than in the previous three cases and included part of the anterior lobe of the cerebellum, however (Fig. 1D), became more tractable. All these animals showed a transient neurological deficit. The two monkeys with neocerebellar lesions (Fig. 1C) remained as aggressive after surgery as they had been before, despite the fact that both were ataxic throughout the study period.

The difference in effect between a fairly complete midline lesion and one which was laterally placed was clearly seen in the two adult isolation-reared animals (Ding and Dong). Ding, whose lesions included culmen, lobulus simplex, pyramis, paramedian lobule, uvula, and flocculonodular

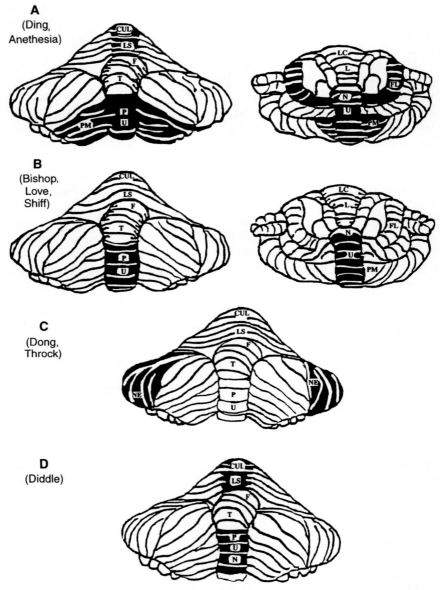

FIG. 1. Cerebellar lesions. CUL, culmen; F, folium vermis; FL, flocculus; L, lingula; LC, lobulus centralis; PM, paramedian lobule; LS, lobulus simplex; N, nodulus; NE, neocerebellum; P, pyramis; T, tuber; U, uvula. (Modified from Berman *et al.*, 1978; with permission.)

lobe, was immediately tamed. He could be handled easily. When placed in the observation chamber with Dong (neocerebellar lesion) during the first postoperative month, little interaction between the two animals occurred. During this period, though, Dong was grossly incapacitated. His gait was extremely unsteady and there was marked ataxia, making it difficult for the animal to even feed itself. Three months after surgery, although still ataxic and unable to climb, Dong pursued Ding with fury. Ding would no longer remain on the ground once Dong entered the cage. Similarly, none of the laboratory personnel could be persuaded to enter the cage with him. Dong's neurological deficit gradually subsided, but his aggressiveness increased and has, in fact, persisted to the present time. Another highly aggressive feral male, slightly larger and heavier than Dong, was placed in the encounter chamber with him. A violent fight ensued, from which it was difficult to extricate the animals. Although Dong emerged from this encounter a decided second best, two additional confrontations resulted in similar pugnacity.

The number of animals used in this study was too few to reach unequivocal conclusions, but there does appear to be a differential effect on aggressive behavior between midline and neocerebellar lesions in the adult animal. The taming seen following midline cerebellar lesions was persistent and dramatic whereas no such effect resulted from neocerebellar lesions. The altered behavior of the two juvenile animals is difficult to interpret within the same framework, however, as a beneficial result seemed to occur in each, despite the fact that one had a vermian and the other a neocerebellar lesion. It should be noted, nevertheless, that the greater socializing effect was apparent in the animal with the midline vermian lesion.

Precise correlation of behavior with the lesions must of course await histological documentation. It is possible that the deep cerebellar nuclei may have been injured at the time of surgery or subsequently as a result of interference with the blood supply. Notwithstanding these reservations, the present study suggests that vermian lesions of the cerebellum result in the modification of aggressive behavior in adult rhesus monkeys and that this alteration in behavior is unrelated to any neurological deficit.

It is tempting to speculate that there might be a subgroup of individuals who suffer from chronic violent aggressive behavior as a consequence of their early somatosensory depravation. If so, it may be fruitful for investigators to consider obtaining neuroimaging studies of the cerebellum with the aim of introducing new therapies for intractable cases including, as a last resort, resection of selected parts of the vermis. Of course further research for replication and verification is needed before such a radical approach could be considered.

Acknowledgment

The original publication was supported in part by National Science Foundation Grant GB30920.

References

Anand, B. K., Malhotra, C. L., Singh, B., and Dua, S. (1959). Cerebellar projections to limbic system. *J. Neurophysiol.* **22,** 451–458.
Berman, A. J. (1970a). Somatosensory-cerebellar lesions and behavior. *In* "Neural-Behavioral Ontogeny of Violent-Aggressive and Autistic-Depressive Disorders" (J. W. Prescott, chairman). Symposium presented at Third Annual Winter Conference on Brain Research, Snowmass-at-Aspen, Colorado.
Berman, A. J. (1970b). Cerebellar decortication and the modification of aggressive behavior. *In* "Maternal–Social Deprivation as Functional Somatosensory Deafferentation in the Abnormal Development of the Brain and Behavior" (A. H. Riesen, chairman). Symposium presented at the 78th Annual Convention of the American Psychological Association, Miami, FL.
Berman, A. J. (1971). Cerebellar decortication and the modification of abnormal behavior in isolation-reared rhesus monkeys. *In* "Neurobiological Perspectives on Parental, Social and Sensory Deprivations" (J. W. Prescott, chairman). Harlow Memorial Symposium presented at the 79th Annual Convention of the American Psychological Association, Hawaii.
Berman, A. J., Berman, D., and Prescott, J. W. (1978). The effect of cerebellar lesions on emotional behavior in the rhesus monkey. *In* "The Cerebellum, Epilepsy and Behavior" (I. S. Cooper, M. Riklan, and R. S. Snider, eds.), pp. 277–284. Plenum Press, New York.
Butter, C. M., Snyder, D. R., and McDonald, J. A. (1970). Effects of orbital frontal lesions on aversive and aggressive behaviors in rhesus monkeys. *J. Comp. Physiol. Psychol.* **72,** 132–144.
Chambers, W. W. (1947). Electrical stimulation of the interior of the cerebellum in the cat. *Am. J. Anat.* **80,** 55–94.
Chambers, W. W., and Sprague, J. M. (1955). Functional localization in the cerebellum. 1. Organization in longitudinal cortico-nuclear zones and their contribution to the control of posture both extrapyramidal and pyramidal. *J. Comp. Neurol.* **103,** 105–129.
Clark, S. L. (1939). Responses following electrical stimulation of the cerebellar cortex in the normal cat. *J. Neurophysiol.* **2,** 1936.
Fraiberg, S., and Freedman, D. (1964). Studies in the ego development of the congenitally blind child. *Psychoanal. Study Child.* **19,** 113–169.
Freedman, D. A. (1968). The influence of congenital and perinatal sensory deprivations on later development. *Psychonomics* **9,** 272–277.
Glees, P., Cole, J., Whitty, C. W. M., and Cairns, H. (1950). The effects of lesions in the cingular gyrus and adjacent areas in monkeys. *J. Neurosurg. Psychiat.* **13,** 178–190.
Harlow, H. F., and Harlow, M. K. (1962). The effect of rearing conditions on behavior. *Bull. Meninger Clin.* **26,** 213–224.
Heath, R. G. (1972a). Physiologic basis of emotional expression: Evoked potential and mirror focus studies in rhesus monkeys. *Biol. Psychiat.* **5,** 15–31.

Heath, R. G. (1972b). Electroencephalographic studies in isolation-reared monkeys with behavioral impairment. *Dis. Nerv. Syst.* **33,** 157–163.

Klein, G. W. (1962). Blindness and isolation. *Psychoanal. Study Child.* **17,** 82–93.

Kluver, H., and Bucy, P. C. (1939). Preliminary analysis of functions of temporal lobes in monkeys. *Arch. Neurol. Psychiat.* **42,** 979–1000.

Leiner, H. C., Leiner, A. L., and Dow, R. S. (1993). Cognitive and language functions of the human cerebellum. *Trends Neurosci.* **16,** 444–447.

Mason, W. A. (1968). "Early Social Deprivation in the Non-Human Primate: Implications for Human Behavior in Environmental Influences." The Rockefeller University Press and Russell Sage Foundation, New York.

Peters, M. (1969). "A Cerebellar Role in Behavior." Ph.D. thesis, University of Western Ontario.

Peters, M., and Monjan, A. A. (1971). Behavior after cerebellar lesions in cats and monkeys. *Physiol. Behav.* **6,** 205–206.

Prescott, J. W. (1971). Early somatosensory depreviation as an ontogenetic process in the abnormal development of the brain and behavior. In "Medical Primatology." Karger, Basel.

Pribram, K. H., and Bagshaw, M. (1953). Further analysis of the temporal lobe syndrome utilizing fronto-temporal ablations. *J. Comp. Neurol.* **99,** 347–375.

Schmahmann, J. D. (1991). An emerging concept: The cerebellar contribution to higher function. *Arch. Neurol.* **48,** 1178–1187.

Schmahmann, J. D. (1996). From movement to thought: Anatomic substrates of the cerebellar contribution to cognitive processing. *Hum. Brain Mapp.* **4,** 174–198.

Sprague, J. M., and Chambers, W. W. (1959). An analysis of cerebellar function in the cat as revealed by its partial and complete destruction and its interaction with cerebral cortex. *Arch. Ital. Biol.* **97,** 68–88.

Ward, A. A. (1948). The cingular gyrus: Area 24. *J. Neurophysiol.* **11,** 12–23.

Weiskrantz, L. (1956). Behavioral changes associated with ablation of the amygdaloid complex in monkeys. *J. Comp. Physiol. Psychol.* **49,** 381–391.

AUTONOMIC AND VASOMOTOR REGULATION

Donald J. Reis and Eugene V. Golanov

Division of Neurobiology, Department of Neurology and Neuroscience, Cornell University Medical College, New York, New York 10021

I. Introduction
II. Regulation of Cerebral Blood Flow and Metabolism by the Fastigial Nucleus
    A. Effect of Fastigial Nucleus Stimulation on the Systemic Circulation: Fastigial Pressor and Depressor Responses
    B. Regulation of the Cerebral Circulation
III. Neuroprotection Evoked from the Fastigial Nucleus
    A. Neuroprotection from the Fastigial Nucleus in Focal Cerebral Ischemia
    B. Salvage is Independent of Changes in Blood Flow: Conditioned Neuroprotection
    C. Alternate Mechanisms of Neurogenic Neuroprotection
IV. Conclusions
    References

The cerebellum not only modulates the systemic circulation, but also profoundly influences cerebral blood flow (rCBF) and metabolism (rCGU), and initiates long-term protection of the brain from ischemia. Electrical stimulation of the rostral ventral pole of the fastigial nucleus (FN), elevates arterial pressure (AP), releases vasoactive hormones, elicits consummatory behavioral and other autonomic events and site specifically elevates rCBF independently of changes in rCGU. Cerebral vasodilation results from the antidromic excitation of axons of brain stem neurons which innervate cerebellum and, through their collaterals, neurons in the rostral ventrolateral reticular nucleus (RVL). RVL neurons initiate cerebral vasodilation over polysynaptic vasodilator pathways which engage a population of vasodilator neurons in the cerebral cortex. In contrast, intrinsic neurons of FN, when excited, elicit widespread reductions in rCGU and, secondarily, rCBF, along with sympathetic inhibition. Electrical stimulation of FN can reduce the volume of a focal cerebral infarction produced by occlusion of the middle cerebral artery by 50%. This central neurogenic neuroprotection is long lasting (weeks) and is not due to changes in rCBF or rCGU. Rather, it appears to reflect alterations in neuronal excitability and/or downregulation of inflammatory responses in cerebral vessels. The FN, therefore, appears to be involved in widespread autonomic, metabolic, and behavioral control, independent of motor control. The findings imply that the FN receives inputs from neurons, probably widely represented in the central autonomic core, which may provide continuing information processing of autonomic and behavioral states. The cerebellum may also widely modulate the state of cortical reactivity to ischemia, hypoxia, and possibly other neurodegenerative events.

I. Introduction

That the cerebellum can regulate the systemic circulation has been recognized for over a century (see Dow and Moruzzi, 1958; Ito, 1984). Numerous studies have focused on how cerebellar stimulation or lesions may regulate the resting and reflex control of arterial pressure (AP), control heart rate, and influence sympathetic and vagal nerve discharge. Other investigations have elaborated on how the cerebellum influences the performance of various viscera and how such control may integrate with behavior. Although almost all of the early studies examined the interaction of the cerebellum with the systemic circulation, none dealt with the possibility that the cerebellum can influence the cerebral circulation and, as a consequence, influence the response of the brain to ischemia.

As a result, this chapter focuses on the interaction of the cerebellum with the cerebral circulation. Specifically, the role of a subarea of the cerebellar fastigial nucleus (FN), its rostral and ventromedial pole, on cerebral blood flow (rCBF), cerebral metabolism (as reflected by regional cerebral glucose utilization, rCGU), and the influence of the region on modifying the expression of focal cerebral ischemia are examined. Investigations have revealed that this region of FN can profoundly modify rCBF, rCGU, and can, most surprisingly, provide long-lasting protection against focal ischemic infarctions.

II. Regulation of Cerebral Blood Flow and Metabolism by the Fastigial Nucleus

A. Effects of Fastigial Nucleus Stimulation on the Systemic Circulation: Fastigial Pressor and Depressor Responses

That electrical stimulation of the cerebellar FN could potently elevate arterial pressure (AP) and modestly elevate heart rate (HR) (Fig. 1) was discovered independently by Miura and Reis (1969) and Achari and Downman (1969) in anesthetized cat. The response, the fastigial pressor response (FPR) (Miura and Reis, 1969), has been replicated in a range of species, including rat (e.g., Nakai, et al., 1982), rabbit (Reis et al., 1982; Bradley et

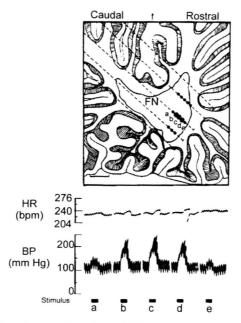

FIG. 1. Localization of optimal loci for fastigial pressor response in a sagittal section of cerebellum in anesthetized cat. (Top) Electrode tracks penetrating the cerebellum. The cerebellum was stimulated at dashed sites along the track. Positive sites are shown by solid circles. Small and large circles represent weak and more powerful pressor responses, respectively. Positive points were identified by stimulation at 200-μm steps with a 12-sec stimulus train (50 cycles/sec, 0.2 ma). (Bottom) Points along tracks identified by letters are represented by appropriate polygraph tracings below each cross section showing heart rate (HR, beats/min) and blood pressure (BP, mmHg). Note that maximal responsive sites are localized in the rostral ventromedial portion of fastigial nucleus (FN). (From Miura and Reis, 1969.)

al., 1987b), dog (Dormer and Stone, 1976), monkey (McKee et al., 1976), and human (Elisevich and Redekop, 1991; Hirano et al., 1993).

Elevations in AP in the FPR result from a sympathetically mediated increase in peripheral resistance resulting from a differentiated activation of spinal preganglionic sympathetic neurons. In cat the response is associated with an increase in resistance and a reduction in blood flow in the mesenteric, renal, and femoral arteries (Doba and Reis, 1972; Martner, 1975a). In intact animals the elevations of AP elicited by a brief stimulus train are graded, appear within 1–2 sec, are sustained, and recover almost immediately when stimulation is terminated (Achari and Downman, 1969; Miura and Reis, 1969). The active site for the response within the FN is highly restricted to the rostral ventrolateral quadrant (Miura and Reis,

1969; Takahashi *et al.*, 1995) (Fig. 1). Vasoconstriction is coupled with suppression of the baroreceptor reflex (Lisander and Martner, 1971; Del Bo *et al.*, 1984), a feature that explains the potency of the rise of blood pressure. Stimulation of the FN also releases adrenaline and noradrenaline from the adrenal medulla (Del Bo *et al.*, 1983a), renin from the kidney (Manning *et al.*, 1985), and arginine vasopressin (AVP) from the posterior pituitary (Del Bo *et al.*, 1983b). The pattern of the neuronal and hormonal responses to stimulation replicates that of the orthostatic reflex, i.e., the antigravity reflex sympathetic response excitation initiated by assumption of an upright posture (Doba and Reis, 1972).

Within the FN, electrical stimulation excites local neurons and also axons projecting to, through, or adjacent to the FN. These may arise from Purkinje cells of the cerebellar cortex, from other deep nuclei, or from fibers entering the cerebellum from brain stem nuclei. Axons may be excited orthodromically or antidromically; in the case of the latter, stimulation will concurrently excite collateral branches of the same neuron to excite all areas sharing its innervation.

To determine which elements in FN are responsible for the FPR, studies have been undertaken to compare the reponses to electrical stimulation of the FN with those produced by microinjection, at the same site, of an excitatory amino acid.

In contrast to electrical stimulation, microinjection of L-glutamate or other excitatory amino acids, agents that only excite perikarya, into the FN fail to elevate AP (Chida *et al.*, 1986; Bradley *et al.*, 1987a; Henry and Connor, 1989). Rather, L-glutamate or, more effectively, its more stable analog (e.g., homocysteic or kainic acids) dose dependently and site specifically lower AP and heart rate (Chida *et al.*, 1986) to evoke a fastigial depressor response (FDR) (Figs. 2 and 3). When FN neurons are selectively destroyed by excitotoxins, the chemically induced depressor response is abolished, while the pressor response persists (Chida *et al.*, 1989a) (Fig. 2).

Thus the FPR and FDR are initiated by different elements within the cerebellum. The FPR results from the excitation of axons innervating and/or passing through the nucleus, whereas the FDR results from the excitation of intrinsic fastigial neurons. Since the FPR persists after the destruction of intrinsic neurons of FN and following chronic ablation of the midline cerebellar cortex (K. Chida *et al.*, unpublished observations), the authors conclude that the FPR is generated by antidromic excitation of the axons of brain stem neurons which innervate the cerebellum. Despite the fact that electrical stimulation also excites local neurons, the finding also indicates that the pressor response is prepotent and overwhelms the sympathoinhibitory response.

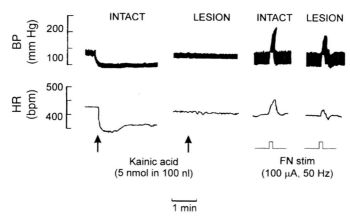

FIG. 2. Effects of microinjection of kainic acid (KA) into, or electrical stimulation of, the fastigial nucleus (FN) before (intact), and 5 days after (lesion), an excitotoxic lesion of FN in anesthetized rats. Note that excitotoxic lesions of FN block the depressor response to KA, but not the pressor responses to electrical stimulation. (Modified from Chida *et al.*, 1989a.)

The brain stem area critical for the elevations and reductions in AP elicited from the chemical stimulation of the FN is the rostral ventrolateral reticular nucleus (RVL). Bilateral lesions of this region abolish pressor and depressor reponses (Fig. 3) (McAllen, 1985; Dormer *et al.*, 1986; Chida *et al.*, 1990a). The implications of this finding are discussed next.

B. REGULATION OF THE CEREBRAL CIRCULATION

In their study in anesthetized cat, Doba and Reis (1972) first noted that while electrical stimulation of the FN constricted vessels and reduced regional blood flow through most vascular beds, it increased flow in the common carotid artery and increased oxygenation of the cerebral cortex. These observations suggested that FN stimulation might elevate rCBF. Subsequent investigations by the authors and others in which rCBF has been measured by electromagnetic flowmetry (Doba and Reis, 1972), H_2 clearance (Reis *et al.*, 1982), isotope dilution (Kety method) in tissue homogenates (Nakai *et al.*, 1982) or tissue sections (Nakai *et al.*, 1983), or with laser Doppler flowmetry (LDF) (Iadecola and Reis, 1990; Golanov and Reis, 1995) have confirmed and extended this inference in rat (Iadecola and Reis, 1990; Golanov and Reis, 1995), cat (Talman *et al.*, 1991), and monkey (McKee *et al.*, 1976; Goadsby and Lambert, 1989).

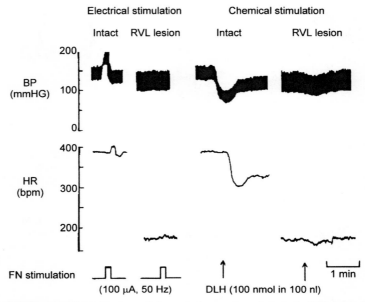

FIG. 3. Effects of bilateral lesions of the rostral ventrolateral medulla (RVL) on electrical- and chemical (D,L- homocysteic acid, DHL)-evoked changes of arterial pressure (AP) and heart rate (HR) elicited by electrical stimulation of the fastigial nucleus (FN). Electrical (left) or chemical (right) stimulation of the FN in RVL-intact animals produced fastigial pressor or depressor responses, respectively. A bilateral lesion of RVL blocked changes in AP and HR evoked by electrical or chemical stimulation of FN. (From Chida et al., 1990b.)

1. *Effects of Electrical Stimulation on Regional Cerebral Blood Flow (rCBF) and Metabolism*

Electrical simulation of the FN in animals anesthetized with suitable amounts of chloralose or halothane and under conditions in which blood gases and AP are stabilized potently elevates rCBF (Fig. 4). rCBF is elevated throughout the entire central nervous system (CNS) (Nakai et al., 1983), including spinal cord (Nakai et al., 1982), albeit with regional variability. The time course of the changes in rCBF differ from those of AP, however. The elevations in rCBF, as measured by LDF, rise gradually, decline during the stimulus epoch, and recover over minutes (Iadecola and Reis, 1990; Yamamoto et al., 1993; Golanov and Reis, 1995). In contrast, the changes in AP are stimulus locked, sustained (Miura and Reis, 1969), and decline slowly after stimulation stops (Yamamoto et al., 1993). The elevations in rCBF are not uniform throughout the brain. Maximal elevations, sometimes nearly 200% of control, appear in the cerebral cortex, particularly in frontal

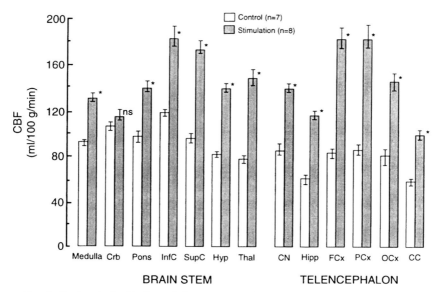

FIG. 4. Regional distribution of changes in cerebral blood flow (rCBF) measured by [^{14}C]-iodoantipyrine in brain homogenates (expressed as ml/100 g × min) elicited by electrical stimulation of fastigial nucleus (FN) in anesthetized, paralyzed rats. Note that FN stimulation increases rCBF in all brain regions except cerebellum, with maximal responses occurring in the cerebral cortex. Crb, cerebellum; InfC, inferior colliculus; SupC, superior colliculus; Hyp, hypothalamus; Thal, thalamus; CN, caudate putamen; Hipp, hippocampus; FCx, PCx, OCx, frontal, parietal, and occipital cortices; CC, corpus callosum. (*$p < 0.005$). (From Reis et al., 1991.)

areas (Nakai et al., 1983) (Fig. 4). In general, the smallest changes occur within white matter.

The most striking feature of the cerebrovascular vasodilation elicited from FN is that, in most regions, it is not associated with proportional changes in rCGU (Fig. 5). This is most pronounced with all areas of the cerebral cortex in which, despite the potent increase in rCBF, there is no metabolic activation. As a result, rCBF and rCGU are uncoupled. In other regions, mostly mono- or polysynaptically linked to FN, some elevations in rCGU can be detected. However, in most, the elevations in rCGU are disproportionally small as compared with the changes in rCBF. It is only within areas of thalamus monosynaptically innervated by FN efferents (Andrezik et al., 1984; Person et al., 1986) that rCGU and rCBF are proportionally elevated, as expected, when primary pathways are excited (Sokoloff et al., 1977).

The uncoupling of flow and metabolism within the cerebral cortex by FN stimulation is a most important finding, for it challenges a traditional

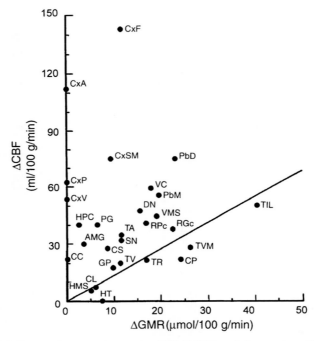

FIG. 5. Relationship between changes in rCBF (ΔCBF) and changes in regional cerebral glucose metabolic rate (ΔGMR) in different brain areas of the rat in response to electrical stimulation of the fastigial nucleus. The straight line represents the relation between rCBF and rGMR observed in unstimulated rats. In most brain regions, changes in ΔCBF are disproportional to changes in ΔGMR, so that most of the points lie away from the proportionality line. AMG, amygdala; CC, corpus callosum; CL, inferior colliculus; CP, caudate putamen; CS, superior colliculus; CxA, auditory cortex; CxF, frontal cortex; CxP, parietal cortex; CxSM, sensory motor cortex; CxV, visual cortex; DN, cerebellar dentate nucleus; GP, globus pallidus; HMS, cerebellar hemisphere; HPC, hippocampus; HT, hypothalamus; PbD, dorsal parabrachial nucleus; PbM, medial parabrachial nucleus; PG, pontine gray matter; RGc, gigantocellular reticular nucleus; RPc, parvocellular reticular nucleus; SN, substantia nigra; TA, thalamic anterior nucleus; TIL, thalamic intralaminar nuclei (centromedial and paracentral); TR, thalamic reticular nucleus; TV, thalamic ventral nucleus; TVM, thalamus ventromedial nucleus; VC, vestibular complex; VMS, cerebellar vermis. (From Reis *et al.*, 1989.)

view that increased neuronal activity is inextricably associated with proportional elevations in rCGU and rCBF, i.e., flow and metabolism are coupled (e.g., Edvinsson *et al.*, 1993; Heistad and Kontos, 1983; Sokoloff *et al.*, 1977; Lou *et al.*, 1987). Coupling, moreover, assumes that the elevations in rCBF are a consequence of metabolism and mediated by release of vasoactive metabolites from discharging neurons and/or associated glia. (Clarke and Sokoloff, 1994). The authors' discovery that rCBF could rise markedly and independently from metabolism with cerebellar stimulation indicated that the brain contains networks that are dedicated to regulating its own blood flow.

The elevations in rCBF elicited from FN appear mediated in their entirety by the excitation of neural pathways entirely contained within the CNS. The response persists after bilateral cervical sympathectomy, adrenalectomy, or transection of the spinal cord at the first cervical segment, indicating that sympathetic neuroeffector mechanisms are not involved (Nakai et al., 1982). They also persist after the pharmacological blockade of cranial parasympathetic vasodilator pathways (Reis et al., 1982; Nakai et al., 1983) or transection of the chorda tympani. However, the response is abolished within the cerebral cortex ipsilateral, but not contralateral or caudal, to a unilateral electrolytic lesion of the basal forebrain (Iadecola et al., 1983), indicating that it depends on a projection, probably extrathalamic, traversing the medial forebrain bundle.

2. Effects of Chemical Stimulation on rCBF and Metabolism

As with AP, chemical stimulation of FN elicits a cerebrovascular response differing in polarity from that evoked by electrical stimulation. Microinjection of excitatory amino acids dose dependently reduces rCBF throughout the brain in association with a parallel reduction in rCGU (Fig. 6) (Chida et al., 1989b). The fact that the relationships across the brain between rCBF and rCGU remain tightly correlated after stimulation suggests that the initial response is a downregulation of metabolism.

Following excitotoxic lesions of FN, the fall of rCBF to chemical stimulation disappears, but the elevation in response to electrical stimulation persists (Fig. 2). These observations indicate that at least two mechanism that influence rCBF are represented in FN. One elevates rCBF without changing rCGU and results from the excitation of local axons, presumably collaterals of brain stem neurons. The other reduces rCBF by inhibiting metabolism and results from the stimulation of intrinsic neurons. The findings also indicate that rCBF and rCGU are not maximally reduced, even under anesthesia, but can be further reduced by neural excitation. Hence, like AP, brain metabolism and blood flow are *tonically* maintained neurogenically. The region of brain that tonically maintains rCGU and rCGF is not known. It is not the RVL (Underwood et al., 1994), the region tonically maintaining AP (Reis et al., 1996), as bilateral lesions fail to modify rCBF and rCGU if AP is maintained (Underwood, 1988; Underwood et al., 1994).

Perhaps, most remarkably, the studies indicate that this small area of cerebellum, when excited, can reduce metabolism throughout the brain. The meaning of this control and its behavioral concomitant is unknown.

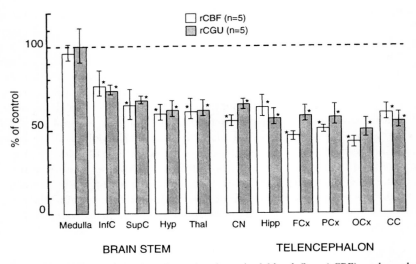

FIG. 6. Proportional reductions in regional cerebral blood flow (rCBF) and cerebral glucose utilization (rCGU) (expressed as percentage of vehicle-injected controls) elicited by the microinjection of kainic acid into fastigial nucleus (FN) in all brain regions, except the medulla, in anesthetized rats. Reductions are greatest in the cerebral cortex. InfC, inferior colliculus; SupC, superior colliculus; Hyp, hypothalamus; Thal, thalamus; CN, caudate putamen; Hipp, hippocampus; FCx, PCx, OCx, frontal, parietal, and occipital cortices; CC, corpus callosum ($*p < 0.05$). (From Reis et al., 1991.)

3. Pathways: Role of the Rostral Ventrolateral Reticular Nucleus

The pathways that mediate systemic responses and cerebral circulations to electrical or chemical stimulation of FN are barely understood. However, it is evident that a crucial relay is localized within the RVL. Bilateral lesions within, or adjacent to, the nucleus will abolish all the vascular effects of electrical or chemical stimulation of FN (Dormer et al., 1989; Chida et al., 1989, 1990a,b) (Fig. 3).

Other evidence supports this contention. First, electrical and chemical stimulation of RVL simulates the systemic and cerebrovascular responses to FN, initiating, with comparable dynamics, a nearly global increase in rCBF that is not associated with changes in rCGU (Underwood, 1988; Underwood et al., 1992; Golanov and Reis, 1996b). Second, stimulation of FN and RVL with single shocks also evokes an identical electrophysiological event in the cerebral cortex, an evoked cortical potential followed by a single brief wave of elevated rCBF, the cerebral burst wave response (Golanov and Reis, 1995) (Fig. 7). Third, although the configurations of the response are identical, the latencies to events are longer from FN than RVL, consistent with the presence of a projection of FN through RVL.

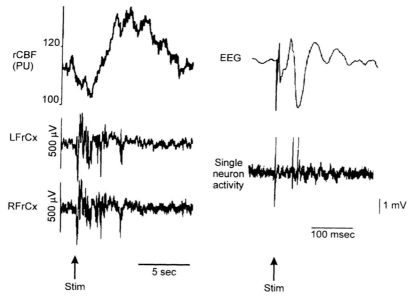

FIG. 7. Burst-cerebrovascular wave complexes elicited by electrical stimulation of fastigial nucleus (FN) with single shocks. Cortical cerebral blood flow (rCBF) was measured by laser Doppler flowmetry (expressed in perfusion units, PU) and EEG was recorded conventionally bilaterally. (Left) Unilateral stimulation of right FN (0.5 msec, 100 μA) triggered identical bursts in right and left frontal cortices (RFrCx and LFrCx, respectively) followed by a transient increase in cortical rCBF. Field cortical electrical potential in response to single-pulse stimulation of FN. (Right) Field activity of the initial characteristic triphasic potential at higher time resolution (upper train) and simultaneously recorded activity of a cortical vasodilator neuron (lower trace) (see text). (Modified from Golanov and Reis, 1996.)

The RVL appears to play a crucial role in generating much of the sympathoexcitation and cerebrovascular vasodilation to hypoxia and brain stem ischemia (Underwood et al., 1994; see Reis et al., 1994; Golanov and Reis, 1996b). Neurons of the RVL are themselves oxygen detectors and are directly excited by cyanide, hypoxia, and/or ischemia (Sun et al., 1992; Sun and Reis, 1994a,b). Moreover, lesions of the area reduce, by up to one-half, the vasodilation elicited by hypoxemia (Underwood et al., 1994; Golanov and Reis, 1996b). The effect is selective, for comparable lesions do not affect the cerebral vasodilation in response to hypercarbia (Underwood et al., 1994; Golanov and Reis, 1996b). The findings therefore suggest that the cerebellum may play a crucial role in influencing cerebral circulatory responses to hypoxia and ischemia. It is not certain whether the RVL neurons generating cerebrovascular vasodilation correspond to the reticulospinal sympathoexcitatory neurons. Evidence shows that they be distributed more caudally (Golanov and Reis, 1996c). Within FN the areas regulating sympathetic tone and rCBF may also differ (Mraovitch et al., 1986).

Two major questions are posed by these studies. First, by what pathway(s) does information antidromically initiated in FN reach RVL, and what is the intracerebral pathway to an important target, the cerebral cortex? Second, what are the cellular and molecular mechanism within a target by which afferent neuronal signals are transduced into a vascular event?

The site of the perikarya of the purported collateralized brain stem neurons which innervate FN and RVL and which are antidromically excited from FN is unknown. However, several brain stem nuclei can be considered candidates by the criteria that they innervate cerebellum and RVL. These include elements in the parabrachial and pedunculopontine nuclei, the periaquaductal gray, several hypothalamic nuclei, and some nuclei of the intralaminar complex of the thalamus, all of which innervate both FN and RVL (Dietrichs, 1985; Dietrichs and Haines, 1985; Dietrichs et al., 1994; Otake et al., 1994). Whether the same neurons innervate FN and RVL, however, has yet to be demonstrated.

The pathway from RVL which dilates vessels in the cerebral cortex is also unknown. As there are no direct projections from RVL to the cortex (Reis et al., 1994), it must include, at least, subcortical synapses. Given the widespread effects on rCBF, it is likely that the cortical afferent system from RVL must project diffusely. The potential relays would be subdivisions of the midline-intralaminar thalamic complex, regions of the lateral hypothalamus/zona incerta complex, and nuclei of the basal forebrain, areas (with the possible exception of the latter) that are part of the hypothalamo-cerebellar network of Dietrichs and Haines (Dietrichs, 1985; Dietrichs and Haines, 1985; Dietrichs et al., 1994).

The mechanism by which an afferent neuronal signal is transduced within the target to increase rCBF is also uncertain. It has been proposed (Reis and Iadacola, 1991; Golanov and Reis, 1996a) that the initial element is an interposed neuron in the cortex that couples the vasodilation with the vascular event. In support is evidence that the excitotoxic destruction of cortical neurons preserving afferent fibers abolishes the evoked vasodilation from FN without altering the resting levels of blood flow (Iadecola et al., 1987; Arneric et al., 1987). A candidate population of such neurons has been identified (Golanov and Reis, 1996a) in deep cortical layers, the activity of which is locked to the burst-wave complexes (Fig. 7). These comprise no more than 5% of spontaneously active cortical cells, discharge invariably about 1.5 sec in advance of waves of vasodilation elicited from FN and RVL, and are excited by single electrical pulses of each. When rCBF is elevated during a stimulus train, their activity parallels the vascular events. The paucity of these neurons can explain why the vasodilation elicited from FN or RVL is not associated with changes in rCGU, as their contribution to overall cortical metabolism is probably negligible.

The cellular events leading from the activity of these vasodilator neurons to the relaxation of blood vessels is not known. It is probably not a product of metabolism such as K^+, given the brief latency for the increase in rCBF following neural activation and the fact that elevations of the ion measured extracellularly are not sufficient to account for flow changes (Iadecola and Kraig, 1991). Whereas generation of nitric oxide may contribute to the response (Iadecola *et al.*, 1993), its contribution is modest (Golanov and Reis, 1995).

III. Neuroprotection Evoked from the Fastigial Nucleus

A. Neuroprotection from the Fastigial Nucleus in Focal Cerebral Ischemia

Electrical stimulation of the FN increases rCBF, but not rCGU. Such treatment, the authors reasoned, might improve the chances of some ischemic neurons to survive. This rationale evolved from new knowledge of the behavior of focal cerebral ischemia (e.g., Siesjö, 1992; Hossmann, 1994; Obrenovitch, 1995).

Focal ischemic infarctions, produced in experimental animals by the occlusion of large arteries [in practice, usually the middle cerebral artery (MCA)], is not homogeneous, but consists of two principal components, an ischemic core and an ischemic penumbra. In the core, rCBF and rCGU are at lowest values and all elements are destined to die (Siesjö, 1992). In the penumbra, rCBF is only partially reduced, whereas rGCU is elevated, resulting in "misery perfusion" (Baron *et al.*, 1981). It is hypothesized that this relative ischemia of the penumbra leads to a cascade of cellular and molecular events involving the gradual failure of energy-dependent ion pumps and transporters (Siesjö, 1992; Hossmann, 1994), and a consequent overflow of neurotransmitters (Matsumoto *et al.*, 1993; Takagi *et al.*, 1993), most notably L-glutamate, which, by actions at NMDA receptors (Hartley *et al.*, 1993; Choi and Hartley, 1993) and by depolarization (Mayer and Westbrook, 1987; Duchen, 1990), facilitates the augmented accumulation of toxic amounts of intracellular Ca^{2+}. As a consequence, neurons of the penumbra gradually die.

In the authors' studies (Underwood *et al.*, 1989; Reis *et al.*, 1989, 1991; Yamamoto *et al.*, 1993; Golanov *et al.*, 1996), the effects of FN stimulation on lesion size induced by permanent occlusion of the MCA were investigated. Rats of the spontaneously hypertensive (SHR) or Sprague–Dawley (SD) strains, anesthetized with halothane or fluothane and with AP and blood gases rigorously controlled, were studied. Electrodes were inserted into the area of FN from which a FPR could be evoked (active site). Controls consisted of rats in which electrodes were inserted into FN, but not stimulated (sham stimulated), or in which areas of cerebellum from which no changes in AP are elicited (e.g., dentate nucleus) were stimulated. The FN was stimulated for 1 hr and, immediately thereafter, the anesthesia discontinued. Animals were returned to their cages, and 24 hr later the brains were processed and the distribution and size of the infarctions determined.

Electrical stimulation of the FN for 1 hr immediately followed by MCA ligation reduces the volume of the ischemic infarction by 40–50% (Golanov et al., 1996; Yamamoto et al., 1993; Zhang and Iadecola, 1992, 1993; Reis et al., 1989, 1991) (Fig. 8). The area of salvage surrounds the infarction and, in general, corresponds to the ischemic penumbra. Such treatment salvages brain in SHRs and also rats of the SD strain (Golanov et al., 1996; Yamamoto et al., 1993; Zhang and Iadecola, 1992, 1993). Sham stimulation or stimulation of extrafastigial sites has no effects (Reis et al., 1989).

The study therefore indicates that excitation of a pathway represented in the FN could salvage part of an ischemic infarction from injury.

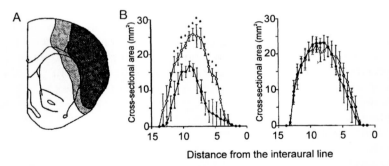

FIG. 8. Stimulation of fastigial nucleus (FN) is neuroprotective. (A) Cross section of rat brain showing a composite of the extent of lesions in seven rats resulting from occlusion of the middle cerebral artery (MCA) 24 hr earlier with (dark shading) and without (dark+light shading) 1 hr of FN stimulation just before occlusion. Dark shading corresponds to the irretrievable (core) zone. Light shading corresponds to the retrievable (penumbral zone) zone. (B) Cross-sectional areas of lesions plotted along the rostral–caudal axis in groups (five to seven) with real (●) or sham (○) stimulation of FN (left) or of the rostral ventrolateral medulla (RVL) (right). Note that stimulation of FN, but not RVL, reduces the lesion volume in a region corresponding to the ischemic penumbra. ($*p < 0.01$; $n = 5-7$). (Adapted from Yamamoto et al., 1993.)

B. Salvage is Independent of Changes in Blood Flow: Conditioned Neuroprotection

The authors' original premise was that FN stimulation might promote neuronal salvage by increasing blood flow, but not metabolism, within the ischemic penumbra. However, this mechanism has been discarded on the basis of three observations.

First, when rCBF is measured by LDF over the area of salvage, stimulation of FN does not elevate rCBF (Yamamoto *et al.*, 1993) (Fig. 9). Moreover, because occlusion of the MCA evokes comparable reductions of rCBF in stimulated and nonstimulated rats, salvage cannot be attributed to differences in the magnitude of ischemia. Second, when rCBF and rCGU are simultaneously measured by autoradiography (Golanov *et al.*, 1996), FN stimulation not only fails to elevate rCBF, but also does not alter rCGU in the salvaged zone. This observation indicates that salvage cannot result from a readjustment of the disproportion between the increased metabolic need and diminished rCBF, i.e., misery perfusion (Baron *et al.*, 1981). Third, electrical simulation of the RVL, the region mediating the elevations in rCBF from FN (Chida *et al.*, 1990a,b) and from which electrical and chemical simulation comparably elevates rCBF, but not rCGU (Underwood *et al.*, 1992; Golanov and Reis, 1996), fails to protect (Yamamoto *et al.*, 1993). The finding indicates that the pathways mediating neuroprotection and the elevation of rCBF are not identical. Finally, the protective effects of FN stimulation are prolonged. One hour of FN stimulation will reduce infarct volume by 40–50% when the MCA is occluded as long as 10 days after FN stimulation (Fig. 10) (Reis *et al.*, 1994). This finding, in the face of the fact that rCBF returns to control values within minutes following termination of the stimulus (Yamamoto *et al.*, 1993), indicates that FN stimulation conditions neuroprotection (Fig. 9).

C. Alternate Mechanisms of Neurogenic Neuroprotection

1. *Reduced Neuronal Excitability*

The mechanism(s) accounting for the conditional neuroprotection elicited from FN is unknown. However, two possibilities have been investigated. One is that conditional stimulation of the FN reduces neuronal excitability and, as a consequence, prevents the expression of the peri-infarction depolarizing waves (PIDs) that appear immediately after MCA occlusion within the ischemic penumbra (Mies *et al.*, 1993, 1994; Nedergaard and Astrup, 1986). The relevance of PIDs to infarct volume is suggested by the fact that their frequency immediately after an infarction directly correlates with

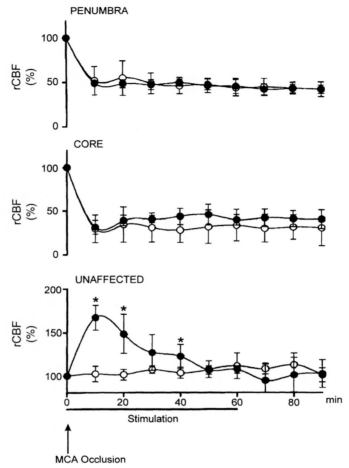

FIG. 9. Stimulation of the fastigial nucleus (FN) does not increase cerebral blood flow (rCBF) in areas of salvage. Changes in rCBF were measured by laser Doppler flowmetry in anesthetized spontaneously hypertensive rats recorded over the ischemic penumbra (upper trace), ischemic core (middle trace), and in the unaffected contralateral hemisphere (lower trace). In one group ($n = 5$), the middle cerebral artery (MCA) was occluded. In the other group ($n = 5$), coincident with MCA occlusion, the FN was stimulated for 60 min and recordings were made for an additional 45 min. Note that the reduction in rCBF in core and penumbra is not affected by FN stimulation. In contrast, rCBF is elevated in the unaffected cortex, although the elevation is not sustained. In stimulated rats there was a significant reduction in lesion volume, despite a comparable reduction in rCBF in the penumbra, indicating that salvage cannot be attributed to differences in the ischemic stimulus. (○, MCA occlusion alone; ●, MCA occlusion combined with FN stimulation). (*$p < 0.05$). (Modified from Yamamoto et al., 1993.)

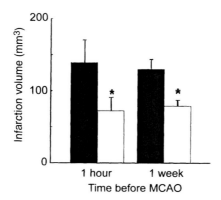

FIG. 10. Prolonged neuroprotection by stimulation of the fastigial nucleus (FN). The middle cerebral artery was occluded (MCAO) in 7 rats after 1 hr of sham (solid bars), or after electrical stimulation (open bars) 1 week prior to MCA occlusion. Lesion volumes were computed 24 hr later. Note that the neuroprotection elicited by FN stimulation is comparable whether MCAO is occluded concurrently or 1 week later) ($*p < 0.05$).

lesion size and, when their expression is suppressed, infarct volumes are reduced (Mies et al., 1994). The deleterious effects of PIDs on neurons within the ischemic penumbra have been attributed to the metabolic cost of repolarization in the face of a compromised blood supply. In support of our hypothesis is the fact that cerebellar stimulation can reduce seizure frequency and cortical excitability (Hablitz and Rea, 1976; Manzoni et al., 1967, 1968).

That this mechanism may be relevant has been suggested by the finding (Golanov and Reis, 1997) that 1 hr of conditional stimulation of the FN, but not of the dentate nucleus (DN), significantly increases the latency and reduces the numbers of PIDs which appear when the MCA is occluded immediately after FN stimulation. Moreover, conditional stimulation of the FN 72 hr prior to occlusion of the MCA will completely abolish the expression of PIDs over the 3 hr of observation (Fig. 11). FN-dependant reduction in PIDs expression is reversed by prior intracerebroventricular administration of the preferential K_{ATP}-channel blocker glibenclamide. This is consistent with the possibility that FN stimulation may elicit the prolonged opening of K^+ channels, possibly of the K_{ATP} subclass, leading, thereby, to hyperpolarization of neurons and diminished excitability. Consistent with this hypothesis are the facts that FN, but not DN, stimulation elevates the thresholds for eliciting waves of cortical spreading depression (Golanov and Reis, 1997) and that, as Heurteaux et al. (1995) have demonstrated, the neuroprotection afforded against global ischemic degeneration elicited by ischemic preconditioning is also apparently dependent on K_{ATP}-channel

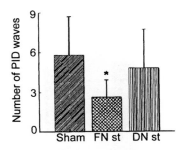

FIG. 11. Stimulation of fastigial nucleus (FN) inhibits expression of peri-infarction depolarizing waves (PIDs). Rats were anesthetized, and the FN or, as control, the dentate nucleus (DN), a region not neuroprotective, was stimulated for 1 hr followed by MCA occlusion. Cortical DC potentials were recorded for 3 hr and the numbers of PIDs were measured. Note that FN, but not DN, stimulation significantly ($p < 0.05$; $n = 5$ in each group) reduces the numbers of PIDs.

opening. This is consistent with observations that cerebellar stimulation may elevate seizure thresholds in experimental animals (Hablitz and Rea, 1976) and reduce seizure susceptibility in humans (Levy and Auchterlonie, 1979).

2. *Impaired Inflammatory Reactivity*

A second mechanism that may contribute to conditioned neuroprotection is modification of the immunoreactivity of brain in response to ischemic injury (see Reis *et al.*, 1997). Focal and global ischemia trigger a cascade of inflammatory reactions in brain that serve, as in most tissues, to isolate and destroy the damaged area and to begin the reparative processes. Two classes of cells are central to the development of postischemic inflammation in brain: vascular endothelial cells and microglia. In response to tissue injury, the microvascular endothelium almost immediately expresses a group of molecules, including the intercellular cell adhesion molecule (ICAM-1) and the vascular cell adhesion molecule (VCAM-1) (Meerschaert and Furie, 1995; Schroeter *et al.*, 1994) which promote adherence and the subsequent migration of leukocytes and monocytes/macrophages into brain (Meerschaert and Furie, 1995; Reis *et al.*, 1995). These infiltrating cells can release cytokines, nitric oxide (NO), and free oxygen radicals, contributing to cell death and enhancement of the inflammatory responses. In addition, vascular cells express the inducible isoform of nitric oxide synthase (iNOS) to generate NO, which may promote vasodilation and cellular necrosis (Relton and Rothwell, 1992; Iadecola *et al.*, 1996). The inflammatory cascade may exacerbate ischemic damage. Blockade of such proinflammatory cytokines as interleukin 1 (IL-1), inhibition of migration of circulating leukocytes or monocytes into ischemic brain, or inhibition of iNOS activity can decrease (up to 50%) the size of the infarctions (Xie *et al.*, 1994; Baeuerle and Henkel, 1994; Iadecola *et al.*, 1995; Zhang *et al.*, 1994; Yamasaki *et al.*, 1995; Rothwell and Relton, 1993).

That stimulation of the FN may suppress ischemic damage by inhibiting inflammatory reactions is supported by the observation that stimulation of the FN 48 hr prior to MCA occlusion reduces the expression of iNOS mRNA in brain microvessels and the infiltration of macrophages into the territory that is salvaged (Reis *et al.*, 1995). The observation, however, poses a problem: Does the conditional stimulus itself inhibit expression of the proinflammatory molecules? Alternatively, is the blunted response secondary to the reduced necrosis in the zone?

To address these questions, the authors have been studying the immune reactivity of microvessels isolated from brain and studied *ex vivo*. The authors have discovered that the incubation of vessels obtained from untreated rats with IL-1β (20 ng/ml) induces the transient expression of iNOS mRNAs. In contrast, the induction of iNOS by IL-1 was reduced (by 50%) in cerebral vessels isolated from rats receiving comparable FN stimulation 48–72 hr before. These results suggest that FN stimulation may reduce the immunoreactivity of cerebral microvessels (Reis *et al.*, 1997; Galea *et al.*, 1997).

These studies demonstrate that excitation of the FN can activate networks in brain that may initiate profound and long-lasting changes in the reactivity of the brain to ischemic injury. These neuroprotective networks may, when excited, alter the excitability of cortical, and possibly other, neurons and also modulate the reactivity of cerebral microvessels to inflammatory stimuli.

IV. Conclusions

It is evident that the vasopressor and cerebrovascular responses of the FPR as elicited by electrical stimulation of the FN result from the stimulation of axons traversing the region. However, there is anatomical specificity: these and most other autonomic and behavioral responses are evoked only from the rostral ventral pole of the nucleus. This region of the cerebellum appears to have actions that are exclusively related to autonomic and behavioral functions as stimulation in unanesthetized animals does not have evident motor effects (e.g., Martner, 1975a; Lisander and Martner, 1975; Dormer and Stone, 1976; Reis *et al.*, 1973) and lesions do not result in motor disabilities (e.g., Giannazzo *et al.*, 1969; Reis *et al.*, 1973; Dormer and Stone, 1976; Fish *et al.*, 1979).

The fact that FPR and elevated rCBF persist after ablation of the cerebellar cortex and destruction of intrinsic FN neurons indicates that the responses arise from excitation of the axons of brain stem neurons that innervate cerebellum. Abolition of all vascular responses by lesions of RVL indicates that this nucleus is interposed in the pathway generating vasodilation in the cerebral cortex. The RVL, however, is excited indirectly, as it is not directly innervated by FN (e.g., Homma *et al.*, 1995), and hence the cerebrovascular vasodilation as well as the hypertension probably results from the excitation of RVL through collateral branches of the same brain stem neurons projecting to the cerebellum. It is likely that the other autonomic and behavioral responses elicited by the electrical stimulation of FN, including the modification of resting and reflex functions of the gastrointestinal tract (Lisander and Martner, 1974; Martner, 1975a,b,c), respiration (Lutherer and Williams, 1986; Lutherer *et al.*, 1989), the induction of aggressive and/or consummatory behaviors (Martner, 1975a; Lisander and Martner, 1975; Dormer and Stone, 1976; Reis *et al.*, 1973), and EEG synchronization (Fig. 12) (Underwood *et al.*, 1989), are also antidromically evoked.

The observations, however, are of theoretical interest with respect to understanding the role of the cerebellum in processing visceral activity. The wide range of autonomic, endocrine, and behavioral responses to electrical stimulation of FN implies that stimulation engages a number of central autonomic nuclei. For example, the release of AVP (Del Bo *et al.*, 1983a,b), the induction of feeding behavior, and the hypertension imply that networks in different hypothalamic nuclei and ventral medulla are simultaneously activated. The morphological basis for such coactivation undoubtedly relates to the fact that nuclear regions of the brain stem

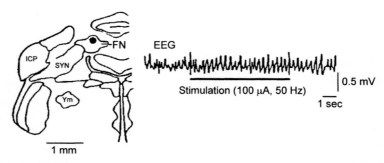

FIG. 12. Effect of electrical stimulation of fastigial nucleus cortical EEG in an anesthetized rat. Note pronounced synchronization of EEG evoked by FN stimulation that outlasts stimulation. (Modified from Underwood *et al.*, 1989).

engaged in autonomic control are highly interconnected, each with the other by parallel and recurrent collateral branch interactions (Otake *et al.*, 1994; Homma *et al.*, 1995). Thus hypothalamus, amygdala, nucleus solitarii, periaquaductal gray, parabrachial nucleus, and visceral thalamus innervate each other in complex reciprocal interactions, often through collaterals (e.g., Otake *et al.*, 1994). The observations that neurons in many of these regions innervate cerebellum, including FN, and that some of these neurons are collateralized provide an anatomical foundation for these interactions (Dietrichs, 1985; Dietrichs and Haines, 1985; Dietrichs *et al.*, 1994). The authors propose that the responses elicited by electrical stimulation of afferents in or near the FN arise from collateralized neurons of this central autonomic core to initiate responses along the network with particular visceral, endocrine, or behavioral responses reflecting activity in specific nuclei.

An implication of this concept is that the cerebellum receives information from many elements of the central autonomic core, including such regions as amygdala, hypothalamus, periaquaductal gray, and parabrachial and solitary nuclei. Data would include not only feedback signals from viscera transmitted over primary visceral afferents (e.g., vagal or baroreceptor nerves), but also information about how visceral inputs are being processed in higher centers into patterns of autonomic and associated behaviors. The formulation therefore suggests a neuronal mechanism whereby the cerebellum is informed of input and output signals from the central autonomic core. Such information may be of importance in the coupling of cardiovascular performance to motor behaviors and in information relating to the elaboration of emotive processes.

It is in this context that the effects of FN on elevating rCBF can be considered. The demonstration that excitation of the FN elevates rCBF independently of metabolism is of importance in its own right by showing that systems in brain can elevate rCBF without metabolic cost. Such a system would be most useful in preparing the brain for possible action, i.e., by increasing the supply not only of oxygen and glucose, but of other metabolic substrates as well. The fact that chemical stimulation of the FN inhibits sympathetic activity and reduces rCBF secondarily to a global reduction in cerebral metabolism indicates that intrinsic neurons of the rostral FN have autonomic actions of their own. This finding raises the possibility that these neurons may be mediators of the action of Purkinje cells in the cerebellar cortex, the excitation of which can excite or inhibit AP and respiration (Bradley *et al.*, 1987a,b). The antisympathetic action of FN neurons also raises the possibility that they are part of a negative feedback loop through which signals associated with the activity of sympathoexcitatory systems of the brain stem signal a sympathoinhibitory outflow from cerebellum. The finding that chemical excitation of FN can generate a widespread reduction in metabolism and secondarily rCBF is intriguing and reveals a heretofore unknown aspect of cerebellar physiology. Its biological relevance, pathways, and transmitters, however, are unknown.

That stimulation of the FN may protect the brain from ischemic injury also represents a new and heretofore unrecognized function of the cerebellum and/or the circuits innervating it. Although we have yet to firmly establish the pathways involved, evidence shows that the response arises from intrinsic neurons of FN, not from antidromic excitation of axons of passage. Notably, salvage evoked from FN is abolished by excitotoxic lesions of the FN, but not of adjacent dentate nucleus (Glickstein et al., 1996). Comparably, stimulation of RVL is not neuroprotective, indicating that the pathways involved in regulating rCBF and protecting the brain are different. The fact that FN stimulation can protect the brain from ischemia for over 10 days (Reis et al., 1994) suggests that cerebellar networks can initiate long-lived and global changes in brain function that are only revealed in the setting of ischemia. Perhaps in some way salvage relates to the functions of FN in metabolic regulation.

The biological significance of the neuroprotection evoked from FN is not evident. It has been suggested that part of the function of the RVL, conceivably modulated by the cerebellum, may be to initiate circulatory reflex reponses that would serve to protect the brain from hypoxia (Reis et al., 1994; Golanov and Reis, 1996b). These would be comparable to the diving reflex of aquatic vertebrates (Blix and Folkow, 1983) and would consist of massive peripheral vasoconstriction and redistribution of blood to brain. Although such adjustments provide relative protection of brain in acute situations, including brain stem ischemia, other neuronal mechanisms might generate long-term molecular changes, assuring long-term protection.

References

Achari, N. K., and Downman, C. B. (1969). Autonomic responses evoked by stimulation of fastigial nuclei in the anaesthetized cat. *J. Physiol. (Lond.)* **204,** 130P.

Andrezik, J. A., Dormer, K. J., Foreman, R. D., and Person, R. J. (1984). Fastigial nucleus projections to the brain stem in beagles: Pathways for autonomic regulation. *Neuroscience* **11,** 497–507.

Arneric, S. P., Iadecola, C., Underwood, M. D., and Reis, D. J. (1987). Local cholinergic mechanisms participate in the increase in cortical cerebral blood flow elicited by electrical stimulation of the festigial nucleus in the rat. *Brain Res.* **411,** 212–225.

Baeuerle, P. A., and Henkel, T. (1994). Function and activation of nf-kappa b in the immune system. *Annu. Rev. Immunol.* **12,** 141–179.

Baron, J. C., Bousser, M. G., Rey, A., Guillard, A., Comar, D., and Castaigne, P. (1981). Reversal of focal "misery-perfusion syndrome" by extra-intracranial arterial bypass in hemodynamic cerebral ischemia: A case study with ^{15}O positron emission tomography. *Stroke* **12**, 454–459.

Blix, A. S., and Folkow, B. (1983). Cardiovascular adjustments to diving in mammals and birds. *In* "Cardiovascular System" (E. M. Renkin and C. C. Michel, eds.), pp. 917–945. American Physiological Society, Bethesda, MD.

Bradley, D. J., Pascoe, J. P., Paton, J. F., and Spyer, K. M. (1987a). Cardiovascular and respiratory responses evoked from the posterior cerebellar cortex and fastigial nucleus in the cat. *J. Physiol. (Lond.)* **393**, 107–121.

Bradley, D. J., Paton, J. F., and Spyer, K. M. (1987b). Cardiovascular responses evoked from the fastigial region of the cerebellum in anaesthetized and decerebrate rabbits. *J. Physiol. (Lond.)* **392**, 475–491.

Chida, K., Iadecola, C., Underwood, M. D., and Reis, D. J. (1986). A novel vasodepressor response elicited from the rat cerebellar fastigial nucleus: The fastigial depressor response. *Brain Res.* **370**, 378–382.

Chida, K., Iadecola, C., and Reis, D. J. (1989a). Differences in selective cardiovascular characteristic of vasopressor responses elicited from the cerebellar fastigial nucleus and the rostral ventrolateral medulla in rats. *Ther. Res.* **10**, 37–55.

Chida, K., Iadecola, C., and Reis, D. J. (1989b). Global reduction in cerebral blood flow and metabolism elicited from intrinsic neurons of fastigial nucleus. *Brain Res.* **500**, 177–192.

Chida, K., Iadecola, C., and Reis, D. J. (1990a). Lesions of rostral ventrolateral medulla abolish some cardio- and cerebrovascular components of the cerebellar fastigial pressor and depressor responses. *Brain Res.* **508**, 93–104.

Chida, K., Underwood, M. D., Miyagawa, M., Kawamura, H., Iadecola, C., Takasu, T., and Reis, D. J. (1990b). Participation of the rostral ventrolateral medulla in the cerebral blood flow of rats: Effects of stimualtion and lesions on systemic and cerebral circulations. *Ther. Res.* **11**, 77–85.

Choi, D. W., and Hartley, D. M. (1993). Calcium and glutamate-induced cortical neuronal death. *Res. Publ. Assoc. Res. Nerv. Ment. Dis. Mol. Cell. Approach. Treat. Neurol. Dis.* **71**, 23–34.

Clarke, D. D., and Sokoloff, L. (1994). Circulation and energy metabolism of the brain. *In* "Basic Neurochemistry" (G. J. Siegel, B. W. Agranoff, R. W. Albers, and P. B. Molinoff, eds.), pp. 645–680. Raven Press, New York.

Del Bo, A., Ross, C. A., Pardal, J. F., Saavedra, J. M., and Reis, D. J. (1983a). Fastigial stimulation in rats releases adrenomedullary catecholamines. *Am. J. Physiol.* **244**, R801–R809.

Del Bo, A., Sved, A. F., and Reis, D. J. (1983b). Fastigial stimulation releases vasopressin in amounts that elevate arterial pressure. *Am. J. Physiol.* **244**, H687–H694.

Del Bo, A., Sved, A. F., and Reis, D. J. (1983c). Pressor effects of vasopressin released by fastigial nucleus stimulation in rat. *J. Hypertension (Suppl. 2)*, 237–239.

Del Bo, A., Sved, A. F., and Reis, D. J. (1984). Fastigial nucleus stimulation and concurrent activation of cardiovascular receptors; differentiate effects on arterial pressure, heart rate and vasopressin release. *J. Hypertens. (Suppl.2)*, S49–S51.

Dietrichs, E. (1983). Cerebellar nuclear afferents from the lateral reticular nucleus in the cat. *Brain Res.* **288**, 320–324.

Dietrichs, E. (1985). Divergent axon collaterals to cerebellum and amygdala from neurons in the parabrachial nucleus locus coeruleus and some adjacent nuclei: A fluorescent double labelling study using rhodamine labelled latex microspheres and fast blue as retrograde tracers. *Anat. Embryol.* **172**, 75–82.

Dietrichs, E., Haines, D. E. (1985). Observations on the cerebello-hypothalamic projection, with comments on non-somatic cerebellar circuits. *Arch. Ital. Biol.* **123**, 133–139.

Dietrichs, E., Haines, D. E., Roste, G. K., and Roste, L. S. (1994). Hypothalamocerebellar and cerebellohypothalamic projections: Circuits for regulating nonsomatic cerebellar activity? *Histol. Histopathol.* **9**, 603–614.

Dietrichs, E., and Walberg, F. (1985). The cerebellar nucleo-olivary and olivo-cerebellar nuclear projections in the cat as studied with anterograde and retrograde transport in the same animal after implantation of crystalline WGA-HRP. II. The fastigial nucleus. *Anat. Embryol. (Berl.)* **173**, 253–261.

Doba, N., and Reis, D. J. (1972). Changes in regional blood flow and cardiodynamics evoked by electrical stimulation of the fastigial nucleus in the cat and their similarity to orthostatic reflexes. *J. Physiol. (Lond.)* **227**, 729–747.

Dormer, K. J., Andrezik, J. A., Person, R. J., Braggio, J. T., and Foreman, R. D. (1986). Fastigial nucleus cardiovascular response and brain stem lesions in the beagle. *Am. J. Physiol.* **250**, H231–H239.

Dormer, K. J., Person, R. J., Andrezik, J. A., Foreman, R. D., and Braggio, J. P. (1989). Ventrolateral medullary lesions and fastigial cardiovascular response in beagles. *Am. J. Physiol.* **256**, H1200–H1208.

Dormer, K. J., and Stone, H. L. (1976). Cerebellar pressor response in the dog. *J. Appl. Physiol.* **41**, 574–580.

Dow, R. S., and Moruzzi, G. (1958). "The Physiology and Pathology of the Cerebellum," pp. 1–675. University of Minnesota Press, Minneapolis.

Duchen, M. R. (1990). Effects of metabolic inhibition on the membrane properties of isolated mouse primary sensory neurones. *J. Physiol. (Lond.)* **424**, 387–409.

Edvinsson, L., MacKenzie, E. T., and McCulloch, J. (1993). "Cerebral Blood Flow and Metabolism," pp. 1–683. Raven Press, New York.

Elisevich, K., and Redekop, G. (1991). The fastigial pressor response: Case report. *J. Neurosurg.* **74**, 147–151.

Fish, B. S., Baisden, R. H., and Woodruff, M. L. (1979). Cerebellar nuclear lesions in rats: Subsequent avoidance behavior and ascending anatomical connections. *Brain Res.* **166**, 27–38.

Galea, E., Golanov, E. V., Feinstein, D. L., Kobylarz, K., and Reis, D. J. (1997). Cerebellar stimulation suppresses expression of inducible nitric oxide synthase in cerebral microvessesls in response to focal ischemia. *J. Cereb. Blood Flow Metab.*, in press.

Giannazzo, E., Manzoni, T., Raffaele, R., Sapienza, S., and Urbano, A. (1969). Effects of chronic fastigial lesions on the sleep-wakefulness rhythm in the cat. *Arch. Ital. Biol.* **107**, 1–18.

Glickstein, S., Golanov, E. V., Kobylarz, K., and Reis, D. J. (1996). Protection against focal ischemic infarction elicited by stimulation of the cerebellar fastigial nucleus results from excitation of intrinsic neurons. *Soc. Neurosci. Abst.* **22**, 716 (Abstract).

Goadsby, P. J., and Lambert, G. A. (1989). Electrical stimulation of the fastigial nucleus increases total cerebral blood flow in the monkey. *Neurosci. Lett.* **107**, 141–144.

Golanov, E. V., and Reis, D. J. (1994). Nitric oxide and prostanoids participate in cerebral vasodilation elicited by electrical stimulation of rostral ventrolateral medulla. *J. Cereb. Blood Flow Metab.* **14**, 492–502.

Golanov, E. V., and Reis, D. J. (1995). Vasodilation evoked from medulla and cerebellum is coupled to bursts of cortical EEG activity in rats. *Am. J. Physiol.* **268**, R454–R467.

Golanov, E. V., and Reis, D. J. (1996a), Cerebral cortical neurons with activity linked to central neurogenic spontaneous and evoked elevations in cerebral blood flow. *Neurosci. Lett.* **209**, 101–104.

Golanov, E. V., and Reis, D. J. (1996b). Oxygen sensitive neurons of the rostral ventrolateral medulla contribute to hypoxic cerebral vasodilation. *J. Physiol. (Lond.)* **495**, 201–216.

Golanov, E. V., and Reis, D. J. (1996c). Different neurons of the rostral ventrolateral medulla (RVL) regulate sympathetic activity and regional cerebral blood flow (rCBF) and EEG. *FASEB J.* **10,** A305 (Abstract).
Golanov, E. V., and Reis, D. J. (1997). Neuroprotective electrical stimulation of the cerebellar fastigial nucleus supresses peri-infarction depolarizing waves. *J. Cereb. Blood Flow Metab.*, in press.
Golanov, E. V., Yamamoto, S., and Reis, D. J. (1996). Electrical stimulation of cerebellar fastigial nucleus fails to rematch blood flow and metabolism in focal ischemic infarctions. *Neurosci. Lett.* **210,** 181–184.
Hablitz, J. J., and Rea, G. (1976). Cerebellar nuclear stimulation in generalized penicillin epilepsy. *Brain Res. Bull.* **1,** 599–601.
Hartley, D. M., Kurth, M. C., Bjerkness, L., Weiss, J. H., and Choi, D. W. (1993). Glutamate receptor induced $^{45}Ca^{2+}$ accumulation in cortical cell culture correlates with subsequent neuronal degeneration. *J. Neurosci.* **13,** 1993–2000.
Heistad, D. D., and Kontos, H. A. (1983). Cerebral circulation. *In* "Handbook of Physiology." (J. T. Shepherd and F. M. Abboud, eds.), Vol. III, pp. 137–182. American Physiological Society, Bethesda, MD.
Henry, R. T., and Connor, J. D. (1989). Axons of passage may be responsible for fastigial nucleus pressor response. *Am. J. Physiol.* **257,** R1436–R1440.
Heurteaux, C., Lauritzen, I., Widmann, C., and Lazdunski, M. (1995). Essential role of adenosine, adenosine A1 receptors, and ATP-sensitive K^- channels in cerebral ischemic preconditioning. *Proc. Natl. Acad. Sci. USA* **92,** 4666–4670.
Hirano, T., Kuchiwaki, H., Yoshida, K., Furuse, M., Taniguchi, K., and Inao, S. (1993). Fastigial pressor response observed during an operation on a patient with cerebellar bleeding: An anatomical review and clinical significance. *Neurosurgery* **32,** 675–677.
Homma, Y., Nonaka, S., Matsuyama, K., and Mori, S. (1995). Fastigiofugal projection to the brainstem nuclei in the cat: An anterograde PHA-l tracing study. *Neurosci. Res.* **23,** 89–102.
Hossmann, K. A. (1994). Viability thresholds and the penumbra of focal ischemia. *Ann. Neurol.* **36,** 557–565.
Iadecola, C., Arneric, S. P., Baker, H. D., Tucker, L. W., and Reis, D. J. (1987). Role of local neurons in cerebrocortical vasodilation elicited from cerebellum. *Am. J. Physiol.* **252,** R1082–R1091.
Iadecola, C., and Kraig, R. P. (1991). Focal elevations in neocortical interstitial K^+ produced by stimulation of the fastigial nucleus in rat. *Brain Res.* **563,** 273–277.
Iadecola, C., Mraovitch, S., Meeley, M. P., and Reis, D. J. (1983). Lesions of the basal forebrain in rat selectively impair the cortical vasodilation elicited from cerebellar fastigial nucleus. *Brain Res.* **279,** 41–52.
Iadecola, C., and Reis, D. J. (1990). Continuous monitoring of cerebrocortical blood flow during stimulation of the cerebellar fastigial nucleus: A study by laser-Doppler flowmetry. *J. Cereb. Blood Flow Metab.* **10,** 608–617.
Iadecola, C., Zhang, F. Y., Casey, R., Clark, H. B., and Ross, M. E. (1996). Inducible nitric oxide synthase gene expression in vascular cells after transient focal cerebral ischemia. *Stroke* **27,** 1373–1380.
Iadecola, C., Zhang, F. Y., and Xu, X. H. (1993). Role of nitric oxide synthase-containing vascular nerves in cerebrovasodilation elicited from cerebellum. *Am. J. Physiol.* **264,** R738–R746.
Iadecola, C., Zhang, F. Y., and Xu, X. H. (1995). Inhibition of inducible nitric oxide synthase ameliorates cerebral ischemic damage. *Am. J. Physiol.* **37,** R286–R292.
Ito, M. (1984). "The Cerebellum and Neural Control," pp. 1–580. Raven Press, New York.

Levy, L. F., and Auchterlonie, W. C. (1979). Chronic cerebellar stimulation in the treatment of epilepsy. *Epilepsia* **20,** 235–245.

Lisander, B., and Martner, J. (1971). Interaction between the fastigial pressor response and the baroreceptor reflex. *Acta Physiol. Scand.* **83,** 505–514.

Lisander, B., and Martner, J. (1974). Influences on gastrointestinal and bladder motility by the fastigial nucleus. *Acta Physiol. Scand.* **90,** 792–794.

Lisander, B., and Martner, J. (1975). Integrated somatomotor, cardiovascular and gastrointestinal adjustments induced from the cerebellar fastigial nucleus. *Acta Physiol. Scand.* **94,** 358–367.

Lou, H. C., Edvinsson, L., and MacKenzie, E. T. (1987). The concept of coupling blood flow to brain function: Revision required? *Ann. Neurol.* **22,** 289–297.

Lutherer, L. O., and Williams, J. L. (1986). Stimulating fastigial nucleus pressor region elicits patterned respiratory responses. *Am. J. Physiol.* **250,** R418–R426.

Lutherer, L. O., Williams, J. L., and Everse, S. J. (1989). Neurons of the rostral fastigial nucleus are responsive to cardiovascular and respiratory challenges. *J. Auton. Nerv. Syst.* **27,** 101–111.

Manning, J. W., Hartle, D. K., Ammons, W. S., and Koyama, S. (1985). The median preoptic area in cardiovascular reflex activity. *J. Auton. Nerv. Syst.* **12,** 239–249.

Manzoni, T., Sapienza, S., and Urbano, A. (1967). Electrocortical influences of the fastigial nucleus in chronically implanted, unrestrained cats. *Brain Res.* **4,** 375–377.

Manzoni, T., Sapienza, S., and Urbano, A. (1968). EEG and behavioural sleep-like effects induced by fastigial nucleus in unrestrained, unanaesthetized cat. *Arch. Ital. Biol.* **106,** 61–72.

Martner, J. (1975a). Cerebellar influences on autonomic mechanisms: An experimental study in the cat with special reference to the fastigial nucleus. *Acta Physiol. Scand. Suppl.* **425,** 1–42.

Martner, J. (1975b). Influences on colonic and small intestinal motility by the cerebellar fastigial nucleus. *Acta Physiol. Scand.* **94,** 82–94.

Martner, J. (1975c). Influences on the defecation and micturition reflexes by the cerebellar fastigial nucleus. *Acta Physiol. Scand.* **94,** 95–104.

Matsumoto, K., Graf, R., Rosner, G., Taguchi, J., and Heiss, W. D. (1993). Elevation of neuroactive substances in the cortex of cats during prolonged focal ischemia. *J. Cereb. Blood Flow Metab.* **13,** 586–594.

Mayer, M. L., and Westbrook, G. L. (1987). The physiology of excitatory amino acids in the vertebrate central nervous system. *Prog. Neurobiol.* **28,** 197–276.

McAllen, R. M. (1985). Mediation of the fastigial pressor response and a somatosympathetic reflex by ventral medullary neurones in the cat. *J. Physiol.* **368,** 423–433.

McKee, J. C., Denn, M. J., and Stone, H. L. (1976). Neurogenic cerebral vasodilation from electrical stimulation of the cerebellum in the monkey. *Stroke* **7,** 179–186.

Meerschaert, J., and Furie, M. B. (1995). The adhesion molecules used by monocytes for migration across endothelium include cd11a/cd18, cd11b/cd18, and vla-4 on monocytes and icam-1, vcam-1, and other ligands on endothelium. *J. Immunol.* **154,** 4099–4112.

Mies, G., Iijima, T., and Hossmann, K. A. (1993). Correlation between peri-infarct DC shifts and ischaemic neuronal damage in rat. *Neuroreport* **4,** 709–711.

Mies, G., Kohno, K., and Hossmann, K. A. (1994). Prevention of periinfarct direct current shifts with glutamate antagonist NBQX following occlusion of the middle cerebral artery in the rat. *J. Cereb. Blood Flow Metab.* **14,** 802–807.

Miura, M., and Reis, D. J. (1969). Cerebellum: A pressor response elicited from the fastigial nucleus and its efferent pathway in brainstem. *Brain Res.* **13,** 595–599.

Mraovitch, S., Pinard, E., and Seylaz, J. (1986). Two neural mechanisms in rat fastigial nucleus regulating systemic and cerebral circulation. *Am. J. Physiol.* **251,** H153–H163.

Nakai, M., Iadecola, C., and Reis, D. J. (1982). Global cerebral vasodilation by stimulation of rat fastigial cerebellar nucleus. *Am. J. Physiol.* **243**, H226–H235.

Nakai, M., Iadecola, C., Ruggiero, D. A., Tucker, L. W., and Reis, D. J. (1983). Electrical stimulation of cerebellar fastigial nucleus increases cerebral cortical blood flow without change in local metabolism: Evidence for an intrinsic system in brain for primary vasodilation. *Brain Res.* **260**, 35–49.

Nedergaard, M., and Astrup, J. (1986). Infarct rim: Effect of hyperglycemia on direct current potential and [^{14}C]-2-deoxyglucose phosphorylation. *J. Cereb. Blood Flow Metab.* **6**, 607–615.

Obrenovitch, T. P. (1995). The ischaemic penumbra: Twenty years on. *Cerebrovasc. Brain Metab. Rev.* **7**, 297–323.

Otake, K., Reis, D. J., and Ruggiero, D. A. (1994). Afferents to the midline thalamus issue collaterals to the nucleus tractus solitarii: An anatomical basis for thalamic and visceral reflex integration. *J. Neurosci.* **14**, 5694–5707.

Person, R. J., Andrezik, J. A., Dormer, K. J., and Foreman, R. D. (1986). Fastigial nucleus projections in the midbrain and thalamus in dogs. *Neurosci.* **18**, 105–120.

Reis, D. J., Berger, S. B., Underwood, M. D., and Khayata, M. (1991). Electrical stimulation of cerebellar fastigial nucleus reduces ischemic infarction elicited by middle cerebral artery occlusion in rat. *J. Cereb. Blood Flow Metab.* **11**, 810–818.

Reis, D. J., Doba, N., and Nathan, M. A. (1973). Predatory attack, grooming and consummatory behavior evoked by electrical stimulation of cerebellar nuclei in cat. *Science* **182**, 845–847.

Reis, D. J., Feinstein, D., Galea, E., and Golanov, E. V. (1997). Central neurogenic neuroprotection: Protection of brain from focal ischemia by cerebellar stimulation. *Fundam. Clin. Pharmacol.*, in press.

Reis, D. J., Golanov, E. V., Ruggiero, D. A., and Sun, M.-K. (1994). Sympatho-excitatory neurons of the rostral medulla are oxygen sensors and essential elements in the tonic and reflex control of the systemic and cerebral circulations. *J. Hypertens.* **12**, S159–S180.

Reis, D. J., Golanov, E. V., Yamamoto, S., Kobylarz, K., and Prabhakar, V. (1994). Conditioned neuroprotection from ischemic infarction elicited by electrical stimulation of the cerebellar fastigial nucleus (FN) in rat. *Soc. Neurosci. Abst.* **20**, 1480 (Abstract).

Reis, D. J., and Iadecola, C. (1991). Intrinsic central neural regulation of cerebral blood flow and metabolism in relation to volume transmission. In "Volume Transmission in the Brain: New Aspects in Electrical and Chemical Communication" (K. Fuxe and L. Agnati, eds.), pp. 523–558. Raven Press, New York.

Reis, D. J., Iadecola, C., MacKenzie, E. T., Mori, M., Nakai, M., and Tucker, L. W. (1982). Primary and metabolically coupled cerebrovascular dilation elicited by stimulation of two intrinsic systems of brain. In "Cerebral Blood Flow: Effects of Nerves and Neurotransmitters" (D. D. Heistad and M. L. Marcus, eds.), pp. 475–484. Elsevier, Amsterdam.

Reis, D. J., Kobylarz, K., Feinstein, D. L., Galea, E., and Golanov, E. V. (1995). Fastigial nucleus stimulation conditions neuroprotection and inhibits expression of inducible nitric oxide synthase gene after focal ischemia. *J. Cereb. Blood Flow Metab.* **15**, S91.

Reis, D. J., Underwood, M. D., Berger, S. B., Khayata, M., and Zaiens, N. I. (1989). Fastigial nucleus stimulation reduces the volume of cerebral infarction produced by occlusion of the middle cerebral artery in rat. In "Neurotransmission and Cerebrovascular Function I" (J. Seylaz and E. T. MacKenzie, eds.), pp. 401–404. Elsevier, Amsterdam.

Reis, D. J., Morrison, S., and Ruggiero, D. A. (1988). The C1 area of the brainstem in tonic and reflex control of blood pressure. *Hypertension* **11** (Suppl.), I8–I13.

Relton, J. K., and Rothwell, N. J. (1992). Interleukin-1 receptor antagonist inhibits ischaemic and excitotoxic neuronal damage in the rat. *Brain Res. Bull.* **29**, 243–246.

Roste, G. K., and Dietrichs, E. (1988). Cerebellar cortical and nuclear afferents from the Edinger-Westphal nucleus in the cat. *Anat. Embryol. (Berl.)* **178,** 59–65.

Rothwell, N. J., and Relton, J. K. (1993). Involvement of interleukin-1 and lipocortin-1 in ischaemic brain damage. *Cerebrovasc. Brain Metab. Rev.* **5,** 178–198.

Schroeter, M., Jander, S., Witte, O. W., and Stoll, G. (1994). Local immune responses in the rat cerebral cortex after middle cerebral artery occlusion. *J. Neuroimmunol.* **55,** 195–203.

Siesjö, B. K. (1992). Pathophysiology and treatment of focal cerebral ischemia. I. Pathophysiology. *J. Neurosurg.* **77,** 169–184.

Sokoloff, L., Reivich, M., Kennedy, C., Des Rosiers, M. H., Patlak, C. S., Pettigrew, K. D., Sakurada, O., and Shinohara, M. (1977). The [^{14}C]deoxyglucose method for the measurement of local cerebral glucose utilisation: Theory, procedure, and normal values in the conscious and anesthetized albino rat. *J. Neurochem.* **28,** 897–916.

Sun, M.-K., Jeske, I. T, and Reis, D. J. (1992). Cyanide excites medullary sympathoexcitatory neurons in rats. *Am. J. Physiol.* **262,** R182–R189.

Sun, M.-K., and Reis, D. J. (1994a). Hypoxia selectively excites vasomotor neurons of rostral ventrolateral medulla in rats. *Am. J. Physiol.* **266,** R245–R256.

Sun, M.-K., and Reis, D. J. (1994b). Hypoxia-activated Ca^{2+} currents in pacemaker neurones of rat rostral ventrolateral medulla *in vitro*. *J. Physiol. (Lond.)* **476,** 101–116.

Takagi, K., Ginsberg, M. D., Globus, M. Y. T., Dietrich, W. D., Martinez, E., Kraydieh, S., and Busto, R. (1993). Changes in amino acid neurotransmitters and cerebral blood flow in the ischemic penumbral region following middle cerebral artery occlusion in the rat: Correlation with histopathology. *J. Cereb. Blood Flow Metab.* **13,** 575–585.

Takahashi, S., Crane, A. M., Jehle, J., Cook, M., Kennedy, C., and Sokoloff, L. (1995). Role of the cerebellar fastigial nucleus in the physiological regulation of cerebral blood flow. *J. Cereb. Blood Flow Metab.* **15,** 128–142.

Talman, W. T., Dragon, D. M., Heistad, D. D., and Ohta, H. (1991). Cerebrovascular effects produced by electrical stimulation of fastigial nucleus. *Am. J. Physiol.* **261,** H707–H713.

Underwood, M. D. (1988). "Control of the Cerebral Circulation and Metabolism by the Rostral Ventrolateral Medulla: Possible Role in the Cerebrovascular Response to Hypoxia." Doctoral thesis, Cornell University, New York.

Underwood, M. D., Berger, S. B., Khayata, M., and Reis, D. J. (1989). Fastigial nucleus stimulation reduces the volume of cerebral infarction produced by occlusion of the middle cerebral artery in rat. *J. Cereb. Blood Flow Metab.* **9,** S32.

Underwood, M. D., Iadecola, C., and Reis, D. J. (1987). Neurons in C1 area of rostral ventrolateral medulla mediate global cerebrovascular responses to hypoxia but not hypercarbia. *J. Cereb. Blood Flow Metab.* **7,** S226.

Underwood, M. D., Iadecola, C., and Reis, D. J. (1983). Stimulation of brain areas increasing cerebral cortical blood flow and/or metabolism does not activate the electroencephalogram. *J. Cereb. Blood Flow Metab.* **3,** S214–S215.

Underwood, M. D., Iadecola, C., and Reis, D. J. (1994). Lesions of the rostral ventrolateral medulla reduce the cerebrovascular response to hypoxia. *Brain Res.* **635,** 217–223.

Underwood, M. D., Iadecola, C., Sved, A. F., and Reis, D. J. (1992). Stimulation of C1 area neurons globally increases regional cerebral blood flow but not metabolism. *J. Cereb. Blood Flow Metab.* **12,** 844–855.

Xie, Q. W., Kashiwabara, Y., and Nathan, C. (1994). Role of transcription factor nf-kappa b/rel in induction of nitric oxide synthase. *J. Biol. Chem.* **269,** 4705-4708.

Yamamoto, S., Golanov, E. V., and Reis, D. J. (1993). Reductions in focal ischemic infarctions elicited from cerebellar fastigial nucleus do not result from elevations in cerebral blood flow. *J. Cereb. Blood Flow Metab.* **13,** 1020–1024.

Yamasaki, Y., Matsuura, N., Shozuhara, H., Onodera, H., Itoyama, Y., and Kogure, K. (1995). Interleukin-1 as a pathogenetic mediator of ischemic brain damage in rats. *Stroke* **26,** 676–680.

Zhang, F. G., and Iadecola, C. (1992). Stimulation of the fastigial nucleus enhances EEG recovery and reduces tissue damage after focal cerebral ischemia. *J. Cereb. Blood Flow Metab.* **12,** 962–970.

Zhang, F. Y., and Iadecola, C. (1993). Fastigial stimulation increases ischemic blood flow and reduces brain damage after focal ischemia. *J. Cereb. Blood Flow Metab.* **13,** 1013–1019.

Zhang, R. L., Chopp, M., Li, Y., Zaloga, C., Jiang, N., Jones, M. L., Miyasaka, M., and Ward, P. A. (1994). Anti-icam-1 antibody reduces ischemic cell damage after transient middle cerebral artery occlusion in the rat. *Neurology* **44,** 1747–1751.

ASSOCIATIVE LEARNING

Richard F. Thompson, Shaowen Bao, Lu Chen, Benjamin D. Cipriano,
Jeffrey S. Grethe, Jeansok J. Kim, Judith K. Thompson, Jo Anne Tracy,
Martha S. Weninger, and David J. Krupa

Neuroscience Program, University of Southern California, Los Angeles, California 90089

I. Introduction
II. The Conditioned Response Pathway
III. The Unconditioned Stimulus Pathway
IV. The Conditioned Stimulus Pathway
V. The Cerebellum and the Reflex Eyeblink Response
VI. Purkinje Neuron Activity
VII. Cerebellar Cortical Lesions
VIII. Decerebration
IX. Locus of the Long-Term Memory Trace
X. The Issues of Performance
XI. Putative Noncerebellar Sites of Memory Storage
XII. Neural Substrate of the Error-Correcting Algorithm in Classical Conditioning
XIII. Supervised Learning and the Cerebellum
XIV. Putative Mechanisms
XV. Cerebellar Involvement in Other Forms of Memory
    References

This chapter reviews evidence demonstrating the essential role of the cerebellum and its associated circuitry in the learning and memory of classical conditioning of discrete behavioral responses (e.g., eyeblink, limb flexion, head turn). It now seems conclusive that the memory traces for this basic category of associative learning are formed and stored in the cerebellum. Lesion, neuronal recording, electrical microstimulation, and anatomical procedures have been used to identify the essential conditioned stimulus (CS) circuit, including the pontine mossy fiber projections to the cerebellum; the essential unconditioned stimulus (US) reinforcing or teaching circuit, including neurons in the inferior olive (dorsal accessory olive) projecting to the cerebellum as climbing fibers; and the essential conditioned response (CR) circuit, including the interpositus nucleus, its projection via the superior cerebellar peduncle to the magnocellular red nucleus, and rubral projections to premotor and motor nuclei. Each major component of the eyeblink CR circuit was reversibly inactivated both in trained animals and over the course of training. In all cases in trained animals, inactivation abolished the CR (and the UR as well when motor nuclei were inactivated). When animals were trained during inactivation (and not exhibiting CRs) and then tested without inactivation, animals with inactivation of the motor nuclei, red nucleus, and superior peduncle had fully learned, whereas animals with inactivation of a very localized region of the cerebellum (anterior interpositus and overlying cortex) had not learned at all. Consequently, the memory traces are formed and stored in the cerebellum. Several alternative possibilities are considered and ruled out. Both the cerebellar cortex and the interpositus nucleus are involved in the memory storage process, suggesting that a phenomenon-like long-term depression (LTD) is involved in the cerebellar cortex and long-term potentiation (LTP) is involved in the interpositus. The experimental findings reviewed in this chapter provide perhaps the first conclusive evidence for the localization of a basic form of memory storage to a particular brain region, namely the cerebellum, and indicate that the cerebellum is indeed a cognitive machine.

I. Introduction

The cerebellum has long been a favored structure for modeling a neuronal learning system, dating from the classic papers of Marr (1969) and Albus (1971). Our empirical work to date on the classical conditioning of discrete behavioral responses has been guided by these models and the related views of Eccles (1977) and Ito (1984); our results constitute a remarkable verification of the spirit of these theories (see also Thach et al., 1992). The highly simplified schematic block diagram of Fig. 1 can serve to summarize our overall results to date and is a much simplified version of our current qualitative working model of the role of the cerebellum in basic *delay* classical conditioning of discrete responses. [Laterality is not shown; the critical region of the cerebellum is ipsilateral to the trained eye (or limb), whereas the critical regions of the pontine nuclei, red nucleus, and inferior olive are contralateral]. Unless otherwise noted, data all refer to the basic delay eyeblink CR (Gormezano et al., 1983; standard conditions are 350-msec tone CS, coterminating with 100-msec corneal airpuff US). We simply note here that with more complex paradigms (e.g., trace conditioning) the hippocampal system also plays a critical role (Kim et al., 1995; Moyer et al., 1990; Solomon et al., 1986b).

II. The Conditioned Response Pathway

We first showed that neurons in the cerebellar interpositus nucleus and in the cerebellar cortex respond to the CS and US and develop amplitude-

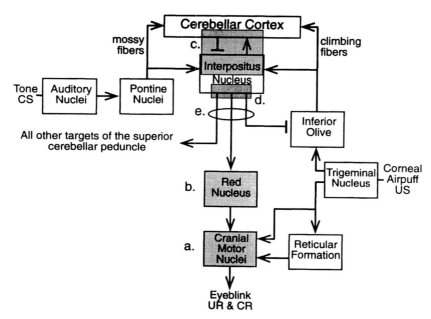

FIG. 1. Simplified schematic of the essential brain circuitry involved in eyeblink conditioning. Shadowed boxes represent areas that have been reversibly inactivated during training. (a) Inactivation of the motor nuclei including facial (seventh) and accessory (sixth). (b) Inactivation of magnocellular red nucleus. (c) Inactivation of dorsal aspects of the interpositus nucleus and overlying cerebellar cortex. (d) Inactivation of ventral interpositus and of white matter ventral to the interpositus. (e) Inactivation of the superior cerebellar peduncle (scp) after it exits the cerebellar nuclei.

time course "models" of the learned behavioral response that precede and predict the occurrence and form of the CR within trials and over the trials of training using classical conditioning of the eyeblink response in rabbits (Foy *et al.*, 1984; McCormick *et al.*, 1982a; McCormick and Thompson, 1984a,b), a result that has been repeatedly replicated (e.g., Berthier and Moore, 1990; Steinmetz, 1990; Tracy *et al.*, 1991). A study with human subjects determining regional cerebral glucose metabolism with positron emission tomography demonstrated that similiar regions of the cerebellum become activated as a result of learning in humans as in rabbits (Logan and Grafton, 1995). Electrical stimulation in the critical region of the interpositus elicits eyeblink responses in naive animals; the circuit is hard wired from interpositus to behavior (Chapman *et al.*, 1988). McCormick *et al.* (1982a) first showed that lesions of the interpositus abolished the CR and had no effect on the UR. This result has been replicated in approximately 15 subsequent studies in several mammalian species (see Clark et al., 1984; Lavond et al., 1985; Steinmetz et al., 1989, 1992; Thompson, 1990; Yeo et al., 1985a). Similarly, lesions of the superior cerebellar peduncle and red nucleus can abolish the CR with no effect on the UR (McCormick et al., 1982b; Rosenfeld et al., 1985; Rosenfeld and Moore, 1983). Importantly, microinfusion of nanomolar amounts of $GABA_A$ antagonists in the critical region of cerebellar cortex (HVI), interpositus, or red nucleus completely and reversibly abolish the CR with no effect at all on the UR in a dose-dependent manner (Haley et al., 1988; Mamounas et al., 1987). Appropriate cerebellar lesions in humans completely prevent learning of the eyeblink CR and have no effect on the UR (Daum et al., 1993; Lye et al., 1988; Solomon et al., 1989). The possibility that effective interpositus lesions abolish the CR by small effects on the UR (the "performance" argument) has been decisively ruled out, as will be detailed later.

III. The Unconditioned Stimulus Pathway

Lesions of the critical region of the inferior olive, the face representation in the dorsal accessory olive (DAO), completely prevent learning if made before training and result in extinction/abolition of the CR if made after training [McCormick et al., 1985; Mintz et al., 1994; Voneida et al., 1990 (limb flexion); Yeo et al., 1986]. Unit activity in this critical DAO region does not respond to auditory stimuli (CS), responds only to US onset, shows no learning-related activity, and decreases as animals learn (Sears and Steinmetz, 1991). Electrical microstimulation of this region serves as a very effective US (Mauk et al., 1986; Steinmetz et al., 1989; Thompson, 1989), as does stimulation of cerebellar white matter (Swain et al., 1992). All these data argue that the DAO climbing fiber system is the essential US reinforcing pathway for the learning of discrete responses (Thompson, 1989). To our knowledge, this is the only system in the brain, other than reflex afferents, where the exact response elicited by electrical stimulation can be conditioned to any neutral stimulus. Current evidence suggests that this system, together with the GABAergic projection from interpositus to DAO (Nelson and Mugnoini, 1987), functions as the error-correcting algorithm in classical conditioning (see later).

IV. The Conditioned Stimulus Pathway

The pontine nuclei send axons directly to the cerebellar cortex and interpositus nucleus (Mihaillof, 1993; Shinoda et al., 1992; Steinmetz and

Sengelaub, 1992; Thompson *et al.*, 1985, 1991). The pontine nuclei in turn receive projections from auditory, visual, and somatosensory systems (Brodal 1978; Glickstein *et al.*, 1980; Mower *et al.*, 1979; Schmahmann and Pandya, 1989, 1991, 1993; Tusa and Ungerleider, 1988). Several regions of the pontine nuclei exhibit short latency evoked unit responses to auditory stimuli (Aitkin and Boyd, 1978; Steinmetz *et al.*, 1987). Appropriate lesions of the pontine nuclei can abolish the CR established to a tone CS, but not a light CS, i.e., can be selective for CS modality (interpositus lesions abolish the CR to all modalities of CS) (Steinmetz *et al.*, 1987). Lesions of the regions of the pons receiving projections from the auditory cortex abolish the CR established with electrical stimulation of auditory cortex as a CS (Knowlton and Thompson, 1992). Infusion of lidocaine into the pontine nuclei in trained animals reversibly attenuates the CR (tone CS) (Knowlton and Thompson, 1988). Extensive lesions of the middle cerebellar peduncle (mcp), which conveys mossy fibers from the pontine nuclei and other sources to the cerebellum, abolish the CR to all modalities of CS (Lewis *et al.*, 1987; Solomon *et al.*, 1986a).

Electrical stimulation of the pontine nuclei serves as a "supernormal" CS, yielding more rapid learning than does a tone or light CS (Steinmetz *et al.*, 1986; Tracy, 1995). With a pontine stimulation CS, lesion of the middle cerebellar peduncle abolishes the CR, thus ruling out the possibility that the pontine CS is activating noncerebellar pathways, e.g., by stimulating fibers of passage or antidromic activation of sensory afferents (Solomon *et al.*, 1986a). Stimulation of the middle cerebellar peduncle itself is an effective CS, and lesion of the interpositus nucleus abolishes the CR established with a pontine or middle peduncle stimulation CS (Steinmetz *et al.*, 1986). When animals are trained using electrical stimulation of the pontine nuclei as a CS (corneal airpuff US), some animals show immediate and complete transfer of the behavioral CR and of the learning-induced neural responses in the interpositus nucleus to a tone CS. These results suggest that the pontine stimulus and tone must activate a large number of memory circuit elements (neurons) in common (Steinmetz, 1990).

Some authors have made the rather astonishing assertion that no evidence exists for the presence of sensory related responses to conditioning stimuli in the cerebellum, i.e., that auditory and visual information do not project to the cerebellum (Gruart and Delgado-Garcia, 1994; Llinas and Walsh, 1993). Beginning with the classic work of Snyder *et al.* (1978), there is extensive literature on sensory projections to the cerebellum (Ito, 1984). In the present context, our own studies and those from other laboratories demonstrate beyond question that the auditory (and visual) stimuli used as CSs activate a number of Purkinje neurons in the cerebellar cortex and many neurons in the interpositus nucleus in naive animals before any training has been given (Berg and Thompson, 1995; Foy *et al.*, 1984, 1992; Foy and Thompson, 1986; Tracy, 1995). There are extensive auditory and visual projections to neurons in the cerebellar cortex and nuclei. Indeed, we and others have recorded Purkinje and interpositus neurons that show convergence of auditory CS and somatosensory (corneal airpuff) US projections prior to any training (e.g., Krupa, 1993; Thompson, 1990; Tracy, 1995).

V. The Cerebellum and the Reflex Eyeblink Response

It has been argued by some authors that the cerebellum is involved in the performance of the reflex eyeblink response and that its role in the conditioned eyeblink response is somehow secondary to this (e.g., Gruart and Delgado-Garcia, 1994). At least in so far as classical conditioning of the nictitating membrane (NM) extension response is concerned, it seems unlikely that the cerebellum plays any direct role in the reflex response. The mean onset latency of the NM reflex response to our standard 3 psi corneal airpuff US is 53 msec (Tracy, 1995) and the peak latency is about 100 msec. However, the mean onset latency of the same NM response to electrical stimulation of the anterior interpositus with parameters that yield a 50% maximum amplitude response (roughly equivalent to the reflex response) is 100 msec. The minimum latency for activation of the cerebellum by the corneal airpuff US is about 20 msec (Tracy, 1995). As a result, the reflex eyeblink response occurs, peaks, and declines before any direct influence by the cerebellum is possible. Tonic modulating influences are of course possible.

VI. Purkinje Neuron Activity

Many Purkinje neurons, particularly in HVI, are responsive to the tone CS and the corneal air puff US in naive animals, as just noted. Before training, many Purkinje neurons that are responsive to the tone CS show variable increases in simple spike frequency (parallel fiber activation) in the CS period. After training, many show learning-induced decreases in simple spike frequency in the CS period; however, a significant number show the opposite effect, and still other patterns of learning-induced responses are found (Berthier and Moore, 1986; Donegan *et al.*, 1985; Foy and Thompson, 1986; Foy *et al.*, 1992; Gould and Steinmetz, 1996; Thompson, 1990). Using electrical stimulation of the forelimb as the CS and periorbital shock as the US, Hesslow and Ivarsson (1994) recorded from Purkinje neurons in the ferret cerebellum. They reported that Purkinje neurons responded with weak increases or decreases (simple spike) to CS alone trials before training. In trained animals, a number of Purkinje neurons showed decreases, sometimes profound, in simple spike discharges in the CS period. In the rabbit eyeblink conditioning preparation, before training, Purkinje neurons that are influenced by the corneal air puff consistently show an evoked complex spike to US onset (climbing fiber activation). In trained animals, this US-evoked complex spike is virtually absent on paired CS–US trials when the animal gives a CR, but is present and normal on US alone test trials (Foy and Thompson, 1986; Krupa et al., 1991; Sears and Steinmetz, 1991; Thompson, 1990) (see Section III).

An interesting aspect of classical conditioning of discrete behavioral reponses (e.g., eyeblink) is the dramatic and almost linear decrease in the ability to learn (e.g., errors or trials to criterion) as a function of age, which has been documented in an important series of studies by Woodruff-Pak and associates (e.g., Woodruff-Pak and Jaeger, 1996; Woodruff-Pak and Thompson, 1988; Woodruff-Pak, this volume). Importantly, there is a close correlation between the decreased number of cerebellar Purkinje neurons and the increased difficulty in learning (Woodruff-Pak et al., 1990).

VII. Cerebellar Cortical Lesions

There is a growing consensus that cortical lesions limited to lobule HVI can impair but not abolish the delay eyeblink CR. It is extremely difficult, however, to completely remove the bottom of the sulcus of HVI without damaging the anterior interpositus, which lies directly underneath. Very large cortical lesions that include the anterior as well as the posterior lobes (intermediate and lateral zones) can massively and permanently impair acquisition and retention of the delay CR and, in some instances, prevent acquisition (e.g., Lavond et al., 1987, 1993; Logan, 1991; Yeo, 1991; Yeo et al., 1985b). Importantly, large cortical lesions abolish the adaptive timing of the CR (Logan, 1991; Perrett et al., 1993) and prevent the acquisition of conditioned inhibition (Logan, 1991). In one study, we explored the possible role of the cerebellar cortex and interpositus in retention of the trace CR (Woodruff-Pak et al., 1985). Interpositus lesions abolished the trace CR. Cortical lesions (posterior lobe) caused a transient decrease but recovery of the CR. Note that the anterior lobe was not lesioned here.

Experimentally, it is virtually impossible to remove the entire cerebellar cortex without damaging the nuclei. In a current study, the mutant Purkinje cell degeneration (pcd) mouse strain was used (Chen *et al.*, 1996). In this mutant, Purkinje neurons (and all other neurons studied) are normal throughout pre- and perinatal development. At about 2–4 weeks postnatal, the Purkinje neurons in the cerebellar cortex degenerate and disappear (Landis and Mullen, 1978). For a period of 2 to 3 months after this time, other neuronal structures appear relatively normal (Goldowitz and Eisenman, 1992). Thus, during this period of young adulthood, the animals have a complete functional decortication of the cerebellum.

We first showed that appropriate lesions of the interpositus nucleus in the wild-type control mice (normal cerebellum) completely prevented learning of the conditioned eyeblink response, as with all other mammals studied. So the cerebellum is completely necessary for learning. We then trained pcd mice. pcd mice learned very slowly and to a much lower level than the wild-type controls, but showed normal extinction with subsequent training to the CS alone. Thus the cerebellar cortex plays a critically important role in normal learning (of discrete behavioral responses), but some degree of learning is possible without the cerebellar cortex. The fact that extinction appears normal raises questions about a report claiming that lesions of the anterior lobe prevent extinction (Perrett and Mauk, 1995).

VIII. Decerebration

We reported that if normal animals are trained in the delay CR and are acutely decerebrated, they retained the CR as long as the red nucleus (and cerebellum) was not damaged (Mauk and Thompson, 1987). Although Kelly *et al.* (1991) claimed to have established CRs in acute decerebrate, decerebellate animals, they did not run necessary controls for sensitization (see Skelton *et al.*, 1988) and used a very short intertrial interval, which Nordholm *et al.* (1991) showed does not yield any learning at all in normal animals. It is therefore quite likely that Kelly *et al.* (1991) were dealing with nonassociative sensitization (see Nordholm *et al.*, 1991). Yeo (1991) used standard training procedures and reported that cerebellar lesions completely abolished CRs established in the acute decerebrate animal, thus directly contradicting Kelly *et al.* (1991). In a similiar vein, Hesslow (1994) reported that stimulation of the eyeblink regions of cerebellar cortex (c1 and c3 zones defined by climbing fiber projections from the periorbital region and by direct stimulation) in trained, high decerebrate cats profoundly inhibited the eyeblink CR but had little effect on the UR (also further evidence against the "performance" argument). Consequently, it seems clear that the cerebellum is necessary for retention of the eyeblink CR in the decerebrate preparation.

IX. Locus of the Long-Term Memory Trace

Overall, the results described to this point demonstrate conclusively that the cerebellum is necessary for learning, retention, and expression of classical conditioning of the eyeblink and other discrete responses (e.g., limb flexion, head turn). The next and more critical issue concerns the locus of the memory traces. The following evidence demonstrates conclusively that the long-term memory traces for this type of learning are formed and stored in the cerebellum.

A new approach to the problem of localizing memory traces in the brain, namely the use of methods of reversible inactivation, together with the recording of neural activity, has been developed. Reversible inactivation methods (e.g., using drugs or cooling) per se have existed for some time and have been used very effectively to produce temporary lesions (e.g., Mink and Thach, 1991). This method has been applied systematically to the major structures and pathways in the cerebellar–brain stem circuit identified as the essential (necessary and sufficient) circuit for classical conditioning of discrete responses (Fig. 1) during performance and acquisition of the CR.

First, the logic of the approach. Consider two structures (A and B) that are connected directly to one another, each of which when inactivated completely prevents expression of the behavioral CR. Further, predictive learning-induced neuronal "*models*" of the behavioral CR are present in both structures after training. For argument, assume that inactivation of these structures has no effect on performance of the UR. Inactivation of A abolishes the learning-induced neuronal response in B, as well as the behavioral CR, whereas inactivation of B abolishes the behavioral CR but does not abolish the learning-induced neuronal response in A. This argues that the memory trace circuit projects from A to B and not B to A. Just these results have been obtained for the interpositus nucleus (A) and the red nucleus (B) (Chapman *et al.*, 1990; Clark and Lavond, 1993). The next step is to inactivate each of these structures during acquisition training. In immediate postinactivation training, assume that animals show no savings (i.e., they take the same number of trials to learn as do naive animals) and must learn as if naive when A was inactivated. However, when B was inactivated, they show immediate asymptotic learning postinactivation. Consequently, the memory must be formed at or beyond A but before B. Further, because there were no savings when A was inactivated, no significant part of the memory could be formed in structures before A in the circuit. Because learning was complete during training with inactivation of B, no significant part of the memory could be formed in structures beyond B in the circuit. As shown in this chapter, these are exactly the results obtained with the anterior interpositus nucleus (A) and the magnocellular red nucleus (B).

A word about methods of inactivation. Infusion of a low dose of muscimol (e.g., 0.7 nmol) over a period of 1 to 2 min inactivates neuron cell bodies (but not axons) for a period of 2–4 hr. It activates $GABA_A$ receptors for this long period of time, thus hyperpolarizing the neurons. Lidocaine acts on sodium channels, thus inactivating both cells and fibers. The duration of action is very brief (a few minutes) so it must be infused continuously. Relatively high doses are necessary, and the site of effective action can be very localized. Tetrodotoxin (TTX) is a much more effective sodium channel blocker and can yield inactivation for a period of hours with a very low dose infused over a minute or two. It also inactivates both cells and fibers. The exact distribution of muscimol in the brain can be determined by the use of radiolabeled (tritiated) mucimol. It is more difficult to determine the effective distribution of radiolabeled lidocaine because it must be infused continuously. (Radiolabeled TTX is not currently available commercially.) Finally, reversible cooling, as developed by Lavond and associates (see Clark et al., 1992), can inactivate a relatively localized region of brain tissue for a period up to hours and has the advantage that it can be turned on or off in a matter of seconds. The exact region inactivated cannot be determined but can be estimated computationally and/or with neuronal recording.

As noted earlier, Fig. 1 shows in a highly simplified schematic form the essential memory trace circuit for classical conditioning of discrete responses based on the lesion, recording and stimulation evidence described previously. Interneuron circuits are not shown, only net excitatory or inhibitory actions of projection pathways. Other pathways, known and unknown, may also of course be involved. Many uncertainties still exist, e.g., concerning details of sensory-specific patterns of projection to pontine nuclei and cerebellum (CS pathways), details of red nucleus projections to premotor and motor nuclei (CR pathway), and the possible roles of recurrent circuits.

Several parts of the circuit have been reversibly inactivated for the duration of training (eyeblink conditioning) in *naive* animals, indicated by shadings labeled a, b, c, d, and encircled e in Fig. 1. Motor nuclei essential for generating UR and CR (primarily seventh and accessory sixth and adjacent neural tissues) were inactivated by the infusion of muscimol (6 days) or cooling (5 days) during standard tone–airpuff training (a in Fig. 1) (Krupa et al., 1996; Thompson et al., 1993; Zhang and Lavond, 1991). The animals showed no CRs and no URs during this inactivation training; indeed, performance was completely abolished. As shown in Fig. 2, the region of inactivation in the brain stem with muscimol was quite large, including the facial (seventh) nucleus, the accessory abducens nucleus, and an extensive surrounding region of reticular formation. The motor paralysis on the side of infusion was complete—the external eyelids were flaccid, the left ear hung down, and no vibrissae movements occurred—and lasted at least 3 hr, followed by full recovery. [Importantly, when the animals were unrestrained, airpuff to the paralyzed (trained) eye caused the animal to jerk its head away, showing that sensory processing of the US occurred]. However, the animals exhibited asymptotic CR performance and normal UR performance from the very beginning of postinactivation training (see Fig. 3). Thus, performance of the CR and UR are completely unnecessary for normal learning, and the motor nuclei and adjacent inactivated tissue make no contribution at all to formation of the memory trace—they are completely efferent from the trace.

Inactivation of the magnocellular red nucleus is indicated by b in Fig. 1. Inactivation by low doses of muscimol for 6 days of training had no effect on the UR but completely prevented expression of the CR (Krupa et al., 1993). Animals showed asymptotic learned performance of the CR from the beginning of postinactivation training (see Fig. 4). Training during cooling of the magnocellular red nucleus gave identical results—animals learned during cooling, as evidenced in postinactivation training, but did not express CRs at all during inactivation training (Clark and Lavond, 1993). However, cooling did impair performance of the UR (but the animals learned normally), yet another line of evidence against the "performance" argument (see Fig. 5). Consequently, the red nucleus must be efferent from the memory trace.

Inactivation of the dorsal anterior interpositus and overlying cortex (c in Fig. 1) by low doses of muscimol (6 days), by lidocaine (3 days, 6 days), and by cooling (5 days) resulted in no expression of CRs during inactivation training and in no evidence of any learning during inactivation training (Clark et al., 1992; Krupa et al., 1993; Nordholm et al., 1993). In subsequent postinactivation training, animals learned normally as though completely naive; they showed no savings at all relative to noninactivated control animals (see Figs. 4–6). None of the methods of inactivation had any effect at all on the performance of the UR on US alone trials. In one study (Nordholm et al., 1993), cerebellar lidocaine infusions effective in abolishing the CR and preventing learning were subsequently tested on US alone trials over a wide range of US intensities and had no effect on performance

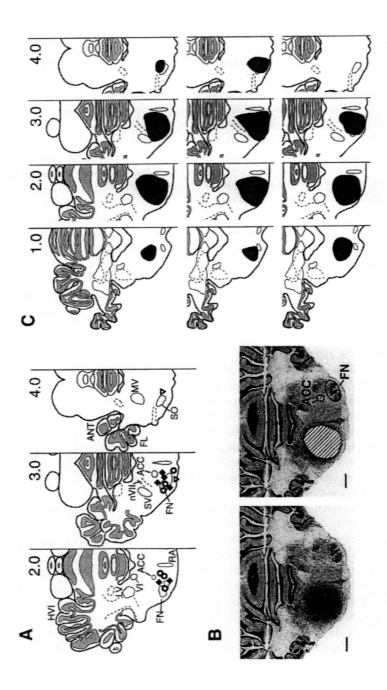

of the UR; indeed, URs were numerically larger with lidocaine inactivation of CRs than with saline control infusions that had no effect on CRs (see Fig. 7). The distribution of [^{3}H]muscimol completely effective in preventing learning included the anterior dorsal interpositus and overlying cortex of lobule HVI, a volume approximately 2% of the total volume of the cerebellum (Krupa et al., 1993). The region of the cerebellum essential for learning this task is extremely localized (see Fig. 6).

Welsh and Harvey (1991) gave rabbits extensive training to a light CS, gave them transfer training to a tone CS with interspersed light CSs (one session) with lidocaine infusion in the anterior interpositus, and trained them to a tone without infusion. Control animals were treated identically except they were given saline infusions during the transfer training session. They reported that CRs to the light CS were prevented during lidocaine infusion but that the animals exhibited virtually asymptotic performance to the tone CS in postinfusion training. As noted earlier, if *naive* animals are given tone CS training during lidocaine infusion in the cerebellum (3 days or 6 days), they do not learn at all and learn subsequently with no savings. One possible explanation of the Welsh and Harvey (1991) results is that substantial transfer of training occurred; indeed, their control animals showed very substantial transfer compared to *naive* animals, i.e., animals with no prior training (see Schreurs and Kehoe, 1987). However, they did not run the control essential to evaluate transfer (see later). Cannula location may also be a factor. When lidocaine was infused in the white matter ventral to the interpositus to inactivate the efferent projections from the interpositus (see later and d in Fig. 1), normal learning occurred, although no CRs were expressed during infusion training (Nordholm et al., 1993),

---

FIG. 2. (A) Cannulae locations for each of the rabbits with motor nuclei infusions of muscimol. ▲, effective muscimol infusion sites; ○, controls; and △, ineffective placements. Numerals above each section represent distances (in millimeters) rostral to lambda in the stereotaxic plane. (B) (left) Photomicrograph of a coronal section through the motor nuclei (MN) of one rabbit infused with [^{3}H]muscimol showing the extent of [^{3}H]muscimol diffusion. The drug diffused throughout the accessory abducens nucleus (ACC), facial nucleus (FN), and surrounding reticular formation. There was no evidence of diffusion into the abducens nucleus. (Right) Same section as on left; hatched region represents the maximal extent of muscimol diffusion. Scale bar is 2.0 mm. (C) Extent of muscimol diffusion (shaded regions, shown on standard sections) for each of the rabbits infused with [^{3}H]muscimol. In each case, the drug diffused throughout the ACC, FN, and surrounding reticular formation. The pattern of diffusion extended (in the rostral/caudal plane) from the level of the inferior olive (section 1.0) to the level of the superior olive (section 4.0). ANT, anterior lobe; FL, flocculus; HVI, hemispheric lobule VI; LV, lateral vestibular nucleus; MV, motor trigeminal nucleus; nVII, seventh nerve; RA, raphe nucleus; SO, superior olive; SV, sensory trigeminal; VI, abducens nucleus (from Krupa et al., 1996).

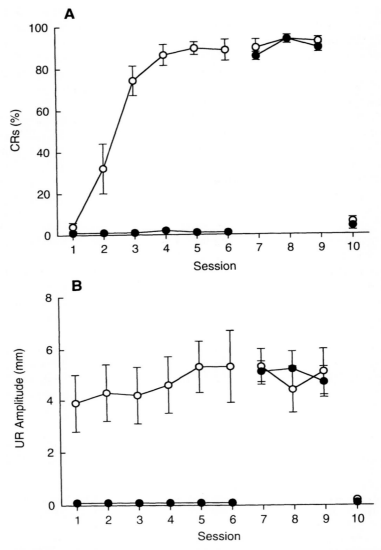

FIG. 3. Mean percentage (CRs and URs) for each training session with motor nuclei infusions of muscimol. One hour prior to sessions 1 to 6 and session 10, one group of rabbits (muscimol, $n = 6$, ●) was infused with 3.5 nmol muscimol (in 0.4 ml saline) into the motor nucleus (MN). The other group (control, $n = 6$, ○) was infused with saline vehicle. No infusions were administered prior to sessions 7 to 9. (A) Mean (± SEM) percentage CRs for each group. Rabbits infused with saline vehicle (sessions 1 to 6) learned the CR normally, reaching asymptotic levels of CRs by the end of session 3. In marked contrast, rabbits infused

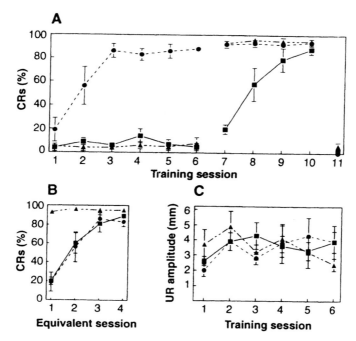

FIG. 4. Effect of muscimol infusion in the cerebellum and red nucleus on CRs and URs. (A) All animals received an infusion before training on sessions 1 to 6. The cerebellar group (■, $n = 6$) received muscimol infusions into the ipsilateral lateral cerebellum, the red nucleus group (▲, $n = 6$) received muscimol in the contralateral red nucleus, and the saline group (●, $n = 6$) received 1 μl of saline vehicle into the ipsilateral lateral cerebellum. No infusions were administered on days 7 to 10. All animals received muscimol infusions before session 11. Data are expressed as percentage CRs averaged over all animals in each group for each training session. (B) Percentage CRs for sessions 1 to 4 of the saline group and sessions 7 to 10 of the cerebellar and red nucleus groups. (C) UR amplitudes on airpuff-only test trials during the six sessions in which infusions were administered. There were no significant differences between groups on these days. All data points are means ± SEM. Symbols are the same for all charts (from Krupa et al., 1993).

with muscimol performed no CRs at all during these sessions. These rabbits performed the CR at asymptotic levels from the start of training on session 7 (no infusion). They had fully learned the CR during the previous inactivation sessions. Muscimol infusions prior to session 10 completely abolished the CR in all rabbits. (B) Mean (± SEM) UR amplitude (measured on US alone trials) for each group. Muscimol infusions prior to sessions 1 to 6 completely prevented performance of the UR. UR performance on sessions 7 to 9 (no infusions) did not differ from that of the controls. Infusion of muscimol on session 10 completely abolished the UR in all of the animals (from Krupa et al., 1996).

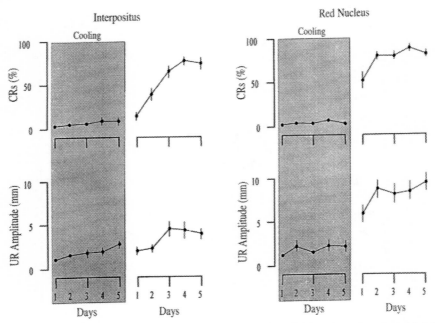

FIG. 5. Effect of cooling the anterior interpositus nucleus (left) or the red nucleus (right) during training of the conditioned eyeblink response (CRs) in two groups of rabbits. Animals were given 5 days of training while the structure was cooled (shading), followed by 5 days of training without cooling. Cooling either structure completely prevented expression of CRs during training. Following removal of cooling, animals that had interpositus cooling during training learned nothing and then learned as though naive. In complete contrast, animals with prior red nucleus cooling learned during cooling, as seen after cooling was removed. Note that the interpositus cooling had no effect on the UR but cooling of the red nucleus substantially impaired performance of the UR (from Clark and Lavond, 1993; Clark et al., 1992).

a result analogous to the result reported by Welsh and Harvey (1991; see Fig. 7). In any event, all studies where *naive* animals were trained during inactivation of the anterior interpositus and overlying cortex (cooling, muscimol, lidocaine) agree in showing that no learning at all occurs during inactivation training.

Cipriano *et al.* (1995) completed an exact replication of the Welsh and Harvey (1991) paradigm but used muscimol instead of lidocaine. The literature and the authors own experience suggest that the region inactivated by the infusion of lidocaine can be very localized (see later). Further, because of its short duration it must be infused continuously; transient recovery could invalidate the procedure used by Welsh and Harvey (1991).

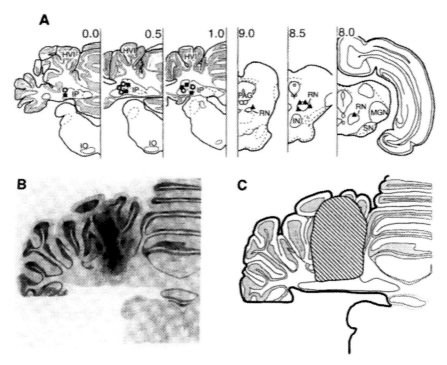

FIG. 6. Localization of inactivated regions in the cerebellum and red nucleus for data in Fig. 4 (muscimol infusions). (A) Locations of all cannulae tips. ■, cerebellar muscimol cannulae; ○, saline controls; and ▲, red nucleus cannulae. Numbers above the first three sections are distances (in millimeters) rostral to lambda, whereas numbers above the last three are distances caudal from the bregma skull sutures. HVI, hemispheric lobule VI; IN, interpeduncular nucleus; IP, interpositus nucleus; IO, inferior olive; MGN, medial geniculate nucleus; PAG, periaqueductal grey; RN; red nucleus; SN, substantia nigra. (B) Digitized image of an autoradiograph showing the greatest extent of [^{3}H]muscimol diffusion in the lateral cerebellum. Labeling encompasses dorsal aspects of the anterior interpositus and the overlying cortex, including lobule HVI. In no instance was any labeling found outside the cerebellum. The autoradiograph is shown superimposed on the Nissl-stained section from which it was exposed. (C) Outline drawing of micrograph in B. Hatched area delineates the maximal extent of [^{3}H]muscimol diffusion (from Krupa et al., 1993).

Cipriano et al. (1995) ran the saline control group as in Welsh and Harvey (1991) and also an additional control group essential to determine the amount of transfer in the experimental animals independent of inactivation, namely a group placed in the training apparatus but given no stimuli. As expected, saline control animals showed substantial transfer relative to unstimulated controls. Even more important, the animals given muscimol infusion during the day of tone training showed no signs of having learned

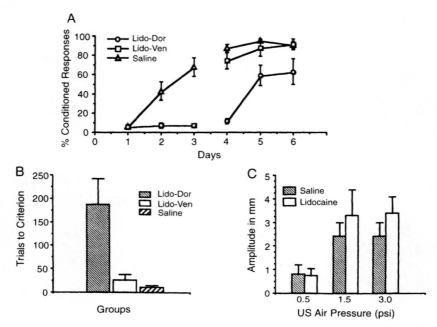

FIG. 7. (A) Effect of lidocaine infusions in the cerebellum targeted to the dorsal or ventral anterior interpositus nucleus. Dorsal (Lido-Dor) and ventral (Lido-Ven) versus saline. Lidocaine-infused animals showed no conditioned responses during infusion, dorsal and ventral group data combined. In postinfusion training animals with ventral cannula placements showed virtually asymptotic learning, whereas those with dorsal placements learned as if naive. (B) Mean postinfusion trials to criterion for the two lidocaine infusion groups and the saline control group. (The number of trials to criterion for the dorsal group was significantly greater than the number of trials for the saline and ventral groups, which did not differ from each other.) (C) Performance on unconditioned stimulus-alone trials at 0.5, 1.5, and 3.0 psi, with dorsal infusions of saline and effective infusions of lidocaine. (There were no differences between unconditioned reflex response amplitudes with saline and lidocaine infusions. Bars indicate standard errors.) from Nordholm et al., 1993.

to tone on the following day—their performance was identical to the performance of the saline control animals on the first day of tone training, i.e., simply the degree of transfer of training shown by the saline controls relative to the unstimulated controls.

Consequently, muscimol gives results opposite to those reported by Welsh and Harvey (1991) but is completely consistent with all other inactivation studies. No learning at all occurs during the day of tone training with muscimol infusion. It could reasonably be concluded that Welsh and Harvey (1991) were inactivating the output from the interpositus with their lidocaine infusions, just as Nordholm *et al.* (1993) demonstrated with ventral interpositus cannulae placements in naive animals.

As just noted, Nordholm *et al.* (1993) showed that infusion of lidocaine in the ventral interpositus (d in Fig. 1) prevented expression but not learning of the CR (Fig. 7). This presumably inactivated white matter conveying information from the cerebellum to other brain regions. Because lidocaine has a very localized region of action, it is quite possible that these ventral interpositus infusions did not inactivate all the efferent projections from the cerebellum. Thus, it is possible that efferent fibers from the interpositus not inactivated by lidocaine project to structure X (e.g., in the thalamus?), critical for formation of the memory trace, and X in turn projects back to the cerebellum. The only way this could occur is if the lidocaine infusion ventral to the interpositus does not inactivate these efferent fibers because the animals learn. However, this hypothetical pathway can play no role at all in the expression of the CR because the ventral infusions, although not preventing learning, completely prevent expression of the CR. This hypothesis is most unlikely but logically possible.

The critical test of this hypothesis is to inactivate essentially all of the output from the cerebellum, i.e., the superior cerebellar peduncle (scp), during training (e in Fig. 1). Results are shown in Fig. 8 (Krupa and Thompson, 1995). TTX was infused in the scp ipsilateral to the critical cerebellar hemisphere before any significant numbers of axons exited the scp. This inactivates both descending and ascending efferent projections of the cerebellar hemisphere. The animals exhibited motor symptoms consistent with this effect. TTX infusion in the scp completely prevented expression of the CR (with no effect on the UR) for the 6 days of training. On the seventh day, TTX was not infused and the animals showed asymptotic learned performance of the CR. Control animals were infused with TTX in the scp for 6 days but not trained, and then trained, and showed normal learning.

Collectively, these data strongly support the hypothesis that the memory trace is formed and stored in a localized region of the cerebellum (anterior interpositus and overlying cortex). Indeed we can conceive of no rational alternative. Inactivation of this region (c) during training completely prevents learning, but inactivation of the output pathway from the region (d and e) and its necessary (for the CR) efferent target, the red nucleus (b), does not prevent learning at all. In no case do the drug inactivations have any effect at all on the performance of the reflex response on US alone trials. If even part of the essential memory trace was formed prior to the cerebellum in the essential circuit, then the animals would have to show savings following cerebellar inactivation training and they show none at all. Similarly, if part of the essential memory trace was formed in the

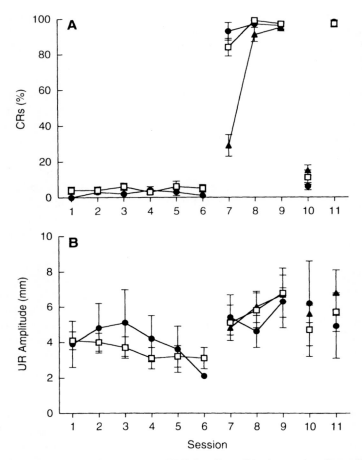

FIG. 8. (A) Percentage CRs (mean ± SEM) for all conditioning sessions from all animals with effective TTX infusion cannulae placements in the superior cerebellar peduncle (scp). TTX was infused into the scp of each animal prior to sessions 1–6 and session 10. No infusions were administered prior to sessions 7–9. Muscimol was infused prior to session 11. Animals trained with an auditory white noise as CS (□, $n = 6$) or with electrical microstimulation of the LRN as CS (●, $n = 4$) performed no significant number of CRs during the first six infusion sessions. On session 7, the first session without infusion, these animals performed the CR at asymptotic levels from the start of training; they had fully learned the CR during the previous six inactivation sessions. Controls (▲, $n = 6$) were infused with TTX and restrained, but presented with no stimuli during sessions 1–6. These animals performed significantly fewer CRs on session 7, their first conditioning session with the auditory CS, and subsequently learned the CR on following sessions. TTX infusions prior to session 10 completely abolished the previously acquired CR in all rabbits. Infusion of muscimol prior to session 11 had no effect on the CR in any rabbit. (B) UR amplitude (mean ± SEM) on airpuff-alone test trials. TTX infusions prior to sessions 1–6 resulted in UR amplitudes significantly lower than URs on sessions 7–9 in which no infusions were administered. Infusion of TTX prior to session 10 or muscimol prior to session 11 had no significant effect on UR amplitudes compared with UR amplitudes on session 9 in which no infusions were administered. Symbols are as in A (from Krupa and Thompson, 1995).

red nucleus or other efferent targets of the interpositus (e.g., brain stem), then animals could not show asymptotic CR performance following red nucleus (b), interpositus efferent (d), or scp (e) inactivation training, but they do.

X. The Issue of Performance

In view of the fact that appropriate lesions of the anterior interpositus nucleus completely and permanently abolish the behavioral CR with no effect on the UR, the argument some have made (Welsh and Harvey, 1989) that the lesion abolition of the CR is due to effects on the UR is most puzzling. These authors claimed that effective lesions of the interpositus nucleus (abolishing the CR) had a very small effect on the UR at very low US intensities in US alone trials. Actually, they did not do this experiment. Instead, they compared effective vs noneffective lesion animals postlesion and selected only some of these animals for comparison. This is not an appropriate comparison—one can obtain any result one wishes by doing this because the interanimal variability in UR performance is considerable. Steinmetz *et al.* (1992) did the experiment by comparing the same animals before and after the effective interpositus lesion. Lesions effective in abolishing the CR completely had no persisting effects on any measurable property of the UR over a wide range of intensities. Welsh and Harvey (1989) asserted that "when one attempts to equate the CS and the UCS (unconditioned stimulus) as response-eliciting stimuli, the deficits in the CR and the UCR (unconditioned response) become more alike" (p. 309). The only way to evaluate this statement is to equate the CS and the US in terms of response elicitation prior to lesion and then determine the effect of the lesion on the two responses that were "psychophysically equivalent" prior to lesion. Welsh and Harvey (1989) did not make this comparison and did not provide any information on the properties of the URs prior to lesion in their animals. When this is done, e.g., when the intensity of the US is reduced so that the UR (US alone trials) is matched in amplitude and percentage response to the CR before lesion, the interpositus lesion abolishes the CR but still has no effect at all on the prelesion equivalent UR (Steinmetz *et al.*, 1992). Furthermore, using data generated by Welsh and Harvey (1989), Steinmetz *et al.* (1992) matched CR amplitudes and UR amplitudes (at low US intensity) postlesion, comparing their "effective" and "ineffective" lesion groups, and found that the CR was abolished and the UR unchanged.

All these studies used a standard US intensity for training (e.g., ≥3 psi), which typically yields a UR (US alone trials) larger in amplitude than the CR. Ivkovich *et al.* (1993) trained animals with a low intensity US just suprathreshold to establish learning. Under these conditions the CR (CS alone trials) and the UR (US alone trials) are equivalent in amplitude (the CR is numerically but not statistically larger than the UR). Interpositus lesions completely abolished the CR and had no effect at all on the UR.

There is in fact a double dissociation in terms of various brain lesion effects on the CR and the UR in eyeblink conditioning. Appropriate partial lesions of the motor nuclei involved in generating the CR and the UR cause immediate abolition of both the CR and the UR; however, with postoperative training the CR recovers almost to the preoperative level but the UR shows little recovery (Disterhoft *et al.*, 1985; Steinmetz *et al.*, 1992). Large lesions of appropriate regions of the cerebellar cortex that markedly impair or abolish performance of the CR (see later) result in an increase in the amplitude of the UR (Logan, 1991; Yeo, 1991).

Collectively, this evidence argues persuasively against the "performance" hypothesis that interpositus lesion abolition of the CR is somehow a result of nonexistant lesion effects on the UR.

XI. Putative Noncerebellar Sites of Memory Storage

Despite all the evidence just reviewed, some authors appear reluctant to deport from the hypothesis that the cerebellum cannot possibly be the site of memory storage for classical conditioning of the eyeblink and other discrete responses (Bloedel, 1992; Bloedel and Bracha, 1995; Llinas and Welsh, 1993; Welsh and Harvey, 1989, 1991). This contrary view holds that the memory trace is somehow stored somewhere in the brain stem. In fact, these authors have not been able to account for the evidence supporting the cerebellar hypothesis just reviewed. The primary evidence that they cite is the fact that learning-induced neuronal models of the behavioral CR that precede and predict the occurrence and form of the CR (just as does the neuronal model in the interpositus nucleus) are present in several regions of the brain, including the red nucleus and a region bordering the trigeminal nucleus. As discussed earlier, the red nucleus cannot possibly be the site of memory storage.

With respect to the trigeminal nucleus, neurons in this region show evoked responses to tone CS onset and corneal airpuff US onset and develop a learning-induced neuronal model of the behavioral CR (Clark and Lavond, 1996; Richards *et al.*, 1991). Large lesions of the trigeminal nucleus will certainly prevent learning by blocking US activation of the brain. The fact that training during inactivation of the critical region of the cerebellum (c in Fig. 1) completely prevents learning of the CR would seem to rule out any significant aspect of memory formation in the trigeminal nucleus, i.e., if any significant learning occurred in the trigeminal nucleus, there would have to be savings following training with cerebellar inactivation and there is no savings at all.

Even more decisive evidence against trigeminal memory storage has been shown by Clark and Lavond (1996). Reversible inactivation of the anterior interpositus (by cooling) in trained animals abolishes the behavioral CR and the learning induced neuronal model of the CR in the trigeminal nucleus, but does not affect the trigeminal CS- and US-evoked neuronal responses. Exactly the same is true with cooling of the red nucleus, i.e., inactivation of the red nucleus abolishes performance of the behavioral CR but does not prevent learning at all. Because inactivating the red nucleus also abolished the learning induced neuronal model in the trigeminal nucleus, the trigeminal nucleus cannot be the site of memory storage. Instead, it is clear that the learning induced neuronal model in the trigeminal nucleus is relayed there from the interpositus nucleus via the red nucleus. The fact that this neuronal model is relayed to the trigeminal nucleus, the first US sensory relay nucleus, is of interest as it may play some sort of modulatory role in learning and performance, but it is clearly not a site of memory storage.

To date, in every site in the brain exhibiting a learning induced neuronal model of the learned behavioral CR that has been studied with inactivation of the interpositus nucleus, this inactivation abolished the learning induced neuronal model (see earlier and L. Chen, S. Bao and R. F. Thompson, unpublished observations; D. G. Lavond and R. E. Clark, personal communication; Sears et al., 1996). The conclusion seems inescapable: the long-term memory trace for classical conditioning of discrete behavioral responses is formed and stored in the cerebellum and this information is then relayed from the interpositus nucleus to other brain regions.

XII. Neural Substrate of the Error-Correcting Algorithm in Classical Conditioning

Evidence reviewed earlier argues for cerebellar localization of the basic memory trace for the delay CR. However, various recurrent pathways also appear to play critical roles in certain aspects of delay learning. One example concerns the error-correcting algorithm in classical conditioning. The basic notion was quantified in the Rescorla–Wagner equation for learning, namely that the associative strength between the CSs and the CR added on a given trial is maximal at the beginning of training and decreases in proportion to the developing strength of the CSs–CR. With asymptotic learning, additional training adds no additional associative strength (Rescorla and Wagner, 1972). This error-correcting algorithm is termed the delta rule in connectionist level cognitive models of learning and memory (Rumelhart and McClelland, 1986). In both instances, this error-correcting algorithm serves to reduce the discrepancy between a neural or computational model of the real world and the real world. In classical conditioning, this notion accounts very nicely for the phenomenon of blocking (Kamin, 1969). In brief, if animals are trained to asymptote on CS_1 and then given additional training on $CS_1 + CS_2$, they later exhibit no learning to CS_2 whereas animals trained from the beginning on $CS_1 + CS_2$ show substantial learning to each stimulus.

It was suggested earlier that the direct GABAergic inhibitory projection from the interpositus nucleus to the inferior olive (Andersson and Hesslow, 1986; Nelson and Mugnoini, 1987) may serve this function in classical conditioning of discrete responses (Donegan et al., 1989; Thompson 1989). At the physiological level there is supporting evidence for this assertion. As noted earlier, dorsal accessory olive (DAO) neurons that are influenced by the US (corneal airpuff) consistently show evoked unit responses to US onset on paired trials before training, but not after training, on trials where CRs occur. As expected, the same result holds for Purkinje neurons (complex spikes) influenced by the US. In well-trained animals, US alone presentations consistently evoke complex spikes, but on paired trials where the CR occur, the US does not evoke complex spikes. We have seen this result consistently in all Purkinje neurons activated by the US alone (Krupa, 1993). When recording from such Purkinje neurons in the trained animal, infusion of the $GABA_A$ antagonist picrotoxin into the DAO reinstates the US-evoked complex spikes to US onset on paired trials where CRs occur (J. J. Kim, D. J. Krupa, and R. F. Thompson, unpublished observation). Thus, blocking GABA inhibition at the DAO disinhibits the US-evoked activity in DAO neurons, climbing fibers, and Purkinje neurons by presumably blocking the direct interpositus to DAO GABAergic inhibitory pathway.

Evidence now shows that this system plays a key role in the behavioral phenomenon of blocking in this form of learning (Kim et al., 1992; J. J. Kim and R. F. Thompson, unpublished observations). Animals are trained on CS_1 to asymptote and are then given extensive $CS_1 + CS_2$ training while a $GABA_A$ antagonist picrotoxin is infused in the dorsal accessory olive. Blocking control animals are treated identically but are infused with vehicle. Control animals for the blocking effect are not given initial CS_1 training. The results of subsequent training to CS_2 are that picrotoxin infused animals show no blocking; they perform identically to the control animals given no initial CS_1 training. In contrast, vehicle control infusion animals show substantial blocking. Thus, the prevention of GABA inhibition in the inferior olive precludes blocking. This form of inhibitory feedback circuitry may act in all brain systems involved in learning and memory to subserve error correction, i.e., the delta rule.

XIII. Supervised Learning and the Cerebellum

Knudsen (1994) identifies what he terms supervised learning in the brain. In supervised learning, information from one network of neurons acts as an instructive signal to influence the pattern of connectivity in another network. He uses the cerebellar networks in classical conditioning (the authors' work); the neuronal networks subserving adaptation of the vestibulo-ocular reflex (VOR) (Ito, 1984; Lisberger, 1988); midbrain networks involved in gaze control in *Xenopus* (e.g., Gaze et al., 1970; Udin, 1985); and calibration of the auditory space map in the barn owl (e.g., Knudsen and Knudsen, 1990) as parallel examples. In his words: "Supervised and unsupervised forms of learning cooperate during development, and in adulthood, to optimize the integrative mechanisms of the brain for the individual. Early in the development of a network, basic patterns of connectivity form by the interaction of genetic determinants and unsupervised forms of learning (Brown et al., 1991; Miller, 1989). This ability of networks to self-organize allows the brain to develop extensively before an animal is even born. As the brain begins to process experience-driven activity, however, supervised learning begins to exert its effects, adjusting and calibrating the original patterns of connectivity. In adulthood, supervised and unsupervised forms of learning continue to adjust connectivity patterns in many networks, altering them adaptively in response to changes in sensory or motor capacities and/or environment that may occur throughout the lifetime of the animal.

"The powerful influence that supervised learning can exert on patterns of connectivity in the brain is shown by many examples. In the case of classical conditioning, strong connections are established between networks of neurons that are not functionally coupled prior to conditioning (McCormick and Thompson, 1984b; Pavlov, 1927; Thompson, 1990). In the development of binocular receptive fields in the optic tectum of the *Xenopus*, the pattern of activity from the contralateral eye instructs the anatomical pattern of tectal connections representing input from the ipsilateral eye (Udin, 1989; Udin and Keating, 1981). In the case of the auditory space map in the owl, connections between pre- and postsynaptic neurons are strengthened if the information conveyed by the auditory inputs is consistent with the visual instructive signal and are weakened if the information conveyed by the inputs is inconsistent with the instructive signal (Brainard and Knudsen, 1993). Thus every aspect of the auditory space map, including its orientation, position, and topography, is under the control of supervised learning.

"The adjustments made by supervised learning in the patterns of connectivity are precise and customized for the individual. The gain of the VOR in the monkey is maintained near a value of 1, despite differences in the sensitivity of the vestibular end organs or the strength of the extraocular muscles that may occur among individuals and within an individual as it ages (Lisberger, 1988)" (Knudsen, 1994, pp. 3994–3995).

Knudsen notes in concluding that processes of plasticity such as long-term potentiation (LTP) and long-term depression (LTD) may be sufficient mechanisms to instantiate all forms of supervised learning.

XIV. Putative Mechanisms

The localization of memory traces for the classical conditioning of discrete responses to the cerebellum is now sufficiently established that a focus on cerebellar cellular mechanisms of plasticity is warranted. Indeed, because much of the essential circuitry has been identified, this system may provide the first instance in the mammalian brain where the entire circuitry from memory trace to learned behavior, the "read out" of memory and the neuronal content of the memory store, can be analyzed (Thompson and Krupa, 1994).

Many of the Purkinje neurons exhibiting learning-related changes show decreases in simple spike responses in the CS period. This would result in the disinhibition of interpositus neurons, consistent with a mechanism of LTD (see earlier). Current evidence suggests that the glutamate activation of AMPA and metabotropic receptors, together with increased intracellular calcium (normally by climbing fiber activation), yields the persisting decrease in AMPA receptor function of LTD (see also Ito and Karachot, 1990; Linden and Conner, 1991). LTD has been proposed as the mechanism underlying synaptic plasticity in the flocculus that subserves adaptation of the VOR (e.g., Ito, 1989).

In behavioral studies, the classical conditioning of discrete responses (e.g., eyeblink, head turn, etc.) evoked by stimulation of the DAO-climbing fibers occurs when paired with stimulation of the mossy fibers as a CS

(Steinmetz et al., 1989), the exact procedure used in the initial studies of LTD (Ito et al., 1982). However, the temporal properties of LTD and classical conditioning appear to differ. Most studies of LTD have used near-simultaneous activation of parallel and climbing fibers. In classical conditioning, mossy-parallel fiber stimulation must precede climbing fiber stimulation by about 100 msec, and best learning occurs with an interval of about 250 msec; simultaneous activation results in extinction of the CR so induced (Steinmetz et al., 1989). As it turns out, virtually all current studies of LTD have used the cerebellar slice or tissue culture and have used GABA antagonists.

Parallel fiber–Purkinje cell field potentials in the cerebellar slice were used to assess the temporal properties of LTD (Chen and Thompson, 1995). The simultaneous stimulation of parallel fibers and climbing fibers (100 pairings) yielded LTD in the presence of bicuculline, as in other studies, but did not yield LTD in the absence of bicuculline. If parallel fiber stimulation preceded climbing fiber stimulation by 250 msec, robust LTD developed in the absence of bicuculline, suggesting that GABA, i.e., inhibitory interneurons, may play a key role in determining the temporal properties of the synaptic plasticity underlying LTD. This possibility is consistent with current views of the mechanisms of LTD, e.g., G protein activation of intracellular cascades (Ito, 1993; Ito and Karachot, 1990; Linden and Connor, 1991). It is perhaps relevant that very large cerebellar cortical lesions severely disrupt the adaptive timing of the conditioned eyeblink response (see earlier; Logan, 1991; Perrett et al., 1993). Interestingly, if a sufficient number of pairings of parallel and climbing fibers is given (600), both simultaneous and 250-msec parallel fiber precedence result in LTD in the absence of GABA antagonists. So the authors' results are not necessarily inconsistent with the studies from other laboratories (Ito, 1993).

Studies using "gene knockout" preparations have strengthened the argument for LTD as a key mechanism of memory storage in cerebellar cortex. Thus, mice that lack the metabotropic glutamate receptor (mGluR1) show marked impairments in cerebellar cortical LTD and eyeblink conditioning (Aiba et al., 1994). They also show generalized motor impairments, i.e., some degree of ataxia, as do the pcd mice (see earlier). Interestingly, current studies present evidence supporting the view that LTD is more important for learning (eyeblink conditioning) than for motor coordination. Thus, using protein kinase C γ knockout mutant mouse, Kano et al. (1995) showed that Purkinje neurons in adult animals maintained the perinatal condition of more than one climbing fiber per neuron (wild-type adults have only one climbing fiber per Purkinje neuron). Chen et al. (1995) showed that this mutant exhibited normal LTD but impaired motor coordination (due, presumably, to the multiple climbing fiber innervation of Purkinje neurons). In striking contrast, these animals learned the conditioned eyeblink response more rapidly than did wild-type controls!

Just the opposite result holds for a quite different mutant, namely the glial fibrillary acidic protein (GFAP) knockout mouse (Shibuki *et al.*, 1996) in which the cerebellar cortex appears to be anatomically normal. Cerebellar cortical LTD and eyeblink conditioning are markedly deficient in these animals (their performance is very similar to that of the pcd mice) but they do not show any impairments at all in motor coordination or general motor behavior!

The Shibuki *et al.* (1996) study is important in another regard as well. GFAP is not present in neurons, only in glial cells. In the cerebellum it is normally present in substantial amounts in the Bergman glia that surround the parallel fiber and climbing fiber–Purkinje neuron dendrite synapses. Although the Bergman glia appear morphologically normal in the GFAP knockout, they have no GFAP. Thus an abnormality limited to glial cells markedly impairs a form of synaptic plasticity (LTD) and a form of basic associative learning and memory. This may be the first direct evidence for a key role of glia in processes of learning and memory.

Several lines of evidence therefore support the hypothesis that a process of LTD in the cerebellar cortex is a mechanism involved in memory storage in classical conditioning of discrete behavioral responses. Similarly, several lines of evidence support such a role for cerebellar cortical LTD in adaptation of the VOR [see earlier discussion and Ito (1984) for detailed discussions]. However, the fact that some degree of learning occurs in the pcd mouse (see earlier), a preparation that functionally has a complete cerebellar decortication, argues that some degree of plasticity must occur in the interpositus nucleus. There is just one paper in the literature (Racine *et al.*, 1986) reporting that tetanus of the white matter yields LTP in the interpositus nucleus. Much work remains to be done in exploring possible mechanisms of plasticity and memory storage in the cerebellum. The fact that GABA agonists and antagonists infused in the cerebellum have such profound effects on the conditioned response (Krupa *et al.*, 1993; Mamounas *et al.*, 1987) at least raises the possibility that GABAergic processes may be involved in cerebellar memory storage.

Insulin-like growth factor I (IGF-I) released from the climbing fibers onto Purkinje neurons has been shown to modulate glutamate-induced GABA release by Purkinje neurons (Castro-Alamancos and Torres-Aleman, 1993). These authors further investigated whether this IGF-I may play a role in learning (Castro-Almancos and Torres-Aleman, 1994). Injection of an IGF-I antisense oligonucleotide in the inferior olive resulted in the complete prevention of eyeblink conditioning in freely moving rats. This blockage was reversible and recovered when the levels of cerebellar IGF-I returned to normal values. After the conditioned eyeblink response had been learned, however, subsequent injection of the IGF-I antisense oligonucleotide had no effect on the learned response, indicating that olivocerebellar IGF-I is essential for learning but not retention. This may constitute further evidence implicating GABAergic processes in cerebellar memory, consistent with earlier studies identifying the IO-climbing fiber system as the essential US reinforcing or teaching system (see earlier discussion and Thompson, 1989).

XV. Cerebellar Involvement in Other Forms of Memory

This chapter has focused on the essential role of the cerebellum in classical conditioning of discrete behavioral responses, a basic form of associative learning and memory. This is the clearest and most decisive evidence for the localization of a memory trace to a particular brain region in mammals (cerebellum) that exists at present. A closely related and increasingly definitive literature supports the view that the cerebellum learns complex, multijoint movements. Thus, Thach et al. (1992) have proposed a model of cerebellar function that suggests that the job of the cerebellum is, among other things, to coordinate elements of movement in its downstream targets and to adjust old movement synergies while learning new ones. These authors suggested that beams of Purkinje cells connected by long parallel fibers could link actions of different body parts represented within each cerebellar nucleus and exert control across nuclei into coordinated multijointed movements. The model Thach et al. (1992) proposed suggests that the cerebellum is involved not only in coordinating multijointed tasks, but that it also learns new tasks through an activity-dependent modification of parallel fiber–Purkinje cell synapses (Thach et al., 1992; Gilbert and Thach, 1977). Evidence in support of this hypothesis includes Purkinje cell recordings, which reveal patterns of activity consistent with a causal involvement in the modification of the coordination between eye position and hand/arm movement (Keating and Thach, 1990), and studies in which focal lesions by microinjection of muscimol severely impair the adaptation of hand/eye coordination without affecting the performance of the task (Keating and Thach, 1991). It is known that the conditioned eyeblink response is a highly coordinated activation of several muscle groups.

There is growing evidence that the cerebellum is critically involved in many other forms of learning, memory and cognition, as documented in this volume. This chapter notes just a few learning examples. Steinmetz and associates (1993) made use of a lever-pressing instrumental response in rats. Animals were trained to press for a food reward and to avoid a shock. Cerebellar lesions abolished the learned lever press avoidance response but did not impair the same lever press response for food. It would seem that the cerebellum is necessary for learning discrete responses to deal with aversive events in both classical and instrumental contingencies. Supple and Leaton (1990a,b) have shown that the cerebellar vermis is necessary for classical conditioning of the heart rate in both restrained and freely moving rats.

In recent years, the hippocampus has become the sine qua non structure for spatial learning and memory in rodents (e.g., Morris et al., 1986; O'Keefe and Nadel, 1978). However, it appears that the cerebellum may also play a critical role in spatial learning and performance, a role that extends well beyond motor coordination. Some years ago Altman and associates found that rats with cerebellar cell loss following early postnatal X irradiation were severly impaired in maze performance based on spatial cues (Pellegrino and Altman, 1979). Lalonde and associates (1990) have made use of a range of mutant mice with cerebellar defects (staggerer, weaver, pcd, lurcher) and also used cerebellar lesions to explore learning impairments. In general, they find significant impairments in tasks involving spatial learning and memory (for a review of this important work see Lalonde and Botez, 1990; Lalonde, this volume).

The Morris water maze has become the quintessential task for detecting hippocampal lesion deficits in spatial learning and memory (Morris 1984). In a most important study, Goodlett et al. (1992) trained pcd mice in the Morris water maze (see earlier discussion for a description of this mutant). Results were striking. The mutants showed massive impairments on this task in distal cue spatial navigation, both in terms of learning and in terms of expressing biases on probe trials. In contrast, they showed normal performance on the proximal cue visual guidance task, thus demonstrating that "the massive spatial navigation deficit was not due simply to motor dysfunction" (Goodlett et al., 1992). The pcd mutants showed clear Purkinje cell loss without significant depletion of hippocampal neurons, but their deficits in spatial learning and memory were actually more severe than those typically seen following hippocampal lesions!

Finally, there is an important and growing body of literature concerned with the role of the cerebellum in timing (Ivry, this volume; Keele and Ivry, 1990). These authors found that the accuracy of timing motor responses correlated across various motor effectors, and with the acuity of judging differences in intervals between tone pairs, suggesting that a localized neural system may underlie timing. Further, the critical timing site (in humans) appears to be the lateral cerebellum, where damage impairs motor and perceptual timing and the perception of visual velocity. These are the same regions critical for classical conditioning.

As Keele and Ivry (1990) note, classical conditioning of discrete adaptive behavioral responses like the eyeblink and limb flexion involves very precise timing, requiring detailed temporal computation. In well-trained animals, the peak of the conditioned eyeblink response occurs within tens of milliseconds of the onset of the US over a wide range of CS–US onset intervals (Smith, 1968; Smith et al., 1969). This adaptive timing of the CR in classical conditioning is one of the most important unsolved mysteries regarding brain substrates of memory. The neural mechanisms that yield this adaptive timing may provide keys to the functions of the cerebellum and to the nature of memory storage processes in the brain.

The evidence reviewed here and in the other chapters of this volume make it abundantly clear that the cerebellum is not simply a structure subserving motor coordination, but plays critical roles in learning and memory storage, in spatial learning and memory, in timing in verbal associative memory in humans (See e.g., Fiez et al., 1992), and, more generally, in cognitive processes (see, e.g., Schmahmann, this volume).

Acknowledgments

The research reported here was supported in part by research grants from the National Science Foundation (IBN-9215069), the National Institute of Health (AG05142), the National Institute of Mental Health (MH52194), the Office of Naval Research (N00014-95-1-1152), and the Sanyo Co., Ltd.

References

Aiba, A., Kano, M., Chen, C., Stanton, M. E., Fox, G. D., Herrup, K., Zwingman, T. A., and Tonegawa, S. (1994). Deficient cerebellar long-term depression and impaired motor learning in mGluR1 mutant mice. *Cell* **79**, 377–388.
Aitkin, L. M., and Boyd, J. (1978). Acoustic input to the lateral pontine nuclei. *Hear. Res.* **1**, 67–77.
Albus, J. S. (1971). A theory of cerebellar function. *Math. Biosci.* **10**, 25–61.

Andersson, G., and Hesslow, G. (1986). Evidence for an inhibitory action by cerebellar nuclear cells on the inferior olive. *Neurosci. Lett. (Suppl)*, **26**, S231.

Berg, M., and Thompson, R. F. (1995). Neural responses of the cerebellar deep nuclei in the naive and well trained rabbit following NM conditioning. *Soc. Neurosci. Abstr.* **21**, 1221.

Berthier, N. E., and Moore, J. W. (1986). Cerebellar Purkinje cell activity related to the classical conditioned nictitating membrane response. *Exp. Brain Res.* **63**, 341–350.

Berthier, N. E., and Moore, J. W. (1990). Activity of deep cerebellar nuclear cells during classical conditioning of nictitating membrane extension in rabbits. *Exp. Brain Res.* **83**, 44–54.

Bloedel, J. R. (1992). Functional heterogenity with structural homogenity: How does the cerebellum operate? *Behav. Brain Sci.* **15**, 666–78.

Bloedel, J. R., and Bracha, V. (1995). On the cerebellum, cutaneomuscular reflexes, movement control and the elusive engrams of memory. *Behav. Brain Res.* **68**, 1–44.

Brainard, M. S., and Knudsen, E. I. (1993). Experience-dependent plasticity in the inferior colliculus: A site for visual calibration of the neural representation of auditory space in the barn owl. *J. Neurosci.* **13**, 4589–4608.

Brodal, P. (1978). The corticopontine projection in the rhesus monkey. Origin and principles of organization. *Brain* **101**, 251–283.

Brown, T. H., Zador, A. M., Mainen, Z. F., and Claiborne, B. J. (1991). Hebbian modifications in hippocampal neurons. *In* "LTP: A Debate of Current Issues" (M. Baudry and J. Davis, eds.), pp. 357–389. MIT, Cambridge, MA.

Castro-Alamancos, M. A., and Torres-Aleman, I. (1993). Long-term depression of glutamate-induced γ-aminobutyric acid release in cerebellum by insulin-like growth factor I. *Proc. Nat. Acad. Sci. USA* **90**, 7386–7390.

Castro-Almancos, M. A., and Torres-Aleman, I. (1994). Learning of the eyeblink response is impaired by an antisense insulin-like growth factor I oligonucleotide. *Proc. Nat. Acad. Sci. USA* **91**, 10203–10207.

Chapman, P. F., Steinmetz, J. E., Sears, L. L., and Thompson, R. F. (1990). Effects of lidocaine injection in the interpositus nucleus and red nucleus on conditioned behavioral and neuronal responses. *Brain Res.* **537**, 140–156.

Chapman, P. F., Steinmetz, J. E., and Thompson, R. F. (1988). Classical conditioning does not occur when direct stimulation of the red nucleus or cerebellar nuclei is the unconditioned stimulus. *Brain Res.* **442**, 97–104.

Chen, C., Masanobu, K., Abeliovich, A., Chen, L., Bao, S., Kim, J. J., Hashimoto, K., Thompson, R. F., and Tonegawa, S. (1995). Impaired motor coordination correlates with persistant multiple climbing fiber innervation in PKCγ mutant mice. *Cell* **83**, 1233–1242.

Chen, C., and Thompson, R. F. (1995). Temporal specificity of long-term depression in parallel fiber-Purkinje synapses in rat cerebellar slice. *Learn. Memory* **2**, 185–198.

Chen, L., Bao, S., Lockard, J. M., Kim, J. J., and Thompson, R. F. (1996). Impaired classical eyeblink conditioning in cerebellar lesioned and Purkinje cell degeneration (pcd) mutant mice. *J. Neurosci.* **16**, 2829–2838.

Cipriano, B. D., Krupa, D. J., Almanza, W., and Thompson, R. F. (1995). Inactivation of the interpositus nucleus prevents transfer of the rabbit's classically conditioned eyeblink response from a light to a tone CS. *Soc. Neurosci. Abstr.* **21**, 1221.

Clark, G. A., McCormick, D. A., Lavond, D. G., and Thompson, R. F. (1984). Effects of lesions of cerebellar nuclei on conditioned behavioral and hippocampal neuronal response. *Brain Res.* **291**, 125–136.

Clark, R. E., and Lavond, D. G. (1993). Reversible lesions of the red nucleus during acquisition and retention of a classically conditioned behavior in rabbit. *Behav. Neurosci.* **107**, 264–270.

Clark, R. E., and Lavond, D. G. (1996). Neural unit activity in the trigeminal complex with interpositus or red nucleus inactivation during classical eyeblink conditioning. *Behav. Neuro.* **110**, 1–9.

Clark, R. E., Zhang, A. A., and Lavond, D. G. (1992). Reversible lesions of the cerebellar interpositus nucleus during acquisition and retention of a classically conditioned behavior. *Behav. Neurosci.* **106**, 879–888.

Daum, I., Schugens, M. M., Ackermann, H., Lutzenberger, W., Dichgans, J., and Birbaumer, N. (1993). Classical conditioning after cerebellar lesions in human. *Behav. Neurosci.* **107**, 748–756.

Disterhoft, J. F., Quinn, K. J., Weiss, C., and Shipley, M. T. (1985). Accessory abducens nucleus and conditioned eye retraction/nicitating membrane extensions in rabbit. *J. Neurosci.* **5**, 941–950.

Donegan, N. H., Gluck, M. A., and Thompson, R. F. (1989). Integrating behavioral and biological models of classical conditioning, *In* "Psychology of Learning and Motivation" (R. D. Hawkins and G. H. Bower, eds.), Vol. 23, pp. 109–156. Academic Press, New York.

Donegan, N. H., Foy, M. R., and Thompson, R. F. (1985). Neuronal responses of the rabbit cerebellar cortex during performance of the classically conditioned eyelid response. *Soc. Neurosci Abstr.* **11**, 835.

Eccles, J. C. (1977). An instruction-selection theory of learning in the cerebellar cortex. *Brain Res.* **127**, 327–352.

Fiez, J. A., Peterson, S. E., Cheney, M. K., and Raichle, M. E. (1992). Impaired non-motor learning and error detection associated with cerebellar damage. *Brain* **115**, 155–178.

Foy, M. R., Krupa, D. J., Tracy, J., and Thompson, R. F. (1992). Analysis of single unit recordings from cerebellar cortex of classically conditioned rabbits. *Soc. Neurosci. Abstr.* **18**, 1215.

Foy, M. R., and Thompson, R. F. (1986). Single unit analysis of Purkinje cell discharge in classically conditioned and untrained rabbits. *Soc. Neurosci. Abstr.* **10**, 122.

Foy, M. R., Steinmetz, J. E., and Thompson, R. F. (1984). Single unit analysis of cerebellum during classically conditioned eyelid response. *Soc. Neurosci. Abstr.* **12**, 518.

Gaze, R. M., Keating, M. J., Szekely, G., and Beazley, L. (1970). Binocular interaction in the formation of specific intertectal neuronal connexions. *Proc. R. Soc. London Biol.* **175**, 107–147.

Gilbert, P. F. C., and Thach, W. T. (1977). Purkinje cell activity during motor learning. *Brain Res.* **128**, 309–328.

Glickstein, M., Cohen, J. L., Dixon, B., Gibson, A., Hollins, M., Labossiere, E., and Robinson, F. (1980). Corticopontine visual projections in macaque monkeys. *J. Comp. Neurol.* **190**, 209–229.

Goldowitz, D., and Eisenman, L. M. (1992). Genetic mutations affecting murine cerebellar structure and function. *In* "Genetically Defined Animal Models of Neurobehavioral Dysfunctions" (P. Driscoll, ed.), pp. 66–88. Birkhauser, Boston.

Goodlett, C. R., Hamre, K. M., and West, J. R. (1992). Dissociation of spatial navigation and visual guidance performance in Purkinje cell degeneration (pcd) mutant mice. *Behav. Brain Res.* **47**, 129–141.

Gormezano, I., Kehoe, E. J., and Marshall-Goodell, B. S. (1983). Twenty years of classical conditioning research with the rabbit. *In* "Progress in Physiological Psychology" (J. M. Sprague and A. N. Epstein, eds.), pp. 197–275. Academic Press, New York.

Gould, T. J., and Steinmetz, J. E. (1996). Changes in rabbit cerebellar cortical and interpositus nucleus activity during acquisition, extinction, and backward classical eyelid conditioning. *Neurobiol. Learn. Memory* **65**, 17–34.

Gruart, A., and Delgado-Garcia, J. M. (1994). Discharge of identified deep cerebellar nuclei neurons related to eyeblinks in the alert cat. *Neuroscience* **61**, 665–681.

Haley, D. A., Thompson, R. F., and Madden, J., IV. (1988). Pharmacological analysis of the magnocellular red nucleus during classical conditioning of the rabbit nictitating membrane response. *Brain Res.* **454,** 131–139.
Hesslow, G. (1994). Inhibition of classically conditioned eyeblink responses by stimulation of the cerebellar cortex in the decerebrate cat. *J. Physiol.* **476,** 245–256.
Hesslow, G., and Ivarsson, M. (1994). Suppression of cerebellar Purkinje cells during conditioned responses in ferrets. *NeuroReport* **5,** 649–653.
Ito, M. (1984). "The Cerebellum and Neural Control." Appleton Century-Crofts, New York.
Ito, M. (1989). Long-term depression. *Ann. Rev. Neurosci.* **12,** 85–102.
Ito, M. (1993). Cerebellar mechanisms of long-term depression. *In* "Synaptic Plasticity: Molecular and Functional Aspects" (M. Baudry, J. L. Davis and R. F. Thompson, eds.), pp. 117–146. MIT Press, Cambridge.
Ito, M., and Karachot, L. (1990). Messenger mediating long-term desensitization in cerebellar Purkinje cells. *Neuroreport* **1,** 129–132.
Ito, M., Sukurai, M., and Tongroach, P. (1982). Climbing fibre induced depression of both mossy fibre responsiveness and glutamate sensitivity of cerebellar Purkinje cells. *J. Physiol. London,* **324,** 113–134.
Ivkovich, D., Lockard, J. M., and Thompson, R. F. (1993). Interpositus lesion abolition of the eyeblink CR is not due to effects on performance. *Behav. Neurosci.* **107,** 530–532.
Kamin, L. J. (1969). Predictability, surprise, attention and conditioning. *In* "Punishment and Aversive Behavior" (R. M. Church, ed.), pp. 279–296. Appleton-Century-Crofts, New York.
Kano, M., Hashimoto, K., Chen, C., Abeliovich, A., Aiba, A., Kurihara, H., Watanabe, M., Inoue, Y., and Tonegawa, S. (1995) Impaired synapse elimination during cerebellar development in PKC_γ mutant mice. *Cell* **83,** 1223–1231.
Keating, J. G., and Thach, W. T. (1990). Cerebellar motor learning: quantitation of movement adaptation and performance in rhesus monkeys and humans implicates cortex as the site of adaptation. *Soc. Neurosci. Abstr.* **16,** 762.
Keating, J. G., and Thach, W. T. (1991). The cerebellar cortical area required for adaptation of monkey's "jump" task is lateral, localized, and small. *Soc. Neurosci. Abstr.* **17,** 1381.
Keele, S. W., and Ivry, R. B. (1990). Does the cerebellum provide a common computation for diverse tasks: A timing hypothesis. *In* "The Development and Neural Bases of Higher Cognitive Functions" (A. Diamond, ed.), pp. 179–211. New York Academy of Sciences Press, New York.
Kelly, T. M., Zuo, C., and Bloedel, J. R. (1991). Classical conditioning of the eyeblink reflex in the decerebrate-decerebellate rabbit. *Behav. Brain Res.* **38,** 7–18.
Kim, J. J., Clark, R. E., and Thompson, R. F. (1995). Hippocampectomy impairs the memory of recently, but not remotely, aquired trace eyeblink conditioned responses. *Behav. Neurosci.* **109,** 195–203.
Kim, J. J., Krupa, D. J., and Thompson, R.F. (1992). Intra-olivary infusions of picrotoxin prevent "blocking" of rabbit conditioned eyeblink response. *Soc. Neurosci. Abstr.* **18,** 1562.
Knowlton, B., and Thompson, R. F. (1988). Microinjections of local anesthetic into the pontine nuclei reduce the amplitude of the classically conditioned eyeblink response. *Physiol. Behav.* **43,** 855–857.
Knowlton, B., and Thompson, R. F. (1992). Conditioning using a cerebral cortical CS is dependent on the cerebellum and brainstem circuitry. *Behav. Neurosci.* **106,** 509–517.
Knudsen, E. I., (1994). Supervised learning in the brain. *J. Neurosci.* **14,** 3985–3997.
Knudsen, E. L., and Knudsen P.F. (1990). Sensitive and critical periods for visual calibration of sound localization by barn owls. *J. Neurosci.* **10,** 222–232.
Krupa, D. J. (1993). "Localization of the Essential Memory Trace for a Classically Conditioned Behavior." Unpublished doctoral dissertation, University of Southern California.

Krupa, D. J., Thompson, J. K., and Thompson, R. F. (1993). Localization of a memory trace in the mammalian brain. *Science* **260**, 989–991.

Krupa, D. J., and Thompson, R. F. (1995). Inactivation of the superior cerebellar peduncle blocks expression but not aquisition of the rabbit's classically conditioned eyeblink response. *Proc. Nat. Acad. Sci. USA* **92**, 5097–5101.

Krupa, D. J., Weiss, C., and Thompson, R. F. (1991). Air puff evoked Purkinje cell complex spike activity is diminished during conditioned responses in eyeblink conditioned rabbits. *Soc. Neurosci. Abstr.* **17**, 322.

Krupa, D. J., Weng, J., and Thompson, R. F. (1996). Inactivation of brainstem motor nuclei blocks expression but not acquisition of the rabbit's classically conditioned eyeblink response. *Behav. Neurosci.* **110**, 1–9.

Lalonde, R., and Botez, M. I. (1990). The cerebellum and learning processes in animals. *Brain Res. Rev.* **15**, 325–332.

Landis, S. C., and Mullen, R. J. (1978). The development and degeneration of Purkinje cells in pcd mutant mice. *J. Comp. Neurol.* **177**, 125–144.

Lavond, D. G., Hambree, T. L., and Thompson, R. F. (1985). Effects of kainic acid lesions of the cerebellar interpositus nucleus on eyelid conditioning in the rabbit. *Brain Res.* **326**, 179–182.

Lavond, D. G., Kim, J. J., and Thompson, R. F. (1993). Mammalian brain substrates of aversive classical conditioning. *Annu. Rev. Psychol.* **44**, 317–342.

Lavond, D. G., Steinmetz, J. E., Yokaitis, M. H., and Thompson, R. F. (1987). Reacquisition of classical conditioning after removal of cerebellar cortex. *Exp. Brain Rese.* **67**, 69–593.

Lewis, J. L., LoTurco, J. J., and Solomon, P. R. (1987). Lesions of the middle cerebellar peduncle disrupt acquisition and retention of the rabbit's classically conditioned nictitating membrane response. *Behav. Neurosci.*, **101**, 151–157.

Linden, D. J., and Connor, J. A. (1991). Participation of postsynaptic PKC in cerebellar long-term depression in culture. *Science* **254**, 656–659.

Lisberger, S. G. (1988). The neural basis for learning of simple motor skills. *Science* **242**, 728–735.

Llinas, R., and Welsh, J. P. (1993). On the cerebellum and motor learning. *Curr. Opin. Neurobiol.* **3**, 958–965.

Logan, C. G. (1991). "Cerebellar Cortical Involvement in Excitatory and Inhibitory Classical Conditioning." Unpublished doctoral dissertation, Stanford University.

Logan, C. G., and Grafton, S. T. (1995). Functional anatomy of human eyeblink conditioning determined with regional cerebral glucose metabolism and positron-emission tomography. *Proc. Natl. Acad. Sci. USA* **92**, 7500–7504.

Lye, R. H., O'Boyle, D. J., Ramsden, R. F., and Schady, W. (1988). Effects of a unilateral cerebellar lesion on the acquisition of eye-blink conditioning in man. *J. Physiol.* **403**, 58.

Mamounas, L. A., Thompson, R. F., and Madden, J. IV (1987). Cerebellar GABAergic processes: Evidence for critical involvement in a form of simple associative learning in the rabbit. *Proc. Nat. Acad. Sci.* **84**, 2101–2105.

Marr, D. (1969). A theory of cerebellar cortex. *J. Physiol.* **202**, 437–470.

Mauk, M. D., and Thompson, R. F. (1987). Retention of clasically conditioned eyelid responses following acute decerebration. *Brain Res.* **403**, 89–95.

Mauk, M. D., Steinmetz, J. E., and Thompson, R. F. (1986). Classical conditioning using stimulation of the inferior olive as the unconditioned stimulus. *Proc. Natl. Acad. Sci. USA* **83**, 5349–5353.

McCormick, D. A., Clark, G. A., Lavond, D. G., and Thompson, R. F. (1982a). Initial localization of the memory trace for a basic form of learning. *Proc. Nat. Acad. Sci. USA* **79**, 2731–2742.

McCormick, D. A., Guyer, P. E., and Thompson, R. F. (1982b). Superior cerebellar peduncle lesions selectively abolish the ipsilateral classically conditioned nictitating membrane/eyelid response of the rabbit. *Brain Res.* **244**, 347–350.

McCormick, D. A., Steinmetz, J. E., and Thompson, R. F. (1985). Lesions of the inferior olivary complex cause extinction of the classically conditioned eyeblink response. *Brain Res.* **359**, 120–130.

McCormick, D. A., and Thompson, R. F. (1984a). Cerebellum: Essential involvement in the classically conditioned eyelid response. *Science* **223**, 296–299.

McCormick, D. A., and Thompson, R. F. (1984b). Neuronal responses for the rabbit cerebellum during acquisition and performance of a classically conditioned nictitating membrane eyelid response. *J. Neur.* **4**, 2811–2822.

Mihailoff, G. A. (1993). Cerebellar nuclear projections from the basilar pontine nuclei and nucleus reticularis tegmenti pontis as demonstrated with PHA-L tracing in the rat. *J. Comp. Neurol.* **330**, 130–146.

Miller, K. D., (1989). Correlation-based models of neural development. In "Neuroscience and Connectionist Theory" (M. A.Gluck and D. E. Rumelhart, eds.), pp.267–352. Erlbaum., Hillsdale, NJ.

Mink, J. W., and Thach, W. T. (1991). Basal ganglia motor control. III. Pallidal ablation: Normal reaction time, muscle contraction, and slow movement. *J. Neurophysiol.* **65**, 330–351.

Mintz, M., Lavond, D. G., Zhang, A. A., Yun, Y., and Thompson, R. F. (1994). Unilateral inferior olive NMDA lesion leads to unilateral deficit in acquisition of NMR classical conditioning. *Behav. Neural Biol.* **61**, 218–224.

Morris, R. G. M. (1984). Developments of a water maze procedure for studying spatial learning in the rat. *J. Neurosci. Meth.* **11**, 47–60.

Morris, R. G. M., Anderson, E., Lynch, G. S., and Baudry, M. (1986). Selective impairment of learning and blockade of long-term potentiation by an N-methyl-D-asparate receptor antagonist, AP5. *Nature* **319**, 774–775.

Mower, G., Gibson, A., and Glickstein, M. (1979). Tectopontine pathway in the cat: Laminar distribution of cells of origin and visual properties of target cells in dorsolateral pontine nucleus. *J. Neurophysiol.* **42**, 1–15.

Moyer, J. R., Jr., Deyo, R. A., and Disterhoft, J. F. (1990). Hippocampectomy disrupts trace eyeblink conditioning in rabbits. *Behav. Neurosci.* **104**, 243–252.

Nelson, B., and Mugnoini, E. (1987). GABAergic innervation of the inferior olivary complex and experimental evidence for its origin. In "The Olivocerebellar System in Motor Control" (P. Strata, ed.). Springer-Verlag, New York.

Nordholm, A. F., Lavond, D. G., and Thompson, R. F. (1991). Are eyeblink responses to tone in the decerebrate, decerebellate rabbit conditioned responses? *Behav. Brain Res.* **44**, 27–34.

Nordholm, A. F., Thompson, J. K., Dersarkissian, C., and Thompson, R. F. (1993). Lidocaine infusion in a critical region of cerebellum completely prevents learning of the conditioned eyeblink response. *Behav. Neurosci.* **107**, 882–886.

O'Keefe, J., and Nadel, L. (1978). "The Hippocampus as a Cognitive Map." Oxford University Press, London.

Pavlov, I. P. (1927). "Conditioned reflexes: An Investigation of the Physiological Activity of the Cerebral Cortex (G. V. Anrep, trans). Oxford University Press, London.

Pellegrino, L. J., and Altman, J. (1979). Effects of differential interference with postnatal cerebellar neurogenesis on motor performance, activity level and maze learning of rats: A developmental study. *J. Comp. Physiol. Psychol.* **93**, 1–33.

Perrett, S. P., and Mauk, M. D. (1995). Extinction of conditioned eyelid responses requires the anterior lobe of cerebellar cortex. *J. Neurosci* **15**, 2074–2080.

Perrett, S. P., Ruiz, B. P.,and Mauk, M. D. (1993). Cerebellar cortex lesions disrupt the timing of conditioned eyelid responses. *J. Neurosci.* **13,** 1708–1718.

Racine, R. J., Wilson, D. A., Gingell, R., and Sutherland, D. (1986). Long-term potentiation in the interpositus and vestibular nuclei in the rat. *Exp. Brain Res.* **63,** 158–162.

Rescorla, R., and Wagner, A. (1972). A theory of Pavlovian conditioning: Variations in the effectiveness of reinforcement. *In* "Classical Conditioning II: Current Research and Theory" (A. Black and W. Prokasy, eds.). Appleton-Century-Crofts, New York.

Richards, W. G., Ricciardi, T. N., and Moore, J. W. (1991). Activity of spinal trigeminal pars oralis and adjacent reticular formation units during differential conditioning of the rabbit nictitating membrane response. *Behav. Brain Res.* **44,** 195–204.

Rosenfield, M. E., and Moore, J. W. (1983). Red nucleus lesions disrupt the classically conditioned nictitating membrane response in rabbit. *Behav. Brain Res.* **10,** 393–398.

Rosenfield, M. D., Dovydaitis, A., and Moore, J. W. (1985). Brachium conjunctivum and rubrobulbar tract: Brainstem projections of red nucleus essential for the conditioned nictitating membrane response. *Physiol. Behav.* **34,** 751–759.

Rumelhart, D. E., and McClelland, J. L. (1986). "Parallel Distributed Processing: Explorations in the Microstructure of Cognition." MIT Press, Cambridge, MA.

Schmahmann, J. D. and Pandya, D. N. (1989). Anatomical investigation of projections to the basis pontis from posterior parietal association cortices in rhesus monkey. *J. Comp. Neurol.* **289,** 53–73.

Schmahmann, J. D., and Pandya, D. N. (1991). Projections to the basis pontis from the superior temporal sulcus and superior temporal region in the rhesus monkey. *J. Comp. Neurol.* **308,** 224–248.

Schmahmann, J. D., and Pandya, D. N. (1993). Prelunate, occipitotemporal, and parahippocampal projections to the basis pontis in rhesus monkey. *J. Comp. Neurol.* **337,** 94–112.

Schreurs, B. G., and Kehoe, E. J. (1987). Cross-model transfer as a function of initial training level in classical conditioning with the rabbit. *Anim. Learn. Behav.* **15,** 47–54.

Sears, L. L., Logue, S. F., and Steinmetz, J. E. (1996). Involvement of the ventrolateral thalamic nucleus in rabbit classical eyeblink conditioning. *Behav. Brain Res.* **74,** 105–117.

Sears, L. L., and Steinmetz, J. E. (1991). Dorsal accessory inferior olive activity diminishes during acquisition of the rabbit classically conditioned eyelid response. *Brain Res.* **545,** 114–122.

Shibuki, K., Gomi, H., Chen, C., Bao, S., Kim, J. J., Wakatsuki, H., Fujisaki, T., Fujimoto, K., Ikeda, T., Chen, C., Thompson, R. F., and Itohara, S. (1996). Deficient cerebellar long-term depression, impaired eyeblink conditioning and normal motor coordination in GFAP mutant mice. *Neuron* **16,** 587–599.

Shinoda, Y., Suguichi, Y., Futami, T., and Izawa, R. (1992). Axon collaterals of mossy fibers from the pontine nucleus in the cerebellar dentate nucleus. *J. Neurophysiol.* **67,** 547–560.

Skelton, R. W., Mauk, M. D., and Thompson, R. F. (1988). Cerebellar nucleus lesions dissociate alpha conditioning from alpha responses in rabbit. *Psychobiology* **16,** 126–134.

Smith, M. C. (1968). CS-US interval and US intensity in classical conditioning of the rabbit's nictitating membrane response. *J. Comp. Physiol. Psychol.* **66,** 679–687.

Smith, M. C., Coleman, S. R., and Gormezano, I. (1969). Classical conditioning of the rabbit's nictitating membrane response at backward, simultaneous, and forward CS-US intervals. *J. Comp. Physiol. Psychol.* **69,** 226–231.

Snyder, R. L., Faull, R. L. M., and Mehler, W. R. (1978). A comparative study of the neurons of origin of the spinocerebellar afferents in the rat, cat and squirrel monkey based on the retrograde transport of horseradish peroxidase. *J. Comp. Neurol.* **181,** 833–852.

Solomon, P. R., Lewis, J. L., LoTurco, J. J., Steinmetz, J. E., and Thompson, R. F.(1986a). The role of the middle cerebellar peduncle in acquisition and retention of the rabbit's classically conditioned nictitating membrane response. *Bull. Psychon. Soc.* **24,** 75–78.

Solomon, P. R., Stowe, G. T., and Pendlbeury, W. W. (1989). Disrupted eyelid conditioning in a patient with damage to cerebellar afferents. *Behav. Neurosci.* **103**, 898–902.
Solomon, P. R., Vander Schaaf, E. R., Thompson, R. F., and Weisz, D. J. (1986b). Hippocampus and trace conditioning of the rabbit's classically conditioned nictitating membrane response. *Behav. Neuro.* **100**, 729–744.
Steinmetz, J. E. (1990). Neuronal activity in the rabbit interpositus nucleus during classical NM-conditioning with a pontine-nucleus-stimulation CS. *Psychol. Sci.* **1**, 378–382.
Steinmetz, J. E., and Sengelaub, D. R. (1992). Possible conditioned stimulus pathway for classical eyelid conditioning in rabbits. *Behav. Neural Biol.* **57**, 103–115.
Steinmetz, J. E., Lavond, D. G., Ivkovich, D., Logan, C. G., and Thompson, R. F. (1992). Disruption of classical eyelid conditioning after cerebellar lesions: Damage to a memory trace system or a simple performance deficit? *J. Neurosci.* **12**, 4403–4426.
Steinmetz, J. E., Lavond, D. G., and Thompson, R. F. (1989). Classical conditioning in rabbits using pontine nucleus stimulation as a conditioned stimulus and inferior olive stimulation as an unconditioned stimulus. *Synapse* **3**, 225–232.
Steinmetz, J. E., Logan, C. G., Rosen, D. J., Thompson, J. K., Lavond, D. G., and Thompson, R. F. (1987). Initial localization of the acoustic conditioned stimulus projection system to the cerebellum essential for classical eyelid conditioning. *Proc. Natl. Acad. Sci. USA* **84**, 3531–3535.
Steinmetz, J. E., Logue, S. F., and Miller D. P. (1993). Using signaled bar pressing tasks to study the neural substrates of appetitive and aversive learning in rats: Behavioral manipulations and cerebellar lesions. *Behav. Neurosci.* **107**, 941–954.
Steinmetz, J. E., Rosen, D. J., Chapman, P. F., Lavond, D. G., and Thompson, R. F.(1986). Classical conditioning of the rabbit eyelid response with a mossy fiber stimulation CS. I. Pontine nuclei and middle cerebellar peduncle stimulation. *Behav. Neurosci.* **100**, 871–880.
Supple, W. F., Jr., and Leaton, R.N. (1990a). Lesions of the cerebellar vermis and cerebellar hemispheres: Effects on heart rate conditioning in rats. *Behav. Neurosci.* **104**, 934–947.
Supple, W. F., Jr., and Leaton, R. N. (1990b). Cerebellar vermis: Essential for classical conditioned bradycardia in rats. *Brain Res.* **509**, 17–23.
Swain, R. A., Shinkman, P. G., Nordholm, A. F., and Thompson, R. F. (1992). Cerebellar stimulation as an unconditioned stimulus in classical conditioning. *Behav. Neurosci.* **106**, 739–750.
Thach, W. T., Goodkin, H. G., and Keating, J. G. (1992). The cerebellum and the adaptive coordination of movement. *Annu. Rev. Neurosci.* **15**, 403–442.
Thompson, J. K., Krupa, D. J., Weng, J., and Thompson, R. F. (1993). Inactivation of motor nuclei blocks expression but not acquisition of rabbit's classically conditioned eyeblink response. *Soc. Neurosci. Abstr.* **19**, 999.
Thompson, J. K., Lavond, D. G., and Thompson, R.F. (1985). Cerebellar interpositus/dentate nuclei afferents seen with retrograde fluorescent tracers in the rabbit. *Soc. Neurosci. Abstr.* **11**, 1112.
Thompson, J. K., Spangler, W. J., and Thompson, R. F. (1991). Differential projections of pontine nuclei to interpositus nucleus and lobule HVI. *Soc. Neurosci. Abstr.* **17**, 871.
Thompson, R. F. (1990). Neural mechanisms of classical conditioning in mammals. *Phil. Trans. R. Soc. Lond. B* **329**, 161–170.
Thompson, R. F., and Krupa, D. J. (1994). Organization of memory traces in the mammalian brain. *Annu. Rev. Neurosci.* **17**, 519–549.
Thompson, R. F. (1989), Role of inferior olive in classical conditioning. In "The Olivocerebellar System in Motor Control" (P. Strata, ed.). Springer-Verlag, New York.
Tracy, J. (1995). "Brain and Behavior Correlates in Classical Conditioning of the Rabbit Eyeblink Response." Unpublished doctoral dissertation, The University of Southern California.

Tracy, J., Weiss, C., and Thompson, R. F. (1991). Single unit recordings of somatosensory and auditory evoked responses in the anterior interpositus nucleus in the naive rabbit. *Soc. Neurosci. Abstr.* **17,** 322.

Tusa, R. J., and Ungerleider, L. G. (1988). Fiber pathways of cortical areas mediating smooth pursuit eye movements in monkeys. *Ann. Neurol.* **23,** 174–183.

Udin, S. B. (1985). The role of visual experience in the formation of binocular projections in frogs. *Cell. Mol. Neurobiol.* **5,** 85–102.

Udin, S. B. (1989). The development of the nucleus isthmi in *Xenopus*. II. Branching patterns of contralaterally projecting isthmotectal axons during maturation of binocular maps. *Vis. Neurosci.* **2,** 153–163.

Udin, S. B., and Keating, M. J. (1981). Plasticity in a central nervous pathway in *Xenopus*: Anatomical changes in the isthmotectal projection after larval eye rotation. *J. Comp. Neurol.* **203,** 575–594.

Voneida, T., Christie, D., Boganski, R., and Chopko, B. (1990). Changes in instrumentally and classically conditioned limb-flexion responses following inferior olivary lesions and olivocerebellar tractotomy in the cat. *J. Neurosci.* **10,** 3583–3593.

Welsh, J. P., and Harvey, J. A. (1989). Cerebellar lesions and the nictitating membrane reflex: Performance deficits of the conditioned and unconditioned response. *J. Neurosci.* **9,** 299–311.

Welsh, J. P., and Harvey, J. A. (1991). Pavlovian conditoning in the rabbit during inactivation of the interpositus nucleus. *J. Physiol.* **444,** 459–480.

Woodruff-Pak, D. S., Cronholm, J. F., and Sheffield, J. B. (1990). Purkinje cell number related to rate of eyeblink classical conditioning. *NeuroReport* **1,** 165–168.

Woodruff-Pak, D. S., and Jaeger, M. (1996). Differential effects of aging on memory systems: Simple classical conditioning compared to paired associate learning. Submitted for publication.

Woodruff-Pak, D. S., Lavond, D. G., and Thompson, R. F. (1985). Trace conditioning: Abolished by cerebellar nuclear lesions but not lateral cerebellar cortex aspirations. *Brain Res.* **348,** 249–260.

Woodruff-Pak, D. S., and Thompson, R. F. (1988). Classical conditioning of the eyeblink response in the delay paradigm in adults aged 18-83 years. *Psychol. Aging* **3,** 219–229.

Yeo, C. H. (1991). Cerebellum and classical conditioning of motor response. *Ann. N.Y. Acad. Sci.* **627,** 292–304.

Yeo, C. H., Hardiman, M. J., and Glickstein, M. (1985a). Classical conditioning of the nictitating membrane response of the rabbit. I. Lesions of the cerebellar nuclei. *Exp. Brain Res.* **60,** 87–98.

Yeo, C. H., Hardiman, M. J., and Glickstein, M. (1985b). Classical conditioning of the nictitating membrane response of the rabbit. II. Lesions of the cerebellar cortex. *Exp. Brain Res.* **60,** 99–113.

Yeo, C. H., Hardiman, M. J., and Glickstein, M. (1986). Classical conditioning of the nictitating membrane response of the rabbit. IV. Lesions of the inferior olive. *Exp. Brain Res.* **63,** 81–92.

Zhang, A. A., and Lavond, D. G. (1991). Effects of reversible lesions of reticular or facial neurons during eyeblink conditioning. *Soc. Neurosci. Abstr.* **17,** 869.

VISUOSPATIAL ABILITIES

Robert Lalonde

Neurology Service, Unit of Behavioral Neurology, Neurobiology, and Neuropsychology,
Hôtel-Dieu Hospital Research Center, and Departments of Medicine and Psychology and
Neuroscience Research Center, University of Montréal, Montréal, Quebec, Canada H2W 1T8

I. Introduction
II. Evaluation of Spatial Learning
   A. Morris Maze
   B. Radial Maze
   C. Spatial Alternation
   D. Hole Board
III. Conclusions
   References

The importance of the hippocampus and its anatomical connections, including the medial septum, thalamic nuclei, and neocortical regions in many spatial tasks including the Morris water maze, has been emphasized. Studies in mutant mice with cerebellar atrophy and in rats with electrolytic lesions of the cerebellum have indicated that the cerebellum has a role in visuospatial and visuomotor processes in the Morris maze. Directional deficits in the water have also been noted in rats whose cerebellum was exposed to X-rays during different developmental stages. Cerebellar interactions with the superior colliculus, the hippocampus, and the neocortex via thalamic nuclei are suggested to be the basis of the cerebellar modulation of directional sense in maze tests.

I. Introduction

The purpose of this chapter is to present available data concerning the effects of cerebellar damage on maze tests in animals. The role of the hippocampus in visuospatial memory has been well characterized. More recently, it has been noted that some of the spatial deficits found in rats with hippocampal lesions can also be found in animals with cerebellar lesions. It is proposed that this result is due to cerebello–hippocampal interactions, together with cerebellar modulation of other subcortical and neocortical regions. An emphasis is placed on the Morris water maze because this paradigm offers the possibility of dissociating spatial defects from visuomotor defects.

II. Evaluation of Spatial Learning

The two maze tasks most frequently employed in the evaluation of the neurobiological bases of spatial learning in rodents are the Morris water maze (Morris *et al.*, 1982) and the radial maze (Olton and Papas, 1979). During spatial learning in the Morris water maze, rats or mice are placed in a pool containing a platform hidden from view. Two frequently used measures are the distance swum and escape latencies before reaching the platform. During sensorimotor testing in the same maze, the platform is raised above water level and is thereby visible from all areas of the pool. In the radial maze, rats (or more rarely mice) are placed in the center of a maze containing eight or more arms spread out like spokes in a wheel. To be able to retrieve a food pellet, the rat must enter each arm once. A second entry into the same arm within a trial is not rewarded and is counted as an error. In one version of the task, food is placed at the end of four of the arms but not at the remaining four. Entries into food-baited arms previously visited are tabulated as errors of working memory, whereas entries into nonbaited arms are tabulated as errors of reference memory (Crusio *et al.*, 1993). The hidden platform task of the Morris maze is considered to be a reference memory task because of the unchanging position of the platform.

A. Morris Maze

1. *Methodological Considerations*

When deficits in spatial orientation occur, distance swum and escape latencies are increased. High escape latencies in the absence of longer path lengths are an indication of slower swim speed, as reported for a spinocerebellar mutant (Lalonde *et al.*, 1993). During some manipulations, e.g., morphine administration, both values are high; subsequent testing revealed that this was due to a decrease in motivation rather than to a spatial defect (McNamara and Skelton, 1992).

In many reports, the animals are first evaluated in the hidden platform condition and then in the visible platform condition. Naive animals learn that only one escape platform is available and learn to locate that platform on the basis of extramaze cues. At the beginning of training, the animals tend to explore mostly the perimeter of the basin. Once they abandon that strategy, normal animals acquire the task rapidly. During visible platform performance, their previous training may help them abandon the wall-hugging strategy, guiding them more effectively toward the goal, located in a new area of the pool. The animals must learn to avoid perseverative responses toward the previously located hidden platform. When first evaluated in the visible platform task, the animals often explore the perimeter of the basin as well. When the location of the invisible platform is changed, they must learn not to perseverate in swimming toward the previously located visible platform.

A methodological difference between studies that may be of some importance is the presence or absence of corrected trials when the cutoff time is exceeded. In some studies, the rat is placed on the platform (corrected trials), whereas in others the rat is removed from the pool (uncorrected trials). It has been argued that this methodological difference explains the presence or absence of a lesion effect (de Bruin *et al.*, 1994).

2. *Hippocampus*

The existence of hippocampal place cells (Foster *et al.*, 1989; Sharp *et al.*, 1995) and the impairments found after bilateral lesions of the hippocampus on a 14-unit T-maze (Jucker *et al.*, 1990) and on a single T-maze (Wan *et al.*, 1994) have underlined the importance of this brain region in spatial orientation. In comparison to a group with cortical lesions overlying the hippocampus and a control group (sham or no surgery), bilateral aspiration lesions of the hippocampus in rats increased distance traveled and escape latencies during acquisition of the hidden platform task (Morris *et al.*, 1982). The visible platform task was minimally affected. Although the impairment was statistically significant, this appeared to be mostly due to high escape latencies during the first trial block. In a probe test, during which the platform was removed from the pool, control and cortically lesioned rats had higher crossings in the region where the platform was previously located than rats with hippocampal lesions, an indication that the hippocampal group had a spatial memory defect and not a loss in motivation. Sutherland and Rudy (1988) examined the effects of bilateral colchicine-induced lesions of the hippocampus on the visible platform task interspersed with hidden platform trials. During either condition, the platform was located in the same region of the pool. The distance traveled was not measured. The lesioned rats had higher escape latencies in the visible platform task only at the beginning of training. Higher escape latencies in the hidden platform condition occurred for the final two trial blocks.

Identical platform positioning in the two tasks does not eliminate the spatial defect caused by hippocampal lesions. McDonald and White (1994) obtained similar results in the same paradigm after lesions of the fornix. Impairment of the place-learning task was found after lesions of the hippocampus by electrolysis and by administration of colchicine, kainate (Sutherland et al., 1982, 1983), and quisqualate (Marston et al., 1993).

Kolb et al. (1984) evaluated the effects of unilateral hippocampal lesions on spatial learning in the Morris maze. Left- or right-sided lesions of the hippocampus augmented escape latencies and the heading angle toward the hidden platform. When the platform was removed, the time spent in the previously correct quadrant was higher in the control group than in either lesioned group. Deficits of spatial orientation were also observed in hemidecorticate rats. Severe impairments in place learning were displayed by totally decorticate rats (Whishaw and Kolb, 1984). These rats were also impaired during cued learning, but not to the same extent. Because hippocampal lesions cause deficits in place learning whereas visuomotor coordination is spared or minimally impaired, the question arises as to which brain regions afferent or efferent to the hippocampus are important for spatially mediated information. Important afferent regions include the entorhinal cortex, the medial septum, and the amygdala.

3. Hippocampal Afferents

Schenk and Morris (1985) evaluated two lesioned groups: one with damage to the entorhinal cortex, the subiculum, and the pre- and parasubiculum, and the other with damage to the entorhinal cortex with less extensive damage to the other areas. A corrected trial procedure was used. In comparison to a control group comprising sham and unoperated animals, the two experimental groups had longer path lengths and higher escape latencies during the acquisition of place learning. The terminal performance of cued learning was not impaired by the lesions (initial performance values were not presented). Using aspiration, Nagahara et al. (1995) lesioned the entorhinal cortex touching the perirhinal cortex border and evaluated place learning with the correction procedure. The escape latencies of lesioned rats were higher than those of sham controls during acquisition of the hidden platform task. During probe trials, the lesioned group spent less time in the correct quadrant of the pool. There was no group difference in swimming speed during the probe trials nor was there a difference in visible platform performance. The importance of the perforant path linking the entorhinal cortex to the hippocampus was underlined by Skelton and McNamara (1992), who reported spatial learning deficits in rats with knife cuts to this structure. Electrolytic lesions of the perirhinal cortex increased escape latencies and the heading angle toward the platform during acquisition of place learning but not during visuomotor performance (Wiig and Bilkey, 1994). No difference was detected for swim distance. The lesioned group had fewer platform crossings during probe trials, indicating a loss in memory. The noncorrected trial procedure was used. However, Kolb *et al.* (1994), also using the noncorrection procedure, found no deficit in place learning after lesions of the posterior temporal cortex that included much of the perirhinal cortex. A number of reports exist concerning the effects of medial septal lesions on acquisition of place learning. Electrolytic lesions of the septum and the diagonal band (Segal *et al.*, 1989) or only the medial septum (Decker *et al.*, 1992; Kelsey and Landry, 1988; Miyamoto *et al.*, 1987; Sutherland and Rodroguez, 1989) impaired acquisition of the hidden platform condition in the Morris task. Visible platform performance (Sutherland and Rodriguez, 1989) and performance with suspended lamp cues (Kelsey and Landry, 1988) were not affected by medial septal lesions. Spatial deficits have been found with neurotoxin-induced lesions (sparing axons of passage) of the medial septum and diagonal band of Broca (Decker *et al.*, 1992; Hagan *et al.*, 1988; Marston *et al.*, 1993; McAlonan *et al.*, 1995), but not with colchicine-induced cell shrinking of the medial septum (Barone *et al.*, 1991). Baxter *et al.* (1995) lesioned the medial septum and the diagonal band of Broca more selectively with the use of 192 IgG-saporin, an immunotoxin of cholinergic neurons. With the correction procedure, rats with damage to the medial septum and diagonal band of Broca were not impaired. These results show that the medial septum is involved in spatial learning, and hippocampal afferents are not exclusively cholinergic.

In contrast to the impairments reported in rats with entorhinal or medial septal damage, electrolytic lesions of the amygdala did not slow down the acquisition of place learning (Sutherland and McDonald, 1990). In the same experiment, hippocampal lesions caused by an ibotenate–colchicine solution or combined lesions of the hippocampus and amygdala increased hidden platform escape latencies.

4. *Hippocampal Efferents*

The hippocampal efferent system includes the association cortex of the frontal lobe and the parietal lobe (Rosene and Van Hoesen, 1977), anterior thalamus, mammillary bodies (Aggleton *et al.*, 1986), and brain stem precerebellar nuclei, namely the inferior olive and pontine nuclei (Saint-Cyr and Woodward, 1980). Kolb *et al.* (1982) evaluated rats with either medial frontal cortex or dorsomedial thalamic lesions during the acquisition of place learning. Whereas rats with thalamic lesions performed like controls, rats with medial frontal lesions had higher escape latencies and a higher heading angle toward the platform. The visible platform condition was not assessed. Sutherland *et al.* (1982) and Kolb *et al.* (1983, 1994) confirmed the deleterious effects of medial frontal lesions on place learning. An absence of a deficit on visible platform training after medial frontal lesions, whether fixed in one position or varied randomly from one trial to the next, was reported by Kolb et al. (1994). Sutherland et al. (1988) compared the effects of suction ablation of the anterior or posterior cingulate cortex and colchicine-induced degeneration of the hippocampus to sham-operated controls on acquisition of place learning. Higher escape latencies were observed in all three lesioned groups. Swimming speed was not affected. There was no assessment of visible platform performance. de Bruin et al. (1994) found an absence of a place-learning defect in rats with medial frontal lesions or smaller lesions encompassing either ventral or dorsal parts. A visible platform deficit in the early part of training was found in all three lesioned groups. This task comprised shifting of platform locations from an identical starting point and was interpreted as a perseveration rather than as a visuomotor problem. They list methological differences between their study and previous ones and conclude that the most important factor explaining the lack of a spatial deficit was the presence of corrected trials. In the Kolb et al. (1982, 1983, 1994) and the Sutherland et al. (1982, 1988) studies, whenever a rat failed to find the hidden platform within the allowed time, the animal was retrieved from the water. The corrected trial procedure permits early recognition of the position of the platform. In the noncorrected procedure, an animal unable to reach the platform learns only where the platform is not and not where the platform is.

In contrast, lesions of the parietal cortex impaired acquisition of the hidden platform condition with both the corrected and the noncorrected trial procedure. Kolb et al. (1983, 1994), using the noncorrected trial procedure, reported that aspiration lesions of the posterior parietal cortex (area 7) in rats increased escape latencies and the heading angle toward the submerged platform. The visible platform condition was not assessed. Using aspiration, DiMattia and Kesner (1988) lesioned a larger section of the parietal lobe, comprising areas 5, 7, 39, 40, and 43. Lesions of the parietal lobe and the hippocampus increased escape latencies and the heading angle toward the submerged platform. A corrected trial procedure was adopted. There was no assessment of the visible platform condition. Crowne et al. (1992), using the noncorrected trial procedure, measured the effects of unilateral aspiration of area 7. In comparison to sham-operated rats, these authors found that lesions of the right parietal lobe increased escape latencies and a target–nontarget quadrant ratio but not the heading angle toward the invisible platform. Left-sided lesions had no effect on any measure. There was no assessment of the visible platform condition. No study has yet undertaken the effects of parietal lesions on visible platform training. The vulnerability of the right parietal lobe on spatial functions in humans had often been emphasized (Botez et al., 1985). The results of this study indicate that hemispheric lateralization of spatial functions in the parietal lobe is not limited to humans.

Sutherland and Rodriguez (1989) evaluated the effects of fornix lesions and its target structures: the anterior thalamus, mamillary bodies, and nucleus accumbens. The latter structure also receives projections from the amygdala. Lesions of the fornix, the anterior thalamus, the mammillary bodies, and the nucleus accumbens lengthened escape latencies in the hidden platform condition. None of these lesions impaired visible platform performance except for the first of four trial blocks among rats with mammillary body lesions. Segal *et al.* (1989) also reported deficits of spatial localization in rats with knife cuts of the fornix.

Table I summarizes the results of lesioning hippocampal afferent and efferent regions. If studies using the noncorrection procedure are included, all hippocampal afferent and efferent regions disrupt place learning except the amygdala. A selective place-learning deficit with a minimal or no deficit in visuomotor coordination has been demonstrated for subcortical structures (the fornix, the septum, the anterior thalamus, the mamillary bodies, and the nucleus accumbens) but not as yet for neocortical structures (prefrontal and parietal cortices). If studies using the correction procedure are included, a place-learning deficit is found after lesions of two main afferent areas, the medial septum and the entorhinal cortex and one efferent area, the parietal cortex, but not the prefrontal cortex.

TABLE I
Effects of Lesioning Hippocampal Afferent and Efferent Regions on Place Learning in the Morris Water Maze with or without the Correction Procedure

Deficit	No deficit
Correction procedure	
Entorhinal cortex	Medial frontal cortex
Medial septum	
Parietal cortex	
Noncorrection procedure	
Medial septum	Amygdala
Perirhinal cortex	Posterior temporal cortex
Parietal cortex	
Medial frontal cortex	Dorsomedial thalamus
Cingulate cortex	
Anterior thalamus	
Mammillary bodies	

5. Precerebellar Nuclei

In addition to hippocampal efferents at levels higher than the midbrain, it is worth exploring brain stem hippocampal efferents projecting to the cerebellum. Electrical stimulation of the fornix alters Purkinje cell activity via climbing fiber and mossy fiber pathways (Saint-Cyr and Woodward, 1980). To the author's knowledge, there is only one report concerning the effects of brain stem precerebellar nuclei lesions on the Morris maze task. Dahhaoui *et al.* (1992b) injected 3-acetylpyridine in rats, causing degeneration of inferior olive cells, the single source of climbing fibers, and then assessed place learning. The rats began a trial from the same pool region, and not from all four cardinal points. A learning criterion of five consecutive trials with escape latencies lower than 9 sec was adopted. Rats with inferior olive lesions took more trials before reaching criterion than saline-injected controls. Visible platform performance was not assessed. This study introduces the possibility of a role for a hippocampo–olivoponto–cerebellar system on spatial orientation or visuomotor performance. Moroever, it is known that hippocampal place cells are dependent on the preparedness of movement (Foster *et al.*, 1989). In conditions of restraint, the spatially selective discharge of pyramidal cells was abolished. Their spatial firing patterns were influenced by visuomotor and vestibular information (Sharp *et al.*, 1995). Visuomotor information may be received from the neocortex and from the cerebellum. The midline cerebellum may provide the hippocampus with the sensorimotor integration necessary to execute spatially mediated behavior (Lalonde and Botez, 1990).

6. Cerebellum

Stein (1986) and Stein and Glickstein (1992) have hypothesized that the neocortico–ponto–cerebellar pathway is involved in the visual guidance of movements. Such considerations have been based on the control of limb and eye movements rather than on whole body movements of animals performing spatial tasks. Attempts have been made to answer the question as to whether cerebellar damage causes deficits in the whole body guidance of movements in water mazes.

Neuropsychological studies have indicated that cerebellar damage affects spatial abilities even in tasks not requiring movement (Botez *et al.*, 1989; Botez-Marquard and Botez, 1993, Botez-Marquard and Routhier, 1995; Fehrenbach *et al.*, 1984; Kish *et al.*, 1988).

a. Mutant Mouse Models: Pathology. The effects of cerebellar damage on water maze learning have been evaluated in five types of mutant mice, with cerebellar degeneration arising during different developmental stages (Table II). These mutants were tested in the water maze at 1–2 months of

TABLE II
NEUROPATHOLOGY OF CEREBELLAR MUTANT MICE AT 1 OR 2
MONTHS OF AGE

Mutant	Cerebellar atrophy	Extracerebellar atrophy
Lurcher	Granule, Purkinje	Inferior olive
Staggerer	Granule, Purkinje	Inferior olive
Weaver	Granule	Substantia nigra pars compacta
pcd	Purkinje	Inferior olive
Hot-foot	Purkinje cell dendritic spines	Unknown

age. At 1 month of age, in lurcher (Caddy and Biscoe, 1979) and in staggerer (Herrup and Mullen, 1979a) mutant mice, there is a massive degeneration of cerebellar granule and Purkinje cells. Abnormalities in the internal and the external granular layers have been detected in lurcher mutants as early as postnatal day 3 (Heckroth, 1992). There is also maldevelopment of basket cell processes around Purkinje cells (Heckroth, 1992). The staggerer mutation has been linked to the chromosome 9 gene site of the neural cell adhesion molecule (D'Eustachio and Davisson, 1993). A decrease of inferior olive neurons has been described in adult lurcher mutants (Heckroth and Eisenman, 1991) and in staggerer mutants as early as 10 days of age (Shojaeian et al., 1985) and in the adult (Blatt and Eisenman, 1985b). It has been determined by studies in chimeric mice that the degeneration of inferior olive cells and of cerebellar granule cells in staggerer mutants is an indirect cause of the staggerer gene action on Purkinje cells (Herrup, 1983; Herrup and Mullen, 1979b; Zanjani et al., 1990). Similarly, the loss of cerebellar granule cells and of inferior olive cells in lurchers is considered to be a secondary consequence of Purkinje cell degeneration (Wetts and Herrup, 1982ab). Nevertheless, there is a loss of granule cell number on postnatal day 4, prior to the beginning of Purkinje cell deneneration on day 8 (Caddy and Biscoe, 1979). Heckroth (1992) reports that the external granular layer of lurcher mutants on days 8 and 9 is approximately half as thick as that of controls. He postulates that the early granule cell loss is the result of a lack of trophic influences provided by Purkinje cells on granule cell development. In lurcher mutants, a 20% drop in deep nuclei numbers occurs at 3 months of age, possibly by means of an anterograde transsynaptic mechanism (Heckroth, 1994), as shown to be the case in Purkinje cell degeneration (pcd) mutants (Triarhou et al., 1987).

pcd mutant mice lose nearly all Purkinje cells at the end of the first postnatal month (Mullen et al., 1976). Cerebellar granule cell degeneration occurs later (Triarhou et al., 1985). Neuronal counts in the cerebellar deep nuclei are normal during the third postnatal week and reduced during the tenth month (Triarhou et al., 1987). There is a more gradual loss of retinal photoreceptors than of Purkinje cells in pcd mutants, attaining approximately 25% by 2 months of age (Mullen and LaVail, 1975). Retinal degeneration was not observed in staggerer or weaver mutants. Inferior olive cell numbers are decreased during the third postnatal week (Ghetti et al., 1987) and later stabilize (Triarhou and Ghetti, 1991). The loss of thalamic cells (O'Gorman and Sidman, 1985) and the loss of mitral cells in the olfactory bulb (Greer and Shepherd, 1982) do not occur until 2 months of age.

At 1 and 2 months of age, hot-foot mutant mice have abnormal spines on Purkinje cell dendrites, together with slight granule cell necrosis (Guastavino et al., 1990). It remains to be determined whether cerebellar afferents are altered in this mutation.

At the end of the first month, weaver mutant mice are characterized by the degeneration of cerebellar granule cells (Hirano and Dembitzer, 1973; Sotelo and Changeux, 1974). Mouse chimera studies indicate that a site of gene action is the cerebellar granule cell (Goldowitz and Mullen, 1982). There is also degeneration of substantia nigra cells during developmental stages in weaver mutants (Triarhou et al., 1988), leading to a reduction in dorsal striatal dopamine concentrations, a result not observed in staggerer and pcd mutants (Roffler-Tarlov and Graybiel, 1984, 1986, 1987). The protein content is reduced in the dorsal striatum, but not in the olfactory tubercle and nucleus accumbens (Roffler-Tarlov and Grabiel, 1984, 1987). A reduced number of tyrosine-hydroxylase immunoreactive neurons in the substantia nigra occurs at postnatal days 20 and 90 in weaver mutants. A reduction of cell numbers also occurs in retrorubral and ventral tegmental areas, but only on day 90 (Triarhou et al., 1988). A reduction of striatal dopamine uptake and an increase of striatal serotonin uptake occur in weaver mutants (Stotz et al., 1994). In addition to the reduction of dopamine uptake sites in the weaver striatum, there is a reduction of D1 receptors (Panagopoulos et al., 1993). However, D2 sites are increased in the dorsolateral part (Kaseda et al., 1987). The number of cerebellar dopamine uptake sites is diminished in weaver mutants, together with D1 and D2 receptor sites (Panagopoulos et al., 1993). There is no degeneration of inferior olive neurons, probably due to the relative sparing of Purkinje cells (Blatt and Eisenman, 1985a).

There is no evidence of neocortical or hippocampal maldevelopment in any of the mutants. The electrophysiological properties of CA1 hippocampal pyramidal cells are normal in staggerer mutants (Fournier and

Crepel, 1984) but have not been tested in other cerebellar mutants. All mutants have lower body weights in comparison to littermate controls, possibly because of a diminished ability to feed themselves during developmental stages. This may be the reason why the weight of some brain regions outside the cerebellum, such as the telencephalon, is lower in nervous, weaver, staggerer, and pcd mutants (Goodlett *et al.*, 1992; Miret-Duvaux *et al.*, 1990; Ohsugi *et al.*, 1986). However, despite a lower body weight, adult staggerer mutants eat more than normal mice of the same C57BL/6J strain (Guastavino *et al.*, 1991). This result is in contrast to cerebellectomized rats, whose food intake and body weight are lower than those of controls and then recover (Jacquart *et al.*, 1989; Mahler *et al.*, 1993). Some behavioral deficits observed in cerebellar mutants may be caused by alterations in feeding patterns during developmental stages, but this possibility is difficult to assess.

b. Mutant Mouse Models: Experimental Observations. Lurcher (Lc/+) mutant mice were compared to normal (+/+) mice of the same background strain (C3H) in a rectangular water basin containing either a hidden or a visible platform (Lalonde *et al.*, 1988). The number of quadrant entries and escape latencies was higher in lurcher mutants than in normal mice under both conditions. Because of the presence of the retinal degeneration gene on the C3H background, the mice were switched to the C57BL/6 background. Visible platform escape latencies of lurcher mutants were still higher. Jackson Laboratory subsequently switched the genetic background to B6CBACa-A^{w-J}/A. When the order of presentation was changed (visible platform followed by the invisible platform) and the task simplified by starting the mice from a single point, the number of quadrant entries and escape latencies was still higher in lurcher mutants under both conditions (Lalonde and Thifault, 1994). There was no correlation in either mutants or controls between hidden platform escape latencies and latencies before falling in a motor coordination test (a coat hanger comprising a suspended thin horizontal bar made of steel and flanked by two diagonal side bars). The absence of an intertest correlation in the mutants is an indication of the presence of distinct physiopathological processes underlying navigational as opposed to motor coordination abilities.

In the same rectangular water basin, weaver mutant mice (wv/wv) were compared to normal mice (+/?) of the same background strain (B6CBA) in the visible platform task, with the task being simplified by starting the mice from a single point (Lalonde and Botez, 1986). Weaver mutants had higher escape latencies than normal mice. In a water maze spatial alternation test, weaver mutants committed more errors than controls, an indication of the directional nature of their navigational deficit. Weaver mutants have cell losses in the substantia nigra (Triarhou *et al.*, 1988) and a reduction in the concentrations of striatal dopamine (Roffler-Tarlov and Graybiel, 1986). Because dopamine depletion in the striatum leads to a visuomotor deficit (Whishaw and Dunnett, 1985), this depletion, together with the cerebellar granule cell loss (Hirano and Dembitzer, 1973; Sotelo and Changeux, 1974), probably contributes to the impairment of navigational abilities displayed by weaver mutants.

Staggerer (sg/sg) mutant mice were compared to normal (+/?) mice of the same background strain (C57BL/6J) under hidden and visible platform conditions in the same water basin as tested in the other mutants (Lalonde, 1987). Quadrant entries and escape latencies were higher in staggerer mutants when the platform was submerged but not when it was visible. In a cerebellar mutant, this pattern reproduces a hippocampal-like deficit. However, Bensoula et al. (1995) compared staggerer mutants to normal mice in two other water maze tests. In the first test, mice were placed in a circular basin starting from a single point where they could reach a visible platform found at the opposite end positioned against the wall. The escape latencies were higher in staggerer mutants. In the second test (water escape pole-climbing test), mice were placed in a small circular basin containing a pole in the center. The water escape latencies of staggerer mutants and of hot-foot mutants were higher than those of normal mice (Fig. 1). When staggerer mutants first perform the invisible platform task followed by the visible platform task, they show a hippocampal-type dissociation. However, when first exposed to a condition where they could see a platform or a pole, they were impaired. In the water escape pole-climbing test, considerable wall-hugging perseveration was displayed by the staggerer mutants. This was not observed during visible platform testing when the animals had first performed the invisible platform task. These observations underline the importance of the order of presentation of the tests in some pathological conditions.

Goodlett et al. (1992) compared pcd mutant mice to normal (+/?) mice of the same background strain (C57BL/6J) at three age levels (30, 50, and 110 days old) first during visible platform training and then during hidden platform training in the circular Morris maze. During visible platform trials, the mice began a trial from different points and the platform location changed from one trial to the next. During hidden platform trials, the mice began a trial from different points but the platform location was fixed. At 30 days of age, there was no difference in path lengths among the groups during visible platform training. However, in the invisible platform condition, the path lengths of pcd mutants were higher. At 50 days of age, the pcd mutants had longer path lengths in the visible platform task during the first four trial blocks but not during the fifth. Distance traveled in the hidden platform trials was higher in pcd mutants. When the mice were

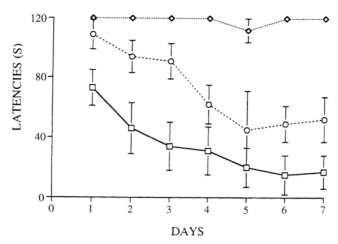

FIG. 1. Latencies before reaching a pole in 5-month old staggerer ($\Diamond$, $n = 7$), hot-foot ($\bigcirc$, $n = 9$), and normal mice ($\square$, $n = 9$) of the C57BL/6J strain controlled for gender. A plastic basin measuring 16 cm in diameter contained water (25°C) at a depth of 9.5 cm. A 4-mm-diameter pole, made of metal rough in texture in order to provide firm gripping, was placed in the center of the basin. The mice could climb the pole and escape from the water. There were two trials per day (cutoff: 60 sec per trial) for 7 consecutive days with an intertrial interval of approximately 30 min. The results showed a significant group × day interaction [$F(12,132) = 6.6$, $P < 0.001$], as normal mice and hot-foot mutants improved over days ($P < 0.001$), whereas staggerer mutants did not ($P > 0.1$). On the final day of training, normal mice outperformed both hot-foot mutants and staggerer mutants ($P < 0.01$).

retested in the visible platform condition, no group difference emerged. At 110 days of age, longer path lengths were displayed by pcd mutants on the third trial block but not on the first two or final three trial blocks. A deficit occurred during hidden platform trials but not during a visible platform trial block at the end of testing. At all age levels, pcd mutants were more impaired in the invisible platform condition than in the visible platform condition. These results are consistent with the hypothesis that the cerebellum is involved in spatial orientation. The presence of retinal degeneration in this mutant (Mullen and LaVail, 1975), although more slowly progressing than the cerebellar pathology, may reduce the perception of distal cues, thereby causing a deficit. As the authors point out, the number of photoreceptors is normal at 30 days of age and lower by 30% at 50 and 50% at 110 days of age. Therefore, 1-month old mutants show a hippocampal-type deficit that cannot be explained by retinal degeneration.

In summary (Table III), water maze testing in cerebellar mutant mice indicates the importance of the cerebellum in the visuomotor coordination necessary to navigate toward a visible platform. This was especially evident

TABLE III
WATER MAZE PERFORMANCE IN 1- TO 2-MONTH-OLD
CEREBELLAR MUTANT MICE

Mutant	Hidden platform	Visible platform
Lurcher	Deficit	Deficit
Weaver	Deficit	Deficit
Staggerer	Deficit	No deficit
pcd	Deficit	No deficit

in lurcher mutant mice, characterized by the massive degeneration of cerebellar granule and Purkinje cells in the absence of known basal ganglia, limbic, or neocortical abnormalities. These results are consistent with the hypothesis (Stein, 1986; Stein and Glickstein, 1992) that the cerebellum is involved in the visual guidance of movements. The involvement of the cerebellum in higher-level spatial processing is indicated by the selective vulnerability of young pcd mutants in the invisible platform condition (Goodlett et al., 1992). A more selective deficit in pcd mutants than lurcher mutants may be due to the higher degree of cerebellar pathology in the latter. Staggerer mutants display a mixed form of navigational deficits depending on methodological factors, but are probably more similar to lurcher mutants than pcd mutants, reflecting their morphological similarities.

c. *Anatomic Correlates.* It is possible that the cerebellum exerts an influence on both spatial orientation and visuomotor coordination. The nature of the deficit displayed by a brain-damaged animal may depend on the extent and regional distribution of the cerebellar pathology. For example, midline as opposed to lateral cerebellar lesions may cause a different profile in navigational abilities. This hypothesis was tested in rats with lesions of either the midline cerebellum, comprising the vermis and the fastigial nucleus, or the lateral cerebellum, comprising the hemispheres and dentate (Joyal et al., 1996). Quadrant entries and escape latencies were higher in rats with midline cerebellar lesions but not in rats with lateral cerebellar lesions during visible platform performance. In contrast, the lateral cerebellar group had higher quadrant entries and escape latencies than a sham-operated group during the hidden platform task. These behavioral differences may be explained by the different anatomical connections of the midline as opposed to the lateral cerebellum.

The midline cerebellum has reciprocal connections with the vestibular system (Walberg, 1972), the superior colliculus (Brodal and Bjaalie, 1992; Carpenter and Batton, 1982; Hirai et al., 1982; May et al., 1990; Saint-Cyr and Courville, 1982; Yamada and Noda, 1987), and the lateral geniculate nucleus of the thalamus (Graybiel, 1974). The fastigial efferent system includes the frontal eye field via thalamic relay nuclei (Kyuhou and Kawaguchi, 1987). Connections with the superior colliculus, frontal eye fields, and lateral geniculate are involved in eye movements in primates, but their functions in rodents remain to be determined. Vermal Purkinje cells receive hippocampal input via brain stem nuclei (Saint-Cyr and Woodward, 1980). In turn, fastigial stimulation alters hippocampal and septal neurons via a multisynaptic pathway (Heath et al., 1978; Newman and Reza, 1979). These anatomical connections indicate that the midline cerebellum may integrate visual and vestibular signals in such a way as to permit effective visuomotor coordination. Visual ambient guidance is crucially dependent on the superior colliculus. During water maze testing, lesions of the superior colliculus cause visuomotor deficits (Lines and Milner, 1985). Vestibular stimulation caused by a rotating platform impaired spatial orientation in the Morris maze, underlining the importance of vestibular cues during oriented swimming movements (Semenov and Bures, 1989). Mice with genetic lesions of the vestibular system sink beneath the water level (Douglas et al., 1979). This is not the case with midline cerebellar lesions.

Other brain regions important in sensorimotor integration may be involved in different aspects of visual guidance. For example, lesions of the striatum or dopamine depletion in the same area impaired visible platform performance in the Morris maze (Whishaw and Dunnett, 1985; Whishaw et al., 1987). Lesioned animals perseverate in wall hugging, but with repeated training abandon this strategy (McDonald and White, 1994). It remains to be determined whether rats with midline cerebellar lesions can reach normal values with repeated training on the visible platform. Striatal damage causes the release of perseverative, inappropriate strategies (Whishaw et al., 1987). The dorsal striatum receives vestibular input and may process vestibular information during food-motivated maze tasks (Potegal, 1982) and during guided swimming movements. Midline cerebellar damage may decrease the animal's ability to use visual and vestibular stimuli during guided swimming movements. Hippocampal pyramidal cells may use the sensorimotor integration provided by the midline cerebellum for navigational place learning.

Lesions of the fastigial nucleus impaired acquisition of the hidden platform task but not visible platform performance (Joyal et al., 1996). Thus, visible platform performance is affected by midline cerebellar lesions but not by fastigial lesions. These results imply that some degree of sensorimotor integration for swimming toward a visible goal occurs at the level of the vermis and the flocculonodular lobe. The lateral cerebellum receives neocortical input from association parts of the frontal and parietal lobes via rostral pontine nuclei (Brodal, 1978; Schmahmann and Pandya, 1989,

1995) and sends information back to these areas via thalamic nuclei (Kakei and Shinoda, 1990; Middleton and Strick, 1994; Sasaki, 1979; Sasaki et al., 1979; Schmahmann and Pandya, 1990; Wannier et al., 1992; Yamamoto and Oka, 1993). In addition to VA-VL thalamic nuclei, the cerebellum projects to the dorsomedial and intralaminar nuclei (Rouiller et al., 1994), the former innervating frontal association cortex and the latter widespread areas of the neocortex. The VA-VL thalamic nuclei are involved in cerebellar input to the parietal cortex (Kakei and Shinoda, 1990; Sasaki, 1979; Sasaki et al., 1979; Schmahmann and Pandya, 1990; Wannier et al., 1992; Yamamoto and Oka, 1993). In addition to the parieto–ponto–cerebellar pathway, a parieto–rubro–olivo–cerebellar pathway exists (Oka et al., 1979). The lateral cerebellum also sends input to the tectum (Hirai et al., 1982; May et al., 1990). The parietal cortex is important in navigational abilities (Crowne et al., 1992; Kolb et al., 1983; DiMattia and Kesner, 1988). The role of the medial frontal cortex is also important (Kolb et al., 1983, 1984; Sutherland et al., 1982), but is dependent on methodological factors (de Bruin et al., 1994).

It has been proposed that the neocortico–ponto–cerebellar pathway controls the visual guidance of limb and eye movements (Stein, 1986; Stein and Glickstein, 1992). This theory may be extended to navigational abilities. Lesions of the lateral cerebellum in rats impaired the hidden platform task but not the visible platform task, an indication of a deficit in spatial orientation and not in sensorimotor guidance (Joyal et al., 1996). On the basis of this study, it may be hypothesized that the cerebellum is implicated in the cognitive processes required to control whole body movements in space. Fastigial and dentate efferents to dopamine cells in the substantia nigra (Snider et al., 1976) may also influence the sensorimotor coordination necessary for visual guidance, as the depletion of striatal dopamine by means of 6-hydroxydopamine caused a visuomotor coordination deficit in the Morris maze (Whishaw and Dunnett, 1985). Ibotenate-induced lesions of the striatum also caused this deficit (Whishaw et al., 1987).

B. RADIAL MAZE

In the Morris maze, the effects of brain lesions are often measured during acquisition of the task. In the radial maze, performance of the already acquired task is measured. Animals must first be trained to retrieve the food at the end of the maze arms before lesioning. As with the Morris maze (McDonald and White, 1994), fornix lesions impair radial maze performance (Packard et al., 1989). The performance of spatial working and reference memory correlates positively with the size of intra- and infrapyramidal fibers in the hippocampus (Crusio et al., 1993).

There is only one report concerning the effects of cerebellar damage on radial maze performance. Goldowitz and Koch (1982) trained staggerer mutants and normal littermates to retrieve food pellets at the end of an eight-arm radial maze. Staggerer mutants had a higher number of working memory errors. These results are consistent with the hypothesis of a role for the cerebellum in spatial learning.

C. Spatial Alternation

The left–right spatial alternation test is a simpler version of the radial maze test, with two instead of eight choices. Impairments of this task have been observed after hippocampal lesions in rats (Wan et al., 1994). Because cerebellar lesions cause Morris maze deficits, there is interest in determining to what extent these lesions affect other spatial tasks. Pellegrino and Altman (1979) X-irradiated the cerebellum during different developmental stages in the rat and assessed the acquisition of single and double water maze alternation tasks. An increase in errors was found in both single and double alternation tasks among irradiated rats. There was no group difference in swimming speed. Dissociations were found between different behavioral tests depending on the length and timing of the irradiations. Late irradiated rats were impaired in the water maze but not in motor coordination tests. When the lower molecular layer developed abnormally, motor coordination deficits were prominent. When the upper molecular layer developed abnormally, water maze deficits were prominent. The authors hypothesized that the early forming parallel fiber system is involved in motor coordination whereas the late forming parallel fiber system is involved in the programming of action.

The role of the cerebellum in water maze alternation has been investigated in cerebellar mutant mice. When positioned at one point of a rectangular basin, the escape platform could be found on one side and when positioned at the opposite end, the same platform was found on the other side. More than one error could be commited per trial whenever a mouse strayed from the platform without turning toward the correct side. Both staggerer (Lalonde, 1987) and weaver (Lalonde and Botez, 1986) mutants committed more errors than their respective controls. A second task comprised route finding in a Z-maze configuration requiring successive left–right turns (Filali et al., 1996; Lalonde et al., 1996). All three cerebellar mutants tested (staggerer, hot-foot, lurcher) committed more errors and had higher escape latencies than their respective controls. The higher number of errors is a further indication of the directional nature of the navigational impairments in mice with hereditary cerebellar atrophy.

D. HOLE BOARD

There is one report concerning the role of the cerebellum during spatial learning in a hole board (Dahhaoui *et al.,* 1992a). Rats were cerebellectomized either before or after training in a hole board, requiring the retrieval of a single pellet in a 16-hole matrix. Cerebellectomized rats committed more errors during acquisition and performance of this task. A similar food-searching task was sensitive to hippocampal lesions (Oades and Isaacson, 1978).

III. Conclusions

Cerebellar damage in animals leads to visuomotor deficits. Depending on the degree or regional specificity of the lesion, cerebellar injury results in disorders of spatial orientation. These results concur with clinical studies showing impairments in visuospatial organization in patients with cerebellar atrophy. Further models of cerebellar atrophy have yet be tested in the Morris maze, together with the elaboration of additional tests able to distinguish spatial orientation from visuomotor coordination.

References

Aggleton, J. P., Desimone, R., and Mishkin, M. (1986). The origin, course, and termination of the hippocampothalamic projections in the macaque. *J. Comp. Neurol.* **243,** 409–421.

Barone, S., Jr., Nanry, K. P., Mundy, W. R., McGinty, J. F., and Tilson, H. A. (1991). Spatial learning deficits are not solely due to cholinergic deficits following medial septal lesions with colchicine. *Psychobiology* **19,** 41–50.

Baxter, M. G., Bucci, D. J., Gorman, L. K., Wiley, R. G., and Gallagher, M. (1995). Selective immunotoxic lesions of basal forebrain cholinergic cells: Effects on learning and memory in rats. *Behav. Neurosci.* **109,** 714–722.

Bensoula, A. N., Guastavino, J.-M., Lalonde, R., Portet, R., Bertin, R., and Krafft, B. (1995). Spatial navigation of staggerer and normal mice during juvenile and adult stages. *Physiol. Behav.* **58,** 823–825.

Blatt, G. J., and Eisenman, L. M. (1985a). A qualitative and quantitative light microscopic study of the inferior olivary complex of normal, reeler, and weaver mutant mice. *J. Comp. Neurol.* **232,** 117–128.

Blatt, G. J., and Eisenman, L. M. (1985b). A qualitative and quantitative light microscopic study of the inferior olivary complex in the adult staggerer mutant mouse. *J. Neurogenet.* **2,** 51–66.

Botez, M. I., Botez, T., and Olivier, M. (1985). Parietal lobe syndromes. *In* "Handbook of Clinical Neurology" (P. J. Vinken, B. W. Bruyn, and H. L. Klawans, eds.), Vol. 45, pp. 63–85. Elsevier, Amsterdam.

Botez, M. I., Botez-Marquard, T., Elie, R., and Attig, E. (1989). Role of the cerebellum in complex human behavior. *Ital. J. Neurol. Sci.* **10,** 291–300.

Botez-Marquard, T., and Botez, M. I. (1993). Cognitive behavior in heredodegenerative ataxias. *Eur. Neurol.* **33,** 351–357.

Botez-Marquard, T., and Routhier, I. (1995). Reaction time and intelligence in patients with olivopontocebellar atrophy. *Neuropsychiat. Neuropsychol. Behav. Neurol.* **8,** 168–175.

Brodal, P. (1978). The corticopontine projection in the rhesus monkey: Origin and principles of organization. *Brain* **101,** 251–283.

Brodal, P., and Bjaalie, J. G. (1992). Organization of the pontine nuclei. *Neurosci. Res.* **13,** 83–118.

Caddy, K. W. T., and Biscoe, T. J. (1979). Structural and quantitative studies on the normal C3H and lurcher mutant mouse. *Phil. Trans. R. Soc. Lond. Ser. B* **287,** 167–201.

Carpenter, M. B., and Batton, R. R., III (1982). Connections of the fastigial nucleus in the cat and monkey. *Exp. Brain Res.* (Suppl. 6), 250–295.

Crowne, D. P., Notovny, M. F., Maier, S. E., and Vitols, R. (1992). Effects of unilateral parietal lesions on spatial localization in the rat. *Behav. Neurosci.* **106,** 808–819.

Crusio, W. E., Schwegler, H., and Brust, I. (1993). Covariations between hippocampal mossy fibers and working and reference memory in spatial and non-spatial radial maze tasks in mice. *Eur. J. Neurosci.* **5,** 1413–1420.

Dahhaoui, M., Lannou, J., Stelz, T., Caston, J., and Guastavino, J. M. (1992a). Role of the cerebellum in spatial orientation in the rat. *Behav. Neur. Biol.* **58,** 180–189.

Dahhaoui, M., Stelz, T., and Caston, J. (1992b). Effects of lesion of the inferior olivary complex by 3-acetylpyridine on learning and memory in the rat. *J. Comp. Physiol. A* **171,** 657–664.

de Bruin, J. P. C., Sanchez-Santed, F., Heinsbroek, R. P. W., and Donker, A. (1994). A behavioral analysis of rats with damage to the medial prefrontal cortex using the morris water maze: Evidence for behavioural flexibility but not for impaired spatial navigation. *Brain Res.* **652,** 323–333.

Decker, M. W., Radek, R. J., Majchrzak, M. J., and Anderson, D. J. (1992). Differential effects of medial septal lesions on spatial-memory tasks. *Psychobiology* **20,** 9–17.

D'Eustachio, P., and Davisson, M. T. (1993). Resolution of the staggerer (sg) mutation from the neural cell adhesion molecule locus (Ncam) on mouse chromosome 9. *Mammal. Genet.* **4,** 278–280.

DiMattia, B. D., and Kesner, R. P. (1988). Spatial cognitive maps: Differential role of parietal cortex and hippocampal formation. *Behav. Neurosci.* **102,** 471–480.

Douglas, R. J., Clark, G. M., Erway, L. C., Hubbard, D. G., and Wright, C. G. (1979). Effects of genetic vestibular defects on behavior related to spatial orientation and emotionality. *J. Comp. Physiol. Psychol.* **93,** 467–480.

Fehrenbach, R. A., Wallesch, C.-W., and Claus, D. (1984). Neuropsychologic findings in Friedreich's ataxia. *Arch. Neurol.* **41,** 306–308.

Filali, M., Lalonde, R., Bensoula, A. N., Guastavino, J.-M., and Lestienne, F. (1996). Spontaneous alternation, motor activity, and spatial learning in hot-foot mutant mice. *J. Comp. Physiol. A* **178,** 101–104.

Foster, T. C., Castro, C. A., and McNaughton, B. L. (1989). Spatial selectivity of rat hippocampal neurons: Dependence on preparedness for movement. *Science* **244,** 1580–1582.

Fournier, E., and Crepel, F. (1984). Electrophysiological properties of in vitro hippocampal pyramidal cells from normal and staggerer mutant mice. *Brain Res.* **311,** 87–96.

Ghetti, B., Norton, J., and Triarhou, L. C. (1987). Nerve cell atrophy and loss in the inferior olivary complex of "Purkinje cell degeneration" mutant mice. *J. Comp. Neurol.* **260,** 409–422.

Glickstein, M., Stein, J., and King, R. A. (1972). Visual input to the pontine nuclei. *Science* **178,** 1110–1111.

Goldowitz, D., and Koch, J. (1986). Performance of normal and neurological mutant mice on radial arm maze and active avoidance tasks. *Behav. Neur. Biol.* **46,** 216–226.

Goldowitz, D., and Mullen, R. J. (1982). Granule cell as a site of gene action in the weaver mouse cerebellum: Evidence from heterozygous mutant chimeras. *J. Neurosci.* **2,** 1474–1485.

Goodlett, C. R., Hamre, K. M., and West, J. R. (1992). Dissociation of spatial navigation and visual guidance in Purkinje cell degeneration (pcd) mutant mice. *Behav. Brain Res.* **47,** 129–141.

Graybiel, A. M. (1974). Visuo-cerebellar and cerebello-visual connections involving the ventral lateral geniculate nucleus. *Exp. Brain Res.* **20,** 303–306.

Greer, C. A., and Shepherd, G. M. (1982). Mitral cell degeneration and sensory function in the neurological mutant mouse Purkinje cell degeneration. *Brain Res.* **235,** 156–161.

Guastavino, J.-M., Bertin, R., and Portet, R. (1991). Effects of the rearing temperature on the temporal feeding pattern of the staggerer mutant mouse. *Physiol. Behav.* **49,** 405–409.

Guastavino, J.-M., Sotelo, C., and Damez-Kinselle, I. (1990). Hot-foot murine mutation: Behavioral effects and neuroanatomical alterations. *Brain Res.* **523,** 199–210.

Hagan, J. J., Salamone, J. D., Simpson, J., Iversen, S. D., and Morris, R. G. M. (1988). Place navigation in rats is impaired by lesions of medial septum and diagonal band but not nucleus basalis magnocellularis. *Behav. Brain Res.* **27,** 9–20.

Heath, R. G., Dempesy, C. W., Fontana, C. J., and Myers, W. A. (1978). Cerebellar stimulation: Effects on septal region, hippocampus, and amygdala of cats and rats. *Biol. Psychiat.* **13,** 501–529.

Heckroth, J. A. (1992). Development of glutamic acid decarboxylase-immunoreactive elements in the cerebellar cortex of normal and lurcher mutant mice. *J. Comp. Neurol.* **315,** 85–97.

Heckroth, J. A. (1994). Quantitative morphological analysis of the cerebellar nuclei in normal and lurcher mutant mice. I. Morphology and cell number. *J. Comp. Neurol.* **343,** 173–182.

Heckroth, J. A., and Eisenman, L. M. (1991). Olivary morphology and olivocerebellar topography in adult lurcher mutant mice. *J. Comp. Neurol.* **312,** 641–651.

Herrup, K. (1983). Role of staggerer gene in determining cell number in cerebellar cortex. II. Granule cell death and persistence of the external granule cell layer in young mouse chimeras. *Dev. Brain Res.* **12,** 271–283.

Herrup, K., and Mullen, R. J. (1979a). Regional variation and absence of large neurons in the cerebellum of the staggerer mouse. *Brain Res.* **172,** 1–12.

Herrup, K., and Mullen, R. J. (1979b). Staggerer chimeras: Intrinsic nature of Purkinje cell defects and implications for normal cerebellar development. *Brain Res.* **178,** 443–457.

Hirai, T., Onodera, S., and Kawamura, K. (1982). Cerebellotectal projections in cats with horseradish peroxidase of tritiated amino acids axonal transport. *Exp. Brain Res.* **48,** 1–12.

Hirano, A., and Dembitzer, H. M. (1973). Cerebellar alterations in the weaver mouse. *J. Cell Biol.* **56,** 478–486.

Jacquart, G., Mahler, P., and Strazielle, C. (1989). Effects of cerebellectomy on food intake, growth, and fecal lipid level in the rat. *Nutrit. Rep. Int.* **39,** 879–888.

Joyal, C. C., Meyer, C., Jacquart, G., Mahler, P., Caston, J., and Lalonde, R. (1996). Effects of midline and lateral cerebellar lesions on motor and non-motor learning in rats. *Brain Res.,* **739,** 1–11.

Jucker, M., Kametani, H., Bresnahan, E. L., and Ingram, D. K. (1990). Parietal cortex lesions do not impair retention performance of rats in a 14-unit T-maze unless hippocampal damage is present. *Physiol. Behav.* **47,** 207–212.

Kakei, S., and Shinoda, Y. (1990). Parietal projection of thalamocortical fibers from the ventroanterior-ventrolateral complex of the cat thalamus. *Neurosci. Lett.* **117,** 280–294.

Kaseda, Y., Ghetti, B., Low, W. C., Richter, J. A., and Simon J. R. (1987). Dopamine D2 receptors increase in the dorsolateral striatum of weaver mutant mice. *Brain Res.* **422,** 178–181.

Kelsey, J. E., and Landry, B. A. (1988). Medial septal lesions disrupt spatial ability in rats. *Behav. Neurosci.* **102,** 289–293.

Kish, S. J., El-Awar, M., Schut, L., Leach, L., Oscar-Berman, M., and Freedman, M. (1988). Cognitive deficits in olivopontocerebellar atrophy: Implications for the cholinergic hypothesis of Alzheimer's dementia. *Ann. Neurol.* **24,** 200–206.

Kolb, B., Burhmann, K., McDonald, R., and Sutherland, R. (1994). Dissociation of the medial prefrontal, posterior parietal, and posterior temporal cortex for spatial navigation and recognition memory in the rat. *Cerebr. Cortex* **6,** 664–680.

Kolb, B., Macintosh, A., Whishaw, I. Q., and Sutherland, R. J. (1984). Evidence for anatomical but not functional asymmetry in the hemidecorticate rat. *Behav. Neurosci.* **98,** 44–58.

Kolb, B., Pittman, K., Sutherland, R. J., and Whishaw, I. Q. (1982). Dissociation of the contributions of the prefrontal cortex and dorsomedial thalamic nucleus to spatially guided behavior in the rat. *Behav. Brain Res.* **6,** 365–378.

Kolb, B., Sutherland, R. J., and Whishaw, I. Q. (1983). A comparison of the contributions of the frontal and parietal association cortex to spatial localization in rats. *Behav. Neurosci.* **97,** 13–27.

Kyuhou, S.-I., and Kawaguchi, S. (1987). Cerebellocerebral projection from the fastigial nucleus onto the frontal eye field and anterior ectosylvian visual area in the cat. *J. Comp. Neurol.* **259,** 571–590.

Lalonde, R. (1987). Exploration and spatial learning in staggerer mutant mice. *J. Neurogenet.* **4,** 285–292.

Lalonde, R., and Botez, M. I. (1986). Navigational deficits in weaver mutant mice. *Brain Res.* **398,** 175–177.

Lalonde, R., and Botez, M. I. (1990). The cerebellum and learning processes in animals. *Brain Res. Rev.* **15,** 325–332.

Lalonde, R., Filali, M., Bensoula, A. N., Monnier, C., and Guastavino, J.-M. (1996). Spatial learning in a Z-maze by cerebellar mutant mice. *Physiol. Behav.* **59,** 83–86.

Lalonde, R., Joyal, C. C., and Côté, C. (1993). Swimming activity in dystonia musculorum mutant mice. *Physiol. Behav.* **54,** 119–120.

Lalonde, R., Lamarre, Y., and Smith, A. M. (1988). Does the mutant mouse lurcher have deficits in spatially oriented behaviours? *Brain Res.* **455,** 24–30.

Lalonde, R., and Thifault, S. (1994). Absence of an association between motor coordination and spatial orientation in lurcher mutant mice. *Beh. Gen.* **24,** 497–501.

Lines, C. R., and Milner, A. D. (1985). A deficit in ambient visual guidance following superior colliculus lesions in rats. *Behav. Neurosci.* **99,** 707–716.

Mahler, P., Guastavino, J.-M., Jacquart, G., and Strazielle. C. (1993). An unexpected role of the cerebellum: Involvement in nutritional organization. *Physiol. Behav.* **54,** 1063–1067.

Marston, H. M., Everitt, B. J., and Robbins, T. W. (1993). Comparative effects of excitotoxic lesions of the hippocampus and septum/diagonal band on conditional visual discrimination and spatial learning. *Neuropsychologia* **31,** 1099–1118.

May, P. J., Hartwich-Young, R., Nelson, J., Sparks, D. L., and Porter, J. D. (1990). Cerebellotectal pathways in the macaque: Implications for collicular generation of saccades. *Neuroscience* **36,** 305–324.

McAlonan, G. M., Wilkinson, L. S., Robbins, T. W., and Everitt, B. J. (1995). The effects of AMPA-induced lesions of the septo-hippocampal cholinergic projection on aversive conditioning to explicit and contextual cues and spatial learning in the water maze. *Eur. J. Neurosci.* **7,** 282–292.

McDonald, R. J., and White, N. M. (1994). Parallel information processing in the water maze: evidence for independent memory systems involving dorsal striatum and hippocampus. *Behav. Neur. Biol.* **61,** 260–270.

McNamara, R. K., and Skelton, R. W. (1992). Pharmacological dissociation between the spatial learning deficits produced by morphine and diazepam. *Psychopharmacology* **108,** 147–152.

Middleton, F. A., and Strick, P. L. (1994). Anatomical evidence for cerebellar and basal ganglia involvement in higher cognitive function. *Science* **266,** 458–461.

Miret-Duvaux, O., Frederic, F., Simon, D., Guenet, J.-L., Hanauer, A., Delhaye-Bouchaud, N., and Mariani, J. (1990). Glutamate dehydrogenase in cerebellar mutant mice: Gene localization and enzyme activity in different tissues. *J. Neurochem.* **54,** 23–29.

Miyamoto, M., Kato, J., Narumi, S., and Nagaoka, A. (1987). Characteristics of memory impairment following lesioning of the basal forebrain and medial septal nucleus in rats. *Brain Res.* **419,** 19–31.

Morris, R. G. M., Garrud, P., Rawlins, J. N. P., and O'Keefe, J. (1982). Place navigation impaired in rats with hippocampal lesions. *Nature* **297,** 681–683.

Mullen, R. J., and LaVail, M. M. (1975). Two new types of retinal degeneration in cerebellar mutant mice. *Nature* **258,** 528–530.

Mullen, R. J., Eicher, E. M., and Sidman, R. L. (1976). Purkinje cell degeneration: A new neurological mutation in the mouse. *Proc. Natl. Acad. Sci. USA* **73,** 208–212.

Nagahara, A. H., Otto, T., and Gallagher, M. (1995). Entorhinal-perirrhinal lesions impair performance of rats on two versions of place learning in the Morris water maze. *Behav. Neurosci.* **109,** 3–9.

Newman, P. P., and Reza, H. (1979). Functional relationships between the hippocampus and the cerebellum: An electrophysiological study of the cat. *J. Physiol. (Lond.)* **287,** 405–426.

Oades, R. D., and Isaacson, R. L. (1978). The development of food search behavior by rats: The effects of hippocampal damage and haloperidol. *Behav. Biol.* **24,** 327–337.

O'Gorman, S., and Sidman, R. L. (1985). Degeneration of thalamic neurons in "Purkinje cell degeneration" mutant mice. I. Distribution of neuron loss. *J. Comp. Neurol.* **234,** 277–297.

Ohsugi, K., Adachi, K., and Ando, K. (1986). Serotonin metabolism in the CNS in cerebellar ataxic mice. *Experientia* **42,** 1245–1247.

Oka, H., Jinnai, K., and Yamamoto, T. (1979). The parieto-rubro-olivary pathway in the cat. *Exp. Brain Res.* **37,** 115–125.

Olton, D. S., and Papas, B. C. (1979). Spatial memory and hippocampal function. *Neuropsychologia* **17,** 669–682.

Packard, M. G., Hirsh, R., and White, N. M. (1989). Differential effects of fornix and caudate nucleus lesions on two radial maze tasks: Evidence for multiple memory systems. *J. Neurosci.* **9,** 1465–1472.

Panagopoulos, N. T., Matsokis, N. A., and Valcana, T. (1993). Cerebellar and striatal dopamine receptors: Effects of reeler and weaver murine mutations. *J. Neurosci. Res.* **35,** 499–506.

Pellegrino, L. J., and Altman, J. (1979). Effects of differential interference with postnatal cerebellar neurogenesis on motor performance, activity level, and maze learning of rats: A developmental study. *J. Comp. Physiol. Psychol.* **93,** 1–33.

Potegal, M. (1982). Vestibular and neostriatal contributions to spatial orientation. *In* "Spatial Abilities: Development and Physiological Foundations" (M. Potegal, ed.), pp. 361–387. Academic Press, New York.

Roffler-Tarlov, S., and Graybiel, S. (1984). Weaver mutation has differential effects on the dopamine-containing innervation of the limbic and nonlimbic striatum. *Nature* **307,** 62–66.

Roffler-Tarlov, S., and Graybiel, A. M. (1986). Expression of the weaver gene in dopamine-containing neural systems is dose-dependent and affects both striatal and nonstriatal regions. *J. Neurosci.* **6,** 3319–3330.

Roffler-Tarlov, S., and Graybiel, A. M. (1987). The postnatal development of the dopamine-containing innervation of dorsal and ventral striatum: Effects of the weaver gene. *J. Neurosci.* **7,** 2364–2372.

Rosene, D. L., and Van Hoesen, G. W. (1977). Hippocampal efferents reach widespread areas of cerebral cortex and amydgdala in the rhesus monkey. *Science* **198,** 315–317.

Rouiller, E. M., Liang, F., Barbalian, A., Moret, V., and Wiesendanger, M. (1994). Cerebellothalamocortical and pallidothalamocortical projections to the primary and supplementary motor cortical areas: A multiple tracing study in macaque monkeys. *J. Comp. Neurol.* **345,** 183–213.

Saint-Cyr, J. A., and Courville, J. (1980). Projections from the motor cortex, midbrain, and vestibular nuclei to the inferior olive in the cat: anatomical organization and functional correlates. *In* "The Inferior Olivary Nucleus: Anatomy and Physiology" (J. Courville, M. De Montigny, and Y. Lamarre, eds.), pp. 97–124. Raven Press, New York.

Saint-Cyr, J. A., and Courville, J. (1982). Descending projections to the inferior olive from the mesencephalon and superior colliculus in the cat. *Exp. Brain Res.* **45,** 333–348.

Saint-Cyr, J. A., and Woodward, D. J. (1980). Activation of mossy and climbing fiber pathways to the cerebellar cortex by stimulation of the fornix in the rat. *Exp. Brain Res.* **40,** 1–12.

Sasaki, K. (1979). Cerebrocerebellar interconnections in cats and monkeys. *In* "Cerebrocerebellar Interactions" (J. Massion, and K. Sasaki, eds.), pp 105–124. Elsevier, Amsterdam.

Sasaki, K., Jinnai, K., Gemba, H., Hashimoto, S., and Mizuno, N. (1979). Projection of the cerebellar dentate nucleus onto the frontal association cortex in monkeys. *Exp. Brain Res.* **37,** 193–198.

Schenk, F., and Morris, R. G. M. (1985). Dissociation between components of spatial memory in rats after recovery from the effects of retrohippocampal lesions. *Exp. Brain Res.* **58,** 11–28.

Schmahmann, J. D., and Pandya, D. N. (1989). Anatomical investigation of projections to the basis pontis from posterior parietal association cortices in rhesus monkey. *J. Comp. Neurol.* **289,** 53–73.

Schmahmann, J. D., and Pandya, D. N. (1990). Anatomical investigation of projections from thalamus to the posterior parietal association cortices in rhesus monkey. *J. Comp. Neurol.* **295,** 299–326.

Schmahmann, J. D., and Pandya, D. N. (1995). Prefrontal cortex projections to the basilar pons in rhesus monkey: Implications for the cerebellar contribution to higher function. *Neurosci. Lett.* **199,** 175–178.

Segal, M., Greenberger, V., and Pearl, E. (1989). Septal transplants ameliorate spatial deficits and restore cholinergic functions in rats with a damaged septo-hippocampal connection. *Brain Res.* **500,** 139–148.

Semenov, L. V., and Bures, J. (1989). Vestibular stimulation disrupts acquisition of place navigation in the Morris water tank task. *Behav. Neur. Biol.* **51,** 346–363.

Sharp, P. E., Blair, H. T., Etkin, D., and Tzanetos, D. B. (1995). Influences of vestibular and visual motion information on the spatial firing patterns of hippocampal place cells. *J. Neurosci.* **15,** 173–189.

Shojaeian, H., Delhaye-Bouchaud, N., and Mariani, J. (1985). Decreased number of cells in the inferior olivary nucleus of the developing staggerer mouse. *Dev. Brain Res.* **21,** 141–146.

Skelton, R. W., and McNamara, R. K. (1992). Bilateral knife cuts to the perforant path disrupt learning in the morris water maze. *Hippocampus* **2**, 73–80.

Snider, R. S., Maiti, A., and Snider, S. R. (1976). Cerebellar pathways to ventral midbrain and nigra. *Exp. Neurol.* **53**, 714–728.

Sotelo, C., and Changeux, J. P. (1974). Bergmann fibers and granule cell migration in the cerebellum of homozygous weaver mutant mouse. *Brain Res.* **77**, 484–491.

Stein, J. F. (1986). Role of the cerebellum in the visual guidance of movement. *Nature* **323**, 217–221.

Stein, J. F., and Glickstein, M. (1992). Role of the cerebellum in visual guidance of movement. *Physiol. Rev.* **72**, 967–1017.

Stotz, E. H., Palacios, J. M., Landwehrmeyer, B., Norton, J., Ghetti, B., Simon, J. R., and Triarhou, L. C. (1994). Alterations in dopamine and serotonin uptake systems in the striatum of the weaver mutant mouse. *J. Neur. Trans. (Gen.)* **97**, 51–64.

Sutherland, R. J., Kolb, B., and Whishaw, I. Q. (1982). Spatial mapping: Definitive disruption by hippocampal or medial frontal cortical damage in the rat. *Neurosci. Lett.* **31**, 271–276.

Sutherland, R. J., and McDonald, R. J. (1990). Hippocampus, amygdala, and memory deficits in rats. *Behav. Brain Res.* **37**, 57–79.

Sutherland, R. J., and Rodriguez, A. J. (1989). The role of the fornix/fimbria and some related subcortical structures in place learning and memory. *Behav. Brain Res.* **32**, 265–277.

Sutherland, R. J., and Rudy, J. W. (1988). Place learning in the Morris place navigation task is impaired by damage to the hippocampal formation even if the temporal demands are reduced. *Psychobiology* **16**, 157–163.

Sutherland, R. J., Whishaw, I. Q., and Kolb, B. (1983). A behavioural analysis of spatial localization following electrolytic, kainate- or colchicine-induced damage to the hippocampal formation in the rat. *Behav. Brain Res.* **7**, 133–153.

Sutherland, R. J., Whishaw, I. Q., and Kolb, B. (1988). Contributions of cingulate cortex to two forms of spatial learning and memory. *J. Neurosci.* **8**, 1863–1872.

Triarhou, L. C., and Ghetti, B. (1991). Stabilisation of neurone number in the inferior olivary complex of aged "Purkinje cell degeneration" mutant mice. *Acta Neuropathol.* **81**, 597–602.

Triarhou, L. C., Norton, J., Alyea, C., and Ghetti, B. (1985). A quantitative study of the granule cells in the Purkinje cell degeneration (pcd) mutant. *Ann. Neurol.* **18**, 146.

Triarhou, L. C., Norton, J., and Ghetti, B. (1987). Anterograde transsynaptic degeneration in the deep cerebellar nuclei of Purkinje cell degeneration (pcd) mutant mice. *Exp. Brain Res.* **66**, 577–588.

Triarhou, L. C., Norton, J., and Ghetti, B. (1988). Mesencephalic dopamine cell deficit involves areas A8, A9 and A10 in weaver mutant mice. *Exp. Brain Res.* **70**, 256–265.

Walberg, F. (1972). Cerebellovestibular relations: Anatomy. *Progr. Brain Res.* **37**, 361–376.

Wan, R.-Q., Pang, K., and Olton, D. S. (1994). Hippocampal and amygdaloid involvement in nonspatial and spatial working memory in rats: Effects of delay and interference. *Behav. Neurosci.* **108**, 866–882.

Wannier, T., Kakei, S., and Shinoda, Y. (1992). Two modes of cerebellar input to the parietal cortex in the cat. *Exp. Brain Res.* **90**, 241–252.

Wetts, R., and Herrup, K. (1982a). Interaction of granule, Purkinje and inferior olivary neurons in Lurcher chimaeric mice: Qualitative studies. *J. Embryol. Exp. Morphol.* **68**, 87–98.

Wetts, R., and Herrup, K. (1982b). Interaction of granule, Purkinje and inferior olivary neurons in Lurcher chimaeric mice. II. Granule cell death. *Brain Res.* **250**, 358–362.

Whishaw, I. Q., and Dunnett, S. B. (1985). Dopamine depletion, stimulation or blockade in the rat disrupts spatial navigation and locomotion dependent upon beacon or distal cues. *Behav. Brain Res.* **18**, 11–29.

Whishaw, I. Q., and Kolb, B. (1984). Decortication abolishes place but not cue learning in rats. *Behav. Brain Res.* **11,** 123–134.

Whishaw, I. Q., Mittleman, G., Bunch, S. T., and Dunnett, S. B. (1987). Impairments in the acquisition, retention and selection of spatial navigational strategies after medial caudate-putamen lesions in rats. *Behav. Brain Res.* **24,** 125–138.

Wiig, K. A., and Bilkey, D. K. (1994). The effects of perirhinal cortical lesions on spatial reference memory in the rat. *Behav. Brain Res.* **63,** 101–109.

Yamada, J., and Noda, H. (1987). Afferent and efferent connections of the oculomotor cerebellar vermis in the macaque monkey. *J. Comp. Neurol.* **265,** 224–241.

Yamamoto, T., and Oka, H. (1993). The mode of cerebellar activation of pyramidal neurons in the cat parietal cortex (areas 5 and 7): An intracellular HRP study. *Neurosci. Lett.* **18,** 129–142.

Zanjani, H. S., Mariani, J., and Herrup, K. (1990). Cell loss in the inferior olive of the staggerer mutant mouse is an indirect effect of the gene. *J. Neurogenet.* **6,** 229–241.

SPATIAL EVENT PROCESSING

Marco Molinari,* Laura Petrosini,†,‡ and Liliana G. Grammaldo*

†Department of Psychology, University of Rome "La Sapienza," *Institute of Neurology, Catholic University of Rome, and ‡Institute of Psychology, CNR, Rome, Italy

I. Morris Water Maze (MWM) for Spatial Function Studies
II. Searching Strategies
III. Neural Structures for Spatial Event Processing
IV. MWM Performances of Hemicerebellectomized Rats
V. Acquisition vs Retention
VI. Cerebellar Contribution to Spatial Event Processing
VII. Cerebellum and Spatial Procedure Development
References

The present review advances experimental evidence on the cerebellar involvement in spatial data processing. In particular, data on Morris water maze (MWM) performances of hemicerebellectomized (HCbed) rats indicate a specific cerebellar role within the procedural aspects of spatial functions. In MWM testing, HCbed animals are impaired in developing efficient exploration strategies and display only old and rather ineffective ways for acquiring spatial information, such as peripheral circling around the pool. This behavior is not exhibited if spatial mapping abilities are preoperatively acquired. Thus, MWM experimental data point toward a procedural deficit that specifically impairs the acquisition phase. The characteristics of the cerebellar involvement in affecting the procedures needed for spatial data management are discussed in the light of recent theories on spatial data processing and on cerebellar timing and ordering functions.

I. Morris Water Maze (MWM) for Spatial Function Studies

Since its introduction in 1981, the Morris water maze has been considered a powerful tool for analyzing the spatial abilities of "good swimmer" animals (Morris, 1981). In fact, it has become *the* tool for investigating spatial learning, because it is easy to use and has an extremely modifiable paradigm. This latter feature has allowed its use in various experimental settings to test which strategies an animal might use in solving spatial tasks and to clarify the contributions of various neural structures and neurochemical systems (Brandeis *et al.*, 1989) to spatial data processing.

II. Searching Strategies

The classic place learning version of the MWM requires the animal to locate a hidden platform. Because the animal starts from various positions in this paradigm, the spatial relations between the platform location and the extramaze cues (windows, laboratory furniture, shelves, etc.) relative to the animal change in every run. In this setting the only way the animal can directly locate the platform is to use a spatial map that contains the absolute relationships between the platform location and the extramaze cues. This finding strategy has been defined as "place strategy" and it depends on the ability of the animal to develop and use a true "allocentric spatial map" (Whishaw and Tomie, 1987).

Spatial abilities do not rely only on "place" functions, and various approaches can be used to solve a spatial task. By modifying the MWM paradigm (e.g., by adding some trials with the platform raised above the water level and thus visible) spatial strategies can be isolated and analyzed separately from place strategies. Specifically, normal rats can learn to reach a location by repeating a specific sequence of movements (praxic strategy), covering the same distances and angles. This strategy can be employed when the starting point and the target always maintain the same positions relative to each other. Furthermore, rats can reach the target, either visible or hidden, by employing a taxic strategy. In this case the animal approaches some specific cues that are proximally or distally associated with the target. Data from several studies indicate that normal rats use praxic, taxic, and place strategies, either separately or concurrently, to solve spatial tasks (Schenk and Morris, 1985; Whishaw *et al.*, 1987). Studies of surgical and neurochemical lesions indicate that various neuronal circuits are rather specific in affecting the different aspects of spatial processing.

Data from numerous studies indicate a severe and lasting place-navigational impairment in tasks requiring mapping strategy following hippocampal ablation. These convincing data offer a coherent framework supporting the notion of a cognitive mapping function for the hippocampus (Morris *et al.*, 1982; Fenton and Bures, 1993; Peinado-Manzano, 1990). Furthermore, spatial mapping deficits observed after frontal, orbital, or parietal cortical damage suggest that all of these structures form an integrated system for learning spatial representations of various aspects of the environment. The frontal cortex is particularly implicated when relationships among exteroceptive cues must be learned, whereas the parietal cortex appears to be important in the processing of allocentric spatial information (DiMattia and Kesner, 1988a,b; Save et al., 1992). Two studies (McDonald and White, 1994; Packard et al., 1994) using lesions of, and local amphetamine injections in, the striatum provided some evidence for a "cue learning" function of the caudoputamen. These results not only add a new structure to the cortical and subcortical regions involved in spatial learning but also advance the hypothesis of the existence of an egocentric navigational control system for movements in extrapersonal space, as opposed to the allocentric spatial relations system of the hippocampus and neocortex.

III. Neural Structures for Spatial Event Processing

Clinical and experimental findings suggest the possibility that cerebellar networks are included among the structures involved in spatial processing. After years of detailed studies on the anatomical and physiological properties of cerebellar networks in the motor domain, an increasing number of findings now suggest that the computational properties of the cerebellum may play a role not only in motor learning but also in cognitive processing of adaptive behaviors (Bracke-Tolkmitt et al., 1989; Lalonde and Botez, 1990; Sanes et al., 1990; Schmahmann, 1991; Akshoomoff and Courchesne, 1992; Ivry and Baldo,1992; Ito, 1993; Leiner et al., 1993; Daum et al., 1993) and, in particular, in the capacity to cope with spatial demands (Appollonio et al., 1993; Grafman et al., 1992, Petrosini et al. 1996).

Impairment in spatial learning in cerebellar patients (Wallesch and Horn, 1990; El-Awar et al., 1991) as well as in cerebellar mutant mice (Goodlett et al., 1992) has been described, supporting the hypothesis of a cerebellar involvement in cognitive spatial operations. In particular, weaver mutant mice, which have massive losses of cerebellar granule cells, do not reach the platform, even when visible, by a direct heading and do not show latencies as low as controls (Lalonde and Botez, 1986). Staggerer mutant mice, which lose Purkinje cells, granule cells, and inferior olive neurons, are impaired when the platform is invisible but not when the platform is visible (Lalonde, 1987). Pcd mutant mice, in which Purkinje cells are selectively degenerated, perform well in the proximal cue task but not in the distal cue version of MWM (Goodlett et al., 1992). Taken together, these findings support the conclusion that the cerebellum contributes to spatial information processing. The difficulty in defining the precise neuroanatomical extent of central nervous system damage in atrophic cerebellar patients and in mouse strains (Schmidt et al., 1982; Triarhou and Ghetti, 1986;

Gupta *et al.*, 1987; Triarhou *et al.*, 1988; Daum and Akermann, chapter 19 of this volume) however, also makes it difficult to exclude the involvement of extracerebellar structures with certainty. Furthermore, the many patterns of spatial function reported to be affected by cerebellar circuits complicate attempts to define the cerebellar role in spatial operations.

We have recently attempted to define the cerebellar contribution to spatial event processing by studying spatial abilities in rats with surgically induced cerebellar lesions (Petrosini *et al.*, 1996). Rats with cerebellar lesions have difficulty maintaining symmetrical posture and coordinated locomotion. These severe motor impairments may overwhelm any true spatial disabilities in dry land testing and/or they may disrupt the motivational aspects of performance. Consequently, we performed the experiments using a behavioral analysis in the MWM. Furthermore, because bilateral cerebellar lesions profoundly affect motor coordination and impede not only dry land performances but also swimming, the approach we chose was hemicerebellectomy (HCb). We analyzed which spatial abilities could be influenced by cerebellar circuits by testing hemicerebellectomized (HCbed) rats in various MWM paradigms that investigated different aspects of spatial data management.

IV. MWM Performances of Hemicerebellectomized Rats

The most impressive and stable deficit exhibited by HCbed rats in solving all MWM paradigms employed, which demanded either processing of distal extramaze cues, such as the place version of the task, or utilization of proximal intramaze cues, such as the cued platform version, was their inability to develop efficient searching behavior. Instead of searching for the platform, they simply swam to the periphery of the pool, displaying compulsive and irrelevant circling (Fig.1). Despite their inability to display any useful exploratory searching or to avoid peripheral circling even when the platform was visible, during repeated cue trials, HCbed animals became progressively more competent in finding the platform location. This improvement was achieved without developing a new strategy: At the beginning of each trial, HCbed animals invariably tended to swim only at the periphery of the pool; then, after a certain delay, they abruptly interrupted peripheral circling and, with a good pointing angle and a rather direct route, swam toward the escape platform (Figs. 1 and 2). This spatial competence was also maintained when the rats had only place strategies to rely on, e.g., when the platform was again hidden and the starting points were

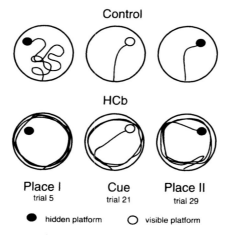

Fig. 1. Searching patterns of a control and a hemicerebellectomized (HCb) animal during successive representative trials with a hidden (Place I), visible (Cue), and again hidden (Place II) platform. Control animal exhibits searching behavior around the pool in the early trial and then a direct approach toward the platform in the successive phases. The HCbed rat displays peripheral searching in all phases of the task, with successful findings only in the Cue and the Place II phases (redrawn from Petrosini et al., 1996).

changed sequentially. However, the animals showed no further improvements.

Evidence thus suggests that HCbed rats can develop spatial maps only in cue paradigms, but are able to use these maps in place paradigms. This may indicate that HCbed rats build and store a spatial map by associating intra- and extramaze cues, but they are able to use the spatial map when only extramaze information is available. In our setting, intramaze information is available only during the cue phase, whereas extramaze information is accessible in both cue and place phases. Thus, according to this interpretation, it can be argued that extramaze information does not provide informative power sufficient for building a spatial map; informative power can be achieved only by simultaneous use of intra- and extramaze cues. Conversely, these extramaze data are sufficient for recall of the spatial template acquired during the cue. In other words, HCbed animals can solve place spatial tasks and plan and execute the appropriate motor repertoire only if a spatial map has already been developed during cue trials.

However, as seen previously, the use of the acquired place map is not immediate, as even in the later trials of the MWM paradigm, HCbed rats continued to swim at the periphery of the pool at the beginning of each trial. Evidently they need time to inhibit the persistent peripheral circling.

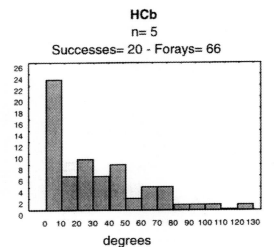

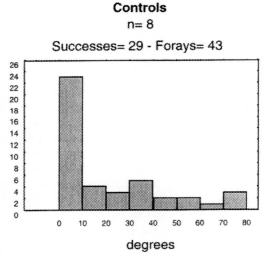

FIG. 2. Heading errors evaluated at the moment of detachment (foray) from peripheral circling (0° on abscissae indicates correct pointing toward the platform) during the last four trials of Place II of a MWM paradigm in which starting points have been randomly changed. Note that in both controls and HCb animals the highest number of forays ranged from 0 to 10°, suggesting the employment of a rather precise place map.

Once they abandon it, however, their direct trajectory to the platform indicates the presence of a quite well-defined spatial map (Figs. 1–3).

The delay in abandoning peripheral circling and the inability to develop efficient searching can be due to difficulties at various levels. HCbed rats may not be able to search either because they are unable to inhibit compulsive behaviors (not searching because of circling) or because their spatial processing system is so impaired that effective searching strategies cannot be performed and only ineffective strategies can be used (circling because of not searching).

Analysis of the strategies employed by HCbed rats in the different MWM tasks, as well as the results obtained in a T-maze paradigm where circling is impeded, indicated that circling per se is not the main factor affecting exploration strategies. Further evidence was obtained by pretraining animals on the MWM, lesioning the right hemicerebellum, and then testing the animals' ability to retain the spatial information acquired preoperatively. Under these conditions no circling develops and spatial performances before and after HCb are indistinguishable. Thus, HCb does not directly induce circling in the MWM paradigm.

If circling is not a compulsive motor schema induced by the cerebellar lesion, why does it develop? By analyzing the results obtained in the MWM paradigm based on a prolonged cue phase, we can formulate a hypothesis. In this paradigm HCbed rats showed improved performances only when the platform was visible. Although hiding the platform did not worsen the performance levels already attained, it was extremely effective in stopping any further improvement. We can hypothesize that, by means of the "trick" of prolonging the cue phase, repetitive association of intra- and extramaze data allows the rats to store place information in an increasingly precise map. As details are added to the stored map, the amount of extramaze information needed to locate the platform correctly in the place paradigm

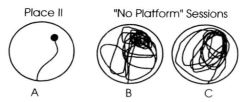

FIG. 3. Acquisition of spatial knowledge in the presence of a HCb at the end of intensive Cue training can be demonstrated by the swimming trajectories of a lesioned animal when tested without the escape platform. (A) Direct finding with the platform hidden. (B and C) Typical searching during two successive trials with the escape platform removed ("no-platform" sessions). The persistent swimming in the location previously occupied by the escape platform clearly indicates that the animal acquired platform location information.

progressively decreases and the need for peripheral swimming diminishes. Thus, in a situation in which more efficient searching strategies are prevented, circling represents a possible exploration strategy and thus becomes a behavior for acquiring spatial data.

In line with this interpretation, the peripheral circling displayed by HCbed animals represents the use of an old function, abandoned long ago by normal animals. In fact, the peripheral searching behavior mimics that exhibited in the MWM by intact developing animals. Peripheral searching characterizes specific developmental windows (17th–24th postnatal days) in which cerebellar maturation is still incomplete and after which the behavior disappears. Thus, HCb disrupts complex and sophisticated spatial abilities and induces the reappearance of rudimentary searching behaviors present in the early phases of development (Rudy et al., 1987). The disappearance of "new" complex strategies coupled with the reappearance of "old" less effective ones has been described for motor behavior following cerebellar lesions. As demonstrated elsewhere (Molinari et al., 1990; Molinari and Petrosini, 1993), adult HCbed rats display locomotor strategies of limb alternation and pivoting maneuvres to change direction, strategies present in intact animals only in early developmental stages and later abandoned.

Taken together, the results reviewed in this chapter provide evidence that cerebellar networks must be included among the neuronal structures involved in spatial data processing. In fact, HCb leads to severe impairment in coping with spatial information in all phases of MWM testing. Within the general inability to cope with the spatial MWM task, some specific features of the deficit induced by the cerebellar lesion can be detected. As previously reported, spatial knowledge of a given environment can be gained through praxic, taxic, and place strategies. Under our testing conditions, no use of praxic strategies was apparent, as no fixed path was ever noted. In addition, taxic strategies, although generally considered the easiest to use, were seldom employed. In fact, even during cue trials, HCbed rats never displayed immediate direct swimming toward the platform. The possibility that the observed behavior may be due to delayed activation of taxic strategies cannot be completely ruled out. However, according to our interpretation of circling behavior, it is likely that during circling in cue trials HCbed rats are building a spatial cognitive map. As this map develops, HCbed rats abandon circling progressively earlier and escape directly onto the platform, demonstrating the actual acquisition and use of a place strategy. The rats appears to use this strategy even when the most obvious approach would be a taxic one, i.e., in the presence of a visible platform. In all testing phases HCbed rats seem to rely essentially on place strategies that, not by chance, are certainly under hippocampal jurisdiction (Morris et al., 1982; Schenk and Morris, 1985).

V. Acquisition vs Retention

Thus, after a HCb, only place strategies are spared. Furthermore, various aspects of reported experiments (Petrosini et al., 1996) demonstrate that HCbed animals are deficient in using the procedures needed for spatial processing. HCbed animals display an almost complete lack of exploration strategies in any MWM phase. In the retention phase, which can only be solved using place strategies, HCbed rats do not display significant performance decays. In T-maze paradigm solving, which is considered to rely more on procedural components than on locale learning (Sutherland et al., 1983; Goodlett et al.,1988; Rawlings and Deacon, 1993), however, rats are more impaired than in the MWM. All successful findings of the platform are achieved without any fixed approach.

In summary, HCbed animals are impaired in executing complex and effective exploration behaviors. They can rely only on old and rather ineffective ways of acquiring spatial information that allow slow acquisition of spatial relations only when both proximal and distal cues are available. Finally, HCbed rats do not seem to be impaired in the utilization of a spatial map once it has been acquired; in fact, both preoperative training and enhanced cue training allow the rat to build a spatial map. Thus, all experimental data point to a procedural deficit that specifically impairs the acquisition phase.

VI. Cerebellar Contribution to Spatial Event Processing

Interestingly, mirrored behavior following large or selective hippocampal lesions has been described. In this case lesioned animals are able to learn a cueing procedure with a visible platform, but they are severely impaired in the place version of the task. These two opposite observations emphasize fractionation of the spatial memory system of the sort postulated by Cohen and Squire (1980). In line with the findings of Schenk and Morris (1985) obtained with entorhinal lesions, we propose that the spatial system continues to operate even in the presence of a cerebellar deficit, but it utilizes only a few "pieces" of the complex mechanism acting in normal subjects.

VII. Cerebellum and Spatial Procedure Development

These behavioral differences observed after hippocampal and cerebellar lesions can be also related to the well-known taxonomy of learning as occurring in explicit (declarative) and implicit (procedural) systems (Squire, 1992). Let us clarify the meaning of the words "declarative" and "procedural" with regard to spatial memory in experimental settings. As stated by Schenk and Morris (1985), "the representation of knowledge in a form which describes the position of the escape platform in relation to other cues in the environment" defines declarative spatial memory, whereas "the representation of stimulus–response habits necessary to guide the animal to the correct location" constitutes procedural spatial memory. Thus, according to this schema of learning systems, hippocampal lesions affect the declarative aspects of spatial memory while sparing the procedural aspects. Conversely, cerebellar lesions affect the procedural components while sparing the declarative components.

In this framework, the cerebellar ability to control procedural aspects, which are prerequisite for any cognitive demand of spatial functions, indicates that the cerebellum may act as a trigger for the progression of steps leading to more elaborate abstractions. However, the possibility that an impairment, even though milder, of the declarative aspects of spatial learning could be present cannot be dismissed. In fact, the spatial performances of HCbed rats are never comparable to those of controls, and these rats are unable to learn a new platform location when only extramaze cues are available. Thus, it is difficult to determine whether these deficits of declarative components are related to specific difficulties in applying the correct procedures, or depend on a malfunctioning of declarative aspect processing, or both.

It is also possible that when cerebellar lesions impair procedural learning, the other central noncerebellar structures, which mediate computation of place representation, might be less effective because they remain either unused or less experienced, since, in order to be called into action, cerebellar-dependent mechanisms must be bypassed.

Several authors have emphasized the role of cerebellar circuits in coordinating and timing sequential events, for example, for the motor domain (Braitenberg et al., 1997), and attempts have also been made to extend such a cerebellar property to cognitive functions (Ivry and Baldo, 1992; Silveri et al.,1994). A timing and ordering cerebellar function within the spatial data processing system can also be hypothesized. A normal animal carries out a rather long chain of events in analyzing and learning a novel environment. As first step, the animal recognizes the novelty of the environment and is motivated to explore it. For effective searching behavior to develop, all irrelevant competing responses must be inhibited. Subsequently, effective searching strategies are planned and executed. The task solution, which must convey a sufficient reward value, is found and recognized as the correct escape. From the second trial on, the environment no longer represents a complete novelty, and the animal progressively adopts more efficient searching to develop a navigational strategy as direct as possible toward the reward target. These sequential events are under the control of different areas of the brain and require precise synchronization. Functional coordination of different cognitive modules has been proposed as a possible contribution of the cerebellum to cognition (Ito, 1993) and it is conceivable that, also in the spatial domain, complex and scattered functions require the timing and sequence controlling cerebellar ability.

To conclude, we stress some aspects of the relationship between the cerebellum and the basal ganglia within the framework of the contribution of "motor structures" to cognition. Parkinsonian patients have been described as having some frontal cognitive deficits, similar to the "mild frontal lobe syndrome" reported in some studies on cerebellar patients. Although the "cognitive" pattern of patients with basal ganglia and cerebellar deficits seems to be quite distinct, it is interesting to note that some of the spatial deficits reported in experimental models of basal ganglia lesions (McDonald and White, 1994; Packard et al., 1994) are similar to those observed in HCbed rats (Petrosini et al., 1996). This correlation between basal ganglia and cerebellum is also sustained by the disynaptic connection that the deep cerebellar nuclei establish with the caudate through the anterior intralaminar nuclei (Bentivoglio et al., 1988; Minciacchi et al., 1991; Parent and Hazrati, 1995).

In this framework the existence of an egocentric navigational control system, under cerebellar and basal ganglia control, opposed to the well-known allocentric spatial relation system, under hippocampal and neocortical control, represents a stimulating working hypothesis.

Acknowledgments

This research was supported by MURST and CNR grants to MM and LP.

References

Akshoomoff, N. A., and Courchesne, E. (1992). A new role for the cerebellum in cognitive operations. *Behav. Neurosci.* **106,** 731–738.

Appollonio, I. M., Grafman, J., Schwartz, V., Massaquoi, S., and Hallett, M. (1993). Memory in patients with cerebellar degeneration. *Neurology* **43**, 1536–1544.

Bentivoglio, M., Minciacchi, D., Molinari, M., Granato, A., Spreafico, R., and Macchi, G. (1988). The intrinsic and extrinsic organization of the thalamic intralaminar nuclei. In " Cellular Thalamic Mechanisms" (M. Bentivoglio and R. Spreafico, eds.), pp. 221–237. Elsevier, Amsterdam.

Bracke-Tolkmitt, R., Linden, A., Canavan, A. G. M., Rockstroh, R., Scholz, E,. Wessel, K., and Diener, H. C. (1989). The cerebellum contributes to mental skills. *Behav. Neurosci.* **103**, 442–446.

Braitenberg, V., Heck, D., and Sultan, F. (1997). Detection and generation of sequences as a key to cerebellar function: Experiments and theory. *Behav. Brain Sci.* in press.

Brandeis, R., Brandys, Y.,and Yehuda, S. (1989). The use of the Morris water maze in the study of memory learning. *Int. J. Neurosci.* **48**, 29–69.

Cohen, N. J., and Squire, L. R. (1980). Preserved learning and retention of pattern analyzing skill in amnesia: Dissociation of knowing how and knowing that. *Science* **210**, 207–209.

Daum, I., Ackermann, H., Schugens, M. M., Reimold, C., Dichgans, J., and Birbmauer, N. (1993). The cerebellum and cognitive functions in humans. *Behav. Neurosci.* **107**, 411–419.

DiMattia, B. V., and Kesner, R. P. (1988a). Role of the posterior parietal association cortex in the processing of spatial event information. *Behav. Neurosci.* **102**, 397–403.

DiMattia, B. V., and Kesner, R. P. (1988b). Spatial cognitive maps: Differential role of parietal cortex and hippocampal formation. *Behav. Neurosci.* **102**, 471–480.

El-Awar, M., Kish, S., Oscar-Berman, M., Robitaille, Y., Schut, L., and Freedman, M. (1991). Selective delayed alternation deficits in dominantly inherited olivopontocerebellar atrophy. *Brain Cogn.* **16**, 121–129.

Fenton, A. A., and Bures, J. (1993). Place navigation in rats with unilateral tetrodotoxin inactivation of the dorsal hippocampus: Place but not procedural learning can be lateralized to one hippocampus. *Behav. Neurosci.* **107**, 552–564.

Goodlett, C. R., Hamre, K. M., and West, J. R. (1992). Dissociation of spatial navigation and visual guidance performance in Purkinje cell degeneration (pcd) mutant mice. *Behav. Brain Res.* **47**, 129–141.

Goodlett, C. R., Nonneman, A. J., Valentino, M. L., and West, J. R.(1988). Constraint on water maze spatial learning in rats: Implications for behavioral studies of brain damage and recovery of function. *Behav. Brain Res.* **28**, 275–286.

Grafman, J., Litvan, I., Massaquoi, S., Stewart, M., Sirigu, A., and Hallett, M. (1992). Cognitive planning deficit in patients with cerebellar atrophy. *Neurology* **42**, 1493–1496.

Gupta, M., Felten, D. L., and Ghetti, B. (1987). Selective loss of monoaminergic neurons in weaver mutant mice: An immunocytochemical study. *Brain Res.* **402**, 379–382.

Ito, M. (1993). Movement and thought: Identical control mechanisms by the cerebellum. *Trends Neurosci.* **16**, 448–450.

Ivry, R. B., and Baldo, J. V. (1992). Is the cerebellum involved in learning and cognition? *Curr. Opin. Neurobiol.* **2**, 212–216.

Lalonde, R. (1987). Exploration and spatial learning in staggerer mutant mice. *J. Neurogen.* **4**, 285–291.

Lalonde, R., and Botez, M. I. (1986). Navigational deficits in weaver mutant mice. *Brain Res.* **398**, 175–177.

Lalonde, R., and Botez, M. I. (1990). The cerebellum and learning processes in animals. *Brain Res. Rev.* **15**, 325–332.

Leiner, H. C., Leiner, A. L., and Dow, R. S. (1993). Cognitive and language functions of the human cerebellum. *Trends Neurosci.* **16**, 444–447.

McDonald, R. J., and White, N. M. (1994). Parallel information processing in the water maze: Evidence for independent memory systems involving dorsal striatum and hippocampus. *Behav. Neural Biol.* **61,** 260–270.

Minciacchi, D., Granato, A., Antonini, A., Sbriccoli, A., and Macchi, G. (1991). A procedure for the simultaneous visualisation of two anterograde and different retrograde fluorescent tracers: Application to the study of the afferent-efferent organisation of thalamic anterior intralaminar nuclei. *J. Neurosci. Meth.* **38,** 183–191.

Molinari, M., and Petrosini, L. (1993). Hemicerebellectomy and motor behaviour in rats. III. Kinematics of recovered spontaneous locomotion after lesions at different developmental stages. *Behav. Brain Res.* **54,** 43–55.

Molinari, M., Petrosini, L., and Dell'Anna, M. E. (1992). Interrelationships between neuroplasticity and recovery of function after cerebellar lesion in rats. *Brain Dysfunct.* **5,** 169–183.

Molinari, M., Petrosini, L., and Gremoli, T. (1990). Hemicerebellectomy and motor behaviour in rats. II. Effects of cerebellar lesion performed at different developmental stages. *Exp. Brain Res.* **82,** 483–492.

Morris, R. G. M. (1981). Spatial localization does not require the presence of local cues. *Learn. Motivat.* **12,** 239–260.

Morris, R. G. M., Garrud, P., Rawlings, J. N. P., and O'Keefe, J. (1982). Place navigation impaired in rats with hippocampal lesions. *Nature* **297,** 681–683.

Packard, M. G., Cahill, L., and Mc Gaugh J. L. (1994). Amygdala modulation of hippocampal-dependent and caudate nucleus-dependent memory processes. *Proc. Natl. Acad. Sci. USA* **91,** 8477–8481.

Parent, A., and Hazrati, L. N. (1995). Functional anatomy of the basal ganglia. II. The place of subthalamic nucleus and external pallidum in basal ganglia circuitry. *Brain Res. Rev.* **20,** 128–154.

Pascual-Leone, A., Grafman, J., Clark, K., Stewart, B. A., and Massaquoi, S. (1993). Procedural learning in Parkinson's disease and cerebellar degeneration. *Ann. Neurol.* **34,** 594–602.

Peinado-Manzano, M. A. (1990). The role of amygdala and the hippocampus in working memory for spatial and non-spatial information. *Behav. Brain Res.* **38,** 117–134.

Petrosini, L., Molinari, M., and Dell'Anna, M. E. (1997). Cerebellar contribution to spatial event processing: Morris water maze and T-maze. Submitted for publication.

Petrosini, L., Molinari, M., and Gremoli, T. (1990). Hemicerebellectomy and motor behaviour in rats. I. Development of motor function after neonatal lesion. *Exp. Brain Res.* **82,** 472–482.

Rawlings, J. N. P., and Deacon, R. M. J. (1993). Further developments of maze procedures. *In* "Behavioral Neurosciences: A Practical Approach" (A. Sahgal, ed), Vol. 1, pp. 95–106. Oxford University Press, Oxford.

Rudy, J. W., Stadler-Morris, S., and Albert, P. (1987). Ontogeny of spatial navigation behaviors in the rat: Dissociation of "proximal" and "distal" -cue-based behaviors. *Behav. Neurosci.* **101,** 62–73.

Sanes, J. N., Dimitrov, B., and Hallett, M. (1990). Motor learning in patients with cerebellar dysfunction. *Brain* **113,** 103–120.

Save, E., Buhot, M. C., Foreman, N., and Thinus-Blanc, C. (1992). Exploratory activity and response to a spatial change in rats with hippocampal or posterior parietal cortical lesions. *Behav. Brain Res.* **47,** 113–127.

Schenk, F., and Morris, R. G. M. (1985). Dissociation between components of spatial memory in rats after recovery from the effects of retrohippocampal lesions. *Exp. Brain Res.* **58,** 11–28.

Schmahmann, J. D. (1991). An emerging concept: The cerebellar contribution to higher function. *Arch. Neurol.* **48,** 1178–1187.

Schmidt, M. J., Sawyer, B. D., Perry, K. W., Fuller, R. W., Foreman, M. M., and Ghetti, B. (1982). Dopamine deficiency in the weaver mutant mouse. *J. Neurosci.* **2,** 376–380.

Silveri, M. C., Leggio, M. G., and Molinari, M. (1994). The cerebellum contributes to linguistic production: A case of agrammatic speech following a right cerebellar lesion. *Neurology* **44,** 2047–2050.

Squire, L. R. (1992). Declarative and nondeclarative memory: Multiple brain systems supporting learning and memory. *J. Cogn. Neurosci.* **4,** 232–243.

Sutherland, R. J., Whishaw, I. Q., and Kolb, B. (1983). A behavioural analysis of spatial localization following electrolytic, kainate -or colchicine-induced damage to the hippocampal formation in the rat. *Behav. Brain Res.* **7,** 133–153.

Triarhou, L. C., and Ghetti, B. (1986). Monoaminergic nerve terminals in the cerebellar cortex of Purkinje cell degeneration mutant mice: Fine structural integrity and modification of cellular environs following loss of Purkinje and granule cells. *Neuroscience* **18,** 795–807.

Triarhou, L. C., Norton, J., and Ghetti, B. (1988). Mesencephalic dopamine cell deficit involves areas A8, A9 and A10 in weaver mutant mice. *Exp. Brain Res.* **70,** 256–265.

Wallesch, C.-W., and Horn, A. (1990). Long-term effects of cerebellar pathology on cognitive functions. *Brain Cogn.* **14,** 19–25.

Whishaw, I. Q., and Tomie, J. A. (1987). Cholinergic receptor blockade produces impairments in a sensorimotor subsystem for place navigation in the rat: Evidence from sensory, motor and acquisition tests in a swimming pool. *Behav. Neurosci.* **101,** 603–616.

SECTION IV
FUNCTIONAL NEUROIMAGING STUDIES

LINGUISTIC PROCESSING

Julie A. Fiez* and Marcus E. Raichle†

*Department of Psychology, University of Pittsburgh, and the Center for the Neural Basis of Cognition, University of Pittsburgh and Carnegie Mellon University, Pittsburgh, Pennsylvania 15260 and †Department of Neurology and Neurological Surgery, and the Mallinckrodt Institute of Radiology, Division of Radiological Sciences, Washington University School of Medicine, St. Louis, Missouri 63110

I. Introduction
II. Articulatory Processes
III. Selection and Production of Verbal Responses
IV. Verbal Learning
V. Future Directions
References

I. Introduction

Since the mid-1980s, a number of functional imaging studies in normal subjects and lesion-behavior studies in patients with cerebellar damage have provided evidence that the cerebellum contributes to nonmotor aspects of language function. In most cases, these studies were not explicitly designed to study the cerebellum and language because historically there has been little evidence that the cerebellum plays any part in language perception, comprehension, or production apart from controlling the motoric aspects of articulation. As a consequence, our understanding of cerebellar contributions to language is still immature and evolving.

There are, however, several general areas which provide a foundation for understanding when, why, and how the cerebellum might be important for language. This chapter focuses on the role of the cerebellum in overt and covert articulation, the selection and production of verbal responses, and the learning of specific verbal associations. In discussing these areas, results from two different methodologies will be emphasized: functional imaging and the lesion method. Functional imaging studies using positron emission tomography (PET), and more recently functional magnetic resonance imaging (fMRI), have been important in this effort because they provide the only relatively noninvasive means of monitoring neuronal activity in the normal human cerebellum by measuring associated changes in local blood flow and blood oxygenation. Unfortunately, the field of view of most PET scanners is smaller in the axial dimension (i.e., dorsal to ventral) than the typical human brain. Thus, it is a limitation of many functional imaging studies that the cerebellum is not fully imaged (or in some cases is excluded entirely). Studies in which cerebellar activation is not found should therefore be treated cautiously, as failure may result from incompletely sampling a physiologically real blood flow change in the cerebellum.

In addressing the potential limitations of functional imaging studies, an important approach has been to evaluate the performance of patients with cerebellar atrophy and focal lesions (caused by strokes or tumors). By studying the behavior of such subjects on tasks associated with cerebellar activation in normal subjects, insight can be gained into whether the cerebellum is necessary for normal task performance. Of course, there are limitations to the lesion method as well. For instance, the locations of cerebellar damage will differ across subjects and thus variability may be introduced into the results, and compensatory mechanisms invoked by an injury (Buckner *et al.*, 1996) may obscure the role of the cerebellum in the normal brain.

This chapter focuses on results that converge across studies and methodologies. Such convergence makes it difficult to dismiss the results on the basis of the limitations specific to each methodology, or idiosyncrasies in a particular subject group. Another theme emphasized is that many language tasks which appear to involve the cerebellum share similarities to nonlanguage tasks which also involve the cerebellum. In trying to understand the contributions of the cerebellum to language, the use of labels such as "language," "motor," and "nonmotor" may create artificial boundaries between tasks; an alternative approach is to consider how a single computational algorithm may underlie a wide variety of tasks.

II. Articulatory Processes

There is general and long-standing agreement that the cerebellum is involved in speech output. Impairments in speech production associated with cerebellar damage have traditionally been viewed as disorders of articulation rather than of language processing, and hence these impairments have not typically been addressed in reviews of the nonmotor functions of the cerebellum. However, evidence now shows that the neural substrates of overt articulation overlap with those involved in covert (silent) articulation. Because covert articulation is thought to play a fundamental role in a wide variety of language tasks (e.g., see Baddeley, 1986; Besner, 1987; Liberman and Mattingly, 1985), any involvement of the cerebellum in creating or maintaining articulatory representations becomes important to understand. For this reason, this chapter begins with a brief review of the data which implicate the cerebellum in overt articulation.

Dysarthria, a disturbance in the muscular control of speech, is a common symptom of cerebellar damage (Ackermann et al., 1992; Amarenco et al., 1991b; Barth et al., 1993). Dysarthria associated with damage to the cerebellum (termed ataxic dysarthria) has a particular set of features, including slurred speech and abnormal rhythm (Brown et al., 1970; Kent et al., 1979). As the name suggests, ataxic dysarthria is often characterized as a specific subtype of the movement disorders associated with cerebellar damage. For instance, comparisons have been made between rhythmic oscillations (tremors) in both speech and posture (Ackermann and Ziegler, 1991). Dysarthria occurs most frequently following damage to superior anterior vermal and paravermal cerebellar regions within the distribution of the superior cerebellar artery (Ackermann et al., 1992; Amarenco et al., 1991a,b; Barth et al., 1993; Lechtenberg and Gilman, 1978); in contrast, patients with damage to more posterior and lateral regions limited to the territory of the posterior inferior cerebellar artery do not typically present with dysarthria (Ackermann et al., 1992; Amarenco et al., 1990; Barth et al., 1993, 1994). Some evidence also shows that dysarthria occurs more frequently following left rather than right cerebellar damage (Amarenco et al., 1991a; Lechtenberg and Gilman, 1978).

Neuroimaging studies have provided results that are in general agreement with this previous lesion work. Relatively simple speech production tasks (such as reading aloud words) produce activation in the cerebellum when compared to control tasks that do not involve a verbal output (such as passively viewing words) (Petersen et al., 1989; Petrides et al., 1993b, 1995). These activations most commonly localize to medial portions of the superior cerebellum (see Fig. 1), although in some cases additional activation has been found more laterally, particularly with slower presentation rates. (As previously noted, poor axial sampling in many studies may bias the results toward finding changes in the superior portion of the cerebellum.) Similar changes in the medial cerebellum are found when subjects merely move their mouth (Fox et al., 1985), indicating that at least some of the cerebellar changes found during word reading and repetition reflect relatively low-level motoric aspects of speech production.

Although evidence that the cerebellum contributes to overt speech can be explained in terms of the control and coordination of a motor output, growing evidence shows that the cerebellum is involved in the generation of articulatory representations which can exist independently from any measurable motor output (covert articulation). Much of the research that has focused on covert articulation has come from studies of working memory. Working memory has commonly been defined as the ability to maintain and manipulate information "on-line" (e.g., see Baddeley, 1986; Goldman-Rakic, 1994), and it is necessary to perform many common tasks, such as rehearsing a telephone number (Miller, 1956), solving a mathematical equation (Logie *et al.*, 1994), imagining the rotation of an object (Corballis and Sidey, 1993), and integrating information across sentence clauses (Just and Carpenter, 1992). Behavioral studies in both normal subjects and patient populations have provided a large body of data which suggests verbal information is maintained using an articulatory rehearsal system (e.g., see Baddeley, 1990; Colle and Welsh, 1976; Murray, 1968).

The central importance of working memory to many different aspects of cognition has stimulated a number of functional imaging studies of both verbal and nonverbal working memory. One of the first studies was reported by Paulesu *et al.* (1993) and was later replicated by the same authors using functional magnetic resonance imaging (Paulesu *et al.*, 1995). For one task in the study reported by Paulesu *et al.* (1993), subjects performed multiple trials in which five sample stimuli were presented at a rate of 1 per sec, followed by a probe stimulus after a 2-sec delay; subjects then pressed a key to indicate whether the probe stimulus matched any of the previously presented sample stimuli. Letters were used as the stimuli for a verbal version of the task, and Korean letters were used as stimuli for a nonverbal version of the task. Comparison of the two task conditions revealed that the cerebellum, left frontal operculum (at or near Brodmann areas 44 and 45), and the supplementary motor area (SMA) were significantly more active in the verbal than in the nonverbal condition. Similar results were reported by Awh *et al.* (1996) for two different verbal working memory tasks. For one task, subjects viewed four target letters presented around a central crosshair for 200 msec, followed by a probe letter after a 3-sec delay; subjects indicated whether the probe letter matched any of the target letters by pressing a mouse button. For a second task, subjects were presented with letters at a rate of one every 3 sec and pressed a mouse button in response to each letter to indicate whether the presented letter

---

FIG. 1. Cerebellar activation found during two different tasks. Four transverse cerebellar sections (taken 12 to 20 mm below a transverse plane through the anterior and posterior commissures) are shown. The top row shows the medial activation found when subjects are asked to read aloud nouns, as compared to a fixation control task. The bottom row shows the additional lateral activation found when subjects are asked to generate verbs in response to visually presented nouns, as compared to the noun-reading task (Raichle *et al.*, 1994).

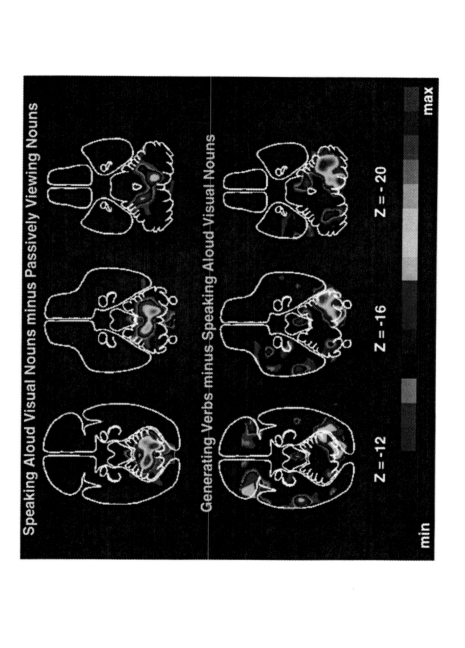

was identical to the letter presented two trials back. Significant activation was found in the cerebellum, SMA, and the left frontal operculum for both tasks when they were compared to control conditions with similar perceptual and motor components. In a third study (Fiez *et al.*, 1996), cerebellar, SMA, and frontal opercular activation was found when subjects simply attempted to remember five verbal items. The items were presented immediately prior to beginning each 40-sec scan and were recalled after each scan terminated. During each scan, the encoding, retrieval, and decision components were thus minimized and there was no measurable motor output. In contrast to these results from verbal working memory studies, cerebellar, frontal opercular, and SMA activation has not been found during object and spatial working memory tasks, except in a few instances in which the stimuli (such as faces) could be encoded verbally (Courtney *et al.*, 1996; Owen *et al.*, 1996; Petrides *et al.*, 1993a; Smith *et al.*, 1995). Figure 2 summarizes the locations of significant cerebellar activation reported across these studies.

In all of the verbal working memory studies discussed earlier, the authors concluded that the cerebellar, SMA, and frontal opercular changes were likely to form a neural substrate for the covert articulatory rehearsal of

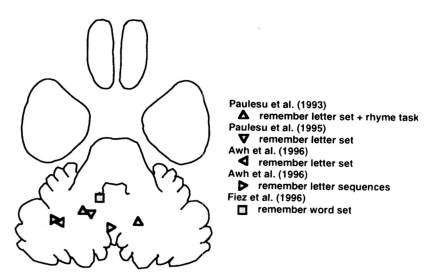

FIG. 2. Significant cerebellar activation found across verbal working memory studies. The locations of significant regions of cerebellar activation found across four different verbal working memory studies are shown by gray-filled symbols. All foci are plotted onto a transverse cerebellar section 26 mm below a transverse plane through the anterior and posterior commissures.

verbal information. However, this does not mean that the processing required for covert articulation is diffusely distributed among these three areas. On the contrary, evidence shows that the contributions made by each area are dissociable, i.e., each area performs unique operations, which contribute in specific ways to the complex phenomena of "inner speech." Support for such specificity comes in part from the fact that damage to these areas produces clearly different impairments in overt speech. For instance, damage to the cerebellum may produce impairments in speech production, but as discussed earlier, these impairments have generally been considered to be purely articulatory in nature. However, damage to the left frontal operculum is strongly associated with Broca's aphasia, a central disturbance of language function characterized by agrammatical speech and reduced fluency, but relatively preserved comprehension. Broca's aphasics, despite their reduced fluency of speech, are generally able to read and repeat single words without the types of articulatory problems associated with dysarthria (Benson, 1979; Damasio, 1992).

Further evidence about the roles of the cerebellum, SMA, and frontal operculum comes from the activation of these areas during different language tasks. Activation in the left frontal operculum has been found in both verbal working memory studies (Awh et al., 1996; Fiez et al., 1996; Paulesu et al., 1993), and studies of phonological processing (Demonet et al., 1992, 1994; Fiez et al., 1995; Zatorre et al., 1992). Importantly, these changes in frontal opercular activation during phonological tasks have been found, in several instances, without concomitant changes in the cerebellum or the SMA (Demonet et al., 1992, 1994; Zatorre et al., 1992). One interpretation of these results is that the left frontal operculum is involved in some types of high-level acoustic/articulatory processes which contribute not only to specific aspects of subvocal rehearsal, but also to a wider range of phonological, articulatory, and acoustic tasks. However, activation in the frontal operculum area has not been found during some simple speech production tasks (such as reading or repeating single words), even though strong cerebellar and SMA activation was reported (Petersen et al., 1989).

Another means of exploring the relationship between overt and covert articulation, and between different articulatory rehearsal processes, has been the investigation of verbal working memory deficits in subjects with various types of speech disorders. Although insights have been gained from these studies, for the most part they have examined subject groups with congenital speech impairments or with focal damage in a mixture of cortical locations; the results are thus of limited value in attempting to associate specific types of impairments with specific brain regions. An exception to this work, however, has come from the investigation of a small number of subjects with anarthria (mutism) caused by frontal opercular or pontine lesions (which should disrupt cortical input into the cerebellum) (Cubelli and Nichelli, 1992; Cubelli *et al.*, 1993; Sala *et al.*, 1991; Vallar and Cappa, 1987). These studies examined the effects of two different factors on anarthric subjects' word span (the number of words the subjects could store in working memory at any given time). The first factor was word length: normal subjects typically recall more short words than long words, presumably because more time is required to covertly articulate longer words (Baddeley, 1986; Baddeley *et al.*, 1975). Pontine subjects failed to show any effect of word length with either auditorily or visually presented words, whereas the frontal opercular subjects showed the normal effect of word length only with auditorily presented stimuli (Cubelli and Nichelli, 1992; Sala *et al.*, 1991).

The second factor examined was phonological similarity: normal subjects typically recall fewer words if those words all sound alike (e.g., hat, cat, pat, etc.), presumably because the words are stored as phonological representations and interference results if these representations are all similar. Although auditorily presented words are thought to gain immediate access to this "phonological store," evidence shows that visually presented words must first be recoded from orthographic to phonological representations using an articulatory process (Baddeley, 1986; Baddeley *et al.*, Vallar, 1984; Conrad and Hull, 1968). Pontine subjects showed an effect of phonological similarity with both visually and auditorily presented words. However, frontal opercular subjects showed an effect of phonological similarity only with auditorily presented stimuli (Cubelli and Nichelli, 1992; Cubelli *et al.*, Pentore, 1993; Sala *et al.*, Wynn, 1991; Vallar and Cappa, 1987). Based on the differential effects of word length and phonological similarity found in the pontine and opercular anarthic subjects, Cubelli and Nichelli (1992) concluded that pontine anarthric subjects are incapable of engaging in an articulatory rehearsal process, whereas frontal opercular anarthric subjects are incapable of recoding visual information into an articulatory form.

Taken as a whole, evidence shows that the frontal operculum, on the one hand, and the cerebellum and the SMA, on the other hand, contribute in very different ways to both overt and covert speech production tasks. Potentially, the frontal operculum may contribute to an articulatory process that is associated with specific aspects of phonological and acoustic analysis, whereas the cerebellum may contribute to an articulatory process that is more motoric in nature. Further support for this hypothesis comes from other types of motor imagery studies. In one of the earliest reports on this subject, Decety *et al.* (1990) found greater activation within the cerebellum when subjects imagined performing movements with a tennis racquet than when they merely rested with their eyes closed. In another study, Fox *et al.* (1987) asked subjects to perform actual and imagined hand movements in separate scans. Relative to a simple eyes-closed rest condition, greater activation was found in the SMA (Fox et al., 1987) and the cerebellum (unpublished data) for both the covert and the overt versions of the task. More recently, Parsons et al. (1995) reported significant cerebellar and SMA activation during a condition in which subjects attempted to discriminate between line drawings of left and right hands shown from different perspectives. Based on data from behavioral studies, it appears that subjects make these judgments by imagining they are moving their own hand to match the orientation of the stimulus hand. These results suggest that the cerebellum and the SMA are involved in the generation of both verbal and nonverbal high-level internal motor representations, in addition to their commonly accepted involvement in overt motor tasks.

A remaining question is whether the contributions of the cerebellum and the SMA to covert and overt articulation tasks are identical. Behavioral results have shown that relatively simple articulation tasks, such as repeatedly saying the word "the," significantly reduce subjects' working memory spans (Levy, 1971; Murray, 1968). This has led to the hypothesis that there are at least some processes which are common to both covert and overt articulation, as both types of tasks cannot be performed concurrently without interference. As discussed earlier, both the cerebellum and the SMA are involved in both covert and overt articulation, and thus they are the most likely neural substrates for this hypothesized common process. However, this conclusion should be treated cautiously, for several reasons. First, functionally distinct regions within the cerebellum and the SMA may be used for overt versus covert speech (Buckner et al., 1996; Picard and Strick, 1996); this issue has yet to be examined in a carefully designed study. Second, in two studies of verbal working memory, subjects were also asked to perform an articulatory rehearsal task that did not impose a significant memory load [silently count from one to five repeatedly (Fiez et al., 1996) and silently repeat a single letter (Awh et al., 1996)]. In both cases, greater SMA and cerebellar activation was found in a verbal working memory task than in the silent rehearsal task, thus indicating that the amount of cerebellar and SMA activation may be affected by the memory load imposed by a task.

III. Selection and Production of Verbal Responses

Evidence that the cerebellum contributes to covert articulation indicates that the cerebellum is involved in more than the production of overt movement. This section turns toward other verbal production tasks which suggest theories of cerebellar function need to be expanded even further.

One of the first examples of cerebellar activation during a clearly cognitive task came from a PET study of language processing reported by Petersen *et al.* (1989). As part of this study, subjects were asked to think of and say aloud appropriate verbs for presented nouns. During the control scan, subjects were asked to merely read aloud or repeat auditorily presented nouns. Unexpectedly, activation of an area within the right lateral cerebellum was found when subjects generated verbs. This activation was found with both visual and auditory presentation of the words and was clearly distinct from more medial cerebellar regions found during both verb generation and noun repetition (see Fig. 1). In addition to the cerebellar activation, a set of left-lateralized frontal cortical regions was activated during the verb generation task, but not the noun reading and repetition tasks. This correspondence (the cerebellar hemispheres receive contralateral cortical input), as well as the lack of lateral cerebellar activation during simpler verbal output tasks, made it difficult to account for the cerebellar activation on a purely motor basis (Petersen *et al.*, 1989).

Cerebellar activation during a verb generation task has now been found in a number of different studies, despite variations on the original task design reported by Petersen *et al.* (1989). Studies by Raichle *et al.* (1994) and Grabowski *et al.* (1996) are most similar to those reported by Petersen *et al.* (1989). Subjects in these studies were asked to produce verbs in response to visually presented nouns, and this task was compared to noun reading; however, each study used different (but overlapping) sets of nouns and the presentation parameters (such as duration and presentation rate) also varied. In a study by Martin *et al.* (1995), subjects were shown pictures of objects and asked to generate appropriate verbs (e.g., "peel" in response to a picture of a banana), in addition to generating verbs in response to visually presented nouns. For all but one of these verb generation studies, right lateral cerebellar activation was found when the verb generation task was compared to a simple verbal output task (see Fig. 3). It should be noted that in some cases a separate focus of activation was also found more medially (Grabowski *et al.*, 1996; Petersen *et al.*, 1989), suggesting multiple cerebellar regions may be activated by the task. However, for the purposes of this chapter, attention will focus on the right lateral response which is consistent across studies.

In interpreting these results, one issue is whether the right cerebellar activation is specific to verb generation or whether it results from a more general component of the task. This question is particularly important because dissociations in the ability of brain-damaged subjects to name pictures of actions versus certain types of objects (such as animals) have

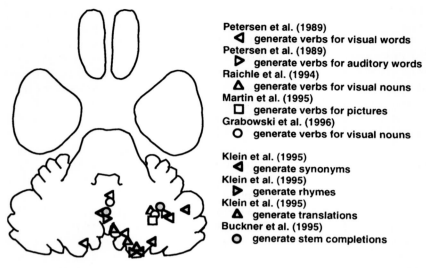

FIG. 3. Lateral cerebellar activation across verb generation tasks. The locations of significant cerebellar activation found across five different verb generation studies are shown by black-outlined symbols, and the foci of activation found in several other verb generation tasks are indicated by gray-filled symbols. In all but one case (Buckner *et al.*, 1995), the generation tasks were compared to a word-reading or object-naming task. All foci are plotted onto a transverse cerebellar section 26 mm below a transverse plane through the anterior and posterior commissures.

been reported (Damasio and Tranel, 1993; McCarthy and Warrington, 1985; Miceli *et al.*, 1988). To account for these dissociations, it has been theorized that the retrieval of word forms denoting actions involves neural systems involved in representing movement through space and time, whereas the retrieval of word forms denoting some objects involves neural systems involved in the perceptual analysis of visual objects (Damasio *et al.*, 1990; Damasio and Tranel, 1993). The results of the imaging study by Martin *et al.* (1995) bear directly on this hypothesis. When subjects generated verbs in response to pictures of objects, activation was found not only in the right lateral cerebellum, but also in the left middle temporal cortex, near an area associated with motion perception. In contrast, when subjects were asked to generate color words (e.g., say "yellow" in response to a picture of a banana), significant changes were found in the left inferior temporal cortex near an area associated with color perception, but not in the right cerebellum or the left middle temporal cortex. (Other regions, such as the left prefrontal cortex and the anterior cingulate, were activated during both word generation tasks). As discussed by Martin *et al.* (1995), these results provide support for the hypothesis that different neural systems are involved in the retrieval of different types of knowledge, and that the retrieval of verbs may specifically involve systems associated with the perception of motion (Damasio et al., 1990; Damasio and Tranel, 1993). The results are also consistent with neuroanatomical evidence that the cerebellum is interconnected with cortical areas primarily involved in visuospatial and visual motion analysis to a much greater degree than those involved in feature detection and object analysis (Schmahmann, 1991; Schmahmann and Pandya, 1995).

Results from other imaging studies, however, argue against the hypothesis that the cerebellum is only involved in generation tasks which involve some analysis of movement. First, and most importantly, other types of generation tasks produce significant cerebellar activation. In a study by Klein et al. (1995), bilingual subjects were presented, in separate scans, with English and French words and were asked to perform three different generation tasks with each word type: generate a synonym, generate a translation of the word, and generate a rhyming word. Compared to a simple word-reading task, significantly greater right cerebellar activation was found for all of the word generation conditions. In a study by Buckner et al. (1995), subjects were asked to generate complete words from three letter word stems (e.g., produce "COUGAR" in response to "COU"). Compared to a simple visual fixation condition, greater right lateral cerebellar activation was found in the stem completion task. Finally, right lateral cerebellar activation has not been found in several task conditions which required the perception or analysis of visual motion (Corbetta et al., 1991; Dupont et al., 1994).

Consistent with the results from imaging studies, subjects with cerebellar damage appear to have deficits on a variety of word retrieval and production tasks. In a study directly motivated by the results of Petersen et al. (1989), Fiez and colleagues (1992) evaluated the performance of a single subject (RC1) with a large region of right cerebellar damage caused by a stroke. RC1 was notable for his absence of any significant motor impairments and for his average to above-average performance on standard tests used to evaluate cognitive functions. When asked to generate verbs in response to nouns, RC1 produced an abnormally high number of nonverb associates for the presented nouns. For instance, whereas normal subjects usually respond to the noun "razor" with the verb "shave," RC1 said "sharp." These errors were particularly striking because they contrasted so markedly with his extremely fluent and grammatically correct conversational skills, as well as his normal performance on standard measures of language function.

In order to explore whether this deficit was specific to verb generation, RC1 was asked to perform four other generation tasks: (1) generate a category label for an exemplar, (2) generate an attribute typically associated with a presented noun, (3) generate a synonym in response to a presented word, and (4) generate a word which begins with the same initial sound as the presented word. On all of these tasks except the phoneme generation task, RC1 also produced a large number of incorrect answers. For example, when asked to generate a synonym to "garbage" he responded with "old," when asked to generate a category label for "cookie" he responded with "eat," and when asked to generate an attribute for "butterfly" he responded with "insect" (Fiez *et al.*, 1992). Other studies have explored the ability of subjects with cerebellar damage to perform word fluency tasks which share similarities to the generation tasks explored with functional imaging. For these fluency tasks, subjects are typically given a single letter, such as "F", and are asked to produce as many words as possible which begin with the target letter within some specified time interval (usually a minute). In several different studies, it has been found that subjects with cerebellar damage produce fewer words than control subjects (Akshoomoff *et al.*, 1992; Appollonio *et al.*, 1993; Leggio *et al.*, 1995). The types of errors produced in these tasks have not been reported.

Although the results from both imaging and lesion-behavior studies indicate that the right lateral cerebellum is involved in word selection and production tasks besides verb generation, a number of important issues remain. One issue is that the location of right lateral cerebellar activation found for verb generation tasks is remarkably consistent, but the location of activation found for other types of generation tasks appears more variable (see Fig. 3). It is unclear whether this variability reflects methodological factors (such as differences in subject anatomy, axial sampling, and anatomical normalization techniques used by different groups) or whether different types of generation tasks might produce activation in specific subregions within the cerebellum. Additional studies will be necessary to resolve this question definitively. A second issue arises from the fact that the right lateral cerebellum is not necessarily used any time a word must be retrieved or produced: simple reading and repetition tasks do not typically produce activation in the right lateral posterior cerebellum; subjects with cerebellar damage can produce fluent speech output; some generation tasks (such as color word generation) have failed to produce significant cerebellar activation; and following repeated practice with the same set of items even the verb generation task produces significantly less cerebellar activation (discussed later).

Some further clues about the type of verbal selection tasks which may particularly involve the cerebellum come from a more detailed analysis of the errors made by RC1 [the subject with cerebellar damage reported by Fiez *et al.* (1992)]. Except in a few instances, RC1 made no explicit comments that his responses were incorrect; using forced-choice response selection paradigms, converging evidence showed that RC1 was impaired in his ability to detect errors. Interestingly, although the incorrect responses made by RC1 across the verb, synonym, attribute, and category generation tasks were inappropriate for the given rule (e.g., in the verb generation tasks the errors were nonverb responses), they were in all cases semantic associates of the presented stimuli. Another potentially important piece of information is that RC1 made fewer errors when there was a context which constrained his response. For example, when RC1 was asked to respond to a noun not with a single verb response, but instead with a simple sentence (e.g., "You (*answer*) with it"), his production of incorrect responses decreased dramatically. One explanation for the failure of Martin *et al.* (1995) to find significant cerebellar activation in normal subjects during a color word generation task is that the task itself provides similar constraints on the possible responses. Not only are color words a more limited set of items, they are exemplars of a single conceptual category (unlike "synonyms" or "attributes"). These results may indicate that the cerebellum is particularly important when a task requires a verbal response to be selected between strongly activated associates, without other types of knowledge that can easily and quickly be implemented to constrain the possible responses. In these cases, the cerebellum may help in the implementation of "internal constraints" based on a rule not discernible from the stimuli.

An implication of this interpretation is that once again the cerebellar contributions to a language task are similar to the contributions it makes to nonverbal tasks. Studies using a variety of methods have provided evidence that the cerebellum can act to detect errors which arise when some previously learned stimulus–response association produces an error in a novel situation. For example, the vestibular–ocular reflex (VOR) helps stabilize visual images on the retina by producing compensatory eye movements in response to vestibular input caused by head movement. When an imperfect compensatory eye movement occurs, the result is a displacement of an image on the retina (retinal slip) (Ito, 1982; Miles and Lisberger, 1981). Analysis of the VOR circuitry suggests climbing fiber discharges alter Purkinje cell discharge, which in turn helps compensate for VOR errors (Stone and Lisberger, 1990).

IV. Verbal Learning

Interestingly, for a number of motor-related tasks, such as the vestibular–ocular reflex just discussed, error information is used to correct for perturbations in an ongoing motor behavior and is also associated with a learned adaptation to a consistent change in some stimulus–response relationship. This section examines evidence that the cerebellum may be involved in similar types of learning in the verbal domain, beginning once again with an examination of the verb generation task.

As might be expected, when initial performance of the verb generation task was compared to a simple noun reading or repetition task, differences were found not only in the cerebellum, but also in a number of cortical regions (Petersen *et al.*, 1989). However, a further examination of the results suggested that the two tasks do not add in a simple hierarchical fashion. In the reading and repetition conditions, activation was found in Sylvian-insular areas, in addition to motor output regions such as the primary motor cortex. In the verb generation condition, activation was found in left frontal, anterior cingulate, and right cerebellar areas, but not in Sylvian-insular areas (Fiez *et al.*, 1996; Raichle *et al.*, 1994). To explain these findings, Raichle *et al.* (1994) hypothesized (1) that subjects must use an effortful strategy which involves the cerebellum, and the prefrontal and anterior cingulate cortices, to select correct responses in the verb generation task, (2) but subjects can use a relatively automatic strategy, involving regions in Sylvian-insular cortex, to perform the noun reading task, and (3) the different strategies are associated with the activation of different regions. As a test of this hypothesis, subjects were asked to practice the task for 10 min with the same list of verbs. Following practice, it was found that subjects performed the verb generation task more automatically (their responses were quicker, more accurate, and stereotyped). These behavioral changes were correlated with striking functional changes: activation decreased in the cerebellum and prefrontal cortex, but increased in the insula. Thus, following practice, subjects appeared to shift strategies, and the regions of activation during verb generation became nearly indistinguishable from those found during noun reading (Raichle *et al.*, 1994). When subjects were given a novel list of words following this practice period, most of these changes appeared to be item specific because the pattern of activation returned to that found in the initial (naive) performance of the verb generation task.

Similar activation differences have been found for a variety of nonverbal tasks, even when these tasks have been compared to control conditions with similar sensory input and motor output. For instance, a region of left lateral cerebellar activation was found when subjects traced through a cutout maze pattern on the basis of tactile guidance, but not when they merely traced through a cut-out square pattern (van Mier *et al.*, 1995). In another study, greater cerebellar activation was found bilaterally when subjects attempted to learn a new sequence of key presses than when they pressed keys in a previously learned sequence (Jenkins et al., 1994). These tasks are obviously very different from the verb generation task and, not surprisingly, the specific locations of cortical and cerebellar activations differ across studies. However, the maze and key press tasks share some underlying behavioral similarities to the verb generation task. All three tasks are initially effortful and correct responses are self-determined (typically through trial and error), but performance on the tasks becomes more automatic with practice. For all three tasks, these behavioral changes have been associated with functional changes. In particular, cerebellar activation has been found to decrease following practice, and shifts have been found in the pattern of cortical activation.

These results suggest that there is a powerful drive to automate many different types of tasks that are performed repeatedly. The hypothesis that there are at least two different mechanisms available for the selection and production of responses has a long history in the psychological literature. William James (1890) distinguished between "ideo-motor" acts, "wherever movement follows unhesitatingly and immediately the notion of it in the mind," and "willed" acts, where "an additional conscious element in the shape of a fiat, mandate, or expressed consent" is necessary. Nearly a century later, Reason and Mycielska (1982) stated a similar concept: "we have two modes of directing our actions, rather like that in a modern aircraft. That is, we can either operate directly (analogous to having our hands on the controls), or we can switch in the automatic pilot." Many different models of skill learning and the development of automaticity have been developed (e.g., see Anderson, 1982; MacKay, 1982; Schneider and Shiffrin, 1977). Although these models differ in specifics, they contain general features that are consistent with PET results. Many of these models postulate that skilled and unskilled performances rely on different cognitive processes. Another feature is that stimulus–response pairings develop through repeated practice, so that once initiated a response can be carried through to completion without further control.

Results from behavioral studies in patients with cerebellar damage provide important evidence that the cerebellum plays a critical role in certain types of practice-dependent learning. The subject RC1 (Fiez et al., 1992) not only tended to produce atypical responses on the verb generation task, he also failed to learn the task normally with repeated exposure to the same set of items. RC1's learning of synonym, category, and attribute generation tasks was also impaired relative to normal controls. Others have also reported that cerebellar patients have difficulty learning new verbal associations. For instance, Bracke-Tolkmitt et al. (1989) found that a group of patients with cerebellar damage were significantly worse than a matched control group at learning random associations between six words and six colors.

Learning impairments in subjects with cerebellar lesions have also been found for cognitive tasks outside of the verbal domain, as would be expected on the basis of the functional imaging results. For instance, RC1's ability to perform and learn the Tower of Toronto puzzle was evaluated. For this puzzle, four colored disks are aligned on a starting peg, and the goal is to move them to one of two other pegs, with the following constraints: (1) only one disk can be moved at a time and (2) a darker disk cannot be placed on top of a lighter disk. Relative to control subjects, RC1 required significantly more moves to solve the puzzle and failed to learn the solution principles with repeated practice (Fiez et al., 1992). Using a more difficult version of a tower puzzle (the Tower of Hanoi), Grafman et al. (1992) also found deficits in the ability of subjects with cerebellar damage to solve the puzzle. Both tower tasks are conceptually similar to the pegboard task evaluated by Kim and colleagues (1994) using fMRI. In this study, greater activation was found bilaterally in the dentate nuclei when normal subjects attempted to solve a pegboard puzzle with particular rules for moving the pegs than when they only had to move the pegs from one adjacent hole to another (Kim et al., 1994).

Prior to imaging and lesion studies in human, there was already an accumulation of evidence in nonhuman primates and other animals which implicated the cerebellum in certain types of motor learning and associative conditioning. For instance, plasticity of the vestibular–occular reflex is lost, or severely reduced, following lesions of the cerebellar flocculus in rabbits and monkeys (Ito et al., 1974; Lisberger et al., 1984). Performance in a manipulandum task which requires a precise motor output against a known load is disrupted by substitution of a novel load. Adaptation to such load changes is disrupted by cerebellar lesions in humans and monkeys (Gilbert and Thach, 1977). Lesions of the interpositus nucleus appear to abolish both reacquisition and retention of a conditioned eye-blink response, while the unconditioned response (such as an eye-blink in response to an air puff) appears largely intact (Lavond et al., 1990; Yeo et al., 1985).

Results such as these have provided strong support for the Marr (1969)/Albus (1971) hypothesis that the architecture of the cerebellum would allow it to act as a modifiable pattern recognition system instrumental in the development of motor skills. One key feature of this hypothesized system was that previously learned stimulus response associations could be modified by practice, with correct responses learned through trial and error. Seen from this perspective, a cognitive task such as the verb generation task can also be viewed as similar to a motor learning task, since with practice the subjects learn to produce, quickly and accurately, a specific response to each visually presented item. This underlying similarity has led to the suggestion that the cerebellum is involved in a similar manner in learning both cognitive and motor tasks, and that the types of computations theorized by Marr (1969) and Albus (1971) should be extended beyond the motor domain (Leiner *et al.,* 1991; Thach, 1997).

V. Future Directions

This chapter placed emphasis on discussing those areas of research in which the results converge across methodologies, and similar findings have been found by different investigators. This focus was chosen to make the important point that studies which have implicated the cerebellum in aspects of language processing are reliable: the convergence of results makes it difficult to dismiss the findings on the basis of the limitations specific to each methodology or to idiosyncrasies in a particular subject group. However, it is important to note that valid criticisms about the results have been raised and that some investigators have failed to replicate previous findings (for critical reviews, see Daum and Ackermann, this volume; Filipek, 1995; Glickstein, 1993). A number of factors, such as lesion location, may account for the discrepancies in the literature.

It is also important to note that by concentrating on a relatively small number of topics, other interesting areas of research were not discussed. For instance, there have been documented instances of agrammatism (Silveri *et al.,* 1994) and mutism (Aguiar *et al.,* 1995) following cerebellar damage. Another line of research has provided evidence for specific regions of vermal hyperplasia in subjects with William's syndrome, a rare developmental disorder (Bellugi *et al.,* 1990). Subjects with William's syndrome have extremely impaired visual–spatial skills, but their linguistic skills are surprisingly preserved (Bellugi *et al.,* 1990). As research into the role of the cerebellum in language continues, it is likely that further developments in these and other areas will emerge.

Finally, although there is a growing consensus that the cerebellum contributes to some aspects of language processing, it is less clear exactly how the cerebellum does so. One approach to understanding the cerebellum is based on an analysis of different tasks to which the cerebellum contributes. As illustrated by several of the examples discussed earlier, such analyses may reveal unexpected parallels between tasks, which in turn provide insight into the specific contributions of the cerebellum. However, many questions remain. For instance, a major issue is whether the existing results can be accounted for by a single theoretical framework or whether the cerebellum performs many different functions. Results from both imaging and lesion-behavior studies suggest that different types of language tasks may involve different cerebellar regions (e.g., compare Figs. 2 and 3). However, it remains possible that the same computational algorithm within each region is performed on different types of information. Attempts to resolve issues of this type should form the basis for many interesting future studies of the cerebellar contributions to language.

References

Ackermann, H., Vogel, M., Petersen, D., and Poremba, M. (1992). Speech deficits in ischaemic cerebellar lesions. *J. Neurol.* **239,** 223–227.
Ackermann, H., and Ziegler, W. (1991). Cerebellar voice tremor: An acoustic analysis. *J. Neurol. Neurosurg. Psychiat.* **54,** 74–76.
Aguiar, P. H., Plese, J. P. P., Ciquini, O., and Marino, R. (1995). Transient mutism following a posterior fossa approach to cerebellar tumors in children: A critical review of the literature. *Child Nerv. Syst.* **11,** 306–310.
Akshoomoff, N. A., Courchesne, E., Press, G., and Iragui, V. (1992). Contribution of the cerebellum to neuropsychology functioning: Evidence from a case of cerebellar degeneration disorder. *Neuropsychologia* **30,** 315–328.
Albus, J. S. (1971). A theory of cerebellar function. *Math. Biosci.* **10,** 25–61.
Amarenco, P., Chevrie-Muller, C., Roullet, E., and Bousser, M. G. (1991a). Paravermal infarct and isolated cerebellar dysarthria. *Ann. Neurol.* **30,** 211–213.
Amarenco, P., Roullet, E., Goujon, C., Cheron, F., Hauw, J.-J., and Bousser, M.-G. (1991b). Infarction in the anterior rostral cerebellum (the territory of the lateral branch of the superior cerebellar artery). *Neurology* **41,** 253–258.
Amarenco, P., Roullet, E., Hommel, M., Chaine, P., and Marteau, R. (1990). Infarction in the territory of the medial branch of the posterior inferior cerebellar artery. *J. Neurol. Neurosurg. Psychiat.* **53,** 731–735.
Anderson, J. R. (1982). Acquisition of cognitive skill. *Psychol. Rev.* **89,** 369–406.
Appollonio, I. M., Grafman, J., Schwartz, W., Massaquoi, S., and Hallet, M. (1993). Memory in patients with cerebellar degeneration. *Neurology* **43,** 1536–1544.
Awh, E., Jonides, J., Smith, E. E., Schumacker, E. H., Koeppe, R., and Katz, S. (1996). Dissociation of storage and rehearsal in verbal working memory: Evidence from PET. *Psychol. Sci.* **7,** 25–31.
Baddeley, A. (1986). "Working Memory." Clarendon Press, Oxford.
Baddeley, A. (1990). "Human Memory: Theory and Practice." Allyn and Bacon, Needham Heights, MA.
Baddeley, A. D., Lewis, V. J., and Vallar, G. (1984). Exploring the articulatory loop. *Q. J. Exp. Psychol.* **36,** 233–252.
Baddeley, A. D., Thomson, N., and Buchanan, M. (1975). Word length and the structure of short-term memory. *J. Verb. Learn. Verb. Behav.* **14,** 575–589.
Barth, A., Bogousslavsky, J., and Regli, F. (1993). The clinical and topographic spectrum of cerebellar infarcts: A clinical-magnetic resonance imaging correlation study. *Ann. Neurol.* **33,** 451–456.

Barth, A., Bogousslavsky, J., and Regli, F. (1994). Infarcts in the territory of the lateral branch of the posterior inferior cerebellar artery. *J. Neurol. Neurosurg. Psychiat.* **57**, 1073-1076.

Bellugi, U., Bihrle, A., Jernigan, T., Trauner, D., and Doherty, S. (1990). Neuropsychological, neurological, and neuroanatomical profile of Williams syndrome. *Am. J. Med. Genet. Suppl.* **6**, 115-125.

Benson, D. F. (1979). "Aphasia, Alexia, and Agraphia." Churchill Livingstone, New York.

Besner, D. (1987). Phonology, lexical access in reading, and articulatory suppression: A critical review. *Q. J. Exp. Psychol.* **39A**, 467-478.

Bracke-Tolkmitt, R., Linden, A., Canavan, A. G. M., Rockstroh, B., Scholz, E., Wessel, K., and Diener, H. C. (1989). The cerebellum contributes to mental skills. *Behav. Neurosci.* **103**, 442-446.

Brown, J. R., Darley, F. L., and Aronson, A. E. (1970). Ataxic dysarthria. *Intl. J. Neurol.* **7**, 302-318.

Buckner, R. L., Corbetta, M., Schatz, J., Raichle, M. E., and Petersen, S. E. (1996). Preserved speech abilities and compensation following prefrontal damage. *Proc. Natl. Acad. Sci. USA* **93**, 1249-1253.

Buckner, R. L., Petersen, S. E., Ojemann, J. G., Miezin, F. M., Squire, L. R., and Raichle, M. E. (1995). Functional anatomical studies of explicit and implicit memory retrieval tasks. *J. Neurosci.* **15**, 12-29.

Buckner, R.L., Raichle, M. E., Miezin, F. M., and Petersen, S. E. (1996). Functional anatomic studies of memory retrieval for auditory words and visual pictures. *J. Neurosci.* **16**, 6219-6235.

Colle, H. A., and Welsh, A. (1976). Acoustic masking in primary memory. *J. Verb. Learn. Verb. Behav.* **15**, 17-32.

Conrad, R., and Hull, A. J. (1968). Information, acoustic confusion and memory span. *Br. J. Psychol.* **55**, 429-432.

Corballis, M. C., and Sidey, S. (1993). Effects of concurrent memory load on visual-field differences in mental rotation. *Neuropsychologia* **31**, 183-197.

Corbetta, M., Miezin, F. M., Dobmeyer, S., Shulman, G. L., and Petersen, S. E. (1991). Selective and divided attention during visual discriminations of shape, color, and speed: Functional anatomy by positron emission tomography. *J. Neurosci.* **11**, 2383-2402.

Courtney, S. M., Ungerleider, L. G., Keil, K., and Haxby, J. V. (1996). Object and spatial visual working memory activate separate neural systems in human cortex. *Cerebr. Cortex* **6**, 39-49.

Cubelli, R., and Nichelli, P. (1992). Inner speech in anarthria: Neuropsychological evidence of differential effects of cerebral lesions on subvocal articulation. *J. Clin. Exp. Neuropsychol.* **14**, 499-517.

Cubelli, R., Nichelli, P., and Pentore, R. (1993). Anarthria impairs subvocal counting. *Percept. Motor Skills* **77**, 971-978.

Damasio, A. R. (1992). Aphasia. *N. Engl. J. Med.* **326**, 531-539.

Damasio, A. R., Damasio, H., Tranel, D., and Brandt, J. P. (1990). Neural regionalization of knowledge access: Preliminary evidence. *Cold Spring Harbor Symp. Quant. Biol.* **55**, 1039-1047.

Damasio, A. R., and Tranel, D. (1993). Nouns and verbs are retrieved with differently distributed neural systems. *Proc. Natl. Acad. Sci. USA* **90**, 4957-4960.

Decety, J., Sjöholm, H., Ryding, E., Stenberg, G., and Ingvar, D. H. (1990). The cerebellum participates in mental activity: Tomographic measurements of regional cerebral blood flow. *Brain Res.* **535**, 313-317.

Demonet, J.-F., Chollet, R., Ramsay, S., Cardebat, D., Nespoulous, J.-L., Wise, R., Rascol, A., and Frackowiak, R. (1992). The anatomy of phonological and semantic processing in normal subjects. *Brain* **115**, 1753-1768.

Demonet, J.-F., Price, C., Wise, R., and Frackowiak, R. S. J. (1994). A PET study of cognitive strategies in normal subjects during language tasks: Influence of phonetic ambiguity and sequence processing on phoneme monitoring. *Brain* **117,** 671–682.

Dupont, P., Orban, G. A., De Bruyn, B., Verbruggen, A., and Mortelmans, L. (1994). Many areas in the human brain respond to visual motion. *J. Neurophysiol.* **72,** 1420–1424.

Fiez, J. A., Petersen, S. E., Cheney, M. K., and Raichle, M. E. (1992). Impaired nonmotor learning and error detection associated with cerebellar damage: A single-case study. *Brain* **115,** 155–178.

Fiez, J. A., Raichle, M. E., and Petersen, S. E. (1996). Use of positron emission tomography to identify two pathways used for verbal response selection. *In* "Developmental Dyslexia: Neural, Cognitive, and Genetic Mechanisms" (C. Chase, G. Rosen, and G. Sherman, eds.), pp. 227-258. York Press, Timonium, MD.

Fiez, J. A., Raife, E. A., Balota, D. A., Schwarz, J. P., Raichle, M. E., and Petersen, S. E. (1996). A positron emission tomography study of the short-term maintenance of verbal information. *J. Neurosci.* **16,** 808–822.

Fiez, J. A., Tallal, P., Raichle, M. E., Miezin, F. M., Katz, W. F., and Petersen, S. E. (1995). PET studies of auditory and phonological processing: Effects of stimulus characteristics and task demands. *J. Cogn. Neurosci.* **7,** 357–375.

Filipek, P. A. (1995). Quantitative magnetic resonance imaging in autism: The cerebellar vermis. *Curr. Opin. Neurol.* **8,** 134–138.

Fox, P. T., Pardo, J. V., Petersen, S. E., and Raichle, M. E. (1987). Supplementary motor and premotor responses to actual and imagined hand movements with positron emission tomography. *Soc. Neurosci. Abstracts* **13,** 1433.

Fox, P. T., Raichle, M. E., and Thach, W. T. (1985). Functional mapping of the human cerebellum with positron emission tomography. *Proc. Natl. Acad. Sci. USA* **82,** 7462–7466.

Gilbert, P. F. C., and Thach, W. T. (1977). Purkinje cell activity during motor learning. *Brain Res.* **128,** 309–328.

Glickstein, M. (1993). Motor skills but not cognitive tasks. *Trends Neurosci.* **16,** 450–451.

Goldman-Rakic, P. S. (1994). The issue of memory in the study of prefrontal function. *In* "Motor and Cognitive Functions of the Prefrontal Cortex" (A.-M. Thierry, J. Glowinski, P. S. Goldman-Rakic, and Y. Christen, eds.), pp. 112–121. Springer-Verlag, New York.

Grabowski, T. J., Frank, R. J., Brown, C. K., Damasio, H., Boles-Ponto, L. L., Watkins, G. L., and Hichwa, R. D. (1996). Reliability of PET activation across statistical methods, subject groups, and sample sizes. *Hum. Brain Mapp.* 23–46.

Grafman, J., Litvan, I., Massaquoi, S., Stewart, M., Sirigu, A., and Hallet, M. (1992). Cognitive planning deficit in patients with cerebellar atrophy. *Neurology* **42,** 1493–1496.

Ito, M. (1982). Cerebellar control of the vestibulo-ocular reflex: Around the flocculus hypothesis. *Ann. Rev. Neurosci.* **5,** 275–296.

Ito, M., Shida, T., Yagi, N., and Yamamoto, M. (1974). Visual influence on rabbit horizontal vestibulo-ocular reflex presumably effected via the cerebellar flocculus. *Brain Res.* **65,** 170–174.

James, W. (1890). "Principles of Psychology," Vol. 2. Henry-Holt and Co., New York.

Jenkins, I. H., Brooks, D. J., Nixon, P. D., Frackowiak, R. S. J., and Passingham, R. E. (1994). Motor sequence learning: A study with positron emission tomography. *J. Neurosci.* **14,** 3775–3790.

Just, M. A., and Carpenter, P. A. (1992). A capacity theory of comprehension: Individual differences in working memory. *Psychol. Rev.* **99,** 122–149.

Kent, R. D., Netsell, R., and Abbs, J. H. (1979). Acoustic characteristics of dysarthria associated with cerebellar disease. *J. Speech Hearing Res.* **22,** 627–648.

Kim, S.-G., Ugurbil, K., and Strick, P. L. (1994). Activation of a cerebellar output nucleus during cognitive processing. *Science* **265**, 949–951.

Klein, D., Milner, B., Zatorre, R. J., Meyer, E., and Evans, A. C. (1995). The neural substrates underlying word generation: A bilingual functional-imaging study. *Proc. Natl. Acad. Sci. USA* **92**, 2899–2903.

Lavond, D. G., Logan, C. G., Sohn, J. H., Garner, W. D. A., and Kanzawa, S. A. (1990). Lesions of the cerebellar interpositus nucleus abolish both nictitating membrane and eyelid EMG conditioned responses. *Brain Res.* **514**, 238–248.

Lechtenberg, R., and Gilman, S. (1978). Speech disorders in cerebellar disease. *Ann. Neurol.* **3**, 285–290.

Leggio, M. G., Solida, A., Silveri, M. C., Gainotti, G., and Molinari, M. (1995). Verbal fluency impairments in patients with cerebellar lesions. *Soc. Neurosci. Abstracts* **21**, 917.

Leiner, H. C., Leiner, A. L., and Dow, R. S. (1986). Does the cerebellum contribute to mental skills? *Behav. Neurosci.* **100**, 443–454.

Levy, B. A. (1971). The role of articulation in auditory and visual short-term memory. *J. Verb. Learn. Verb. Behav.* **10**, 123–132.

Liberman, A. M., and Mattingly, I. G. (1985). The motor theory of speech perception revised. *Cognition* **21**, 1–36.

Lisberger, S. G., Miles, F. A., and Zee, D. S. (1984). Signals used to compute errors in monkey vestibuloocular reflex: Possible role of flocculus. *J. Neurophysiol.* **52**, 1140–1153.

Logie, R. H., Gilhooly, K. J., and Wynn, V. (1994). Counting on working memory in arithmetic problem solving. *Memory Cognit.* **22**, 395–410.

MacKay, D. G. (1982). The problems of flexibility, fluency, and speed-accuracy trade-off in skilled behavior. *Psychol. Rev.* **89**, 483–506.

Marr, D. (1969). A theory of cerebellar cortex. *J. Physiol.* **202**, 437–470.

Martin, A., Haxby, J. V., Lalonde, F. M., Wiggs, C. L., and Ungerleider, L. G. (1995). Discrete cortical regions associated with knowledge of color and knowledge of action. *Science* **270**, 102–105.

McCarthy, R., and Warrington, E. K. (1985). Category specificity in an agrammatic patient: The relative impairment of verb retrieval and comprehension. *Neuropsychologia* **23**, 709–727.

Miceli, G., Silveri, M. C., Nocentini, U., and Caramazza, A. (1988). Patterns of dissociations in comprehension and production of nouns and verbs. *Aphasiology* **2**, 351–358.

Miles, F. A., and Lisberger, S. G. (1981). Plasticity in the vestibulo-ocular reflex: A new hypothesis. *Annu. Rev. Neurosci.* **4**, 273–299.

Miller, G. A. (1956). The magical number seven, plus or minus two: Some limits on our capacity for processing information. *Psychol. Rev.* **63**, 81–97.

Murray, D. J. (1968). Articulations and acoustic confusability in short-term memory. *J. Exp. Psychol.* **78**, 679–684.

Owen, A. M., Evans, A. C., and Petrides, M. (1996). Evidence for a two-stage model of spatial working memory processing within the lateral frontal cortex: A positron emission tomography study. *Cerebr. Cortex* **6**, 31–38.

Parsons, L. M., Fox, P. T., Downs, J. H., Glass, T., Hirsch, T. B., Martin, C. C., Jerabek, P. A., and Lancaster, J. L. (1995). Use of implicit motor imagery for visual shape discrmination as revealed by PET. *Science* **375**, 54–58.

Paulesu, E., Frith, C. D., and Frackowiak, R. S. J. (1993). The neural correlates of the verbal component of working memory. *Nature* **362**, 342–345.

Paulesu, E., Connelly, A., Frith, C. D., Friston, J. J., Heather, J., Meyers, R., Gadian, D. G., and Frackowiak, R. S. J. (1995). Functional MRI correlations with positron emission tomography: Initial experience using a cognitive activation paradigm on verbal working memory. *Neuroimag. Clin. North Am.* **5**, 207–225.

Petersen, S. E., Fox, P. T., Posner, M. I., Mintun, M., and Raichle, M. E. (1989). Positron emission tomographic studies of the processing of single words. *J. Cognit. Neurosci.* **1**, 153–170.

Petrides, M., Alivisatos, B., and Evans, A. C. (1995). Functional activation of the human ventrolateral frontal cortex during mnemonic retrieval of verbal information. *Proc. Natl. Acad. Sci. USA* **92**, 5803–5807.

Petrides, M., Alivisatos, B., Evans, A. C., and Meyer, E. (1993a). Dissociation of human mid-dorsolateral from posterior dorsolateral frontal cortex in memory processing. *Proc. Natl. Acad. Sci. USA* **90**, 873–877.

Petrides, M., Alivisatos, B., Meyer, E., and Evans, A. C. (1993b). Functional activation of the human frontal cortex during the performance of verbal working memory tasks. *Proc. Natl. Acad. Sci. USA* **90**, 878–882.

Picard, N., and Strick, P. L. (1996). Motor areas of the medial wall: A review of their location and functional activation. *Cerebr. Cortex* **6**, 342–353.

Raichle, M. E., Fiez, J. A., Videen, T. O., MacLeod, A. K., Pardo, J. V., Fox, P. T., and Petersen, S. E. (1994). Practice-related changes in human brain functional anatomy during nonmotor learning. *Cerebr. Cortex* **4**, 8–26.

Reason, J., and Mycielska, K. (1982). "Absent-minded? The Psychology of Mental Lapses and Everyday Errors." Prentice-Hall, Englewood Cliffs, NJ.

Sala, S. D., Logie, R. H., Marchetti, C., and Wynn, V. (1991). Case studies in working memory: A case for single cases? *Cortex* **27**, 169–191.

Schmahmann, J. D. (1991). An emerging concept: The cerebellar contribution to higher function. *Arch. Neurol.* **48**, 1178–1187.

Schmahmann, J. D., and Pandya, D. N. (1995). Prefrontal cortex projections to the basilar pons in rhesus monkey: Implications for the cerebellar contribution to higher function. *Neurosci. Lett.* **199**, 175–178.

Schneider, W., and Shiffrin, R. M. (1977). Controlled and automatic human information processing. I. Detection, search, and attention. *Psycholog. Rev.* **84**, 1–53.

Silveri, M. C., Leggio, M. G., and Molinari, M. (1994). The cerebellum contributes to linguistic production: A case of agrammatic speech following a right cerebellar lesion. *Neurology* **44**, 2047–2050.

Smith, E. E., Jonides, J., Koeppe, R. A., Awh, E., Schumacher, E. H., and Minoshima, S. (1995). Spatial versus object working memory: PET investigations. *J. Cogn. Neurosci.* **7**, 337–356.

Stone, L. S., and Lisberger, S. G. (1990). Visual responses of Purkinje cells in the cerebellar flocculus during smooth-pursuit eye movements in monkeys. II. Complex spikes. *J. Neurophysiol.* **63**, 1262–1275.

Thach, W. T. (1997). On the specific role of the cerebellum in motor learning and cognition: Clues from PET activation and lesion studies in man. *Behav. Brain Sci.,* in press.

Vallar, G., and Cappa, S. F. (1987). Articulation and verbal short-term memory: Evidence from anarthria. *Cogn. Neuropsychol.* **4**, 55–78.

van Mier, H., Tempel, L., Perlmutter, J., Raichle, M., and Petersen, S. (1995). Generalization of practice-related effects in motor learning using the dominant and non-dominant hand measured by PET. *Soc. Neurosci. Abstracts* **21**, 1441.

Yeo, C. H., Hardiman, M. J., and Glickstein, M. (1985). Classical conditioning of the nictitating membrane response of the rabbit. I. Lesions of the cerebellar nuclei. *Exp. Brain Res.* **60**, 87–98.

Zatorre, R. J., Evans, A. C., Meyer, E., and Gjedde, A. (1992). Lateralization of phonetic and pitch discrimination in speech processing. *Science* **256**, 846–849.

SENSORY AND COGNITIVE FUNCTIONS

Lawrence M. Parsons and Peter T. Fox

Research Imaging Center
University of Texas Health Science Center at San Antonio
San Antonio, Texas 78284

I. Introduction
II. Early Studies of Cerebellar Function in Cognition
III. Dissociating Perceptual/Cognitive and Somatomotor Functions within Cerebellar Regions
IV. Double Dissociation of Cerebellar Function and Motor Processing
V. Implications for Hypotheses about Cerebellar Function
References

New neuroimaging studies provide striking evidence that the cerebellum is intensely and selectively active during sensory and cognitive tasks, even in the absence of explicit or implicit motor behavior. Focal activity is observed in the lateral cerebellar hemispheres during the processing of auditory, visual, cutaneous, spatial, and tactile information, and in anterior medial cerebellar regions during somatomotor behavior. Moreover, a double dissociation exists between (a) cerebellar activity and sensory processing and (b) motor behavior and activity in known motor areas in the cerebral cortex. These findings contradict the classical motor coordination theory of cerebellar function but are predicted by, or are at least consistent with, new alternative theories.

I. Introduction

Two classical tenets, that (1) the cerebellum does not participate in mental processes such as cognition and perception and that (2) the cerebellum only serves to support fine motor control, have become increasingly challenged by results in neuropsychology and in functional neuroimaging of healthy humans. This chapter discusses challenges to these tenets from neuroimaging studies of sensory, perceptual, and cognitive processing, excluding studies of linguistic processing and skill learning which are covered elsewhere in this volume (see chapters by J. A. Fiez and M. E. Raichle and by J. Doyon). This chapter discusses some basic features of neuroimaging, reviews principal empirical data, and concludes with a discussion evaluating how these data fit current proposals of cerebellar function.

The functional neuroimaging studies discussed here employed either positron emission tomography (PET), which detects regional cerebral blood flow (rCBF) (Fox et al., 1984, 1989), or functional magnetic resonance imaging (fMRI), which is sensitive to changes in blood oxygenation (Kwong et al., 1992; Ogawa et al., 1992). Both measures can be used as indirect correlates of neural activity (Bandettini et al., 1993; Fox et al., 1988; Orrison et al., 1995; Ramsey et al., 1996; Roland, 1993; Shulman et al., 1993) and each yields data typically consistent with, or complementary to, data from both lesion-deficit and electrophysiological studies of neural function (e.g., Puce et al., 1995; Wang et al., 1996).

Results from functional neuroimaging studies, a methodology only a decade old (Posner and Raichle, 1994; Prichard and Rosen, 1994), appear to be having a considerable impact on hypotheses of cerebellar function. The impact is by virtue of increasing the variety of data implicating cerebellar function in nonmotor behaviors and reflects two important features of PET and fMRI, as compared to neurophysiology and neurology. First, neuroimaging allows for paradigms in which subjects (healthy humans) can be trained to perform more intricately controlled, and more cognitive, tasks than is possible with other animals. Second, whereas researchers in neurology and neuropsychology must wait for accidents of nature to yield "pure" cases of selective brain damage, a neuroimaging study can be designed to efficiently localize components of the neural substrate of a mental process and be conducted in a relatively short amount of time.

For various reasons, however, there are at present just a few functional neuroimaging studies examining cerebellar involvement in cognition and perception. Many studies are constrained by their imaging instruments (either PET or fMRI) to have a limited field of view of subjects' brains, often excluding all or much of the cerebellum. In addition, many studies use regions of interest (ROI) analyses in brain areas other than cerebellum and do not collect or analyze data outside of those regions. Furthermore, among those studies that have observed and analyzed activation in the cerebellum, often the researchers were not prepared to interpret unanticipated cerebellar activation and, consequently, deemphasized it in reporting their findings. Finally, in the Talairach and Tournoux (1988) atlas, which is the stereotaxic coordinate space used in nearly all published reports for standardized reference of neuroimaging activation, the cerebellum is represented in poor and imprecise anatomical detail.

Fortunately, there will likely be many more neuroimaging studies evaluating the role of the cerebellum in cognitive and perceptual tasks. New instruments with wider fields of view will soon be more widely disseminated.

Stereotaxic coordinate atlases that include more precise representations of cerebellum will soon be available (e.g., Schmahmann *et al.*, 1996). There is likely to be heightened interest in cerebellar participation in nonmotor processing as a result of publications such as the present volume. Moreover, the total number of neuroimaging studies reported is doubling biennially, and there is an increasing sophistication of experimental designs, higher resolution measurement instruments, and more powerful analysis methods (e.g., Bailey, 1997; Fox *et al.*, 1997).

II. Early Study of Cerebellar Function in Cognition

An early attempt to use neuroimaging methods to evaluate the possible role of the cerebellum in nonmotor sensory or cognitive activity was the 1994 study by Kim and colleagues. This experiment used fMRI to assess whether deep lateral cerebellar nuclei, which are the only output for the large cerebellar hemispheres, are active during cognitive activity. Healthy subjects performed two tasks that were assumed to have comparable motor activity but different extents of cognitive processing. In the control task, subjects performed a visually guided task in which they reached for and grasped a peg in a hole, then placed the peg in the next in a series of holes, eventually moving four pegs from one end of the series to the other. In the experimental task, the same subjects performed the same visually guided peg movement, but were required to move the pegs under the constraints of three rules. This task was presumed to elicit cognitive activity (i.e., problem solving) as subjects decided in which sequence the pegs should be moved in order to comply with the rules. The extent of activity imaged in a single transverse plane containing the dentate nuclei was three to four times greater during the experimental task than during the control task. Because there was comparable overt motor behavior in the two tasks, the authors concluded that the increase in dentate activity was associated with cognitive processing.

A plausible alternative interpretation of the increase in dentate activity in the Kim *et al.* (1994) study is that the increased activity may have occurred if subjects used a strategy of imagining moving their hands to rearrange the pegs as an aid in planning a sequence of peg movements. Because subjects were not allowed to move any peg backwards in the experimental task and because future moves depended on prior moves, sequences of moves had to be planned to complete the task. It is well documented in cognitive science that in solving physical problems similar to this task, subjects often imagine possible manipulations and configurations (Anderson, 1995; Gentner and Stevens, 1983; Johnson-Laird and Byrne, 1991; Shepard and Cooper, 1982). If the Kim et al. (1994) subjects used this strategy, then there would be a considerable amount of implicit motor behavior during the experimental task and none during the control. A PET study by Parsons et al. (1995) showed that when subjects imagine reaching into a visually presented target hand orientation (Parsons, 1987a, 1994), even in the absence of any motor behavior, there is strong activity in the cerebellum, including the lateral hemispheres, and in many motor areas in the cerebral cortex. [Other neuroimaging studies have also detected cerebellar activation during various imagined or observed motor performance, e.g., Decety et al. (1990, 1994) and Ryding et al. (1993).] Without data that unconfound implicit motor behavior and any other, more abstract, problem-solving cognitions that may be present, the most accurate conclusion from Kim et al. (1994) data is probably that deep cerebellar lateral nuclei are activated by imagined motor behavior. Thus, these early data do not appear to demonstrate cerebellar activation for cognitive processing unrelated to motor control.

III. Dissociating Perceptual/Cognitive and Somatomotor Functions within Cerebellar Regions

Another approach is to use an experimental design capable of yielding a regional dissociation within the cerebellum between somatomotor processing and perceptual/cognitive processing. This strategy is exemplified by three studies, each employing different experimental designs. Each study implicates posterior lateral cerebellar regions in perceptual processing (of either visual or auditory information) and implicates medial and anterior cerebellar regions in somatomotor processing.

The first of these studies (P. T. Fox, L. M. Parsons, D. A. Hodges, and J. S. Sergent, manuscript in preparation) used PET to investigate the neural substrate of the representation of music, a complex rule-based communication informed by harmonic structure, melody, rhythm, timbre, emotional expression, and poetic semantics. This experiment recorded the brain activation of expert pianists performing either music (the third movement of Bach's Italian concerto) or scales. Both performances were bimanual, executed from memory (without reading a score or seeing the keyboard or hands), and required each hand to play comparable sequences of notes. By contrasting brain states in scales and Bach, we hoped to outline the neural substrate of music representation. We found that the scales and Bach conditions both produced equally extensive bilateral activation in premotor and motor cortex and in anterior medial cerebellar areas that correspond to sensory-motor representations of each hand. More importantly, however, the laterality of activations produced by scales and Bach were dramatically different in other brain areas. Although strong multifocal activations were observed only during scales in the left auditory temporal cortex [Brodmann area (BA) 22], comparable activations were present only during Bach in the right auditory temporal cortex (BA 21). These areas of auditory cortex were the only cortical activations showing a double dissociation between hemisphere and performed task, but they mirror an observed pattern of dissociation between the cerebellum and tasks, wherein intense focal activations were observed in an intermediate lateral posterior area in the right cerebellum for scales only and in a homologous area in the left cerebellum for Bach only. Because right cerebellum has its primary connectivity with the left cerebral cortex (Carpenter, 1991), we hypothesize that activation in the right auditory cortex during Bach, is related to that in the left cerebellum. (Indeed, across eight subjects and three trials, there was a +0.77 correlation between activation in the right auditory areas and that in the left lateral cerebellar ones.) Likewise, we hypothesize that activation in the left auditory cortex during scales, is related to that in the right cerebellum (which may be an auditory cerebellar area).

Overall, these PET data suggest that lateral cerebellar areas selectively support the auditory and cognitive (but nonmotor) representations of music and scales. At least one other neuroimaging study has observed cerebellar activation (also with PET) during an auditory processing task (Griffiths *et al.*, 1994). In this task, subjects attended either to a moving sound or to a stationary sound. Subjects made no motor responses during these tasks. Cerebellar activation was present only during the moving sound task, suggesting that the cerebellum supports auditory–spatial information processing.

The second experiment using a regional dissociation approach is an fMRI study of visual selective attention and motor tasks (Allen *et al.*, 1996). Subjects performed (1) a visual selective attention condition which possessed no motor behavior component, (2) a motor behavior condition which possessed no selective visual attention, or (3) both the visual selective attention and motor response tasks together. Prior to the experiment, researchers selected two ROIs in a coronal image plane passed through the cerebellum. An anterior ROI in the right cerebellar hemisphere was deemed likely to be associated with somatomotor processing involving the right hand, and another ROI in the superior region of the posterior (left) cerebellum was assumed to be involved in visual attention. The authors' predictions were confirmed. During the motor (no selective visual attention) condition, the anterior ROI was activated but the superior posterior ROI was not. During the visual attention (no motor response) condition the superior posterior ROI was activated but the anterior ROI was not.

During the conjoint condition, both of these regions were active to about the same extent as in their separate conditions. Researchers concluded that the cerebellum is involved in selective attention operations that are anatomically differentiated from cerebellar somatomotor activations.

A similar conclusion was reached by Shulman *et al.* (1997) who performed a meta-analysis of PET investigations of various visual information processing tasks, some with motor components, some without motor components. The analysis located activated regions which generalized across tasks, while isolating other regions that were differentially sensitive to selected task components. Left medial cerebellar areas, which were active for somatomotor processes, were dissociated from right lateral cerebellar areas, which were active for nonmotor, apparently visual processing functions. The latter regions were active and unmodulated by the presence or absence of motor responses during either active or passive conditions. In addition, increases in these areas were sensitive to experimental variables that held the motor response constant. These observations provide more evidence in favor of a regional dissociation within the cerebellum between somatomotor processing in medial and anterior regions and perceptual or cognitive processing in posterior and lateral areas.

IV. Double Dissociation of Cerebellar Functions and Motor Processing

Strong evidence of cerebellar participation in perceptual and cognitive tasks without motor components is revealed by a PET study of the mental rotation of abstract objects (Parsons and Fox, 1995). Subjects discriminated between identical and mirror image pairs of simple Shepard–Metzler objects visually presented (Shepard and Metzler, 1971; Parsons, 1997). Each subject performed two tasks (see Fig. 1): (1) when the objects in a pair were separated by a rotation about their long central segment and (2) when corresponding pairs were not separated by a rotation. Subjects made only covert judgment responses. Activations specific to the mental rotation process were dissociated from activation for encoding, comparison, and judgment processes by subtracting (2) from (1). Observed activations specifically associated with mental rotation were the strongest and most extensive in the cerebellum (see Fig. 2), e.g., net cerebellar activation was 4.5 times that for parietal cortex, 2 times that for occipital cortex, and 3.5 times that for temporal cortex. Cerebellar activations in superior vermis, deep nuclei, and inferior and superior lateral areas were observed bilaterally but were twice as strong on the right. Our subjects made no gross movements, apart from eye movements, activations for which were subtracted

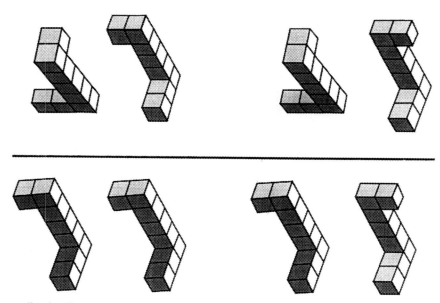

FIG. 1. (Top) Examples of the mental rotation task stimuli: A pair with an identical shape and a pair with a different (mirror image) shape. (Bottom) Examples of the shape judgment control task stimuli: A pair with an identical shape and a pair with a different (mirror image) shape. From Parsons and Fox (1995).

out by the statistical contrast with the shape judgment control task (which elicited similar eye movements). Indeed, there was no significant rotation-specific activation detected in areas known to be involved in the execution of eye movements (e.g., the frontal eye fields or the superior colliculus). In addition, strong evidence from psychophysical studies shows that observers do not perform this task by implicit motor activity such as imagining manipulating the stimulus with one's hand (Parsons, 1987a,b,c, 1994). In fact, no significant activation was detected in the other areas involved either in imagined or implicit body movement or in planning or executing motor behavior. Thus, these data show that the cerebellum can participate intensely in a cognitive and perceptual task, without the presence of motor behavior and without the participation of the known motor areas in the cerebral cortex. These findings suggest that the cerebellum supports the visual spatial processing performed during mental rotation, but they do not indicate specifically what function it performs (see discussion later).

Thus, four studies (P. T. Fox *et al.*, manuscript in preparation; Allen *et al.*, 1996; Shulman *et al.*, 1997; Parsons and Fox, 1995), particularly the latter, indicate a single dissociation between (a) cerebellar activity (specifi-

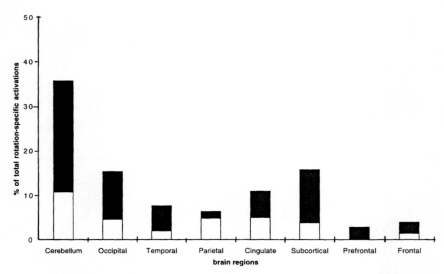

FIG. 2. The percentage of rotation-specific activation detected in a whole brain, pixel-based analysis in left (□) and right (■) cerebellum and cortical areas. From Parsons and Fox (1995).

cally, in the lateral hemispheres) and (b) motor behavior and activity in motor areas in the cerebral cortex. These studies also suggest a positive association between (c) cerebellar activity and (d) perceptual or cognitive processing.

The strongest test yet of such possibilities was conducted in an fMRI study by Gao et al. (1996). This study monitored activity in the deep lateral cerebellar (dentate) nucleus in humans performing tasks involving both passive and active sensory tasks with their fingers. The dentate nucleus provides the sole output for the large lateral cerebral hemispheres of the cerebellum and its activity is most often associated with finger movements. This experiment was designed to test the implications of the hypothesis that the cerebellum monitors and adjusts sensory acquisition to optimize processing in the rest of the brain and is not engaged by motor control per se (Bower, 1995, 1997). This hypothesis was suggested by results in electrophysiology in rat cerebellum during tactile tasks (Bower and Kassell, 1990). Four implications of this hypothesis were tested in humans: (1) dentate nuclei should respond to sensory stimuli even when there are no accompanying overt finger movements; (2) finger movements not

FIG. 3. Functional MRI (color) overlaid on anatomical MRI (gray) showing dentate activations ($P < 0.05$) for (A) cutaneous stimulations, (B) cutaneous discrimination, (C) grasp object, and (D) grasp object discrimination tasks. From Gao et al. (1996), with permission.

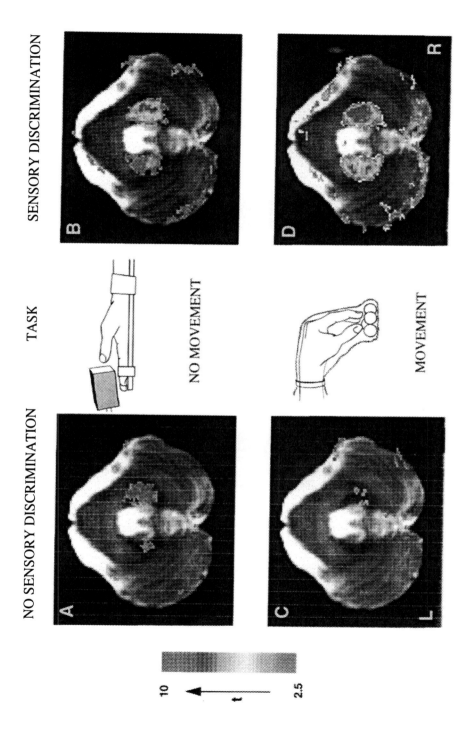

associated with tactile sensory discrimination should not induce substantial dentate activation; (3) the requirement to make a sensory discrimination with the fingers should induce an increase in dentate activation, with or without accompanying finger movements; and (4) the dentate should be most strongly activated when there is the most opportunity to modulate the acquisition of sensory data, i.e., when the sensory discrimination involves the active repositioning of tactile sensory surfaces through finger movements.

Subjects performed each of four tasks (see Fig. 3). In the cutaneous stimulation task, they experienced sandpaper rubbed against the immobilized pads of the fingers of each hand. In the cutaneous discrimination task, subjects were asked to actively compare (without making an overt response) whether the coarseness of the sandpaper on one hand matched that on the other. In the grasp objects task, they used each hand to repeatedly reach for, grasp, raise, and then drop an object (one of four irregularly shaped spheres). In the grasp objects discrimination task, they used each hand independently to repeatedly grasp one object with one hand, while using the other hand to grasp another object, and noticed (without making an overt response) whether the two objects had the same shape or not. In neither task could a subject see the objects being manipulated or discriminated. During these tasks, a transverse plane containing the dentate nucleus was imaged in each subject.

Each of the four implications just described were confirmed. The cerebellar output nuclei (dentate) showed significant task-induced blood flow increases when participants experienced cutaneous stimulation alone (Fig. 3A). Thus, there are responses in the dentate to purely sensory stimuli, i.e., those unaccompanied by overt motor behavior of the type classically associated with dentate activation. This finding confirms PET results that show cerebellar activation during hand vibration (Fox et al., 1985). When the same stimuli were presented under identical conditions, but a covert discrimination of the cutaneous stimuli was required, the dentate nuclei exhibited activation more than twice as intense and three times as extensive (Fig. 3B). The enhanced activity could be a consequence of the cerebellar connectivity with the prefrontal cortex which supports the working memory processes that may be necessary for comparing and discriminating sensory experiences (Baddeley, 1986; Goldman-Rakic, 1995; Middleton and Strick, 1994; Schmahmann and Pandya, 1995). The activations observed in these two cutaneous tasks are thus consistent with the hypothesis that the cerebellum is engaged during the acquisition and processing of sensory information and is even more strongly engaged during discrimination.

In the second pair of tasks, we compared cerebellar activation in a sensory discrimination task requiring rapid, coordinated movements of the fingers to a control task with similar finger movements but no discrimination. The control task produced very slight, statistically insignificant dentate nucleus activation (Fig. 3C). The slight activation likely reflected cutaneous stimulation of the fingers that touched the stimuli. The lack of activation confirms that rapid, coordinated, fine finger movements, in the absence of a sensory discrimination, do not detectably engage the dentate nucleus. This slight activation is consistent with the slight activation in the Kim *et al.* (1994) control condition which also required fine motor control. This finding provides evidence that the primary role of the cerebellum is not coordination of motor behavior for its own sake. By far the strongest and most extensive activation for any of the tasks studied here was observed when subjects made a covert discrimination of object shape using their fingers (Fig. 3D). The striking difference between the degree of dentate activation in the two grasping tasks provides strong evidence of cerebellar involvement in sensory discrimination. The latter finding is consistent with earlier indications of posterior cerebellar activity during tactile exploration tasks (Seitz *et al.*, 1991).

A follow-up PET study imaging the whole brain during these four tasks (L. M. Parsons, J. M. Bower, J. Xiong, and P. T. Fox, manuscript in preparation) showed that the anterior cerebellar regions were active during the grasping tasks. We predicted this activity on the basis of the hypothesized role of the cerebellum in the acquisition of proprioceptive information. In addition, we found, as expected, that these regions were equally active during the grasping tasks whether or not there was a task requirement to discriminate. In somatosensory areas of the cerebral cortex during the two grasping tasks, there was, as predicted, greater activity when there was a requirement to discriminate object shape than when there was no such requirement. There was no detectable activity in anterior cerebellar regions during the two cutaneous tasks. Thus, results from these last two studies indicate that the lateral cerebellar hemispheres are active to the degree that a sensory discrimination is required. Also, independently, the anterior cerebellar regions are active only to the extent to which there is motor behavior, and the activity is hypothesized to be due to sensory acquisition of proprioceptive information accompanying motor behavior.

The findings from these two experiments are not inconsistent with data from studies that implicate the cerebellum in motor behavior. In both neurophysiology studies of awake animals and other neuroimaging studies, the sensory and motor components of task performance have not been well dissociated. Determining whether a brain area has a motor or sensory function is a subtle problem. Motor behavior is guided by the ongoing sensory acquisition of information about the object toward which action is directed, and continuously updated sensory data are necessary for accurate, coordinated, and smooth motor behavior. The two studies just described appear to unconfound sensory and motor components in behavior. Because earlier imaging and animal studies have not decoupled sensory and motor processing, it is quite plausible that cerebellar activity in the other studies (e.g., Ellerman *et al.*, 1994) was also due to the acquisition of proprioceptive or tactile information rather than to motor behavior. Additional studies separating the influences of these two factors are necessary to resolve this issue.

The results of these studies provide the second part of a double dissociation. Data described earlier show that there is a single dissociation indicating that the cerebellum may be intensely active, without the presence of motor behavior or activity in known motor areas in the cerebral cortex. Current data (Gao *et al.*, 1996; L. M. Parsons *et al.*, manuscript in preparation) show the oppositely directed single dissociation: that fine motor behavior can occur without the participation of the lateral hemispheres of the cerebellum. In combination then, there is double dissociation between (1) cerebellar processing and sensory processing and (2) motor behavior and cerebral cortical motor system activity. Furthermore, these data strongly implicate the cerebellum in sensory acquisition.

V. Implications for Hypotheses about Cerebellar Function

To summarize, the neuroimaging results reviewed in this chapter indicate that regions in the posterior lateral cerebellum are active selectively during tasks involving auditory, auditory–spatial, visual, visual–spatial, cutaneous, and tactile–spatial information processing. The results also indicate that anterior and medial cerebellar areas are active during explicit and implicit somatomotor behaviors. Other neuroimaging studies, which have not been reviewed, report cerebellar activation in support of other cognitive processes, such as semantic association (Martin *et al.*, 1995; Mellet *et al.*, 1995; Petersen *et al.*, 1989), attention (Le and Hu, 1996), working memory (Desmond *et al.*, 1995; Klingberg *et al.*, 1995), and verbal learning and memory (Andreasen *et al.*, 1995; Grasby *et al.*, 1993). There are also initial indications from neuroimaging studies that vermal cerebellar activation is associated with emotional states such as panic, sadness, depression, and fear (Bench *et al.*, 1992; Dolan *et al.*, 1992; George *et al.*, 1995; Mayberg *et al.*, 1995; Reiman *et al.*, 1989). It is likely that these mental conditions have few or no motor components but possess various sensory and cognitive components (Ekman and Davidson, 1994).

Thus, neuroimaging data showing cerebellar activation during nonmotor processing are quite numerous and varied. However, because the findings do not indicate precisely what function(s) the cerebellum performs, we shall consider the data reviewed here in light of the various current hypotheses about cerebellar function (excluding those pertaining to linguistic processing and skill learning, which are discussed elsewhere in this volume).

These hypotheses can be very briefly described as follows. (1) The cerebellum is dedicated only to fine motor control (e.g., Ghez, 1991; Houk and Wise, 1995; Ito, 1984; Llinas and Sotelo, 1992; Thach *et al.*, 1992). (2) The cerebellum models dynamic change in objects formed by perceptual and cognitive processes (Ito, 1993). (3) The cerebellum tracks dynamic changes in perceived states of objects (Paulin, 1993). (4) The cerebellum performs timer functions, operating multiple, independent timer processes for each potential response being prepared for in a situation (Ivry, 1996). (5) The cerebellum binds interval-timing and sequence-coding processes to other forms of encoding and representation (Grafman, 1996). (6) The cerebellum monitors and adjusts sensory acquisition (Bower, 1997). (7) The cerebellum modulates attention by learning to predict and prepare for imminent information acquisition, analysis, or action (Courchesne *et al.*, 1994). (8) The cerebellum modulates cerebral cortical processes, damping oscillation, maintaining function steadily around a homeostatic baseline, and smoothing performance (Schmahmann, 1996).

The classical motor control theory of cerebellar function (1) is contradicted by the neuroimaging data reviewed here. The apparent falsification of that theory signals a new era in the understanding of cerebellar physiology. It is perhaps to be expected that the new alternative hypotheses (2)–(8) are still in the early stages of development, are expressed in general terms, and have common features. It is also not surprising that the hypotheses have not yet been tested by neuroimaging or other studies in ways designed to discriminate among them nor do any neuroimaging data fortuitously invalidate any of the new hypotheses. Those hypotheses that postulate cerebellar functions that occur in many different tasks, such as attention, sensory acquisition control, event timing, event sequencing, and process modulation, appear to be consistent with cerebellar activation during the sort of sensory and cognitive tasks discussed here. However, certain neuroimaging findings reviewed in this chapter seem to be more closely described by other hypotheses proposing more specialized functions. For example, the predominance of cerebellar activation during the operation of mentally rotating abstract objects, which may involve kinematic modeling, seems more closely described by hypotheses (2) and (3).

On balance, the sensory acquisition controller hypothesis (6) appears to be the most promising current alternative because of its power to predict unexpected neuroimaging results, its firm neurobiological support, and its parsimony (for details, see Bower, 1997). This proposal is an example of a hypothesis which assumes a single general function for all cerebellar regions rather than assuming multiple functions (e.g., Paulin, 1993; Bloedel, 1992). An important argument for a single general function is that the structure of the cerebellum is uniform, almost crystalline (Palay and Chan-Palay, 1974), and in general, structure is known to have a strong influence on computational function (e.g., Shepherd, 1989). This theory assumes that cerebellar circuitry performs the same function everywhere, but that the kind of information on which the computation is performed varies with regional variation in efferent and afferent projections. For example, the information operated on could vary in the sense modality involved, as it is known that the cerebellum has reciprocal projections with nearly every sensory system (Brodal, 1978, 1981; Welker, 1987).

In conclusion, we note that the scientific understanding of the principles of distributed computation is still rudimentary. As we deepen our understanding of these principles, hypotheses about cerebellar function are likely to be substantially different in form, and to have quite different supporting arguments, than at present. At the same time, because known facts, explicit hypotheses, and effective research strategies from cerebellar neurobiology, neuroimaging, neurology, and experimental psychology are becoming better integrated, we look forward to a more comprehensive and accurate basis of knowledge about the cerebellum. This knowledge will surely have significant implications for our understanding of brain function in general and for our practice of clinical neuroscience.

References

Allen, G., Courchesne, E., Buxton, R. B., and Wong, E. C. (1996). Dissociation of attention and motor operations in the cerebellum. In "Proceedings of the Third Annual meeting of the Cognitive Neuroscience Society," p. 28.
Anderson, J. R. (1995). "Cognitive Psychology and Its Implications," 4th Ed. Freeman, New York.
Andreasen, N. C., O'Leary, D. S., Arndt, S., Cizadlo, T., Hurtig, R., Rezai, K., Watkins, G. L. Ponto, L. L., and Hichwa, R. D. (1995). Short-term and long-term verbal memory: A positron emission tomography study. Proc. Nat. Acad. Sci. USA **92,** 5111–5115.
Baddeley, A. (1986). "Working Memory." Oxford University Press, New York.
Bailey, D. (1997). Recent trends in PET camera designs. In "PET: Critical Assessment of Recent Trends" (B. Gulyas and H. W. Muller-Gartner, eds.). Kluwer Academic Press, Dordrecht, The Netherlands.

Bandettini, P. A., Jesmanowicz, A., Wong, E. C., and Hyde, J. S. (1993). Processing strategies for time-course data sets in functional MRI of the human brain. *Magn. Reson. Med.* **30,** 161–173.

Bench, C. J., Friston, K. J., Brown, R. G., Scott, L. C., Frackowiak, R. S. J., and Dolan, R. J. (1992). The anatomy of melancholia: Focal abnormalities of cerebral blood flow in major depression. *Psychol. Med.* **22,** 607–615.

Bloedel, J. (1992). Functional heterogeneity with structural homogeneity: How does the cerebellum operate? *Behav. Brain Sci.* **15,** 666–678.

Bower, J. M. (1995). The cerebellum as a sensory acquisition controller. *Hum. Brain Mapp.* **2,** 255–256.

Bower, J. M. (1997). Is the cerebellum sensory for motor's sake, or motor for sensory's sake: The view from the whiskers of a rat? *Progr. Brain Res.* **114,** 483–516.

Bower, J. M., and Kassel, J. (1990). Variability in tactile projection patterns to cerebellar folia crus-IIa of the Norway rat. *J. Comp. Neurol.* **302,** 768–778.

Brodal, A. (1981). "Neurological Anatomy in Relation to Clinical Medicine." Oxford University Press, New York.

Brodal, P. (1978). The corticopontine projection in the rhesus monkey: Origin and principles of organization. *Brain* **101,** 251–283.

Carpenter, M. B. (1991). "Core Text of Neuroanatomy," 4th Ed. Williams and Wilkins, Baltimore, MD.

Courchesne, E., Townsend, J., Akshoomoff, N. A., Saitoh, O., Yeung-Courchesne, R., Lincoln, A. J., James, H. E., Haas, R. H., Schreibman, L., and Lau, L. (1994). Impairment in shifting attention in autistic and cerebellar patients. *Behav. Neurosci.* **108,** 848–865.

Decety, J., Sjohom, H., Ryding, E., Stenberg, G., and Ingvar, D. H. (1990). The cerebellum participates in mental activity; Tomographic measurements of regional cerebral blood flow. *Brain Res.* **535,** 313–317.

Decety, J., Perani, D., Jeannerod, M., Bettinardi, V., Tadary, B., Woods, R., Mazziotta, J. C., and Fazio, F. (1994). Mapping motor representations with PET. *Nature* **371,** 600–602.

Desmond, J. E., Gabrieli, J. D. E., Ginier, B. I., Demb, J. B., Wagner, A. D., Enzmann, D. R., and Glover, G. H. (1995). A functional MRI (fMRI) study of cerebellum during motor and working memory tasks. *Soc. Neurosci. Abstr.* 1210.

Dolan, R. J., Bench, C. J., Scott, R. G., Friston, K. J., and Frackowiak, R. S. J. (1992). Regional cerebral blood flow abnormalities in depressed patients with cognitive impairment. *J. Neurol. Neurosurg. Psychiat.* **55,** 768–773.

Ekman, P., and Davidson, R. J. (1994). "The Nature of Emotion." Oxford University Press, New York.

Ellerman, J. M., Flament, D., Kim, S.-G., Fu, Q-G., Merkle, H., Ebner, T. J., Ugurbil, J. (1994). Spatial patterns of functional activation of the cerebellum investigated using high-field (4-t) MRI. *Magn. Reson. Imag. Biomed.* **7,** 63–68.

Fox, P. T., Lancaster, J. L., and Friston, K. J. (1997). "Mapping and Modeling the Human Brain." Wiley, New York.

Fox, P. T., Mintun, M. A., Raichle, M. E., and Herscovitch, P. (1984). A non-invasive approach to quantitative functional brain mapping with $H_2^{15}O$ and positron emission tomography. *J. Cerebr. Blood Flow Metab.* **4,** 329–333.

Fox, P. T., Mintun, M. A., Reiman, E. A., and Raichle, M. E. (1989). Enhanced detection of focal brain responses using intersubject averaging and distribution of subtracted PET images. *J. Cerebr. Blood Flow Metab.* **8,** 642–653.

Fox, P. T., Raichle, M. E., Mintun, M. A., and Dence, C. (1988). Nonoxidative glucose consumption during focal physiological neural activity. *Science* **241,** 462–464.

Fox, P. T., Raichle, M. E., and Thach, W. T. (1985). Functional mapping of the human cerebellum with positron emission tomography. *Proc. Natl. Acad. Sci. USA* **82,** 7462–7466.
Flourens, P. (1824). Recherches experimentales sur les proprietes et les fonctions du systeme nerveux, dans les animaux vertegres. Cervot, Paris.
Gao, J.-H., Parsons, L. M., Bower, J. M., Xiong, J., Li, J., and Fox, P. T. (1996). Cerebellum implicated in sensory acquisition and discrimination rather than motor control. *Science* **272,** 545–547.
Gentner, D., and Stevens, A. (1983). "Mental Models." Erlbaum, Mahwah, NJ.
George, M. S., Ketter, T. A., Parekh, P. I., Horwitz, B., Hersocovitch, P. and Post, R. M. (1995). Brain activity during transient sadness and happiness in healthy women. *Am. J. Psychiat.* **152,** 341–351.
Ghez, C. (1991). The cerebellum. *In* "Principles of Neural Sciences" (E. R. Kandel, J. H. Schwartz, and T. M. Jessell, eds.), 3rd Ed., pp. 626–646. Elsevier, New York.
Goldman-Rakic, P. S. (1995). Toward a circuit model of working memory and the guidance of voluntary motor action. *In* "Models of Information processing in the Basal Ganglia" (J. C. Houk, J. L. Davis, and D. G. Beiser, eds.), pp. 131–148. Massachusetts institute of Technology Press, Cambridge, MA.
Grafman, J. (1996). Does the cerebellum contribute to complex cognitive processing? *In* "Symposium at the Annual Meeting of the International Society for Behavioral Neuroscience." Albequerque, NM.
Grasby, P. M., Firth, C. D., Friston, K. J., Bench, C., Frackowiak, R. S. J., and Dolan, R. J. (1993). Functional mapping of brain areas implicated in auditory-verbal function. *Brain* **116,** 1–20.
Griffiths, T. D., Bench, J. C., and Frackowiak, R. S. (1994). Human cortical areas selectively activated by apparent sound movement. *Curr. Biol.* **4,** 892–895.
Houk, J. C., and Wise, S. P. (1995). Distributed modular architectures linking basal ganglia, cerebellum, and cerebral-cortex: Their role in planning and controlling action. *Cereb. Cortex* **5,** 95–110.
Ito, M. (1984). "The Cerebellum and Neural Control." Appleton-Century-Crofts, New York.
Ito, M. (1993). Movement and thought: Identical control mechanisms by the cerebellum. *Trends Neurosci.* **16,** 448–450.
Ivry, R. B. (1996). The timing hypothesis. *In* "Symposium at the Annual meeting of the International Society for Behavioral Neuroscience," Albequerque, NM.
Johnson-Laird, P. N., and Byrne, R. M. J. (1991). "Deduction." Erlbaum, Mahwah, NJ.
Kim, S.-G., Ugurbil, and Strick, P. L. (1994). Activation of a cerebellar output nucleus during cognitive processing. *Science* **265,** 949–951.
Klingberg, T., Roland, P. E., and Kawashima, R. (1995). The neural correlates of the central executive function during working memory: A PET study. *Hum. Brain Mapp. Suppl.* **1,** 414.
Kwong, K. K., Belliveau, J. W., Chesler, D. A., Goldberg, I. E., Weiskoff, R. M., Poncelet, B. P., Kennedy, D. N., Hoppel, B. E., Cohen, M. S., and Turner, R. E. (1992). Dynamic magnetic resonance imaging of the human brain activity during primary sensory stimulation. *Proc. Natl. Acad. USA* **89,** 5951–5955.
Le, T. H., and Hu, X. (1996). Involvement of the cerebellum in intramodality attention shifting. *Neuroimage* **3,** 246.
Llinas, R., and Sotelo, C. (1992). "The Cerebellum Revisited. Springer-Verlag, New York.
Martin, A., Haxby, J. V., LaLonde, F. M., Wiggs, C. L., and Ungerleider, L. G. (1995). Discrete cortical regions associated with knowledge of color and knowledge of action. *Science* **270,** 102–105.
Mayberg, H. S., Liotti, M., Jerabek, P. A., Martin, C. C., and Fox, P. T. (1995). Induced sadness: A PET model of depression. *Hum. Brain Mapp. Suppl.* **1,** 396.

Mellet, E., Crivello, F., Tzourio, N., Joliot, M., Petit, L., Laurier, L., Denis, M., and Mazoyer, B. (1995). Construction of mental images based on verbal description: Functional neuroanatomy with PET. *Hum. Brain Mapp. Suppl.* **1,** 273.

Middleton, F. A., and Strick, P. L. (1994). Anatomical evidence for cerebellar and basal ganglia involvement in higher cognitive function. *Science* **266,** 458–461.

Ogawa, S., Menon, R. S., Tank, D. W., Kim, S.-G., Merkle, H., Ellermann, J. M., and Ugurbil, K. (1992). Functional brain mapping by blood oxygenation level-dependent contrast magnetic resonance imaging: A comparison of signal characteristics with a biophysical model. *Biophys. J.* **64,** 309–397.

Orrison, W. W., Lewine, J. D., Sanders, J. A., and Hartshorne, M. F. (1995). "Functional Brain Imaging." Mosby, St. Louis, MO.

Palay, S. L., and Chan-Palay, V. (1974). "Cerebellar Cortex: Cytology and Organization." Springer, Berlin.

Parsons, L. M. (1987a). Imagined spatial transformation of one's hands and feet. *Cogn. Psychol.* **19,** 176–241.

Parsons, L. M. (1987b). Imagined spatial transformation of one's body. *J. Exp. Psychol. Gen.* **116,** 172–191.

Parsons, L. M. (1987b). Visual discrimination of abstract mirror-reflected three-dimensional objects at many orientations. *Percept. Psychophys.* **42,** 49–59.

Parsons, L. M. (1994). Temporal and kinematic properties of motor behavior reflected in mentally simulated action. *J. Exp. Psychol. Hum. Percept. Perform.* **20,** 709–730.

Parsons, L. M. (1997). New psychophysical constraints on theories of how an object's orientation and rotation are mentally represented. Submitted for publication.

Parsons, L. M., Bower, J. M., Xiong, J., and Fox, P. T. (in preparation). Neuroimaging studies of whole brain activity during active and passive sensory tasks: Evidence for sensory acquisition control functions localized within cerebellum.

Parsons, L. M., and Fox, P. T. (1995). Neural basis of mental rotation. *Soc. Neurosci. Abst.* **21,** 272.

Parsons, L. M., Fox, P. T., Downs, J. H., Glass, T., Hirsch, T. B., Martin, C. C., Jerabek, P. A., and Lancaster, J. L. (1995). Use of implicit motor imagery for visual shape discrimination as revealed by PET. *Nature* **375,** 54–59.

Paulin, M. G. (1993). The role of the cerebellum in motor control and perception. *Brain Behav. Evol.* **41,** 39–50.

Petersen, S. E., Fox, P. T., Posner, M. I., Mintun, M., and Raichle, M. E. (1989). Positron emission tomographic studies of the processing of single words. *J. Cogn. Neurosci.* **1,** 153–170.

Posner, M. I., and Raichle, M. E. (1994). "Images of Mind." Freeman, New York.

Prichard, J. W., and Rosen, B. R. (1994). Functional study of the brain by NMR. *J. Cerebr. Blood Flow Metab.* **14,** 365–372.

Puce, A., Constable, R. T., Luby, M. L., McCarthy, G., Nobre, A. C., Spencer, D. D., Gore, J. C., and Allison, T. (1995). Functional magnetic resonance imaging of sensory and motor cortex: Comparison with electrophysiological localization, *J. Neurosurg.* **83,** 262–270.

Ramsey, N. F., Kirkby, B. S., Van Gelderen, P., Berman, K. F., Duyn, J. H., Frank, J. A., Mattay, V. S., Van Horn, J. D., Esposito, G., Mooner, C. T. W., and Weinberger, D. W. (1996). Functional mapping of human sensorimotor cortex with 3D BOLD fMRI correlates highly with $H_2^{15}O$ PET rCBF. *J. Cerebr. Blood Flow Metab.* **16,** 755–764.

Reiman, E. M., Raichle, M. E., Robins, E., Mintun, M. A., Fusselman, M. J., Fox, P. T., Price, J. L., and Hackman, K. A. (1989). Neuroanatomical correlates of a lactate-induced anxiety attack. *Arch. Gen. Psychiat.* **24,** 493–500.

Roland, P. E. (1993). "Brain Activation." Wiley-Liss, New York.

Ryding, E., Decety, J., Sjohom, H., Stenberg, G., and Ingvar, D. H. (1993). Motor imagery activates the cerebellum regionally: SPECT rCBF study with ^{99m}Tc-HMPAO. *Cogn. Brain Res.* **1,** 94–99.

Schmahmann, J. D. (1996). From movement to thought: Anatomic substates of the cerebellar contribution to cognitive processing. *Hum. Brain Mapp.* **4,** 174–198.

Schmahmann, J. D., Doyon, J., Makris, N., Petrides, M., Holmes, C., Evans, A., and Kennedy, D. (1996). An MRI atlas of the human cerebellum in Talairach sapce. *Neuroimage* **3,** 122.

Schmahmann, J. D., and Pandya, D. N. (1995). Prefrontal cortex projections to the basilar pons: Implications for the cerebellar contribution to higher function. *Neurosci. Lett.* **199,** 175–178.

Seitz, R. J., Roland, P. E., Bohm, C., Greitz, T., and Stoneelander, S. (1991). Somatosensory discrimination of shape: Tactile exploration and cerebral activation. *Eur. J. Neurosci.* **3,** 481–492.

Shepard, R. N., and Cooper, L. A. (1982). "Mental Images and Their Transformations." MIT Press, Cambridge, MA.

Shepard, R. N., and Metzler, J. (1971). Mental rotation of three-dimensional objects. *Science* **171,** 701–703.

Shepherd, G. (1989). The significance of real neuron architectures for neural network simulations. *In* "Computational Neuroscience." (E. Schwartz, ed.), pp. 80–96. MIT Press, Cambridge, MA.

Shulman, G. L., Corbetta, M., Buckner, R. L., Fiez, J. A., Miezin, F. M., Raichle, M. E., and Petersen, S. E. (1997). Common blood flow changes across visual tasks. I. Increases in subcortical structures and cerebellum, but not in non-visual cortex. *J. Cogn. Neurosci.,* in press.

Shulman, R. G., Blamire, A. M., Rothmann, D. L., and McCarthy, G. (1993). Nuclear magnetic resonance imaging and spectroscopy of human brain function. *Proc. Natl. Acad. Sci. USA* **90,** 3127–3133.

Talairach, J., and Tournoux, P. (1988). "Co-planar Stereotaxtic Atlas of the Human Brain. Thieme Medical, New York.

Thach, W. T., Goodkin, H. P., and Keating, J. G. (1992). The cerebellum and the adaptive coordination of movement. *Annu. Rev. Neurosci.* **15,** 403–442.

Wang, G., Tanaka, K., and Tanifuji, M. (1996). Optical imaging of functional organization in the monkey inferotemporal cortex. *Science* **272,** 1665–1668.

Welker, W. (1987). Spatial organization of somatosensory projections to granule cell cerebellar cortex: Functional and connectional implications of fractured somatotopy. *In* "New Concepts in Cerebellar Neurobiology" (J. S. King, ed.), pp. 239–280. Liss, New York.

SKILL LEARNING

Julien Doyon

Department of Psychology, and Rehabilitation Research Group, Francois-Charon Centre,
Laval University, Quebec City, Quebec, Canada G1M 2S8

I. Introduction
II. Conceptual Framework
III. Motor and Visuomotor Skill Learning and the Cerebellum
IV. Discussion
    A. Skill-Learning Paradigms
    B. Cognitive Processes
    References

This chapter reviews recent experiments that have examined the functional neuroanatomy of motor and visuomotor skill learning using brain imaging techniques such as single photon emission computed tomography, positron emission tomography, and functional magnetic resonance imaging. Special attention has been given to the cerebral blood flow changes in the cerebellum that are associated with the acquisition of these skills, although localizations of other activated regions (cortical and subcortical) are also included. The cognitive processes involved in different skill acquisition paradigms are discussed with particular reference to the learning stages at which subjects were scanned. This approach examines the conditions that are likely to produce cerebellar activation and helps us understand the role of the cerebellum in acquiring skilled behaviors.

I. Introduction

Considerable evidence from neurophysiological studies and from lesion studies in animals and humans suggests that the cerebellum plays a crucial role in the learning of various types of motor and nonmotor abilities (for reviews see Bloedel, 1992; Fiez, 1996; Ito, 1993; Leiner *et al.*, 1986, 1993, 1995; Schmahmann, 1991, 1996; Thach *et al.*, 1992; Thompson and Krupa, 1994). In recent years, research using new quantitative measures of cerebral blood flow (CBF) with single photon emission computed tomography (SPECT) and positron emission tomography (PET) or of changes in blood oxygenation levels with functional magnetic resonance imaging (fMRI) has not only confirmed this notion, but has also extended further our understanding of the neural substrate involved in the acquisition of different skills. Despite these findings, however, the nature of the cerebellar contribution to the acquisition of skilled behaviors in humans still remains unclear, as different patterns of findings have been obtained in studies using similar experimental paradigms and the same imaging technique. For example, while some investigators have reported an increase in blood flow in the cerebellum with learning of a sequence of movements (Doyon et al., 1996), others have reported either a decrease (Friston et al., 1992; Jenkins et al., 1994) or no change of activity (Grafton et al., 1995) in this structure.

This chapter reviews experiments that have examined the functional neuroanatomy involved in the acquisition of a skill using brain imaging techniques (see Buckner and Tulving, 1995; Daum and Ackermann, 1995; Grafton, 1995; Karni, 1997 for reviews that have included some of the studies described later in the chapter). Emphasis has been placed on the results of studies that have focused on motor and visuomotor skills, as other investigations which have looked at the neural substrate involved in the incremental learning of cognitive abilities (Kim et al., 1994; Petersen et al., 1989; Raichle et al., 1994) are described elsewhere in this volume. Special attention has also been given to the CBF changes in the cerebellum that are associated with learning, although localization of other activated regions (cortical and subcortical) are reported as well. In accordance with the view of Bloedel and Bracha (this volume) that the motor–nonmotor and the motor–cognitive distinctions may not be completely adequate to categorize the functions of the cerebellum in the acquisition and performance of skilled behaviors, a more "process-oriented" approach is presented. Finally, the discussion focuses primarily on the cognitive processes involved in different skill acquisition paradigms and the learning stages at which subjects were scanned in order to try to elucidate the conditions that are apt to elicit a cerebellar contribution and to help us understand further the role of the cerebellum in skill learning.

II. Conceptual Framework

Skill learning (also called "procedural memory") refers to the capacity to acquire an ability through practice (e.g., Squire, 1992). Acquisition of such an ability can be purely implicit, as it does not require conscious recognition or retrieval of a prior event. Numerous types of skills have been assessed in experimental situations using a variety of learning paradigms. For example, motor and visuomotor skills have been tested using tasks that necessitate the learning of a repeated sequence of finger or limb movements, the ability to maintain contact between a metal stylus and a small target located on a disk that can be adjusted to rotate at different velocities (rotor pursuit test), or the capacity to draw figures through the reflection of a mirror (mirror-drawing test). Visuoperceptual skills have been measured with tests in which the subjects are required to read words that have been mirror inverted (mirror reading task) or to discriminate a visual target texture from a uniform background. Finally, cognitive skill learning has been evaluated using problem-solving paradigms such as the insanity task, in which subjects are asked to solve a pegboard puzzle following a specific set of rules. Learning on these types of tasks is usually observed through test performance and is measured by a reduction in reaction time, a decrease in the number of errors, and/or a reduction in the number of trials to reach criterion. Such changes in performance are gradual and a function of the amount of practice [although some performance gains are time dependent and require several hours during waking or sleep states to evolve, due possibly to consolidation of the learning, see Karni and Sagi (1993) and Karni et al. (1994)]. Furthermore, the slope of the learning curve is dependent on the stage at which the subjects are performing the task. Rapid changes are universally observed at the beginning of the acquisition process and are then followed by slow improvement, which may take days and sometimes weeks before the performance reaches asymptote (for a review see Karni, 1997).

Very little theoretical attention has been given to the procedural memory system (see Moscovitch et al., 1993); consequently, our understanding of the mnemonic processes involved in acquiring a skill is limited. Nevertheless, models of skill acquisition (e.g., Anderson, 1990; Fitts, 1962) have been proposed, and thus can be used to help interpret the results of imaging studies that have tried to determine the neural circuits implicated in learning skills. According to such models, the learning process would follow three steps. First, in the early stages, the subject would acquire (implicitly or explicitly) the knowledge necessary to perform a particular skill-learning task. Conscious attention and mental effort would be required at this stage. Second, with practice on the task, the subject would enter into a "composition stage" in which serially executed productions would be combined into a single production. In that phase, the subject would experience a significant speedup in the time needed to execute the task and a large reduction in the number of errors. Again, mental effort and attention would be required, although to a lesser extent. Finally, with extended practice, the subject would reach a "proceduralization phase" in which the execution of the task has become automatic, hence reducing the effort and attention to be directed to perform the skill, as well as the load on the working memory capacity. It is important to note that the amount of practice needed to progress from one stage to the other, and to achieve automaticity, can vary among subjects and skill-learning tasks.

Based on this brief review of the concepts that characterize the memory system involved in learning skills, one would expect different results in functional imaging studies to emerge depending not only the type of paradigm used, but also on the specific stage of acquisition attained at the time of scanning. Indeed, distinct patterns of activation should be anticipated when subjects are tested at the beginning of the learning process versus when they are scanned after they have achieved some level of automatization of the skill. Different profiles should also be obtained when one examines changes in CBF that are associated with the encoding (i.e., incremental acquisition) of an ability compared to when one explores blood flow activity during the retrieval of a well-learned skill. Furthermore, differences should be found when subjects are learning a skill implicitly versus when they can develop and use explicit strategies to perform the skill-learning task or have acquired explicit knowledge of the task before practice begins.

The next sections discuss the results of brain imaging studies that have looked at CBF changes using SPECT, PET, or fMRI techniques in motor-related skills in order to review (1) the type of paradigms with which a change in blood flow in the cerebellum has been observed, and (2) the nature of the learning processes that have elicited an increase or decrease in cerebellar activity.

III. Motor and Visuomotor Skill Learning and the Cerebellum

CBF changes in the cerebellar cortex and/or nuclei associated with learning have been observed in a number of motor and visuomotor skill-learning paradigms, as well as in different stages of the learning process. Lang and colleagues (1988) were the first to report a cerebellar activation that was related to the acquisition of a visuomotor skill using the SPECT Tc-99m HMPAO technique. The performance of healthy volunteer subjects was tested on two versions of a visual tracking task: one in which the movements of the target on the screen and that of the hand were reversed (i.e., when the target moved to the right, the subject had to produce a movement to the left, etc.) and another in which the target and hand movements were not inverted. In the inverted tracking condition, the subjects reduced their number of errors by 28% on average. Because no other details regarding their performance were given, it is difficult to determine the level of learning that the subjects achieved during the scanning session. When the blood flow changes in the two experimental conditions were contrasted, hence subtracting away the visual and motoric aspects involved in the task, increased activity associated with learning in the inverted condition was observed in the left cerebellum, the midfrontal gyri, the frontomedial region, including the supplementary motor area (SMA), and the right caudate and putamen. The activation in the left cerebellum involved the entire hemisphere, as more precise localization was not further defined.

In a series of well-controlled PET activation studies, Grafton and colleagues (1992, 1994) examined the functional neuroanatomy, and the role of the cerebellum in particular, of visuomotor skill learning using another type of visual tracking task: the rotor pursuit test. In the first study, Grafton *et al.* (1992) scanned a group of normal control subjects while they were learning to keep a stylus on a disk that was set to rotate at 60 rotations/min with their right arm. Periods of practice, consisting of seven trials of 20 sec each, were given after every scan. Subjects were also scanned twice in a control condition in which they were required to keep the stylus immobile in the center of the rotating disk and to follow the movement of the target with their eyes only. All subjects showed an improvement in performance on the rotor pursuit task; the mean time on target increased from 5 to 68% over the course of the four learning scans. Analyses of the functional data were carried out using three separate analysis of variance (ANOVA) with weighted comparisons of means by linear contrasts. The first examined the movement effect by comparing the mean changes in CBF during the four learning scans to those of the two control conditions. The second measured the effects of acquiring the visuomotor skill by comparing the blood flow changes seen in each of the performance scans. The third evaluated the time effect by comparing the change of activity between the two control scans. Simple motor execution of the task activated several cortical (M1, SMA) and subcortical areas (putamen, substantia nigra), as well as the middle and right parasagittal zones of the cerebellum extending from anterior lobe to the inferior vermal region. In contrast, learning of the skill produced changes in a subset (M1, SMA, and pulvinar) of these motor-related structures, but not in the cerebellum. The authors acknowledged that the lack of significant CBF change in the cerebellum may reflect the fact that the limited field of view of the scanner precluded full visualization of the inferior portion of the cerebellar hemispheres that are thought to be important in learning a skill (Leiner *et al.*, 1993). Grafton and colleagues (1992) also suggested that the apparent inactivity in the cerebellum resulted from the fact that subjects were scanned while still in the early phases of the acquisition process, thus automatization of the skill had not yet been achieved.

In a second study, Grafton *et al.* (1994) repeated the experiment just reported with the rotor pursuit test in order to determine whether the cerebellum contributes to this type of visuomotor skill learning and to examine the effects of extended practice (automaticity) on this task. The learning and control conditions during scanning were identical to those of the first study, except that the subjects were tested and scanned on two separate days: (a) while the subjects were learning the rotor pursuit task (day 1) and (b) after an extensive period of practice was given such that they could achieve some level of automatization of the skill (day 2). Similar to the results of the first study, learning to execute the rotor pursuit task produced activations in motor-related areas such as the contralateral M1 and SMA bilaterally. This time, however, increased activity associated with acquisition of this skill was also seen in the ipsilateral anterior cerebellum and parasagittal vermal area, as well as in the cingulate and inferior parietal regions. Activity in the left (but not the right) anterior cerebellar region was also correlated with the speed of learning. In contrast, on day 2, increased activity related to the automaticity of the task was seen bilaterally in the parietal cortex, in the inferior left premotor area, and in the left putamen. No change in blood flow was observed in the cerebellum. The authors concluded that visuomotor skill learning is mediated by a neural network including cortical and cerebellar areas, whereas automaticity of the skill involves other structures such as the striatum.

Blood flow changes in the cerebellum have also been reported during learning of a motor maze in two studies by van Mier and colleagues (1994, 1995). In the first study, a group of right-handed control subjects was asked to move a pen continuously and as quickly as possible in a clockwise direction with their dominant hand through a cutout maze with their eyes closed. The subjects were scanned in the following experimental conditions: at rest, while performing the maze for the first time (naive condition), after having practiced the same maze for a period of 10 min (practice condition), and while tracing a new maze (novel condition). Practice-related effects produced a shift of activity from the right premotor cortex, right parietal areas, and left cerebellum to the supplementary motor cortex (SMA). When subjects traced a novel maze, right premotor, right parietal, and left cerebellar regions were activated again. Interestingly, a very similar pattern of findings (reduced CBF in right premotor, right parietal, and left cerebellum) was also observed in a second study (van Mier *et al.,* 1995), in which another group of subjects were scanned under the same experimental conditions, except that tracing of the cutout maze and square was done with the left (instead of the right) hand in a counterclockwise direction. The authors concluded that the modulation in the right cerebral and left cerebellar regions was independent of the hand used and suggested that these areas were not involved in motor performance per se, but were instead processing abstract information about the skill-learning task. They also proposed that the cerebellum is most likely important for processing the temporal aspects of the task, such as when and where to accelerate and decelerate their movements to produce the maze task in a smooth fashion.

Seitz and colleagues (1994) studied the functional contribution of the cerebellum during performance and learning of unilateral two-dimensional trajectorial movements with PET. In this study, the subjects were required either to write the letter "r" or to draw two ideograms at different heights on a digitizer board with the right hand. The scans were performed while the subjects were writing the letter as quickly or as accurately as possible, while they were drawing the ideograms for the first time, and after practice tracing the ideograms for a period of 15 min. Analyses of the blood flow data were performed by comparing each of the experimental tasks to the baseline condition in which the subjects were simply asked to keep the stylus immobile on a blank writing field and by looking at the speed at which the letters and ideograms were written during each scan. Subjects were slowest to produce the movements when drawing the ideograms for the first time; at this stage, increased activity was observed in the right dentate nucleus, cerebellar vermis, and left primary motor cortex. As their performance improved after practice, additional activations were reported in the right lateral and anterior cerebellum, as well as in the left SMA and right premotor cortex. Accurate writing of the letter "r" produced changes of rCBF in the right premotor and right parietal areas, whereas fast writing of this letter elicited a significant increase of activity in several contralateral motor regions (primary motor cortex, SMA, premotor cortex) as well as activations in the right putamen and left pontine nuclei. Importantly, however, the level of rCBF in the dentate nucleus was no longer different from that in the baseline condition. Based on their findings, Seitz *et al.* (1994) concluded that the acquisition of new trajectorial movements involves the cerebellar cortex and dentate nucleus, the latter being particularly implicated in scaling the velocity of new movements. In contrast, they proposed that overlearned movements engage the contralateral motor cortex and the premotor regions bilaterally, thus supporting the idea that the movement trajectories could be stored in both motor and premotor cortices.

Repeated sequence tests have been used in most studies of motor and visuomotor skill acquisition. Typically in these studies, subjects are asked to produce a sequence of movements that they have been taught explicitly prior to scanning (Schlaug *et al.*, 1994; Seitz *et al.*, 1990), to discover a particular sequence by trial and error (Jenkins *et al.*, 1994), or to follow the display of visual stimuli on a screen within which a repeating sequence has been embedded (Doyon *et al.*, 1996; Grafton *et al.*, 1995). The motor responses usually involve finger–thumb opposition movements or movements of the whole arm.

In 1990, Seitz and colleagues were the first to demonstrate a modulation of blood flow in the cerebellum as subjects learned a self-paced complex sequence of finger-to-thumb opposition movements with the right hand. The subjects were given explicit knowledge of the sequence of finger movements to produce before the scanning session began. They were scanned at three levels of the acquisition process (initial learning, advanced learning, and skilled performance), and the changes in blood flow in each phase were compared to those in a rest condition. The level of performance on the task was measured by assessing the speed of the movements and the number of errors. Periods of practice of the motor sequence lasting 20–40 min each were included between scans, and results of the kinematic and electromyographic analyses (Seitz and Roland, 1992) showed that the finger movements became significantly more rapid and smoother after practice than in the initial learning phase. At the beginning of learning, increased activity was found in motor-related areas, including the right anterior lobe of the cerebellum. As the subjects' ability to execute the sequence of movements improved with practice, there was an increase in blood flow in the striatum, whereas the ipsilateral increase of rCBF in the cerebellum remained stable across the different learning stages. Because the frequency of finger movements increased progressively with practice, however, the authors suggested that circuits involving both the cerebellum and the striatum play an important role in the learning of a motor sequence.

In the series of studies by Seitz and colleagues (Schlaug *et al.*, 1994; Seitz *et al.*, 1990; Seitz and Roland, 1992), subjects were required to produce the motor sequence as quickly as possible. As mentioned earlier, the frequency of finger movements increased progressively from the initial to the skilled performance, hence making it difficult to dissociate the motoric aspect of the task from learning of the sequence per se. Consequently, some investigators have used sensory–auditory guidance to ensure that the rate of finger movements would be the same for all subjects in each scan. For example, Friston *et al.* (1992) used PET to identify the cerebral structures involved in the execution of sequential finger-to-thumb opposition movements with each digit (2–5) in turn. The subjects were scanned a total of six times divided into three pairs of the following experimental conditions: (a) repetitive motor task and (b) rest. No measure of the subject's performance was recorded, thus making it difficult to determine whether there was any improvement over time. Using the statistical parametric mapping (SPM) approach, main motor activation effects were seen bilaterally in the cerebellum (right more than left) and in the left sensorimotor motor cortex. Other peaks of activation were observed in the premotor cortex (left more than right), left putamen, left lateral thalamus, and cerebellar nuclei. In contrast, attenuation of activation with practice (adaptation) was found only in the lateral cerebellar cortex and in the medial region at the level of the deep cerebellar nuclei. This adaptation mechanism was thought to reflect the consequences of long-term changes in synaptic excitability, possibly due to the phenomenon of long-term depression described by Ito (e.g., 1993), and was presented as evidence that the cerebellum is involved in the early stages of the acquisition process. This finding should be interpreted with caution however, as the interaction was due not only to the reduced rCBF level observed after practice in the right cerebellar region, but also to an upward drift of blood flow activity in the three rest conditions. Further, it should be noted that the finger-opposition sequence task used in this study was the simplest possible, hence requiring minimal learning. Thus, it is difficult to relate these findings to studies in which a more complex sequence of finger movements was required.

Jenkins and colleagues (1994) examined the functional neuroanatomy of motor skill learning using another version of the finger sequence task in which the subjects were required to learn sequences of movements by trial and error. The authors aimed at identifying the pattern of changes in blood flow that can be observed when subjects are performing a sequence that was well learned before scanning (retrieval of the prelearned condition) versus when they are learning new sequences by trial and error (encoding of the new learning condition). These two conditions were compared to a rest condition. The subjects' level of automatization in performing the well-learned sequence was tested during the last trial of the practice session using a dual-task paradigm in which the subjects had, simultaneously, to repeat strings of digits in the same order as they were presented and to execute the motor sequence. As in the study by Friston *et al.* (1992), the rate of movements in both learning conditions was kept constant using a pacing tone. Contrary to other experiments, the extent of the cerebellum was fully imaged using a new method of data acquisition in which half of the subjects were positioned low in the scanner to see the entire cerebellum, while the other half was positioned high to see the SMA and other regions of the dorsolateral prefrontal cortex. The results showed that all subjects had achieved some level of automatization of the practiced sequence, as they were capable of repeating strings that were five digits long without any error at the same time as they were performing the sequence. In general, when the prelearned condition was compared to the rest condition, peaks of activations were observed in the cerebral motor associated structures [i.e., left sensorimotor, lateral premotor, and parietal (area 40) regions, SMA, anterior cingulate, and left putamen, as well as ventral thalamus and parietal cortex (area 7) bilaterally] and in the cerebellar hemispheres bilaterally including the anterior and posterior lobes of the vermis and cerebellar nuclei. When the new learning condition was compared to rest condition, similar cortical and subcortical regions were activated. However, there were additional activations in the prefrontal cortex, whereas activations in the cerebellum were more extensive. Finally, when the blood flow activity in the prelearned condition was subtracted from that of the new learning condition (this subtraction constituting a better measure of the learning per se, without motoric confounds), activations were reported in the cerebellar vermis, cortex and nuclei bilaterally, and the prefrontal cortex (in the vicinity of areas 9, 10, and 46) bilaterally, as well as in other subcortical and cortical areas (medial thalamus, red nuclei, anterior cingulate, bilateral parietal areas 7 and 40, and bilateral premotor cortex). The authors concluded that the prefrontal cortex is engaged in the learning of a new sequence of movements, especially in a task in which the subjects are required to use problem-solving strategies to acquire the motor skill, but that this region of cortex is no longer activated when the skill has become automatic. In contrast to the prefrontal cortex, the cerebellum was significantly activated during both encoding and retrieval of a sequence of movements, and therefore Jenkins *et al.* (1994) suggested that the cerebellum plays a critical role not only in the initial learning phase of a motor sequence, but in the automatization stage as well. These findings were replicated and further extended in a series of studies (Passingham *et al.*, 1995) designed to differentiate between neural networks that could mediate some of the cognitive components elicited during the trial and error learning of a sequence of finger movements (e.g., decision making, level of attention, and mental rehearsal of the motor sequence). Indeed, this study showed that the dorsal prefrontal and anterior cingulate cortices were reactivated when subjects were asked to pay attention to the movements they had to execute in the prelearned sequence condition. This suggested that these cortical areas play a critical role in attention to action. Passingham *et al.* (1995) also noted an increase in blood flow in the caudate nucleus and cerebellum when comparing the changes in rCBF during learning of a new sequence to that of a free-selection condition. In this condition, subjects had to pay attention to and make decisions regarding the movements to be made, but they were not required to mentally rehearse or to learn any sequence. This was a more appropriate control than the rest condition used in the initial study of Jenkins *et al.* (1994). Consistent with their previous conclusions, the authors indicated that these structures are involved in the early stages of motor learning. Unfortunately, however, the activity in the free-selection condition was not subtracted from that in the prelearned condition. It was therefore not possible to determine whether the cerebellum is still critical in the automatization phase of finger sequence learning per se when confounding cognitive processes have been controlled for.

The latter issue was addressed by Doyon and colleagues (1996) who demonstrated that the cerebellum (and the striatum) is involved in the automatization stage of the acquisition process of a visuomotor skill. The authors used a computerized, touch-screen version of the repeated sequence test developed by Nissen and Bullemer (1987), which consists of a visual reaction-time task with a fixed embedded sequence of finger movements. Subjects were scanned (total of six times) in two stages of the implicit learning of a visuomotor sequence, two types of control conditions (perceptual and random sequence), and two declarative memory conditions. Responses were made with the right index finger, but also involved movements of the arm and shoulder. This design allowed the authors to identify the specific patterns of activation that are associated with the implicit acquisition of both novel (after 10 presentations of an embedded sequence) and highly trained sequences of movements (after 160 presentations of another embedded sequence) and to compare these directly with those observed when subjects are provided with explicit (declarative) knowledge of these two sequences. Also, a random sequence condition was used in which subjects were exposed to the same visual stimuli and required to give the same motor response, except that stimuli were presented at random instead of in a repeating sequence. This permitted the authors to examine the CBF changes associated with sequence acquisition per se and not motor performance. Individual trials were presented with a fixed interstimulus interval (ISI) of 800 msec so that the number of responses made during the scanning period (60 sec) was controlled. Substraction of the random condition from the highly trained condition revealed specific areas of activation in the right ventral striatum and dentate nucleus of the cerebellum (see Fig. 1). Blood flow changes in the right hemisphere were also seen in the medial posterior parietal and prestriate regions, as well as in the anterior cingulate cortex. In contrast, once the subjects had acquired explicit knowledge of the embedded sequence in the highly trained condition, increased CBF activity was observed only in the right mid-ventrolateral frontal area. These findings confirmed that both the cerebellum and the striatum are involved in the implicit acquisition of a visuomotor skill, especially in advanced stages of the learning process, whereas the ventrolateral prefrontal cortex contributes preferentially to the declarative aspect of this task. A similar pattern of findings has been reported by Rauch and colleagues (1995) who used PET to investigate the functional anatomy of both implicit and explicit skill learning. Foci of activations during early phases of implicit learning were found in motor-associated areas in the right hemisphere, including a peak in the vicinity of the right ventral striatal activation, consistent with the observations of Doyon *et al.* (1996). After subjects had acquired explicit knowledge of the sequence, changes in activity were observed in the ventrolateral prefrontal cortex and the vermal cerebellar region. No cerebellar activation was found in the implicit condition, although as the authors acknowledge (Rauch *et al.*, 1995), the limited field of view of the camera excluded the inferior portions of the cerebellum.

Grafton and colleagues (1995) further investigated the neural substrate that mediates both implicit and explicit learning of a visuomotor sequence using a version of the repeated sequence task developed by Nissen and Bullemer (1987), with or without a dual-task paradigm. Subjects underwent 12 scans during which visual stimuli either followed a six-element repeating sequence or were presented at random. Half the scans were performed while subjects were executing an attentional interference test (dual-task condition) in which they were required to monitor a stream of audible pure tones and to keep track of the low-pitched (50 Hz) tones that were presented among distracters (1000-Hz tones). None of the subjects developed awareness of the sequence; this condition permitted the identification of the structures that are involved in the implicit learning of a motor sequence. The remaining six scans were executed as the subjects performed the visuomotor skill learning task without attentional interference (single-task condition). Seven of the 12 subjects demonstrated explicit knowledge of the sequence in this condition, hence allowing identification of the cerebral structures important in this type of learning process. In both the dual- and the single-task conditions, subjects were given seven blocks of 84 trials each where stimuli were presented at random, eight blocks in which the same sequence was repeating, and two other blocks of random trials. The scans in both conditions were performed in blocks 2 and 5 (random sequence) blocks 8, 11, and 15 (repeating sequence), and on block 17

FIG. 1. Merged PET–MRI sections illustrating CBF increases averaged for all 14 subjects in the highly learned sequence minus random sequence condition. This subtraction yielded focal changes in blood flow shown as *t*-statistic images; the range is coded by the color scale. The subject's left is on the left side in these sections. The sagittal image in this figure, taken at coordinate $x = +12$ (right hemisphere), illustrates the significant CBF increases observed in the ventral striatum and the dentate nucleus of the cerebellum. Both of these peaks are also illustrated below on coronal sections: (a) ventral striatum ($y = +5$) and (b) dentate nucleus of the cerebellum ($y = -59$). Significant blood flow changes were also seen in the medial posterior parietal and prestriate regions, as well as in the anterior cingulate cortex (from Doyon *et al.*, 1996).

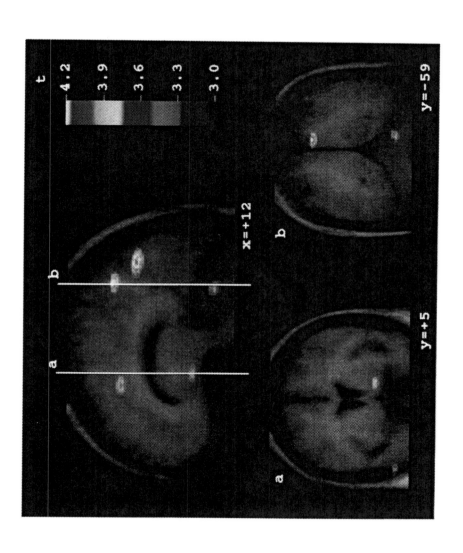

(random). The trials were presented using a fixed ISI of 1500 msec, hence controlling for the speed of movements during learning. A multivariate repeated measures analysis of variance (MANOVA) was used to look at longitudinal changes of rCBF that correlated with changes in reaction time during the three repeating sequence scans. Contrary to the results of several groups of researchers (e.g., Doyon *et al.*, 1996, Friston *et al.*, 1992; Jenkins *et al.*, 1994; Passingham *et al.*, 1995), no cerebellar activation was found in either version (implicit and explicit) of the skill-learning task. However, this discrepancy may be explained by the following reasons. First, the use of a dual-task paradigm with the repeating sequence test may constitute a unique, and fundamentally different, way to explore the functional anatomy of implicit skill learning. This may reveal an anatomical substrate quite distinct from that utilized when no interference is elicited by a distracting task. Second, such an inconsistency may be related to the differences in the learning stages at which blood flow data were acquired in the two studies. In the Grafton *et al.* (1995) experiment, subjects were still in the early stages of the implicit learning process when performing the sequence in the dual-task condition as only eight blocks of 84 trials were administered. It should be noted that no cerebellar activation was observed in the study of Doyon *et al* (1996) when the random condition was subtracted from the new learning condition (after only 10 presentations of the sequence). However, a significant peak of activity was found in the cerebellum when the random condition was subtracted from the highly learned sequence condition (i.e., after 160 presentations of the sequence, hence allowing subjects to achieve some level of automatization of the skill). Finally, failure to observe cerebellar activation may result from the fact that the blood flow changes observed in the repeating sequence scans were not compared directly to those of a random sequence, nor to those of a rest control condition. This contrasts with the study of Doyon *et al.* (1996), in which significant cerebellar activation was observed when the pattern of blood flow activity in the highly learned sequence was contrasted to that in the random sequence condition.

The role of the cerebellum in motor learning has been investigated with fMRI (4T scanner) during performance of a step-tracking task. Subjects were required to make center-out movements to eight targets displayed one at a time on a projection screen by superimposing a cursor onto the target locations using a joystick (Flament *et al.*, 1994, 1995). In the first study (Flament *et al.*, 1994), normal control subjects were tested under three different visuomotor relationships between the joystick and the cursor: (a) normal joystick/cursor relationship (control condition) where a movement from the cursor in one direction produced movement of the cursor in the same direction; (b) random changes in joystick/cursor relationship (random condition); and (c) reversed but constant joystick/cursor relationship (learning condition). In this last condition, movements from the joystick in one direction always produced movements of the cursor in the opposite direction. In general, functional data showed an inverse relation between the level of activation in the cerebellum and the subjects' performance. In the random condition, subjects did not show any improvement in their ability to move the cursor on the target and the level of cerebellar activity remained higher than in the control condition. As subjects improved their performance with practice on the learning task, however, the intensity and area of activation in both cerebellar hemispheres declined to levels that were equal (or sometimes lower) than those observed in the control task. It is interesting to note that one of the seven subjects did not demonstrate the same level of learning as the others and that the level of cerebellar activation remained high in that subject. Because reduced levels of activity were seen in the cerebellum after subjects had acquired the skill, Flament and colleagues (1994) concluded that the cerebellum does not seem to be the site of storage of the learned process. In contrast, a subsequent fMRI study by Flament and colleagues (1995) demonstrated that learning to execute the joystick task in the reversed condition is associated with increased activations in the primary motor cortex, SMA, premotor region, and lentiform nucleus, suggesting that these structures may be involved in the storage of this kind of learning.

IV. Discussion

Studies that have examined the circuitry mediating the acquisition of motor skilled behaviors in humans using brain imaging techniques were reviewed in order to assess the role that the cerebellum plays in this type of memory. More specifically, the goal of this chapter was to identify the skill-learning paradigms that have produced an activation in the cerebellum and to determine the nature of the cognitive processes during which cerebellar changes in blood flow have been elicited. This section discusses these two subjects in turn and points out some of the issues that need to be addressed in future research.

A. SKILL-LEARNING PARADIGMS

It is apparent that CBF modulations in both the cortex and/or nuclei of the cerebellum have been observed in a variety of skill-learning paradigms, suggesting that this structure is critical for the acquisition of skills in motor and visuomotor modalities. Indeed, activations in the cerebellum have been observed in almost every study of the dynamic changes in rCBF that occur with the acquisition of a skill. Some exceptions have been reported (Grafton *et al.,* 1992, 1995; Rauch *et al.,* 1995), but in two of these three studies, the limited field of view of the PET camera excluded the ventral and lateral cerebellar regions that have been shown to be activated in several types of skill-learning tasks.

It has been suggested that blood flow changes in the cerebellum are observed only in situations in which the experimental condition is more difficult than the control task (Bower, 1995). Based on their series of studies in the somatosensory system, Bower and colleagues proposed that the cerebellum "coordinates the acquisition of sensory data, rather than coordinating movements per se" (Bower, 1992; Bower and Kassel, 1990). According to this view, increased cerebellar activity should be observed in tasks requiring the greatest integration of sensory input. Furthermore, with automatization of skilled behaviors, the level of cerebellar activity should decrease because less accurate sensory processing would then be necessary. This interpretation accounts for several studies in which experimental tasks that are more demanding than the control conditions were used (e.g., Kim *et al.,* 1994; Grafton *et al.,* 1992, 1994; Raichle *et al.,* 1994), but it cannot easily account for the entire set of data reported in skill-learning studies. For example, in the experiment by Doyon and colleagues (1996) using the repeated sequence task, increased cerebellar activity was observed when subtracting the "random" from the "highly learned sequence" condition. Because the random condition is more difficult than the sequence task (as demonstrated from the reaction time data and the subjects' subjective accounts of their experience in performing the two types of tasks), this finding is not consistent with Bower's (1995) explanation of cerebellar activations seen in PET studies.

B. COGNITIVE PROCESSES

CBF changes in the cerebellum have been reported in conditions requiring a variety of cognitive processes, as well as at different phases of the acquisition process. The cerebellum was found to be involved not only in the encoding of "motor programs" necessary to execute visual tracking (Grafton *et al.,* 1994; Flament *et al.,* 1994, 1995), motor maze (van Mier *et al.,* 1994, 1995), two-dimensional trajectorial movement (Seitz *et al.,* 1994), and sequence tasks (Doyon *et al.,* 1996; Jenkins *et al.,* 1994; Passingham *et al.,* 1995; Rao *et al.,* 1995; Rauch *et al.,* 1995; Schlaug *et al.,* 1994; Seitz *et al.*, 1990), but it was also found to contribute to the retrieval of learned sequences of movements (Friston *et al.*, 1992; Jenkins *et al.*, 1994) as well as of verbal semantic information (e.g., Petersen *et al.*, 1988, 1989; Raichle *et al.*, 1994). Based on studies that used a repeating sequence paradigm, it has also been shown that the cerebellum is active during both implicit learning (Doyon *et al.*, 1996) and when subjects are practicing a motor sequence for which they have complete explicit knowledge (Friston *et al.*, 1992; Grafton *et al.*, 1995; Schlaug *et al.*, 1994; Seitz *et al.*, 1990). Other investigators have reported that the cerebellar contribution extends to skills for which subjects need to utilize problem-solving abilities to find a repeating sequence of finger movements by trial and error (Jenkins *et al.*, 1994; Passingham *et al.*, 1995) or to elucidate the solution to a complex pegboard puzzle (Kim *et al.*, 1994).

Blood flow changes in the cerebellum have also been seen at different stages of the acquisition of a skill. In fact, modulations of activity in this structure have been observed in the early stages of the learning process, i.e., during the phase in which subjects show quick improvement in performance of the task (e.g., Friston *et al.*, 1992; Grafton *et al.*, 1994; Kim *et al.*, 1994; Petersen *et al.*, 1988, 1989; Seitz *et al.*, 1990, 1994; van Mier *et al.*, 1994, 1995). This learning stage has been conceptualized as the "composition phase" (Anderson, 1990; Fitts, 1962). In addition, differential cerebellar activity was found somewhat later in the acquisition process, i.e., during the "proceduralization phase," at which time subjects have achieved a certain level of automatization of the skill and are still improving, but at a much slower rate (Doyon *et al.*, 1996; Jenkins *et al.*, 1994; Seitz *et al.*, 1990).

Thus the evidence reviewed earlier is consistent with the notion that the cerebellum is critical for the acquisition of several types of skills under numerous cognitive conditions. Such an interpretation of human imaging data is in agreement with work in both animals (e.g., Shimansky *et al.*, 1995; see also Bloedel, 1992; J. R. Bloedel and V. Bracha, this volume; Thach *et al.*, 1992) and humans (e.g., Doyon *et al.*, 1997; Pascual-Leone *et al.*, 1993; Sanes *et al.*, 1990), in which a variety of skill-learning impairments have been observed following a lesion in the cerebellum. Much less is known regarding the nature of the dynamic changes in cerebellar activity that can be expected during the acquisition of skills. Some investigators have reported a reduction of activity following varying amounts of practice on a skill-learning task (Flament *et al.*, 1994; Friston *et al.*, 1992; Raichle *et al.*, 1994; van Mier *et al.*, 1994, 1995), suggesting that the cerebellum is mainly involved in the early stages of the learning process. Conversely, others have found an increase in cerebellar activity after practice (Doyon *et al.*, 1996; Grafton *et al.*, 1994; Seitz *et al.*, 1994), suggesting that this structure may be implicated in the automatization phase of a skill. It is possible that such a disparity in results may be due to methodological differences between studies. There is, however, an alternative interpretation to be found within the framework of the phases of skill learning. The importance of this view is that it provides several strong predictions and it also accounts for the observations to date. A number of studies (Flament *et al.*, 1994, 1995; Grafton *et al.*, 1994; Raichle *et al.*, 1994; Seitz *et al.*, 1994) suggest that the discrepant findings may not result from differences in the type of learning paradigm, nor of the cognitive processes involved in acquiring a skill, but rather that the differences in cerebellar blood flow changes may be due to the stage of learning during which the subjects' performance was scanned. The profile that emerges from these studies is that the cerebellum is active in the early "composition" and "proceduralization" phases of learning, but that its level of activation drops significantly when subjects have achieved an asymptotic level of performance or when subjects are performing a task that is overlearned. At the same time, this reduction in cerebellar activity appears to be coupled to an increase in CBF in specific cortical and/or subcortical regions. For example, Grafton and colleagues (1994) demonstrated that the subjects' gain in performance on the rotor pursuit task in day 1 was correlated to an increase of CBF in the cerebellum, whereas significant changes were observed in other cortical areas and the striatum, but not in the cerebellum, after these subjects had received additional practice and achieved an asymptotic level of performance on day 2. Using fMRI, Flament *et al.* (1994) reported that the intensity and area of activation in the cerebellum returned to baseline levels when subjects learned to perform in both the reversed and the normal experimental conditions of a joystick-tracking task. However, in a further experiment, the same group of researchers (Flament *et al.*, 1995) found an increase in activation compared to baseline in motor cortical regions (e.g., SMA and primary and premotor regions) as performance improved. Finally, the results of other studies suggest that this pattern of hemodynamic changes between the cerebellum and the cortical regions is not limited only to the learning of motor related skills, but it extends also to other modalities and other types of nondeclarative learning such as the classically conditioned eyeblink response (CR). For example, Raichle and colleagues (1994) reported that as subjects are learning to generate an appropriate verb for a noun (Petersen *et al.*, 1989) the initial activation seen in the cerebellum disappears when subjects become stereotyped in their responses after 15 min of practice. Cerebellar activation is then replaced instead by an increase in rCBF in the insular area. Similarly, in a study designed to determine whether the cerebellum is important for the retention of CR responses that are acquired naturally prior to the occurrence of a lesion, Bracha *et al.*, (1996) have shown that patients with cerebellar lesions produce normal anticipatory, eyeblink responses to a ball that is quickly approaching their forehead. These results suggest that the cerebellum is not essential for the retention of CR responses and that it is not critical for storing this type of learning.

The reason for the reported lack of activity in the cerebellum after subjects have reached optimal level of performance is unknown. However, some explanations can be considered. First, from a methodological point of view, it is possible that the cerebellum continues to be active, even in the most advanced phases of learning, but the level of neuronal activity in the learned task does not statistically exceed that in the control condition because the latter elicits a high cerebellar contribution on its own (e.g., Raichle et al., 1994). This relativity effect cannot explain all the results though, as an absence of activation has been seen with different types of controls [e.g., rest (Seitz et al., 1994)] and when no direct comparison to a control task was made, i.e., using a regression method of analysis (Grafton et al., 1994). From a physiological viewpoint, it is conceivable that the failure to detect cerebellar activation in the latest phases of learning may reflect the fact that less cells are necessary to perform the task (e.g., because of better synchronization). This new, more efficient level of activity would therefore not be detected by the PET camera or by the fMRI sequences presently available because of a lack of sensitivity. Alternatively, the overall pattern of blood flow changes occurring as a function of skill (i.e., absence of rCBF in the cerebellum, coupled with an increase of blood flow in cortical and subcortical regions) may indicate that the cerebellum is involved in early learning phases of a skill, but the "neuronal representation" (or program, engram, etc.) of the learning per se would not reside within the cerebellum. Instead, this "representation" of the skill would be mediated by a distributed cortical and subcortical system (possibly involving the striatum) that would depend on the type of task and cognitive processes involved. Contrary to the classical "take over" view of the cerebellum during learning (e.g., Sanes et al., 1990; Stein, 1986), this would imply that the cerebellum [in relation with cerebral cortical structures (e.g., Schmahmann 1991, 1996; Leiner et al., 1993; Middleton and Strick, 1994; Schmahmann and Pandya, this volume)] would be important for the acquisition of skilled behaviors. When subjects have learned the ability and have practiced it until it becomes automatic and overlearned, however, the cerebral cortical circuitry, either on its own or with help from subcortical structures, would then be sufficient to produce and retain the learned behavior.

This cerebellar to cortical and subcortical shift in control after the skill has become automatic and overlearned is consistent with experimental evidence and theoretical models from several investigators who have studied the functions of the cerebellum using a neuroanatomical and neurophysiological approach (e.g., for reviews see Bloedel, 1992; J. R. Bloedel and V.

Bracha, this volume; Houck, 1991). Indeed, such a notion is in accord with Bloedel's view (Bloedel, 1992, 1993; Bloedel and Bracha, this volume) that the cerebellum and the cognitive processes that it performs are "involved in" skill learning, but that this structure does not constitute a "storage site" for the memory trace of the skill. This proposal also agrees with the adjustable pattern generator model of Houk (1991), which suggests that the cerebellar output via the Purkinje cells would be implicated in the control of movements by influencing selectively the activity of premotor cortical circuitry, but that with repetition, this control of movements during the learning of motor habits would transfer to, and be carried out automatically by, the premotor networks. Finally, it should be noted that this notion is consistent with the results of a fMRI study by Karni and colleagues (1995, 1997) who have shown that the overleaning (i.e., after 4 weeks of daily practice) of a simple sequence of finger movements was associated with an enlargement in spatial extent of the activation within the primary motor cortical hand area.

The cortical and subcortical "take over" hypothesis of learning is still conjectural, as several questions remain unanswered. For example, it is unknown whether the same pattern of hemodynamic change in cerebellar, and cerebral cortical and subcortical regions can be seen across a variety of learning skills or whether it is task dependent. With the exception of the study of Grafton and colleagues (1994) subjects in other studies (e.g., Doyon et al., 1996; Grafton et al., 1992; Seitz et al., 1994) were not given enough practice to reach an asymptotic level of performance or to attain an overlearning phase, and thus very little information regarding the levels of rCBF in those stages is yet available. More research in which several measures of CBF are taken at different phases of the acquisition process, and in which the task is practiced sufficiently to achieve the overlearning phase (e.g., Karni et al., 1995), will be necessary to assess the extent of variations in the level of neuronal activity during a complete learning cycle. With the advent of (1) new three-dimensional PET cameras that are more sensitive, hence enabling a significant reduction in the amount of radioactive tracer injected on each scan and thus a greater number of repetitions of the same conditions, and (2) fMRI which can be used to test the same subjects repetitively until a novel skill is fully automatized, it is hoped that such experimental designs will be implemented more frequently in future studies.

Acknowledgments

I thank Drs. Avi Karni, Jeremy Schmahmann, and Viviane Sziklas as well as Philip Jackson, Martin Lafleur, Robert Laforce, Jr., and Virginia Penhune for their constructive criticisms on an earlier version of this chapter. This work was supported by a grant (OGPIN-012) from the Natural Sciences and Engineering Research Council of Canada.

References

Anderson, J. R. (1990). "Cognitive Psychology and Its Implications," 3rd Ed. Freeman, New York.
Bloedel, J. R. (1992). Functional heterogeneity with structural homogeneity: How does the cerebellum operate? *Behav. Brain Sci.* **15,** 666–678.
Bloedel, J. R. (1993). "Involvement in" versus "storage of." *Trends Neurosci.* **16,** 451–452.
Bower, J. M. (1992). Is the cerebellum a motor control device? *Behav. Brain Sci.* **15,** 714–715.
Bower, J. M. (1995). The cerebellum as sensory acquisition controller. *Hum. Brain Mapp.* **2,** 255–256.
Bower, J. M., and Kassel, J. (1990). Variability in tactile projection patterns to cerebellar folia Crus IIA in the normal rat. *J. Comp. Neurol.* **302,** 768–778.
Bracha, V., Wunderlich, D. A., Zhao, L., Brachova, L., and Bloedel, J. R. (1996). Is the human cerebellum required for the storage of conditioned eyeblink memory traces? *Soc. Neurosci. Abst.* **22,** 280.
Buckner, R. L., and Tulving, E. (1995). Neuroimaging studies of memory: theory and recent PET results. *In* "Handbook of Neuropsychology" (F. Boller and J. Grafman, eds.), Vol. 10, pp. 439–466. Elsevier, New York.
Daum, I., and Ackermann, H. (1995). Cerebellar contributions to cognition. *Behav. Brain Res.* **67,** 201–210.
Doyon, J., Gaudreau, D., Laforce, R., Jr., Castonguay, M., Bédard, P. J., Bédard, F., and Bouchard, J-P. (1997). Role of the striatum, cerebellum and frontal lobes in the learning of a visuomotor sequence. *Brain Cogn.*
Doyon, J., Owen, A. M., Petrides, M., Sziklas, V., and Evans, A. C. (1996). Functional anatomy of visuomotor skill learning in human subjects examined with positron emission tomography. *Eur. J. Neurosci.* **8,** 637–648.
Fiez, J. A. (1996). Cerebellar contributions to cognition. *Neuron* **16,** 13–15.
Fitts, P. (1962). Factors in complex skill training. *In* "Training Research and Education" (R. Glaser, ed.). University of Pittsburgh Press, Pittsburgh.
Flament, D., Ellermann, J., Ugurbil, K., and Ebner, T. J. (1994). Functional magnetic resonance imaging (fMRI) of cerebellar activation while learning to correct for visuo-motor errors. *Soc. Neurosci. Abstr.* **20,** 20.
Flament, D., Lee, J. H., Ugurbil, K., and Ebner, T. J. (1995). Changes in motor cortical and subcortical activity during the acquisition of motor skill, investigated using functional MRI (4T, echo planar imaging). *Soc. Neurosci. Abstr.* **21,** 1422.
Friston, K. J., Frith, C. D., Passingham, R. E., Liddle, P. F., and Frackowiak, R. S. J. (1992). Motor practice and neuropsychological adaptation in the cerebellum: A positron tomography study. *Proc. R. Soc. Lond. Ser. B Biol. Sci.* **248,** 223–228.
Grafton, S. T. (1995). PET imaging of human motor performance and learning. *In* "Handbook of Neuropsychology" (F. Boller and J. Grafman, eds.), pp. 405–422. Elsevier, New York.
Grafton, S. T., Hazeltine, E., and Ivry, R. E. (1995). Functional mapping of sequence learning in normal humans. *J. Cogn. Neurosci.* **7,** 497–510.
Grafton, S. T., Mazziotta, J. C., Presty, S., Friston, K. J., Frackowiak, R. S. J., and Phelps, M. E. (1992). Functional anatomy of human procedural learning determined with regional cerebral blood flow and PET. *J. Neurosci.* **12,** 2542–2548.

Grafton, S. T., Woods, R. P., and Mike, T. (1994). Functional imaging of procedural motor learning: Relating cerebral blood flow with individual subject performance. *Hum. Brain Mapp.* **1,** 221–234.

Houk, J. C. (1991). Outline for a theory of motor learning. *In* "Tutorials in Motor Neuroscience" (G. E. Stelmach, ed.). Kluwer Academic Publishers.

Ito, M. (1993). A new physiological concept on cerebellum. *Rev. Neurol.* **146,** 564–569.

Jenkins, I. H., Brooks, D. J., Nixon, P. D., Frackowiak, R. S. J., and Passingham, R. E. (1994). Motor sequence learning: A study with positron emmission tomography. *J. Neurosci.* **14,** 3775–3790.

Karni, A. (1997). The acquisition of perceptual and motor skills: A memory system in the adult human cortex. *Behav. Brain Res.*, in press.

Karni, A., Meyer, G., Jezzard, P., Adams, M., Turner, R., and Ungerleider, L. G. (1995). Functional MRI evidence for adult motor cortex plasticity during skill learning. *Nature* **377,** 155–158.

Karni, A., and Sagi, D. (1993). The time course of learning a visual skill. *Nature* **365,** 250–252.

Karni, A., Tanne, D., Rubenstein, B. S., Askenasy, J. J. M., and Sagi, D. (1994). Dependence on REM sleep of overnight improvement of a perceptual skill. *Science* **265,** 679–682.

Kim, S.-G., Ugurbil, K., and Strick, P. L. (1994). Activation of a cerebellar output nucleus during cognitive processing. *Science* **265,** 949–954.

Lang, W., Lang, M., Podreka, I., Steiner, M., Uhl, F., Suess, E., Muller, C., and Deecke, L. (1988). DC-potential shifts and regional cerebral blood flow reveal frontal cortex involvement in human visuomotor learning. *Exp. Brain Res.* **71,** 353–364.

Leiner, H. C., Leiner, A. L., and Dow, R. S. (1986). Does the cerebellum contribute to mental skills? *Behav. Neurosci.* **100,** 443–454.

Leiner, H. C., Leiner, A. L., and Dow, R. S. (1993). Cognitive and language functions of the human cerebellum. *Trends Neurosci.* **16,** 444–447.

Leiner, H. C., Leiner, A. L., and Dow, R. S. (1995). The underestimated cerebellum. *Hum. Brain Mapp.* **2,** 244–254.

Middleton, F. A., and Strick, P. L. (1994). Anatomical evidence for cerebellar and basal ganglia involvement in higher cognitive function. *Science* **266,** 458–461.

Moscovitch, M., Vriezen, E., and Goshen-Gottstein, Y. (1993). Implicit tests of memory in patients with focal lesions or degenerative brain disorders. *In* "Handbook of Neuropsychology" (F. Boller and J. Grafman, eds.), pp. 133–173. Elsevier, New York.

Nissen, M. J., and Bullemer, P. (1987). Attentional requirements of learning: Evidence from performance measures. *Cogn. Psychol.* **19,** 1–32.

Pascual-Leone, A., Grafman, J., Clark, K., Stewart, M., Massaquoi, S., and Hallet, M. (1993). Procedural learning in Parkinson's disease and cerebellar degeneration. *Ann. Neurol.* **34,** 594–602.

Passingham, R. E., Jueptner, M., Frith, C., Brooks, D. J., and Frackowiak, R. S. J. (1995). An analysis of motor learning. *Hum. Brain Mapp. Suppl.* **1,** 410.

Petersen, S. E., Fox, P. T., Posner, M. I., Mintun, M., and Raichle, M. E. (1988). Positron emission tomographic studies of the cortical anatomy of single-word processing. *Nature* **331,** 585–589.

Petersen, S. E., Fox, P. T., Posner, M. I., Mintun, M., and Raichle, M. E. (1989). Positron emission tomographic studies of the processing of single words. *J. Cogn. Neurosci.* **1,** 153–170.

Raichle, M. E., Fiez, J. A., Videen, T. O., Macleod, A.-M. K., Pardo, J. V., Fox, P. T., and Petersen, S. E. (1994). Practice-related changes in human brain functional anatomy during nonmotor learning. *Cereb. Cortex* **4,** 8–26.

Rao, S. M., Harrington, D. L., Haaland, K. Y., Bobholz, J. A., Binder, J. R., Hammeke, T. A., Frost, J. A., Myklebust, B. M., Jacobson, R. D., Bandettini, P. A., and Hyde, J. S. (1995). Functional MRI correlates of cognitive-motor learning. *Hum. Brain Mapp. Suppl.* **1,** 412.

Rauch, S. L., Savage, C. R., Halle, D. B., Curran, T., Alpert, N. M., Kendrick, A., Fischman, A. J., and Kosslyn, S. M. (1995). A PET investigation of implicit and explicit sequence learning. *Hum. Brain Mapp.* **3,** 271–286.

Sanes, J. N., Dimitrov, B., and Hallet, M. (1990). Motor learning in patients with cerebellar dysfunction. *Brain* **113,** 103–120.

Schlaug, G., Knorr, U., and Seitz, R. J. (1994). Inter-subject variability of cerebral activations in acquiring a motor skill: A study with positron emission tomography. *Exp. Brain Res.* **98,** 523–534.

Schmahmann, J. D. (1991). An emerging concept: The cerebellar contribution to higher function. *Arch. Neurol.* **48,** 1178–1187.

Schmahmann, J. D. (1996). From movement to thought: Anatomic substrates of the cerebellar contribution to cognitive processing. *Hum. Brain Mapp.* **4,** 174–198.

Seitz, R. J., Canavan, A. G. M., Yaguez, L., Herzog, H., Tellmann, L., Knorr, U., Yanxiong, H., and Homberg, V. (1994). Successive roles of the cerebellum and premotor cortices in trajectorial learning. *NeuroReport* **5,** 2541–2544.

Seitz, R. J., and Roland, E. (1992). Learning of sequential finger movements in man: A combined kinematic and positron emission tomography (PET) study. *Eur. J. Neurosci.* **4,** 154–165.

Seitz, R. J., Roland, E., Bohm, C., Greitz, T., and Stone-Elander, S. (1990). Motor learning in man: A positron emission tomographic study. *NeuroReport* **1,** 57–60.

Shimansky, Y., Wang, J.-J., Bracha, V., and Bloedel, J. R. (1995). Cerebellar inactivation abolishes the capability of cats to compensate for unexpected but not expected perturbations of a reach movement. *Soc. Neurosci. Abstr.* **21,** 914.

Squire, L. R. (1992). Declarative and nondeclarative memory: Multiple brain systems supporting learning and memory. *J. Cogn. Neurosci.* **4,** 232–243.

Stein, J. F. (1986). Role of the cerebellum in the visual guidance of movement. *Nature* **323,** 217–221.

Thach, W. T., Goodkin, H. P., and Keating, J. G. (1992). The cerebellum and the adaptive coordination of movement. *Annu. Rev. Neurosci.* **15,** 403–442.

Thompson, R. F., and Krupa, D. J. (1994). Organization of memory traces in the mammalian brain. *Annu. Rev. Neurosci.* **17,** 519–549.

van Mier, H., Petersen, S. E., Tempel, L. W., Perlmutter, J. S., Snyder, A. Z., and Raichle, M. E. (1994). Practice related changes in a continuous motor task measured by PET. *Soc. Neurosci. Abstr.* **20,** 361.

van Mier, H., Tempel, L. W., Perlmutter, J. S., Raichle, M. E., and Petersen, S. E. (1995). Generalization of practice-related effects in motor learning using the dominant and nondominant hand measured by PET. *Soc. Neurosci. Abstr.* **21,** 1441.

SECTION V

CLINICAL AND NEUROPSYCHOLOGICAL OBSERVATIONS

EXECUTIVE FUNCTION AND MOTOR SKILL LEARNING

Mark Hallett* and Jordan Grafman†

*Human Motor Control Section and †Cognitive Neuroscience Section, Medical Neurology Branch, National Institute of Neurological Disorders and Stroke, National Institutes of Health, Bethesda, Maryland 20892

I. Introduction
II. Motor Learning
   A. Motor Adaptation Learning
   B. Motor Skill Learning
   C. Functional Neuroimaging in Motor Learning
   D. Sequence Learning with the Serial Reaction Time Test
   E. Conclusions on Motor Learning
III. Executive Function
   A. Performance on "Tower-Type" Planning Tasks
   B. The Cerebellum and Retrieval of Information from Memory
   C. Controlled Attention and the Cerebellum
   D. Temporal Order Processing
   E. Time Estimation and the Cerebellum
   F. Functional Neuroimaging in Executive Function
   G. Hypothesized Roles of the Cerebellum in Cognitive Processing: A Reprise
IV. Conclusions
   References

I. Introduction

Accumulating evidence shows that the cerebellum participates in various aspects of motor learning and cognition. Evidence from behavioral studies of patients with cerebellar atrophy implies that the cerebellum plays a role in visuomotor learning and adaptation, planning, strategic thinking, time processing, and associative learning. Evidence from studies using functional neuroimaging supports this implication and substantiates the hypothesis that the cerebellum acts in concert with other structures as part of a frontal-subcortical system devoted to the storage and organization of timed sequential behaviors. The role of the cerebellum in timed sequential cognitive processing may be analogous to its role in motor processing and suggests a mechanism by which cognitive events become sequenced and temporally labeled.

II. Motor Learning

The cerebellum is clearly involved in the coordination of movement, and the additional concept that the cerebellum might be involved in motor learning was promoted by Marr (1969), who theorized that learning could occur by cellular interactions in the cerebellar cortex. The theory was that the climbing fibers modified the responses of the Purkinje cells to mossy fiber input. Albus (1971) refined the theory, suggesting that climbing fiber activity should decrease Purkinje cell excitability, and Ito (1989) produced recordings of cerebellar cellular activity that supported this concept. While modification of cellular behavior is an appropriate substrate for motor learning, it is still unclear what aspect of motor learning might go on in the cerebellum.

Motor learning itself is a complex phenomenon with many different components. One aspect can be defined as a change in motor performance with practice. Other aspects include increasing the repertoire of motor behavior and maintaining a new behavior over a period of time. Even considering only a change in motor performance, several different phenomena are likely. A distinction is made between motor adaptation learning and motor skill learning (Sanes *et al.,* 1990). It is probably easiest to make this distinction by referring to the concept of an operating characteristic. An operating characteristic is a descriptor of a set of movements that relate different movement variables to each other. It describes the current state of the capability of the motor system. Generally, a change in one variable will affect another. The best known operating characteristic of motor performance is Fitts' law, which relates movement speed and accuracy (Fitts, 1954). Movement from point A to point B can be made at various speeds, and each speed is associated with accuracy. Slower speeds are more accurate, whereas faster speeds are less accurate. Another operating characteristic is the gain associated with a visuomotor tracking task. For a particular visual stimulus, there is an associated movement. With a change in gain, the appropriate movement may be smaller or larger.

Motor adaptation learning can be defined as a change in motor performance without a change in the operating characteristic. In a point-to-point movement, a faster movement with a predictable decrease in accuracy is a change in performance, but not a change in the operating characteristic. This is not necessarily just a trivial change in performance. Learning the new speed may require considerable practice, but, if the learning is associated with a decrease in accuracy, it does not indicate a new capability of the motor system. Likewise, a change in visuomotor gain by itself does not indicate anything more than a change in the point of working on the operating characteristic.

Motor skill learning can be defined as a change in motor performance with a change in the operating characteristic. It indicates a new capability of the motor system. If a point-to-point movement is made both fast and with great accuracy, there is an apparent violation of Fitts' law. This means that there is a new operating characteristic in effect. In many circumstances, this performance would be clearly recognized as a new skill. Skill learning probably cannot stand alone separate from adaptation learning. In satisfying a new motor requirement, it is not enough to achieve a new operating characteristic, but it probably is also necessary to find the correct place to work on the operating characteristic.

The distinction between motor adaptation learning and motor skill learning may be analogous in some circumstances to the distinction made by Brooks *et al.* (1983) between "what to do" and "how to do." In learning a new complex movement, it is first necessary to understand the requirements of the task and to develop a strategy or gross kinematic plan that is at least an appropriate form of a response. This is called "what to do." Then it is necessary to refine the plan to produce better performance. This is called "how to do." The setting up of a new motor plan, including a unique sequence of actions, would likely be a new skill, while the refinement of the behavior would likely be achieved by adaptation.

Therefore, when assessing a situation for motor learning of the type characterized by a change in performance with practice, it is important to determine whether there is a change in the operating characteristic. In some circumstances, the learning will not demonstrate such a change, and it can be considered mainly adaptation learning. In other circumstances, when the operating characteristic is changed, the situation may be dominated by skill learning, but it is likely that adaptation learning is also occurring.

A. Motor Adaptation Learning

Perhaps the classic example of adaptation learning is the change in gain of the vestibulo-ocular reflex. The gain of the vestibulo-ocular reflex refers to the magnitude of eye movement resulting from head movement. The change in gain depends on the environmental circumstances. The amount of eye movement with a specific head movement depends on the working point on the operating characteristic defined by the gain. Selecting the appropriate gain is learning, but no apparent skill is acquired when the gain changes. Adaptation of the vestibulo-ocular reflex requires participation of the cerebellum and associated brain stem structures (Lisberger, 1988).

Eye blink conditioning is recognized as a form of motor learning and could be argued to fit the proposed definition for adaptation learning. Blinking of the eyelid to a conditioned stimulus may or may not occur. Whether or not it will occur, and how much, could be described by a single operating characteristic. At any one time, the amount of "conditioning" could be the working point on the operating characteristic. Thompson (1990) found that eye blink conditioning in animals seems to require an intact cerebellum, at least for the expression and timing of the response. Topka et al. (1993) studied eye blink conditioning in humans to determine if the intact cerebellum is required for eye blink conditioning in humans. They employed a classical delay conditioning paradigm in five patients with pure cerebellar cortical atrophy and in seven patients with olivopontocerebellar atrophy. The patients' results were compared with the results from neurologically healthy volunteers of the same sex and similar age. The two groups of patients had similar abnormalities in the acquisition of the conditioned response and produced fewer conditioned responses than the control subjects in any given block of trials. Many of the patients' conditioned responses were inappropriately timed with respect to the conditioned stimulus. Such results have also been found by others (Lye et al., 1988; Daum et al., 1993a,b).

Adaptation to lateral displacement of vision, as produced by prism glasses, has been used to assess learning of a visuomotor task (Weiner et al., 1983). Pointing to a target is a clear example of the visual system directing the motor system. When prism glasses are used, there is at first a mismatch between where the target is seen and where the pointing is directed. With experience, normal subjects adjust to the distortion and begin to point correctly. Correct pointing can be a product of a true change in the visuomotor coordination or an intellectual decision to point in a direction other than where the target appears to be located. When the glasses are removed, typically the subject initially points in the direction opposite that pointed to when the glasses were worn. In the naive subject, this is an excellent measure of true change in the visuomotor task because there is no reason for making an intellectual decision to point other than in the direction where the target appears to be located. With additional experience, the subjects return to correct performance. This type of motor learning fits the definition of adaptation well. With a stimulus, pointing could be anywhere. Choosing the correct visuomotor coordination to fit the current environmental situation is a type of adaptation learning. Patients with cerebellar damage show poor or no adaptation (Weiner et al., 1983) (Fig. 1), but patients with damage elsewhere in the brain, including the basal ganglia and different regions

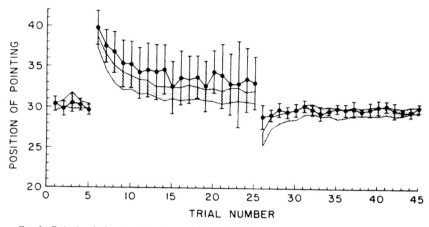

FIG. 1. Pointing behavior of patients with cerebellar lesions compared with normal subjects on 45 trials. Trials 6 to 25 were performed with prism glasses that shifted vision to the right. On trial 26, the first trial after the prism glasses were removed, the patients pointed almost exactly as they did on trial 5, before the prism glasses were worn. Thus, no adaptation occurred. The shaded area indicates the normal range (mean ± 2 SE). The filled circles and error bars indicate the mean ± 2 SE for the patients. From Weiner et al. (1983), with permission.

of the cortex, and patients with verbal memory deficits all show appropriate adaptation.

Another task that can test adaptation learning is one that includes a change in the visuomotor gain. An example is making movements of the elbow by matching targets on a computer screen. If the gain of the elbow with respect to the display on the computer screen is changed, then the amount of movement needed to match the targets will change. This simple gain change nicely fits the definition of adaptation motor learning. In the normal circumstance after a change in gain, there would be an error that would gradually be reduced with continued practice. Deuschl et al. (1996) used this task in studying 10 patients with cerebellar damage from degenerative diseases and 10 normal subjects. They measured the rate of adaptation by fitting to a curve the amplitudes of successive movements during the learning. Patients showed much slower learning than normal subjects.

All these studies suggest that the cerebellum plays a part in adaptation learning. The cerebellum is traditionally thought to play an important role in motor coordination. Adaptation, the planning of "how" to make the movement, or refinement of behavior may all be thought of as improving coordination. Adaptation learning may be considered an aspect of coordination.

B. MOTOR SKILL LEARNING

Complex, multijoint arm movement tasks, such as throwing a ball or playing the piano, are typically considered skills. The ability to sequence all the component movements correctly, smoothly, and in the appropriate amount of time is clearly difficult and appears to increase the behavioral repertoire. As such tasks are learned, they can be accomplished more quickly and more accurately. This violates Fitts' law and establishes a new operating characteristic. Hence, by definition, such learning would be skill learning.

Pursuing a possible role of the cerebellum in skill learning, Sanes *et al.* (1990) studied patients with cerebellar degeneration and normal subjects who traced polygons both with direct vision and while watching their performance in a mirror. The subjects were asked to perform the traces as quickly and as accurately as possible. Both groups performed faster in the direct vision task than in the mirror vision task, but accuracy deteriorated. Hence, the performance was in keeping with Fitts' law, and skill learning could not be shown. In the mirror vision task, normal subjects showed improvement in both time and accuracy, an indication that they had developed new skill. The performance of patients with cerebellar cortical atrophy was deficient in this task, which led to the conclusion that the cerebellum may play a role in skill learning as well as adaptation learning. There were several problems in the interpretation of the mirror vision experiment. First, the performance of patients with oliovopontocerebellar atrophy was normal, and in fact was faster than that of the normal subjects from the beginning. The authors interpreted this result as possibly being due to deficient visuomotor coordination in the patients, giving rise to less of a mismatch in visual and proprioceptive guidance, which confused the normal subjects. Second, there was clearly an important adaptation component to the task because there was altered visuomotor transformation. Thus, an abnormality might be due to the important adaptation component.

Topka *et al.* (1991) studied skill learning in 18 patients with cerebellar degeneration and in 15 normal subjects who performed multijoint arm movements on a data tablet, generating a trajectory connecting five via points in a given sequence. Subjects were asked to increase their accuracy but maintain constant movement time. A more accurate performance in the same amount of time would indicate a new operating characteristic and the development of skill learning. The subjects performed 100 trials with a movement time of approximately 3500 msec (relatively slowly), and then performed another 100 trials as quickly as possible. In the slower task, both groups were successful in keeping time and both improved relative accuracy at about the same rate. In the faster task, normal subjects performed more rapidly than the patients, yet they improved at a faster rate than the patients. A difference in learning rate with the faster task was confirmed by comparing the performance of a few normal subjects and patients who executed the faster task first (before the slower task). These results made some facts clear. First, patients with cerebellar damage were able to learn new skills, by our definition of skill learning. Second, their ability to learn appeared to be speed related because it was worse when a fast speed was required. Massaquoi and Hallett (1996) showed that the coordination deficits of patients with cerebellar damage are also speed dependent. The patients did much better with slow movements than with fast movements. Thus, their ability to refine movement variables needed for an adaptation component would likely be better with slow movements than with fast ones. The authors concluded that the deficits of the cerebellar patients in this task could be related to an adaptation component.

Given that patients with cerebellar damage are able to learn a new skill, it is logical to want to identify mechanisms related to skill learning. The cerebral motor cortex is clearly involved in movement, and studies have examined it as well as other cortical regions. Hallett et al. (1993) described studies on the plasticity of the motor cortex. Pascual-Leone et al. (1995a) used transcranial magnetic stimulation to map the cortical motor areas targeting the forearm finger flexor and extensor muscles in normal subjects who were learning a one-handed, five-finger exercise on an electronic keyboard. The task was paced by a metronome so that improvement in accuracy could identify skill learning. The keyboard was connected by a musical instrument digital interface to a personal computer so that the times of key presses could be measured. Subjects practiced the task for 2 hr daily. Their ability to keep accurate time with the metronome improved, and the number of errors was reduced. The corticomotor output maps targeting the fingers used to perform this task expanded in association with error reduction and improved task performance. It is not unreasonable to consider the motor cortex as a relevant site for motor skill learning. It is clearly involved in movement, and cortical cells have complex patterns of connectivity, including variable influences on multiple muscles within a body part. Long-term potentiation has been demonstrated in the motor cortex (Iriki et al., 1989).

C. FUNCTIONAL NEUROIMAGING IN MOTOR LEARNING

With the use of cerebral blood flow as a marker for neuronal activity, it is possible to image areas of the brain that are active during different tasks. Positron emission tomography coupled with methods for measuring cerebral blood flow is the standard, and functional magnetic resonance imaging (fMRI) based on deoxyhemoglobin has been introduced. In movement tasks, various brain regions are activated, depending on the task. The primary motor cortex is almost always activated to some extent, although because of problems of resolution it has been often difficult to separate primary motor cortex from premotor cortex or primary sensory cortex. Many studies using functional neuroimaging have concentrated on motor learning. The results are often somewhat confusing because of the different techniques and experimental paradigms used.

Seitz and Roland (1992) studied learning of a complex finger tapping sequence, and data were reanalyzed subsequently by Schlaug *et al.* (1994). The only region that was consistently active in all subjects with learning of the task was the contralateral sensorimotor region. Unfortunately, learning was accompanied by an increase in movement frequency, which also causes an increase in activation in the contralateral sensorimotor region (Sadato *et al.,* 1996). Cerebellar activation remained constant, and the authors speculated that this may have indicated a relative decline because an increase in activation may have been expected with the more rapid movement rate. Friston *et al.* (1992) studied practice of a simple finger movement sequence, and the only change observed with learning was a decrease of activity in the cerebellum. Seitz *et al.* (1994) studied learning of hand trajectories. With learning, activation of the ipsilateral dentate nucleus declined while the premotor cortical regions bilaterally were increasingly activated.

Grafton *et al.* (1992, 1994) studied learning of a pursuit rotor task. The task was to keep a stylus on a target that was moving on a rotating disc. The experiment is appealing because motor behavior is continuous, but with practice the ability to stay on target is much improved. One problem, however, is that the motor strategy and patterns of muscle activation may well change during the learning. These studies showed clear activation of the contralateral sensorimotor region during learning. Other regions, including the supplementary motor area, thalamus, contralateral cingulate area, and precuneate cortex, were also active. Activation of the ipsilateral anterior cerebellum increased with learning, although the authors speculated that it might have declined eventually if they had carried out the experiment further.

Jenkins *et al.* (1994) studied learning of a sequence of key presses by trial and error using auditory feedback. This behavior was compared with a sequence of key presses that was already learned. They found equal activation of primary motor cortex with both tasks. The prefrontal, premotor, and parietal cortices and the cerebellum, all bilaterally, were more active with learning.

Flament *et al.* (1994) used fMRI of the cerebellum during studies of motor learning with different visuomotor relationships. The task clearly had an important adaptation learning component. The cerebellum showed early activity that declined with learning.

Another study using fMRI (Karni *et al.*, 1995) focused attention on the contralateral primary motor cortex and utilized the experimental paradigm of finger-tapping sequences. Two sequences were compared, one which was in the process of being learned and another that was already learned. Although the learned sequence could have been performed faster, both sequences were performed at the same rate, which was paced by an auditory stimulus. Hence, motor activity was well matched. As the motor task was learned, more area of the motor cortex was activated. Within the same session, repetitions of the same sequence at first activated a progressively smaller region of motor cortex, but as a sequence was learned, the region became progressively larger.

In most of these studies, cerebellar activation was evident in the learning phase and declined when the movement was learned. This certainly indicates that the cerebellum has a role in learning. The pattern is consistent with many possible roles, but clearly adaptation learning is one. The decline of cerebellar activation when the movement is learned negates the idea that the cerebellum stores the movement and is in some way responsible for the automatic running of the motor program when it is well learned. However, the contralateral primary motor cortex and other cortical regions are also involved and may contribute to aspects of skill learning.

D. Sequence Learning with the Serial Reaction Time Test

Many of the studies of motor learning are complicated, and it is difficult to separate out the different facets of the process. One facet is learning the order of a number of components of a complex movement with sequential elements. The serial reaction time test (SRTT) appears to be a good paradigm for studying motor learning of sequences (Nissen and Bullemer, 1987; Pascual-Leone *et al.*, 1995b). The ability to carry out sequences of motor actions is clearly a critical part of most complex tasks, and the SRTT should be helpful in understanding this aspect of learning. The task is a choice reaction time with typically four possible responses. The responses can be carried out by key presses with four different fingers. A visual stimulus indicates which is the appropriate response. The completion of one response triggers the next stimulus. Each movement is simple and separate from the others so that the movement aspect of this task is different (and easier) than other tasks such as finger tapping or piano playing. The naive subject is unaware that the stimuli are a repeating sequence. With practice at this task, the responses become faster, even though the subject has no conscious recognition that the sequence is repetitive. This is called implicit learning. With continuing practice and improvement, the subject recognizes that there is a sequence, but may not be able to specify what it is. Now, knowledge is becoming explicit. With even more practice, the subject can specify the sequence, and the task has become declarative as well as procedural. Performance becomes even better at this stage, but the subject's strategy can change because he or she can anticipate the stimuli.

Thus, the SRTT appears to assess two processes relating to the sequencing of motor behavior while factoring out elements of motor coordination. As such, it might be considered a test of some components of motor skill learning.

Wachs *et al.* (1994) studied the intermanual transfer of implicit learning of the SRTT. After a few blocks of training with one hand, the subject performed subsequent blocks with the other hand. Four groups of normal subjects were studied, one under each condition: (1) random sequence, (2) a new sequence, (3) parallel image of the original sequence, and (4) mirror image of the original sequence. Only group 4 showed a carryover effect from the original learning. This result suggests that what is stored as implicit learning is a specific sequence of motor outputs and not a spatial pattern.

Implicit learning is impaired in patients with cerebellar degeneration, Parkinson's disease, Huntington's disease, and progressive supranuclear palsy (Grafman *et al.*, 1990; Willingham, 1992; Pascual-Leone *et al.*, 1993). Patients with cerebellar degeneration, in particular, are severely affected (Fig. 2). The performance of these patients was characterized not only by a lack of improvement in reaction time, but also by a deficiency in development of explicit knowledge (Pascual-Leone *et al.*, 1993). Moreover, giving the patients information about the sequence in advance (explicit knowledge) did not help improve reaction time. Pascual-Leone *et al.* (1996) reported that transient disruption of the dorsolateral prefrontal cortex with repetitive TMS impaired implicit learning of the SRTT. However, implicit learning was preserved in patients with temporal lobe lesions and in patients with short-term declarative memory disturbances, as in most patients with Alzheimer's disease.

To study the involvement of the primary motor cortex in implicit learning, Pascual-Leone *et al.* (1994a,b) mapped the motor cortex with TMS on the side of the head contralateral to the hands of normal subjects performing the SRTT. Mapping was done at intervals while the subjects were at rest between performing blocks of the SRTT. The map gradually enlarged during the implicit and explicit learning phases, but as soon as full explicit

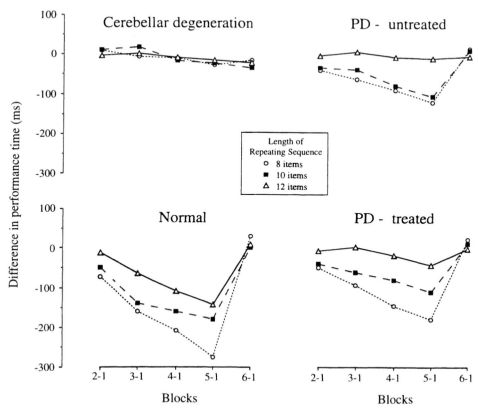

FIG. 2. Mean differences in response times (msec) between block 1 (random) and the subsequent blocks (2–5, repeating sequence; 6, random sequence) in 30 normal volunteers, 20 untreated and treated patients with Parkinson's disease (PD), and 15 patients with cerebellar degeneration, according to the length of the repeating sequence. See text and Pascual-Leone et al. (1993) for details.

learning was achieved, the map returned to its baseline size. This finding suggests an important role of the primary motor cortex in the SRTT.

Grafton et al. (1995) used PET to study learning during the SRTT. Two situations were imaged. In one, a second distracting task was performed at the same time as the SRTT. Such distraction does not interfere with implicit learning, but makes explicit learning much less likely. Hence, active brain regions are likely to reflect implicit learning. In the second situation, there was no other task, and subjects were scanned in the explicit learning phase. In the implicit learning situation, the contralateral primary motor cortex, supplementary motor area, and putamen were active. Involvement of the basal ganglia is consistent with the finding that patients with Parkinson's disease have difficulty with the SRTT. In the explicit learning situation, the dorsolateral prefrontal cortex ipsilaterally, the premotor cortex, and the parietal cortex bilaterally were active. This suggests that different structures are active in implicit and explicit learning. Heightened cerebellar activation was not found in either stage of learning.

In summary of the studies of the SRTT, it appears that multiple structures in the brain are involved, but the involvement might come at different stages of learning. The primary motor cortex appears to play a role in implicit learning, and the premotor and parietal cortical areas play a role in explicit learning. The latter conclusion is supported by the clinical finding that damage of premotor and parietal areas might lead to apraxia; this finding might be interpreted as a deficiency of motor memories for complex movements. The cerebellum also appears to be important in the learning of movement sequences given the results in the patients with cerebellar degeneration, but the nature of the role is not clear.

E. Conclusions on Motor Learning

Motor learning is a complex phenomenon with many components. Depending on the particular task, different anatomical structures are involved. It would be an oversimplification to say that only one part of the brain is involved with any task; it is more likely that a network is functional. However, it is possible to identify some aspects where particular structures play a major role. The cerebellum takes the principal part in adaptation learning. In skill learning, however, the cerebellar role is smaller, and cortical structures, including the motor cortex, are important. Skill learning has many facets and likely engages large portions of the brain. To the extent that sequencing is important, the cerebellum appears to have an important role.

III. Executive Function

During most of this century, it was thought that the cerebellum was simply an organ of movement (and concerned primarily with adaptive and skilled movements). However, since the mid-1980s, the results from a modest number of behavioral and functional neuroimaging studies have suggested that the cerebellum is also involved in nonmotor cognitive functions (Leiner *et al.*, 1986, 1993; Botez *et al.*, 1989; Keele and Ivry, 1990;

Schmahmann, 1991; Daum et al., 1993a). In particular, it has been thought that the cerebellum makes a fundamental contribution to so-called executive functions, such as the planning, reasoning, and thinking functions associated with the human prefrontal cortex (Shallice, 1982; Grafman, 1989). The exact cognitive contribution of the cerebellum is currently undetermined, although the kinds of tasks on which patients with cerebellar damage fail are limited in scope. Therefore, by defining the damaged cerebellum's cognitive sphere of influence, we may be able to identify some of its possible roles in cognition. The following neuropsychological studies relied almost exclusively on the study of patients with cerebellar atrophy, and we believe that the results have general relevance for understanding the role of the cerebellum in cognition.

A. Performance on "Tower-Type" Planning Tasks

Grafman et al. (1992) hypothesized that patients with cerebellar atrophy may have difficulty with tasks that require cognitive (as opposed to motor) sequencing. They found that the patients had difficulty solving problems on the Tower of Hanoi task (which requires a sequence of disk moves across three pegs to achieve a goal state) that could not be accounted for by motor impairment, age, educational level, level of dementia, depression, visuomotor procedural learning, verbal memory, or verbal fluency. Furthermore, compared with a control group, a subgroup of patients with cerebellar atrophy had a significant increase in planning time before they made their initial move (with no increase in between-move pause time), reinforcing the idea that the locus of their deficit on this task was a cognitive planning problem that was independent of motor control problems. Different initial (but similar goal) states were used for every problem, thereby maximizing the use of different routes to a solution, so that learning a particular pattern of moves would not be relevant to performance on the task. Because the subjects were always able to view the current position of the disks on the three pegs (minimizing the memory demands of the task), the authors suggested that a likely locus of the patients' failure on this task was the specific demands of cognitive planning or the need for implementation of a cognitive sequence.

In addition to patients with cerebellar atrophy, patients with frontal lobe lesions or frontal lobe dysfunction due to subcortical disease also failed to perform normally on the Tower of Hanoi task (Goel and Grafman, 1995). Patients with frontal lobe lesions performed poorly on the Tower of London task (Shallice, 1982). Patients with Parkinson's disease showed similar impairment on the Tower of Toronto task (Saint-Cyr et al., 1988).

The reason that patients with frontal lobe dysfunction fail on tower-type tasks is unclear, although assembling a sequential series of events or actions into a coherent "plan" is apparently difficult for patients with frontal lobe, basal ganglia, or cerebellar lesions (Grafman, 1995). The tower task is sufficiently complex, however, that in addition to planning, a variety of other cognitive processes are necessary to perform the task, including counterintuitive reasoning processes, visual imagery, working memory, and basic perceptual processes. An alternative view is that patients with frontal lobe dysfunction are unable to identify or plan for counterintuitive moves contained within a sequence of moves required to reach a goal (Goel and Grafman, 1995).

In either case, cognitive planning could be viewed as a cognitive analogue to complex motor procedures that require a series of individual movements to function as a unitary sequence (Pascual-Leone et al., 1993). Given that both cognitive and motor processes are required to perform most tasks, it is likely that a set of neural components making up a distributed visuomotor processing system would work together with the relative intensity of each component's activity, depending on the ongoing demands of a task. For example, motor initiation may depend on the supplementary and premotor cortex, and execution of motor sequences may depend on the cerebellum for the timing of movements and on the motor cortex for the targeting of muscles used in the task, whereas cognitive sequences may depend on the prefrontal cortex for the initiation of cognitive plans and on the cerebellum for the on-line timing and sequencing of the individual symbolic events that make up the cognitive plan or action.

Thus, cerebellar dysfunction disrupts a visuomotor neural system that includes subcortical structures and the prefrontal cortex concerned with both motor and cognitive activity (Kim et al., 1994; Middleton and Strick, 1994; Schmahmann and Pandya, 1995). Lesions of the prefrontal cortex would hypothetically affect the activation of plans and actions, subcortical structures such as the thalamus and basal ganglia might allow for their automatic execution (Willingham, 1992), and the cerebellum would ensure their correct sequencing and timing. In the study of Pascual-Leone et al. (1993), cognitive sequencing problems could have affected patients with cerebellar atrophy either when they tried to develop a planning strategy before actual disk movement or during their on-line cognitive processing. A breakdown in sequencing and timing could lead to on-line errors in selecting the correct solution path. The more novel the planning problem, the greater the likelihood of a planning failure, because a well-established memory (which might include sequencing and decision patterns) for the plan would not exist to help compensate for on-line timing and sequence encoding difficulty. Within this explanatory framework, sequencing refers to the order of events, whereas timing refers to the relative temporal placement of one event to another. Sequence and timing processes can operate independently or interact.

B. THE CEREBELLUM AND RETRIEVAL OF INFORMATION FROM MEMORY

Convergent evidence based on the performance of patients with brain lesions and functional neuroimaging studies in normal subjects has led investigators to conclude that the prefrontal cortex plays an important role in memory retrieval (Lalonde and Botez, 1990; Fiez et al., 1992; Ivry and Baldo, 1992; Appollonio et al., 1993; Roland, 1993; Raichle et al., 1994; Fiez, 1996). In particular, it is thought that the effortful strategic control over retrieval (consciously searching through memory for an item while being aware of using a particular strategy such as category of knowledge or temporal sequence to identify the item) requires the participation of the prefrontal cortex. If the cerebellum, in conjunction with the prefrontal cortex, helps to code timing and sequential relations between stimuli and actions, then patients with cerebellar damage may have trouble encoding or retrieving information from memory because timing information or sequential coding is required for either the encoded knowledge or the retrieval search itself.

To test this general hypothesis, Appollonio et al. (1993) conducted a comprehensive assessment of both implicit and explicit learning and memory processes in patients with cerebellar atrophy. Compared with the performance of normal subjects on free-recall measures, the patients' performance was impaired even though they displayed normal performance on cued recall and recognition memory (suggesting that the patients had at least partially intact encoding operations). The authors thought that the patients might have had more difficulty with the elaborate and effortful operations required for intentional encoding and the search strategy procedures used to process previously presented material than with recognition processes. The dissociation between free-recall and recognition has often been used as a marker to distinguish "subcortical dementia" from cortical dementia. However, an effortful memory deficit is rather nonspecific because it occurs in almost all patients with brain disease (e.g., Weingartner et al., 1984). Both cerebellar patients and normal subjects performed normally on measures of automatic memory (e.g., remembering how often a stimulus was presented) (Appollonio et al., 1993). Automatic operations are usually involved in encoding and retrieving the surface features of a stimulus, and they demand less sustained concentration and fewer cognitive resources (Weingartner et al., 1984).

The study by Weingartner et al. (1984) of patients with untreated Parkinson's disease showed a pattern of results almost identical to that observed by Appollonio et al. (1993) in patients with cerebellar atrophy, and this clinical picture was considered to provide support for the hypothesis that automatic and effort-demanding processes are conducted through different mechanisms. The results of Appollonio et al. (1993) are consistent with this hypothesis. Furthermore, the hypothesis is confirmed by the fact that the patients with cerebellar atrophy performed similarly to normal subjects on verbal and visual implicit memory tasks which, by definition, do not require the allocation of any conscious cognitive resources for recall.

When Appollonio et al. (1993) examined two other aspects of information processing—the mental scanning rate of a limited amount of verbal information held in short-term memory and the rapid retrieval of verbal knowledge from long-term memory—only the latter was consistently impaired within and across different tasks. They analyzed the number of words spoken every 15 sec on a word fluency task. Because the patients' and normal subjects' scores were not significantly different during the first 15 sec of word production, it is unlikely that the patients' diminished verbal fluency was due simply to motor speech problems (they may fatigue faster, but the effects of fatigue on cognitive processing have not been documented in patients with cerebellar atrophy). A profound impairment in verbal fluency was reported in a single case study of a patient with an idiopathic cerebellar degenerative disorder (Fiez et al., 1992; Fiez, 1996).

Estes (1974) pointed out that successful performance on word fluency tests depends on the subject's ability to organize his or her output in terms of clusters of meaningfully related words. As a result, word fluency tests require not only motor execution and intact access to semantic memory, but also planning and organizational abilities, which appear to be disrupted in patients with cerebellar atrophy. In support of this idea, Appollonio et al. (1993) found that the performance of patients with cerebellar atrophy on the initiation and perseveration subscale of the Mattis dementia rating scale (Mattis, 1976) accounted for the results obtained in the effortful free-recall and word fluency tasks. That is, after covarying for the initiation and perseveration score, the patients' word fluency and free-recall performances were no longer different from those of normal subjects. A possible explanation is that the initiation and perseveration subscale of the Mattis dementia rating scale, the free-recall task, and the word fluency task share at least one common "executive" process (e.g., goal attainment), which, when dysfunctional, is responsible for the patients' impaired performance (Duncan et al., 1996). Initiation and perseveration errors have traditionally been associated with prefrontal lobe dysfunction. However, caution is advised before simply attributing the cognitive deficits of cerebellar patients to prefrontal lobe dysfunction. Rather, the possibility that cerebellar dysfunction disrupts a specific emergent property (e.g., on-line timing and sequencing of cognitive events) of the functioning of a complex neural network that includes the prefrontal regions should also be taken into consideration. These results, however, confirm the general hypothesis that patients with cerebellar atrophy have no impairment on memory tasks, except when intentional and sustained effort is required.

C. Controlled Attention and the Cerebellum

Dimitrov et al. (1996) studied the performance of patients with cerebellar atrophy on various tests of attention. The patients performed a selective attention test as quickly and as accurately as normal subjects, and so appeared as sensitive as normal subjects to the various within-attentional task manipulations. The patients' response times on a spatial attention task were significantly slowed compared with those of normal subjects, but their performance was as accurate as that of normal subjects. There were no significant group × condition response time or accuracy interactions on the spatial attention task. Because this task was selectively administered to patients with spinocerebellar atrophy, the results may not be generalizable to patients with cerebellar atrophy.

These results indicate that visual processes concerned with selective attention are spared in patients with cerebellar atrophy. Although the patients with spinocerebellar atrophy had generally slower response times on the spatial attention task, the pattern of their response times across task conditions was similar to that of normal subjects. This finding is in contrast to some reports that patients with cerebellar damage are impaired when shifting attention (Akshoomoff and Courchesne, 1992; Akshoomoff et al., 1992). It also suggests that problems of timing or sequencing in patients with cerebellar atrophy are not simply attributable to impaired attention during encoding.

D. Temporal Order Processing

Given the findings of Grafman et al. (1992) on the Tower of Hanoi task, Dimitrov et al. (1996) were surprised to observe that patients with cerebellar atrophy generally had no difficulty on a verbal temporal order retrieval task. In this task, the patients were asked to place all the verbal items in the order in which they had been presented to them earlier. The patients were as accurate as normal subjects in reproducing the exact position of stimuli in a list, but if they could not find the exact position of an item, they were less effective in placing it in an adjacent position (i.e., close to its original location). If the patients had deficits in visuomotor and cognitive timing and sequencing, the impairment was in dynamic (i.e., online) timing or sequencing rather than in self-paced retrieval of sequence order (i.e., after the memory of the sequence or time has been stored). Perhaps on-line sequential placement of stimuli is an "all or none" phenomenon, so that if stimulus sequence placement is "missed" because of increased variation in event-time onset and duration coding, the eventual placement of that stimulus in a sequence will be more variable than normal. Thus, recall of stimulus order will be poorer for stimuli originally encoded incorrectly within a sequence by the cerebellum. This admittedly post-hoc explanation needs to be tested experimentally.

E. TIME ESTIMATION AND THE CEREBELLUM

Patients with cerebellar atrophy are impaired in both motor and perceptual timing. Early studies (Ivry and Keele, 1989) argued that patients with cerebellar atrophy are impaired in making perceptual discrimination when there are small differences in time duration. For example, patients performed worse than normal subjects on tasks designed to estimate kinaesthetic duration and velocity perception (Grill *et al.*, 1994). These findings suggest that the cerebellum may be involved in processing sensory signals that are relevant for motor control and conscious perception. However, the threshold procedure used in some early studies (Taylor and Creelman, 1967; Pentland, 1980) could not control for the effects of attentional deficits on performance or for any particular bias in the decision criterion.

Nichelli *et al.* (1996) showed that patients with cerebellar atrophy are impaired in discriminating intervals of 100 to 600 msec and that their impairment cannot be explained by attention deficits (i.e., normal subjects and patients showed similar reliability in estimation). In that study, several different timing tasks were administered to patients and normal subjects.

1. Time Production Task

In a time production task, the subject was seated in front of a computer screen. At the beginning of the task, a stopwatch 10 cm in diameter appeared in the center of the screen. The subject's task was to get the hand of the stopwatch moving at its regular rate by tapping on a computer key at 1 tap per second. The computer recorded the interval between each key tap. To avoid any run-in and run-out effect, the first five and last five intervals were eliminated from the data analysis. For the remaining 50 intervals, the relative error and the coefficient of variation were calculated for each subject's performance. Measures of variability in rhythmic tapping tasks confound two processes: the timekeeper, which determines when a response should be emitted, and the implementation system, which executes the program. According to the model of Wing and Kristofferson (1973), the tapping system operates in an open-loop mode (i.e., the timekeeper and the implementation system behave as independent random variables with normal variance). Therefore, the total variance of intertapping intervals were decomposed into separable estimates of the timekeeper (clock delay) and the implementation variability (motor delay).

2. Bisection Tasks

Bisection tasks consisted of a time bisection task and a spatial bisection task. The time bisection task consisted of two conditions: short-interval discrimination and long-interval discrimination. In short-interval discrimination, the standard short interval was 100 msec and the standard long interval was 900 msec. Subjects were asked to classify nine different durations (100–900 msec in 100-msec increments) as being more similar to the standard short interval or to the standard long interval. In long-interval discrimination, the standard short interval was 8 sec and the standard long interval was 32 sec. Subjects were asked to classify seven different durations (8–32 sec in 4-sec increments) as being more similar to the standard short interval or to the standard long interval. To avoid time estimates obtained by overt counting, numbers were presented in the center of the screen during the intervals and subjects had to read them aloud.

In the spatial bisection task, the condition was line length discrimination, in which the stimulus was a horizontal line displayed in the center of the screen. The standard short line was 6 mm and the standard long line was 54 mm. Subjects were asked to classify nine different lengths (6–54 mm in 6-mm increments).

The number of times a subject classified an interval (or a line) as "long" was plotted against stimulus duration. The proportion of long responses were analyzed by iterative least-square fitting to an unbiased logistic regression for interval duration. Then, based on each individual function, the following were calculated: the bisection point (i.e., the duration classified as long on 50% of trials), the difference limen (i.e., half the difference between the duration classified as long on 75% of trials and the duration classified as long on 25% of trials), and the Weber ratio (i.e., the difference limen divided by the bisection point). Lower bisection points indicated that the subject tended to classify short stimuli as more similar to the long stimuli rather than to the short standard, whereas higher bisection points suggested the opposite tendency. According to the Weber law, the difference limen is expected to vary as a function of the bisection point. Thus, the ability of the system to discriminate time intervals is better measured by the value of the Weber ratio. Finally, a measure of the subjects' precision (i.e., consistency over trials) was obtained from logistic regression. Precision was defined as the complement to 1 of the ratio between unexplained and total variance.

The results indicated that whenever the stimulus to classify encompassed a slightly longer interval, the patients showed a leftward shift of the bisection function compared with normal subjects. Nichelli et al. (1996) interpreted this shift as meaning that patients with cerebellar atrophy perceived the durations to be closer to each other than normal subjects perceived them to be. The patients' leftward shift was unaccompanied by changes in reliability of classification. Curiously enough, in discriminating stimuli that were closer to each other (as in a 100- to 325-msec discrimination task), patients with cerebellar atrophy performed as well as normal subjects (Nichelli et al., 1996). As noted earlier, patients' difficulty in time estimation occurred at a longer interval (100–600 msec). An explanation for this finding is to localize the patients' deficits either at the level of a system that gates the impulses of a pacemaker into an accumulator or within the accumulator itself. Consequently, errors in classifying temporal intervals would be likely to occur when errors in gating the pacemaker's pulses accumulate (i.e., only after a duration >325 msec).

In the Nichelli et al. (1996) study, patients with cerebellar atrophy also showed a profound deficit in discriminating long intervals (i.e., intervals in the range of seconds). However, in this case, consistency of performance was also affected. It follows that this impairment may not be due to a specific timing component, but related to deficits in sustained attention or strategy. Mangels et al. (1994) (also see Nichelli et al., 1995) demonstrated defective perception of long intervals in patients with frontal lobe lesions. Evidence also shows that cerebellar patients fail in many so-called frontal tasks [see earlier and Leiner et al. (1993) for a review]. The neural basis for the participation of the cerebellum in cognitive tasks is often attributed to its extended connections with the frontal lobe. The same may be true for its putative role in discriminating long intervals. In agreement with this hypothesis, Nichelli et al. (1996) found that consistency of the estimates in long-interval discrimination in patients with cerebellar atrophy was significantly related to their performance on the Wisconsin card sorting test.

Additional observations by Nichelli et al. (1996) of increased variability in rhythmic tapping by cerebellar patients confirmed earlier findings of Ivry and Keele (1989) and Ivry (1993). Both motor delay and clock delay estimates were affected in patients with cerebellar atrophy. However, only clock delay was significantly related to measures of perceptual timing. These results can be interpreted as providing evidence for a common timekeeping mechanism that is used in both execution and perceptual functions involving time-related decisions.

Braitenberg (1967) proposed, and then rejected (Fahle and Braitenberg, 1984), a theory that the cerebellum may function as an internal clock. The theory was based on different delay lines that may provide signals for as long as 200 msec. However, many models (Pellionisz and Llinas, 1982; Fahle and Braitenberg, 1984) include time as part of the computation process that is carried out by the cerebellum in order to anticipate joint positions to be achieved in the course of a movement. Could it be that this same computational timing process is also done to anticipate the sequential position of cognitive events to be perceived or expressed in the course of strategic implementation or the expression of a "script" or "schema" [a series of events sequentially organized into a routine or meaningful activity, such as the events that define the activity of "eating at a restaurant" (Grafman, 1995)]?

F. Functional Neuroimaging in Executive Function

Functional neuroimaging can be used to provide convergent evidence about the role of a neural structure in a particular process. Not suprisingly, many studies have confirmed that the cerebellum is activated in conjunction with certain motor processes and even during complex motor learning (see earlier). Almost all studies of simple perceptual, attentional, or recognition processes find no prominent cerebellar contribution. However, several studies are of relevance to the thesis that the cerebellum is involved in certain forms of cognitive processing (Decety et al., 1990; Leiner et al., 1993). These studies (Decety et al., 1990; Jenkins and Frackowiak, 1993; Roland, 1993; Kim et al., 1994; Raichle et al., 1994; Fiez, 1996; Gao et al., 1996; Molchan, manuscript in preparation) have demonstrated cerebellar activation during classical conditioning, associative learning, motor preparation and imagery, and retrieval of stored information. Almost all of these processes were shown to activate not only the cerebellum but also a large set of neural regions that form part of a distributed network that includes the prefrontal and frontal cortices, thalamus, and basal ganglia. Although the cerebellum may be activated during a functional neuroimaging task, it remains a challenge to determine its unique role within a distributed network of neural structures. Given the neuropsychological data reviewed earlier, it would not be suprising if the explanation for the cerebellum's activation in various neuroimaging tasks involved its sequence and timing functions.

G. Hypothesized Roles of the Cerebellum in Cognitive Processing: A Reprise

We propose that the cerebellum assists in the binding together of discrete events into linked sequences of specified durations. In lower species, the discrete events must have originally been individuated movements, thus leading to the cerebellum's crucial role in the programming and storage of movement patterns. We, and others, have shown that a motor sequence pattern can be learned only if the cerebellum is able to contribute to that learning by virtue of its timing and sequence activity. We suspect that as the prefrontal cortex evolved and expanded to account for 30% of the entire cerebral cortex, it became capable of storing cognitive events in much the same way that the posterior frontal cortex may store motor patterns (Grafman, 1995), i.e., cognitive events could be stored as a structured event complex or, in other words, a linked sequence of events (Grafman, 1995). The same timing and sequence mechanism that the cerebellum provided for motor processes through its linkage to the thalamus and motor cortex may have similarly become responsible for the timing and sequencing of cognitive events through a parallel indirect linkage to the prefrontal cortex. We propose that these functions allowed individuated cognitive events to be sequentially bound to each other and appropriately timed to sequentially emerge when retrieved for execution. Because the expression of cognitive ideas and events is on a longer time scale than the expression of motor programs, even an indirect connection of the prefrontal cortex to the cerebellum might be sufficient for timing purposes at that level of knowledge.

If this framework is plausible, then the cerebellum should be primarily involved in cognitive activities that require the linkage of events (e.g., associative learning, planning, implementation of strategic thinking, and controlled processes) over time. Future studies of imagery capacity in cerebellar patients may help constrain its role in cognition (e.g., Decety *et al.*, 1990). Because imaging ideas and events activate brain regions that are also activated during execution, and because imaging can involve a series of images, the cerebellum should also be active in the retrieval and linkage of images. Because imagery is restricted to internal representations without requiring overt motor output or sensory input processes, it should be easier to specify the cognitive role of the cerebellum in certain activities.

IV. Conclusions

Notable among brain structures for its beautiful neural architecture, the cerebellum has achieved a prominent status among the assembly of neural structures devoted to motor processes. Thus, it was somewhat suprising and counterintuitive to suspect that it may also have a role in cognitive processing. Perhaps its late emergence was associated with our paucity of knowledge, since the early 1980s, on the role of the prefrontal cortex and basal ganglia in cognition. With this new knowledge came a reexamination of the role of many neural structures, including the cerebellum, in behavior. The behavioral evidence from our studies on patients with cerebellar atrophy implied that the cerebellum plays a role in visuomotor learning and adaptation, planning, strategic thinking, time processing, and associative learning. Functional neuroimaging has supported this inference and has substantiated the belief that the cerebellum acts in concert with other structures as part of a frontal-subcortical system devoted to the storage and organization of timed sequential behaviors. The role of the cerebellum in timed sequential cognitive processing may be analogous to its role in motor processing and suggests a mechanism by which cognitive events become sequenced and temporally labeled.

Acknowledgment

Substantial portions of the text on motor learning were taken from Hallett *et al.* (1996).

References

Akshoomoff, N. A., and Courchesne, E. (1992). A new role for the cerebellum in cognitive operations. *Behav. Neurosci.* **106,** 731–738.
Akshoomoff, N. A., Courchesne, E., Press, G. A., and Iragui, V. (1992). Contribution of the cerebellum to neuropsychological functioning: Evidence from a case of cerebellar degenerative disorder. *Neuropsychologia* **30,** 315–328.
Albus, J. S. (1971). A theory of cerebellar function. *Math. Biosci.* **10,** 25–61.
Appollonio, I. M., Grafman, J., Schwartz, V., Massaquoi, S., and Hallett, M. (1993). Memory in patients with cerebellar degeneration. *Neurology* **43,** 1536–1544.
Botez, M. I., Botez, T., Elie, R., and Attig, E. (1989). Role of the cerebellum in complex human behavior. *Ital. J. Neurol. Sci.* **10,** 291–300.
Braitenberg, V. (1967). Is the cerebellar cortex a biological clock in the millisecond range? *Prog. Brain Res.* **25,** 334–346.
Brooks, V. B., Kennedy, P. R., and Ross, H. G. (1983). Movement programming depends on understanding of behavioral requirements. *Physiol. Behav.* **31,** 561–563.
Daum, I., Ackermann, H., Schugens, M. M., Reimold, C., Dichgans, J., and Birbaumer, N. (1993a). The cerebellum and cognitive functions in humans. *Behav. Neurosci.* **107,** 411–419.
Daum, I., Schugens, M. M., Ackermann, H., Lutzenberger, W., Dichgans, J. and Birbaumer, N. (1993b). Classical conditioning after cerebellar lesions in humans. *Behav. Neurosci.* **107,** 748–756.

Decety, J., Sjoholm, H., Ryding, E., Stenberg, G., and Ingvar, D. H. (1990). The cerebellum participates in mental activity: Tomographic measurements of regional cerebral blood flow. *Brain Res.* **535**, 313–317.

Deuschl, G., Toro, C., Zeffiro, T., Massaquoi, S., and Hallett, M. (1996). Adaptation motor learning of arm movements in patients with cerebellar diseases. *J. Neurol. Neurosurg. Psychiat.* **60**, 515–519.

Dimitrov, M., Grafman, J., Kosseff, P., Wachs, J., Alway, D., Higgins, J., Litvan, I., and Lou, J.-S. (1996). Preserved cognitive processes in cerebellar degeneration. *Behav. Brain Res.* **79**, 131–135.

Duncan, J., Emslie, H., Williams, P., Johnson, R., and Freer, C. (1996). Intelligence and the frontal lobe: The organization of goal-directed behavior. *Cogn. Psychol.* **30**, 257–303.

Estes, W. (1974). Learning theory and intelligence. *Am. Psychol.* **29**, 740–749.

Fahle, M., and Braitenberg, V. (1984). Some quantitative aspects of cerebellar anatomy as a guide to speculation on cerebellar functions. *In* "Cerebellar Functions" (J. Bloedel, J. Dichgans, and W. Precht, eds.), pp. 186–200. Springer-Verlag, Berlin.

Fiez, J.A. (1996). Cerebellar contributions to cognition. *Neuron* **16**, 13–15.

Fiez, J. A., Petersen, S. E., Cheney, M. K., and Raichle, M. E. (1992). Impaired non-motor learning and error detection associated with cerebellar damage: A single case study. *Brain* **115**, 155–178.

Fitts, P. M. (1954). The information capacity of the human motor system controlling the amplitude of movement. *J. Exp. Psychol.* **47**, 381–391.

Flament, D., Ellermann, J., Ugurbil, K., and Ebner, T. J. (1994). Functional magnetic resonance imaging (fMRI) of cerebellar activation while learning to correct for visuomotor errors. *Soc. Neurosci. Abstr.* **20**, 20.

Friston, K. J., Frith, C. D., Passingham, R. E., Liddle, P. F., and Frackowiak, R. S. J. (1992). Motor practice and neurophysiological adaptation in the cerebellum: A positron tomography study. *Proc. R. Soc. Lond. B Biol. Sci.* **248**, 223–228.

Gao, J. H., Parsons, L. M., Bower, J. M., Xiong, J., Li, J., and Fox, P. T. (1996). Cerebellum implicated in sensory acquisition and discrimination rather than motor control. *Science* **272**(5261), 545–547.

Goel, V., and Grafman, J. (1995). Are the frontal lobes implicated in "planning" functions? Interpreting data from the Tower of Hanoi. *Neuropsychologia* **33**, 623–642.

Grafman, J. (1989). Plans, actions, and mental sets: Managerial knowledge units in the frontal lobes. *In* "Integrating Theory and Practice in Clinical Neuropsychology" (E. Perecman, ed.), pp. 93–138. Lawrence Erlbaum Associates, Hillsdale, NJ.

Grafman, J. (1995). Similarities and distinctions among models of prefrontal cortical functions. *Ann. N.Y. Acad. Sci.* **769**, 337–368.

Grafman, J., Litvan, I., Massaquoi, S., Stewart, M., Sirigu, A., and Hallett, M. (1992). Cognitive planning deficit in patients with cerebellar atrophy. *Neurology* **42**, 1493–1496.

Grafman, J., Weingartner, H., Newhouse, P., Sunderland, T., Thompsen-Putnam, K., Lalonde, F., and Litvan, I. (1990). Implicit learning in patients with Alzheimer's disease. *Pharmacopsychiatry* **23**, 94–101.

Grafton, S. T., Hazeltine, E., and Ivry, R. (1995). Functional mapping of sequence learning in normal humans. *J. Cogn. Neurosci.* **7**, 497–510.

Grafton, S. T., Mazziotta, J. C., Presty, S., Friston, K. J., Frackowiak, R. S. J., and Phelps, M. E. (1992). Functional anatomy of human procedural learning determined with regional cerebral blood flow and PET. *J. Neurosci.* **12**, 2542–2548.

Grafton, S. T., Woods, R. P., and Tyszka, M. (1994). Functional imaging of procedural motor learning: Relating cerebral blood flow with individual subject performance. *Hum. Brain Map.* **1**, 221–234.

Grill, S.E., Hallett, M., Marcus, C., and McShane, L. (1994). Disturbances of kinaesthesia in patients with cerebellar disorders. *Brain* **117,** 1433–1447.

Hallett, M., Cohen, L. G., Pascual-Leone, A., Brasil-Neto, J., Wassermann, E. M., and Cammarota, A. N. (1993). Plasticity of the human motor cortex. *In* "Spasticity: Mechanisms and Management" (A. F. Thilmann, D. J. Burke, and W. Z. Rymer, eds.), pp. 67–81. Springer-Verlag, Berlin.

Hallett, M., Pascual-Leone, A., and Topka, H. (1996). Adaptation and skill. *In* "Acquisition of Motor Behavior in Vertebrates" (J. R. Bloedel, T. J. Ebner, and S. P Wise, eds.), pp. 289–301. MIT Press, Cambridge, MA.

Iriki, A., Pavlides, C., Keller, A., and Asanuma, H. (1989). Long-term potentiation of motor cortex. *Science* **245,** 1385–1387.

Ito, M. (1989). Long-term depression. *Annu. Rev. Neurosci.* **12,** 85–102.

Ivry, R. (1993). Cerebellar involvement in the explicit representation of temporal information. *Ann. N.Y. Acad. Sci.* **682,** 214–230.

Ivry, R. B., and Baldo, J. V. (1992). Is the cerebellum involved in learning and cognition? *Curr. Opin. Neurobiol.* **2,** 212–216.

Ivry, R.L., and Keele, SW. (1989). Timing functions of the cerebellum. *J. Cogn. Neurosci.* **1,** 136–152.

Jenkins, I. H., Brooks, D. J., Nixon, P. D., Frackowiak, R. S. J., and Passingham, R. E. (1994). Motor sequence learning: A study with positron emission tomography. *J. Neurosci.* **14,** 3775–3790.

Jenkins, I. H., and Frackowiak, R. S. (1993). Functional studies of the human cerebellum with positron emission tomography. *Rev. Neurol. (Paris)* **149,** 647–653.

Karni, A., Meyer, G., Jezzard, P., Adams, M., Turner, R., and Ungerleider, L. G. (1995). Functional MRI evidence for adult motor cortex plasticity during motor skill learning. *Nature* **377,** 155–158.

Keele, S. W., and Ivry, R. (1990). Does the cerebellum provide a common computation for diverse tasks? A timing hypothesis. *Ann. N.Y. Acad. Sci.* **608,** 179–207.

Kim, S. G., Ugurbil, K., and Strick, P. L. (1994). Activation of a cerebellar output nucleus during cognitive processing. *Science* **265**(5174), 949–951.

Lalonde, R., and Botez, M. I. (1990). The cerebellum and learning processes in animals. *Brain Res. Rev.* **15,** 325–332.

Leiner, H. C., Leiner, A. L., and Dow, R. S. (1986). Does the cerebellum contribute to mental skills? *Behav. Neurosci.* **100,** 443–454.

Leiner, H. C., Leiner, A. L., and Dow, R. S. (1993). Cognitive and language functions of the human cerebellum. *Trends Neurosci.* **16,** 444–447.

Lisberger, S. G. (1988). The neural basis for learning of simple motor skills. *Science* **242,** 728–735.

Lye, R. H., O'Boyle, D. J., Ramsden, R. T., and Schady, W. (1988). Effects of a unilateral cerebellar lesion on the aquisition of eyeblink-conditioning in man. *J. Physiol. (Lond.)* **403,** 58.

Mangels, J. A., Ivry, R. B., and Helmuth, L. L. (1994). The perception of short and long intervals in patients with frontal lobe, cerebellar, or basal ganglia lesions. *In* "Proceedings of the Cognitive Neuroscience Society Inaugural Meeting," p. 120. San Francisco, CA.

Marr, D. (1969). A theory of cerebellar cortex. *J. Physiol. (Lond.)* **202,** 437–470.

Massaquoi, S., and Hallett, M. (1996). Kinematics of initiating a two-joint arm movement in patients with cerebellar ataxia. *Can. J. Neurol. Sci.* **23,** 3–14.

Mattis, S. (1976). Mental status examination for organic mental syndrome in the elderly patient. *In* "Geriatric Psychiatry" (L. Bellak and T. B. Karasu, eds.). Grune and Stratton, New York.

Middleton, F. A., and Strick, P. L. (1994). Anatomical evidence for cerebellar and basal ganglia involvement in higher cognitive function. *Science* **266**(5184), 458–461.

Nichelli, P., Alway, D., and Grafman, J. (1996). Perceptual and motor timing in cerebellar degeneration. *Neuropsychologia* 34, 863-871.

Nichelli, P., Clark, K., Hollnagel, C., and Grafman, J. (1995). Duration processing after frontal lobe lesions. *Ann. N.Y. Acad.Sci.* **769**, 183–190.

Nissen, M. J., and Bullemer, P. (1987). Attentional requirements of learning: Evidence from performance measures. *Cogn. Psychol.* **19**, 1–32.

Pascual-Leone, A., Dang, N., Cohen, L. G., Brasil-Neto, J. P., Cammarota, A., and Hallett, M. (1995a). Modulation of muscle responses evoked by transcranial magnetic stimulation during the acquisition of new fine motor skills. *J. Neurophysiol.* **74**, 1037–1045.

Pascual-Leone, A., Grafman, J., Clark, K., Stewart, M., Massaquoi, S., Lou, J. S., and Hallett, M. (1993). Procedural learning in Parkinson's disease and cerebellar degeneration. *Ann. Neurol.* **34**, 594–602.

Pascual-Leone, A., Grafman, J., and Hallett, M. (1994a). Modulation of cortical motor output maps during development of implicit and explicit knowledge. *Science* 263, 1287-1289.

Pascual-Leone, A., Grafman, J., and Hallett, M. (1994b). Transcranial magnetic stimulation in the study of human cognitive function. *In* "New Horizons in Neuropsychology" (M. Sugishita, ed.), pp. 93–100. Elsevier, Amsterdam.

Pascual-Leone, A., Grafman, J., and Hallett, M. (1995b). Procedural learning and prefrontal cortex. *Ann. N.Y. Acad.Sci.* **769**, 61–70.

Pascual-Leone, A., Wassermann, E. M., Grafman, J., and Hallett, M. (1996). The role of the dorsolateral prefrontal cortex in implicit procedural learning. *Exp. Brain Res.* **107**, 479–485.

Pellionisz, A., and Llinas, R. (1982). Space-time representation in the brain: The cerebellum as a predictive space-time metric tensor. *Neuroscience* **7**, 2949–2970.

Pentland, A. (1980). Maximum likelihood estimation: The best PEST. *Percept. Psychophys.* **28**, 377–379.

Raichle, M. E., Fiez, J. A., Videen, T. O., MacLeod, A. M., Pardo, J. V., Fox, P. T., and Petersen, S. E. (1994). Practice-related changes in human brain functional anatomy during nonmotor learning. *Cereb. Cortex* **4**, 8–26.

Roland, P. E. (1993). Partition of the human cerebellum in sensory-motor activities, learning and cognition. *Can. J. Neurol. Sci.* **20**, S75–S77.

Sadato, N., Ibañez, V., Deiber, M.-P., Campbell, G., Leonardo, M., and Hallett, M. (1996). Frequency-dependent changes of regional cerebral blood flow during finger movements. *J. Cereb. Blood Flow Metab.* **16**, 23–33.

Saint-Cyr, J. A., Taylor, A. E., and Lang, A. E. (1988). Procedural learning and neostriatal dysfunction in man. *Brain* **111**, 941–959.

Sanes, J. N., Dimitrov, B., and Hallett, M. (1990). Motor learning in patients with cerebellar dysfunction. *Brain* **113**, 103–120.

Schlaug, G., Knorr, U., and Seitz, R. J. (1994). Inter-subject variability of cerebral activations in acquiring a motor skill: A study with positron emission tomography. *Exp. Brain Res.* **98**, 523–534.

Schmahmann, J. D. (1991). An emerging concept: The cerebellar contribution to higher function. *Arch. Neurol.* **48**, 1178–1187.

Schmahmann, J. D., and Pandya, D. N. (1995). Prefrontal cortex projections to the basilar pons in rhesus monkey: Implications for the cerebellar contribution to higher function. *Neurosci. Lett.* **199**, 175–178.

Seitz, R. J., and Roland, P. E. (1992). Learning of sequential finger movements in man: A combined kinematic and positron emission tomography (PET) study. *Eur. J. Neurosci.* **4**, 154–165.

Seitz, R. J., Canavan, A. G., Yaguez, L., Herzog, H., Tellmann, L., Knorr, U., Huang, Y., and Homberg, V. (1994). Successive roles of the cerebellum and premotor cortices in trajectorial learning. *NeuroReport* **5**, 2541-2544.

Shallice, T. (1982). Specific impairments of planning. *Philos. Trans. R. Soc. Lond. B Biol. Sci.* **298**, 199-209.

Taylor, M., and Creelman, C. (1967). PEST: Efficient estimates of probability functions. *J. Acoust. Soc. Am.* **41**, 782-787.

Thompson, R. F. (1990). Neural mechanisms of classical conditioning in mammals. *Philos. Trans. R. Soc. Lond. B Biol. Soc.* **329**, 161-170.

Topka, H., Massaquoi, S. G., Zeffiro, T., and Hallett, M. (1991). Learning of arm trajectory formation in patients with cerebellar deficits. *Soc. Neurosci. Abst.* **17**, 1381.

Topka, H., Valls-Solé, J., Massaquoi, S., and Hallett, M. (1993). Deficit in classical conditioning in patients with cerebellar degeneration. *Brain* **116**, 961-969.

Wachs, J., Pascual-Leone, A., Grafman, J., and Hallett, M. (1994). Intermanual transfer of implicit knowledge of sequential finger movements. *Neurology* **44**(Suppl. 2), A329.

Weiner, M. J., Hallett, M., and Funkenstein, H. H. (1983). Adaptation to lateral displacement of vision in patients with lesions of the central nervous system. *Neurology* **33**, 766-772.

Weingartner, H., Burns, S., Diebel, R., and LeWitt, P. A. (1984). Cognitive impairments in Parkinson's disease: Distinguishing between effort-demanding and automatic cognitive processes. *Psychiatry Res.* **11**, 223-235.

Willingham, D. B. (1992). Systems of motor skill. In "Neuropsychology of Memory" (L. R. Squire and N. Butters, eds.), pp. 166-178. Guilford Press, New York.

Wing, A., and Kristofferson, A. (1973). Response delays and the timing of discrete motor responses. *Percept. Psychophys.* **14**, 5-12.

VERBAL FLUENCY AND AGRAMMATISM

Marco Molinari, Maria G. Leggio, and Maria C. Silveri

Institute of Neurology, Experimental Neurology Laboratory, Neuropsychology Service,
Catholic University, 00168 Rome, Italy

I. Historical Background
II. Dysarthria
III. Verbal Fluency
IV. Fluency Strategies
V. Agrammatism
VI. Dysgraphia
VII. Hypothesis for a Cerebellar Role in Language
References

Since the beginning of this century it has been documented that cerebellar lesions induce speech deficits but these were thought to result from lack of motor coordination in the muscular activity needed for phonation. The pure motor nature of the cerebellum has been challenged on different grounds, and cerebellar activation has been documented in language-related tasks independently from motor activity. This chapter reviews the available evidence in favor of a cerebellar contribution to linguistic processing, focusing mainly on clinical observations in patients. The clinical findings are discussed in the light of recent theories on cerebellar functions.

I. Historical Background

The cognitive abilities of patients with cerebellar lesions have classically been considered to be unaffected. In recent years, this view has been challenged on different grounds. Evolutionary and neuroanatomical evidence and functional studies in normal subjects, as well as clinical observations, are converging in assigning a cognitive role to the cerebellum. However, criticisms of the alleged importance of the cerebellum in cognition are not lacking, and the sceptical view also has its proponents (see Daum and Ackermann, this volume).

This chapter deals with available evidence suggesting that the cerebellum plays a role in linguistic processing and it focuses mainly on clinical observations of patients. The rapidly growing literature on functional activation studies in healthy subjects, which provides additional support for the importance of the cerebellum in linguistic processing, is reviewed in this volume by Fiez and Raichle and thus will be cited here only when appropriate for discussion.

II. Dysarthria

Speech deficits in cerebellar patients have long been considered to be limited to the motor aspects of language, namely to a lack of motor coordination of the laryngeal muscles that causes the so-called ataxic dysarthria characteristic of cerebellar speech. The cerebellar site responsible for dysarthria has been localized in the paravermal region of the anterior cerebellum (Ackerman et al., 1992; Gilman et al., 1981). Most researchers believe that the contribution of the cerebellum to language is limited to the execution of the motor repertoire required for phonation. Suggestions about the functional relationship between cerebellar damage and speech also derive from the poorly understood phenomenon of mutism after posterior fossa surgery in children (Ammirati et al. 1989; Pollack et al., 1995; Rekate et al., 1985). This transitory speech block after recovery from anaesthesia has been related to dysarthria mechanisms, possibly due to cerebellar damage (Van Dongen et al., 1994). Nevertheless, the complete mutism observed and its hypothesized origin in lesions of the lateral cerebellum (Fraioli and Guidetti, 1975) indicate the possible involvement of cerebellar structures different from the ones responsible for dysarthria.

Support for a possible cerebellar involvement in language at a "higher" level than the simple articulatory one comes from neuroanatomical (Bauman and Kemper, 1985; Bauman et al., this volume) and neuroimaging (Courchesne et al., 1988) studies of infantile autism. Gross deficits in language development and speech patterns characterized by echolalia, metaphorical language, and pronominal reversal are often present in these children in which a reduction in cerebellar posterior vermis volume has also been reported (Courchesne et al., 1994).

Further correlations between cerebellar volume and language abilities have also been adduced by comparing speech performances and cerebellar volume in different genetic disorders. In fact, although children affected by Joubert or Down's syndrome present vermal hypoplasia associated with reduced verbal abilities, children affected by William's syndrome do not present cerebellar anomalies and, despite general retardation in mental abilities, display a good preservation of linguistic skills (Joubert et al., 1969;

Jerningan *et al.*, 1993). These observations prompted Leiner *et al.* (1993) to hypothesize a cerebellar role in language, namely that cerebellar circuits could provide the correct analysis of the sensory information needed to correctly produce language.

It should be noted, however, that despite the theory proposed by Leiner *et al.* (1993), clinical observations in diffuse central nervous system developmental pathologies do not allow more than the generic conclusion that some relationship exists between cerebellum and speech, as many of these patients may have other lesions that could cause the language disorders. In other words, these studies do not necessarily infer a possible relationship between cerebellum and linguistic competence, as the linguistic deficits in these pathologies have to be considered within the context of a general cognitive impairment. Observations of single patients with acquired cerebellar lesions allow for more specific hypotheses about the relationship between cerebellum and defined aspects of linguistic function, particularly word fluency and sentence production.

III. Verbal Fluency

Verbal fluency is the capacity to generate lists of words according to a given rule that may be either a letter of the alphabet (e.g., retrieval of words that begin with the letter F) or a semantic category (e.g., retrieval of words from the semantic category of "animals"). Verbal fluency performances on both semantic categories and letters tasks are generally considered to depend on both verbal and executive, namely "frontal" abilities (Ramier and Hecaen, 1970; Milner, 1964). Despite data suggesting a substantial contribution of temporal regions (Parks *et al.*, 1988) to word fluency, this task is considered a useful test in exploring mostly frontal lobe functioning (McCarthy and Warrington, 1990).

Frontal cortices, including nonmotor prefrontal areas, are reciprocally interconnected with the cerebellum (Middleton and Strick, 1994; Schmahmann and Pandya, 1995) and, based on clinical evidence, the possibility has been advanced that cerebellar patients might present some sort of "mild frontal syndrome" (Botez *et al.*, 1985; Grafman *et al.*, 1992). Verbal fluency testing has been used by different groups to address the question of the "frontal" abilities in cerebellar patients and conflicting results have been reported.

Verbal fluency deficits in both semantic category and letters have been reported in two studies on patients with degenerative cerebellar pathologies (Akshoomoff *et al.*, 1992; Appollonio *et al.*, 1993). However, these findings were not confirmed by Helmuth and Ivry (1994) in which atrophic and focal cerebellar patients were tested on a semantically cued attention shifting task (Posner's cueing task), a verbal fluency task, and a verbal discrimination task. In all tasks, cerebellar patients presented relatively spared performances despite the severity of their motor impairment. In a well-documented single case study, Fiez et al. (1992) reported difficulties in different verb generation tasks after a right lateral cerebellar lesion. Although Petersen et al. (1989) demonstrated a specific cerebellar activation during linguistic tasks using positron emission tomography (PET), they hypothesized that no specific language-related deficits are present after right cerebellar damage and that the verbal performance impairments observed are related to a general inability in practice-related learning and in error detection tasks.

Leggio et al. (1995) studied word fluency ability in three groups of patients with cerebellar damage, one group with cerebellar atrophy, and two groups with focal lesions confined to the right or the left cerebellar hemisphere. Both letter and semantic fluency were evaluated. The results are summarized in Fig. 1. All groups performed at a lower level than matched controls, and the difference was more evident for letter fluency

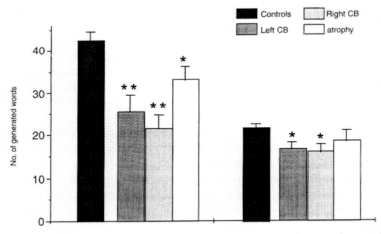

FIG. 1. The total number of generated words (verbal fluency) in letter and semantic tasks in controls and in three cerebellar patient groups. A 4 × 2 two-way ANOVA (group × fluency) revealed significant effects of group ($F_{3,51} = 9.79$; $P = 0.000033$) and fluency ($F_{1,51} = 78.06$; $P < 0.000001$). Interaction was also significant ($F_{3,51} = 8.90$; $P = 0.000076$). Asterisks indicate one-way ANOVA significance vs controls. *$P < 0.05$; **$P < 0.0005$.

than semantic fluency. Interestingly, some differences also emerged among the three groups of patients. Atrophic patients, in whom the cerebellar lesion mainly involved the vermal and paravermal regions and dysarthria and motor impairments were greatest, performed better than patients with focal lesions. In fact, whereas focal patients were significantly impaired in both letter and semantic fluency, atrophic patients differed significantly from controls only in letter fluency, and semantic fluency was only slightly inferior to the control group. Although the worst performance of the cerebellar patients vs the controls could be interpreted as a possible nonspecific effect of brain damage, the quantitative difference among the three groups of patients with cerebellar lesions indicates some specific role of the cerebellum in this ability.

The evidence that patients with left lesions perform better than patients with right lesions is of particular interest. Indeed, whereas the difference does not reach statistical significance, this tendency is consistent with previous findings attributing a prevalence of the right cerebellar hemisphere in linguistic production, probably by means of cross connections with the language dominant left cerebral hemisphere (Petersen et al., 1989).

The better word fluency of cerebellar atrophy patients compared to focal patients clearly indicates that word fluency reduction is not related to motor deficits, i.e., dysarthria. In fact, atrophic patients, with the most affected motor abilities, have the least affected word fluency.

Further considerations arise from the different anatomical distribution of the lesions of focal and atrophic patients. Lesions were mainly vermal and paravermal in the atrophy group, but were almost completely confined to the lateral part of the hemispheres in the focal group. PET findings indicate that the cerebellar hemispheres are specifically involved in cognitive functions whereas motor tasks induce activation of the vermis and the paravermal regions (Jenkins et al., 1994). Our findings, which demonstrate an association of medial lesions with a prevalence of motor deficits and lateral lesions with linguistic deficits, provide further support for these functional studies. Nevertheless, it must also be stressed that the nature (vascular and neoplastic vs degenerative) and the time course (abrupt vs progressive) of cerebellar damage may play some role in producing either quantitative or qualitative clinical differences. Because of their nature, focal lesions are more likely to produce massive damage of the neural substrate and thus more clear-cut modular deficits than atrophic lesions, which are more likely to be associated with diffuse cognitive impairment. Thus, at least some of the differences between atrophic and focal patients might be related to the pathology of the lesion more than to its anatomical distribution. A solution to this problem might be found by comparing patients with vascular lesions in the medial and lateral cerebellum, respectively.

IV. Fluency Strategies

A qualitative analysis of subjects' responses in word fluency revealed further differences among cerebellar groups. It has been argued that generating a list of words according to a given rule, whatever the criterion requested, is normally achieved by clustering words according to semantic or phonemic criteria (Lezak, 1995). Clustering is differentially defined according to the verbal fluency task examined. In *letter fluency,* two or more successive words are produced that share not only the same first but also the same second phoneme (e.g., freddo [cold] frase [sentence]). Two or more successive words that rhyme with each other (e.g., allòro [laurel] adòro [adore]) are considered as a phonemic cluster, whereas any two successive words sharing the same semantic category (e.g., ananas [pineapple] arancia [orange]) or that are two forms of a verb (e.g., amare [to love] amatore [lover]) are classified as semantic clusters. *Semantic fluency* phonemic clusters include two successive words beginning with the same phoneme (e.g., canarino [canary] condor [condor]) or which rhyme (e.g., pappagallo [parrot] gallo [cock]), whereas semantic clusters are represented by two successive words from the same subcategory (e.g., falco [hawk] aquila [eagle] subcategory wild birds of the broad bird category). In normal subjects, phonemic and semantic clustering are adopted concurrently, although to a different extent, in both semantic and letter fluency. However, clustering strategies can be differentially affected by brain lesions. For instance, semantic clustering has been reported to be selectively affected in Alzhemeir's disease patients and in patients with damage to the right posterior cortical structures (Rosen, 1980; Laine, 1988; Laine and Neimi, 1988), whereas no semantic clustering impairment has been reported after left anterior lesions (Laine and Neimi, 1988). Thus, cluster analysis would allow investigating the strategies adopted by the subject in the retrieval process.

Different studies have suggested that the cerebellum might be involved in strategy planning and in controlling the procedural aspects of different cognitive functions (Salmon and Butters, 1995; Petrosini *et al.,* 1996). Clustering analysis could shed some light on the strategic and procedural aspects of word fluency. Clustering was analyzed in word lists produced by the three groups of patients in letter and semantic fluency. Semantic and phonemic clustering strategies were identified and quantified according to the procedure described by Raskin *et al.* (1992), which was briefly indicated earlier. In agreement with previous reports (Raskin *et al.*, 1992), the strategy of semantic clustering (considering semantic clustering in semantic and letter tasks) predominates over phonemic clustering in controls as well as in the three cerebellar groups (Fig. 2). Semantic clustering was well preserved in all cerebellar patients, both in letter and in semantic tasks, and no significant differences were detected between cerebellar patients and controls (Fig. 2). However, all three cerebellar groups presented a deficit in clustering words phonologically in both letter and semantic tasks (Fig. 2). Thus, all cerebellar groups present a pattern opposite to that of Alzheimer patients, with a specific impairment of phonological clustering. This evidence provides further support for the hypothesis that cerebellar lesions do not generically affect the word fluency process, but play a specific role in the strategic processes involved in word retrieval.

At present, it is difficult to further clarify why cerebellar damage affects phonological clustering. However, it is tempting to speculate that the cere-

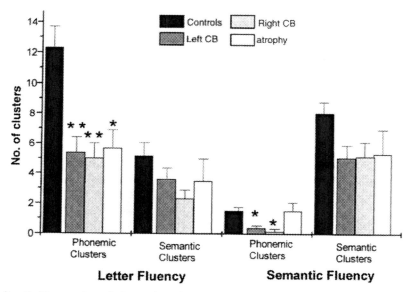

FIG. 2. The number of phonemic and semantic clusters on letter and semantic tasks by control and cerebellar patient groups. A 4 × 4 two-way ANOVA (group × cluster type) revealed significant effects of group ($F_{3,51} = 7.17$; $P = .0.000416$) and of cluster type ($F_{3,153} = 24.32$; $P < 0.000001$). Interaction was also significant ($F_{9,153} = 3.24$; $P = 0.001233$). Asterisks indicate one-way ANOVA significance vs controls. *$P < 0.05$; **$P < 0.005$.

bellar phonemic clustering deficit might depend on difficulties in coupling phonological motor output with the auditory sensory feedback required for phonological grouping. Although this is only a working hypothesis, it fits nicely with the proposed role of the cerebellum as a central coordinator for different interactive central functions (Silveri et al., 1994), as well as with the emphasized importance of the cerebellum in sensory discrimination (Gao et al., 1996).

V. Agrammatism

Further evidence supporting cerebellar involvement in language comes from two recent reports of specific linguistic deficits, namely agrammatism, in patients with focal cerebellar lesions (Silveri et al., 1994; Zettin et al., 1995). In the case reported by Silveri et al. (1994), the linguistic deficit was very specific. No impairments were detected on an extensive battery that explored general intelligence, orientation, memory, visuospatial skills, praxis, and frontal lobe functions. Language examination was normal for all parameters, including sentence comprehension, with the notable exception of dysarthria and agrammatic speech. A sample of the patient's spontaneous speech is reported in Fig. 3.

One of the main criticisms raised against the existence of cognitive deficits in cerebellar patients derives from the difficulty of precisely characterizing the anatomical extent of the lesion and thus reliably correlating the deficits observed to a lack of cerebellar functioning (Daum and Ackermann, 1995). In the agrammatic patient reported by Silveri et al. (1994), the correlation between cerebellar lesion and cognitive symptoms was consistent, not only because of the neuroradiological evidence of cerebellar damage in the absence of supratentorial lesions, but also because the evolution of the motor cerebellar symptoms (right side dysmetria and adiadochokinesia, bilateral gaze evoked horizontal nystagmus) paralleled the observed cognitive linguistic impairment. A further consideration of the relationship between cerebellar lesions and cognitive functioning derives from the time course of cerebral blood flow (rCBF) alterations in this patient. Consistent with the reported inverse cerebrocerebellar diaschisis following a cerebellar infarct (Broich et al., 1987), a diffuse rCBF reduction in the left hemisphere was present immediately after the onset of agrammatic speech. Four months later, motor deficits were milder and speech was only occasionally agrammatic. A new single photon emission tomography demonstrated a reduction of the left hemisphere hypoperfusion. Interestingly, rCBF reduction was

1.   ho sentito
2. Io guardavo   [la] televisione. Nel momento dopo, subito dopo, sentire   [che]
3. I was watching   television. One moment after, immediately after feel(inf)

1.   andava . Avevo   potevo
2. metà   [del corpo] non andare. Avere   un attacco, non potere parlare.
3. one half   not go(inf). Have(inf) an attack, not able to (inf) speak.

2. Sopra   c'era mia moglie che dormiva perchè era mezzanotte. Io tutto a un tratto
3. Upstairs my wife was sleeping   because it was midnight. I   suddenly

1.   sono alzato   sono andato   sono fatto
2. [mi] alzare,   tutto a un tratto andare   giù per terra.   Non [mi] fare
3.   stand up(inf) suddenly   fall(inf)   down on the ground. Not
       do(inf)

1.   parlavo   ho vomitato
2. niente   perchè c'era il tappeto.   Non parlare, poi vomitare   tutta la
   notte.
3. nothing because there was the carpet. Not speak(inf) then vomit(inf) all night
   long.

1.   ho rovesciato   ha praticato
2. Io rovesciare.   E' venuto mio figlio, subito una iniezione   [mi] pratica,   e   le
3. I   vomit (inf)   My son came,   immediately an injection   he does   and
   the

1.   ho aspettato   entrassi
2. flebo. Io aspettare   che qui al Policlinico   entra
3 drips. I   wait(inf.)   that here at the Policlinic is admitted

1.   Ho aspettato   fosse
2. che [ci fosse] vuoto   il posto. Aspettare   che qui [ci] stare   il posto vuoto,
   [che mi]
3. that   free   the bed. Wait(inf.)   that here   be(inf) the bed   free

1. dessero
2. dare   il posto vuoto.
3. give(inf.) the bed free.

FIG. 3. A sample of spontaneous speech from an agrammatic patient reported by Silveri et al. (1994). Patient's description of his history of illness; Line 1, correct target words; line 2, patient's speech (underlined); and line 3, word-by-word English translation of the patient's speech. inf, infinitive form; omissions are shown in brackets.

more persistent in the left temporal region, which is one of the regions where grammatical processing is hypothesized to occur.

It seems reductive to consider the observed cognitive deficit as depending on a general diaschisis effect related to a lack of cerebellar input to the left hemisphere. Furthemore, because the neurobiological phenomena that induce diaschisis, as well as the clinical implication of the diaschisis-related changes in rCBF, are still a matter of discussion (Meyer et al., 1995), to refer to a functional phenomenon as due to diaschisis does not add useful information.

Zettin et al. (1995) reported a similar patient. In their presentation they stressed the possibility that agrammatism in cerebellar-damaged patients may be due to compensatory mechanisms. This interpretation of agrammatism is not new. Influenced by the simplicity of the agrammatic speech appearing as a form of "telegraphic communication" and by the evidence that agrammatism does not occur in isolation from other speech impairments, early researchers suggested that agrammatism could be the result of a compensatory mechanism, i.e., the use of a simplified emergency language that allows for an economy of effort (Isserlin, 1922). The "economy of efforts" hypothesis has been criticized by different authors. Luria (1970) went well beyond the simplistic theory of effort economy, suggesting that agrammatism refers to a disturbance of the dynamic aspects of language that prevents the development of the "dynamic schemata of sentences." More recently, the theoretical framework for syntactic production has been under close scrutiny, and different hypotheses have been advanced (for a discussion of agrammatism see Miceli et al., 1989; Badeker and Caramazza, 1985; Grodzinsky, 1991, Caplan, 1991).

Criticism of the economy of efforts theory is essentially based on two points. First, verbal production in agrammatic patients is not always simpler than the correct form, e.g., in agrammatism, verbs are often produced in the infinitive, even when the inflexed form would be shorter. Second, there is clear evidence that patients with severe articulatory disorders may not be affected by speech agrammatism (Ackerman et al., 1992). Both observations are also valid in interpreting agrammatism after cerebellar lesions. Therefore, if agrammatism after a cerebellar lesion is not a trick to compensate for motor difficulties in speech, how does it develop?

In line with the observation of Luria (1970) on the dynamic aspects of agrammatism and the known role of the cerebellum in timing motor sequences (Braitenberg et al., 1996), a different hypothesis on the importance of the cerebellar computational properties for cognitive sentence processing can be advanced. As stressed by Caplan (1995), different linguistic functions pertaining to different neuronal circuits must be mastered to complete syntactic processing. Such complex interarea coordination requires a precise temporal and sequential organization. Nespoulous et al. (1988) suggested that agrammatism may be the result of a patient's adaptation to a deficit lying outside the mental linguistic system. Thus, difficulties in the syntactic constructions of sentences may derive from lesions of structures outside the language-specific cortical areas. To put grammatical morphemes into accordance, they must be available in the working memory during the application of syntactic rules. If the latter process moves slowly or is not on a parallel with the working memory availability of morphemes, the sentence construction will disintegrate. The cerebellum has all the credentials needed to serve as the interarea functional coordinator to produce grammatically correct sentences. In fact, timing and cerebellum have been associated by different experimental and clinical evidence (see Ivry, this volume) and, at least for the motor domain, the modularity of the cerebellar circuits has been associated with the ability to generate functional sequences (Ito, 1982; Braitenberg et al., 1997). Taken together, these data indicate that the cerebellum might provide the correct timing needed for an optimal functional interrelationship among the different functional modules needed for the cognitive processing of words and sentences.

VI. Dysgraphia

Support for this hypothesis of the cerebellum as a cognitive controller outside the main cognitive processor but needed for intermodule coordination derives from an observation of peripheral afferent dysgraphia in a patient affected by cerebellar atrophy (Silveri et al., 1996). According to the most recent theories on dysgraphias, peripheral afferent dysgraphia is characterized by two different groups of deficits. The first group, the so-called neglect-related features, is represented by the tendency to write on the right-hand side of the page and by the difficulty of maintaining horizontal lines. Omissions and repetitions of strokes and letters form the second group of dysgraphia deficits (feedback-related features), which have been related to defects of visual and proprioceptive feedback during writing movements. Interestingly, in the case reported, in addition to handwriting difficulties consistent with movement dysmetria, some high-order problems with the characteristics of the feedback-related features of the peripheral afferent dysgraphia were present. Because no sensorial or praxic deficits were present, it has been suggested that the lack of effectiveness of the sensorial feedback, similar to the hypothesis advanced for interpreting peripheral dysgraphia after right hemispheric lesions, can be due to a complex disorder of sensorial feedback related to a form of *inattention* to the feedback. In the case of cerebellar lesions, this *inattention* may be due to the uncoupling of motor planning and proprioceptive feedback. Again this event may be a consequence of the lack of cerebellar ordering and timing control over cortical areas.

VII. Hypothesis for a Cerebellar Role in Language

In concluding this chapter, the authors would like to stress that in all the linguistic problems reported—verbal fluency, phonemic clustering, and agrammatism—it is possible to localize the functional deficit outside the main linguistic processor system. This hypothesis is consistent with current pathophysiological interpretations of speech production. Furthermore, in all three language-related deficits reported, an uncoupling of sensory feedback and motor output can be hypothesized, giving support to the recently advanced theory that implicates the cerebellum in sensory acquisition and discrimination rather than in motor control (Gao *et al.*, 1996).

However, the cerebellar circuits represent a highly apt system for performing ordering and timing tasks (Braitenberg *et al.*, 1997), and the vast interconnection with the cerebral cortex (see J. D. Schmahmann and D. N. Pandya, this volume) suggests that this function may also be effective in coordinating cortical functions. Along these lines, it cannot be excluded that the cerebellar-dependent cognitive deficits may be a consequence of a lack of coordination and optimization of cortico-cortical functional interrelations physiologically under cerebellar jurisdiction.

Acknowledgments

This research was supported by MURST and CNR grants.

References

Ackerman, H., Vogel, M., Petersen, D., and Poremba M. (1992). Speech defcts in ischaemic cerebellar lesions. *J. Neurol.* **239,** 223–227.
Akshoomoff, N. A., Courchesne, E., Press, G. A., and Iragui, V. (1992). Contribution of the cerebellum to neuropsychological functioning: Evidence from a case of cerebellar degenerative disorder. *Neuropsychologia* **30**(4), 315–328.
Ammirati, M., Mirzai, S., and Samii, M. (1989). Transient mutism following removal of a cerebellar tumor: A case report and review of the literature. *Child. Nerv. Syst.* **5,** 12–14.

Apollonio, I. M., Grafman, J., Schwartz, M. S., Massaquoi, S., and Hallett, M. (1993). Memory in patients with cerebellar degeneration. *Neurology* **43,** 1536–1544.

Badecker, W., and Caramazza, A. (1985). On considerations of method and theory governing the use of clinical categories in neurolinguistics and cognitive neuropsychology: The case against agrammatism. *Cognition* **20,** 97–125.

Bauman, M., and Kemper, T. L. (1985). Histoanatomic observations of the brain in early infantile autism. *Neurology* **35,** 866–874.

Botez, M. I., Gravel, J., Attig, E., and Vezina, J.-L. (1985). Reversible chronic cerebellar ataxia after phenytoin intoxication: Possible role of cerebellum in cognitive thought. *Neurology* **35,** 1152–1157.

Braitenberg, V., Hack, D., and Sultan, F. (1997). The detection and generation of sequences as a key to cerebellar function: Experiments and theory. *Behav. Brain Sci.*, in press.

Broich, K., Hartmann, A., Biersack, H. J., and Horn, R. (1987). Crossed cerebello-cerebral diaschisis in a patient with cerebellar infarction. *Neurosci. Lett.* **83,** 7–12.

Caplan, D. (1991). Agrammatism is a theoretically coherent aphasic category. *Brain Lang.* **40,** 274–281.

Caplan, D. (1995). The cognitive neuroscience of syntactic processing. *In* "The Cognitive Neurosciences" (M. S. Gazzaniga, ed.), pp. 871–879. MIT Press, Cambridge/London.

Courchesne, E., Townsend, J., and Saitoh, O. (1994). The brain in infantile autism: Posterior fossa structures are abormal. *Neurology* **44,** 214–223.

Courchesne, E., Yueng-Couchesne, R., Press, G. A., Hesselink, J. R., and Jernigan, T. L. (1988). Hypopalsia of cerebellar vermal lobules VI and VII in autism. *N. Engl. J. Med.* **318,** 1349–1354.

Daum, I., and Ackerman, H. (1995). Cerebellar contributions to cognition. *Behav. Brain Res.* **67,** 201–201.

Fiez, J. A., Petersen, S. E., Cheney, M. K., and Raichle, M. E. (1992). Impaired non-motor learning and error detection associated with cerebellar damage. *Brain* **115,** 155–178.

Fraioli, B., and Guidetti, B. (1975). Effects of stereotactic lesions of the dentate nucleus of the cerebellum in man. *Appl. Neurophysiol.* **38,** 81–90.

Gao, J.-H., Parsons, L. M., Bower, J. M, Xiong, J., Li, J., and Fox, P. T. (1996). Cerebellum implicated in sensory acquisistion and discrimination rather than motor control. *Science* **272,** 545–547.

Gilman, S., Bloedel, J. R., and Lechtenberg, R., (1981). "Disorders of the Cerebellum." Davis, Phildelphia.

Grafman, J., Litvan, I., Massaquoi, S., Steward, M., Siriqu, A., and Hallett, M. (1992). Cognitive planning deficit in patients with cerebellar atrophy. *Neurology* **42,** 1493–1496.

Grodzinsky, Y. (1991). There is an entity called agrammatic aphasia. *Brain Lang.* **41,** 565–589.

Helmuth, L. L., and Ivry, R. B. (1994). Cognitive deficits following cerebellar lesions in humans: Studies of attention and verbal fluency. *Soc. Neurosci. Abstr.* **20,** 412.12.

Isserlin, M. (1922). Über agrammatismus. *Zeitsch. Gesamte Neurol. Psychiatr.* **75,** 322–410.

Ito, M. (1982). Questions in modeling the cerebellum. *J. Theor. Biol.* **99,** 81–86.

Jenkins, I. H., Brooks, D. J., Nixon, P. D., Frackowiak, R. S. J., and Passingham, R. E. (1994). Motor sequence learning: A study with positron emission tomography. *J. Neurosci.* **14,** 3775–3790.

Jerningan, T. L., Bellugi, U., Sowell, E., Doherty, S., and Hesselink, J. R. (1993). Cerebral morphologic distinctions between Williams and Down syndromes. *Arch. Neurol.* **50,** 186–191.

Joubert, M., Eisenring, J. J., Robb, J. P., and Andermann, F. (1969). Familial agenesis of the cerebellar vermis. *Neurology* **19,** 813–825.

Laine, M. (1988). Correlates of word fluency performance. *In* "Studies in Languages" (P. Koivuselkā-Sallinen and L. Sarajārvi, eds.), Vol. 12, University of Joensuu, Joensuu, Finland.

Laine, M., and Neimi, J. (1988). Word fluency production strategies of neurological patients: Semantic and phonological clustering. *J. Clin. Exp. Neuropsychol.* 10.

Leggio, M. G., Solida, A., Silveri, M. C., G., Gainotti, and Molinari, M. (1995). Verbal fluency impairment in patients with cerebellar lesions. *Soc. Neurosci. Abstr.* 21, 21, 917.

Leiner, H. C., Leiner, A. L., and Dow, R. S. (1993). Cognitive and language functions of the human cerebellum. *Trends Neurosci.* 16, 444–454.

Lezak, M. D. (1995). "Neuropsychological Assessment." Oxford University Press, Oxford.

Luria, A. K. (1970). "Traumatic aphasia." The Hague/Paris Mouton.

McCarthy, R. A., and Warrington, E. K. (1990). "Cognitive Neuropsychology." Academic Press, San Diego.

Meyer, J. S., Obara, K., and Muramatsu, K. (1995). Diaschisis. *Neurol. Res.* 15, 362–366.

Miceli, G., Silveri, M. C., Romani, C., and Caramazza, A. (1989). Variation in the pattern of omissions and substitutions of grammatical morphemes in the spontaneous speech of so-called agrammatic patients. *Brain Lang.* 36, 447–492.

Middleton, F. A., and Strick P. L. (1994). Anatomical evidence for cerebellar and basal ganglia involvement in higher cognitive function. *Science* 266, 458–461.

Milner B. (1964). Some effects of frontal lobectomy in man. *In* "The Frontal Granular Cortex and Behaviour" (J. M. Warren and K. Akert, eds.), pp. 313–33. McGraw-Hill, New York.

Nespoulous, J. L., Dordain, M., Perron, C., Ska, B., Bub, D., Caplan, D., Meheler, J., and Roch Lechours, A. (1988). Agrammatism in sentence production without comprehension deficits: Reduced availability of syntactic structures and/or grammatical morphemes? *Brain Lang.* 33, 273–295.

Parks, R. W., Loewenstein, D. A., Dodrill, K. L., Barker, W. W., Yoshii, F., Chang, J. Y., Emran, A., Apicella, A., Sheramata, W. A., and Duara, R. (1988). Cerebral metabolic effects of verbal fluency test: A PET scan study. *J. Clin. Exp. Neuropsychol.* 10, 565–575.

Petersen, S. E., Fox, P. T, Posner, M. I., Mintun, M., and Raichle, M. E. (1989). Positron emission tomography studies of the processing of single words. *J. Cogn. Neurosci.* 1, 153–170.

Petrosini, L., Molinari, M., and Dell'Anna, M. E. (1996). Cerebellar contribution to spatial event processing: Morris water maze and T-maze. *Eur. J. Neurosci.* 8(8), 101–115.

Pollack, I. F., Polinko, P., Albright, A. L., Towbin, R., and Fritz, C. (1995). Mutism and pseudobulbar symptoms after resection of posterior fossa tumors in children: Incidence and pathophysiology. *Neurosurgery* 37, 885–893.

Ramier, A. M., and Hecaen, S. C. (1970). Role respecto des atteintes frontales et de la lateralisation lesionelle dans deficits de la "fluence verbale." *Rev. Neurol. (Paris)* 123, 17–22.

Raskin, S. A., Sliwinski, M., and Borod, J. C. (1992). Clustering strategies on tasks of verbal fluency in Parkinson's disease. *Neuropsychologia* 1, 95–99.

Rekate, H. L., Grub, R. L., Aram, D. M., Hahn J. F., and Ratcheson, R. A. (1985). Muteness of cerebellar origin. *Arch. Neurol.* 42, 697–698.

Rosen, W. (1980). Verbal fluency in aging and dementia. *J. Clin. Neuropsycol.* 2, 135–146.

Salmon, D. P., and Butters, N. (1995). Neurobiology of skill and habit learning. *Curr. Opin. Neurobiol.* 5, 184–190.

Schmahmann, J. D., and Pandya, D. N. (1995). Prefrontal projections to the basilar pons: Implications for the cerebellar contribution to higher function. *Neurosci. Lett.* 199, 175–178.

Silveri, M. C., Leggio, M. G., and Molinari, M. (1994). The cerebellum contributes to linguistic production: A case of agrammatism of speech following right hemicerebellar lesion. *Neurology* 44, 2047–2050.

Silveri, M. C., Misciagna, S., Leggio, M. G., and Molinari, M. (1997). Spatial dysgraphia and cerebellar lesion: A case report. *Neurology*, in press.

Van Dongen, H. R., Catsman-Berrevoets, C. E., and van Mourik, M. (1994). The syndrome of "cerebellar" mutism and subsequent dysarthria. *Neurology* **44,** 2040–2046.

Zettin, M., Rago, R., Perino, C., Messa, C., Perani, D., and Cappa, S. F. (1995). "Eloquio Agrammatico da Lesione Cerebellare: Un Secondo Caso." Fall meeting of the Neuropsychology Section of the Italian Society of Neurology, October 28, Bologna.

CLASSICAL CONDITIONING

Diana S. Woodruff-Pak

Department of Psychology
Temple University
Philadelphia, Pennsylvania 19122
and
Laboratory of Cognitive Neuroscience
Philadelphia Geriatric Center
Philadelphia, Pennsylvania 19141

I. Introduction
  A. History of Investigations of Neurobiological Substrates of Classical Conditioning
  B. The Rabbit Model System of Nictitating Membrane/Eyeblink Classical Conditioning
II. Eyeblink Classical Conditioning in Patients with Cerebellar Lesions
  A. Initial Case Study Reports
  B. Patients with Cerebellar Lesions Compared to Age-Matched Healthy Participants
  C. Comparisons of Cerebellar Lesions in Humans and Rabbits and Performance on Eyeblink Conditioning
III. Normal Aging and Eyeblink Conditioning: Cerebellar Purkinje Cell Loss
  A. Purkinje Cell Alterations during Normal Aging
  B. Purkinje Cell Correlations with Eyeblink Conditioning
  C. Counts of Interpositus Nucleus Cells and Correlations with Eyeblink Conditioning
IV. Positron Emission Tomography Detection of Cerebellar Involvement in Eyeblink Conditioning
V. Eyeblink Conditioning and Other Neuropsychological Tasks
  A. Predictors of Eyeblink Classical Conditioning
  B. Interference during Dual-Task Conditions
VI. Eyeblink Classical Conditioning in Noncerebellar Lesions, Neurodegenerative Disease, and Other Syndromes
  A. Medial Temporal Lobe Lesions
  B. Huntington's Disease
  C. Alzheimer's Disease
  D. Autism
VII. Summary and Conclusions
  References

Evidence has amassed from research in humans indicating that the cerebellar circuitry serving as the substrate for eyeblink classical conditioning is similar to that in nonhuman primates. In patients with bilateral cerebellar lesions or neurodegenerative cerebellar disease, few conditioned eyeblink responses are produced with either the ipsilesional or the contralesional eye. Cerebellar patients with lateralized lesions, like rabbits with experimentally produced unilateral cerebellar lesions, produce relatively normal conditioned responses (CRs) with the contralesional eye and few or no CRs with the ipsilesional eye. Age-related deficits in eyeblink classical conditioning appear in humans and rabbits in middle age. In normal aging in many species, including humans, there is Purkinje cell loss in cerebellar cortex. In rabbits, the Purkinje cell number correlates highly with the rate of learning, regardless of age. Positron emission tomography imaging of normal young adults during eyeblink conditioning reveals changes in activity in the cerebellum. Timed interval tapping, a task that assesses cerebellar function, also predicts performance on eyeblink conditioning. In dual-task conditions involving simultaneous performance of eyeblink conditioning and timed interval tapping, eyeblink conditioning is impaired. Investigations of patients with lesions or neurodegenerative disease not involving the cerebellum demonstrate that acquisition of CRs is possible, although prolonged in the case of hippocampal cholinergic disruption. Evidence to date suggests that the human analogue of the rabbit interpositus nucleus, the globose nucleus, is essential for the production of the conditioned eyeblink response and that cerebellar cortical Purkinje cells play a role in normal acquisition.

I. Introduction

In the widely used paradigm, classical conditioning of the eyeblink response, perhaps more than in any other form of learning, neuronal mechanisms of memory elucidated in infrahuman mammals apply directly to the human condition. An array of neuroscientific techniques have been used to demonstrate that the structure essential for learning and memory of the conditioned eyeblink response in rabbits is the cerebellum ipsilateral to the conditioned eye. More specifically, the ipsilateral dorsolateral interpositus nucleus is the site of plasticity for learning, with additional plasticity residing in the ipsilateral medial–lateral cerebellar cortex that projects to the interpositus nucleus.

Evidence has amassed from research in humans indicating that the cerebellar circuitry serving as the substrate for eyeblink classical conditioning is similar to that in rabbits. The lines of investigation that suggest this include (i) studies of patients with cerebellar lesions or neurodegenerative cerebellar disease; (ii) changes in eyeblink conditioning with age in normal adults; (iii) positron emission tomography (PET) imaging of normal young adults; (iv) correlations with measures of cerebellar function; and (v) dual-task conditions involving the simultaneous performance of tasks engaging the cerebellum. Additionally, investigations of patients with lesions or neurodegenerative disease not involving the cerebellum have provided further parallels with the rabbit model system of eyeblink classical conditioning. The human analogue of the rabbit interpositus nucleus, the globose nucleus, appears to be the site of plasticity for acquisition of the conditioned eyeblink response. As it does in rabbits, the cerebellar cortex also plays a role in acquisition.

A. History of Investigations of Neurobiological Substrates of Classical Conditioning

The history of the study of the neurobiological substrates of eyeblink classical conditioning is relatively long, beginning at the start of the 20th century. After receiving the Nobel Prize in 1904 for his research on the physiology of digestion in dogs, Ivan Petrovitch Pavlov spent much of the rest of his career formally investigating the phenomenon of classical conditioning (Pavlov, 1927). One of Pavlov's former students, Jerzy Konorski (1948), attempted to recruit neuropsychological and cellular mechanisms for the exploration of conditioning phenomena. In the United States, Ernest Hilgard conducted prototypical studies on human eyeblink conditioning (Hilgard and Campbell, 1936). Hilgard's work established the close correspondence in properties of the conditioned eyeblink response in humans and other animals, suggesting that the underlying neuronal mechanisms of memory storage and retrieval may be the same in all mammals, including humans (Hilgard and Marquis, 1936).

B. The Rabbit Model System of Nictitating Membrane/Eyeblink Classical Conditioning

Much of the general literature on classical conditioning is based on data collected with the human eyeblink conditioning paradigm and in the rabbit nictitating membrane or "third eyelid" paradigm first introduced by Isadore Gormezano (Gormezano et al., 1962; Schneiderman et al., 1962). Evidence has converged from a number of sources to suggest that the cerebellum ipsilateral to the conditioned eye is essential for eyeblink classical conditioning in rabbits and humans.

The most extensive body of literature linking the cerebellum and eyeblink classical conditioning comes from research with animals. This research is eloquently described by its primary instigator and motivator,

Richard F. Thompson *et al.*, this volume, and will be mentioned only briefly here. A variety of techniques including electrophysiological recording of multiple and single units, electrolytic and chemical lesions, physical and chemical reversible lesions, neural stimulation, genetic mutations, and pharmacological manipulation have been used to demonstrate that the dorsolateral interpositus nucleus ipsilateral to the conditioned eye is the essential site for acquisition and retention (Berthier and Moore, 1986, 1990; Chen *et al.*, 1996; Clark *et al.*, 1992; Gould and Steinmetz, 1994; Krupa *et al.*, 1993; Lavond *et al.*, 1985; Lincoln *et al.*, 1982; McCormick *et al.*, 1981; McCormick and Thompson, 1984a,b; Steinmetz *et al.*, 1992; Thompson, 1986, 1990; Yeo *et al.*, 1985). Involvement of the cerebellar cortex has also been demonstrated during normal acquisition, although it may not be essential (Chen *et al.*, 1996; Lavond and Steinmetz, 1989).

II. Eyeblink Classical Conditioning in Patients with Cerebellar Lesions

Whereas it has been possible to investigate neural substrates of eyeblink classical conditioning extensively in animals using invasive lesion and recording techniques, such experimental manipulations are not feasible in humans. Instead, a body of literature is amassing that reports eyeblink classical conditioning performance in patients with focal brain lesions and those with neurodegenerative diseases. These studies consistently indicate that the circuitry for eyeblink classical conditioning is similar in animal and human brains.

A. Initial Case Study Reports

It appears that cerebellar lesions in humans impair or prevent acquisition of the conditioned eyeblink response. The initial evidence for the role of the cerebellum in eyeblink classical conditioning in humans was provided in a case study report by Lye *et al.* (1988). These investigators carried out eyeblink classical conditioning, testing both the right and left eye, in a 62-year-old patient with a spontaneous right cerebellar hemisphere infarction almost 6 years previously. Magnetic resonance imaging confirmed that the lesion was restricted to the right cerebellum. The patient had a normal eyeblink unconditioned response (UR) in both eyes to every corneal airpuff unconditioned stimulus (US), but showed few conditioned responses (CRs) in the right eye in 396 pairings of a tone-conditioned stimulus (CS) and corneal airpuff US. The left eye showed large CRs beginning on the second trial of training (and after 396 paired CS–US presentations to the right eye) and continuing for 35 of 36 CS–US presentations. This pattern of good left eye conditioning and poor right eye conditioning was maintained throughout four further sequential reversals of the eye receiving the US. Neurological, audiometric, and electrophysiological examinations of the patient suggested that poor eyeblink classical conditioning was not attributable solely to primary sensory or motor deficits.

A second case study report was presented by Solomon *et al.* (1989) who observed impaired eyeblink classical conditioning in the 400-msec delay paradigm, relative to control subjects. The 54-year-old patient had cerebellar dysfunction associated with an atrial myxoma (tumor in the heart). The subject presumably suffered multiple strokes over a 3-year period due to emboli from the tumor. The researchers suggested that the lesions interrupted afferents to the cerebellum, although anatomic confirmation of damage to the inferior cerebellar peduncle was lacking and the total extent of deafferentation was unclear. In 100 eyeblink conditioning trials, the patient produced only six CRs and never emitted two consecutive CRs. This 6% CR rate is significantly lower than observed with normal age-matched humans who produce 40–50% CRs. Indeed, when human subjects are presented with explicitly unpaired presentations of the CS and US, i.e., when the tone never precedes the airpuff in a predictable manner, they produce around 6% CRs (Durkin *et al.* 1993; D. S. Woodruff-Pak and M. Jaeger, manuscript submitted for publication). Thus, this patient was performing at baseline as if there was no association at all between the CS and the US.

B. Patients with Cerebellar Lesions Compared to Age-Matched Healthy Participants

Topka *et al.* (1993) investigated eyeblink classical conditioning in a group of 12 patients with cerebellar atrophy using a shock US paired with a tone CS. The eyeblink UR was normal in these patients, but their ability to acquire CRs was severely impaired. Daum *et al.* (1993) assessed eyeblink classical conditioning and electrodermal conditioning in 7 patients with cerebellar degeneration and found severely disrupted eyeblink classical conditioning but normal autonomic conditioning.

Woodruff-Pak *et al.* (1996a) tested six patients with unilateral and seven with bilateral cerebellar lesions. The effects of lesions of various parts of the cerebellum (cerebellar cortex medial versus lateral, cerebellar deep nuclei including globose nucleus) were also examined. Cerebellar timing measures and a measure of declarative memory were assessed to investigate how cerebellar disruption affected these behaviors. Results replicated the findings of Daum *et al.* (1993) and Topka *et al.* (1993) of eyeblink classical conditioning in patients with bilateral cerebellar lesions. Comparisons among six patients with lateralized lesions and age-matched healthy control participants also tested on the left and right eye demonstrated that eyeblink classical conditioning was significantly impaired on the lesioned side but less impaired on the side contralateral to the lesion (see Fig. 1). The eye contralateral to the lesion acquired CRs about as well as that of normal adults. These results parallel observations in ipsilateral cerebellar lesions and eyeblink classical conditioning in rabbits (e.g., Lavond *et al.*, 1985; Woodruff-Pak *et al.*, 1993). Results suggest that the brain circuitry for acquisition of the conditioned eyeblink response, at least in the delay eyeblink classical conditioning paradigm, is similar in rabbits and humans.

C. COMPARISONS OF CEREBELLAR LESIONS IN HUMANS AND RABBITS AND PERFORMANCE ON EYEBLINK CONDITIONING

Rabbits with permanent (Lincoln *et al.*, 1984) or reversible (Clark *et al.*, 1992; Krupa *et al.*, 1993) lesions of interpositus nucleus are unable to

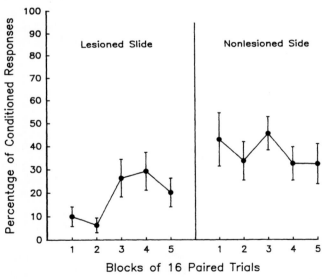

FIG. 1. Percentage of conditioned responses in six patients with unilateral lesions of the cerebellum in two five-block sessions. Cerebellar patients were tested first for five 16-trial blocks using the eye ipsilateral to the lesion. They were then tested for five 16-trial blocks using the eye contralateral to the lesion. Conditioning was carried out in the delay paradigm with a 400-msec interval between the tone-conditioned stimulus and the corneal airpuff unconditioned stimulus.

acquire CRs with the eye ipsilateral to the lesion, but have a high percentage of CRs with the eye contralateral to the lesion. Woodruff-Pak *et al.* (1996a) studied eyeblink classical conditioning in patients with cerebellar pathology comparable to the experimental lesions in rabbits. A 73-year-old man with a left cerebellar aneurysm that destroyed the deep nuclear region, including globose nucleus, produced no CRs with the left eye. In contrast, the patient produced 28% CRs with the right eye which is within the low–normal range for his age (see Fig. 2). Another patient with a unilateral lesion that included a small portion of the deep nuclear region had suffered a cerebellar infarct 2 years before eyeblink classical conditioning testing. This 57-year-old man had only 12% CRs with the eye ipsilateral to the lesion (right eye), but he had 74% CRs with the contralesional eye.

Patients with severe cerebellar degeneration bilaterally were shown to be severely impaired in eyeblink classical conditioning (Woodruff-Pak *et*

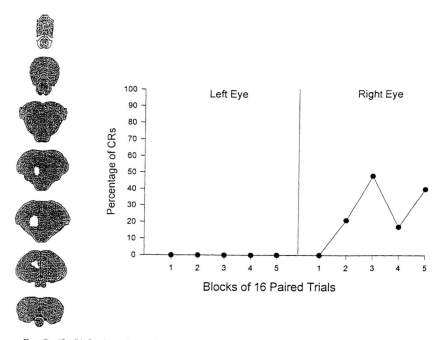

FIG. 2. (Left) Lesion size and extent depicted for a 73-year-old cerebellar patient with an aneurysm in the region including the left globose nucleus. Data are presented on a standardized set of templates. A horizontal series of seven slices of the cerebellum is shown with the lesion in white. The side of the template matches the side of the actual lesion. From top to bottom, the templates run from most superior to most inferior. The globose nucleus is located in the fourth and fifth slices from the top. (Right) Percentage of conditioned responses of this patient in the left (eye ipsilateral to the lesion) and right eye.

al., 1996a). One 60-year-old patient with bilateral atrophy of the cerebellum and alcoholism produced 1% CRs with the left eye and 12% CRs with the right eye (Fig. 3). A 23-year-old man with Holmes disease and severe cerebellar degeneration produced 8% CRs with the right eye and 7% CRs with the left eye. When he was tested a second time on eyeblink classical conditioning, this man performed even more poorly, producing 5% CRs with the right eye and none with the left eye.

McCormick and Thompson (1984a) reported that rabbits with lateral cerebellar cortical lesions do not have impaired eyeblink classical conditioning. A comparable case in humans occurred with a 65-year-old man with a small infarct in the lateral right superior cerebellar hemisphere (Woodruff-Pak *et al.*, 1996a). This patient produced 75% CRs with the right eye and 69% CRs with the left eye (Fig. 4). His performance was comparable to a healthy subject of his age at the high–normal end of the range. His

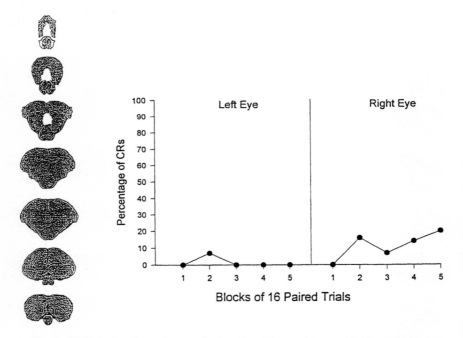

FIG. 3. (Left) Lesion size and extent depicted in a 60-year-old man with bilateral degeneration of cerebellum and alcoholism. Data are presented on a standardized set of templates. A horizontal series of seven slices of the cerebellum is shown with the lesion in white. The side of the template matches the side of the actual lesion. From top to bottom, the templates run from most superior to most inferior. (Right) Percentage of conditioned responses of this patient in the left and right eye.

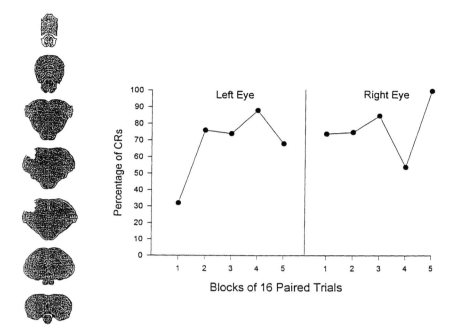

FIG. 4. (Left) Lesion size and extent depicted in a 65-year-old man with a small left cerebellar infarct in the superior hemisphere. In rabbits, lesions in the lateral cerebellar cortex do not impair acquisition or retention of the conditioned eyeblink response. Data are presented on a standardized set of templates. A horizontal series of seven slices of the cerebellum is shown with the lesion in white. The side of the template matches the side of the actual lesion. From top to bottom, the templates run from most superior to most inferior. (Right) Percentage of conditioned responses of this patient in the left and right eye.

data showed a learning curve in the initial conditioning in the left eye, and his high level of performance carried over to the first block of trials in the right eye.

It therefore seems that eyeblink classical conditioning is impaired ipsilaterally in patients with unilateral cerebellar lesions and bilaterally in patients with bilateral cerebellar lesions. Furthermore, there appears to be a dissociation between eyeblink classical conditioning and declarative learning and memory (Woodruff-Pak *et al.*, 1996a). Assessments of immediate and delayed recall on the revised Wechsler Memory Scale (WMS-R) Verbal Paired Associates were similar in cerebellar patients and healthy control subjects. Declarative memory was not impaired in cerebellar patients whose nondeclarative memory (as assessed by eyeblink classical conditioning) was severely impaired.

Timed interval tapping is a task that is impaired in cerebellar patients (see chapter by R. Ivry). This task makes subjects judge intervals by presenting 12 pacing tones and then requiring subjects to tap at the same pace for 31 trials when there is no pacing tone. The six patients with lateralized cerebellar lesions tested on eyeblink conditioning showed greater variability (indicating poorer performance) with the hand ipsilateral to the lesions. In normal control subjects, there was a significant correlation between eyeblink classical conditioning and the clock component but not the motor component of the timed interval tapping score. The clock component assesses variability due to the timing of the response and has been shown to be abnormal in patients with lateral cerebellar cortical lesions (Ivry and Keele, 1989; Ivry *et al.*, 1988). The CR in eyeblink classical conditioning requires precise timing and is optimal when it peaks shortly before US onset. The result that the clock component correlated significantly with percentage of CRs supports the perspective of Keele and Ivry (1990) that the cerebellum plays a role in the precise timing of the CR in eyeblink classical conditioning.

III. Normal Aging and Eyeblink Conditioning: Cerebellar Purkinje Cell Loss

Eyeblink classical conditioning is performed more poorly by aged organisms in four species: humans, cats, rats, and rabbits (for review see Woodruff-Pak, 1990). Studies of eyeblink classical conditioning over the adult human age span documented poorer performance in older subjects (Durkin *et al.*, 1993; Solomon *et al.*, 1989; D. S. Woodruff-Pak and M. Jaeger, manuscript submitted for publication; Woodruff-Pak and Thompson, 1988). Given the generality of the results of age-related impairment in eyeblink conditioning across species, an age-related change occurring across species was sought to account for these findings. Using the rabbit model system to implicate potential causal mechanisms, two cerebellar cellular sites were apparent in which the CS and US information was integrated: Purkinje cells in cerebellar cortex and interpositus nucleus principal cells.

A. Purkinje Cell Alterations During Normal Aging

Loss of Purkinje cells with age has been reported in many mammalian species. Indeed, describing age-related changes in the brain for 47 species of vertebrates, Dayan (1971) stated "there is a generalized loss of neurons most easily detected as fall-out of Purkinje cells from the cerebellar cortex" (p. 37).

Age-related alterations in Purkinje cells are not confined to cell loss. Studies in aged rodents have shown reductions in synaptic contacts between parallel fibers and Purkinje cell dendrites (Glick and Bondareff, 1979; Rogers et al., 1984) and decreases in spine density and in total length of Purkinje cell dendrites (Pentney, 1986). Electrophysiological assessment of Purkinje cells in aged rats showed an increase in cells with abnormal firing patterns (Rogers et al., 1981). These abnormal cells also exhibited poor Nissl staining. Alterations in physiological or biochemical processes may reduce the efficiency of surviving neurons, thus compromising the overall capacity of systems to encode memory traces. Nandy (1981) reported a 44% decrease in the number of Purkinje cells in the left cerebellar cortex of 20-year-old rhesus monkeys as compared to 4-year-old monkeys. The number of granule cells in the same area was relatively equal in the two age groups.

In human cerebellum, Ellis (1920) made a painstaking count of Purkinje cells and found them to decrease with age in adulthood. Hall et al. (1975) reported that up to 25% of the Purkinje cells in the human cerebellum were lost in very old patients. Ninety normal cerebella were assessed in subjects ranging from childhood to >100 years. Wide variation was found within age groups at all ages, but a mean reduction of 2.5% of the Purkinje cells per decade was observed which represents a 25% reduction over the 100-year period of life that was studied. Cerebellar Purkinje cell counts require postmortem analysis. Imaging resolution in humans is not fine enough to reveal presence or absence of individual Purkinje cells. Therefore, investigations of Purkinje cell–eyeblink classical conditioning relationships in living humans are not yet possible.

Thompson's (1986) model for the neural circuitry for eyeblink classical conditioning suggests that Purkinje cells play an important role in learning and retention. If there is pruning of the synaptic tree or the total loss of Purkinje cells in aging, the net effect would be to eliminate some of the central coordination between the mossy fiber–granule cell–parallel fiber CS input and the climbing fiber US input. Reducing the number of Purkinje cells may reduce the degree to which the organism can be conditioned. One means of eliminating Purkinje cells is to aspirate cerebellar cortex. When the cerebellar cortex is aspirated, acquisition ipsilateral to the aspiration is delayed (Lavond and Steinmetz, 1989) and retention is impaired (Lavond et al., 1987). Of course, cerebellar cortical aspiration removes all neural tissue, not just Purkinje cells. More selective elimination of Purkinje cells was achieved by Chen et al. (1996) who tested mutant mice with Purkinje cell degeneration (*pcd*). These mutant mice are born with Purkinje cells, but the cells are lost by the third and fourth postnatal weeks. At the time of testing when the *pcd* mice were adults, histology revealed that they had no cerebellar Purkinje cells. *pcd* mice acquired CRs, but their acquisition was delayed in a manner similar to animals with cerebellar cortical aspirations.

B. PURKINJE CELL CORRELATIONS WITH EYEBLINK CONDITIONING

In rabbits, it has been possible to carry out Purkinje cell counts after the rabbits were trained in eyeblink classical conditioning. To relate behavioral aging data to the phenomenon of Purkinje cell loss, we counted Purkinje cells in Larsell's (1970) area HVI and in vermis in rabbits ranging in age from 3 to 50 months which had been trained in the trace eyeblink classical conditioning paradigm, the paradigm in which the tone and airpuff do not overlap (Woodruff-Pak *et al.*, 1990a). Nissl staining was used. Comparisons of the total number of Purkinje cells counted for each age group revealed a highly significant difference. The older rabbits had fewer cells. Cell counts in the molecular layer of the same animals indicated no age differences, making it unlikely that differential tissue shrinkage was the cause of the age difference in Purkinje cell number. For the group, there was a correlation of -0.79 ($P < 0.005$) between Purkinje cell number and trials to criterion. The fewer Purkinje cells a rabbit had, the longer it took it to condition. Further analysis demonstrated that this relationship was relatively independent of age. A partial correlation was computed removing the variance due to age, and the resulting relationship between Purkinje cell number and trials to criterion was $r = -0.61$ ($P < 0.025$). Thus, when age was statistically held constant, Purkinje cell number and trials to criterion were still related.

Given the results with aging rabbits, we suspected that the Purkinje cell number may account for some of the individual variation in learning in young rabbits. A second experiment used only young rabbits to test the hypothesis that the Purkinje cell number is related to the rate of acquisition of eyeblink classical conditioning (Woodruff-Pak *et al.*, 1990a). Eighteen 3-month-old male rabbits were classically conditioned in the trace paradigm using the behavioral and histological procedures identical to those used in the first study. Identical methods were also used in the Purkinje cell counts. Even among young rabbits in which Purkinje cell variability was more limited, there was a striking relationship between Purkinje cell number and rate of acquisition. The correlation between the Purkinje cell number and trials to criterion for this sample was -0.60 ($P < 0.01$). Whereas the problem of differential tissue staining as a function of age may have affected data in the first experiment, this problem was not present in the second experiment in which all rabbits were aged around 3 months.

In a subsequent study, brains of rabbits ranging in age from 3 months to 7 years were analyzed using a library of monoclonal antibodies (MAbs) and advanced immunohistochemical techniques (J. M. Coffin *et al.*, manuscript submitted for publication). These rabbits were tested in the delay classical conditioning paradigm. Several MAbs specific to neuronal cytoskeletal proteins were identified, including RMD020, DP1, and RM0254. The MAb RMD020 produced the most intense staining and eliminated much of the background artifact. Counting 10 samples of cerebellar cortical tissue per rabbit covering Larsell's hemisphere VI, a wide variation in Purkinje cell number was observed. The correlation between Purkinje cell number and trials to learning criterion was -0.61 ($P < 0.01$). In 13 of the rabbits, 1-year retention data were available, and the correlation between Purkinje cells and retention trials to criterion was -0.69 ($P < 0.01$).

C. COUNTS OF INTERPOSITUS NUCLEUS CELLS AND CORRELATIONS WITH EYEBLINK CONDITIONING

Interpositus nucleus cells in 25 rabbits aged 3 to 85 months were immunostained with the MAb RMD020 (J. M. Coffin *et al.*, manuscript submitted for publication). Because the cells in the deep nucleus are smaller and more difficult to count than cerebellar cortical Purkinje cells, five slides per rabbit representing anterior to posterior interpositus nucleus were photographed. Counts were made by placing a grid on the photomicrographs. Cell counts were related to acquisition and retention of eyeblink classical conditioning. In the case of the interpositus nucleus, correlations with acquisition were not significant, but the correlation with retention was -0.80 ($P < 0.01$). Neither acquisition nor retention was correlated with hippocampal pyramidal cell number in the CA1 region, although there was a 19% loss in pyramidal cells in hippocampus over the age range of 3 to 85 months. Taken together, these results support the hypothesis that loss of Purkinje cells with age impairs both acquisition and retention of eyeblink conditioning in older organisms and that loss of cells in the interpositus nucleus impairs retention.

IV. Positron Emission Tomography Detection of Cerebellar Involvement in Eyeblink Conditioning

Studies of normal young adult subjects performing eyeblink classical conditioning during PET assessment provide another direct line of evidence indicating cerebellar involvement in eyeblink classical conditioning. Three PET investigations of eyeblink classical conditioning concur in their reports of changes in the cerebellum (Blaxton *et al.*, 1996; Logan and Grafton, 1995; Molchan *et al.*, 1994). During associative learning in eyeblink classical conditioning, Logan and Grafton (1995) observed significant increases in activation in inferior cerebellar cortex/deep nuclei, anterior cerebellar vermis, contralateral cerebellar cortex, and in some noncerebellar regions. Molchan *et al.* (1994) and Blaxton *et al.* (1996) reported decreases in cerebellar cortex. Differences among the groups may have occurred as a result of differences in the PET scanning technique and the time during acquisition that scanning occurred. For example, Molchan *et al.* (1994) and Blaxton *et al.* (1996) reported scanning over early periods of acquisition, whereas Logan and Grafton (1995) scanned after subjects had received much more training. Regardless of differences in technique, all three PET groups observed changes in human cerebellum as a function of the acquisition of CRs.

V. Eyeblink Conditioning and Other Neuropsychological Tasks

In healthy adult subjects, a means of examining cerebellar involvement in eyeblink classical conditioning apart from assessment with PET is to use neuropsychological assessment. Neuropsychological measures assess behaviors with relatively well-documented neurological substrates. For example, timed interval tapping is a test assessing cerebellar function. Tests such as the WMS-R Verbal Paired Associates assess declarative memory that has medial temporal lobe substrates. Neuropsychological tests that predict, do not predict, or interfere with eyeblink classical conditioning during dual-task performance have been identified.

A. PREDICTORS OF EYEBLINK CLASSICAL CONDITIONING

Timed interval tapping data provide evidence associating the cerebellum with eyeblink classical conditioning. As mentioned in Section IIC, timed interval tapping is impaired in cerebellar patients. Clock variability is the precision of the timing of the tapping, with greater variability reflecting poorer timing precision. In patients with unilateral cerebellar lesions, clock variability was always greater in the hand ipsilateral to the lesion (Woodruff-Pak *et al.*, 1996a). The hand contralateral to the cerebellar lesion showed less variability.

In 150 adults ranging in age from 20 to 89 years, eyeblink classical conditioning in the 400-msec delay paradigm, timed interval tapping, WMS-R, and other neuropsychological measures were assessed (D. S. Woodruff-Pak and M. Jaeger, manuscript submitted for publication). Four groups of variables were evaluated for their utility in predicting eyeblink classical conditioning performance. Group 1 was a control component composed of age and hearing test variables. Group 2 consisted of the four cerebellar timed interval tapping measures. Group 3 was blink RT. Group 4 included declarative measures assessed by the WMS-R (Visual and Verbal Paired Associates). Multiple regression using the forward selection procedure was carried out on the dependent measure of percentage of CRs. The combined control variables of age and hearing accounted for 28.6% of the variance which was statistically significant. The second highest predictor of eyeblink classical conditioning performance was the timed interval tapping assessment of cerebellar function. This component accounted for an additional 9.0% of the variance and was also statistically significant. Declarative learning and memory measures accounted for an additional 1.9% of the variance, but this increment in explained variance of eyeblink classical conditioning was not significant. Blink reaction time accounted for 0.4% of the variance and was not a significant predictor of eyeblink classical conditioning. Thus, a cerebellar measure accounted for a significant amount of the variance in eyeblink classical conditioning performance after the variance due to age had been removed. Other behavioral assessments, including blink reaction time and declarative memory, did not predict performance on eyeblink conditioning.

B. Interference During Dual-Task Conditions

Another line of evidence relating the cerebellum to the production of CRs in humans also comes from normal young adult subjects. The investigation involved the behavioral assessment of subjects' performance of eyeblink classical conditioning and selected other tasks in dual-task combinations (Papka *et al.*, 1995). Each of the seven groups in this study was composed of 20 young adults, making the sample size 140 participants. One dual-task group performed the timed interval tapping task and eyeblink classical conditioning simultaneously. Timed interval tapping stimuli used to pace the subject were presented visually, and the CS was an auditory tone for the eyeblink classical conditioning task. A second dual-task group performed a declarative memory task (viewing words, pictures, and numbers on a computer screen and recognizing them later in a longer group of words and pictures) while performing eyeblink classical conditioning.

Other dual tasks included eyeblink classical conditioning and choice reaction time and eyeblink classical conditioning and viewing an engaging video (with instruction that memory of the video would be tested at the end of the session). Additional groups were tested with unpaired presentations of tones and airpuffs (explicitly unpaired eyeblink classical conditioning control) and timed interval tapping, unpaired presentations and the declarative memory task, and unpaired presentations and choice reaction time.

In general, the eyeblink classical conditioning/timed interval tapping group performed significantly more poorly than the other paired eyeblink classical conditioning groups, i.e., timed interval tapping impaired the performance of eyeblink classical conditioning. The other eyeblink classical conditioning groups (with the declarative task, the choice reaction time task, or with video viewing) tended to perform more similarly. Eyeblink conditioning was normal in all three of these dual-task groups. These results provided evidence that the selective disruption of eyeblink classical conditioning, when concurrently performed with timed interval tapping, may be attributed to cerebellar involvement in both tasks.

VI. Eyeblink Classical Conditioning in Noncerebellar Lesions, Neurodegenerative Disease, and Other Syndromes

Because it is possible that any form of brain damage might impair eyeblink classical conditioning, the following research is discussed to demonstrate that lesions affecting eyeblink conditioning in humans are specific to the neural circuitry involved in this learning and identified in rabbits. In both rabbits and humans, the cerebellum is essential for acquisition and retention of the conditioned eyeblink response. The hippocampus plays a modulatory role in acquisition in that manipulations of the hippocampal cholinergic system can alter the rate of learning.

A. Medial Temporal Lobe Lesions

In rabbits, the hippocampus is activated during eyeblink classical conditioning, but conditioning in the delay paradigm in which the CS and US overlap can occur in the absence of the hippocampus (Schmaltz and Theios, 1972; Solomon and Moore, 1975). Eyeblink classical conditioning in the delay paradigm also occurs in humans with hippocampal lesions. Classical conditioning has been studied in patients with global amnesia resulting from bilateral lesions of the medial temporal-diencephalic system and in those (presumably without global amnesia) who had unilateral temporal-lobe lesions. In a study of two amnesic patients (one Korsakoff's syndrome and one post encephalitic case), Weiskrantz and Warrington (1979) reported evidence of classical eyeblink conditioning in the delay paradigm, with retention at 10 min and 24 hr. Daum *et al.* (1989) replicated the results of Weiskrantz and Warrington (1979) with three amnesics (one post encephalitic case and two cases with epilepsy of unspecified etiology). This study examined eyeblink conditioning using a delay paradigm with a 720-msec CS–US interval. All three patients had relatively normal conditioning in the delay paradigm but were impaired in discrimination learning (measured by pairing the airpuff with one frequency of tone CS and not with a second different frequency of tone CS) and discrimination reversal (pairing the airpuff with the second frequency of tone CS and not with the CS frequency with which the airpuff was originally paired). In a subsequent study, Daum *et al.* (1991) investigated electrodermal as well as eyeblink conditioning in the delay and discrimination paradigms in 17 normal control subjects and 17 patients who had undergone unilateral en bloc resection of the right or left temporal lobe for the relief of intractable epilepsy. Daum *et al.* (1991) found that the acquisition of CRs in the delay paradigm was comparable in the unilateral hippocampectomized group and the normal control group, but that performance on discrimination conditioning was impaired. Individuals with unilateral hippocampal lesions produced CRs to both tone-conditioned stimuli whereas normal control subjects discriminated by making CRs only to the CS to which the US was paired. Seven amnesic patients, five of whom had radiological evidence of bilateral damage to the hippocampal formation, were tested using a 750-msec delay paradigm. Results revealed normal conditioning in comparison to age-matched, healthy control subjects (Gabrieli *et al.*, 1995). The authors concluded that in humans, as in rabbits, brain structures critical for declarative memory are not essential for the acquisition of elementary CS–US associations.

We tested the thoroughly investigated amnesic patient H.M. who had bilateral removal of medial temporal lobe structures including most of hippocampus. H.M. was 64 years old at the time of the first testing period and 66 when he was retested. We also tested a second amnesic patient with bilateral medial temporal lobe lesions resulting from herpes simplex encephalitis (Woodruff-Pak, 1993). Both amnesic patients were able to acquire CRs in the 400-msec delay and 750-msec trace paradigms. The acquisition of the postencephalitic patient was within the normal range for healthy adults of his age (57 years). H.M.'s acquisition in the trace paradigm was normal, and this result was interpreted to signify that the 750-msec interval gave him a longer time period to respond. In the 400-msec delay paradigm, however, H.M.'s acquisition was slow compared to healthy adults of his age. H.M. has rather extensive degeneration in the cerebellar vermis, probably as the result of long-term treatment with Dilantin. His slow rate of acquisition in the 400-msec delay paradigm may have resulted from impairment of his cerebellar cortex. According to Brooks (1986), the cerebellar cortex helps to program the relative timing and intensity of the action of muscles in the response. Keele and Ivry (1990) hypothesized that the cerebellar cortex provides a critical computation of timing capability required in the performance of a variety of tasks, including eyeblink conditioning. Damage to H.M.'s cerebellar cortex may account for the poor timing of his responses, especially his difficulty in producing CRs in the 400-msec CS–US delay interval.

B. HUNTINGTON'S DISEASE

Huntington's disease (HD) is a progressive degenerative neurological disease resulting in abnormalities of voluntary movement and dementia. The basal ganglia are grossly atrophied, and there is thinning and shrinkage of the cerebral cortex. The hippocampus and hippocampal cholinergic system, however, remain relatively intact, as does the cerebellum. Although the eyeblink reflex is hyposensitive in HD, we anticipated that this lack of excitability could be overcome with an airpuff directed at the cornea. On the basis of what is known about the neural circuitry essential or normally involved in eyeblink classical conditioning, the possibility existed that the pattern of neurodegeneration in HD would not interfere with this type of learning. Thus, it was anticipated that these patients would condition relatively normally.

The performance of seven patients with HD was compared to agematched, healthy participants using the 400-msec delay eyeblink classical conditioning paradigm (Woodruff-Pak and Papka, 1996b). There were no differences in the production of CRs between HD patients and normal control subjects, but the timing of the CR was abnormal in HD. These results indicated that the ability to acquire CRs was normal in HD, but the striatum may have some role in optimizing the timing of the CR.

The neostriatum is not essential for eyeblink classical conditioning in rabbits, although lesions of the caudate nucleus (Powell *et al.*, 1978) and disruption of neostriatal dopamine (Harvey and Gormezano, 1981; Kao and Powell, 1988; Sears and Steinmetz, 1990) impair acquisition. Extracellular recordings of multiple- and single-unit responses in neostriatum of rabbits during eyeblink classical conditioning revealed conditioning-related patterns of neural firing. However, the neostriatal units related to the CR occurred too late to initiate the CR (White et al., 1994). The investigators suggested that the neostriatum may receive CR-related information from other structures, such as the cerebellum, that are critical for the optimal timing of the CR.

C. ALZHEIMER'S DISEASE

It has been our working hypothesis that disruption of the septo-hippocampal cholinergic system delays acquisition of eyeblink classical conditioning in probable Alzheimer's disease (AD) beyond the impairment observed in normal aging (Woodruff-Pak et al., 1989). We based the prediction of grossly impaired eyeblink classical conditioning in AD patients on data collected with the rabbit model system. The rabbit model system demonstrated that the hippocampus can play a modulatory role in conditioning (Berger et al., 1986). Although the hippocampus is not essential for acquisition in the delay classical conditioning paradigm in rabbits, disruption or facilitation of the hippocampus affects the rate of conditioning. Disruption of muscarinic receptors in the septo-hippocampal cholinergic system with scopolamine injections impairs acquisition of CRs (Harvey et al., 1983; Moore et al., 1976; Solomon et al., 1983) and eliminates pyramidal cell activity in conjunction with the CR and UR (Salvatierra and Berry, 1989). Electrical stimulation of the perforant path establishing long-term potentiation in hippocampus causes more rapid classical conditioning in a discrimination paradigm (Berger, 1984). Thus, the animal model for AD using the eyeblink classical conditioning paradigm is an older rabbit (having Purkinje cell loss in cerebellum) with a scopolamine-disrupted hippocampal cholinergic system.

Since we first proposed that patients with probable AD would be impaired on eyeblink classical conditioning based on evidence available from the older rabbit scopolamine-injected model, data have consistently supported that model. In a sample of 20 probable AD patients and 20 healthy age-matched adults, eyeblink conditioning was significantly poorer in AD patients who showed almost no acquisition in a 90-trial session (Woodruff-Pak et al., 1990b). This result was independently replicated in another laboratory (Solomon et al., 1991). This result has been replicated and extended in our laboratory with new samples of patients and normal control subjects (Woodruff-Pak and Papka, 1996a; Woodruff-Pak et al., 1996b). In the later study, we also demonstrated that eyeblink classical conditioning in some cases was effective in differentiating cerebrovascular dementia from probable AD. In addition to working with probable AD and cerebrovascular dementia patients, we extended this work to adults with Down's syndrome over the age of 35 who inevitably develop AD (called DS/AD; Papka et al., 1994; Woodruff-Pak et al., 1994). Patients with DS/AD perform eyeblink classical conditioning similarly to probable AD patients, but young adults with DS perform eyeblink classical conditioning significantly better.

Rabbits with disrupted hippocampal cholinergic systems have delayed acquisition of eyeblink classical conditioning but eventually acquire CRs (Solomon et al., 1983). On this basis, we predicted that if probable AD and DS/AD patients were given enough training trials, they would eventually produce CRs. Probable AD and DS/AD patients were tested with eyeblink classical conditioning for 5 consecutive days, and most of them eventually attained a learning criterion of eight CRs in 9 consecutive trials (Woodruff-Pak et al., 1996c). Solomon et al. (1995) tested probable AD patients in paired tone and corneal airpuff presentations in the 400-msec delay paradigm for four consecutive 70-trial sessions and reported similar results. The neural substrate supporting eyeblink classical conditioning therefore seems to be impaired by probable AD and DS/AD beyond the impairment observed in normal aging, but it is not destroyed.

The fact that probable AD and DS/AD patients eventually acquire CRs argues against lesions of the cerebellum as the cause of the significant impairment of eyeblink classical conditioning in that patient population. We have argued that normal age-related Purkinje cell loss that occurs in many mammals, including humans, may account for the age-related decline in eyeblink classical conditioning. Using immunohistochemical methods with postmortem cerebellar tissue from AD, DS/AD, and normal elderly control brains, we reported that AD neuropathology in the form of β-amyloid plaques in DS/AD cerebellum does not appear to impair Purkinje cells beyond the effects of normal aging (Li et al., 1994). Furthermore, there are only a limited number of plaques in the AD cerebellum and none in the cerebellum in normal aging. Neurofibrillary tangles have not been found in the cerebellum in AD and are rarely present in DS/AD. Given the relative absence of neuropathology that could disrupt eyeblink classical conditioning in the AD and DS/AD cerebellum, it is not likely that cerebellar impairment is solely responsible for impairment of eyeblink classical conditioning in AD and DS/AD.

D. Autism

In autism, there are demonstrated cerebellar abnormalities in cerebellar cortex and deep nuclei (see chapters by M. Bauman et al. and by N. Askhoomoff et al.). Eleven autistic patients with a mean age of 12 years were tested on eyeblink classical conditioning in a 350-msec delay paradigm, and they were compared to 11 age-matched normal control participants (Sears *et al.*, 1994). Learning occurred more rapidly in the autistic subjects and they produced a similar percentage of CRs to the normal controls. However, the timing of the response was abnormal. One of the characteristic features of eyeblink conditioning in the autistic subjects was the shorter, abnormally timed CR onset and CR peak latency. This result is similar to the abnormal timing of CRs that has been reported in rabbits with cerebellar lesions (e.g., Perrett *et al.*, 1993). Abnormal eyeblink classical conditioning in autistic children was interpreted by Sears *et al.* (1994) as evidence for deviant cerebellar–hippocampal interactions because limbic system pathologies have also been reported in autism (e.g., Bauman, 1991; Bauman and Kemper, 1985).

VII. Summary and Conclusions

Using a number of different strategies, research has converged to implicate the human cerebellum as essential in classical conditioning of the eyeblink response. In patients with bilateral cerebellar lesions or neurodegenerative cerebellar disease, CRs are not produced on either side. Cerebellar patients with lateralized lesions, like rabbits with experimentally produced unilateral cerebellar lesions, produce relatively normal CRs with the eye contralateral to the lesion and few or no CRs ipsilateral to the lesion. Age-related deficits in eyeblink classical conditioning appear in humans and rabbits in middle age. In normal aging, there is Purkinje cell loss in the cerebellar cortex. In rabbits, the Purkinje cell number correlates highly with the rate of learning, regardless of age. PET imaging of normal young adults during eyeblink conditioning reveals changes in activity in the cerebellum. Timed interval tapping, a task that assesses cerebellar function, also predicts performance on eyeblink conditioning. In dual-task conditions involving the simultaneous performance of eyeblink conditioning and timed interval tapping, eyeblink conditioning is impaired. Investigations of patients with lesions or neurodegenerative disease not involving the cerebellum demonstrate that acquisition of CRs is possible, although in the case of hippocampal cholinergic disruption, acquisition is prolonged. Evidence to date suggests that the human analogue of the rabbit interpositus nucleus, the globose nucleus, is essential for the production of the conditioned eyeblink response and that cerebellar cortical Purkinje cells play a role in facilitating acquisition.

References

Bauman, M. (1991). Microscopic neuroanatomic abnormalities in autism. *Pediatr. Suppl.* **1,** 791–796.
Bauman, M., and Kemper, T. (1985). Histoanatomic observations of the brain in early infantile autism. *Neurology* **3,** 866–874.
Berger, T. (1984). Long-term potentiation of hippocampal synaptic transmission affects rate of behavioral learning. *Science* **20,** 810–816.
Berger, T., Berry, S. D., and Thompson, R. F. (1986). Role of the hippocampus in classical conditioning of aversive and appetitive behaviors. *In* "The Hippocampus" (R. I. Isaacson and K. H. Pribram, eds.), Vol. IV, pp. 203–239. Plenum, New York.
Berthier, N. E., and Moore, J. (1986). Cerebellar Purkinje cell activity related to the classically conditioned nictitating membrane response. *Exp. Brain Res.* **63,** 341–350.
Berthier, N. E., and Moore, J. (1990). Activity of deep cerebellar nuclei during classical conditioning of nictitating membrane extension in rabbit nictitating response. *Exp. Brain Res.* **83,** 44–54.
Blaxton, T. A., Zeffiro, T. A., Gabrieli, J. D. E., Bookheimer, S. Y., Carrillo, M. C., Theodore, H., and Disterhoft, J. F. (1996). Functional mapping of human learning: A PET activation study of eyeblink conditioning. *J. Neurosci.* **16,** 4032–4040.
Brooks, V. B. (1986). "The Neural Basis of Motor Control." Oxford University Press, New York.
Chen, L., Bao, S., Lockard, J. M., Kim, J. J., and Thompson, R. F. (1996). Impaired classical eyeblink conditioning in cerebellar lesioned and Purkinje cell degeneration (*pcd*) mutant mice. *J. Neurosci.* **16,** 2829–2838.
Clark, R. E., Zhang, A. A., and Lavond, D. G. (1992). Reversible lesions of the cerebellar interpositus nucleus during acquisition and retention of a classically conditioned behavior. *Behav. Neurosci.* **106,** 879–888.
Daum, I., Channon, S., and Canavan, A. G. M. (1989). Classical conditioning in patients with severe memory problems. *J. Neurol. Neurosurg. Psychiat.* **52,** 47–51.
Daum, I., Channon, S., Polkey, C. E., and Gray, J. A. (1991). Classical conditioning after temporal lobe lesions in man: Impairment in conditional discrimination. *Behav. Neurosci.* **105,** 396–408.
Daum, I., Schugens, M. M., Ackermann, H., Lutzenberger, W., Dichgans, J., and Birbaumer, N. (1993). Classical conditioning after cerebellar lesions in humans. *Behav. Neurosci.* **107,** 748–756.
Dayan, A. D. (1971). Comparative neuropathology of ageing: Studies on the brains of 47 species of vertebrates. *Brain* **94,** 31–42.
Durkin, M., Prescott, L., Furchtgott, E., Cantor, J., and Powell, D. A. (1993). Concomitant eyeblink and heart rate classical conditioning in young, middle-aged, and elderly human subjects. *Psychol. Aging* **8,** 71–81.
Ellis, R. S. (1920). Norms for some structural changes in the cerebellum from birth to old age. *J. Comp. Neurol.* **32,** 1–33.
Gabrieli, J. D. E., McGlinchey-Berroth, R., Carrillo, M. C., Gluck, M. A., Cermak, L. S., and Disterhoft, J. F. (1995). Intact delay-eyeblink classical conditioning in amnesia. *Behav. Neurosci.* **109,** 819–827.
Glick, R., and Bondareff, W. (1979). Loss of synapses in the cerebellar cortex of the senescent rat. *J. Gerontol.* **34,** 818–822.
Gormezano, I., Schneiderman, N., Deaux, E., and Fuentes, I. (1962). Nictitating membrane: Classical conditioning and extinction in the albino rabbit. *Science* **138,** 33–34.

Gould, T. J., and Steinmetz, J. E. (1994). Multiple-unit activity from rabbit cerebellar cortex and interpositus nucleus during classical discrimination/reversal eyelid conditioning. *Brain Res.* **652**, 98–106.

Hall, T. C., Miller, K. H., and Corsellis, J. A. N. (1975). Variations in the human Purkinje cell population according to age and sex. *Neuropathol. Appl. Neurobiol.* **1**, 267–292.

Harvey, J. A., and Gormezano, I. (1981). Effects of haloperidol and pimozide on classical conditioning of the rabbit nictitating membrane response. *J. Pharmacol. Exp. Ther.* **218**, 712–719.

Harvey, J. A., Gormezano, I., and Cool-Hauser, V. A. (1983). Effects of scopolamine and methylscopolamine on classical conditioning of the rabbit nictitating membrane response. *J. Pharmacol. Exp. Therap.* **22**, 42–49.

Hilgard, E. R., and Campbell, A. A. (1936). The course of acquisition and retention of conditioned eyelid responses in man. *J. Exp. Psychol.* **19**, 227–247.

Hilgard, E. R., and Marquis, D. G. (1936). Conditioned eyelid responses in monkeys, with a comparison of dog, monkey, and man. *Psychol. Monogr.* **47**, 186–198.

Ivry, R., and Keele, S. W. (1989). Timing functions of the cerebellum. *Cogn. Neurosci.* **1**, 134–150.

Ivry, R., Keele, S. W., and Diener, H. C. (1988). Dissociation of the lateral and medial cerebellum in movement timing and movement execution. *Exp. Brain Res.* **73**, 167–180.

Kao, K.-T., and Powell, D. A. (1988). Lesions of the substantia nigra retard Pavlovian eyeblink but not heart rate conditioning in the rabbit. *Behav. Neurosci.* **102**, 515–525.

Keele, S. W., and Ivry, R. (1990). Does the cerebellum provide a common computation for diverse tasks: A timing hypothesis. *In* "The Development and Neural Bases of Higher Cognitive Functions" (A. Diamond, ed.), pp. 179–211. New York Academy of Sciences Press, New York.

Konorski, J. (1948). "Conditioned Reflexes and Neuron Organization." Cambridge University Press, Cambridge.

Krupa, D. J., Thompson, J. K., and Thompson, R. F. (1993). Localization of a memory trace in the mammalian brain. *Science* **260**, 989–991.

Larsell, O. (1970). Rabbit. *In* "The Comparative Anatomy and Histology of the Cerebellum from Monotremes through Apes" (J. Jansen, ed.). University of Minnesota Press, Minneapolis.

Lavond, D. G., Hembree, T. L., and Thompson, R. F. (1985). Effect of kainic acid lesions of the cerebellar interpositus nucleus on eyelid conditioning in the rabbit. *Brain Res.* **326**, 179–182.

Lavond, D. G., and Steinmetz, J. E. (1989). Acquisition of classical conditioning without cerebellar cortex. *Behav. Brain Res.* **33**, 113–164.

Lavond, D. G., Steinmetz, J. E., Yokaitis, M. H., and Thompson, R. F. (1987). Reacquisition of classical conditioning after removal of cerebellar cortex. *Exp. Brain Res.* **67**, 69–93.

Li, Y.-T., Woodruff-Pak, D. S., and Trojanowski, J. Q. (1994). Amyloid plaques in cerebellar cortex and the integrity of Purkinje cell dendrites. *Neurobiol. Aging* **1**, 1–9.

Lincoln, J. S., McCormick, D. A., and Thompson, R. F. (1982). Ipsilateral cerebellar lesions prevent learning of the classically conditioned nictitating membrane/eyelid response. *Brain Res.* **242**, 190–193.

Logan, C. G., and Grafton, S. T. (1995). Functional anatomy of human eyeblink conditioning determined with regional cerebral glucose metabolism and positron emission tomography. *Proc. Natl. Acad. Sci. USA* **92**, 7500–7504.

Lye, R. H., O'Boyle, D. J., Ramsden, R. T., and Schady, W. (1988). Effects of a unilateral cerebellar lesion on the acquisition of eye-blink conditioning in man. *J. Physiol. (London)* **403**, 58P.

McCormick, D. A., Lavond, D. G., Clark, G. A., Kettner, R. E., Rising, C. E., and Thompson, R. F. (1981). The engram found? Role of the cerebellum in classical conditioning of nictitating membrane and eyelid responses. *Bull. Psychonomic Soc.* **18,** 103-105.

McCormick, D. A., and Thompson, R. F. (1984a). Cerebellum: Essential involvement in the classically conditioned eyelid response. *Science* **223,** 296-299.

McCormick, D. A., and Thompson, R. F. (1984b). Neuronal responses of the rabbit cerebellum during acquisition and performance of a classically conditioned nictitating membrane-eyelid response. *J. Neurosci.* **4,** 2811-2822.

Molchan, S. E., Sunderland, T., McIntosh, A. R., Herscovitch, P., and Schreurs, B. G. (1994). A functional anatomical study of associative learning in humans. *Proc. Natl. Acad. Sci. USA* **91,** 8122-8126.

Moore, J. W., Goodell, N. A., and Solomon, P. R. (1976). Central cholinergic blockage by scopolamine and habituation, classical conditioning, and latent inhibition of the rabbit's nictitating membrane response. *Physiol. Psychol.* **4,** 395-399.

Nandy, K. (1981). Morphological changes in the cerebellar cortex of aging *Macaca nemestrina*. *Neurobiol. Aging* **2,** 61-64.

Papka, M., Ivry, R. B., and Woodruff-Pak, D. S. (1995). Selective disruption of eyeblink classical conditioning by concurrent timed interval tapping. *NeuroReport* **6,** 1493-1497.

Papka, M., Simon, E. W., and Woodruff-Pak, D. S. (1994). A one-year longitudinal investigation of eyeblink classical conditioning and cognitive and behavioral tests in adults with down's syndrome. *Aging Cogn.* **1,** 89-104.

Pavlov, I. P. (1927). "Conditioned Reflexes" (G. V. Anrep, ed.). Oxford University Press, Oxford.

Pentney, R. J. (1986). Qualitative analysis of dendritic networks of Purkinje neurons during aging. *Neurobiol. Aging* **7,** 241-348.

Perrett, S., Ruiz, B., and Mauk, M. (1993). Cerebellar cortex lesions disrupt learning-dependent timing of conditioned eyelid responses. *J. Neurosci.* **13,** 1708-1718.

Powell, D. A., Mankowski, D., and Buchanan, S. (1978). Concomitant heart rate and corneoretinal potential conditioning in the rabbit (*Oryctolagus cuniculus*): Effects of caudate lesions. *Physiol. Behav.* **20,** 143-150.

Rogers, J., Zornetzer, S. F., and Bloom, F. E. (1981). Senescent pathology of cerebellum: Purkinje neurons and their parallel fiber afferents. *Neurobiol. Aging* **2,** 15-21.

Rogers, J., Zornetzer, S. F., Bloom, F. E., and Mervis, R. E. (1984). Senescent microstructural changes in the rat cerebellum. *Brain Res.* **292,** 23-32.

Salvatierra, A. T., and Berry, S. D. (1989). Scopolamine disruption of septo-hippocampal activity and classical conditioning. *Behav. Neurosci.* **103,** 715-721.

Schmaltz, L. W., and Theios, J. (1972). Acquisition and extinction of a classically conditioned response in hippocampectomized rabbits (*Oryctolagus cuniculus*). *J. Comp. Physiol. Psychol.* **79,** 328-333.

Schneiderman, N., Fuentes, I., and Gormezano, I. (1962). Acquisition and extinction of the classically conditioned eyelid response in the albino rabbit. *Science* **136,** 650-652.

Sears, L. L., Finn, P. R., and Steinmetz, J. E. (1994). Abnormal classical eyeblink conditioning in autism. *J. Autism Dev. Disord.* **24,** 737-751.

Sears, L. L., and Steinmetz, J. E. (1990). Haloperidol impairs classically conditioned nictitating membrane responses and conditioning-related cerebellar interpositus nucleus activity in rabbits. *Pharmacol. Biochem. Behav.* **36,** 821-830.

Solomon, P. R., Brett, M., Groccia-Ellison, M., Oyler, C., Tomasi, M., and Pendlebury, W. W. (1995). Classical conditioning in patients with Alzheimer's disease: A multiday study. *Psychol. Aging* **10,** 248-254.

Solomon, P. R., Groccia-Ellison, M., Flynn, D., Mirak, J., Edwards, K. R., Dunehew, A., and Stanton, M. E. (1993). Disruption of human eyeblink conditioning after central cholinergic blockade with scopolamine. *Behav. Neurosci.* **107,** 271–279.

Solomon, P. R., Levine, E., Bein, T., and Pendlebury, W. W. (1991). Disruption of classical conditioning in patients with Alzheimer's disease. *Neurobiol. Aging* **12,** 283–287.

Solomon, P. R., and Moore, J. W. (1975). Latent inhibition and stimulus generalization of the classically conditioned nictitating membrane response in rabbits (*Oryctolagus cuniculus*) following dorsal hippocampal ablations. *J. Comp. Physiol. Psychol.* **69,** 1192–1203.

Solomon, P. R., Pomerleau, D. Bennett, L., James, J., and Morse, D. L. (1989). Acquisition of the classically conditioned eyeblink response in humans over the lifespan. *Psychol. Aging* **4,** 34–41.

Solomon, P. R., Solomon, S. D., VanderSchaaf, E., and Perry, H. E., (1983). Altered activity in the hippocampus is more detrimental to classical conditioning than removing the structure. *Science* **220,** 329–331.

Solomon, P. R., Stowe, G. T., and Pendlebeury, W. W. (1989). Disrupted eyelid conditioning in a patient with damage to cerebellar afferents. *Behav. Neurosci.* **103,** 898–902.

Steinmetz, J. E., Lavond, D. G., Ivkovich, D., Logan, C. G., and Thompson, R. F. (1992). Disruption of classical eyelid conditioning after cerebellar lesions: Damage to a memory trace system or a simple performance deficit? *J. Neurosci.* **12,** 4403–4426.

Steinmetz, J. E., and Thompson, R. F. (1991). Brain substrates of aversive classical conditioning. *In* "Neurobiology of Learning, Emotion, and Affect" (J. Madden IV, ed.), pp. 97–120. Raven Press, New York.

Thompson, R. F. (1986). The neurobiology of learning and memory. *Science* **233,** 941–947.

Thompson, R. F. (1990). Neural mechanisms of classical conditioning in mammals. *Phil. Trans. R. Soc. Lond.* **319,** 161–170.

Thompson, R. F., and Krupa, D. J. (1994). Organization of memory traces in the mammalian brain. *Annu. Rev. Neurosci.* **17,** 519–549.

Topka, H., Valls-Sole, J., Massaquoi, S. G., and Hallett, M. (1993). Deficit in classical conditioning in patients with cerebellar degeneration. *Brain* **116,** 961–969.

Weiskrantz, L., and Warrington, E. K. (1979). Conditioning in amnesic patients. *Neuropsychologia* **17,** 187–194.

White, I. M., Miller, D. P., White, W., Dike, G. L., Rebec, G. V., and Steinmetz, J. E. (1994). Neuronal activity in rabbit neostriatum during classical eyelid conditioning. *Exp. Brain Res.* **99,** 179–190.

Woodruff-Pak, D. S. (1990). Mammalian models of learning, memory, and aging. *In* "Handbook of the Psychology of Aging" (J. E. Birren and K. W. Schaie, eds.), 3rd Ed., pp. 235–257. Academic Press, San Diego.

Woodruff-Pak, D. S. (1993). Eyeblink classical conditioning in H.M.: Delay and trace paradigms *Behav. Neurosci.* **107,** 911–925.

Woodruff-Pak, D. S., Cronholm, J. F., and Sheffield, J. B. (1990a). Purkinje cell number related to rate of eyeblink classical conditioning. *NeuroReport* **1,** 165–168.

Woodruff-Pak, D. S., Finkbiner, R. G., and Katz, I. R. (1989). A model system demonstrating parallels in animal and human aging: Extension to Alzheimer's disease. *In* "Novel Approaches to the Treatment of Alzheimer's Disease" (E. M. Meyer, J. W. Simpkins, and J. Yamamoto, Eds.), (pp. 355–371). Plenum, New York.

Woodruff-Pak, D. S., Finkbiner, R. G., and Sasse, D. K. (1990b). Eyeblink conditioning discriminates Alzheimer's patients from non-demented aged. *NeuroReport* **1,** 45–48.

Woodruff-Pak, D. S., Lavond, D. G., Logan, C. G., Steinmetz, J. E., and Thompson, R. F. (1993). Cerebellar cortical lesions and reacquisition in classical conditioning of the nictitating membrane response in rabbits. *Brain Res.* **608,** 67–77.

Woodruff-Pak, D. S., and Papka, M. (1996a). Alzheimer's disease and eyeblink conditioning: 750 ms trace versus 400 ms delay paradigm. *Neurobiol. Aging* **17**(3), 397–404.

Woodruff-Pak, D. S., and Papka, M. (1996b). Huntington's disease and eyeblink classical conditioning: Normal learning but abnormal timing. *J. Intl. Neuropsychol. Soc.* **2**, 323–334.

Woodruff-Pak, D. S., Papka, M., and Ivry, R. (1996a). Cerebellar involvement in eyeblink classical conditioning in humans. *Neuropsychology* **10**(4), 443–458.

Woodruff-Pak, D. S., Papka, M., Romano, S., and Li, Y.-T. (1996b). Eyeblink classical conditioning in Alzheimer's disease and cerebrovascular dementia. *Neurobiol. Aging* **17**, 505–512.

Woodruff-Pak, D. S., Papka, M., and Simon, E. W. (1994). Down's Syndrome adults aged 35 and older show eyeblink classical conditioning profiles comparable to Alzheimer's disease patients. *Neuropsychology* **8**, 1–11.

Woodruff-Pak, D. S., Romano, S., and Papka, M. (1996c). Training to criterion in eyeblink classical conditioning in Alzheimer's disease, Down's syndrome with Alzheimer's disease, and healthy elderly. *Behav. Neurosci.* **110**(1), 22–29

Woodruff-Pak, D. S., and Thompson, R. F. (1988). Classical conditioning of the eyeblink response in the delay paradigm in adults aged 18-83 years. *Psychol. Aging* **3**, 219–229.

Yeo, C. H., Hardiman, M. J., and Glickstein, M. (1985). Classical conditioning of the nictitating membrane response of the rabbit. I. Lesions of the cerebellar nuclei. *Exp. Brain Res.* **60**, 87–98.

EARLY INFANTILE AUTISM

Margaret L. Bauman
Children's Neurology Service
Massachusetts General Hospital
Boston, Massachusetts 02114

Pauline A. Filipek
Department of Pediatrics
University of California at Irvine
Orange, California 92668

Thomas L. Kemper
Departments of Neurology, and Anatomy and Neurobiology
Boston University School of Medicine
Boston, Massachusetts 02118

I. Introduction
II. Neuroimaging Studies
III. Microscopic Observations in the Cerebellum and Related Olive
IV. Implications of Cerebellar Abnormalities in Autism
V. Conclusion
References

I. Introduction

Autism is a behaviorally defined syndrome, first described by Kanner in 1943. Symptoms become evident by 3 years of age and include atypical social interaction, disordered language and cognitive skills, impaired imaginary play, poor eye contact, and an obsessive insistence on sameness. Perseveration, repetitive, and stereoptypic behavior and a restricted range of interests may be present in some cases. Although not unusual in physical appearance, a significant number of autistic individuals exhibit hypotonia, dyspraxia, and a disordered modulation of sensory input (Rapin, 1994). Some clinical features of autism have been reported in conditions such as tuberous sclerosis, phenylketonuria, and fragile X syndrome but, in most cases, a specific etiology cannot be identified. Although the cause of autism is believed by many to be multifactorial, twin and family studies have suggested evidence for a genetic liability, the mechanism of which remains unknown (Piven and Folstein, 1994).

For many years, autism was believed to be the result of poor parenting and environmental factors. However, research over the past 20 years has provided increasing evidence for a neurological basis for the disorder. A number of regions of the brain have been hypothesized as candidate sites of abnormality, including the basal ganglia (Vilensky et al., 1981), the thalamic nuclei (Coleman, 1979), the vestibular system (Ornitz and Ritvo, 1968), structures of the medial temporal lobe (Boucher and Warrington, 1976; Delong, 1978; Damasio and Maurer, 1978; Maurer and Damasio, 1982; Bauman and Kemper, 1985, 1995; Bachevalier and Merjanian, 1994), and the cerebellum (Ritvo et al., 1986; Courchesne et al., 1988, 1994b; Bauman and Kemper, 1985, 1995). Microscopic analysis of the autistic brain has delineated abnormalities of the limbic system, the cerebellum, and related inferior olivary nucleus (Bauman and Kemper, 1985, 1994, 1995).

II. Neuroimaging Studies

Attempts to radiographically study the brain in autism began with the study of Hauser et al. (1975) in which enlargement of the left temporal horn was reported in 15 of 18 autistic children using pneumoencephalography. Imaging studies expanded dramatically in the mid-1970s following the introduction of computerized tomographic scanning technology. The major focus of these studies relative to autism was related to the delineation of asymmetries of the cerebral hemispheres (Hier et al., 1979) and later observations of ventricular size (Campbell et al., 1982; Rosenbloom et al., 1984; Jacobson et al., 1988).

With the introduction of magnetic resonance imaging (MRI), morphometric analysis of specific brain structures became the focus of neuroimaging research in autism. These studies have largely concentrated on the cerebellum. Using in vivo quantitative MRI techniques, Courchesne et al. (1988) reported finding hypoplasia of vermal lobules VI and VII in 16 male and 2 female autistic subjects, aged 6 to 30 years of age, in comparison with control subjects. The authors concluded that vermal hypoplasia represented a consistent abnormality in most autistic individuals and suggested a relationship between these findings and deficits in attention, sensory modulation, motor, and behavioral initiation, symptoms frequently observed in these patients. The following year, Murakami et al. (1989), reanalyzing data on 9 of these same patients, reported that, using paramidline saggital images, the cumulative slice area measure of the cerebellar hemispheres was 12% smaller in the autistic subjects than in the control group. However, no differences were found in cerebellar width between the two groups. The authors noted that the finding of reduced hemispheric area correlated positively with the area of lobules VI and VII only in the autistic patients.

Subsequently, five similarly designed studies reported no differences in the area of vermal lobules VI and VII measured on midsagittal MRI images (Ritvo and Garber, 1988; Holttum et al., 1992; Piven et al., 1992; Kleinman et al., 1992; Filipek et al., 1992b). A total of 77 male and 10 female autistic subjects, ranging in age from 2 to 53 years, were included in these studies. Different age groups were emphasized in each of these investigations. In addition, the nonautistic comparison groups for each study met distinctly differing inclusionary criteria and were not consistently matched for IQ or socioeconomic status (SES) in each project. Thus, differences in research design among these studies make comparison of results difficult to interpret (Filipek, 1995).

Nowell et al. (1990) studied 53 patients and 32 controls with MRI. The autistic subjects were matched for age but not sex, SES, or IQ and all males were screened for fragile X syndrome. Only 5 (7.6%) of the autistic subjects showed evidence of vermian atrophy and the degree of atrophy, when present, did not appear to correlate with the severity of the patient's autism. The investigators did not find any specific morphologic criteria to be associated with autism and concluded that either vermian hypoplasia was too subtle to be detected on 10-mm images or it was an epiphenonmenon unrelated to autism.

Later, Courchesne et al. (1994b), measuring area on midsagittal MRI images, reported hypoplasia of vermal lobules VI and VII in 43 of 50 autistic subjects and hyperplasia in the remaining 7 patients, resulting in a bimodal distribution. The authors concluded that the combination of these two subtypes within the autistic population was responsible for the failure to delineate abnormalities of mean vermal areas in the other MRI studies. In that same year, Courchesne et al. (1994a) reported on their reanalysis of data from two of the previously reported MRI studies (Piven et al., 1992; Kleinman et al., 1992) as well as measurements from two of their own earlier investigations (Courchesne et al., 1988, 1994b). Collectively, data obtained on 78 autistic subjects were reexamined. Although no significant interstudy differences were found in the area measurements of vermal lobules VI and VII among the autistic subjects, the vermal area measurements for the autistic group as a whole were reported to be significantly smaller than those of normal controls. Plotting the results against "verbal and social intelligence," the authors noted a correlation between the degree of vermal hypoplasia and the degree of "deficit." Further, as in their earlier study, Courchesne *et al.* (1994b) described a bimodal distribution of vermal hypoplasia in 87% and hyperplasia in 13% of the autistic subjects as compared with normal controls.

More recently, Schaefer *et al.* (1996) measured the cerebellar vermis in 125 normal individuals and 102 patients with a variety of neurological disorders. Hypoplasia of lobules VI and VII was noted in neurological conditions unassociated with autistic behavior and was not uniformly seen in autistic subjects. The authors concluded that hypoplasia of lobules VI and VII was a nonspecific finding, that it was not a specific marker for autism, and that cerebellar dysgenesis was unlikely to be solely responsible for the clinical presentation of autistic behavior.

These collective neuroimaging studies of the cerebellum represent muliple differing methods of MRI scanning, with dissimilar slice thicknesses, slice orientation, and position. The anatomy seen on any two-dimensional MRI slice represents the average of the total anatomy existing through the entire thickness of that slice (i.e., "volume averaging"). Therefore, the anatomy on three contiguous 3-mm slices will be more accurate than the anatomy seen on a single 9-mm slice, especially of structures with oblique edges. Some MRI scans also include variably sized gaps between slices (interslice gaps), where the brain is not imaged, whereas others produced thin contiguous slices that image the entire volume of the brain. Because slice orientation and position of a single MRI slice may also contribute to neuroanatomic variability, individual slices selected for unidimensional or two-dimensional measurements may significantly differ across subjects. Added to these differing methodologies is the fact that these imaging studies represent a combination of qualitative and quantitative image analysis methods, without uniform anatomic definitions. It is not surprising, therefore, that the combination of one-dimensional measures of length or width, two-dimensional measures of area on single selected slices, three-dimensional volumes interpolated through interslice gaps, and three-dimensional volumes on thin contiguous slices, all performed with variable anatomic definitions, have produced heterogeneous results that cannot be directly compared (Filipek *et al.* 1992a; Filipek, 1996).

Adding to the variations in methodology between these imaging investigations is the selection of subjects and control groups with differing levels of inclusion criteria, further compounding the difficulty of comparing results between studies. For example, due to the retrospective nature of the Courchesne *et al.* (1994b) reanalysis study, it was not possible for the authors to directly match the collective autistic and control subjects for IQ. Several studies (Courchesne *et al.*, 1988, 1994b; Kleinman *et al.*, 1992) reported verbal IQ for the autistic subjects with scores ranging from 48 to

77. The Courchesne *et al.* (1988, 1994b) studies formally measured verbal IQ (VIQ) on Wechsler intelligence scales whereas the Kleinman *et al.* (1992) study utilized Vineland composite scores. The Vineland is a measure of performance of adaptive behaviors; the composite score for this test is not equivalent to a verbal IQ score. Second, the mean verbal IQ score for the control group cited by Courchesne *et al.* (1994b) is 109, thus indicating a substantial IQ discrepancy between the autistic subjects, whose IQ scores are largely in the mentally retarded range of intellectual ability, and the controls. Yakovlev (1960) observed that, in mental retardation, there is a generalized decrease in brain size involving all structures uniformly. Because whole brain measurements are not included in any of the studies of cerebellar size, it is difficult to know whether the cerebellar findings truly reflect differences relative to the rest of the brain or a more generalized reduction in brain size, possibly related to mental retardation. It is of interest to note that Piven *et al.* (1992) have observed that the presence of vermal hypoplasia of lobules VI and VII appears to be closely correlated with IQ. Thus, it is possible that the IQ discrepancy between autistic and control subjects may be related to the differences in vermal area reported by Courchesne *et al.* (1994b).

Given the lack of uniformity between studies at this time, the significance of MRI observations of the cerebellum in autism remains uncertain and their interpretation should be treated with caution. It is likely that larger numbers of subjects studied against carefully selected age-, sex-, IQ-, and SES-matched controls, using strict uniform methodological criteria, will need to be performed before any definite conclusions can be reached.

III. Microscopic Observations in the Cerebellum and Related Olive

Few neuropathological studies of the brain in autism have been reported and the large majority of these have been primarily devoted to observations of the cortical and subcortical regions of the cerebral hemispheres (Aarkrog, 1968; Darby, 1976; Williams *et al.*, 1980; Coleman *et al.*, 1985). In 1984, histoanatomic abnormalities of the cerebellum and the limbic system were first reported in the brain of a well-documented autistic man (Bauman and Kemper, 1984) and a more detailed description of this case was published the following year (Bauman and Kemper, 1985). Subsequently, Ritvo *et al.* (1986) quantitatively studied the microscopic anatomy of the cerebellar hemispheres and vermis from four autistic subjects and controls. Decreased numbers of Purkinje cells were reported throughout the cerebellum, and the authors speculated on the possible relationship of these findings to some of the clinical features of autism.

Since these initial studies, eight additional brains from well-documented autistic patients have been systematically analyzed using the technique of gapless serial section (Bauman and Kemper, 1995). These cases included six children, two girls and four boys ranging in age from 5 to 12 years, and two young adult males, ages 22 and 28 years, for a total of three young adults. All have been studied in comparison with identically processed age- and sex-matched control material. In addition to an apparent developmental curtailment of the limbic system in all cases, consistent abnormalities have been observed in the cerebellum and related inferior olive.

All brains have shown a marked reduction in the number of Purkinje cells and a variable decrease in granule cells throughout the cerebellar hemispheres (Figs. 1 and 2), with the most significant findings appearing in the posterolateral neocerebellar cortex and adjacent archicerebellar cortex (Arin et al., 1991). In contrast to these dramatic hemispheric abnormalities, no statistically significant differences in the size or number of Purkinje cells have been observed in any area of the vermis (Fig. 3; Table 1) (Bauman and Kemper, 1996). In addition, abnormalities have been observed in the fastigial, globose, and emboliform nuclei in the roof of the cerebellum that appear to differ with the age of the patient. Small pale neurons, which are reduced in number, characterize these nuclei in the three older autistic brains (Fig. 4). In the younger cases, however, these same nerve cells, as well as those of the dentate nucleus, are enlarged in size and present in adequate numbers (Fig. 5) (Bauman and Kemper, 1994).

Areas of the principal inferior olivary nucleus in the brain stem, known to be related to the abnormal cerebellar cortex observed in these autistic brains (Holmes and Stewart, 1908), failed to show retrograde cell loss and atrophy, features invariably seen following perinatal and postnatal Purkinje cell loss in human pathology (Norman, 1940; Greenfield, 1954). The olivary neurons in the three older autistic brains were small and pale but present in adequate numbers (Fig. 6). In contrast, these same neurons were significantly enlarged in size but otherwise normal in appearance and number in all six of the younger autistic brains (Fig. 7). In all of the autistic brains, some of the olivary neurons tended to cluster along the periphery of the olivary convolutions, a pattern reported in several disorders of prenatal origin associated with mental retardation (Sumi, 1980; DeBassio et al., 1985).

IV. Implications of Cerebellar Abnormalities in Autism

The cerebellar abnormalities observed in the autistic brain provide some insight into the timing of onset of the pathologic process in this disorder.

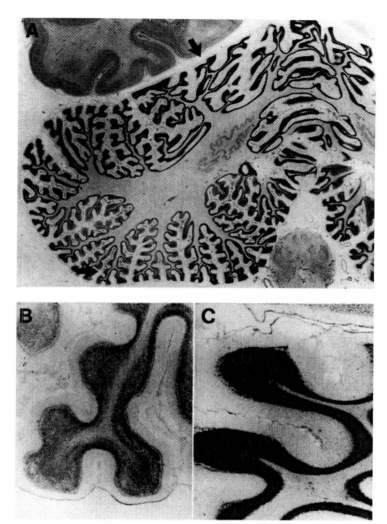

FIG. 1. Nissl-stained section of the cerebellum from the brain of an adult autistic male. Atrophy of the cerebellar cortex can be seen in the inferior and lateral regions of the hemisphere (A). Markedly reduced numbers of Purkinje cells and to a lesser extent, granule cells are evident (B) in compariosn with the grossly normal appearing superior cerebellum (C) (Bauman and Kemper, 1985). Original magnification, ×25.

In all of the autistic brains studied to date, there is a bilaterally symmetrical decrease in the number of Purkinje cells and, to a lesser extent, granule cells throughout the cerebellar hemispheres without significant gliosis. The

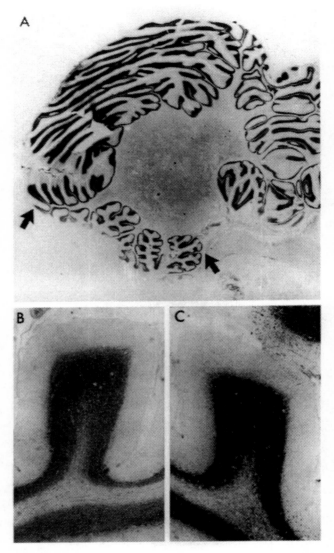

FIG. 2. Nissl-stained section from the brain of an autistic child. As in adult cases, there is atrophy of the inferior and lateral regions of the cerebellar hemispheres (A) associated with a pronounced reduction in the number of Purkinje cells in these same areas (B and C) (Bauman and Kemper, 1994). Original magnification, ×25.

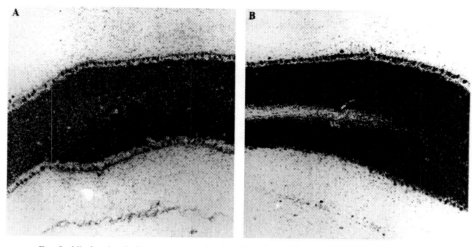

FIG. 3. Nissl-stained photomicrograph of sections through vermal lobule VI from an autistic adult (A) and an age- and sex-matched control (B). Note that the size and number of the Purkinje cells appear to be comparable in both brains. Original magnification, ×63.

absence of glial hyperplasia suggests that these abnormalities were acquired early in development. In animals, cerebellar lesions have shown an associated glial response which becomes less evident as the onset of the lesion develops at progressively earlier ages (Brodal, 1940). Further, the absence

TABLE I
PURKINJE CELL DIAMETER P VALUES IN VERMAL LOBULES I–IX[a]

Vermal lobule	P value
Lingula (I and II)	0.1080
Lingula centralis (III)	0.5000
Culmen (IV and V)	0.2979
Declive (VI)	0.0259
Folium (VIIA)	0.2979
Tuber (VIIA and VIIB)	0.5000
Pyramis (VIII)	0.5000
Uvula (IX)	0.1884
Nodulus (X)	0.1884

[a] There is no statistically significant difference between the size of Purkinje cells in autistic and control subjects in any lobule of the vermis.

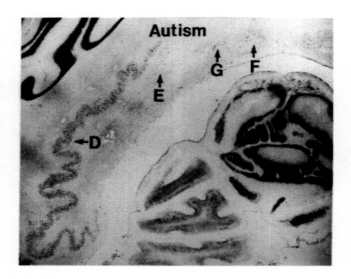
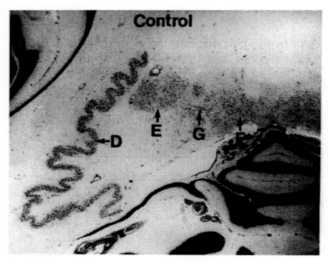

FIG. 4. Nissl-stained photomicrograph of the deep cerebellar nuclei in the brain of an autistic adult male compared to an age- and sex-matched control. The neurons of the fastigial (F), globose (G), and emboliform (E) nuclei are small, pale, and reduced in numbers compared to the control. The dentate nucleus (D) is slightly distorted and the neurons are small but present in adequate numbers (Bauman and Kemper, 1985). Original magnification, ×63.

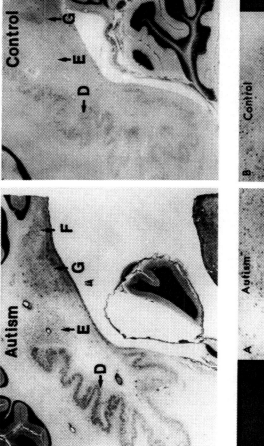
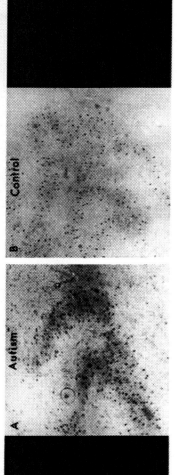

FIG. 5. Nissl-stained photomicrograph of the deep cerebellar nuclei seen in an autistic child. The neurons of these nuclei are enlarged and present in adequate numbers. Similar findings can be seen in the dentate nucleus in the lower panels (Bauman and Kemper, 1994). Original magnification, ×63.

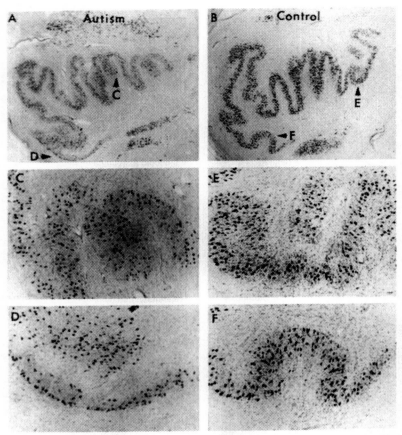

FIG. 6. High-power photomicrographs of the neurons of the inferior olive from the brain of an adult autistic male compared to a normal control. The neurons of the olive are small and pale in comparison with the control but are present in adequate numbers. Note the peripheral distribution of neurons along the inferior loop of this nucleus in the autistic brain (Bauman and Kemper, 1985). Original magnification, ×63.

of a retrograde loss of inferior olivary neurons is a striking observation in these brains. Retrograde loss of olivary neurons has been regularly observed following cerebellar lesions in the immature postnatal and adult animal (Brodal, 1940) and in neonatal and adult humans (Holmes and Stewart, 1908; Norman, 1940; Greenfield, 1954). According to the topographic map of Holmes and Stewart (1908), a lesion in the neocerebellar cortex corresponding in distribution to the abnormalities observed in these autistic brains would result in the loss of olivary neurons in the lateral regions of

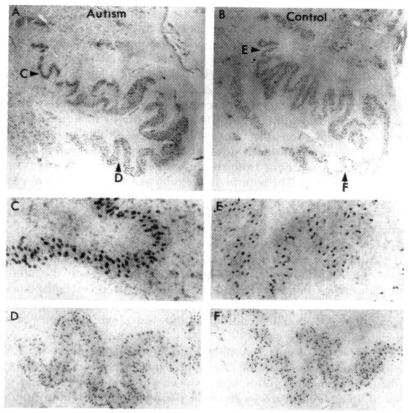

FIG. 7. Nissl-stained photomicrographs of the inferior olive from the brain of an autistic child compared to an age- and sex-matched control. The olivary neurons of the autistic brain are large and plentiful in comparison with the control. As in the adult autistic brain, some of the neurons are disributed peripherally along the inferior loop of this nucleus (Bauman and Kemper, 1994). Original magnification, ×63.

the nucleus. However, in all cases, these neurons, although enlarged in the younger autistic brain and reduced in size in all of the older autistic cerebra, were present in adequate numbers.

The occurrence of retrograde olivary cell loss following cerebellar lesions is believed to be secondary to the tight relationship of the olivary climbing fiber axons to the Purkinje cell dendrites (Eccles et al., 1967). In the fetal monkey, it has been shown that the olivary climbing fibers synapse with the Purkinje cells in a transitory zone located beneath the Purkinje cell layer called the lamina dissecans (Rakic, 1971). Because this zone is no longer evident in the human fetus after 30 weeks of gestation (Rakic and

Sidman, 1970), it is likely that the cerebellar cortical lesion observed in the autistic brain occurred at or before this time. In an analogous situation, expected retrograde cell loss in the medial dorsal nucleus of the thalamus in the rhesus monkey failed to occur following prefrontal lesions prior to, but not after, 160 days of gestation (Goldman and Galkin, 1978).

The abnormalities of the deep cerebellar nuclei show an inconsistent relationship to the findings in the cerebellar cortex and inferior olivary complex. The least involved cerebellar nucleus, the dentate, normally receives direct projections from the most involved cerebellar region, the neocerebellar cortex. Further, one of the most involved nuclei, the fastigial, receives direct projections from the histologically unremarkable vermal cortex (Brodal, 1981). The neurons of the principal inferior olivary nucleus, which are small in adult autistic brains and large in younger brains, are reciprocally related to the dentate nucleus which is histologically normal in adult autistic brains but which shows neuronal enlargement in all of the childhood cases. In contrast, the morphologically unremarkable medial and dorsal accessory olivary nuclei are related to the consistently abnormal globose and emboliform nuclei. With few exceptions, because the efferent projections from the cerebellum derive from the deep cerebellar nuclei, the abnormalities in these nuclei, combined with those in the cerebellar cortex, are in a position to broadly disrupt the function of these projections (Brodal, 1981).

The presence of enlarged, plentiful neurons in the deep cerebellar nuclei and inferior olive in the childhood autistic cases and small cell size in these same nuclear groups, with reduced numbers in the fastigial, globose, and emboliform nuclei in all of the adult autistics, is of interest. It has been shown that advanced myelination is already present in the olivocerebellar tracts in the inferior cerebellar peduncle by 28 weeks of gestation (Flechsig, 1920; Yakovlev and Lecours, 1967). This observation suggests that a functional circuit already exists between the olivary nucleus and the cerebellum at this stage of development. Because the intimate relationship between the Purkinje cells and the inferior olivary nucleus has not yet been established at this time, the prenatal olivary projection is presumed to be to the cerebellar nuclei, and it is likely that this pathway is the primary cerebellar circuit prior to 30 weeks of gestation.

In the infant and adult human, the appearance of a retrograde loss of olivary neurons following cerebellar cortical lesions suggests that the dominant inferior olivary projection in the perinatal and postnatal periods is to the Purkinje cells, with the more primitive circuit remaining as a collateral projection to the cerebellar nuclei. Given these observations, it has been hypothesized that because of the marked early lack of Purkinje cells in the autistic brain, the normal mature olivary to Purkinje cell circuitry cannot be adequately established, resulting in the persistence of the more primitive prenatal connectivity. If this hypothesis is correct, the abnormal retention of the more primitive fetal circuit might account for the presence of "compensatory" neuronal enlargement of the cerebellar nuclei and inferior olive in the childhood autistic brains. Because this fetal circuit was not "designed" to function as the dominant postnatal pathway, it is possible that it cannot be sustained over time, resulting in the later reduction of neuronal size and an eventual loss of neurons in the older autistic patients.

The relationship of the cerebellar findings to those in the forebrain and to the clinical features of autism remains uncertain. Dysfunction of the cerebellum beginning before birth may be associated with few, if any, neurological symptoms (Norman, 1940; Adams et al., 1984). Studies in animals have demonstrated the existence of a direct pathway between the fastigial nucleus and the amygdala and septal nuclei, and a reciprocal circuitry between this nucleus and the hippocampus, suggesting that the cerebellum may play a role in the regulation of emotion and higher cortical thought (Heath and Harper, 1974; Heath et al., 1978). Both animal and human studies have implicated the cerebellum in the regulation of affective behavior (Berman et al., 1974) and functional psychiatric disorders (Heath et al., 1979). Further, based on studies in rabbits, the dentate and interpositus nuclei have been determined to play a role in the elaboration of classical conditioned reflex responses (McCormick and Thompson, 1984).

It has been well established that the cerebellum is involved in the acquisition of some types of motor learning and in the regulation of motor coordination (Bloedel, 1992). It has also been suggested that the cerebellum may play a role in the perception and production of the control of time involving both motor and sensory systems (Ivry and Keele, 1989). Accurate timing requires the ability to predict when an event should occur and the establishment of synchrony between dynamic events. Research has suggested that the lateral cerebellar hemispheres and dentate nuclei play an important role in this function (Holmes, 1939). Additional studies have suggested that the cerebellum may be important in mental imagery, in anticipatory planning (Leiner et al., 1987), and in some aspects of language processing (Peterson et al., 1989). Further, the cerebellum has been implicated in the control of attention, particularly the voluntary shift of selective attention between sensory modalities (i.e., auditory and visual), possibly due to its relationship with the parietal association cortices through connections in the pons (Schmahmann and Pandya, 1989; Courchesne et al., 1992; Akshoomoff and Courchesne, 1992). It has also been suggested that the cerebellum may play a role in cognitive planning, a function that is independent of memory and that is most significant in novel situations (Grafman et al., 1992). More recently, studies in the cebus monkey have established that the dorsolateral prefrontal cortex, believed to be involved in "spatial working memory," is the target of output from the dentate nucleus of the cerebellum (Middleton and Strick, 1994). This relationship to the prefrontal cortex suggests that the cerebellum is involved in the planning and timing of future behavior. In addition to these many functions, the cerebellum also appears to play a role in the regulation of speed, consistency, and appropriateness of mental and cognitive processes, as well as the control and integration of motor and sensory information and activity (Schmahmann, 1991).

Thus, there is mounting evidence for the importance of the cerebellum in the modulation of emotion, behavior, learning, and language, and it is likely that the neuroanatomic abnormalities observed in the cerebellum in autism may contribute to some of the atypical behaviors and disordered information processing characteristic of the syndrome. However, the precise functional significance of these findings, their relationship to the abnormalities noted in the limbic system, and their specific impact on the clinical features of autism remain to be elucidated.

V. Conclusion

Consistent microscopic abnormalities have been observed in the brains of nine autistic individuals which have been confined to the limbic system and cerebellar circuits. The findings in the cerebellum and related inferior olive suggest that the process which resulted in these morphologic changes occurred or had its onset before birth. Although the effect of prenatal abnormalities on cognitive and behavioral development is unknown, it is likely that early dysfunction in these circuits could have a substantial impact on the acquisition and processing of information, and might well contribute to some of the clinical features of autism.

Given the absence of an animal model for research and the limited availability of autopsy material for neuroanatomic study, expanded and more detailed neuroimaging techniques may offer improved opportunities for the *in vivo* study of the brain. Much of the present research in this area has been devoted to measurements of area and volume in specific brain regions, most notably the cerebellum. Although it was initially assumed that differences in the structural size of specific brain regions were a reflection of microscopic neuroanatomic abnormalities, it has become clear that imaging and histologic findings are not necessarily equivalent. In the autistic cerebellum, the histologic findings are most marked in the posterior and lateral regions of the cerebellar hemispheres, which have received little imaging attention. In contrast, the major MRI abnormality reported in the autistic brain has been hypoplasia and hyperplasia of the vermis, a structure that is histologically normal. While there may be methodological factors which have contributed to the inconsistent findings on MRI imaging, it is also possible that the microscopic cerebellar findings may presently be beyond detection by modern imaging techniques. In either case, at this point in time, there appears to be a mismatch between microscopic and imaging studies of the cerebellum in autism and it is therefore prudent to interpret the results of imaging studies in this disorder with caution.

References

Aarkrog, T. (1968). Organic factors in infantile psychoses and borderline psychoses: Retrospective study of 45 cases subjected to pneumoencephalography. *Dan. Med. Bull.* **15,** 283–288.
Adams, J. H., Corselis, J. A. N., and Duchen, L. W. (1984). "Greenfield's Neuropathology." Wiley, New York.
Akshoomoff, N. A., and Courchesne, E. (1992). A new role for the cerebellum in cognitive operations. *Behav. Neurosci.* **106,** 731–738.
Arin, D. M., Bauman, M. L., and Kemper, T. L. (1991). The distribution of Purkinje cell loss in the cerebellum in autism. *Neurology* **41,** 307.
Bachevalier, J., and Merjanian, P. M. (1994). The contribution of medial temporal lobe structures in infantile autism: A neurobehavioral study in primates. *In* "The Neurobiology of Autism" (M. L. Bauman and T. L. Kemper, eds.), pp. 146–169. Johns Hopkins University Press, Baltimore.
Bauman, M. L., and Kemper, T. L. (1984). The brain in infantile autism: A histoanatomic report. *Neurology* **34,** 275.
Bauman, M. L., and Kemper, T. L. (1985). Histoanatomic observations of the brain in early infantile autism. *Neurology* **35,** 866–874.
Bauman, M. L., and Kemper, T. L. (1994). Neuroanatomic observations of the brain in autism. *In* "The Neurobiology of Autism" (M. L. Bauman and T. L. Kemper, eds.), pp. 119–145. Johns Hopkins University Press, Baltimore.
Bauman, M. L., and Kemper, T. L. (1995). Neuroanatomical observations of the brain in autism. *In* "Advances in Biological Psychiatry" (J. Panksepp, ed), pp. 1–26. JAI Press, New York.
Bauman, M. L., and Kemper, T. L. (1996). Observations on the Purkinje cells in the cerebellar vermis in autism. *J. Neuropathol. Exp. Neurol.* **55,** 613.
Berman, A. J., Berman, D., and Prescott, J. W. (1974). The effect of cerebellar lesions on emotional behavior in the rhesus monkey. *In* "The Cerebellum, Epilepsy and Behavior" (I. S. Cooper, M. Riklan, R. S. Snyder, eds.), pp. 277–284. Plenum, New York.
Bloedel, J. R. (1992). Functional heterogeneity with structural homogeneity: How does the cerebellum operate? *Behav. Brain Sci.* **15**(4), 666–678.
Boucher, J., and Warrington, E. K. (1976). Memory deficits in early infantile autism: Some similarities to the amnestic syndrome. *Br. J. Psychol.* **67,** 73–87.
Brodal, A. (1940). Modification of the Gudden method for study of cerebral localization. *Arch. Neurol. Psychiat.* **43,** 46–58.
Brodal, A. (1981). "Neurological Anatomy in relation to Clinical Medicine." Oxford University Press, New York.

Campbell, M. S., Rosenbloom, S., Perry, R., George, A. E., Kricheff, I. I., Anderson, L., Small, A. M., Jenings, S. J. (1982). Computerized axial tomography in young autistic children. *Am. J. Psychiat.* **139,** 510–512.

Coleman, M. (1979). Studies of the autistic syndromes. *In* "Congenital and Acquired Cognitive Disorders" (R. Katzman, ed.), pp. 265–303. Raven Press, New York.

Coleman, P. D., Romano, J., Lapham, L., and Simon, W. (1985). Cell counts in cerebral cortex in an autistic patient. *J. Aut. Dev. Disorder* **15,** 245–255.

Courchesne, E., Akshoomoff, N. A., and Townsend, J. (1992). Recent advances in autism. *In* "Neurobiology of Infantile Autism" (H. Naruse and E. M. Ornitz, eds.), pp. 111–128. Elsevier, Amsterdam.

Courchesne, E., Saitoh, O., Yeung-Couchesne, R., Press, G. A., Lincoln, A. J., Haas, R. H., and Schreibman, L. (1994a). Abnormalities of cerebellar lobules VI and VII in patients with infantile autism: Identification of hypoplastic and hyperplastic subgroups by MRI imaging. *Am. J. Roentgenol.* **162,** 123–130.

Courchesne, E., Townsend, J., and Saitoh, O. (1994b). The brain in infantile autism. *Neurology* **44,** 214–228.

Courchesne, E., Yeung-Couchesne, R., Press, G. A., Hesselink, J. R., and Jernigan, T. L. (1988). Hypoplasia of cerebellar lobules VI and VII in autism. *N. Engl. J. Med.* **318,** 1349–1354.

Damasio, A. R., and Maurer, R. G. (1978). A neurological model for childhood autism. *Arch Neurol* **35,** 777–786.

Darby, J. H. (1976). Neuropathological aspects of psychosis in childhood. *J. Aut. Child. Schizophr.* **6,** 339–352.

DeBassio, W. A., Kemper, T. L., and Knoefel, J. E. (1985). Coffin-Siris syndrome: Neuropathological findings. *Arch. Neurol.* **42,** 350–353.

Delong, G. R. (1978). A neuropsychological interpretation of infantile autism. *In* "Autism" (M. Rutter and E. Schopler, eds.), pp. 207–218. Plenum Press, New York.

Eccles, J. C., Ito, M., and Szentagothai, J. (1967). "The Cerebellum as a Neuronal Machine." Springer, New York.

Filipek, P. A. (1995). Quantitative magnetic resonance imaging in autism: The cerebellar vermis. *Curr. Opin. Neurol.* **8,** 134–138.

Filipek, P. A. (1996). Neuroimaging in autism: The state of the science 1995. *J. Aut. Dev. Disord.* **26**(2), 211–215.

Filipek, P. A., Kennedy, D. N., and Caviness, V. S. (1992a). Neuroimaging in child neuropsychology. *In* "Child Neuropsychology" (I. Rapin and S. Segalowitz, eds.), Vol. 6, pp. 301–329. Elsevier, Amsterdam.

Filipek, P. A., Richelme, C., Kennedy, D. N., Rademacher, J., Pitcher, D. A., Zidel, S., and Caviness, V. S. (1992b). Morphometric analysis of the brain in developmental language disorders and autism. *Ann. Neurol.* **32,** 475.

Flechsig, P. (1920). "Anatomie des Menchlichen Gehirn und Ruchenmachs auf Myelogenetischer Grundlage." George Theime, Leipzig.

Goldman, P. S., and Galkin, T. W. (1978). Prenatal removal of frontal association cortex in the fetal rhesus monkey: Anatomic and functional consequences in postnatal life. *Brain Res.* **152,** 452–485.

Grafman, J., Litvan, I., Massaquoi, S., Stewart, M., Sivigu, A., and Hallet, M. (1992). Cognitive planning deficit in patients with cerebellar atrophy. *Neurology* **42,** 1493–1496.

Greenfield, J. G. (1954). "The Spino-cerebellar Degenerations." Thomas, Springfield, IL.

Hauser, S. L., Delong, G. R., and Rosman, N. P. (1975). Pneumoencephalographic findings in the infantile autism syndrome: A correlation with temporal lobe disease. *Brain* **98,** 667–688.

Heath, R. G., Dempsey, C. W., Fontana, C. J., and Myers, W. A. (1978). Cerebellar stimulation: Effects on septal region, hippocampus and amygdala of cats and rats. *Biol. Psychiat.* **113,** 501–529.

Heath, R. G., Franklin, D. E., and Shraberg, D. (1979). Gross pathology of the cerebellum in patients diagnosed and treated as functional psychiatric disorders. *J. Nerv. Ment. Disord.* **167,** 585–592.

Heath, R. G., and Harper, J. W. (1974). Ascending projections of the cerebellar fastigial nucleus to the hippocampus, amygdala and other temporal lobe sites: Evoked potential and other histologic studies in monkeys and cats. *Exp. Neurol.* **45,** 268–287.

Hier, D. B., LeMay, M., and Rosenberger, P. B. (1979). Autism and unfavorable left-right asymmetries of the brain. *J. Aut. Dev. Disord.* **9,** 153–159.

Holmes, G. (1939). The cerebellum of man. *Brain* **62,** 1–30.

Holmes, G., and Stewart, T. G. (1908). On the connection of the inferior olives with the cerebellum in man. *Brain* **31,** 125–137.

Holttum, J. R., Minshew, N. J., Sanders, R. S., and Phillips, N. E. (1992). Magnetic resonance imaging of the posterior fossa in autism. *Biol. Psychiat.* **32,** 1091–1101.

Ivry, R. B., and Keele, S. W. (1989). Timing functions of the cerebellum. *J. Cogn. Neurosci.* **1,** 136–152.

Jacobson, R., Lecouteur, A., Howlin, P., and Rutter, M. (1988). Selective subcortical abnormalities in autism. *Psychol. Med.* **18,** 39–48.

Kanner, L. (1943). Autistic disturbances of affective contact. Nervous Child **2,** 217–250.

Kleinman, M. D., Neff, S., and Rosman, N. P. (1992). The brain in infantile autism. *Neurology* **42,** 753–760.

Leiner, H. C., Leiner, A. L., and Dow, R. S. (1987). Cerebrocerebellar learning loops in apes and humans. *Ital. J. Neurol. Sci.* **8,** 425–436.

Maurer, R. G., and Damasio, A. R. (1982). Childhood autism from the point of view of behavioral neurology. *J. Aut. Dev. Disord.* **12,** 195–205.

McCormick, D. A., and Thompson, R. F. (1984). Cerebellum: Essential involvement in the classically conditioned eyelid response. *Science* **223,** 296–299.

Middleton, F. A., and Strick, P. L. (1994). Anatomical evidence for cerebellar and basal ganglia involvment in higher cognitive function. *Science* **266,** 458–461.

Murakami, J. W., Courchesne, E., Press, G. A., Young-Courchesne, R., and Hesselink, J. R. (1989). Reduced cerebellar hemisphere size and its relationship to vermal hypoplasia in autism. *Arch. Neurol.* **46,** 689–694.

Norman, R. M. (1940). Cerebellar atrophy associated with etat marbre of the basal ganglia. *J. Neurol. Psychiat.* **3,** 311–318.

Nowell, M. A., Hackney, D. B., Murak, A. S., and Coleman, M. (1990). Varied appearance of autism: Fifty-three patients having the full autistic syndrome. *Magn. Reson. Imag.* **8,** 811–816.

Ornitz, E. M., and Ritvo, E. R. (1968). Neurophysiologic mechanisms underlying perceptual inconstancy in autistic and schizophrenic children. *Arch. Gen. Psychiat.* **19,** 22–27.

Peterson, S. F., Fox, P. T., Posner, M. I., Mintum, M. A., and Raichle, M. E. (1989). Positron emission tomographic studies in the processing of single words. *J. Cogn. Neurosci.* **1,** 153–170.

Piven, J., and Folstein, S. E. (1994). The genetics of autism. In "The Neurobiology of Autism" (M. L. Bauman and Kemper, T. L. eds.), pp. 18–44. Johns Hopkins University Press, Baltimore.

Piven, J., Nehme, E., Simon, P., Pearlson, G., and Folstein, S. E. (1992). Magnetic resonance imaging in autism: Measurement of the cerebellum, pons, and fourth ventricle. *Biol. Psychiat.* **31,** 491–504.

Rakic, P. (1971). Neuron-glia relationship during granule cell migration in developing cerebellar cortex: A Golgi and electron microscopic study in macacus rhesus. *J. Comp. Neurol.* **141,** 282–312.

Rakic, P., and Sidman, R. L. (1970). Histogenesis of the cortical layers in human cerebellum, particularly the lamina dissecans. *J. Comp. Neurol.* **139,** 473–500.

Rapin, I. (1994). Introduction and overview. *In* "The Neurobiology of Autism" (M. L. Bauman and Kemper, T. L. eds.), pp. 1–17. Johns Hopkins University Press, Baltimore.

Ritvo, E. R., Freeman, B. J., Scheibel, A. B., Duong, T., Robinson, H., Guthrie, D., and Ritvo, A. (1986). Lower Purkinje cell counts in the cerebella of four autistic subjects: Initial findings of the UCLA-NSAC autopsy research report. *Am. J. Psychiat.* **146,** 862–866.

Ritvo, E. R., and Garber, J. H. (1988). Cerebellar hypoplasia and autism. *N. Engl. J. Med.* **319,** 1152.

Rosenbloom, S., Campbell, M., and George, A. E. (1984). High resolution CT scanning in infantile autism: A quantitative approach. *J. Am. Acad. Child. Pschiat.* **23,** 72–77.

Schaefer, G. B., Thompson, J. N., Bodensteiner, J. B., McConnell, J. H., Kimberling, W. J., Gay, C. T., Dutton, W. D., Hutchings, D. C., and Gray, S. B. (1996). Hypoplasia of cerebellar veins in neurogenic syndromes. *Ann. Neurol.* **39,** 382–384.

Schmahmann, J. D. (1991). An emerging concept: The cerebellar contribution to higher function. *Arch. Neurol.* **48,** 1178–1187.

Schmahmann, J. D., and Pandya, D. N. (1989). Anatomical investigation of projections to the basis pontis from posterior parietal association cortices in rhesus monkey. *J. Comp. Neurol.* **289,** 53–73.

Sumi, S. M. (1980). Brain malformation in the trisomy 18 syndrome. *Brain* **93,** 821–830.

Vilensky, J. A., Damasio, A. R., and Maurer, R. G. (1981). Gait disturbances in patients with autistic behavior. *Arch. Neurol.* **38,** 646–649.

Williams, R. S., Hauser, S. L., Purpura, D. P., Delong, G. R., and Swisher, C. N. (1980). Autism and mental retardation. *Arch. Neurol.* **37,** 749–753.

Yakovlev, P. I. (1960). Anatomy of the human brain and the problem of mental retardation. *In* "Proceedings of the First International Conference on Mental Retardation" (P. W. Bowman and H. V. Mautner, eds.), pp. 1–43. Grune and Stratton, New York.

Yakovlev, P. I. (1970). Whole brain serial histological sections. *In* "Neuropathology: Methods and Diagnosis" (C. G. Tedeschi, ed.), pp. 371–378. Little Brown, Boston.

Yakovlev, P. I., and Lecours, A. R. (1967). The myelogenetic cycles of regional maturation of the brain. *In* "Regional Development of the Brain in Early Life" (A. Minkowski, ed.), pp. 3–70. Blackwell Sci., Oxford.

OLIVOPONTOCEREBELLAR ATROPHY AND FRIEDREICH'S ATAXIA: NEUROPSYCHOLOGICAL CONSEQUENCES OF BILATERAL VERSUS UNILATERAL CEREBELLAR LESIONS

Thérèse Botez-Marquard and Mihai I. Botez

Neurobehavioral, Neurobiological and Neuropsychology Research Unit,
Hôtel-Dieu de Montréal and University of Montréal,
Montréal, Quebec, Canada H2W 1T8

I. Introduction
II. Neuropsychological and Neurobehavioral Studies of Patients with Bilateral Cerebellar Damage
   A. Epileptic Patients
   B. Olivopontocerebellar Atrophy (OPCA) and Friedreich's Ataxia (FA) Patients
III. Reaction Time (RT) and Movement Time (MT) Assessment in Patients with Bilateral Cerebellar Damage
   A. Rationale and Methods
   B. Simple Visual RT and MT in Epileptics with and without Cerebellar Atrophy
   C. RT and MT in OPCA and FA Patients
   D. RT and MT in Patients with Unilateral Cerebellar Damage
IV. Single Photon Emission Computed Tomography Studies and Neuropsychology of the Cerebellum
V. Neuropsychological Findings in Patients with Unilateral Cerebellar Damage
VI. Negative Findings
VII. Conclusion and Summary
References

This chapter deals with neuropsychological disturbances in patients with bilateral cerebellar damage (BCD), i.e., epileptic patients chronically receiving phenytoin, patients with olivopontocerebellar atrophy (OPCA), and Friedreich's ataxia (FA) versus those with unilateral cerebellar damage (UCD), i.e., patients with cerebellar strokes. BCD patients showed: (i) impaired executive functions in planning and programming of daily activities, elaborating and using structures, and difficulty in abstract thinking, functions that are related to cerebello-frontal loops; (ii) deficits in visuospatial organization for a concrete task and deficient visual-spatial working memory, functions related to cerebello-parietal loops; (iii) lower general intellectual abilities than controls (especially those with OPCA); (iv) difficulties with memory retrieval, diminished global memory quotient, and reduced spatial working memory ability; and (v) slower speed of information processing, as measured by simple and multiple choice reaction time (RT). In UCD patients, neuropsychological and neurobehavioral abilities were deficient for 2–5 months; after this time period, their performances returned to normal. In both BCD and UCD patients, single photon emission computed tomography (SPECT) studies showed different degrees of "reverse" cerebellar → basal ganglia → frontoparietal diaschisis which may underlie permanent or transitory neuropsychological deficits. The relationships among neuropsychological findings, SPECT studies, and chemical neuroanatomy are discussed.

I. Introduction

The cerebellum has long been considered one of the structures underlying the starting mechanism of speech, i.e., speech initiation, maintenance of speech fluency and volume, control of the articulation process, and motor patterning of words (Botez and Carp, 1968; Botez and Barbeau, 1971). In those earlier studies, speech was considered to be the output side of information processing, and as the vehicle of language it required constant "modulation and control" by subcortical mechanisms. In 1978, Watson postulated that the cerebellum may participate in the integration of sensory information, motor skill learning, visual and auditory discrimination, emotion, and motivation control and reinforcement. The possible role of the cerebellum in influencing emotions and psychotic behavior was further developed by Heath *et al.* (1979, 1982) and Schmahmann (1991). The possible correlation of cerebellar damage with some cognitive disorders was suggested by some earlier investigators (Watson, 1978; Fehrenbach *et al.*, 1984), but clear clinical relationships or a physiopathological background was neither postulated nor documented. Botez *et al.* (1985) underscored the role of cerebello-frontal and cerebello-parietal associative loops as anatomophysiological substrata of mild frontal- and parietal-like syndromes encountered in a patient with reversible chronic cerebellar ataxia after phenytoin intoxication. This paper was followed by the report of Leiner *et al.* (1986). Subsequent investigations (Botez, 1992, 1993, 1994) have addressed the following questions: (i) which specific (if indeed they are) neuropsychological disorders and reaction time (RT) deficits are associated with well-delimited bilateral cerebellar damage (BCD) versus those found in patients with unilateral cerebellar damage (UCD); (ii) which neuropsychological measures are not impaired with chronic cerebellar damage, i.e., negative findings; (iii) is there a relationship between anatomic imagery (CT scan and MRI) and the severity of neurobehavioral impairment; (iv) which abnormalities on single photon emission computed tomography (SPECT) with hexamethylpropyleneamine oxide (HMPAO) uptake in cerebellar patients versus controls are correlated with neuropsychological deficits; and (v) are there some neurochemical substrates accompanying the heredodegenerative ataxias and, consequently, are we able to recommend replacement therapy? This chapter presents the results of our investigations to date.

The rationale for studying patients with BCD versus those with UCD arises from two sources: (i) UCD in monkeys is followed by a lengthening of RT which returns to normal values after 20–30 days (Spidalieri *et al.*, 1983); and (ii) our "anecdotal" clinical observations in two young women (19 and 21 years old) with migraine and unilateral cerebellar stroke following contraception pill consumption who were in our care between 1970–1971 and 1996, respectively; subsequent to their strokes, both had high intellectual achievements at university and in everyday life (one became a lawyer and the other became a successful business woman).

II. Neuropsychological and Neurobehavioral Studies of Patients with Bilateral Cerebellar Damage

Two different groups of patients with bilateral cerebellar damage were studied: (i) Epileptics receiving long-term phenytoin (PHT) treatment, who were divided into two subgroups, one with and one without cerebellar atrophy; and (ii) patients with heredodegenerative ataxias, i.e., those with olivopontocerebellar atrophies (OPCA) and those with Friedreich's ataxia (FA).

The inclusion criteria in both studies were (i) men and women aged between 18 and 65 years and (ii) a handgrip strength of at least 5 kg with each hand.

The exclusion criteria were (i) patients showing central (i.e., ventricular dilation) and cortical (i.e., cortical sulci dilation) atrophies according to previously defined radiological measures (Botez *et al.*, 1988); (ii) alcoholics and patients with medical disease in evolution; (iii) patients showing lacunar states on CT scans; (iv) subjects with a deficient visual acuity or oculomotor disturbances affecting smooth reading of small print; (v) patients with severe motor or articulatory disturbances precluding neuropsychological assessment; (vi) all OPCA patients with even mild parkinsonian signs; and (vii) patients with depression.

Appropriate neuropsychological batteries were used, taking into consideration the presence of motor disturbances.

A. Epileptic Patients

Cerebellar atrophy has often been noted in epileptic patients, although not usually in association with clinical manifestations of chronic ataxia

(Hofmann, 1958; Salcman et al., 1978; Koller et al., 1980; Botez et al., 1988). Histologic changes of cerebellar atrophy in epileptics were reported by Spielmeyer (1920) long before the identification of PHT as a major anticonvulsant drug by Merritt and Putman (1938). In our study of 134 PHT-treated and nontreated epileptics versus control subjects, we concluded that cerebellar atrophy in epileptics was significantly correlated with both the length of the illness and the amount of PHT ingested during the patient's lifetime. The number of seizures did not appear to be related to cerebellar atrophy (Botez et al., 1988). In a more recent study of 36 epileptics, cerebellar atrophy was found in PHT-exposed patients with partial epileptic seizures, even in the absence of generalized tonic–clonic seizures (Ney et al., 1994). We agree with the conclusion of Ney et al. (1994) that "whether it is PHT or the seizures that play the primary etiologic role remains unanswered. These factors may be synergistic."

From a group of 55 epileptics with normal CT scans and 49 with pure cerebellar atrophy as detected by radiologic criteria (Botez et al., 1988), 31 epileptics with normal CT scans and 33 epileptics with pure cerebellar atrophy were selected (Botez et al., 1989a). The selection was carried out randomly, i.e., following a tabulation of numbers.

No statistically significant differences were found between groups with regard to age, education, and number of grand mal and other seizures. Neuropsychological assessment revealed lower performances by the atrophic group on the following measures: full IQ scale, verbal IQ scale, performance IQ scales, information, arithmetic, block design, object assembly, digit symbol, and Stroop test forms I and II. No significant differences were observed between the two groups for the remaining five subtests of the Ottawa–Wechsler IQ scale (comprehension, digit span, similarities, picture arrangement, picture completion) as well as for the immediate recall and delayed recall subtests (logical stories, visual memory, verbal association) belonging to the Wechsler memory scale. Analyses of the composite scores of neuropsychological performances suggested that the cerebellum interferes with the following complex behavioral functions: (i) visuospatial organization for a concrete task, as demonstrated by deficient performances on the object assembly, digit symbol, and block design subtests; and (ii) planning and programming of daily activities, as revealed by deficient performance on the Stroop test.

B. Olivopontocerebellar Atrophy (OPCA) and Friedreich's Ataxia (FA) Patients

With OPCA patients, the biochemical and heredoclinical classification of Huang and Plaitakis (1984) was used: (1) recessive OPCA associated with glutamate dehydrogenase deficiency, (2) sporadic OPCA, (3) dominant OPCA ("S" family), (4) dominant OPCA (spinopontine variety), and (5) dominant OPCA associated with slow saccades (Wadia and Swami, 1971). The majority of OPCA patients belonged to a family from Gaspé County with the dominant OPCA form and slow occular saccades (form 5). Genetic studies were undertaken in form 3 and 5 OPCA patients in Dr. Guy Rouleau's laboratory. In form 3 OPCA, the locus is on the short arm of chromosome 6, whereas it is assigned to the long arm of chromosome 12 in form 5 OPCA patients (Lopes-Cendes *et al.*, 1994). No genetic or biochemical studies were done in form 1 and 2 OPCA patients. In these cases, the diagnosis was made exclusively on clinical grounds. FA patients were adults who met all diagnostic criteria of Harding (1988); 85% of them were in wheelchairs.

Fifteen OPCA and 15 FA patients were evaluated with a neuropsychological test battery that was constructed to assess general nonverbal intelligence as well as frontal and parietal lobe functions (Botez-Marquard and Botez, 1993). They were pair-matched with normal controls for age, sex, and education. All tests were carried out in such a way that almost no motor action had to be executed by the patient except for the copy of Rey's complex figure and block design test. In agreement with previous data, OPCA patients showed more severe degrees of cerebellar atrophy on CT scans than FA patients: in the OPCA group, 8 subjects had severe, 5 had moderate, and 2 had mild cerebellar atrophy, whereas in the FA group, these classifications involved 3, 2, and 7 individuals, respectively. In addition, 3 FA patients had normal CT scans.

Both patient groups showed lower performances than their controls on Raven's standard progressive matrices, on Rey's complex figure, and on block design (timed variant). In addition, OPCA patients performed worse than their controls on block design (untimed variant), on the similarities subtest of the Ottawa–Wechsler IQ scale, and on the Hooper visual organization and Trail B test; in contrast, no impairments on these four tests were found in FA patients.

Both groups presented significant deficits as compared with their controls on the following measures: digit span forward, digit span backward, and picture arrangement subtests of the Ottawa–Wechsler IQ scale (Chagnon, 1953).

Cognitive functions seemed to be more impaired in OPCA than in FA, a fact most likely related to the occurrence and severity of radiologically confirmed cerebellar atrophy.

Botez-Marquard *et al.* (1996) investigated whether there is a correlation in OPCA patients between the severity of neuropsychological impairment and the degree of cerebellar atrophy on CT scans. Thirty-two OPCA patients without cortical or central atrophy were studied. In addition to inspection of films by a neuroradiologist (i.e., "subjective" assessment), five neuroradiological measures were used: estimation of brain stem ratio, midbrain ratio, fourth ventricular ratio, brachium pontis ratio, and bicaudate ratio.

The brain stem ratio assessed the following structures: pontine nuclei, medial lemniscus, reticular formation, corticospinal, corticomedullary and corticopontine tracts, facial and trigeminal nuclei, and central tegmental and spinothalamic tracts, as well as superior olivary nuclei. The structures measured by the midbrain ratio were the frontopontine, corticospinal and corticomedullary tracts, the locus niger and origin of the nigrostriatal and striopallidal tracts, the locus ceruleus, red nuclei, lateral and medial lemniscus, and the decussation of the brachium conjonctivum. The structures measured by the fourth ventricle ratio were (i) anteriorly: the central tegmental and longitudinal tracts; (ii) laterally: dentate and vestibular nuclei, brachium pontis, and the restiform body; and (iii) posteriorly: the nodulus of vermis, dentate, and interpositus nuclei. The brachium pontis ratio measured pontocerebellar fibers. The bicaudate cerebroventricular ratio assessed the dimensions of caudate nuclei, the internal and external capsules, and the centrum semiovalis. Practically, the measurements of this ratio did not reveal any significant difference between OPCA patients and control subjects. For each criterion, the patients were divided into two groups: one with mild and the other with moderate to severe atrophy (16 subjects in each group).

Neuropsychological tests assessed verbal memory, learning, retrieval mechanisms, attention, abstract thinking, logical thinking, shifting, planning, organization, visuoperceptive, visuospatial and constructive abilities assessed by Raven's standard progressive matrices, the Wechsler memory scale, the associated verbal learning subtest of the latter scale, auditory-verbal learning of Rey's 15 words (learning abilities, proactive recall, delayed recall, and recognition), and certain Ottawa–Wechsler IQ subtests, i.e., digits forward, digits backward, similarities, picture arrangement as well as the Stroop test, Rey's Oesterreith complex geometric figure, and Hooper's visual organization test.

The relationships between neuroradiological measures and neuropsychological assessment in OPCA patients with mild *versus* those with moderate–severe atrophy revealed cognitive impairment, the extent and severity of which were correlated with the degree of cerebellar atrophy.

The most severe cognitive impairment correlated with the fourth ventricle ratio assessment which measures both neocerebellar and paleocerebellar structures. On 10 tests (global memory quotient, digit backward, similarities, block design, picture arrangement, Raven's standard progressive matrices (score and time), Rey's total recall of 15 words, Rey Oesterreith figure copy, and Stroop test), the differences were significant, whereas in 2 other tests (delayed recall and recognition) they were of borderline significance. The brachium pontis ratio, measuring predominantly corticopontocerebellar fibers, i.e., a pure neocerebellar structure, showed significant differences in 8 tests [global memory quotient, associated words, similarities, Raven's standard progressive matrices (score only), total recall of Rey's 15 words, proactive recall, delayed recall, and recognition]; performance on Rey's figure copy was of borderline significance. Comparison of neuropsychological performances in patients with severe OPCA versus those with mild OPCA revealed significantly lower performances in OPCA patients with severe atrophies on 6, 2, and 4 tests, respectively. These findings of Botez-Marquard et al. (1996) demonstrated that the cerebellum probably participates in neuroanatomic circuits underlying spatial abilities, selective and focused attention, abstract reasoning, and specific memory abilities, consistent with the concept of anatomic and metabolic neocerebellar → basal ganglia → associative cerebral cortical loops. Our observations do not, however, suggest that the neocerebellum and dentate nucleus are exclusively involved in cognition (Leiner et al., 1991; Kim et al., 1994) but rather that other structures (fastigius and vermis) as well as fastigial → limbic loops may also be involved (Botez-Marquard et al., 1996).

In order to further elucidate the possible role of the cerebellum in cognitive behavior, 21 OPCA patients and 21 normal controls with equivalent age, gender, and educational level underwent a neurobehavioral test battery (Arroyo-Anllo and Botez-Marquard, 1996): (i) simple visual and auditory RT; (ii) two tests of motor programming of Luria (1966): simultaneous fist clenching and unclenching, and sequential "fist-edge-palm" movements; (iii) an arithmetic solving problem of Luria (1966); (iv) the meaning of six commonly known proverbs in Quebec; (v) verbal reasoning, i.e., five nonsense sentences modified from the Chapman–Cook speech reading test and eight problems of verbal reasoning; (vi) simple figure copying and delayed recall; (vii) trail making A and B; (viii) visuoperceptive functions, i.e., the battery of Warrington and James (1991); and (ix) a visuospatial short-term memory scan by the Corsi block-tapping task (Milner, 1971).

OPCA patients had lower scores for some tests sensitive to lesions of the frontal lobe, such as hand sequencing, verbal reasoning, and proverb interpretation. In addition, deficits in copying a simple figure and immediate visuospatial memory suggested a mild, parietal-like syndrome. Although immediate memory of the figure was impaired, long-term memory was not. These results are congruent with involvement of the cerebellum in visuospatial working memory.

III. Reaction Time (RT) and Movement Time (MT) Assessment in Patients with Bilateral Cerebellar Damage

A. RATIONALE AND METHODS

RT is recognized as one of the best behavioral measures of central nervous integrity (Birren *et al.*, 1980) and primarly reflects cognitive processing speed (Spirduso *et al.*, 1988). In more general terms, RT evaluates some basic cognitive operations that are engaged in many forms of intellectual behavior (Vernon, 1981).

Botez *et al.* (1991) used RT and MT measurements for objective clinical assessments in ataxic patients before and after clinical trials because no practice effects occur in RT and MT tests (Benton and Blackburn, 1957; Vernon, 1983; Baker *et al.*, 1986).

Simple, forewarned visual and auditory RT and MT were measured separately with the Lafayette apparatus (Model No. 63107) according to the technique of Hamsher and Benton (1977) and described previously (Botez *et al.*, 1989a,b, 1993).

More recently, a microcomputer was programmed to measure multiple-choice RT with progressively more difficult tasks in order to evaluate procedural learning, fast screening memory, and visuospatial functions. Auditory stimuli were delivered by an amplifier interface placed on top of the screen (Botez-Marquard and Routhier, 1995).

Figure 1 illustrates the program from simple visual RT to progressively more difficult multiple-choice RTs.

Specific mental ability tests measure something they have in common. Spearman (cf. Jensen, 1993) termed this "something" the general factor, which he symbolized as "g". Various studies show the relationship between RT variables and a single test that has a high loading "g" factor, such as Raven's progressive matrices (Jensen, 1993). MT is a factor that is uncorrelated with RT and apparently reflects motor speed and dexterity. It has little or no relation to "g" or information processing (Jensen, 1993).

B. SIMPLE VISUAL RT AND MT IN EPILEPTICS WITH AND WITHOUT CEREBELLAR ATROPHY

Epileptics with cerebellar atrophies had significantly lengthened simple visual and auditory RTs when compared to those without cerebellar atrophies (Botez *et al.*, 1989a). MT measurements, however, did not show significant differences between the two groups.

This study confirmed the postulate (Jensen, 1993) that RT and MT are independent measures and concluded that the lengthened RT in epileptic

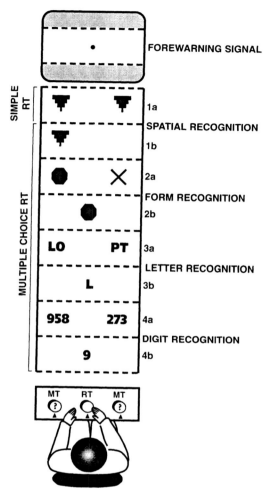

FIG. 1. Simple and multiple-choice reaction times (RT). ·forewarning signal. 1a, simple RT with bilateral visual stimulation. Subject presses the middle key upon appearance of the stimulus and releases key (RT), then moves to the right key with the right hand and to the left key with the left hand. 1b, with appearance of the stimulus, subject moves to the right- or left-sided key. 2a, demonstration of stimulus (1 min). 2b, upon appearance of the stimulus in the central field, subject has to release the key and move to either the left or the right side where the stimulus was initially situated. 3a, 3b, 4a, 4b, same procedure but more stimuli to remember.

patients with cerebellar atrophy could be related to more severe neuropsychological impairment in these patients.

C. RT AND MT IN OPCA AND FA PATIENTS

Simple visual and auditory RT and MT were studied in patients with OPCA and FA, and were found to be significantly lengthened in OPCA patients for visual RT and MT and for auditory RT and MT (calculated for mean RTs and MTs of the left and right hand together); $P < 0.001$ for each of the four parameters. In FA patients, there was also a lengthening of performances in comparison to the controls: visual RT, $P < 0.02$; auditory RT, $P < 0.007$; visual MT, $P < 0.03$ and auditory MT, $P < 0.01$. FA patients sometimes performed better on both visual and auditory RT testing than OPCA patients, in agreement with the lower neuropsychological performances of the OPCA patients (Botez-Marquard and Botez, 1993).

The group of 32 patients with OPCA divided by radiologic criteria into mild–moderate atrophy (16 subjects) and severe cerebellar atrophy (16 cases) was studied with respect to their performance on four RT and four MT (right and left hand) measures (Botez et al., 1993). Very few measures were significantly lengthened in patients with severe atrophy versus those with mild–moderate atrophy as assessed by the brain stem, brachium pontis, and fourth ventricle ratios. In contrast, when comparing RT and MT performances in patients with mild–moderate versus those with severe atrophy, as revealed by the midbrain ratio, seven of eight measures showed a significant lengthening of both RTs and MTs with severe atrophy versus mild–moderate atrophy, whereas the eighth variable was of borderline significance.

The first conclusion of this study is that OPCA patients have significantly slower RTs and MTs than their controls. These findings are in agreement with experimental data in monkeys with cerebellar, i.e., dentate, lesions (Lamarre and Jacks, 1978).

Second, it is possible that this difference in RT and MT performances between OPCA patients with severe versus those with mild–moderate atrophy at the midbrain level is due to neurochemical factors implicated at this level. Among all neuroradiologic ratios, atrophy of the midbrain is the only one that involves dopaminergic, noradrenergic, and glutamatergic structures and pathways (Snider, 1975; Snider et al., 1976; Huang and Plaitakis, 1984).

The multiple-choice RT study of Botez-Marquard and Routhier (1995) examined four levels of difficulty in visual RT and one level in auditory RT. Results showed that OPCA patients were significantly slower ($P < 0.05$)

than the controls on all visual RT tasks (Fig. 2) except for multiple-choice letters (RT IV variant) and the auditory multiple-choice RT two-tone task. Additionally, when relationships were analyzed between different RTs and MTs and Raven's progressive matrices, a visuospatial and concept formation task, OPCA patients presented a rather strong negative correlation coefficient for all RTs: Octagon/cross was an exception, however, and multiple-choice RT (letters) showed only a correlation tendency ($P < 0.08$). Controls had a significant correlation on only one measure: multiple-choice RT (letters). No correlation was noted for all MTs versus Raven in OPCA patients. In control subjects, several correlations were found: Octagon/cross, letters, and the two-tone task.

This study confirmed the previous observations indicating that RT and MT are two different measures. In both groups (OPCA and controls), RT increased as the tasks became increasingly complex, confirming our hypothesis that OPCA patients have screening memory disabilities. Increased MT for either group could reflect either additional decision time or delayed procedural learning. We believe this is an important finding for two reasons: (i) the study showed that RT and MT are two different processes with RT (simple or multiple choice) being a measure of cognitive components whereas MT measures motor performance once the ceiling effect has been reached (Birren et al., 1980; Jensen, 1993); and (ii) the integrity of subcortical structures has been documented as a requisite for

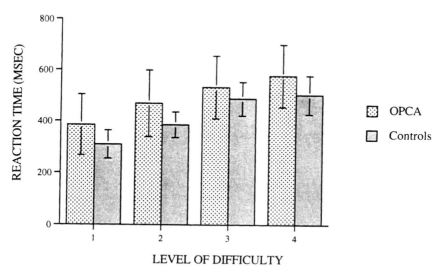

FIG. 2. Multiple-choice RT as a function of complexity in patients with olivopontocerebellar atrophy.

normal RT performance (Cummings and Benson, 1984), and our results are consistent with the observations that the cerebellum, like other subcortical structures (Cummings and Benson, 1984), is also involved in information processing speed, as assessed by RT.

In conclusion, in OPCA patients there was a strong correlation between all RT measurements and Raven's standard progressive matrices. Therefore, in these patients, the slowed speed of information processing (SIP) appears to be associated with visuospatial analogy processing and immediate memory functions.

Conversely, choice RT on tasks that were progressively more and more difficult was carried out with greater difficulty in OPCA patients than in controls. There was a strong relationship between RT performance and intelligence or factor "g", as measured by Raven's standard progressive matrices in OPCA patients; in normal subjects, the correlation was low, in accordance with the observations of Jensen (1993).

Finally, simple visual and auditory RT and MT in patients with lesions at various levels of the central nervous system, including 25 patients with well-delimited BCD, were tested (Botez-Marquard et al., 1989). Patients were individually pair-matched with normal controls for age, sex, and level of education. Simple visual and auditory RT and independent MT tests were correlated with the performance on Raven's standard progressive matrices. BCD patients demonstrated significant diminution of SIP as measured by RT with both hands ($P < 0.0001$), and a significant negative correlation between RT performances and the Raven's test was noted ($P < 0.05$).

D. RT and MT in Patients with Unilateral Cerebellar Damage

Our study included six patients with chronic unilateral cerebellar infarcts. They were individually pair-matched with six control subjects in terms of sex, education level, and age. Average patient age ($X \pm SD$) was 59.5 ± 10.7 versus 58.2 ± 9.15 years in the controls ($P = NS$) whereas patient education level was 12.7 ± 5.1 versus 13.5 ± 5.0 years in the controls ($P = NS$).

Clinical features and SPECT findings in patients with UCD are presented in Table I with the percentage of HMPAO uptake reduction at different levels of the central nervous system. It may be inferred that all patients had cerebello → basal ganglia → cortical diaschisis, as described in previous papers (Botez et al., 1989b, 1990, 1991; Rousseau and Steinling, 1992; Sommezoglu et al., 1993). The Mann–Whitney U test was used for statistical analyses of neuropsychological performances. P values were corrected for equal ranges. RT and MT results showed no significant differences between

TABLE I
Clinical Features, CT Scans, and SPECT Findings in Patients with Chronic Unilateral Cerebellar Infarcts[a]

Patient' initials	Sex	Age (years)	Duration	HKST	FNFT	Medical conditions	CT findings	SPECT Cerebellum	SPECT Basal ganglia	SPECT Parieto-frontal cortex
P.M.	F	63	8 years	++(R)	++(R)	Elevated blood pressure	Right PICA infarct	−40 R	N	−20 LFP
M.N.	M	63	2 years	+++(L)	++(L)	Elevated blood pressure, diabetes	Left PICA infarct	−30 L NR	−15 R NL	−20 R NL
D.C.	F	39	15 years	++(R)	+(R)	Anovulant consumption, migraine, smoker	Right SCA infarct	−20 R NL	−20 L NR	−30 LFP NR
R.J.	M	70	1 year	+(L)	+(L)	Elevated blood pressure, hypercholesterolemia	Left PICA and AICA infarct	−50 L NR	−20 R NL	−15 RFP NL
R.D.	M	61	5 years	++(R)	++(R)	Elevated blood pressure, diabetes	Right PICA and SCA infarct	−90 R NL	−20 L NR	−20 LFP NR
P.L.	F	64	6 months	++(L)	+(L)	Elevated blood pressure	SCA infarct	−20 L	−20 R	−15 RFP

[a]SPECT values are expressed as percentages of reduction in HMPAO uptake (Botez *et al.*, 1991). HKST, heel–knee skin test; FNFT, finger–nose-finger test; −, no deficit; +, mild; ++, moderate; +++, severe deficit; R, right; L, left; LFP, left fronto-parietal; RFP, right fronto-parietal; NR, NL, normal right, normal left. The percentages of reduced HMPAO uptake are expressed for the cerebellum, basal ganglia, and cerebral cortex; PICA, posterior inferior cerebellar artery; AICA, anterior inferior cerebellar artery; SCA, superior cerebellar artery.

patient and control groups. In the presence of cognitive impairment due to chronic UCD, one might expect a significantly lower performance on RT measurements with the hand ipsilateral to the cerebellar lesion, corresponding to the contralateral fronto-parietal area. By employing the Wilcoxon matched pair signed ranks test, RT and MT performances were calculated with the hand ipsilateral to the cerebellar lesion versus the contralateral hand. No significant differences were found for all RT measurements, whereas for MT performances, only one (visual MT-right hand) showed a significantly lengthened value with the ipsilateral hand.

No significant differences between performances of UCD patients versus controls were found on the following tests: digit span forward, digit span backward, similarities, picture arrangement, block design, Rey's complex figure (copy and incidental memory), Raven's standard progressive matrix, Trail B, Rey's word test, Hooper visual organization, and Smedley's handgrip tests.

In four other acute unilateral cerebellar infarcts involving posterior inferior cerebellar or superior cerebellar arteries, frankly lengthened RTs and a low performance on Raven's standard progressive matrices were noted 5 and 10 days, respectively, following the stroke. In one patient, 12 months later, performance was within the normal range compared with the controls.

IV. Single Photon Emission Computed Tomography Studies and Neuropsychology of the Cerebellum

Both PET and SPECT have related classic crossed cerebellar diaschisis to lesions of the contralateral cerebral hemisphere. Interruption of the cerebro-cerebellar loops, i.e., decreased activity of the corticopontocerebellar pathways, is thought to be the most likely mechanism of this remote transneuronal metabolic depression (Feeney and Baron, 1986). Its frequent association with frontal and parietal lesions—the origin of the corticopontine fiber tract—fits this hypothesis.

Using SPECT, Broich et al. (1987) described the first case of crossed cerebello-cerebral diaschisis: in a patient with right cerebellar infarction, they observed hypoperfusion in the left cerebral hemisphere with the most marked reduction in the left premotor region. No mention was made about the possible involvement of the thalamus and/or basal ganglia in this study. Botez et al. (1989b) performed SPECT scans in 4 patients with unilateral cerebellar infarcts. In 3 of these patients, HMPAO uptake was reduced not only in the contralateral fronto-parietal cortex, as shown by Broich et al.

(1987), but also in the contralateral basal ganglia. This phenomenon of crossed cerebellar-cerebral diaschisis in unilateral cerebellar strokes was subsequently confirmed by other investigations (Botez et al., 1990, 1991; Attig et al., 1991; Rousseau and Steinling, 1992; Sommezoglu et al., 1993). Seventeen patients with OPCA and FA were also studied using SPECT; 6 of these patients had reduced HMPAO uptake limited to the cerebellum whereas 11 had cerebello → fronto-parietal reverse diaschisis (Botez et al., 1991). The following conclusions may reasonably be observed from these studies.

Crossed cerebellar-cerebral diaschisis at the subcortical level is associated with remote hypoperfusion mostly in the basal ganglia and not in the thalamus (Botez et al., 1989b, 1990, 1991; Rousseau and Steinling, 1992; Sommezoglu et al., 1993); at the cortical level, diminution of HMPAO uptake is more pronounced in the contralateral frontal lobe and less so in the parietal lobe.

Of the 15 OPCA cases with marked cerebellar hypoperfusion, 11 had reduced HMPAO uptake in the fronto-parietal regions and basal ganglia bilaterally. By comparison, in FA patients, fronto-parietal hypoperfusion was less frequent (in only 33% of cases). The differences in these SPECT results on OPCA versus FA patients are probably explained by the primary localization of lesions: cerebellar lesions in OPCA and afferent cerebellar pathways in FA. These findings are also consistent with the more severe neuropsychological deficits in OPCA than in FA patients.

It is interesting that reduced HMPAO uptake in cerebellar-cortical diaschisis generally spared the thalamus and was found mainly in the basal ganglia. Some anatomical and neurochemical facts may explain this phenomenon. Snider et al. (1976) noted the predominance of fastigial efferents synapsing on cells in the ventral tegmental area in cats and rats, whereas projections from interpositus-dentate nuclei showed a preference for the dopaminergic substantia nigra pars compacta. Projections from the substantia nigra enter the neostriatum and pass to the frontal cortex. The release of dopamine in both caudate nuclei and both substantia nigrae in response to unilateral stimulation of cerebellar nuclei in the cat was demonstrated by Nieoullon et al. (1978).

Although the corticostriate projection has excitatory effects, the nigrostriatal dopaminergic projection has been generally regarded as inhibitory (Côté and Crutcher, 1991; Sommezoglu et al., 1993). On the basis of these anatomical and physiological data, the low blood flow in basal ganglia could be explained by a loss of excitatory inputs from the cortex and/or perhaps from the contralateral dentate nucleus in the cerebellum (via dopaminergic projections). The reduced HMPAO uptake in basal ganglia could therefore be the result of lack of excitation due to reduced inputs from the contralateral dentate nucleus. We can conclude that cerebellar → basal ganglia → cortical diaschisis follows the neurochemical–anatomic pathways.

Reversed cerebello → basal ganglia → cortical diaschisis involves frontal and parietal areas most prominently. This is in accord with anatomic data showing feedforward (Schmahmann and Pandya, 1995) and feedback connections (Middleton and Strick, 1994) between the prefrontal cortex and the cerebellum, as well as cerebellar connections with the parietal lobe (Sasaki et al., 1975; Glickstein et al., 1985; Schmahmann and Pandya, 1989).

V. Neuropsychological Findings in Patients with Unilateral Cerebellar Damage

The neuropsychological test battery used in our six patients had three objectives: (i) to assess general intellectual ability; (ii) to evaluate SIP, as measured by simple visual and auditory RT; and (iii) to examine specific cognitive functions, i.e., frontal- and parietal-like syndromes occurring with cerebellar lesions of other etiologies. The sequence of testing was RT and MT, Raven's standard progressive matrices, Rey's complex figure test, Rey's 15-word test, four subtests of the Ottawa–Wechsler scale, the Hooper test, and the Smedley handgrip test.

Neuropsychological assessments, including RT determinations, did not reveal any impairment in patients with chronic unilateral cerebellar infarcts as compared to their controls. These findings do not concur with those of Wallesch and Horn (1990) in 12 cases involving cerebellar pathology. In that study, it is possible that the neuropsychological results reflected the fact that 11 of their 12 patients had cerebellar tumors and underwent shunting procedures for elevated intracranial pressure and consequent hydrocephalus.

Despite these normal neuropsychological performances, chronic UCD patients nevertheless had cerebello → basal ganglia → cortical reverse diaschisis in SPECT studies (Table I).

A series of experimental findings support our data, showing an absence of lengthened RT or other neuropsychological deficits in patients with chronic UCD. Unilateral experimental dentate lesions in monkeys induced lengthened auditory RT in the immediate postlesion period, which gradually declined over about 20 days postoperatively and approached normal values (Spidalieri et al., 1983).

Amrani et al. (1988) produced unilateral lesions of the right dentate and interpositus nuclei in monkeys and observed severe movement dysfunction of the ipsilateral limb which recovered in 10 to 15 days. After recovery, the left cerebellar nuclei were lesioned, which caused a deficit in the left arm and in the right contralateral arm that had recovered from the previous lesion inflicted on the right side. These experimental findings and our observations on humans converge to the same conclusions: there is some functional relationship between both sides of the cerebellum which could promote compensation for both motor and cognitive behavior in chronic unilateral cerebellar disease. This kind of compensation could be included in the so-called "plasticity" of the nervous system.

We conclude that only bilateral chronic cerebellar lesions are followed by neuropsychological disorders. Are these disorders related only to the lack of compensation due to bilateral cerebellar lesions or are they related both to cerebellar lesions and to metabolic changes (diaschisis) in the basal ganglia, thalamus, and cerebral cortex?

We have observed behavioral and neuropsychological disorders acutely following cerebellar infarction. Neuropsychological testing was performed in a 64-year-old right-handed woman with an MRI-documented infarct in the territory of the *left* superior cerebellar artery. SPECT revealed crossed cerebello → basal ganglia → cortical diaschisis (Botez-Marquard *et al.*, 1994). The first neuropsychological examination took place 2 weeks after the stroke and revealed *right* hemisphere dysfunction manifested by a significant discrepancy between verbal versus performance IQ testing, particularly in subtests of object assembly, picture arrangement, digit symbols, Hooper's visual organization test, and Benton's judgment of line orientation. Frontal-like deficiencies were displayed by low performances in picture arrangement, phonemic verbal fluidity, and the Rey Oesterreith complex figure test. Considerable improvement was noted on all these parameters when the patient was retested 2 years later. RT and MT also improved gradually, and her performances became normal (Table II). Neuroimaging and SPECT studies, however, remained unchanged.

Silveiri *et al.* (1994) reported that a patient with a *right* cerebellar infarct developed agrammatic speech and that SPECT with ^{99m}Tc-HMPAO showed markedly decreased perfusion in the whole *left* cerebral hemisphere. The patient was *not* aphasic and his agrammatism was limited to spontaneous speech. The agrammatism was characterized as a "delay in the process of sentence construction."

We have not encountered aphasic disorders or word-finding difficulties in patients with cerebellar disease. It is difficult to accept such a concept biologically because even in cases of thalamic aphasia, where the lesion is at a higher level of the central nervous system than the cerebellum, the condition is reversible within 3–4 weeks and is associated with thalamocortical diaschisis. Speech comprehension requires the highest level of neural organization in both ontogenesis and phylogenesis. It therefore seems difficult to find such difficulties occurring at lower levels of the central

TABLE II
REACTION TIME (RT) PERFORMANCE VALUES (MEANS AND SD) OF ONE UCD PATIENT AND CONTROLS

Reaction time[a]	January 1991		July 1992		May 1993		Normal values (N = 15)	
	RTs	MTs[b]	RTs	MTs	RTs	MTs	RTs	MTs
SVRT I (msec)								
Right hand (RH)	366 (153)	251 (153)	252 (92)	161 (29)	279 (61)	179 (74)	240 (35)	179 (21)
Left hand (LH)	425 (142)	362 (138)	302 (70)	187 (33)	278 (62)	187 (74)		
MCRT II (msec)								
RH	389 (133)	280 (78)	316 (59)	208 (86)	298 (38)	186 (32)	309 (55)	185 (42)
LH	499 (130)	382 (137)	328 (80)	227 (39)	311 (58)	223 (68)		
MCRT III (msec)								
RH	425 (131)	331 (127)	416 (75)	230 (73)	384 (73)	224 (63)	385 (50)	203 (46)
LH	529 (102)	364 (132)	397 (86)	312 (104)	343 (62)	281 (05)		

[a] SVRT I, simple visual reaction time (arrow left and right side simultaneously); MCRT II, multiple-choice reaction time (arrow on left or right side); and MCRT III, multiple-choice reaction time (octagon left side/cross right side).
[b] Movement times.

nervous system. The case of Silveiri *et al.* (1994) is interesting because: (i) the patient was not aphasic; (ii) the delay in sentence construction could be included in the concept of the role of the cerebellum in modulation and maintenance of speech flow (Botez and Carp, 1968; Botez and Barbeau, 1971); and (iii) the agrammatism disappeared after a few months, despite persistent cerebello-cortical diaschisis, in agreement with previous findings (Botez *et al.*, 1990, 1991; Botez-Marquard *et al.*, 1994).

VI. Negative Findings

We are presently following 140 patients in our outpatient ataxia clinic but we have not observed any language comprehension disorders, long-term memory deficits, word-finding difficulty, or verbal learning deficits. Dysarthria and difficulty in initiation of speech are frequently observed. None of our patients have major frontal syndromes or major psychiatric syndromes such as aggressivity or anxiety; only minor reactive depression is occasionally encountered. No behavioral changes resembling mild autistic behavior have been noted.

VII. Conclusion and Summary

Our findings in BCD patients (OPCA and FA) as well as in epileptics with and without cerebellar atrophies show two groups of syndromes: (i) *impaired executive functions* in planning and programming of daily activities, elaborating and using structures, and difficulty in abstract thinking, functions that are related to cerebello-frontal loops; these findings were confirmed and extended by more sophisticated techniques of El-Awar *et al.* (1991) and Grafman *et al.* (1992); and (ii) a deficit in visuospatial organization for a concrete task and deficient visual–spatial working memory, functions related to cerebello-parietal loops. We have postulated that the role of the cerebellum in mild frontal- and parietal-like syndromes is mainly *indirect*, i.e., modulation occurs through cerebello-frontal and cerebello-parietal pathways, mediated by physiological and neurochemical mechanisms (Botez *et al.*, 1985; Schmahmann, 1991; Schmahmann and Pandya, 1989, 1995; Botez, 1994; Barraquer Bordas, 1994).

The background of mild frontal and mild parietal lobe behavioral syndromes in BCD patients is the consequence, we believe, of cerebello → basal ganglia → cortical diaschisis. The more severe diminution of HMPAO uptake at the cortical level in OPCA patients compared with FA patients seems to be responsible for their greater neuropsychological impairment. In addition to spatial abilities and executive functions, the cerebellum is indirectly involved in selective attention and abstract reasoning abilities (Botez-Marquard et al., 1996).

Memory retrieval difficulties are encountered and the global memory quotient is diminished in BCD patients (although mean scores are still within the normal range) with a reduction of spatial working memory ability. We generally agree with the findings of Appolonio et al. (1993) who suggested that memory in patients with relatively pure cerebellar dysfunction is only partially compromised and that the impairment (at least in part) seems to be secondary to a deficit in executive functions.

BCD patients, especially those with OPCA, had lower general intellectual abilities compared to the controls, which was reflected by lower Raven's standard matrice performances. FA patients took more time to complete Raven and obtained better scores than OPCA subjects who performed faster but scored lower.

RT, which measures SIP, is always lengthened in BCD patients. The lengthening is even more evident on multiple-choice RT with increasingly difficult tasks. Whereas the role of the cerebellum in mild frontal and parietal lobe syndromes is mainly *indirect* through the cerebello-cortical loops, its role in RT performances is *direct*, i.e., the lesion itself is responsible for the lengthening of performance.

Our findings do not support an *exclusive role* of the neocerebellum and of the dentate nucleus in cognition; rather the paleocerebellar structures, particularly the vermis and fastigial limbic loops, could be involved in the control of nonmotor behavior.

Dysfunctions of glutamatergic, dopaminergic, noradrenergic, and perhaps serotoninergic pathways may underlie the impairment of SIP and some other neuropsychological performances in cerebellar disease. The behavioral neurochemistry of cerebellar disease is in its infancy.

In patients with UCD, some neuropsychological performances as well as SIP can be compromised for 3–6 months after the stroke; performances return to normal thereafter despite the persistence of cerebello → basal ganglia → cortical diaschisis for years. In some patients, neuropsychological performances are suggestive of impairment of the contralateral cerebral hemisphere (Botez-Marquard et al., 1994; Silveiri et al., 1994).

Acknowledgments

This study was supported by grants from the Canadian Association of Friedreich's Ataxia and Du Pont Merck Pharmaceutical Company, Wilmington, Delaware. The authors are indebted to Mr. Ovid M. Da Silva for editing this manuscript. Secretarial support was very capably provided by Mrs. Michèle Mathieu.

REFERENCES

Amrani, K., Pellerin, J. P., and Lamarre, Y. (1988). The effect of bilateral lesions of the cerebellar nuclei on motor performance in the monkey. *Soc. Neurosci. Abst.* **13,** 1239.
Apollonio, I. M., Grafman, J., Schwartz, V., Massaquoi, S., and Hallett, M. (1993). Memory in patients with cerebellar degeneration. *Neurology* **43,** 1536–1544.
Arroyo-Anllo, E. M., and Botez-Marquard, T. (1996). Neurobehavioral assessment in olivopontocerebellar atrophy patients. Submitted for publication.
Attig, E., Botez, M. I., Hublet, C. L., Vervonck, C., Jacques, J., and Capon, A. (1991). Diaschisis cérébral croisé par lésion cérébelleuse. *Rev. Neurol. Paris* **147,** 200–207.
Baker, S. J., Maurissen, J. P., and Chrzan, G. J. (1986). Simple reaction time and movement time in human volunteers. *Percept. Motor Skills* **63,** 767–774.
Barraquer Bordas, L. (1994). Aspectos cognitivos de la funcion del cerebelo y reflexiones a que ello nos induce. *Archiv. Neurobiol.* **58,** 267–272.
Benton, A. L., and Blackburn, H. L. (1957). Practice effects in reaction time tasks in brain-injured patients. *J. Abn. Soc. Psychol.* **54,** 109–113.
Birren, J. E., Woods, A. M., and Williams, M. V. (1980). Behavioral slowing with age: Causes, organization, and consequences. *In* "Aging in the 1980s: Psychological Issues" (L. W. Poon, ed.), pp. 293–308. American Psychological Association, Washington, D.C.
Botez, M. I. (1992). The neuropsychology of the cerebellum: An emerging concept. *Arch. Neurol.* **49,** 1229–1230.
Botez, M. I. (1993). Cerebellar cognition. *Neurology* **43,** 10.
Botez, M. I. (1994). The cerebellum. *In* "Encyclopedia of Human Behavior" (S. Ramachandran, ed.), Vol. 1, pp. 549–560. Academic Press, San Diego.
Botez, M. I., Attig, E., and Vézina, J. L. (1988). Cerebellar atrophy in epileptic patients. *Can. J. Neurol. Sci.* **15,** 299–303.
Botez, M. I., and Barbeau, A. (1971). Role of subcortical structures, and particularly of the thalamus, in the mechanisms of speech and language. *Int. J. Neurol.* **8,** 300–320.
Botez, M. I., Botez, T., Elie, R., and Attig, E. (1989a). Role of the cerebellum in complex human behavior. *Ital. J. Neurol. Sci.* **10,** 291–300.
Botez, M. I., Botez, T., Léveillé, J., and Lambert, R. (1990). Cerebello-cerebral diaschisis. *Neurology* **40**(Suppl.1), 173.
Botez, M. I., and Carp, N. (1968). Nouvelles données sur le problème du mécanisme de déclenchement de la parole. *Rev. Roum. Neurol.* **5,** 152–158.
Botez, M. I., Gravel, J., Attig, E., and Vézina, J. L. (1985). Reversible chronic cerebellar ataxia after phenytoin intoxication: Possible role of the cerebellum in cognitive thought. *Neurology* **35,** 1152–1157.
Botez, M. I., Léveillé, J., and Botez, T. (1989b). Role of the cerebellum in cognitive thought; SPECT and neurological findings. *In* "Proceedings of the Thirteen Annual Brain Impairment Conference" (Matheson and H. Newman, eds.), pp. 179–195. The Australian Society for the Study of Brain Impairment, Richmond, Australia.
Botez, M. I., Léveillé, J., Lambert, R., and Botez, T. (1991). Single photon emission computed tomography (SPECT) in cerebellar disease: Cerebello-cerebral diaschisis. *Eur. Neurol.* **31,** 405–421.

Botez, M. I., Pedraza, O. L., Botez-Marquard, T., Vézina, J. L., and Elie, R. (1993). Radiologic correlates of reaction time measurements in olivopontocerebellar atrophy. *Eur. Neurol.* **33,** 304–309.

Botez-Marquard, T., and Botez, M. I. (1992). Unilateral and bilateral cerebellar lesions: Neuropsychological performances. *Neurology* **42**(Suppl. 3), 290.

Botez-Marquard, T., and Botez, M. I. (1993). Cognitive behavior in heredodegenerative ataxias. *Eur. Neurol.* **33,** 351–357.

Botez-Marquard, T., Botez, M.I., Cardu, B., and Léveillé, J. (1989). Speed of information processing and its relationship to intelligence at various levels of the central nervous system. *Neurology* **39**(Suppl. 1), 318.

Botez-Marquard, T., Léveillé, J., and Botez, M. I. (1994). Neuropsychological functioning in unilateral cerebellar damage. *Can. J. Neurol. Sci.* **21,** 353–357.

Botez-Marquard, T., Pedraza, O. L., and Botez, M. I. (1996). Neuroradiological correlates of neuropsychological disorders in olivopontocerebellar atrophy (OPCA). *Eur. J. Neurol.* **3,** 89–97.

Botez-Marquard, T., and Routhier, I. (1995). Reaction time and intelligence in patients with olivopontocerebellar atrophy. *Neuropsychiat. Neuropsychol. Behav. Neurol.* **8,** 168–175.

Broich, K., Hartmann, A., Biersack, H.-J., and Horn, R. (1987). Crossed cerebello-cerebral diaschisis in a patient with cerebellar infarction. *Neurosci. Lett.* **83,** 7–12.

Chagnon, M. (1953). "Manuel et normes de l'échelle d'intelligence Ottawa–Wechsler." Université d'Ottawa, Ottawa.

Côté, L., and Crutcher, M. D. (1991). The basal ganglia. *In* "Principles of Neural Science" (E. R. Kandel *et al.*, eds.), pp. 647–659. Elsevier, New York.

Cummings, J. L., and Benson, D. F. (1984). Subcortical dementia: Review of an emerging concept. *Arch. Neurol.* **41,** 874–879.

El-Awar, M., Kish, S., Oscar-Berman, M., Robitaille, Y., Schut, L., and Freedman, M. (1991). Selective delayed alternation deficits in dominantly inherited olivopontocerebellar atrophy. *Brain Cogn.* **16,** 121–129.

Feeney, D. M., and Baron, J. C. (1986). Diaschisis. *Stroke* **17,** 817–830.

Fehrenbach, R. A., Wallesch, C. W., and Claus, D. (1984). Neuropsychologic findings in Friedreich's ataxia. *Arch. Neurol.* **41,** 306–314.

Glickstein, M., May, J. G., and Mercier, B. E. (1985). Corticopontine projections in the macaque: The distribution of labelled cortical cells after large injections of horseradish peroxidase in the pontine nuclei. *J. Comp. Neurol* **235,** 343–359.

Grafman, J., Litvan, I., Massaquoi, S., Stewart, M., Sirigu, A., and Hallett, M. (1992). Cognitive planning deficit in patients with cerebellar atrophy. *Neurology* **42,** 1493–1496.

Hamsher, K. S., and Benton, A. L. (1977). The reliability of reaction time determinations. *Cortex* **13,** 306–310.

Harding, A. E. (1988). The inherited ataxias. *In* "Advances in Neurology, Molecular Genetics of Neurological and Neuromuscular Disease" (S. Di Donato *et al.*, eds.), Vol. 48, pp. 37–45. Raven Press, New York.

Heath, R. G., Franklin, D. E., and Shraberg, D. (1979). Gross pathology of the cerebellum in patients diagnosed and treated as functional psychiatric disorders. *J. Nerv. Ment. Dis.* **167,** 585–592.

Heath, R. G., Franklin, D. E., Walker, C. F., and Keating, J. W. (1982). Cerebellar vermal atrophy in psychiatric patients. *Biol. Psychiat.* **17,** 569–583.

Hoffmann, W. W. (1958). Cerebellar lesions after parenteral dilantin administration. *Neurology* **8,** 210–214.

Huang, Y. P., and Plaitakis, A. (1984). Morphological changes of olivopontocerebellar atrophy in computed tomography and comments on its pathogenesis. *In* "The Olivopontocerebelar Atrophies" (R. C. Duvoisin, and A. Plaitakis, eds.), pp. 39–85. Raven Press, New York.

Jensen, A. R. (1993). Spearman's g: link between psychometrics and biology. *Ann. N. Y. Acad. Sci.* **702**, 103–129.
Kim, S. G., Ugurbil, K., and Strick, P. L. (1994). Activation of a cerebellar output nucleus during cognitive processing. *Science* **265**, 949–951.
Koller, W. C., Glatt, S. L., and Fox, J. H. (1980). Phenytoin-induced cerebellar degeneration. *Ann. Neurol.* **8**, 203–204.
Lamarre, Y., and Jacks, B. (1978). Involvement of the cerebellum in the initiation of fast ballistic arm movement in the monkey. *Contemp. Clin. Neurophysiol.* **34**(EEG Suppl. 2), 441–447.
Leiner, H. C., Leiner, A. L., and Dow, R. S. (1986). Does the cerebellum contribute to mental skills? *Behav. Neurosci.* **100**, 443–454.
Leiner, H. C., Leiner, A. L., and Dow, R. S. (1991). The human cerebro-cerebellar system: Its computing, cognitive and language skills. *Behav. Brain Res.* **44**, 113–128.
Lopes-Cendes, I., Andermann, E., Attig, E., Cendes, F., Radvany, J., Bosch, S., Wagner, M., Gerstenbrand, F., Andermann, F., and Rouleau, G.A. (1994). Confirmation of the SCA-2 locus as an alternative locus for dominantly inherited spinocerebellar ataxias and refinement of the candidate region. *Am. J. Hum. Genet.* **54**, 774–781.
Luria, A. R. (1966). "Higher Cortical Functions in Man." Basic Book, New York.
Meritt, H. H., and Putnam, T. J. (1938). Sodium diphenyl-bidautoinate in the treatment of convulsive disorders. *J. Am. Med.* **111**, 1068–1073.
Middleton, F. A., and Strick, P. L. (1994). Anatomical evidence for cerebellar and basal ganglia involvement in higher cognitive function. *Science* **266**, 458–461.
Milner, B. (1971). Interhemispheric differences in the localization of psychological processes in man. *Br. Med. Bull.* **27**, 3.
Ney, C. G., Lantos, G., Barr, W. B., and Schaul, N. (1994). Cerebellar atrophy in patients with long-term phenytoin exposure and epilepsy. *Arch. Neurol.* **51**, 767–771.
Nieoullon, A., Cheramy, A., and Glowinski, J. (1978). Release of dopamine in both caudate nuclei and both substantia nigrae in response to unilateral stimulation of cerebellar nuclei in the cat. *Brain Res.* **148**, 143–152.
Rousseau, M., and Steinling, M. (1992). Crossed hemispheric diaschisis in unilateral cerebellar lesions. *Stroke* **23**, 511–514.
Salcman, M., Defendini, R. Corell, J., and Gilman, S. (1978). Neuropathological changes in cerebellar biopsies of epileptic patients. *Ann. Neurol.* **3**, 10–19.
Sasaki, K., Oka, H., Matsuda, Y., Shimono, T., and Mizuno, M. (1975). Electrophysiological studies of the projections from the parietal association area to the cerebellar cortex. *Exp. Brain Res.* **23**, 91–102.
Schmahmann, J. D. (1991). An emerging concept: The cerebellar contribution to higher function. *Arch. Neurol.* **48**, 1178–1187.
Schmahmann, J. D., and Pandya, D. N. (1989). Anatomical investigation of projections to the basis pontis from posterior parietal association cortices in the rhesus monkey. *J. Comp. Neurol.* **289**, 53-73.
Schmahmann, J. D., and Pandya, D. N. (1995). Prefrontal cortex projections to the basilar pons: Implications for the cerebellar contribution to higher function. *Neurosci. Lett.* **199**, 175–178.
Silveiri, M. C., Leggio, M. G., and Molinari, M. (1994). The cerebellum contributes to linguistic production: A case of agrammatic speech following a right cerebellar lesion. *Neurology* **44**, 2047–2050.
Snider, R. S. (1975). A cerebellar-ceruleus pathway. *Brain Res.* **88**, 59–63.
Snider, R. A., Maiti, A., and Snider, S. R. (1976). Cerebellar pathways to ventral midbrain and nigra. *Exp. Neurol.* **53**, 714–728.

Sommezoglu, K., Sperling, B., Henricksen, T., Tfelt-Hansen, P., and Lassen, N. A. (1993). Reduced contralateral hemispheric flow measured by SPECT in cerebellar lesions: Crossed cerebral diaschisis. *Acta Neurol. Scand.* **87,** 275–280.

Spidalieri, G., Busby, L., and Lamarre, Y. (1983). Fast ballistic arm movement triggered by visual, auditory and somesthetic stimuli in the monkey. II. Effects of unilateral dentate lesion on discharge of precentral cortical neurons and reaction time. *J. Neurophysiol.* **50,** 1359–1379.

Spielmeyer, W. (1920). Uber einige Beziehungen zwischen Ganglien zellenveranderungen und gliosen Erscheinungen, besonders am Kleinhirn. *Z. Ges. Neurol. Psychiat.* **54,** 1–38.

Spirduso, W. W., MacRae, H. H., MacRae, P. G., Prewitt, J., and Osborne, L. (1988). Exercise effects on aged motor function. *Ann. N. Y. Acad. Sci.* **515,** 363–375.

Vernon, P. A. (1981). Reaction time and intelligence in the mentally retarded. *Intelligence* **5,** 345–355.

Vernon, P. A. (1983). Speed of information processing and general intelligence. *Intelligence* **7,** 53–70.

Wadia, N. H., and Swami, R. K. (1971). A new form of heredo-familial spinocerebellar degeneration with slow eye movements. *Brain* **94,** 359–374.

Wallesch, C. V., and Horn, A. (1990). Long-term effects of cerebellar pathology on cognitive functions. *Brain Cogn.* **14,** 19–25.

Watson, P. J. (1978). Nonmotor functions of the cerebellum. *Psychol. Bull.* **85,** 944–967.

POSTERIOR FOSSA SYNDROME

Ian F. Pollack

Department of Neurosurgery and University of Pittsburgh Cancer Institute Brain Tumor Center
University of Pittsburgh School of Medicine and Children's Hospital of Pittsburgh
Pittsburgh, Pennsylvania 15213

I. Introduction
II. Patient Population
   A. General Features
   B. Temporal Profile of Mutism and Associated Impairments
   C. Diagnostic Evaluation
   D. Neuropsychological Testing
   E. Illustrative Case
III. Discussion
   A. Clinical Features of the Syndrome
   B. Anatomic Basis for the Posterior Fossa Mutism Syndrome
   C. Anatomic Basis for the Associated Affective Symptoms
   D. Implications
   E. Summary
      References

Transient mutism is a well-recognized sequela of posterior fossa tumor resection in children. A recent review from our institution indicated that 12 of 142 children undergoing such procedures (8.5%) exhibited transient speech impairment, the largest series of such patients reported to date. Each child had a vermian neoplasm that was approached by division of the inferior vermis ($n = 10$) and/or superior vermis ($n = 3$). Seven children had medulloblastomas, three had astrocytomas, and two had ependymomas. None of the affected children had cerebellar hemispheric lesions; in contrast, the incidence among children with vermian neoplasms was 13%. In general, mutism developed 1 to 4 days postoperatively and typically was associated with puzzling neurobehavioral abnormalities. All children had bizarre personality changes, emotional lability, and/or decreased initiation of voluntary movements; nine exhibited poor oral intake; and five had urinary retention. Detailed neuropsychological testing was performed in seven children and confirmed the presence of widespread impairments not only in speech, but also in initiation of other motor activities. These deficits generally resolved during a period of several weeks to months, although two children had residual impairment. Characteristically, affect and oral intake normalized before the speech began to improve. These deficits were noted to correlate with the presence of edema within the brachium pontis bilaterally, although this association was not absolute. The latter observation suggests that there was not a single locus underlying this disorder, but rather that the involved neural pathways may have been impaired at any one of a number of sites within the posterior fossa. Based on the results in our patients and in others described in the literature, we postulate an important role for the cerebellum and/or its afferent and efferent connections in initiating (rather than merely coordinating) speech and other complex motor activities and a potential role for these structures in influencing overall behavior and affective state.

I. Introduction

The development of mutism after resection of large pediatric vermian tumors was first noted anecdotally by Hirsch *et al.* (1979) and Sakai *et al.* (1980) and has remained a source of confusion and consternation for neurosurgeons caring for such children. Studies in our institution (Pollack *et al.*, 1995) and elsewhere (Aguiar *et al.*, 1993; Ammirati *et al.*, 1989; Boratynski and Wocjan, 1993, Catsman-Berrevoets *et al.*, 1992; Cochrane *et al.*, 1994; Dailey *et al.*, 1995; D'Avanzo *et al.*, 1993; Dietze and Mickle, 1990–1991; Ferrante *et al.*, 1990; Gaskill and Marlin, 1991–1992; Herb and Thyen, 1992; Humphreys, 1989; Nagatani *et al.*, 1991; Rekate *et al.*, 1985; Salvati *et al.*, 1991; Van Calenbergh *et al.*, 1995; Van Dongen *et al.*, 1994; Volcan *et al.*, 1986; Wisoff and Epstein, 1984) indicate that the development of mutism in this setting is not rare, occurring in as many as 15% of children with vermian tumors. In addition to the speech impairment, affected patients often exhibit a spectrum of characteristic neurobehavioral deficits (Cochrane *et al.*, 1994; Dailey *et al.*, 1995; Gaskill and Marlin, 1991–1992; Humphreys, 1989; Pollack *et al.*, 1995; Rekate *et al.*, 1985; Von Dongen *et al.*, 1994; Volcan *et al.*, 1986; Wisoff and Epstein, 1984), which may include eating dysfunction, emotional lability, impaired eye opening, incontinence, and, in some instances, decreased initiation of a wide range of voluntary activities. Fortunately, in most cases, these deficits resolve completely during a period of several weeks to months.

Despite the frequency and dramatic nature of this syndrome, the anatomical substrate for the speech and neuropsychiatric dysfunction has remained uncertain. However, the recent availability of high-resolution imaging techniques and the growing recognition of the complex interconnections between the cerebellum and higher cortical centers provide the basis for rational insights into the structural etiology for this phenomenon and potential ways for minimizing the frequency of this syndrome. This chapter reviews our experience with the diagnosis and treatment of such patients in the context of the existing neurosurgical and neuroanatomic literature, presents a hypothesis regarding the anatomical substrates for this unusual syndrome, and discusses ways in which the frequency of this problem may be decreased.

II. Patient Population

A. GENERAL FEATURES

One hundred and forty-two children underwent resection of an infratentorial tumor between 1985 and 1994 at the Children's Hospital of Pittsburgh; 92 had lesions that primarily involved the vermis. Twelve of the children (8.5%) exhibited postoperative mutism in the absence of any obvious cause, such as profound neurological impairment as a result of the tumor, the surgical procedure, or meningitis (Table I). As noted later in the chapter, the spectrum of the syndrome in terms of the duration of the speech impairment and the severity of the associated neurobehavioral manifestations varied considerably among these children. Each of the patients had a tumor of the cerebellar vermis (Figs. 1A and 1B), in nine cases arising from the inferior vermis, in two cases from the superior vermis, and in one case with multiple metastatic nodules, affecting both areas. The incidence of this syndrome was 13.0% among the 92 children with vermian lesions. Seven of the affected patients had a medulloblastoma, 3 had an astrocytoma, and 2 had an ependymoma. The patients ranged in age from 3.5 to 16 years; 4 were girls and 8 were boys. All had normal speech preoperatively.

Each of these children underwent placement of an external ventricular drain as an initial step in the operation. The tumor was then approached using a standard suboccipital craniotomy or craniectomy, performed with the patient in a modified prone (Concorde) position. In each case, an area of the vermis was split vertically to facilitate tumor removal. Somatosensory-evoked potentials and brain stem auditory-evoked potentials were monitored intraoperatively; in no case were significant or lasting abnormalities detected on these studies. Continuous sixth and seventh nerve electromyography was performed intraoperatively on Cases 4 through 12.

Eleven of these children underwent resections that were judged to be complete or nearly complete by the operating surgeon and the postoperative images. In 10 children, the ventriculostomy was "weaned" out in the

TABLE I
Summary of 12 Children with Mutism and Neurobehavioral Symptoms after Resection of Posterior Fossa Tumors[a]

Age (years)/sex	Tumor site	Brain stem invasion	Histology	Postoperative interval until onset of syndrome	Speech impairment	Neurobehavioral symptoms	Poor oral intake	Other symptoms	Time until beginning/completion of speech recovery	Residual deficits
6/F	Vermis	Fourth ventricle floor and R lateral recess	Medulloblastoma	18–24 hr	Mutism	Emotional lability, limited initiation of activities, poor short-term memory	No	R ataxia and tonic posturing, urinary retention	6 weeks/4 months	R ataxia, dysmetria
11/F	Vermis	L lateral recess of fourth ventricle	Ependymoma	24 hr	Mutism	Lay curled up in bed, eyes closed, whined, refused to follow commands	Yes	L sixth and seventh CN paresis, urinary retention	2 weeks/2 months	L sixth nerve paresis
9/M	Vermis	No	Medulloblastoma	24 hr	Mutism	Emotional lability, limited initiation of activities, poor short-term memory	No	Ataxia	7 days/4 weeks	None
10/M	Vermis	Fourth ventricle floor	Medulloblastoma	Immediate	Mutism	Emotional lability, refused to initiate activities	Yes	L sixth and seventh CN paresis, ataxia	3 weeks/6 months	Ataxia, swallowing difficulties
16/M	Vermis	No	Pilocytic astrocytoma	48 hr	Mutism	Lay curled up in bed, eyes closed, whined, refused to follow commands, poor problem-solving and memory	Yes	L hemiparesis, urinary retention	1 month/3 months	L ataxia

Age/Sex	Tumor location	Tumor extension	Tumor type	Onset	Speech	Behavior	Initiated activities?	Neurological findings	Duration mutism/other	Residual
9/M	Vermis	Fourth ventricle floor	Medulloblastoma	<24 hr	Mutism	Refused to initiate activities, difficulty with problem-solving	Yes	L ataxia, neglect	4 weeks/4 months	L ataxia, dysmetria
6/F	Vermis	L middle cerebellar peduncle	Pilocytic astrocytoma	72–96 hr	Mutism	Lay cured up in bed, eyes closed, whined	Yes	Ataxia, sixth CN paresis	10 days/3 weeks	L ataxia
9/M	Vermis	Both middle cerebellar peduncles, R fourth ventricle floor	Ependymoma	48–72 hr	Mutism	Lay curled up in bed, whined, emotional lability	Yes	L ataxia, sixth CN paresis, urinary retention	10 days/2 months	None
8/F	Vermis	L fourth ventricle floor, middle cerebellar peduncle	Medulloblastoma	48–72 hr	Mutism	Refused to initiate activities	Yes	Ataxia, L sixth and seventh CN paresis	7 days/3 weeks	None
5/M	Vermis	R floor of fourth ventricle	Medulloblastoma	72 hr	Mutism	Lay curled up in bed, whined	Yes	R hemiparesis, upgaze paresis, ataxia, urinary retention	14 days/6 weeks	Ataxia, mild dysarthria
3.5/M	Vermis	Lateral superior floor of fourth ventricle; R middle cerebellar peduncle	Astrocytoma	72 hr	Mutism	Lay curled up in bed, whined	No	R ataxia, L hemiparesis	14 days/2 months	Ataxia
4/M	Vermis with mets to both cerebellar hemispheres	Fourth ventricle floor	Medulloblastoma	72 hr	Mutism	Lay curled up in bed, whined	Yes	L arm ataxia, INO	18 days/4 weeks	Ataxia

a Modified from Pollack *et al.* (1995).

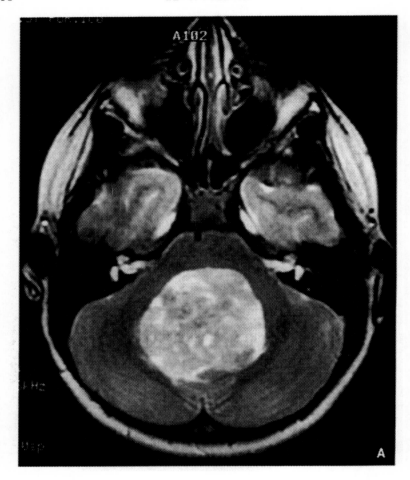

FIG. 1. Axial T2-weighted (A) and sagittal T1-weighted, contrast-enhanced (B) MR images in a child with a large vermian tumor and obstructive hydrocephalus who developed mutism after undergoing a complete tumor resection. (C) A follow-up image 3 months after the operation shows evidence of the vermian incision and persistent fourth ventricular dilatation, but no evidence of focal cerebellar hemisphere or brain stem injury.

immediate postoperative period after it was clear that the patient's cerebrospinal fluid was being adequately absorbed internally. Two children required permanent cerebrospinal fluid diversion. In all cases, the ventricular dilatation was much improved postoperatively in comparison to the preoperative state.

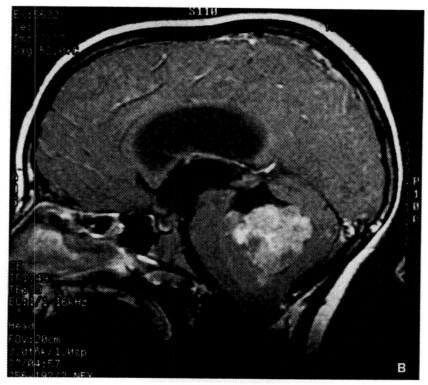

FIG. 1—Continued

B. TEMPORAL PROFILE OF MUTISM AND ASSOCIATED IMPAIRMENTS

Each of our patients awoke uneventfully after the tumor resection with the expected amounts of ataxia and dysmetria. Several had associated sixth and/or seventh nerve paresis. Ten of 12 had relatively normal speech initially, vocalizing a few words or short phrases, and in some cases, communicating in sentences. These children subsequently developed mutism 24 to 96 hr after surgery. Two other children were not observed to speak after awakening from anesthesia, but were otherwise alert.

In general, the mutism was not manifested as an isolated finding, but instead typically occurred in association with obvious neurobehavioral dysfunction. Eleven children exhibited an almost stereotypical response, lying curled up in bed and whining inconsolably, without speaking intelligible words. Three children seemed unable to initiate voluntary eye opening

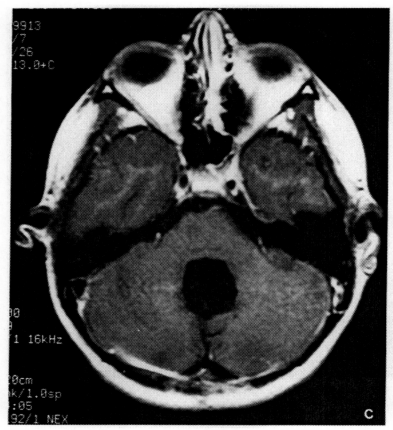

Fig. 1—*Continued*

and lay in bed with their eyes closed. Five patients had urinary retention or incontinence without any apparent urological or pharmacological cause. Nine patients had significant impairment in oral motor coordination and seemed unwilling or unable to eat. Only two of these children had evidence of an impaired gag reflex or abnormal pharyngoesophageal motility on a barium cine-esophagram study; in the other seven patients, no convincing explanation for the eating impairment was apparent. Characteristically, these children had difficulty initiating the chewing and swallowing process, but once food was swallowed, they exhibited no sign of neurogenic dysphagia, such as nasal regurgitation or aspiration. Six patients had associated long-tract signs.

The time course of clinical improvement varied significantly among individual children. In general, affect, oral intake, and urinary function recovered to normal before the speech impairment began to improve. In the hope of hastening the rate of recovery of speech, most of the children were enrolled in a comprehensive program of speech therapy while in the hospital and eight patients who were not talking by the time of discharge were continued in intensive outpatient rehabilitation. Five children began speaking single words within 2 weeks of surgery and the other seven began to speak within 2 months of surgery. All patients regained fluent speech within 4 months of surgery. However, the character of the speech was often strikingly abnormal during the recovery phase. Three of the children began speaking in a whispered voice and four others spoke in a high-pitched "whiny" voice. Several of the children initially began to converse with expletives, in some ways resembling the speech of patients with a recovering Broca's aphasia. The speech pattern in all 12 children also exhibited a dysarthric quality before recovering fully. One child was left with mild residual dysarthria.

C. Diagnostic Evaluation

Because of our underlying concern that these impairments might reflect a remediable structural lesion, such as a posterior fossa hemorrhage, each of these patients underwent a series of diagnostic studies to determine whether there was an obvious cause for the speech and neurobehavioral deficits. On initial review, these studies were surprisingly unrevealing. Eight patients underwent computerized tomography (CT) and eight underwent magnetic resonance imaging (MRI), which in each instance failed to show significant hemorrhage in the resection bed, worsening hydrocephalus, or significant abnormalities elsewhere within the brain. Only one of these patients had radiographically detectable residual tumor. Three children underwent xenon CT cerebral blood flow studies, which showed no focal areas of hypoperfusion. Single photon emission tomography with both ^{99m}Tc-HMPAO and ^{201}Th and positron emission tomography with 18-fluorodeoxyglucose in one patient each were likewise unremarkable.

A more detailed imaging analysis was subsequently performed (Pollack et al., 1995) in the hope of identifying a structural basis for the speech and neuropsychiatric dysfunction. For this study, preoperative and postoperative CT and MRI studies for each of the previously described 12 children, along with two age-matched controls for each patient (with similar tumors but without postoperative mutism), were reviewed in a blinded fashion by two neuroradiologists who were unaware of the patient's history. In all patients, postoperative studies were obtained within 48 hr of operation. The images were graded on a variety of criteria, including lesion location and size, and involvement of cerebellum, cerebellar peduncles, and brain stem by tumor or edema preoperatively and by edema or hemorrhage postoperatively. In addition, the length of the cerebellar incision was measured. Late follow-up studies were also examined for areas of permanent injury.

To our initial disappointment, this extensive review identified no single feature that was pathognomonic for the presence of postoperative mutism. Not unexpectedly, tumor location was not a significant factor that distinguished the two groups because each patient selected for the control group had a vermian lesion as did the 12 with mutism. Maximum lesion diameter in the children with mutism (4.4 ± 1.1 cm) was comparable to that of children without mutism (4.6 ± 1.6 cm). Hydrocephalus was present in the vast majority of patients in both groups. Similarly, postoperative edema within the brachium pontis and/or brachium conjunctivum, characterized by low density on CT or high signal intensity of T2-weighted MRI, which was apparent in most patients in both groups, was not a useful predictor of the presence of mutism ($P > 0.1$, Fisher's exact test). However, we observed that the edema was typically bilateral in children with mutism, but only unilateral in those without mutism. Bilateral edema within these structures was present in 9 of 12 children with mutism vs only 6 of 24 controls, a statistically significant association ($P < 0.01$). Edema within the brain stem was seen in 4 patients with mutism vs 2 controls ($P > 0.1$).

Interestingly, the rostrocaudal length of the cerebellar vermis incision in the patients with mutism (2.9 ± 1.2 cm) was comparable to that of the children without mutism (3.4 ± 1.0 cm). Finally, a review of the late follow-up MRI studies in each child demonstrated no clear-cut foci of injury to account for the mutism. Although several patients in both the mutism and control groups who had received posterior fossa radiotherapy exhibited volume loss within the cerebellum in association with persistent dilatation of the fourth ventricle, no obvious areas of infarction were apparent (Fig. 1C). Thus, the only radiologic factor that clearly correlated with the development of mutism was the presence of bilateral edema within the cerebellar peduncles, but this relationship was far from absolute.

D. Neuropsychological Testing

Detailed neuropsychological testing was performed in seven of these children during the period of recovery from their deficits. In many cases, earlier testing was not possible because the children would not communicate and would only lie in bed and whine. Thus, the testing was typically deferred until after the child had become interactive and cooperative, but before the speech deficits had fully resolved. These studies almost uniformly demonstrated difficulties not only in speech, but also in the initiation and completion of age-appropriate motor activities. Several children also had impairments in recent memory, attention span, and problem-solving ability. The fact that these deficits were apparent even as the speech and affective disorders were resolving implies that the initial neuropsychiatric dysfunction may have reflected a global dysfunction in the initiation of motor activities and in the regulation of affective state. These difficulties ultimately cleared completely in each of the patients. Although several children have residual ataxia or dysmetria, all but two patients were able to resume a full range of normal activities without functional limitations. Each of these children is currently alive without obvious disease, although one child with a large vermian astrocytoma that invaded the brain stem required reoperation for a focal area of progressive tumor within the middle cerebellar peduncle and brain stem.

E. Illustrative Case

A 16-year-old boy presented with a 6-week history of headaches and left dysmetria. CT and MRI showed a hypodense, 4-cm mass within the superior vermis that enhanced uniformly after administration of intravenous contrast medium, and caused mild obstructive hydrocephalus. The tumor was exposed via a supracerebellar trajectory, followed by an incision in the upper one-third of the vermis. A complete resection of the lesion, which proved to be pilocytic astrocytoma, was achieved. Upon awakening from anesthesia, the patient exhibited no neurological deficits and had clear speech. However, by the third postoperative day, he had developed mutism in association with a bizarre, depressed affect and a mild left hemiparesis. He lay curled up in bed with his eyes closed, whining intermittently, refusing to follow commands, with poor oral intake and urinary retention. His speech consisted only of occasional expletives, which were uttered in a high-pitched nasal voice when he was disturbed. CT and MRI showed edema around the resection cavity, within the brachium pontis and brachium conjunctivum bilaterally, and within the rostral pons. A xenon CT/cerebral blood flow study was unremarkable.

The patient's poor oral intake and bizarre personality changes each began to improve during the third postoperative week; his urinary retention resolved shortly thereafter. Neuropsychological testing during the period of recovery demonstrated poor initiation and completion of age-appropriate motor and problem-solving activities, despite improvement in his overall mood. Follow-up testing several months later showed improvement in these deficits. During the interim, his speech improved through a stage of dysarthria to its preoperative baseline. He returned to regular school classes by 3 months postoperatively; his only persistent deficit was mild left ataxia. No adjuvant therapy was administered and he remains progression-free.

III. Discussion

A. Clinical Features of the Syndrome

During the last decade, more than 50 cases of mutism and/or bizarre personality changes have been reported after the removal of posterior fossa mass lesions. These are reviewed in detail in Pollack *et al.* (1995). As with our own cases, these patients have generally been young and have had large midline cerebellar and fourth ventricular tumors that were resected via an inferior vermian incision. However, two of our cases and one of Dietze and Mickle (1990) are distinctive in that a superior vermian incision was employed and the inferior vermis was not traversed. The greater numbers of cases associated with inferior vermian approaches may reflect that the majority of childhood midline posterior fossa tumors, particularly medulloblastomas and ependymomas, are most accessible through the inferior vermis as they grow inferiorly to present at the foramen of Magendie. However, a causal effect of the inferior vermian incision cannot be excluded (Dailey *et al.*, 1995). Similarly, the association with age may be more reflective of a tendency for children to have large midline vermian tumors, which are relatively uncommon in adults, than a true age-related predisposition to postoperative mutism. Adults with an identical syndrome have been reported (Cakir *et al.*, 1994; D'Avanzo *et al.*, 1993; Salvati *et al.*, 1991) and the author has personally cared for an adult patient with a large hemorrhagic metastasis to the vermis who developed postoperative mutism and neuropsychiatric dysfunction in the absence of other objective neurological deficits. Thus, one feature common to all reported cases was the involvement of the midline cerebellum by a large mass lesion that had been resected through an incision in the vermis.

A second feature that is characteristic of this syndrome is that symptoms often develop after an interval of relatively normal functioning in the immediate postoperative period and that the deficits are largely reversible during the first few weeks to months after surgery. Although mutism is the central element in this syndrome, the majority of patients in our series and in several previous reports (Cochrane *et al.*, 1994; Dailey *et al.*, 1995; Gaskill and Marlin 1991; Humphreys, 1989; Pollack *et al.*, 1995; Rekate *et al.*, 1995; Van Dongen *et al.*, 1994; Volcan *et al.*, 1996; Wisoff and Epstein 1984) have had associated neurobehavioral abnormalities that included a combination of emotional lability, poor oral intake, decreased spontaneous initiation of movements, impaired eye opening, and urinary retention. The frequency of this syndrome in our series [12 of 142 patients with posterior fossa tumors (8.5%)] is strikingly similar to the incidences reported by Cochrane *et al.*, [1993; 6 of 105 patients (5.7%)], van Calenbergh *et al.*, [1989; 5 of 63 patients (7.9%)], and Dailey *et al.*, [1995; 9 of 102 patients (8.2%)].

B. ANATOMIC BASIS FOR THE POSTERIOR FOSSA MUTISM SYNDROME

The cerebellum has long been thought to play an important role in *coordinating* speech (Holmes, 1917), based on the observation that both acute and chronic cerebellar hemispheric lesions often result in dysarthria (Ackermann, *et al.*, 1992; Lechtenberg and Gilman, 1978). However, the concept that one or more areas within the cerebellum might actually play a role in *generating* speech is comparatively novel. The clinical experience in our own and other studies of children with mutism, coupled with recent neuroanatomic and physiological data, support the validity of such a role and shed some light on the anatomical substrate for the speech disturbances observed in our patients.

The fact that mutism occurs almost exclusively with midline cerebellar mass lesions that have been resected via an inferior vermian incision, but not with large cerebellar hemispheric tumors that have been resected without splitting the vermis, has been offered as presumptive evidence that injury to the lower vermian region is a crucial element in the pathophysiology of this disorder. In this regard, Dailey *et al.* (1995) noted that the development of the mutism syndrome correlated with the use of a long inferior vermian incision that extended to the superior portion of the caudal vermis.

However, several factors indicate that the anatomical substrate for the overall syndrome of mutism and associated neurobehavioral changes may not be within the inferior vermis per se, but instead may more accurately be localized to adjacent structures. If the caudal cerebellum did indeed contain an area crucial for the initiation of speech, deficits should not only be more common but also more persistent after resection of midline cerebellar tumors. In fact, the majority of patients who undergo resection of an inferior vermian or fourth ventricular tumor through an incision in the inferior vermis experience neither mutism nor even severe dysarthria postoperatively. Furthermore, patients who have undergone complete section of the vermis may manifest little if any postoperative speech impairment

(Lechtenberg and Gilman, 1978), whereas two of our patients, who underwent section of the superior vermis with sparing of the inferior vermis, exhibited prolonged postoperative mutism. This implies that if an anatomical substrate for this syndrome is present within the cerebellum, it does not reside within the inferior vermis, but instead is localized more laterally. This fits with the observation in our series that the length of the vermian incision did not correlate with the development of postoperative mutism.

The fact that symptoms often develop after an interval of 1 to 3 days postoperatively provides additional evidence that the structures responsible for the syndrome do not generally suffer direct injury from intraoperative transection. This indicates that the area responsible for the symptom complex is not directly in the midline because if the midline cerebellar structures themselves played an essential role in speech control, these should be affected immediately, rather than in a delayed fashion.

However, the fact that the postoperative imaging studies that were performed in the children in this series showed no evidence of discrete areas of infarction, hypoperfusion, or decreased metabolic activity within the cerebellar hemispheres, diencephalon, or cerebral cortex indicates that the lesion responsible for this syndrome is not located at a distant site within the brain, but instead resides close to the operative bed within the paramedian portions of the cerebellum or brain stem. In this context, our observation that patients with mutism often had bilateral edema within the cerebellar peduncles, which was seen in a much smaller percentage of unaffected children, is of particular interest. This implies that a critical pathway responsible for initiating speech and other complex voluntary movements travels within this structure. It can also be inferred that injury or edema occurring proximally or distally along such a pathway could induce a similar syndrome. For example, bilateral injury or impairment of the paramedian cerebellar relay nuclei might produce an identical syndrome in the absence of involvement of either the peduncles or brain stem.

The plausibility of this mechanism is supported by the results of stereotactic lesioning studies in which mutism has been reported as a complication of bilateral lesions to the region of the dentate nuclei (Fraoli and Guidetti, 1975; Guidetti and Fraoli, 1977) or interposed nuclei (Siegfried *et al.*, 1970). This would also account for the presence of mutism in the patients in our series without evidence of significant edema within the brain stem or peduncles. This is also supported by positron emission tomography and single photon emission tomography data (Petersen *et al.*, 1989; Ryding *et al.*, 1993), which have demonstrated bilateral cerebellar activation during speech. It is also clear that mutism may result from injury to the afferent and/or efferent projections that transit through the dendate and/or interposed nuclei. This observation fits with neuroanatomic data that show complex reciprocal connections between the cerebellar and cerebral cortices via pathways traversing the cerebellar nuclei and brain stem (Schmahmann, 1991; Sherman and Schmahmann, 1995). These include somatotopically organized connections between the dendate and interposed nuclei and Broca's area and the premotor and supplementary motor cortices via the brachium pontis, brain stem, diencephalon, and cerebellorubrothalamocortical system (Adams and Victor, 1993; Allen and Tsukahara, 1974; Leiner et al., 1986, 1989, 1991, 1993; Orioli and Strick, 1989; Schell and Strick, 1984; Schmahmann, 1991; Schmahmann and Pandya, 1991; Wiesendanger and Wiesendanger, 1985). In this regard, Frim and Ogilvy (1995) noted "cerebellar" mutism in a child who underwent resection of a cavernous angioma within the central pons. Similarly, mutism has been noted in patients with Parkinson's disease who have undergone bilateral stereotactic lesioning of the thalamus, one of the principal targets of the cerebellar efferent pathway to the cerebral cortex (Orioli and Strick, 1989; Schell and Strick, 1984; Wiesendanger and Wiesendanger, 1985).

Thus, the syndrome of "cerebellar" mutism may really represent interruption of one of the critical subcortical components of the speech-generating system that exist proximal to Broca's area. The fact that this pathway can be affected at several sites fits with our observation that there is no single abnormality on imaging that is pathognomonic for this syndrome. The delayed onset of the speech impairment probably reflects the interval until the edema, inflammation, or focal hypoperfusion resulting from manipulation within the operative bed has affected these adjacent structures. Because this syndrome does not occur after unilateral cerebellar exposure, it can also be inferred that bilateral impairment of this pathway is necessary to produce significant deficits. This is consistent with the hypothesis that complex movements involving bilaterally integrated musculature, such as talking, eating, and eye opening, require that both cerebellar hemispheres act together (Brodal, 1981; Carpenter, 1985; Ghez and Fahn, 1985). Recent studies indicate that the cerebellum functions not only to coordinate these activities, but may also be involved in encoding or "memorizing" the stereotyped movements that underly them (Fiez et al., 1992; Lalonde and Botez, 1990; Leiner et al., 1986, 1989, 1991, 1993). As such, the speech and oropharyngeal impairments that are observed in the posterior fossa syndrome may be best described as an oral-pharyngeal apraxia, as noted by Dailey et al. (1995), with impaired initiation of complex learned behaviors by these structures in the absence of oropharyngeal motor or sensory deficits.

Our observation that several patients without postoperative speech dysfunction had evidence of edema within the cerebellar peduncles bilaterally and/or brain stem, whereas others with mutism had comparatively unimpressive postoperative imaging studies, suggests that there is substantial variability between patients in the presence and severity of impairments that result from apparently similar injuries to the cerebellar pathways. This diversity in the pattern of speech deficits following otherwise similar lesions has also been noted by Lechtenberg and Gilman (1978) and Ackermann et al. (1992) in patients sustaining ischemic, neoplastic, and traumatic injuries to the cerebellum.

It is also apparent from our own cases and a review of the literature that the duration of the speech impairment and the spectrum of associated neurological and neurobehavioral abnormalities vary widely among affected patients. In the most limited manifestation of this process, patients have impairment in coordinating the highly complex, bilaterally integrated movements necessary to produce speech. More severe expressions of this syndrome affect not only speech, but also coordination of the oropharyngeal movements necessary for initiating chewing and swallowing (Dailey et al., 1995). In the most severe expression of this syndrome, initiation of a broad spectrum of volitional movements is impaired, including eye opening, voiding, and a variety of other activities. This more global impairment is apparent not only clinically, but also on the basis of neuropsychological testing. The implication of these observations is that isolated cerebellar mutism may involve a comparatively focal bilateral lesion to the dentate and/or interposed nuclei or their afferent or efferent pathways, whereas the more extensive manifestations of oropharyngeal apraxia and global impairment of volitional movements reflect progressively more extensive involvement of the paravermian region and/or its connections. Alternatively, the wide range of manifestations may reflect differential susceptibility of affected patients to lesions within the paravermian structures. Whereas patients with a "high vulnerability" to such lesions might exhibit widespread impairments in the initiation of a broad range of activities, those who were comparatively less vulnerable might exhibit impairment only in the initiation of the extremely complex, bilaterally integrated movements necessary for producing speech.

In accordance with either view, the symptoms that resolve first in patients with the most severe manifestations are usually those that involve global akinesia with decreased initiation of all voluntary activities. Second, eating and urinary incontinence improves. During this stage, patients will often appear to be relatively normal other than their speech difficulties, but on detailed testing, they still manifest subtle impairments in initiation of complex activities. Finally, speech begins to recover, initially with a dysarthric and often bizarre vocal quality, and ultimately returns to normal.

C. Anatomic Basis for the Associated Affective Symptoms

The anatomical basis for the bizarre affective symptoms that are sometimes associated with the mutism syndrome remains puzzling, particularly if one views the cerebellum as strictly involved in planning the execution of motor activities. Although it is conceivable that the affective symptoms simply reflect the patient's extreme frustration and sense of despondency at not being able to communicate verbally or to easily initiate voluntary movements, the stereotypical appearance of the affected patients suggests that a behavioral process is unlikely to be strictly involved and that a pathophysiological process is more likely to be responsible. Although bacterial or aseptic meningitis can also produce personality changes after posterior fossa surgery, and has been suggested in previous studies to be a possible precipitating factor for the development of mutism and neurobehavioral symptoms (Ferrante et al., 1990; Humphreys, 1989; Rekate et al., 1985), this explanation did not account for the symptoms noted in our series. None of the affected patients had evidence of bacterial meningitis or significant cerebrospinal fluid leukocytosis. In addition, high-dose steroids were administered to several patients and were ineffective in reversing either the mutism or behavioral changes, which suggests that these symptoms were not due to aseptic meningitis.

Wisoff and Epstein (1984) postulated that the postoperative "pseudobulbar syndrome," which typically occurred in patients with evidence of edema bilaterally within the cerebellar peduncles and/or brain stem, reflected temporary dysfunction of the brain stem tegmentum. In support of this explanation, unusual personality changes are well-described manifestations of primary brain stem lesions, such as intrinsic brain stem gliomas (Petronio and Edwards, 1989), central pontine myelinolysis (Price and Mesulam, 1987), and vascular insufficiency from proximal basilar artery ischemia (Caplan, 1980).

Although, in light of these observations, it is conceivable that the affective symptoms in our patients reflected injury to the brain stem tegmentum, it is also possible that injury or edema within the cerebellum itself may have played a causative role. In addition to its well-known connections with the motor pathways, the cerebellum may have similar interactions with areas of the brain involved in nonmotor activities (Fiez et al., 1992) and even with regions traditionally viewed as regulating emotional state (Middleton and Stricks, 1994; Schmahmann, 1991, 1994; Schmahmann and Pandya, 1991; Sherman and Schmahmann, 1995). In this regard, Fiez et al. (1992) and Sherman and Schmahmann (1995) have reported on patients who exhibited impaired nonmotor learning after cerebellar infarction or degeneration. Such patients manifested impairments in verbal fluency, planning, and concept formation that are typical of frontal lobe lesions; deficits in visual memory and visuospatial organization and construction, traditionally attributed to temporal and parietal lobe lesions; and affective blunting, characteristic of paralimbic lesions. In many ways, these global cognitive and affective impairments, when occurring in the absence of obvious weakness or sensory loss, resemble the widespread neurobehavioral abnormalities that were grossly apparent in our patients early in their course. Even during the period when these children were beginning to recover clinically, the diffuse extent of their deficits was clearly evident on detailed neuropsychological testing. The observation that anomalies of the cerebellum are often detected in children with autism (M. L. Bauman et al., this volume; Schmahmann, 1994) is also intriguing in this regard.

D. IMPLICATIONS

In light of these observations, we would predict that the incidence of the mutism syndrome after vermian tumor resections could be minimized by limiting manipulation around the tumor bed. Although we have made an effort to minimize the length of the vermian incision, an approach that has also been recommended by Dailey et al. (1995), in the hope of decreasing the degree of edema and/or injury to the midline and paramedian cerebellar structures, this has not eliminated the problem. Two of our 12 cases occurred despite these measures. Because the primary source of injury and/or edema probably results from lateral dissection and removal of tumor that is adherent to or invading the cerebellar peduncles and paramedian cerebellum bilaterally, the only obvious way to avoid this syndrome would be to limit aggressive tumor resection in these regions. However, because the duration of progression-free survival correlates strongly with extent of resection for astrocytoma (Hayostek et al., 1993; Pollack, 1994), ependymoma (Nazar et al., 1990; Pollack, 1994; Vanuytsel et al., 1992), and medulloblastoma (Bourne et al., 1992; Jenkin et al., 1990; Pollack, 1994), the three most common vermian tumors, we would not advocate limiting the lateral tumor resection in order to avoid the potential for mutism, particularly because the mutism–pseudobulbar symptom complex is generally reversible. However, we would recommend that surgeons make a conscious effort to avoid excessive paravermian manipulation, a strategy that has been noted anecdotally to minimize the frequency of this complication. In contrast to the impairments produced by cerebellar dissection, which generally are largely reversible, brain stem injuries from overly aggressive tumor removal are typically irreversible. Thus, extensive brain stem invasion, which occurs in some of these tumors, may indeed pose a practical limit to the extent of resection that can be safely achieved.

E. Summary

Mutism and associated neurobehavioral impairments occur in a small but significant percentage of patients who undergo resection of a vermian tumor. The pattern of symptoms ranges from isolated impairment of speech to oropharyngeal apraxia to global impairment in the initiation of volitional activities. In most cases, this syndrome is self-limited and relatively normal speech is recovered within 3 to 4 months. Although we have made a conscious effort to minimize the extent of the vermian incision, we have still encountered this syndrome in two patients with large lesions exposed through comparatively limited vermian incisions. Thus, this syndrome may not result directly from the incision itself but from edema, inflammation, or focal hypoperfusion around the resection cavity, which may reversibly compromise the functioning of the dentate and interposed nuclei and/or the afferent or efferent connections to this region bilaterally.

Acknowledgment

This work was supported in part by a grant from the National Institutes of Health (KO8NS01810).

References

Ackermann, H., Vogel, M., Petersen, D., and Poremba, M. (1992). Speech deficits in ischaemic cerebellar lesions. *J. Neurol* **239,** 223–227.
Adams, R. D., and Victor, M. (1993). "Principles of Neurology," 5th Ed. pp. 74–82. McGraw-Hill, New York.
Aguiar, P. H., Plese, J. P. P., Ciquini, O., and Marino, R., Jr. (1993). Cerebellar mutism after removal of a vermian medulloblastoma; literature review. *Pediat. Neurosurg.* **19,** 307.
Allen, G. I., and Tsukahara, N. (1974). Cerebrocerebellar communication systems. *Physiol. Rev.* **54,** 957–1006.
Ammirati, M., Mirzai, S., and Samii, M. (1989). Transient mutism following removal of a cerebellar tumor: A case report and review of the literature. *Child. Nerv. Syst.* **5,** 12–14.
Boratynski, W., and Wocjan, J. (1993). Mutism after surgeries with removal of posterior cranial fossa neoplasms. *Neurol. I Neurochir. Polska* **27,** 261–265.

Bourne, J. P., Geyer, R., Berger, M., Griffin, B., and Milstein, J. (1992). The prognostic significance of postoperative residual contrast enhancement on CT scan in pediatric patients with medulloblastoma. *J. Neurooncol.* **14,** 262–270.

Brodal, A. (1981). "Neurological Anatomy in Relation to Clinical Medicine," 3rd Ed. Oxford University Press, New York.

Cakir, Y., Karakisi, D., and Kocanaogullario, O. (1994). Cerebellar mutism in an adult: Case report. *Surg. Neurol.* **41,** 342–344.

Caplan, L. R. (1980). "Top of the basilar" syndrome. *Neurology* **30,** 72.

Carpenter, M. B. (1985) "Core Text of Neuroanatomy," 3rd Ed. Williams and Wilkins, Baltimore.

Catsman-Berrevoets, C. E., van Dongen, H. R., and Zwetsloot, C. P. (1992). Transient loss of speech followed by dysarthria after removal of posterior fossa tumour. *Dev. Med. Child. Neurol.* **34,** 1102–1109.

Cochrane, D. D., Gustavsson, B., Poskitt, K. P., Steinbok, P., and Kestle, J. R. W. (1994). The surgical and natural morbidity of aggressive resection of posterior fossa tumors in childhood. *Pediat. Neurosurg.* **20,** 19–29.

Dailey, A. T., McKhann, G. M., II, and Berger, M. S. (1995). The pathophysiology of oral pharyngeal apraxia and mutism following posterior fossa tumor resection in children. *J. Neurosurg.* **83,** 467–475.

D'Avanzo, R., Scuotto, A., Natale, M., Scotto, P., and Cioffi, F. A. (1993). Transient "cerebellar" mutism in lesions of the mesencephalo-cerebellar region. *Acta Neurol.* **15,** 289–296.

Dietze, D. D., Jr., and Mickle, J. P. (1990–1991). Cerebellar mutism after posterior fossa surgery. *Pediat. Neurosurg.* **16,** 25–31.

Ferrante, L., Mastronardi, L., Acqui, M., and Fortuna, A. (1990). Mutism after posterior fossa surgery in children: Report of three cases. *J. Neurosurg.* **72,** 959–963.

Fiez, J. A., Petersen, S. E., Cheney, M. K., and Raichle, M. F. (1992). Impaired non-motor learning and error detection associated with cerebellar damage: A single case study. *Brain* **115,** 155–178.

Fraoli, B., and Guidetti, B. (1975). Effect of stereotactic lesions of the dentate nucleus of the cerebellum in man. *Appl. Neurophysiol.* **38,** 81–90.

Frim, D. M., and Ogilvy, C. S. (1995). Mutism and cerebellar dysarthria after brainstem surgery: Case report. *Neurosurgery* **36,** 854–857.

Gaskill, S. J., and Marlin, A. E. (1991–1992). Transient eye closure after posterior fossa tumor surgery in children. *Pediat. Neurosurg.* **17,** 196–198.

Ghez, C., and Fahn, S. (1985). The cerebellum. *In* "Principles of Neural Science" (E. R. Kandel, and J. H. Schwartz, eds.), 2nd Ed., pp. 502–522. Elsevier, New York.

Guidetti, B., and Fraoli, B. (1977). Neurosurgical treatment of spasticity and dyskinesias. *Acta Neurochir. Suppl.* **24,** 27–39.

Hayostek, C. J., Shaw, E. G., Scheithauer, B., O'Fallon, J. R., Weiland, T. L., Schomberg, P. J., Kelly, P. J., and Hu, T. C. (1993). Astrocytomas of the cerebellum: A comparative clinicopathological study of pilocytic and diffuse astrocytomas. *Cancer* **72,** 856–869.

Herb, E., and Thyen, U. (1992). Mutism after cerebellar medulloblastoma surgery. *Neuropediatrics* **23,** 144–146.

Hirsch, J. F., Renier, D., Czernichow, P., Benveniste, L., and Pierre-Kahn, A. (1979). Medulloblastoma in childhood: Survival and functional results. *Acta Neurochir.* **48,** 1–15.

Holmes, G. (1917). The symptoms of acute cerebellar injuries due to gunshot injuries. *Brain* **40,** 461–535.

Humphreys, R. P. (1989). Mutism after posterior fossa tumor surgery. *In* "Concepts in Pediatric Neurosurgery" (A. E. Marlin, ed.), pp. 57–64. Karger, Basel.

Jenkin, D., Goddard, K., Armstrong, D., Becker, L., Berry, M., Chan, H., Doherty, M., *et al.* (1990). Posterior fossa medulloblastoma in childhood: Treatment results and a proposal for a new staging system. *Int. J. Radiat. Oncol. Biol. Phys.* **19,** 265–274.

Lalonde, R., and Botez, M. I. (1990). The cerebellum and learning processes in animals. *Brain Res. Rev.* **15,** 325–332.

Lechtenberg, R., and Gilman, S. (1978). Speech disorders in cerebellar disease. *Ann. Neurol.* **3,** 285–290.

Leiner, H. C., Leiner, A. L., and Dow, R. S. (1986). Does the cerebellum contribute to mental skills? *Behav. Neurosci.* **100,** 443–454.

Leiner, H. C., Leiner, A. L., and Dow, R. S. (1989). Reappraising the cerebellum: What does the hindbrain contribute to the forebrain? *Behav. Neurosci.* **103,** 998–1008.

Leiner, H. C., Leiner, A. L., and Dow, R. S. (1991). The human cerebro-cerebellar system: Its computing, cognitive, and language skills. *Behav. Brain Res.* **44,** 113–128.

Leiner, H. C., Leiner, A. L., and Dow, R. S. (1993). Cognitive and language function of the human cerebellum. *Trends Neurosci.* **16,** 444–447.

Middleton, F. A., and Strick, P. L. (1994). Anatomical evidence for cerebellar and basal ganglia involvement in higher cognitive function. *Science* **266,** 458–461.

Nagatani, K., Waga, S., and Nakagawa, Y. (1991). Mutism after removal of a vermian medulloblastoma: Cerebellar mutism. *Surg. Neurol.* **36,** 307–309.

Nazar, G. B., Hoffman, H. J., Becker, L. E., Jenkin, D., Humphreys, R. P., and Hendrick, E. B. (1990). Infratentorial ependymomas in childhood: Prognostic factors and treatment. *J. Neurosurg.* **72,** 408–417.

Orioli, P. J., and Strick, P. L. (1989). Cerebellar connections with the motor cortex and the arcuate premotor area: An analysis employing retrograde transneuronal transport of WGA-HRP. *J. Comp. Neurol.* **288,** 621–626.

Petersen, S. E., Fox, P. T., Posner, M. I., Mintun, M. A., and Raichle, M. E. (1989). Positron emission tomographic studies of the processing of single words. *J. Cogn. Neurosci.* **1,** 153–170.

Petronio, J. A., and Edwards, M. S. B. (1989). Management of brain stem tumors in children. *Contemp. Neurosurg.* **11**(7), 1–6.

Pollack, I. F. (1994). Current concepts: Brain tumors in children. *N. Engl. J. Med.* **331,** 1500–1507.

Pollack, I. F., Polinko, P., Albright, A. L., Towbin, R., and Fitz, C. (1995). Mutism and pseudobulbar symptoms after resection of posterior fossa tumors in children: Incidence and pathophysiology. *Neurosurgery* **37,** 885–893.

Price, B. H., and Mesulam, M. M. (1987). Behavioral manifestations of central pontine myelinolysis. *Arch. Neurol.* **44,** 671–673.

Rekate, H. L., Grubb, R. L., Aram, D. M., Hahn, J. F., and Ratcheson, R. A. (1985). Muteness of cerebellar origin. *Arch. Neurol.* **42,** 697–698.

Ryding, E., Decety, J., Sjoholm, H., Sternberg, G., and Ingvar, D. (1993). Motor imagery activates the cerebellum regionally: A SPECT rCBF study with ^{99m}Tc-HMPAO. *Cogn. Brain Res.* **1,** 94–99.

Sakai, H., Sekino, H., and Nakamura, N. (1980). Three cases of "cerebellar mutism." *Shinkeinaika* **12,** 302–304.

Salvati, M., Missori, P., Lunardi, P., and Orlando, E. R. (1991). Transient cerebellar mutism after posterior fossa surgery in an adult: Case report and review of the literature. *Clin. Neurol. Neurosurg.* **93**(4), 313–316.

Schell, G. R., and Strick, P. L. (1984). The origin of thalamic inputs to the arcuate premotor and supplementary motor areas. *J. Neurosci.* **4,** 539–560.

Schmahmann, J. D. (1991). An emerging concept: The cerebellar contribution to higher function. *Arch. Neurol.* **48,** 1178–1187.
Schmahmann, J. D. (1994). The cerebellum in autism: Clinical and anatomic perspectives. *In* "The Neurobiology of Autism" (M. L. Bauman, and T. L. Kemper, eds.), pp. 195–226. Johns Hopkins University Press, Baltimore.
Schmahmann, J. D., and Pandya, D. N. (1991). Projections to the basis pontis from superior temporal sulcus (STS) in the rhesus monkey. *J. Comp. Neurol.* **308,** 224–248.
Sherman, J. C., and Schmahmann, J. D. (1995). The spectrum of neuropsychological manifestations in patients with cerebellar pathology. *Hum. Brain Mapp. Suppl.* **1,** 361.
Siegfried, J., Esslen, E., Gretener, U., Ketz, E., and Perrett, E. (1970). Functional anatomy of the dentate nucleus in the light of stereotaxic operations. *Confin. Neurol.* **32,** 1–10.
Van Calenbergh, F., Van De Laar, A., Plets, C., Goffin, J., and Casaer, P. (1995). Transient cerebellar mutism after posterior fossa surgery in children. *Neurosurgery* **37,** 894–898.
Van Dongen, H. R., Catsman-Berrevoets, C. E., and van Mourik, M. (1994). The syndrome of "cerebellar" mutism and subsequent dysarthria. *Neurology* **44,** 2040–2046.
Vanuytsel, L. J., Bessell, E. M., Ashley, S. E., Bloom, H. J. G., and Brada, M. (1992). Intracranial ependymoma: Long-term results of a policy of surgery and radiotherapy. *Int. J. Radial. Oncol. Biol. Phys.* **23,** 313–319.
Volcan, I., Cole, G. P., and Johnston, K. A. (1986). A case of muteness of cerebellar origin. *Arch. Neurol.* **43,** 313–314.
Wiesendanger, R., and Wiesendanger, M. (1985). The thalamic connections with medial area 6 (supplementary motor cortex) in the monkey (*Macaca fascicularis*). *Exp. Brain. Res.* **59,** 91–104.
Wisoff, J. H., and Epstein, F. J. (1984). Pseudobulbar palsy after posterior fossa operation in children. *Neurosurgery* **15,** 707–709.

CEREBELLAR COGNITIVE AFFECTIVE SYNDROME

Jeremy D. Schmahmann and Janet C. Sherman

Department of Neurology, Massachusetts General Hospital and Harvard Medical School, Boston, Massachusetts 02114

I. Introduction
II. Patient Selection and Methods of Study
III. Subjects
IV. Results
   A. Elementary Neurologic Examination
   B. Bedside Mental State Testing
   C. Neuropsychological Testing
V. Discussion
   References

There has been persistent uncertainty as to whether lesions of the cerebellum are associated with clinically significant disturbances of behavior and cognition. To address this question, 20 patients with diseases confined to the cerebellum were studied prospectively over a 7-year period and the nature and severity of the changes in neurological and mental function were evaluated. Neurological examination, bedside mental state testing, neuropsychological studies, and anatomic neuroimaging were administered at the time of presentation and during follow-up assessments. Behavioral changes were clinically prominent in patients with lesions involving the posterior lobe of the cerebellum and the vermis and, in some cases, overwhelmed other aspects of the presentation. These changes were characterized by an impairment of working memory, planning, set shifting, verbal fluency, abstract reasoning, and perseveration; visual–spatial disorganization, visual memory deficits, and logical sequencing; and a bland or frankly inappropriate affect. Lesions of the anterior lobe of the cerebellum produced only minor changes in executive and visual–spatial functions. This newly defined clinical entity is called the cerebellar cognitive affective syndrome. The constellation of deficits is suggestive of disruption of the cerebellar modulation of neural circuits that link frontal, parietal, temporal, and limbic cortices with the cerebellum.

I. Introduction

Clinical teaching regarding the effects of cerebellar lesions is that deficits are confined to the realm of voluntary movement, gait, and equilibrium.

Early case reports describing behavioral consequences of cerebellar pathology were anecdotal and have not been pathologically verified (Dow and Moruzzi, 1958; Schmahmann, 1991), and therefore this notion has not been adopted by the clinical neuroscience community. The body of evidence now linking the cerebellum with cognitive processing is derived largely from anatomic, physiologic, theoretical, and functional neuroimaging studies, with some support also derived from neuropsychological evaluations (see chapter by J. D. Schmahmann).

A concern shared by investigators, particularly by clinicians who have considered the possibility of a cerebellar contribution to nonmotor behavior, is that there is a dearth of descriptions of clinically relevant cases. The question legitimately posed is, are deficits detectable only by subtle neuropsychological tests but without clinical consequence sufficient to warrant a major overhaul of current concepts of cerebellar function? This question was therefore specifically addressed by performing neurologic examinations, bedside mental state tests, and neuropsychological evaluations on patients with acute and chronic diseases confined to the cerebellum (J. D. Schmahmann and J. C. Sherman, manuscript in preparation). This chapter summarizes the principal findings of that study, which suggest that there is a predictable pattern of clinically relevant cognitive and behavioral changes in patients with cerebellar lesions.

II. Patient Selection and Methods of Study

Patients with cerebellar pathology were studied prospectively over a 7-year period (July 1989 to August 1996). Patients were not selected specifically for the presence of cognitive disturbance. Each patient received a comprehensive medical and neurologic examination and bedside mental state testing (by JDS). Neuropsychological testing was performed (by JCS) in the majority of cases, using standardized measures, and comparing patients' performance with normative data. Patients were evaluated between 1 week and 6 years from the onset of the illness. Patients were excluded from study if there was evidence of disease outside the cerebellum on clinical examination or neuroimaging, or if there were risk factors likely to affect other brain regions. The mental state examination was conducted according to established clinical methods (Weintraub and Mesulam, 1985; Hodges, 1994). Magnetic resonance imaging (MRI) or computerized axial tomography (CT) was performed according to standard protocols. The determination of the vascular territory of those cases with infarction was derived from the work of Amarenco (1991). Electroencephalographic (EEG) studies were performed on an 18-channel Grass monitor.

III. Subjects

A total of 20 patients were studied. There were 8 women and 12 men, ranging in age from 23 to 74 years (mean 48.2), excluding one 12-year-old boy. The educational level ranged from 9 to 20 years (mean 13.9). Fifteen patients suffered cerebellar infarctions. Strokes involved the posterior inferior cerebellar artery (PICA) territory bilaterally in two patients, the right PICA in five patients, the left PICA in three patients, the right anterior inferior cerebellar artery (AICA) territory in one patient, and one each in the left and right superior cerebellar artery (SCA) territories. Three patients (one with R-PICA and two with L-PICA tellitory strokes) underwent posterior fossa decompression because of threatened hydrocephalus. Neuroimaging performed the day after surgery showed no evidence of hydrocephalus, and the patients were alert and cooperative during subsequent testing. Three patients had postinfectious cerebellitis, three had cerebellar cortical atrophy, and one underwent resection of a low-grade ganglioglioma at the vermis. The EEG was normal in each case at the time of cognitive testing.

IV. Results

A. ELEMENTARY NEUROLOGIC EXAMINATION

Patients with pancerebellar involvement had incoordination of arms and legs; unstable (ataxic) gait; dysarthria; and eye movement abnormalities, including nystagmus, saccadic breakdown of pursuit, hypometric and hypermetric saccades, periodic alternating nystagmus and square wave jerks at rest, and failure to suppress the vestibulo-ocular reflex. Patients with bilateral PICA infarction were motorically greatly compromised as a consequence of incoordination, but showed some improvement over the ensuing months. In those with unilateral infarction in the PICA or SCA territories, motor disturbances involved the limbs, were mild, and resolved within weeks. The patient with cerebellar vermis excision for tumor had a minimally abnormal elementary neurologic examination at the time of a florid behavioral change. Strong bilateral palmar grasp reflexes were noted in some patients with pancerebellar or bilateral cerebellar disease of recent onset.

B. BEDSIDE MENTAL STATE TESTING

All patients were awake, cooperative, and able to give an account of their history, although the level of attention was variable. No patient demonstrated a standard clinical aphasic syndrome, hemispatial neglect, or agnosia. In contrast, the behavior of these patients was abnormal in varying degrees of severity, with features that differed according to lesion site and acuity of onset. Those with bihemispheric infarction, pancerebellar disease, and large PICA lesions were affected in a manner that was clinically obvious. Patients with small lesions in the posterior cerebellum and in the anterior superior cerebellum (AICA or SCA territories) were only mildly involved. Almost without exception, patients demonstrated problems with working memory, planning, set shifting, abstract reasoning, and perseveration. Impairment of verbal fluency was present in virtually every patient; in some cases this was so severe as to resemble the mutism seen in children subsequent to posterior fossa surgery. Fluency was decreased in spontaneous conversation, so that some patients were telegraphic in their speech output or talked very little unless prodded. The response rate with semantic or phonemic fluency was completely outside the range of normal. This was unrelated to dysarthria as some patients wtih minimal dysarthria in the setting of acute lesions performed more poorly than others with severe dysarthria and disease of greater duration.

Visuospatial disintegration, most marked in attempting to draw or copy a diagram, was found in all patients, no matter whether the lesion was acute or chronic, and the dysmetria of the extremities minimal or severe. The sequential approach to the drawing of diagrams and the conceptualization of figures were disorganized. Some patients also demonstrated simultagnosia, particularly those with bilateral or pancerebellar dysfunction.

Naming was impaired in patients with bilateral or pancerebellar lesions, particularly in those of recent onset. Truly agrammatic speech was most notable in those with bilateral acute disease. Elements of abnormal syntactic structure were noted in others, but less prominently. Similarly, prosody was unusual and in some patients quite abnormal, with tone of voice characterized by a high-pitched, whining, childish, and hypophonic quality. Mild difficulties with verbal learning and recall, visual memory, mental calculation, and praxis were noted in some patients, but these were not consistent features in the clinical mental state evaluation.

It was evident from the onset that the most flagrant and defining features of the mental state in this series of patients were problems in the modulation of behavior and personality style. With the notable exception of those patients whose strokes were confined to the anterior lobe, a major component of the clinical evaluation was either flattening of affect or disinhibition. This was manifested as inappropriate behavior, overfamiliarity, flamboyant and impulsive actions, and humorous but flippant comments. Additionally, behavior was regressive and child like in some cases, particularly following acute lesions in the posterior cerebellum that also involved vermis or paravermian structures.

Autonomic changes were the central feature in one patient whose stroke in a medial branch of the right PICA involved the fastigial nucleus and paravermian cortex region. This manifested as spells of hiccuping and coughing which precipitated bradycardia and loss of consciousness.

C. Neuropsychological Testing

Results of the neuropsychological studies provided further evidence of cognitive dysfunction in each of the patients tested. The range and level of deficit varied according to lesion site and duration. The findings were in agreement with the clinical observations that those with bilateral lesions and posterior lobe lesions were most impaired whereas those with the disease confined to the anterior lobe of the cerebellum were least affected. There was a similarity across patients in the type of impairments demonstrated. Patients demonstrated deficits on tasks of executive function, including diminished verbal fluency, problem-solving ability, and set shifting; poor visual construction with evidence of significant difficulty in appreciating organizational structure; difficulties with memory that appeared to be specific to visual material; and disturbed sequential reasoning as demonstrated by inferior performance on a picture-ordering task. In addition, a flattened or inappropriate effect was observed frequently in this population. The presence of these common features resulted in a general lowering of level of intellectual functioning and was suggestive of a definable neurobehavioral syndrome.

V. Discussion

The debate concerning the putative nonmotor functions of the cerebellum has been hampered by the lack of clinical data in patients showing a correlation between cerebellar damage and changes in behavior. The study (J. D. Schmahmann and J. C. Sherman, manuscript in preparation) summarized here documents the existence and nature of clinically relevant behavioral manifestations from lesions of the cerebellum. Findings indicate that these behavioral changes can be diagnosed at the bedside, conform to an identifiable clinical syndrome, and are consistent with the predictions derived from experimental work in anatomy, physiology, and functional neuroimaging.

This syndrome, which the authors call the cerebellar cognitive affective syndrome, is characterized by (1) Disturbances of executive function including poor planning, set shifting including perseveration, abstract reasoning, and verbal fluency; (2) visual–spatial disorganization and impaired visual–spatial memory; (3) personality change characterized by flattening or blunting of affect, disinhibited, inappropriate, or "giddy" behavior reminiscent of "witselzucht"; (4) difficulty with interpreting and producing logical sequences; and (5) language difficulties including dysprosodia, mild anomia, and agrammatism. The net effect of these disturbances in cognitive functioning is a general lowering of overall intellectual function.

The neurobehavioral presentation in these patients was more pronounced and generalized in patients with pancerebellar disorders and in those with acute onset cerebellar disease, but it was also present in a less florid form in those with more chronic or restricted cerebellar pathology. Lesions of the posterior lobe were particularly important in the generation of the disturbed cognitive behaviors, and the vermis was consistently involved in patients with pronounced affective presentations. The anterior lobe seemed to be less prominently involved in the generation of these cognitive and behavioral deficits. The one patient with an autonomic syndrome had a lesion involving the medial posterior lobe, including the fastigial nucleus. Caution is warranted in drawing conclusions regarding the details of the organization of these various functions within the cerebellum because of the small sample size and the heterogeneity of the lesion type in this study. The correlation of each component of the cognitive affective syndrome with the precise region of the cerebellum destroyed will depend on the analysis of a larger group of patients.

Elements of the cerebellar cognitive affective syndrome can be found in previous reports. Patients with cerebellar degeneration have been described to have difficulties with executive, or frontal lobe type functions (Grafman *et al.*, 1992; Appollonio *et al.*, 1993), visual–spatial deficits (Botez *et al.*, 1989), and linguistic difficulties including agrammatism (Silveri *et al.*, 1994). In addition, children subjected to posterior fossa surgery develop mutism and a behavioral syndrome (Pollack *et al.*, 1995) that has some features in common with those described here. The results of the study are compatible with these observations and indeed extend and elaborate upon them.

The cognitive and affective abnormalities described in this chapter are those usually encountered in patients who have disorders of the cerebral hemispheres, particularly of the association areas and paralimbic regions, or of the subcortical areas with which they are interconnected. Thus disturbances of executive function are usually encountered in patients with lesions of the prefrontal cortex; visual–spatial deficits are seen following damage to the parietal lobe; decreased verbal fluency and linguistic processing difficulties are seen in the setting of either frontal or temporal lobe pathology; impaired visual–spatial sequencing accompanies lesions of the right temporal lobe; and changes in affect and motivation commonly reflect disturbances in limbic-related regions in the cingulate and parahippocampal gyri (see Critchley, 1953; Fuster, 1980; Mesulam, 1985). The presence of these cognitive deficits in patients with cerebellar lesions invokes reports of anatomical connections linking the cerebral association areas and paralimbic regions with the cerebellum (Schmahmann, 1996; see chapters by D. E. Haines *et al.*, by F. A. Middleton and P. L. Strick, and by J. D. Schmahmann and D. N. Pandya). Thus there would appear to be a correlation between the neuropsychological and affective manifestations in patients with cerebellar disorders and the anatomically derived hypotheses concerning the putative cerebellar contributions to higher order function (Schmahmann, 1991, 1996).

How could this syndrome have been missed through the past century of advances in clinical neurology and neuroscience? In the early stages of the acute syndrome, the variable attention, withdrawn attitude of the patient, and multiple but specific cognitive disturbances raise the possibility of cerebellar swelling with brain stem compression, hydrocephalus, brain stem infarction, or infarction elsewhere in the hemispheres. Prior to the advent of contemporary neuroimaging (CT/MRI) it was not possible to determine whether these extracerebellar features were present or not. Pathologic confirmation of the cerebellar infarct at some future time after the patient's condition had run its course would not provide the "realtime" anatomic–pathologic correlation necessary to establish this relationship. Thus anecdotal clinical reports were largely dismissed for lack of substantiating proof, or perhaps because it was not possible to exclude concomitant pathology. The clinical–anatomic correlation available now with MRI, particularly with diffusion-weighted MRI that is sensitive within hours to the earliest evidence of ischemic change in brain tissue (Moseley *et al.*, 1995), provides greater confidence in excluding the complications that may occur following cerebellar infarction. Electroencephalography has also been helpful in establishing that the cerebellar cognitive affective syndrome is not merely another form of "encephalopathy" or a nonspecific confusional state.

A larger cohort of patients, studied morphologically in conjunction with a newly available detailed atlas of the human cerebellum (Schmahmann et al., 1996) and with more searching experimental psychology paradigms, will be valuable in exploring this cerebellar cognitive affective syndrome and the mechanisms responsible for its development. In so doing, it may be possible to increase our understanding of cerebellar function, specifically how the cerebellum plays a role in the modulation of cognition and affect.

References

Amarenco, P. (1991). The spectrum of cerebellar infarctions. *Neurology* **41**, 973–979.
Appollonio, I. M., Grafman, J., Schwartz, V., Massaquoi, S., and Hallett, M. (1993). Memory in patients with cerebellar degeneration. *Neurology* **43**, 1536–1544.
Botez, M. I., Botez, T., Elie, R., and Attig, E. (1989). Role of the cerebellum in complex human behavior. *Ital. J. Neurol. Sci.* **10**, 291–300.
Critchley, M. (1953). "The Parietal Lobes." Hafner Press, New York.
Dow, R. S., and Moruzzi G. (1958). The Physiology and Pathology of the Cerebellum. University of Minnesota Press, Minneapolis.
Fuster, J. M. (1980). "The Prefrontal Cortex: Anatomy, Physiology and Neuropsychology of the Frontal Lobe." Raven Press, New York.
Grafman, J., Litvan, I., Massaquoi, S., Stewart, M., Sirigu, A., and Hallett, M. (1992). Cognitive planning deficit in patients with cerebellar atrophy. *Neurology* **42**, 1493–1496.
Hodges, J. R. (1994)."Cognitive Assessment for Clinicians." Oxford Univ. Press, Oxford.
Mesulam, M.-M. (1985). "Principles of Behavioral Neurology." F. A. Davis, Philadelphia.
Moseley, M. E., Butts, K., Yenari, M. A., Marks, M., and de Crespigny, A. (1995). Clinical aspects of DWI. *NMR Biomed.* **8**, 387–396.
Pollack, I. F., Polinko, P., Albright, A. L., Towbin, R., and Fitz, C. (1995). Mutism and pseudobulbar symptoms after resection of posterior fossa tumors in children: Incidence and pathophysiology. *Neurosurgery* **37**, 885–893.
Schmahmann, J. D. (1991). An emerging concept: The cerebellar contribution to higher function. *Arch. Neurol.* **48**, 1178–1187.
Schmahmann, J. D. (1966). From movement of thought: Anatomic substrates of the cerebellar contribution to cognitive processing. *Hum. Brain Mapp.* **4**, 174–198.
Schmahmann, J. D., Doyon, J., Holmes, C., Makris, N., Petrides, M., Kennedy, D. N., and Evans, A. C. (1996). An MRI atlas of the human cerebellum in Talairach space. *NeuroImage* **3**, S122.
Silveri, M. C., Leggio, M. G., and Molinari, M. (1994). The cerebellum contributes to linguistic production: A case of agrammatic speech following a right cerebellar lesion. *Neurology* **44**, 2047–2050.
Weintraub, S., and Mesulam, M.-M. (1985). Mental state assessment of young and elderly adults in behavioral neurology. *In* "Principles of Behavioral Neurology" (M.-M. Mesulam, ed.), Vol. 26, pp. 71–124. F. A. Davis, Philadelphia.

INHERITED CEREBELLAR DISEASES

Claus W. Wallesch and Claudius Bartels

Department of Neurology, Otto-von-Guericke University, 39120 Magdeburg, Germany

I. Introduction
II. Animal Models
III. Human Pathology
   A. Cerebellar Cortical Atrophy
   B. Olivopontocerebellar Atrophy
   C. Friedreich's Disease
IV. Discussion
V. Conclusion
   References

This chapter analyzes the neuropsychological deficits in inherited cerebellar diseases and compares their symptomatology with animal models in which the exact anatomical localization of degeneration is known and limited to the cerebellum. Both animal and human data suggest that cerebellar cortical atrophy affects functions of the frontal lobe system. Olivopontocerebellar atrophy is genetically and clinically inhomogeneous. The dementia syndrome that occurs in a proportion of patients does not seem to be linked with cerebellar dysfunction. Patients suffering from Friedreich's disease have been described as exhibiting cognitive slowing and deficits in spatial tasks. Because other structures are more prominently involved than the cerebellum in this disease, other pathoanatomical correlates may explain the symptomatology.

I. Introduction

The theme of this chapter narrows the general topic of the role of the cerebellum in cognitive functions to a quite specific aspect of cerebellar pathology, namely neuropsychological alterations in subjects suffering from an inherited cerebellar disease. In humans, genetically determined diseases involving the cerebellum broadly fall into three categories: those that affect the cerebellum or parts thereof exclusively [as in cerebellar cortical atrophy (CCA)]; those that include the cerebellum and other anatomical structures, often with individual or pedigree dependent patterns [as in olivopontocerebellar atrophy (OPCA)]; and diseases in which the focus of degeneration is not in the cerebellum but elsewhere in the nervous system [as in Friedreich's disease (FD)]. In the latter two types, possible contributions of affected neural structures outside the cerebellum have to be considered. In all cases, the effects of sensorimotor impairment and handicap on the functional and cognitive status have to be taken into account.

This chapter considers the following pathologies: animal models, cerebellar cortical atrophy, olivopontocerebellar atrophy, and Friedreich's disease.

This chapter does not consider studies in which the type of cerebellar disease was not sufficiently defined and cerebellar pathologies resulting from nongenetic etiology, such as tumor, infarct, inflammatory, demyelinating, and toxic disease. The neuropsychological features of the latter especially seem to be a very promising field of investigation, as some toxins are quite specifically directed against certain components of the cerebellum: organic mercury compounds destroy the granular cells but spare Purkinje cells, whereas the opposite is found with high-dose cytosine arabinoside and phenytoin intoxication. Another interesting model in human pathology is paraneoplastic cerebellar degeneration, which also focally affects the Purkinje cells.

As reviewed by Schmahmann (1991, 1996), a contribution of the cerebellum to cognitive functions is anatomically plausible. The parietal, temporal, and frontal association cortices project to pontine nuclei, the efferent fibers of which terminate in the cerebellar cortex (Brodal, 1979; Brodal and Steen, 1983; Schmahmann and Pandya, 1992, 1995). Cerebellar efferents back to forebrain cortex project via thalamic nuclei to areas 4 and 6, as well as to prefrontal and posterior parietal cortices (Jones, 1985; Schmahmann and Pandya, 1990; Middleton and Strick, 1994; Schmahmann, 1996).

The outlined anatomical connections support the hypothesis that cerebellar disease may affect forebrain cortical, and especially frontal and parietal cognitive functions.

II. Animal Models

A number of mouse mutants have been bred that develop quite specific degenerations of certain cerebellar neuron types within the first weeks of life. In homozygotes of the lurcher mutant, the majority of the Purkinje cell population degenerates (Caddy and Biscoe, 1979). These animals exhibit spatial orientation deficits in a water maze (Lalonde et al., 1988) that do not correlate with motor coordination impairment (Lalonde and Thifault, 1994) and reduced spontaneous alternation in choice situations (Lalonde *et al.*, 1986). Other cerebellar mutants, such as the staggerer, weaver, Purkinje cell degeneration, and nervous mutant mice, show quite specific impairments in spatial navigational and learning tasks and in spontaneous alternation. Goodlett *et al.* (1992) suggest that mouse visual learning based on a single cue is not impaired in the Purkinje cell degeneration, whereas the elaboration of a spatial map is severely deficient. In clinical neuropsychology, a somewhat analogous deficit has been described with prefrontal lesions (Karnath and Wallesch, 1992), and an inability to spontaneously alternate behavioral strategies is also regarded as a symptom of frontal lobe damage (e.g., Fuster, 1980).

Mutant mouse data therefore support the hypothesis that cerebellar hemisphere degeneration affects nonsensorimotor functions that are usually related to the integrity of the frontal and possibly the parietal lobes.

III. Human Pathology

A. CEREBELLAR CORTICAL ATROPHY

According to Harding (1982), pure CCA is rare as an autosomal dominantly inherited disease. A sporadic form is more common (Bahlo *et al.*, 1986).

Clinically, patients exhibit gait and limb ataxia and dysarthria. There is no agreement in the clinical literature with respect to the frequency of dementia. It seems to occur in a subgroup of hereditary patients, but is rare in other populations (Orozco Diaz *et al.*, 1990) and in sporadic forms. The differential diagnosis of OPCA is difficult if not impossible in sporadic cases, and the diagnosis may only be established with certainty after prolonged observation of the clinical course or at autopsy. Furthermore, some well-investigated families of dominant cerebellar ataxia exhibit the neuropathological features of OPCA (Orozco Diaz *et al.*, 1990).

Grafman *et al.* (1992), in a well-controlled investigation in patients with CCA, reported significantly increased planning times in the Tower of Hanoi problem that could not be explained by age, educational level, severity of dementia, memory deficit, response time, or visuomotor procedural learning ability. The authors described a significant increase in premovement planning time, a deficit that has also been found in Parkinson's disease patients (Saint-Cyr *et al.*, 1988). Grafman *et al.* (1992) suggested the presence of a functional link among cerebellum, basal ganglia, and the frontal lobe concerned with specific cognitive planning processes.

A further study of the same group (Pascual-Leone et al., 1993) that specifically addressed procedural learning but included patients suffering from CCA ($N = 10$) and OPCA ($N = 5$) demonstrated a profound deficit in the acquisition of implicit procedural and declarative knowledge in a fixed sequence visuomotor association task. The authors suggest an inability to order events in the time domain as the functional basis for this deficit. In contrast to the majority of other studies in the literature, Pascual-Leone et al. (1993) compared patients not only with normal controls, but also with patients suffering from Parkinson's disease. Although the patient groups were not matched for motor disability, the comparison with normal subjects allowed the detection of dissociations between the profile of impairments in the two diseases.

A third study of the National Institutes of Health (NIH) group addressed memory functions in patients with relatively selective cerebellar degeneration (Appollonio et al., 1993). Of 11 patients, 2 exhibited neuroradiological signs of brain stem involvement and clinically manifested extrapyramidal symptoms. In discussing the problem of whether their patients corresponded to CCA or OPCA, the authors concluded that in any case the pattern of neuropsychological deficits was likely to be cerebellum specific in accordance with clinical and radiographic findings. Appollonio et al. (1993) found an impairment in demanding free recall measures but normal performance in cued recall, recognition, automatic, and implicit memory. The speed of information processing was normal, but verbal fluency (clinically regarded as a symptom of frontal lobe dysfunction) was diminished. In summary, the impairments could be explained by a deficit in executive functions, similar to those in patients with frontal lobe lesions and Parkinson's disease.

The NIH group thus convincingly demonstrated a dyfunction of the executive (in neuropsychological terms) or "frontal lobe system" (in terms of functional neuroanatomy, compare Rosvold and Szwarcbart, 1964) in the patients they investigated. In the absence of neuropathological confirmation, however, it remains a possibility that their patients included some OPCA patients in whom basal ganglia dysfunction may have contributed to the executive impairments.

There certainly is a clinical problem with the diagnosis of CCA in patients in whom there is no available pathology of a family member. In a clinical review, Staal and de Jong (1991) attempted to "convince the reader that if one finds the combination of degenerative disease with isolated infratentorial atrophy on CT, there exists hardly any other possibility than OPCA or the not infrequent combination of OPCA and SND" (Striatonigral degeneration, p. 520).

B. OLIVOPONTOCEREBELLAR ATROPHY

Olivopontocerebellar atrophy occurs in familial and sporadic forms. According to Harding (1984), there are five types of hereditary OPCA, four with dominant and one with recessive inheritance, which show variation with respect to the prominence of symptoms from the affected neurological systems. Dementia occurs in about one quarter of hereditary cases. Nonfamilial OPCA is characterized by a variety of neurological symptoms and signs (e.g., ataxia, dysarthria, pyramidal, and extrapyramidal) that may occur in diverse combinations, among which dementia was included in 6 out of 47 patients in the series of Staal *et al.* (1990).

Pathologically, OPCA is characterized by atrophy of the cerebellar hemispheres, the nuclei and fibers of the pons, the olives, the dentate nuclei, and, in some instances, the basal ganglia, substantia nigra, thalamus, red nucleas, and subthalamic nuclei. The cerebral cortex is affected in a few cases [for reviews, see Eadie (1991) for the sporadic and Harding (1984) for hereditary forms]. Kish *et al.* (1987), however, demonstrated markedly reduced choline acetyltransferase activity in various cortical regions in 17 hereditary OPCA patients from five pedigrees, none of whom exhibited clinical signs of global cognitive impairment. Although the authors conclude that the cortical cholinergic abnormality may not be causally related to cognitive deterioration either in OPCA or in Alzheimer's disease and may not be reflected in clinical abnormality, their findings pose an obstacle to the interpretation of OPCA as an extracortical disease.

An illustrative case of dementia in probable OPCA has been reported by Akshoomoff *et al.* (1992). The authors report that their patient had been an honors student at high school. For a sporadic form of OPCA, he was unusually young at onset (16 years). The electrophysiological investigation suggested that the disease process was not entirely limited to the cerebellum, and the MRI scan exhibited mild pontine atrophy, so a diagnosis of OPCA seems likely. In neuropsychological assessment, the patient exhibited a global cognitive impairment of the subcortical dementia type similar to that found in supranuclear palsy or Huntington's disease.

Kish *et al.* (1988) found neuropsychological deficits otherwise assigned to frontal lobe and basal ganglia dysfunction in patients from a family with OPCA (spinocerebellar ataxia type 1 in the nomenclature used by the authors). Kish *et al.* (1994) extended their first study to include 43 hereditary patients with a wide range of severity of ataxia in an attempt to control for sensorimotor deficit and to detect the possible presence of dementia in a subgroup. The assessment included colored progressive matrices, the Wechsler memory scale, the continuous performance test, the Hooper visual organization test, and the Wisconsin card sorting test. Mildly ataxic patients as a group performed normally or close to normal. Half of the moderately and all of the severely ataxic subjects showed an impairment in the Wisconsin card sorting test that was independent of the level of education and degree of depression. A minority of moderately ataxic patients and the majority of severely ataxic patients also exibited deficits in tasks of attention and memory. The authors conclude that dominantly inherited spinocerebellar ataxia is not homogeneous with respect to cognitive status. A subgroup with greater neurological impairment exhibits deficits of executive functions, and a smaller number shows more widespread cognitive deficits. Previously, Landis *et al.* (1974) had described a subgroup of OPCA patients with deficits of verbal and nonverbal intelligence, memory and frontal system functions. In summary, Kish *et al.* (1994) postulate that damage to the olivopontocerebellar system affects cortical executive functions or that involvement by the disease process of other neuronal systems such as cholinergic pathways (El-Awar *et al.*, 1991) results in executive and other neuropsychological impairment.

Furthermore, these investigations suggest that executive functions generally related to the frontal lobes are affected in a subpopulation of OPCA patients, whereas other cognitive deficits seem less frequent and may differ among families and types of inheritance. The findings of Kish *et al.* are corroborated to some extent by Matthew *et al.* (1993), who were able to demonstrate lower regional frontal and prefrontal metabolic rates on positron emission tomography (PET) in patients with advanced disease and mild cognitive impairment. A review of the PET and single photon emission tomography (SPECT) literature with respect to the diaschisis phenomenon (metabolic depression in areas not directly affected by disease) suggests, however, that the presence of this phenomenon only indicates the presence of anatomical connections but does not predict functional status.

Berent *et al.* (1990) investigated a large series ($N = 39$; 15 hereditary and 24 sporadic) of OPCA patients. They applied the revised Wechsler intelligence scale, verbal fluency, Wechsler memory, and simple and choice reaction times. Their assessment, however, did not include more refined measures of frontal lobe system function such as Wisconsin card sorting. When adjustments for education and motor performance were made, patients did not differ in their performance from normal controls. Four subjects presented with an abnormally low IQ, but history suggested lifelong functioning at such levels. Thus, Berent *et al.* (1990) could not find evidence of dementia in their OPCA sample. They conclude that when dementia is found in OPCA it may be that other systems, such as the cholinergic or the basal ganglia, may account for its presence. The design of their study, however, did not allow for the detection of more specific neuropsychological impairments.

Botez-Marquard and Botez (1993) reported findings from 15 OPCA patients, 12 of whom had inherited disease, who were pair-matched with normal controls for age, sex, level of education, and socioeconomic background. Impairments in the Raven standard progressive matrices, untimed block design and similarities from the Wechsler intelligence scale, and the quantitative analysis of copying the Rey complex figure suggested the presence of a mild parietal-like syndrome. The authors also described longer visual and auditory reaction times. A detailed analysis performed by Botez *et al.* (1993) suggests that the reaction time impairment in OPCA is correlated with midbrain but not with cerebellar atrophy.

A number of other studies suggest the representation of a supramodal clock mechanism in the cerebellar hemispheres (for a review, see Keele and Ivry, 1990). In a group study that included cerebellar atrophy (OPCA cases), stroke, and tumor patients, Ivry and Keele (1989) demonstrated that cerebellar pathology affected not only the timing of motor responses but also the perceptual judgement of duration (intervals between acoustic events).

C. FRIEDREICH'S DISEASE

Friedreich's disease is a heredodegenerative disease with onset in childhood or early adolescence that primarily affects the posterior columns of the spinal cord, dorsal root ganglion cells, the pyramidal, and the spinocerebellar tracts. Degeneration of the cerebellar cortex, dentate nucleus, and superior cerebellar peduncle is characteristic for advanced disease only and does not correlate well with clinical status (Wessel *et al.*, 1989). It has not been established whether the mild to moderate cerebellar atrophy found in FD is an independent feature or a consequence of cerebellar deafferentation.

Friedreich (1863) himself noted no macroscopic abnormality in the cerebellum in his autopsy reports. However, histological abnormalities in the cerebellar cortex affecting both the Purkinje cells and the molecular layer have been reported as a regular feature of FD (Lamarche *et al.*, 1984). Using molecular biological methods, Duclos *et al.* (1993) described a putative transmembrane protein as a possible transcript from the Friedreich gene locus that is prominently expressed in the granular layer of the cerebellum.

Forebrain cortical atrophy accentuated in the frontal and parietal lobes has been described, but is not prominent in the majority of cases (Spiller, 1910; Claus and Aschoff, 1980).

Clinically, FD is characterized mainly by progressive ataxia, dysarthria, impairment of the position sense, absence of lower limb tendon reflexes, foot and spine deformities, and cardiomyopathy. The majority of patients become wheelchair bound and severely handicapped before their 25th year. There are, however, individual and familial exceptions who remain ambulatory for a longer period. The average age of death is in the fourth decade, mainly from cardiac disease. Dementia is rare and seems to occur only in genetically defined subpopulations (Bell and Carmichael, 1939; Fehrenbach et al., 1984).

The literature is not unanimous whether FD leads to neuropsychological abnormalities. Leclercq et al. (1985) found no specific deficit. Hart et al. (1985) and Botez-Marquand and Botez (1993) described cognitive slowing in the absence of other deficits that could not be explained by sensorimotor and coordination impairment.

Fehrenbach et al. (1984) investigated 15 patients with Friedreich's disease with a battery of neuropsychological tests including an abbreviated version of the Wechsler adult intelligence scale (WAIS) that was adapted for the investigation of neurologically impaired patients, Raven's progressive matrices, the Wisconsin card sorting test, and a visuoconstructive task that included no motor performance (mental folding of the two-dimensional layouts of three-dimensional structures). In comparison to matched normal controls, FD patients were impaired in the mental folding task and in the Picture arrangement subtest of the WAIS. Performance in block design was statistically not different from controls when speed credits were not given. Fehrenbach et al. (1984) are cautious in their interpretation and discuss, in addition to a possible cerebellar contribution to cognitive spatial operations, a long-term lack of training with respect to actions in space due to a chronic sensorimotor handicap. The authors emphasize that they found no deficits in verbal functions and no evidence for the presence of dementia in FD.

It is certainly difficult to interpret the neuropsychological results obtained with FD patients as evidence in favor of a role of the cerebellum in cognitive functions. Alternative interpretations outlined by Fehrenbach et al. (1984) are that the deficits in the visuoconstructive domain may be a consequence of lack of training (ataxia, sensory loss, being bound to wheelchair etc.) or a result of damage to extracerebellar gray matter (e.g., parietal cortex in the case of visuoconstructive impairment).

Furthermore, as Friedreich's disease leads indirectly (via thalamic nuclei) to a massive deafferentation of parietal lobe structures from somatosensory input, this could also result in disturbances in parietal lobe function.

IV. Discussion

The results obtained in lurcher and mutant mice suggest that these mouse strains are indeed impaired in functions conventionally related to the frontal and parietal lobes. However, as long as the gene, its product, and the distribution of expression and the function of the protein are not known, it cannot be ruled out that the biological defect in these animals transcends the boundaries of the cerebellum and affects other brain systems physiologically or biochemically. We do not know, for example, the activity of cortical choline acetyltransferase in lurcher mutant mice. The mouse models nevertheless have great potential and their further investigation may contribute to our understanding of the functions of the cerebellum. Cerebellar atrophy mice and, to a similar extent, some toxic lesions are of particular importance as they seem to affect only substructures of the cerebellum. Taken together with the available understanding of the internal wiring of the cerebellum, reasearch could aim at the biological and computational basis of cognitive operations. However, detailed knowledge regarding the extent and mechanism of pathology is a precondition for further speculation.

In human pathology, data concerning the role of the cerebellum as a supramodal clock (Ivry and Keele, 1989, Keele and Ivry, 1990) provide support for a cerebellar role in cognitive operations. These investigators were able to show numerous double dissociations to other pathologies, thus separating the cerebellar deficit from other cerebral pathology. In the terminology of cognitive neuropsychology, the cerebellar clock is a modular function in Fodor's sense: it is peripheral to cognition, operates automatically without conscious insight, and is functionally segregated (Fodor, 1982).

The work of the NIH group on CCA patients suggests that patients suffering from this disease in an advanced state exhibit a deficit of executive functions that is similar to but can be separated from impairments found in Parkinson's disease patients (Pascual-Leone *et al.*, 1993). Pascual-Leone *et al.* (1993) suggest an inability to order events in the time domain as the functional basis and relate the described deficits with functions of the cerebellar clock. From the pathoanatomical point of view, it cannot be ruled out that at least groups of CCA that include sporadic cases are contaminated by OPCA patients.

The investigations of Kish *et al.* (1994) suggest that executive ("frontal") functions are affected in a subpopulation of OPCA patients, whereas other cognitive deficits seem less frequent and may differ among pedigrees and types of inheritance. Correspondingly, the PET study of Matthew *et al.* (1993) found decreased regional frontal and prefrontal metabolic rates during an auditory continuous performance task. These data can hardly support the hypothesis of a cerebellar contribution to frontal lobe function (and are not interpreted by the authors in that way) as (a) a functional role of diaschisis effects as shown by PET and SPECT is not yet established, (b) the findings of Kish *et al.* (1987) suggest functional cortical pathology in OPCA, and (c) the task used is hardly related to specific functions of the frontal lobes. Even if it was agreed upon or established that data do represent diaschisis, then nothing beyond the known anatomical projections is proven. Further conclusions require more detailed knowledge of the gene and its product. That OPCA may include dementia only demonstrates the variability of system atrophies but cannot contribute to our understanding of cerebellar functions.

In Friedreich's disease, cognitive integrity, descreased cognitive speed, and visuoconstructive impairment have been described. The neuropathological literature is divided whether Friedreich's disease affects the cerebellum at all. The findings of Duclos *et al.* (1993) suggest that the gene is expressed in the granular layer and in many other parts of the brain. Although the findings in FD fall into the general pattern of frontal and parietal neuropsychological impairments in hereditary cerebellar disease, more knowledge on molecular biology is required. Alternative explanations for cognitive deficits in hereditary cerebellar disease have already been discussed and can be applied to other human cerebellar diseases as well.

V. Conclusion

A functional link of the cerebellum to frontal and parietal association cortices is anatomically plausible. A critical analysis of the neuropsychological findings in hereditary cerebellar disease shows that these pathologies are far from ideal for giving clinical support to the anatomical plausibility.

In summary on mice and men, animal data on cognitive changes in hereditary cerebellar disease seem promising but are not yet entirely conclusive. Studies on humans with hereditary cerebellar disease are equivocal, as the effects of cerebellar pathology cannot be separated clearly from extracerebellar disease.

References

Akshoomoff, N. A., Courchesne, E., Press, G. A., and Iragui, V. (1992). Contribution of the cerebellum to neuropsychological functioning: Evidence from a case of cerebellar degenerative disorder. *Neuropsychologia* **30,** 315–328.

Appollonio, I. M., Grafman, J., Schwartz, V., Massaquoi, S., and Hallett, M. (1993). Memory in patients with cerebellar degeneration. *Neurology* **43,** 1536–1544.

Bahlo, R. W., Yee, R. D., and Honrubia, V. (1986). Late cortical cerebellar atrophy: Clinical and oculographic features. *Brain* **109,** 159–180.

Bell, J., and Carmichael, E. A. (1939). On hereditary ataxia and spastic paraplegia. *Treat. Hum. Inherit.* **4,** 141–281.

Berent, S., Giordani, B., Gilman, S., Junck, L., Lehtinen, S., Markel, D.S., Boivin, M., Kluin, K. J., Parks, R., and Koeppe, R. A. (1990). Neuropsychological changes in olivopontocerebellar atrophy. *Arch. Neurol.* **47,** 997–1001.

Botez, M. I., Pedraza, O. L., Botez-Marquard, T., Vezina, J. L., and Elie, R. (1993). Radiologic correlates of reaction time measurements in olivopontocerebellar atrophy. *Euro. Neurol.* **33,** 304–309.

Botez-Marquard, T., and Botez, M. I. (1993). Cognitive behavior in heredodegenerative ataxias. *Eur. Neurol.* **33,** 351–357.

Brodal, P. (1979). The pontocerebellar projection in the rhesus monkey: An experimental study with retrograde axonal transport of horseradish peroxidase. *Neuroscience* **4,** 193–208

Brodal, P., and Steen, N. (1983). The corticopontocerebellar pathway to crus I in the cat as studied with anterograde and retrograde transport of horseradish peroxidase. *Brain Res.* **267,** 1–17.

Caddy, K. W. T., and Biscoe, T. J. (1979). Structural and quantitative studies on the normal C3H and lurcher mutant mouse. *Phil. Trans. R. Soc. London (Biol.)* **287,** 167–201.

Claus, D., and Aschoff, J. C. (1980). Computer-Tomografhie bei Atrophien im Bereich der hinteren Schädelgrube. *Arch. Psychiat. Nervenkrank.* **229,** 179–187.

Duclos, F., Boschert, U., Sirugo, G., Mandel, J.-L., Hen, R., amd Koenig, M. (1993). Gene in the region of the Friedreich ataxia locus encodes a putative transmembrane protein expressed in the nervous system. *Proc. Natl. Acad. Sci. USA* **90,** 109–113.

Eadie, M. J. (1991). Olivo-ponto-cerebellar atrophy (Dejerine-Thomas type). *In* "Handbook of Clinical Neurology" (J. B. M. V. de Jong, ed.), Vol. 16/60, pp. 511–518. Elsevier, Rotterdam.

El-Awar, M., Kish, S., Oscar-Berman, M., Robitaille, Y., Schut, L., and Freedman, M. (1991). Selective delayed alternation deficits in dominantly inherited olivopontocerebellar atrophy. *Brain Cogn.* **16,** 121–129.

Fehrenbach, R. A., Wallesch, C. W., and Claus, D. (1984). Neuropsychologic findings in Friedreich's ataxia. *Arch. Neurol.* **41,** 306–308.

Fodor, J. (1982). The Modularity of Mind. MIT Press, Cambridge, MA.

Friedreich, N. (1863). Über degenerative Atrophie der spinalen Hinterstränge. *Virch. Arch. Pathol. Anat.* **26,** 391–419, 433–459; **27,** 1–26.

Fuster, J. M. (1980). "The Prefrontal Cortex: Anatomy, Physiology and Neuropsychology of the Frontal Lobe." Raven Press, New York.

Goodlett, C. R., Hamre, K. M., and West, J. R. (1992). Dissociation of spatial navigation and visual guidance performance in Purkinje cell degeneration (pcd) mutant mice. *Behav. Brain Res.* **47,** 129–141.

Grafman, J., Litvan, I., Massaquoi, S., Stewart, M., Sirigu, A., and Hallett, M. (1992). Cognitive planning deficit in patients with cerebellar atrophy. *Neurology* **42,** 1493–1496.

Harding, A. E. (1982). The clinical features and classification of the late onset autosomal dominant cerebellar ataxias. *Brain* **105,** 1–28.

Harding, A. E. (1984). "The Hereditary Ataxias and Related Disorders." Churchill Livingstone, Edinburgh.

Hart, R. P., Kwentus, J. A., Leshner, R. T., and Frazier, R. (1985). Information processing speed in Friedreich's ataxia. *Ann. Neurol.* **17,** 612–614.

Ivry, R. I., and Keele, W. (1989). Timing functions of the cerebellum. *Cogn. Neurosci.* **1,** 134–150.

Jones, E. G. (1985). "The Thalamus." Plenum, New York.

Karnath, H. O., and Wallesch, C. W. (1992). Inflexibility of mental planning: A characteristic disorder with prefrontal lobe lesions? *Neuropsychologia* **30,** 1011–1016.

Keele, S. W., and Ivry, R. (1990). Does the cerebellum provide a common computation for diverse tasks? *In* "The Development and Neural Bases of Higher Cognitive Functions" (A. Diamond, ed.), Vol. 608, pp. 179–201. New York Academy of Science, New York.

Kish, S. J., Currier, R. D., Schut, L., Perry, T. L., and Morito, C. L. (1987). Brain choline acetyltransferase reduction in dominantly inherited olivopontocerebellar atrophy. *Ann. Neurol.* **22,** 272–275.

Kish, S. J., El-Awar, M., Schut, L., Leach, L., Oscar-Berman, M., and Freedman, M. (1988). Cognitive deficits in olivopontocerebellar atrophy: Implications for the cholinergic hypothesis of Alzheimer's dementia. *Ann. Neurol.* **24,** 200–206.

Kish, S. J., el-Awar, M., Stuss, D., Nobrega, J., Currier, R., Aita, J. F., Schut, L., Zoghbi, H. Y., and Freedman, M. (1994). Neuropsychological test performance in patients with dominantly inherited spinocerebellar ataxia: Relationship to ataxia severity. *Neurology* **44,** 1738–1746.

Lalonde, R., Lamarre, Y., and Smith, A. M. (1988). Does the mutant mouse lurcher have deficits in spatially oriented behaviours? *Brain Res.* **455,** 24–30.

Lalonde, R., Lamarre, Y., Smith, A. M., Botez, M. I. (1986). Spontaneous alternation and habituation in lurcher mutant mice. *Brain Res.* **362,** 161–164.

Lalonde, R., and Thifault, S. (1994). Absence of an association between motor coordination and spatial orientation in lurcher mutant mice. *Behav. Genet.* **24,** 497–501.

Lamarche, J. B., Lemieux, B., and Liu, H. B. (1984). The neuropathology of 'typical' Friedreich's ataxia in Quebec. *Can. J. Neurol. Sci.* **11,** 592–600.

Landis, D. M. D., Rosenberg, R. N., Landis, S. C., Schut, L., and Nyhan, W. L. (1974). Olivopontocerebellar degeneration. *Arch. Neurol.* **31,** 295–307.

Leclercq, M., Harmant, J., and Debarry, T. (1985). Psychometric studies in Friedreich's ataxia. *Acta Neurol. Belg.* **85,** 202–221.

Matthew, E., Nordahl, T., Schut, L., King, A. C., and Cohen, R. (1993). Metabolic and cognitive changes in hereditary ataxia. *J. Neurol. Sci.* **119,** 134–140.

Middleton, F. A., Strick, P. L.(1994). Anatomical evidence for cerebellar and basal ganglia involvement in higher cognitive function. *Science* **266,** 458–461.

Orozco Diaz, G., Nodarse Fleites, A., Cordoves Sagaz, R., and Auburger, G. (1990). Autosomal dominant cerebellar ataxia. *Neurology* **40,** 1369–1375.

Pascual-Leone, A., Grafman, J., Clark, K., Stewart, M., Massaquoi, S., Lou, J.S., and Hallett, M. (1993). Procedural learning in Parkinson's disease and cerebellar degeneration. *Ann. Neurol.* **34,** 594–602.

Rosvold, H. E., and Szwarcbart, M. K. (1964). Neural structures involved in delayed-response performance. In "The Frontal Granular Cortex and Behavior" (J. M. Warren and K. Akert, eds.). McGraw Hill, New York.

Saint-Cyr, J. A., Taylor, A. E., and Lang, A. E. (1988). Procedural learning and neostriatal dysfunction in man. *Brain* **111**, 941–959.

Schmahmann, J. D. (1991). An emerging concept: The cerebellar contribution to higher function. *Arch. Neurol.* **48**, 1178–1187.

Schmahmann, J. D. (1996). From movement to thought: Anatomical substrates of the cerebellar contribution to cognitive processing. *Hum. Brain Map.* **4**, 174–198.

Schmahmann, J. D., and Pandya, D.N. (1990). Anatomical investigation of projections from thalamus to posterior parietal cortex in the rhesus monkey: A WGA-HRP and fluorescent tracer study. *J. Comp. Neurol.* **295**, 299–326.

Schmahmann, J. D., and Pandya, D. N. (1992). Course of the fiber pathways to pons from parasensory association areas in the rhesus monkey. *J. Comp. Neurol.* **326**, 159–179.

Schmahmann, J. D., and Pandya, D. N.(1995). Prefrontal cortex projections to the basilar pons in rhesus monkey: Implications for the cerebellar contribution to higher function. *Neurosci. Lett.* **199**, 175–178.

Spiller, W. G. (1910). Friedreich's ataxia. *J. Nerv. Ment. Dis.* **37**, 411–435.

Staal, A., and de Jong, J. M. B. V. (1991). Non-familial olivopontocerebellar atrophy. In "Handbook of Clinical Neurology" (J. B. M. V. de Jong, ed.), Vol.16/60. pp. 519–536. Elsevier, Rotterdam.

Staal, A., Meerwaldt, J. D., van Dongen, K. J., Mulder, P. G., and Busch, H. F. M. (1990). Non-familial degenerative disease and atrophy of brainstem and cerebellum. Clinical and CT data in 47 patients. *J. Neurol. Sci.* **95**, 259–269.

Wessel, K., Schroth, G., Diener, H. C., Müller-Forell, W., and Dichgans, J. (1989). Significance of MRI-confirmed atrophy of the cranial spinal cord in Friedreich's ataxia. *Eur. Arch. Psychiat. Neurol. Sci.* **238**, 225–230.

NEUROPSYCHOLOGICAL ABNORMALITIES IN CEREBELLAR SYNDROMES—FACT OR FICTION?

Irene Daum* and Hermann Ackermann†

*Institute of Medical Psychology and Behavioral Neurobiology, University of Tübingen, 72074 Tübingen, Germany and †Department of Neurology, University of Tübingen, 72076 Tübingen, Germany

I. Introduction
II. Clinical Observations after Cerebellar Damage
    A. Mental Retardation in Cerebellar Malformations
    B. Dementia in Cerebellar Dysfunction
III. Cerebellar Involvement in Motor Learning
    A. Models of Motor Learning
    B. Motor Adaptation and Habituation
    C. Motor Skill Acquisition
    D. Classical Conditioning of Motor Responses
IV. Cerebellar Involvement in Temporal Processing
V. Cerebellar Involvement in Higher Cognitive Functions
    A. Anatomical Substrates
    B. General Intellectual Abilities
    C. Frontal Lobe Functions
    D. Visuospatial Abilities
    E. Memory and Nonmotor Skill Learning
    F. Language
VI. Methodological Considerations and Possible Directions
References

In recent years, theoretical considerations and a large number of empirical investigations have been published in support of a cerebellar involvement in cognitive processing. This chapter aims at a critical evaluation of the neuropsychological findings from clinical studies of patients with cerebellar syndromes. The discussion will mainly consider data from patients with selective cerebellar dysfunction, as data from patients with combined cerebellar and extracerebellar damage are of limited value for the issue of a cerebellar involvement in cognition. Early clinical observations indicated that degenerative diseases or selective cerebellar lesions did not necessarily give rise to general intellectual impairment such as dementia. Recent neuropsychological evidence based on standardized testing does not yet provide a clear picture. Deficits in motor learning or temporal processing are consistently observed in patients with cerebellar syndromes, while the cerebellum does not appear to be critically involved in general intellectual capacities or memory. Deficits in frontal lobe function, visuospatial processing or nonmotor skill learning have been reported in several studies, but have not been replicated in others. Such discrepancies may relate to a number of methodological problems. Future neuropsychological studies should take such methodological issues into account by using patients with selective cerebellar dysfunction, adequately matched clinical and nonclinical comparison groups, and theory-driven test batteries comprising a wide range of tests.

I. Introduction

In the last century, Rolando (1809) stated on the basis of animal experiments that cerebellar damage leads to an impairment in motor functions without affecting sensory or intellectual functions of the brain. Ferrier (1894) later came to similar conclusions. In the clinical literature, the early detailed patient studies reported by Stewart and Holmes (1904) and Holmes (1917) also concentrated on the disruption of motor functions following lesions to the cerebellum, and cognitive impairments in association with cerebellar damage were generally discussed in relation to extracerebellar dysfunction (Gilman et al., 1981).

The view that the cerebellum is primarily involved in motor control, the regulation of muscular tone, and the coordination of skilled movements has been strongly supported by physiological and pathophysiological data (Adams and Victor, 1989). An early case study by Bond (1895), however, already pointed to a possible involvement of brain stem and cerebellar dysfunction in the development of dementia, and Snider (1950) later called for a broader concept of cerebellar function, emphasizing the cerebellar modulation of the limbic and autonomic system. This issue has received some empirical support by clinical observations of emotional changes in patients with cerebellar atrophy (Snider and Maiti, 1976; Harding, 1984).

In recent years, interest in the possible cerebellar participation in cognition has reemerged. This development was based on a number of theoretical considerations and associated neuropsychological findings. In a series of articles, Leiner et al. (1986, 1989, 1993) have stressed the marked enlargement of the lateral cerebellum and the dentate nucleus in humans and discussed the functional implications with respect to cognition. The reciprocal connections of the cerebellum with different cortical areas may provide the basis for a cerebellar modulation of the higher cognitive functions subserved by these regions (see Schmahmann, 1991). Along similar lines, Ito (1993) argued that movement and thought might be controlled by the same neuronal mechanisms and that the cerebellum might serve as a multipurpose learning machine supporting all kinds of neural control, including the control of cognition. It should be noted, however, that other authors did not share the view that enlargement of the dentate nucleus and two-way cerebellocortical projections indicate a cerebellar involvement in cognitive processing. Bloedel (1993) and Glickstein (1993) have questioned the specificity of the dentate nucleus in humans and the functional implications of cerebellar enlargement with respect to cognition. It has also been argued that interconnections with the cerebral cortex may indicate a cortical influence on cerebellar activity rather than vice versa (Bloedel, 1993).

In parallel to the theoretical contributions in support of a cerebellar role in cognition, a series of neuropsychological studies have been published relating to this issue. This chapter aims at a critical evaluation of the empirical evidence from these clinical studies. Findings from neuroimaging studies or from experimental animal work will only be considered peripherally. The discussion will center on the performance of patients with selective cerebellar dysfunction; most studies in patients with combined cerebellar and extracerebellar damage (e.g., Friedreich's ataxia) will not be included, as they do not allow clearcut conclusions about the specific role of the cerebellum in cognition.

II. Clinical Observations after Cerebellar Damage

A. MENTAL RETARDATION IN CEREBELLAR MALFORMATIONS

Because of its long course of maturation, developmental abnormalities of the cerebellum are not uncommon. The investigation of the general cognitive status of children with cerebellar malformations provides an important source of evidence for the possible role of the cerebellum in cognition. Cerebellar aplasia, i.e., incomplete cerebellar development, is frequently, but not invariably, associated with mental retardation (Cutting, 1976; Macchi and Bentivoglio, 1987). The clinical symptoms related to partial malformations of the vermis or the cerebellar hemispheres show considerable variability. Mental retardation was seen to be linked to varying degrees of impaired coordination (Macchi and Bentivoglio, 1987), whereas other cases showed neither motor nor intellectual impairments during life, and cerebellar malformations were accidentally discovered during autopsy (e.g., Erskine, 1950; Macchi et al., 1964). Partial malformations therefore do not necessarily lead to clinically significant cognitive impairments either.

Children with cerebellar hypoplasia, i.e., reduced cerebellar volume, frequently present with mental retardation. There are, however, often signs of additional extracerebellar morphological alterations in these cases, so that no clear conclusions can be reached with respect to the specific cerebellar contribution (Harding, 1984). Similarly, a study of children with congenital cerebellar atrophy indicated mild cognitive retardation on standardized neuropsychological tests, but the authors stressed that the possible role of extracerebellar dysfunction could not be ruled out (Guzzetta et al., 1993). It is noteworthy that impaired cognition in developmental malformations may partly result from the impaired acquisition of psychomotor skills, and it is difficult to disentangle the motor and nonmotor consequences of congenital or early cerebellar dysfunction.

B. Dementia in Cerebellar Dysfunction

Degenerative cerebellar diseases in adulthood have sometimes been reported to give rise to dementia (Harding, 1984). However, in most of these cases the degenerative process extended beyond the cerebellum. Exceptions are sporadic olivocerebellar atrophy (Eadie, 1975), pure cerebellar atrophy (Dichgans et al., 1989), and the autosomal-dominant cerebellar ataxia III (ADCA III, Harding, 1984). The majority of cases for which postmortem examination confirmed damage restricted to the cerebellum did not present with dementia (for a discussion see Ackermann and Daum, 1995). The presence of progressive cognitive decline in more advanced age, many years after onset of ataxia, could be attributed to generalized cerebrovascular disease (see Ackermann and Daum, 1995). Harding (1981) reported development of dementia in 28% of cases with idiopathic late onset cerebellar ataxia. Again, more than half of these patients had clinical signs of additional extracerebellar dysfunction. The lack of postmortem data in most patients precludes firm conclusions about the contribution of cerebellar dysfunction to the observed cognitive decline.

Ischemia restricted to the cerebellum does not seem to result in dementia or global cognitive impairments (Amarenco, 1991; Amarenco et al., 1993). Memory problems observed after cerebellar infarcts may be due to disruption of the thalamus or the temporal lobes, areas which are also supplied by the posterior circulation (Amarenco, 1991). Such infarcts can also lead to attentional deficits because of dysfunction of the reticular formation (e.g., edema with subsequent brain stem compression).

Similarly, there is little evidence that cerebellar tumors which do not affect neighboring structures lead to dementia. Amici *et al.* (1976) reported "psychological changes" in 37% of patients with cerebellar tumors. Given the nature of the problems (confusional states or psychomotor slowing), they are more likely due to raised intracranial pressure rather than selective cerebellar dysfunction.

In summary, degenerative diseases and selective lesions of the cerebellum do not necessarily give rise to overt general intellectual impairment. However, in most of the investigations just reported, the integrity of higher cognitive functions was evaluated on the basis of clinical observations and was not documented by standard neuropsychological testing. It is therefore possible that the patients may have shown more subtle cognitive deficits which did not necessarily become apparent in everyday life or general clinical observations.

III. Cerebellar Involvement in Motor Learning

A. MODELS OF MOTOR LEARNING

It has long been recognized that the cerebellum plays a major role in motor control, with the lateral regions being involved in movement planning and programming and the intermediate and medial regions being involved in movement execution (Brooks and Thach, 1981; Dichgans and Diener, 1984). There is also convincing evidence that motor learning, changes in skillful movements as a result of experience, may depend on the integrity of the cerebellum (Ito, 1984). Marr (1969) and Albus (1971) have developed models of the basic mechanisms of experience-dependent modifications of Purkinje cells through concurrent activation of climbing fiber and parallel fiber inputs. Ito (1984) described a related cerebellar mechanism which serves to adapt movements to changing environmental demands. The empirical evidence for a cerebellar involvement in different types of motor learning is briefly described in the following sections.

B. MOTOR ADAPTATION AND HABITUATION

The term motor adaptation refers to the exchange of a behavior for another in response to changes in sensory input (see Sanes *et al.*, 1990). Motor adaptation is frequently investigated by means of the vestibulo-ocular reflex (VOR), compensatory eye movements for the stabilization of retinal images if the head is turned or if prisms are used (Stone and Lisberger, 1986). Patients with cerebellar dysfunction are impaired at prism adaptation and other forms of visuomotor adaptation to altered visual environment (Gauthier et al., 1979; Weiner et al., 1983).

Habituation is characterized by response decrements with repeated presentations of the same stimulus and is often investigated using the acoustic startle reflex, i.e., motor responses (such as eyeblinks) to a series of loud noises. Auditory input to the medial cerebellum and efferent projections to the brain stem areas which mediate the primary startle reflex may form the substrate of the cerebellar involvement in habituation (Lalonde and Botez, 1990). Long-term but not short-term habituation of the acoustic startle reflex was found to be severely impaired after damage to the cerebellar vermis in animals (Leaton and Supple, 1986; 1991). As yet, there are no data in human subjects with cerebellar dysfunction.

C. Motor Skill Acquisition

Motor skill acquisition refers to improvements in motor performance as a result of practice; the aim is to perform movements fast and accurately without attentional control (see Sanes et al., 1990). The hypothesis that the cerebellum is involved in the acquisition of such skills is supported by a number of lesion and electrophysiological studies in animals (e.g., Sasaki and Gemba, 1982; Lalonde and Botez, 1986). There is also empirical evidence from clinical investigations, as patients with cerebellar dysfunction had difficulties in acquiring a manual skill (repetitive tracing of a geometrical pattern) (Sanes et al., 1990) or in learning the skillful execution of serial movements (Inhoff et al., 1989). Neuroimaging studies also confirmed the cerebellar participation in motor skill learning, although it remained unclear whether the magnitude of the cerebellar activation was related to learning as such or to the control of rapid sequential finger movements (Seitz et al., 1990; Seitz and Roland, 1992).

D. Classical Conditioning of Motor Responses

Classical conditioning of motor responses is usually investigated within the context of eyeblink conditioning. An acoustic stimulus (conditioned stimulus, CS) is paired with a corneal airpuff (unconditioned stimulus, US) which evokes a reflex blink (Thompson, 1991; Daum and Schugens, 1996). With repeated pairings, eyeblinks occur in the tone-airpuff interval before airpuff onset, thereby constituting conditioned responses (CRs).

Animal work has documented that eyeblink conditioning depends on the integrity of the cerebellum (Yeo, 1991; Lavond et al., 1993). The essential neuronal circuit involves the convergence of somatosensory US information and of auditory CS information in the cerebellum (nuclei and/or cortex), and an efferent pathway from the nucleus interpositus which acts on the brain stem motor nuclei which control eyeblink responses (see Thompson, 1991). This model has received further support from clinical studies. A single case study and three controlled group studies indicated that patients with cerebellar lesions were severely impaired at acquiring classically conditioned eyeblink responses in the presence of intact reflex blinks to the US (Lye et al., 1988; Daum et al., 1993b; Topka et al., 1993; Woodruff-Pak et al., 1996). It should be emphasized that the conditioning deficit of cerebellar patients was limited to motor responses, conditioning of autonomic and electrocortical responses (which were recorded in parallel to eyeblink responses) appeared to be unaffected (Daum et al., 1993b). The involvement of the cerebellum thus seems to be specific to motor conditioning and does not represent a more general associative learning deficit.

IV. Cerebellar Involvement in Temporal Processing

Some of the motor control and motor learning deficits observed after cerebellar lesions have been interpreted within the context of the hypothesis that the cerebellum is necessary for the computation of the timing requirements of motor programs (see Keele and Ivry, 1991). Consistent with this hypothesis, animal studies showed that cerebellar output precedes activity in the motor cortex, with the cerebellocortical signal containing explicit timing information (Sasaki and Gemba, 1984). In clinical studies, key symptoms of cerebellar dysfunction such as dysmetria and dysdiadochkinesia have also been attributed to an impaired ability to time temporal relations between antagonist muscles (Dichgans and Diener, 1984). The impairments in rhythmic tapping observed in patients with cerebellar lesions have also been interpreted in terms of a temporal processing deficit (Ivry et al., 1988) and their deficient speech production may reflect a difficulty in the temporal coordination of the neuromuscular events involved in articulation (Ivry and Gopal, 1992).

According to Keele and Ivry (1991), the cerebellum acts as an internal clock not only in motor control, but during any condition which requires explicit temporal computations. In accordance with this hypothesis, patients with cerebellar lesions were impaired at judging the relative duration of time intervals or the velocity of moving visual stimuli (Ivry and Keele, 1989; Ivry and Diener, 1991). There was, however, no deficit when the discrimination of time intervals relevant for linguistic material was required. Patients with cerebellar lesions showed intact discrimination of voice onset times or vowel durations (Ivry and Gopal, 1992). It has been shown, however, that the discrimination of two spoken words that exclusively differed in durational parameters was severely impaired in patients with bilateral cerebellar damage, whereas unilateral lesion patients performed normally on this task (Ackermann *et al.*, 1996).

V. Cerebellar Involvement in Higher Cognitive Functions

A. Anatomical Substrates

Several reciprocal cerebellocortical pathways have been identified which are thought to serve as the possible neuroanatomical basis for a cerebellar contribution to cognition. The cerebellum receives projections from the posterior parietal cortex, from the superior temporal sulcus, and from frontal regions via corticopontocerebellar pathways, and efferent projections to the same areas emerge from the dentate nucleus (see Schmahmann, 1991). These loops are thought to imply the possible cerebellar modification of information arriving from association cortex, and thereby an involvement in the higher cognitive functions subserved by these cortical regions.

B. General Intellectual Abilities

As degenerative cerebellar diseases have occasionally been reported to give rise to dementia, the investigation of intellectual capacities in cerebellar patients by means of standardized neuropsychological tests is of some interest. The results are far from conclusive. Although some studies reported reduced scores in either the verbal or both the verbal and performance subtests of the Wechsler Adult Intelligence Scale-Revised (e.g., Kish *et al.*, 1988; Bracke-Tolkmitt *et al.*, 1989; Akshoomoff *et al.*, 1992), other studies reported scores in the normal or even the superior range (e.g., Fiez *et al.*, 1992; Appollonio *et al.*, 1993; Daum *et al.*, 1993a). In the latter studies, clinical and neuroradiological data documented that the damage was restricted to the cerebellum whereas there was some evidence of additional extracerebellar involvement in the patients described in the other studies. Similar to the clinical observations of dementia in cerebellar syndromes (see earlier discussion), impairments in general intellectual functions may be contingent upon the presence of extracerebellar damage.

C. Frontal Lobe Functions

The idea of "frontal-like" cognitive impairment in patients with cerebellar dysfunction has received considerable interest, partly because of recent neuroimaging studies which have documented a functional relationship between the cerebellum and frontal cortex (Junck et al., 1988; Kim et al., 1994).

Neuropsychological data with respect to the status of frontal functions in patients with cerebellar lesions are inconsistent. Anticipatory planning as assessed by the Tower of Hanoi test was found to be impaired in cerebellar atrophy patients (Grafman et al., 1992); they solved fewer problems and needed more time to plan their moves. Significant cerebellar activation was also observed during performance of a related task in a magnetic resonance imaging study (Kim et al., 1994).

Verbal fluency tasks require subjects to name as many items of a certain category or words starting with a certain letter within a limited time period, usually 1 min. There is disagreement as to whether cerebellar patients show deficits on such tasks (Akshoomoff et al., 1992; Fiez et al., 1992; Appollonio et al., 1993). An important issue for the interpretation of these data is that cerebellar degeneration often leads to dysarthria with slowed speech tempo (Ackermann and Ziegler, 1992). Such speech motor deficits interfere with verbal fluency tasks, and speech slowing needs to be considered as an alternative interpretation for fluency deficits. A study by Appollonio et al. (1993) indicated difficulties in initiation and evidence of perseveration in verbal as well as written output, lending some credence to the hypothesis of frontal dysfunction in cerebellar patients. Even though cerebellar activation was observed in normal subjects during performance of the Wisconsin card sorting test, a traditional test of frontal lobe function (Berman et al., 1995), cerebellar patients did not show perseverative tendencies or other deficits on this task (Bracke-Tolkmitt et al., 1989; Daum et al., 1993a; Fiez et al., 1992). The severe problems of cerebellar patients in shifting attention between modalities, however, would be consistent with impaired cerebellofrontal interactions, with deficient implementation of the "frontal" plan to change attentional behavior due to cerebellar dysfunction (see Akshoomoff et al., 1992).

D. Visuospatial Abilities

The two-way links between the cerebellum and the parietal cortex have formed the theoretical basis for the hypothesis of the cerebellar involvement in visuospatial organization. Surprisingly, there are only a few empirical studies on visuospatial abilities in cerebellar patients. Evidence supporting the hypothesis stems from findings of deficient visuospatial recall (Bracke-Tolkmitt *et al.*, 1989) and impaired visuospatial manipulations (Wallesch and Horn, 1990) in patients with cerebellar syndromes. Other investigators, however, did not observe visuospatial processing deficits in similar patient samples (Appollonio *et al.*, 1993; Daum *et al.*, 1993a).

E. MEMORY AND NONMOTOR SKILL LEARNING

Cognitive processes which do not clearly relate to a mediation by the two-way cerebellocortical loops have also been investigated. Short-term and long-term declarative memory as well as priming effects were largely found to be intact in patients with cerebellar damage (Bracke-Tolkmitt *et al.*, 1989; Fiez *et al.*, 1992; Appollonio *et al.*, 1993; Daum *et al.*, 1993a). Some theoretical accounts have suggested that the cerebellar participation in skill acquisition may not be limited to motor learning, but may extend into the cognitive learning domain (Leiner *et al.*, 1986). A controlled group study reported normal performance of cerebellar patients on standard perceptual and cognitive skill learning tasks (Daum *et al.*, 1993a), whereas analysis of a single case yielded evidence for an impairment in cognitive skill acquisition (Fiez *et al.*, 1992). The patient needed more trials than controls to solve a puzzle, but the practice-related changes in performance over time were similar which questions the interpretation of data in terms of a genuine learning deficit.

Patients with cerebellar degeneration were also impaired at learning visuomotor associations in a serial reaction time task (Pascual-Leone *et al.*, 1993). As learning was assessed on the basis of changes in reaction times that were not separately analyzed for movement initation and execution components, it is difficult to rule out the possible contribution of motor performance deficits (e.g., differential fatigue effects).

F. LANGUAGE

A long-standing tradition of clinical research did not yield signs of aphasia, such as agrammatism or semantic problems in cerebellar syndromes (see Ackermann and Ziegler, 1992). Apart from a report of agrammatic speech after a right cerebellar infarction (Silveri *et al.*, 1994), there is little evidence for language dysfunction in patients with cerebellar damage. Using standardized neuropsychological tests, Akshoomoff *et al.* (1992) observed mild word finding and naming deficits in a case of cerebellar degenerative disorder, while comprehension, syntax, and reading were unaffected. The patient with a right cerebellar infarct described by Fiez *et al.* (1992) showed intact performance on the different subtests of the Boston diagnostic aphasia exam. His ability to generate words according to different rules was, however, severely impaired, and there appeared to be no evidence of learning with increasing practice. The lack of a learning effect may partly be explained by a deficient initiation of speech motor output due to dysarthria. This hypothesis, however, does not explain the considerable number of errors. Moreover, a positron emission tomography (PET) study yielded a significant activation of the right lateral cerebellum in similar tasks (Petersen *et al.*, 1989) which supports the hypothesis of a possible cerebellar involvement in this regard.

VI. Methodological Considerations and Possible Directions

Currently available clinical and neuropsychological findings in patients with cerebellar syndromes do not yet form a consistent picture. A number of studies have yielded evidence of deficits in frontal lobe function, visuospatial information processing or nonmotor skill learning, whereas other investigations did not replicate these effects. There are at least two explanations which may account for some of the discrepancies in the present results.

The first explanation relates to patient selection and the composition of the clinical groups investigated so far. Many samples included not only patients with selective cerebellar damage, but also comprised patients with additional extracerebellar damage, such as patients with Friedreich's ataxia or olivopontocerebellar atrophy (OPCA). OPCA is known to be associated with generalized cognitive impairments and a mild form of dementia (Kish *et al.*, 1988; Matthew *et al.*, 1993; Botez-Marquard and Botez, 1993). It is likely that the cognitive symptoms in this condition result from brain stem damage or functional changes in the cerebral cortex and frontal areas in particular (Kish *et al.*, 1988; Matthew *et al.*, 1993). As the neuropsychological profile of patients with selective cerebellar damage and patients suffering from OPCA is distinctly different, the pooling of the two patient groups is not advisable (Daum *et al.*, 1993b; Daum and Ackermann, 1995).

A second explanation for the inconsistent picture is related to methodological shortcomings and inadequately matched control or comparison groups in particular. Control subjects should be matched on the basis of present-state IQ, rather than educational level, since the latter measure may be too insensitive for the evaluation of general intellectual abilities which often form the basis for the evaluation of cognitive performance in terms of sparing of function or the presence of deficits (Daum et al., 1993b). Whether there are changes in IQ as a result of acquired cerebellar damage may be judged by the comparison of measures of premorbid and present-state IQ. Negative mood states may also present an alternative explanation for reduced neuropsychological test scores in cerebellar syndromes and should therefore be taken into consideration (Berent *et al.*, 1990).

A further methodological issue relates to the neuropsychological batteries employed so far. Several studies have concentrated on the functional implications of the cerebellofrontal and cerebelloparietal loops (see earlier discussion). A possible explanation for the discrepant results concerning frontal function in cerebellar syndrome may relate to the fact that most studies only considered one or two experimental procedures. However, patients with selective frontal lesions frequently show intact performance on a range of so-called frontal tests, and patients with posterior cortex lesions may be impaired on such tasks (Mayes and Daum, 1996). Given the lack of specificity of many frontal tests, a useful strategy would be to (a) administer a wide range of tests which should include all aspects of cognitive functions which are attributed to the frontal lobes (e.g., problem solving, executive functions, fluency, memory for context, false memory phenomena etc.) and (b) directly compare the performance of patients with cerebellar lesions and patients with frontal lobe lesions on the same tasks. This approach should also involve the investigation of spatial working memory, given the recent anatomical evidence for a projection between the prefrontal area involved in spatial working memory (area 46) and the lateral cerebellum (Middleton and Strick, 1994).

Along similar lines, in order to get a clearer picture of the possible involvement of the cerebellum in visuospatial processing, the different cognitive components involved in such abilities should be considered. For example, distinctions are drawn between visuoperceptual functions (e.g., visual synthesis in pattern recognition), visuomotor functions (e.g,. visuoconstructive tasks as in block design), visuospatial functions as such (e.g., judgment of distance or direction), and visuospatial imagery (e.g., mental rotation) (see Kolb and Whishaw, 1995). The comprehensive analysis of the performance of cerebellar patients with respect to these different components may help develop a more specific hypothesis about cerebellar participation in such abilities. A possible hypothesis would be that visuospatial judgments are only impaired if judgments have to be made relative to the individual's own position in space. It is also surprising that there is no data base yet with regard to many functions which have been shown to be impaired in patients with parietal lobe lesions, such as selective attention for visual stimuli, left–right discrimination, or apraxia (see Kolb and Whishaw, 1995).

Apart from lesion studies, functional neuroimaging techniques provide exciting new opportunities in gaining insight into brain–behavior relationships. But their limitations should not be overlooked. In addition to technical problems associated with earlier attempts to investigate the human cerebellum with PET techniques (see Jenkins and Frackowiak, 1993), paradigms from experimental psychology have to be adjusted or modified to fit in with the restrictions of measurement techniques. It should be taken into consideration how the nature of the cognitive processes is affected by the necessary alterations in experimental procedures. A second issue that is often neglected is that on the basis of neuroimaging data alone, it is very difficult to determine which brain structures are critically involved in what aspects of cognitive function. For example, neuroimaging studies of classical eyeblink conditioning have yielded significant changes in regional blood flow not only in the ipsilateral cerebellum, but also in multiple other sites, including the auditory cortex, the prefrontal and parietal cortices, the cingulate cortex, the neostriatum and the hippocampus (Molchan et al., 1994; Logan and Grafton, 1995). Because lesions in many of these areas do not affect the acquisition of eyeblink conditioning (Daum and Schugens, 1996), these regions may be involved in the processing of some aspects of the task, but they are not necessary for learning to occur. It is therefore important to complement neuroimaging by lesion data and to investigate the differential contribution of different activated brain areas to behavior. It may also be useful to supplement the behavioral data by the recording of event-related potentials during the performance of cognitive tasks, as such data may give additional information about the nature of the impairment and possible compensatory mechanisms (Tachibana et al., 1995).

In summary, neuropsychological studies in patients with cerebellar syndromes provide strong evidence for a cerebellar participation in those cognitive processes that modulate motor behavior, such as motor planning or motor learning. At least some of these problems may relate to a deficient processing of temporal intervals. The picture concerning nonmotor cerebellar functions is less clearcut. The cerebellum does not seem to be critically involved in general intellectual and memory abilities, while the possible contribution to frontal lobe functions and visuospatial processing remains to be specified on the basis of empirical studies using patients with selective cerebellar dysfunction, adequately matched clinical and nonclinical comparison groups, and theory-driven test batteries comprising a wide range of tests.

References

Ackermann, H., and Daum, I. (1995). Kleinhirn und kognition: Psychopathologische, neuropsychologische und neuroradiologische Befunde. *Fortschr. Neurol. Psychiat.* **63,** 30–37.

Ackermann, H., and Ziegler, W. (1992). Die zerebelläre dysarthrie: Eine literaturübersicht. *Fortschr. Neurol. Psychiat.* **60,** 28–40.

Ackermann, H., Gräber, S., Hertrich, I., and Daum, I. (1996). Categorical speech perception in cerebellar disorders. *Brain Lang.,* in press.

Adams, R. D., and Victor, M. (1989). "Principles of Neurology." McGraw Hill, New York.

Akshoomoff, N., Courchesne, E., Press, G., and Iragui, V. (1992). Contribution of the cerebellum to neuropsychological functioning: evidence from a case of cerebellar degenerative disorder. *Neuropsychologia* **30,** 315–328.

Albus, J. S. (1971). A theory of cerebellar function. *Math. Biosci.* **10,** 41–74.

Amarenco, P. (1991). The spectrum of cerebellar infarctions. *Neurology* **41,** 973–979.

Amarenco, P., Rosengart, A., DeWitt, L. D., Pessin, M. S., and Caplan, L. R. (1993). Anterior inferior cerebellar artery territory infarcts. Mechanisms and clinical features. *Arch. Neurol.* **50,** 154–161.

Amici, R., Avanzini, G., and Pacini, L. (1976). "Cerebellar Tumors: Clinical Analysis and Physiopathological Correlations." Karger, Basel.

Appollonio, I. M., Grafman, J., Schwartz, V., Massaquoi, S., and Hallett, M. (1993). Memory in patients with cerebellar degeneration. *Neurology* **43,** 1536–1544.

Berent, S., Giordani, B., Gilman, S., Junck, L., Lehtinen, S., Markel, D. S., Bolvin, M., Kluin, K. J., Parks, R., and Koeppe, R. A. (1990). Neuropsychological changes in olivopontocerebellar atrophy. *Arch. Neurol.* **47,** 997–1001.

Berman, K. F., Ostrem, J. L., Randolph, C., Gold, J., Goldberg, T. E., Coppola, R., Carson, R. E., Herscovitch, P., and Weinberger, D. R. (1995). Physiological activation of a cortical network during performance of the Wisconsin Card Sorting Test: A positron emission tomography study. *Neuropsychologia* **33,** 1027–1046.

Bloedel, J. R. (1993). "Involvement in" versus "storage of". *TINS* **16,** 451–452.

Bond, C. H. (1895). Atrophy and sclerosis of the cerebellum. *J. Ment. Sci.* **41,** 409–420.

Botez-Marquard, T., and Botez, M. I. (1993). Cognitive behavior in heredodegenerative ataxias. *Eur. Neurol.* **33,** 351–357.

Bracke-Tolkmitt, R., Linden, A., Canavan, A. G. M., Rockstroh, B., Scholz, E., Wessel, K., and Diener, H. C. (1989). The cerebellum contributes to mental skills. *Behav. Neurosci.* **103,** 442–446.

Brooks, V., and Thach, W. (1981). Cerebellar control of posture and movement. *In* "Handbook of Physiology: Motor Control" (V. Brooks, ed.), pp. 877–946. American Physiological Society, Washington, DC.

Cutting, J. C. (1976). Chronic mania in childhood: Case report of a possible association with a radiological picture of cerebellar disease. *Psychol. Med.* **6,** 635–642.

Daum, I., and Ackermann, H. (1995). Cerebellar contributions to cognition. *Behav. Brain Res.* **67,** 201–210.

Daum, I., Ackermann, H., Schugens, M. M., Reimold, C., Dichgans, J., and Birbaumer, N. (1993a). The cerebellum and cognitive functions in humans. *Behav. Neurosci.* **107,** 411–419.

Daum, I., and Schugens, M. M. (1996). On the cerebellum and classical conditioning. *Curr. Dir. Psychol. Sci.* **5,** 58–61.

Daum, I., Schugens, M. M., Ackermann, H., Lutzenberger, W., Dichgans, J., and Birbaumer, N. (1993b). Classical conditioning after cerebellar lesions in humans, *Behav. Neurosci.* **107,** 748–756.

Dichgans, J., and Diener, H. C. (1984). Clinical evidence for functional compartmentalisation of the cerebellum. *In* "Cerebellar Functions" (J. Bloedel, J. Dichgans, and W. Precht, eds.), pp. 126–147. Springer, Berlin.

Dichgans, J., Diener, H. C., and Klockgether, T. (1989). Zu den Heredoataxien. *In* "Verhandlungen der Deutschen Gesellschaft für Neurologie" (P. A. Fischer, H. Baas, and W. Enzensberger, eds.), pp. 754–762. Springer, Berlin.

Dow, R. S., and Moruzzi, G. (1958). "The Physiology and Pathology of the Cerebellum." University of Minnesota Press, Minneapolis.

Eadie, M. J. (1975). Cerebello-olivary atrophy (Holmes type). *In* "Handbook of Clinical Neurology. System Disorders and Atrophies" (P.J. Vinken and G.W. Bruyn, eds.), pp. 403–414. Elsevier, Amsterdam.

Erskine, C. A. (1950). Asymptomatic unilateral agenesis of the cerebellum. *Mschr. Psychiat. Neurol.* **119**, 321–339.

Ferrier, D. (1894). Recent work on the cerebellum and its relations, with remarks on the central connexions, and trophic influence of the fifth nerve. *Brain* **17**, 1–26.

Fiez, J. A., Petersen, S. E., Cheney, M. K., and Raichle, M. E. (1992). Impaired non-motor learning and error detection associated with cerebellar damage. *Brain* **115**, 155–173.

Gauthier, G. M., Hofferer, J. M., Hoyt, W. F., and Stark, L. (1979). Visual-motor adaptation: Quantitative demonstration in patients with posterior fossa involvement. *Arch. Neurol.* **36**, 155–160.

Gilman, S., Bloedel, J. R., and Lechtenberg, R. (1981). "Disorders of the Cerebellum." Davis, Philadelphia.

Glickstein, M. (1993). Motor skills but not cognitive tasks. *TINS* **16**, 450–451.

Grafman, J., Litvan, I., Massaquoi, S., Stewart, M., Sirigu, A. and Hallett, M. (1992). Cognitive planning deficit in patients with cerebellar atrophy, *Neurology.* **42**, 1493–1496.

Guzzetta, F., Mercuri, E., Bonanno, S., Longo, M., and Spano, M. (1993). Autosomal recessive congenital cerebellar atrophy: A clinical and neuropsychological study. *Brain Dev.* **15**, 439–445.

Harding, A. E. (1981). Idiopathic late onset cerebellar ataxia: A clinical and genetic study of 36 cases. *J. Neurol. Sci.* **51**, 259–271.

Harding, A. E. (1984). "The Hereditary Ataxias and Related Disorders." Churchill Livingstone, Edinburgh.

Holmes, G. (1917). The symptoms of acute cerebellar injuries due to gunshot injuries. *Brain* **40**, 503–535.

Inhoff, A. W., Diener, H. C., Rafal, R. D., and Ivry, R. (1989). The role of the cerebellar structures in the execution of serial movements. *Brain* **112**, 565–581.

Ito, M. (1984). "The Cerebellum and Neural Control." Raven Press, New York.

Ito, M. (1993). Movement and thought: Identical control mechanisms by the cerebellum. *TINS* **16**, 448–450.

Ivry, R. B., and Diener, H. C. (1991). Impaired velocity perception in patients with lesions to the cerebellum. *J. Cogn. Neurosci.* **3**, 355–366.

Ivry, R. B., and Gopal, H. S. (1992). Speech production and perception in patients with cerebellar lesions. *In* "Attention and Performance XIV: Synergies in Experimental Psychology, Artificial Intelligence and Cognitive Neuroscience" (D. E. Meyer and S. Kornblum, eds.), pp. 771–802. Erlbaum, Hillsdale, NJ.

Ivry, R. B., and Keele, S. W. (1989). Timing functions of the cerebellum. *J. Cogn. Neurosci.* **1**, 136–152.

Ivry, R. B., Keele, S. W., and Diener, H. C. (1988). Dissociation of the lateral and medial cerebellum in movement timing and movement execution. *Exp. Brain Res.* **73**, 167–180.

Jenkins, I. H., and Frackowiak, R. S. (1993). Functional studies of the human cerebellum with positron emission tomography. *Rev. Neurol. (Paris)* **149,** 647–653.

Junck, L., Gilman, S., Rothley, J. J., Betley, A., Koeppe, R., and Hichwa, R. (1988). A relationship between metabolism in frontal lobes and cerebellum in normal subjects studied with PET. *J. Cereb. Blood Flow Metab.* **8,** 774–782.

Keele, S. W., and Ivry, R. (1991). Does the cerebellum provide a common computation for diverse tasks. A timing hypothesis. *Ann. N.Y. Acad. Sci.* **608,** 179–211.

Kim, S. G., Ugurbil, K., and Strick, P. L. (1994). Activation of a cerebellar output nucleus during cognitive processing. *Science* **265,** 949–951.

Kish, S. J., El-Awwar, M., Schut, L., Leach, L., Oscar-Berman, M., and Freedman, M. (1988). Cognitive deficits in olivopontocerebellar atrophy: Implications for the cholinergic hypothesis of Alzheimer's dementia. *Ann. Neurol.* **24,** 200–206.

Kolb, B., and Whishaw, I. Q. (1995). "Fundamentals of Human Neuropsychology." Freeman, New York.

Lalonde, R., and Botez, M. I. (1990). The cerebellum and learning processes in animals. *Brain Res. Rev.* **15,** 325–332.

Lalonde, R., and Botez, M. I. (1986). Navigational deficits in weaver mutant mice. *Brain Res.* **398,** 175–177.

Lavond, D. G., Kim, J. J., and Thompson, R. F. (1993). Mammalian brain substrates of aversive classical conditioning. *Annu. Rev. Psychol.* **44,** 317–342.

Leaton, R. N., and Supple, W. F. (1986). Cerebellar vermis: Essential for long-term habituation of the acoustic startle response. *Science* **232,** 513–515.

Leaton, R. N., and Supple, W. F. (1991). Medial cerebellum and long-term habituation of acoustic startle in rats. *Behav. Neurosci.* **105,** 804–816.

Leiner, H. C., Leiner, A. L., and Dow, R. S. (1986). Does the cerebellum contribute to mental skills? *Behav. Neurosci.* **100,** 443–454.

Leiner, H. C., Leiner, A. L., and Dow, R. S. (1989). Reappraising the cerebellum: What does the hindbrain contribute to the forebrain? *Behav. Neurosci.* **103,** 998–1008.

Leiner, H. C., Leiner, A. L., and Dow, R. S. (1993).Cognitive and language functions of the human cerebellum. *TINS* **16,** 444–447.

Logan, C. G., and Grafton, S. T. (1995). Functional anatomy of human eyeblink conditioning determined with regional cerebral glucose metabolism and positron emission tomography. *Proc. Nat. Acad. Sci. USA* **92,** 7500–7504.

Lye, R. H., O'Boyle, D. J., Ramsden, R. T., and Schady, W. (1988). Effects of a unilateral cerebellar lesion on the acquisition of eyeblink conditioning in man. *J. Physiol. (London)* **403,** 58.

Macchi, G., and Bentivoglio, M. (1987). Agenesis or hypoplasia of cerebellar structures. In "Handbook of Clinical Neurology: Malformations" (P. J. Vinken, G.W. Bruyn, H. L. Klawans, and N. C. Myrianthopoulos, eds.), pp. 175–194. Elsevier, Amsterdam.

Macchi, G., de Biase, G., and Pagnini, P. (1964). Le aplasie e le ipoplasie del cervelletto. *Arch. de Vecchi Anat. Pat.* **54,** 1–45.

Marr, D. (1969). A theory of cerebellar cortex. *J. Physiol. (London)* **202,** 437–470.

Matthew, W., Nordahl, T., Schut, L., King, A. C., and Cohen, R. (1993). Metabolic and cognitive changes in hereditary ataxia. *J. Neurol. Sci.* **119,** 134–140.

Mayes, A. R., and Daum, I. (1996). How specific are the memory and other cognitive deficits caused by frontal lobe lesions? In "Methodology in Frontal and Executive Functions" (P. Rabbitt, ed.). Erlbaum, Hove, in press.

Middleton, F. A., and Strick, P. L. (1994). Anatomical evidence for cerebellar and basal ganglia involvement in higher cognitive function. *Science* **266,** 458–461.

Molchan, S. E., Sunderland, T., McIntosh, A. R., Herscovitch, P., and Schreurs, B. G. (1994). A functional anatomical study of associative learning in humans. *Proc. Nat. Acad. Sci. USA* **91,** 8122-8126.

Pascual-Leone, A., Grafman, J., Clark, K., Stewart, M., Massaquoi, S., Lou, J. S., and Hallett, M. (1993). Procedural learning in Parkinson's disease and cerebellar degeneration. *Ann. Neurol.* **34,** 594-602.

Petersen, S. E., Fox, P. T., Posner, M. I., Mintun, M., and Raichle, M. E. (1989). Positron emission tomographic studies of the processing of single words. *J. Cogn. Neurosci.* **1,** 153-170.

Rolando, L. (1809). Saggio sopra la vera struttura del cervello dell'uomo e delgi animali e sopra le funzioni del sistema nervoso. Cited in Dow and Moruzzi (1958).

Sanes, J. N., Dimitrov, B., and Hallett, M. (1990). Motor learning in patients with cerebellar dysfunction. *Brain* **113,** 103-120.

Sasaki, K., and Gemba, H. (1982). Development and change of cortical field potentials during learning processes of visually initiated hand movements in the monkey. *Exp. Brain Res.* **48,** 429-437.

Sasaki, K., and Gemba, H. (1984). Compensatory motor function of the somatosensory cortex for dysfunction of the motor cortex following cerebellar hemispherectomy in the monkey. *Exp. Brain Res.* **56,** 532-538.

Schmahmann, J. D. (1991). An emerging concept: The cerebellar contribution to higher function. *Arch. Neurol.* **48,** 1178-1187.

Seitz, R. J., and Roland, P. E. (1992). Learning of sequential finger movements in man: A combined kinematic and positron emission tomography (PET) study. *Eur. J. Neurosci.* **4,** 154-165.

Seitz, R. J., Roland, P. E., Bohm, C., Greitz, T., and Stone-Elander, S. (1990). Motor learning in man: A positron emission tomographic study. *NeuroReport* **1,** 57-66.

Silveri, M. C., Leggio, M. G., and Molinari, M. (1994). The cerebellum contributes to linguistic production: A case of agrammatic speech following a right cerebellar lesion. *Neurology* **44,** 2047-2050.

Snider, R. S. (1950). Recent contributions to the anatomy and physiology of the cerebellum. *Arch. Neurol. Psychiat.* **64,** 196-219.

Snider, R. S., and Maiti, A. (1976). Cerebellar contributions to the Papez circuit. *J. Neurosci. Res.* **2,** 133-146.

Stewart, T. G., and Holmes, G. (1904). Symptomatology of cerebellar tumours: A study of forty cases. *Brain* **27,** 522-591.

Stone, L. S., and Lisberger, S. G. (1986). Detection of tracking errors by visual climbing fiber inputs to monkey cerebellar flocculus during pursuit eye movements. *Neurosci. Lett.* **72,** 163-168.

Tachibana, H., Aragane, K., and Sugita, M. (1995). Event-related potentials in patients with cerebellar degeneration: Electrophysiological evidence for cognitive impairment. *Brain Res. Cogn. Brain Res.* **2,** 173-180.

Thompson, R. F. (1991). Are memory traces localized or distributed? *Neuropsychologia* **29,** 571-582.

Topka, H., Valls-Sole, J., Massaquoi, S. G., and Hallett, M. (1993). Deficit in classical conditioning in patients with cerebellar degeneration. *Brain* **116,** 961-969.

Wallesch, C. W., and Horn, A. (1990). Long-term effects of cerebellar pathology on cognitive functions. *Brain Cogn.* **14,** 19-25.

Weiner, M. J., Hallett, M., and Funkenstein, H. H. (1983). Adaptation to lateral displacement of vision in patients with lesions of the central nervous system. *Neurology* **33,** 766-772.

Woodruff-Pak, D. S., Papka, M., and Ivry, R. B. (1996). Cerebellar involvement in eyeblink classical conditioning in humans. *Neuropsychology* **10,** 443-458.

Yeo, C. H. (1991). Cerebellum and classical conditioning of motor responses. *Ann. N.Y. Acad. Sci.* **627,** 292-305.

SECTION VI
THEORETICAL CONSIDERATIONS

CEREBELLAR MICROCOMPLEXES

Masao Ito

Laboratory for Synaptic Function, Frontier Research Program, Institute of Physical and Chemical Research (RIKEN), Wako, Saitama 351-01, Japan

I. Introduction
II. Evolutionary View
III. Roles in Spinal Cord and Brain Stem Functions
IV. Roles in Voluntary Movements
V. Possible Roles in Thought
VI. Cerebellar Microcomplexes
VII. Comments
   References

I. Introduction

Advances in the past three decades in studies of the cerebellum have led to the development of the concept of cerebellar corticonuclear microcomplexes (referred to hereafter as cerebellar microcomplexes) which act as structural and functional units of the cerebellum (Ito, 1984, 1990, 1993). A cerebellar microcomplex receives dual inputs via mossy fibers and climbing fibers (Fig. 1). Mossy fiber signals drive a cerebellar microcomplex, whereas climbing fiber signals represent errors and act to reorganize internal connections in the cerebellar microcomplex. Cerebellar microcomplexes are connected to various systems of the brain and so play diverse roles in central nervous system functions. This situation would be similar to that of a computer chip which can be used for a great many purposes.

This chapter first discusses the possible roles of cerebellar microcomplexes from the viewpoint of evolution and examines how nature has found uses of the cerebellar microcomplexes while functions of the central nervous system evolved step by step up to mental activity. The second point of discussion is the functional capacity of cerebellar microcomplexes. Because their structure appears to be largely uniform throughout the cerebellum, the question of how they contribute to mental activity with the same structural and operational principles as applied to lower nervous system functions arises. These two lines of discussion provide a theoretical basis for

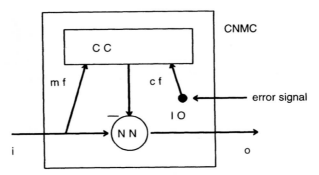

FIG. 1. Simplified structure of a cerebellar corticonuclear microcomplex (CNMC). CC, cerebellar cortex; NN, cerebellar or vestibular nucleus; mf, mossy fiber; cf, climbing fiber; IO, inferior olive; i, input; o, output.

the hypothesis that the cerebellum contributes to mental activity as first suggested by Leiner *et al.* (1986) and is substantiated by new observations on humans as documented in this volume. Even though it is true that presently available experimental evidence is limited (Daum and Ackermann, 1995), these theoretical considerations would help experimenters in their efforts to conduct appropriate experimental testings of the hypothesis.

II. Evolutionary View

Common to all vertebrates from fish to humans, the central nervous system has five basic duties. First, reflexes, both somatic and autonomic, and second, compound movements, such as posture, locomotion, breathing, and saccadic eye movement, are both mediated via neuronal activity in the spinal cord and brain stem. Third, innate behaviors such as food intake, drinking, attack, flight, and sexual behaviors are mediated via neuronal activity in the hypothalamus. The contributions of cerebellar microcomplexes to reflex arcs and compound movements are well documented (see Ito, 1984). Whether cerebellar microcomplexes contribute to innate behaviors is much less certain, but the existence of neuronal connections between the cerebellum and the hypothalamus suggests that they do (see Ito, 1984). The postulated contribution of the cerebellum to shock avoidance behavior (see Lalonde, 1994) could be related to this third type of function.

The fourth duty of the central nervous system is the cerebral sensorimotor function developed in birds and mammals. In middle-sized mammals (rodents, cats, dogs), the sensorimotor cortex occupies major areas of the cerebral neocortex. Contributions of the cerebellum to instrumental learning and voluntary movements mediated by neuronal activity in the sensorimotor cortex have been shown (see Ito, 1984; Thach *et al.*, 1991; Lalonde, 1994). The cerebral association function developed in primates is the fifth duty. It occupies three-fourths of the human cerebral cortex, which accounts for the emergence of mental activities such as thought in humans. In parallel with the evolutionary expansion of the cerebrum, the cerebellum has also expanded, suggesting a contribution of cerebellar microcomplexes to the association cortex function.

As illustrated in Fig. 2, the central nervous system is hierarchically organized with respect to the just-mentioned five executive functions (reflexes, compound movements, innate behavior, cerebral sensorimotor function, and cerebral association function). Other parts of the vertebrate central nervous system are assumed in Fig. 2 to exert regulatory actions on the five executive functions. The limbic system acts as a positive or negative reinforcer to enhance a purposeful innate behavior and to suppress a harmful one in a given situation and is viewed as a system which, by estimating biological values of stimuli, secures survival of individuals and maintenance of species (Fig. 2A). The biological clock and sleep–wakefulness system are also regulatory systems located primarily in the brain stem (Fig. 2D). The function of the basal ganglia has not yet been well specified, but here it is assumed to be stabilization, as suggested by the hyperkinesia in Huntington's chorea or hypokinesia in Parkinsonism (Fig. 2B). The stabilization may be exerted via selection from among many simultaneous activities occurring in many elements of the executive systems. In other words, a particular movement or behavior is promoted by inhibiting competing ones (see Kimura, 1995). A unique function of the cerebellum is assumed to afford adaptiveness in the five executive systems (Fig. 2C), as explained in the following sections.

III. Roles in Spinal Cord and Brain Stem Functions

A cerebellar microcomplex may directly be connected to a reflex arc, as in the case of the vestibulo-ocular reflex (Fig. 3A). A cerebellar microcomplex in a reflex arc is considered to enable adaptive control of the reflex in ever-changing environments. A reflex itself is stereotyped and has no capability for adapting to environmental changes. A cerebellar microcomplex senses errors in the performance of a reflex and accordingly changes command signals flowing across the microcomplex to the reflex arc. The

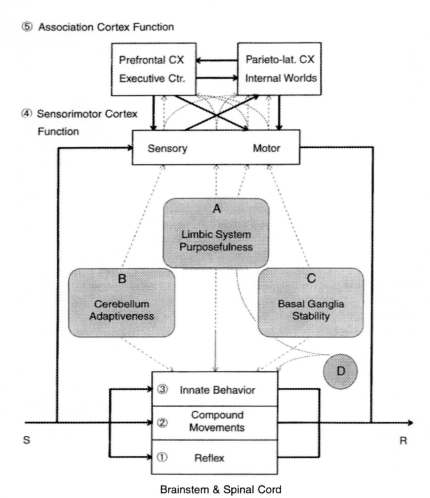

FIG. 2. Hierarchical structure of the vertebrate central nervous system. 1–5, five executive systems. A–D, four regulatory systems. D, sleep–wakefulness centers. S, stimulus; R, response.

changes so induced in the command signals modify the reflex performance toward a minimization of errors. A cerebellar microcomplex could similarly control compound movements adaptively in changing environments. The extent of the contribution of the cerebellum to the innate behavior systems is still unclear, but it is probable that a microcomplex confers adaptiveness on an innate behavior which otherwise is stereotyped. The proposed roles of cerebellar microcomplexes in reflexes or compound movements agree

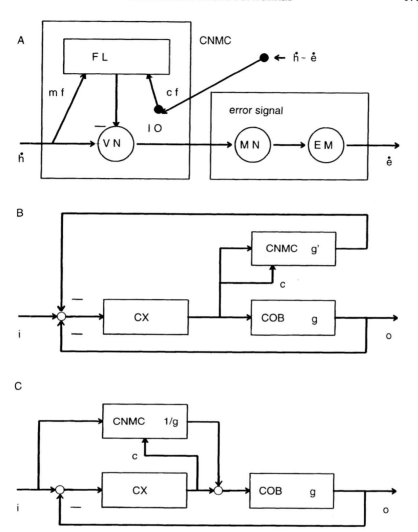

FIG. 3. Three model systems incorporating CNMC. (A) The case of vestibuloocular reflex. FL, flocculus; VN, vestibular nucleus; MN, oculomotor nuclei; EM, extraocular muscles; h, head velocity; e, eye velocity. (B) Dynamics model control. CX, executive motor cortices; COB, control object; g, dynamics of COB; g', model of g represented by CNMC. (C) Inverse dynamics model control.

very well with the modern concept of adaptive control whereby a microcomplex provides an error-driven adaptive control mechanism to reflexes or compound movements which themselves are classic control systems.

IV. Roles in Voluntary Movements

A microcomplex is inserted in a loop originating from and returning from the motor cortex (Fig. 3B). It is suggested that the microcomplex serves as an internal model of an object to be controlled by the motor cortex (Ito, 1970, 1984, 1993). It would be helpful to remember that in engineering systems, an adaptive control system can be used to develop a model representing dynamics properties of an object. This is done by connecting the object and the adaptive control system in parallel. Although they are fed with the same input, differences between their outputs are fed to the adaptive control system as error signals. When the adaptive control system is modified toward minimization of the error signals, it will have the same dynamics as the object. This is the theoretical basis for the postulate that a cerebellar microcomplex forms a model by means of its error-driven adaptive mechanism.

To date, two types of control have been suggested for voluntary movement. First, a cerebellar microcomplex forms an internal model representing the dynamics properties of a musculoskeletal system that is controlled by the motor cortex (Ito, 1970, 1984) (Fig. 3B). After this internal model is formed, the motor cortex performs accurate control in a feedforward manner without referring to the consequences of the actually performed movements, provided that the microcomplex precisely represents the dynamic properties of the musculoskeletal system after adaptation through practice. This dynamics model control, as it may be called, is consistent with the pattern of anatomical connections from the motor cortex to the pontine nucleus, the intermediate cortex of the cerebellum, and back to the motor cortex via the nucleus interpositus and thalamus.

In the second type of control proposed by Kawato et al. (1987) (see also Kawato and Gomi, 1993), a cerebellar microcomplex is assumed to act as a feedforward controller which operates in parallel with the motor cortex which peforms feedback control (Fig. 3C). In this case, the feedforward control is performed perfectly as long as the cerebellar microcomplex functions accurately and represents the inverse dynamics properties of the musculoskeletal system to be controlled. Error signals during this learning may be obtained from the motor cortex (Kawato and Gomi, 1993). This inverse dynamics model control, as it may be called, requires the existence of neuronal connections from a premotor cortex to the pontine nucleus, the cerebellar hemisphere, the dentate nucleus, and a brain stem center of a spinal descending system (possibly a reticulospinal system). Such connections have been shown to exist (see Ito, 1984).

The dynamics model control implies that the motor cortex can perform precise movement control in a feedforward manner using the dynamics model in the internal loop (Fig. 3B). For example, one would be able to hit a golf ball precisely even with one's eyes closed, as if one were hitting it in the cerebellum. Performance of the finger-to-nose test of touching the nose with a finger with the eyes closed can be explained similarly by assuming that one is touching the nose in the cerebellum. It is interesting that a patient with acute cerebellar dysfunction reported that he felt that he had missed his nose when his finger was approaching it (K. Sasaki, personal communication).

With an inverse dynamics model in the cerebellum, however, one would be able to perform a movement automatically without conscious thought because command signals would bypass the motor cortex. This may explain the fact that we can perform a skilled movement unconsciously, except in the initial phase of practice. Coexistence of the two types of control may suggest that two types of motor learning proceed in parallel: a dynamics model and an inverse dynamics model are formed in the cerebellum through repeated practice with external feedback control. However, it is pointed out that once the dynamics model is established, an inverse dynamics model can be formed through practice using the internal feedback through the dynamics model in the cerebellum. One may speculate that if movement were inhibited during such practice, "image training" would occur.

It has been confirmed that firing patterns of Purkinje cells in the ventral paraflocculus represent inverse dynamics of eyeballs, consistent with the hypothesis that the ventral paraflocculus is involved in the inverse dynamics model control of eye movement (Shidara et al., 1995).

V. Possible Roles in Thought

Even though neural mechanisms of thought are not well defined, it is certain that our capability of thought emerged as a consequence of enormous expansion of the association cortex. Its evolutionarily newest portion is divided into two parts: the prefrontal and parietolateral association cortices. The parietolateral association cortex integrates sensory information collected through sensory and perisensory cortices, and in humans includes Wernicke's speech center which encodes verbal concepts. The parietolateral association cortex is the highest level at which our internal world is formed (Fig. 4). The prefrontal association cortex, located anterior to the premotor and motor cortices, is the area where command signals appear to emerge. It contains an area which generates no-go signals to suppress voluntary movements (Sasaki et al., 1993) and is regarded as central ex-

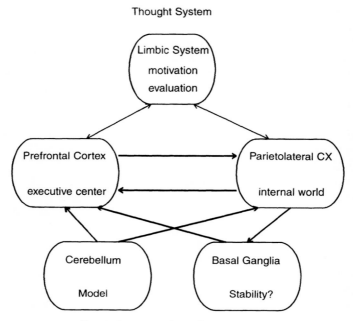

FIG. 4. Structure of the thought system.

ecutive centers of the brain equipped with working memory mechanisms (Goldman-Rakic, 1992; Funahashi and Kubota, 1994). Thought may be affected by the prefrontal cortex acting on the ideas and concepts encoded in the parietolateral cortex under regulatory influences from the limbic system, cerebellum, and basal ganglia (Fig. 4). Simultaneous activation of the prefrontal association cortex and Wernicke's speech center, together with the cerebellum, has been demonstrated to occur in a positron emission tomography study on subjects performing a simple thought task using language (Petersen et al., 1989).

The author's suggestion that the cerebellum is involved in thought is based on an analogy between movement and thought from the viewpoint of control (Ito, 1990, 1993). In voluntary movements, command signals from the motor cortex act on a musculoskeletal system (Figs 3B and 3C), whereas in thought, command signals from the prefrontal association cortex act on ideas or concepts encoded in the parietolateral association cortex. Even though the musculoskeletal system and ideas or concepts are very different in nature, the same control system principle in Figs. 3B and 3C postulated for voluntary movement control may also apply to the thought.

According to the psychological concept of a mental model (Johnson-Laird, 1983), thought may be viewed as a process of manipulating a mental model formed in the parietolateral association cortex by command signals from the prefrontal association cortex. A cerebellar microcomplex may be connected to neuronal circuits involved in thought and may represent a dynamics or an inverse dynamics model of a mental model. In other words, a mental model might be transferred from the parietolateral association cortex to the cerebellar microcomplex during repetition of a thought. By analogy to voluntary movement, one may speculate that formation of a dynamics model in the cerebellum would enable us to think correctly in a feedforward manner, i.e., without the need to check the outcome of the thought. This may be the case when one performs a quick arithmetical calculation. This dynamics model requires a loop connection through a cerebellar microcomplex from the prefrontal association cortex back to the same cortex (as in Fig. 3B). Connections from the dentate nucleus to the prefrontal cortex have been demonstrated in monkeys (Middleton and Strick, 1994). However, an inverse dynamics model in the cerebellum (as in Fig. 3C) would enable us to think automatically without conscious effort. This may apply to the case of speech, for example, which should be consistent with the report that the number of errors in a task of converting nouns to verbs or of identifying the appropriate use of objects decreased during repeated performance of the tasks by normal subjects, but remained constant during the repeated performance of the same tasks by a patient with a cerebellar hemispheric infarction (Fiez et al., 1992).

Experimental testing of these possibilities is difficult due to our ignorance of how mental models are represented in neuronal networks. The manner of representation of words, concepts, or ideas in the cerebral cortex needs to be clarified before we examine a mechanism by which the postulated mental model is copied by cerebellum for dynamics or inverse dynamics model control of thought.

VI. Cerebellar Microcomplexes

The cerebellum can be divided into numerous microcomplexes, each of which consists of a cortical microzone as defined by Oscarsson (1979), a small group of neurons in the cerebellar or vestibular nuclei receiving inhibitory signals from Purkinje cells in that microzone, a small group of inferior olivary neurons supplying their climbing fiber axons to Purkinje cells in that microzone and collaterals of the climbing fiber axons to the nuclear neurons, and small groups of neurons in the precerebellar nuclei of the brain stem or spinal cord supplying mossy fiber afferents to that microzone and collaterals to the nuclear neurons (Fig. 5). Each microcomplex also receives serotonergic and noradrenergic afferents (see Ito, 1984).

Essential assumptions concerning the function of a microcomplex are that (1) stem fibers of mossy fibers arising from precerebellar nuclei convey command signals across the cerebellar or vestibular nuclei to the brain stem or spinal cord, or to the cerebral cortex via the thalamus; (2) the transmission of the command signals across nuclear neurons is modulated by inhibitory signals from Purkinje cells driven by the mossy fiber signals; (3) climbing fibers convey signals representing errors in the consequences induced by the command signals; and (4) error signals of climbing fibers induce long-term depression (LTD) in those parallel fiber-to-Purkinje cell synapses, driven by repeated mossy fiber signals almost in synchrony with climbing fiber signals.

LTD is thus proposed as a core process accounting for the adaptive capability of microcomplexes. Its occurrence has been confirmed, and complex chemical reactions involving receptors, second messengers, protein kinases, and protein phosphatases have been demonstrated to underlie LTD (reviewed by Ito, 1991; Crepel and Audinat, 1991; Linden, 1994). LTD is due to reduced sensitivity of glutamate receptors mediating parallel fiber-to-Purkinje cell transmission, caused by phosphorylation of the glutamate receptors. Antibodies raised against the phosphorylated glutamate receptors thus label synapses undergoing LTD (Nakazawa *et al.*, 1995). Crucial roles of LTD in motor learning are indicated by the fact that reagents blocking LTD prevent adaptation of the vestibuloocular reflex (Nagao and Ito, 1991) and locomotion (Yanagihara and Kondo, 1996). It has also been revealed that movement disorders in certain gene knockout mice are paralleled by diminution of LTD, suggesting causality between them (Chen *et al.*, 1995).

Activity-dependent plastic changes also occur in other types of synapses within a cerebellar microcomplex: potentiation in inhibitory synapses on Purkinje cells (Kano *et al.*, 1992) and a depression in Purkinje cell-derived inhibitory synapses in cerebellar nuclear cells (Morishita and Sastry, 1993). It has been proposed that motor learning is due to plasticity not only in the cerebellar cortex but also in deep cerebellar nuclei, but even if it is so, the plasticity in the cortex must guide the plasticity in the deep nuclei (Raymond *et al.*, 1996). Mossy fiber to granule cell synapses may also exhibit plasticity because of the presence of NMDA receptors in these synapses, but this plasticity may play a role in the developmental organization of the mossy fiber-to-granule cell connections.

The operational principle for a cerebellar microcomplex to serve as an error-driven adaptive element in a neural system is easily understood from the neural diagram shown in Fig. 5. However, clarification of how a cerebellar microcomplex can represent a dynamics or an inverse dynamics model requires a mathematical theory. These models can be represented by functions including variables such as velocity, acceleration, or position in movement control as shown for eyeballs (Shidara et al., 1995), but in thought, they should include variables specific to mental models. Although capabilities of multilayered neuronal networks to approximately represent any function have been shown (e.g., see Funahashi, 1989), the final question arises as to how ideas, concepts, or words are represented by a mental model in the parietolateral cortex and how such a mental model can be transferred to a cerebellar microcomplex.

VII. Comments

Opinion is sometimes divided as to whether cerebellar functions are motor or sensory (Goa et al., 1996) or motor or cognitive. Since outputs of the executive functions 1–4 in Fig. 2 are largely represented by movement, involvement of the cerebellum in these four may lead to the impression that the cerebellum is primarily a motor organ. However, since the output

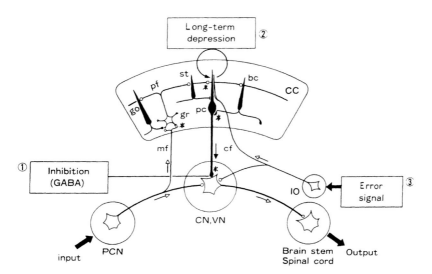

FIG. 5. Detailed structure of CNMC. pc, Purkinje cell; bc, basket cell; st, stellate cell; pf, parallel fiber; go, Golgi cell; gr, granule cell; PCN, precerebellar nuclei; CN, VN, cerebellar and vestibular nuclei. Sites where plasticity is assumed to be present are shown by asterisks.

of functional system 5 is expressed as mental activity, cerebellar contribution to 5 will not be represented by movement; hence it may be described as nonmotor or cognitive. An emphasis must be made, however, that the cerebellum in Fig. 2 is not sensory, motor, or cognitive, but it is designated as a regulatory organ subserving five executive functions. The cerebellum is defined uniquely by its error-driven adaptive control mechanism and the model-building capability based on it, but not by the nature of the executive system to which it is connected.

In analogy to the contribution of the cerebellum to motor activity, its contribution to mental activity may be specified as regulating the speed, consistency, and appropriateness of cognitive processes, with dysfunction leading to a dysmetria of thought (Schmahmann, 1991). This chapter provides theoretical bases for explaining cerebellar symptoms such as dysmetria as being due to impairment of a cerebellar model of musculoskeletal system. A similar explanation applies to mental dysmetria that may occur due to lack of the model which copies a mental model.

References

Chen, C., Kano, M., Abeliovich, A., Chen, L., Bao, S., Kim, J. J., Hashimoto, K., Thompson, R. F., and Tonegawa, S. (1995). Impaired motor coordination correlates with persistent multiple climbing fiber innervation in PKC γ mutant mice. *Cell* **83,** 1233–1242.

Crepel, F., and Audinat, E. (1991). Excitatory amino acid receptors of cerebellar Purkinje cells: Development and plasticity. *Prog. Biophys. Mol. Biol.* **55,** 31–46.

Daum, I., and Ackermann, H. (1995). Cerebellar contributions to cognition. *Behav. Brain Res.* **67,** 201–210.

Fiez, J. A., Petersen, S. E., Cheney, M. K., and Raichle, M. E. (1992). Impaired non-motor learning and error detection associated with cerebellar damage. *Brain* **52,** 203–225.

Funahashi, K. (1989). On the approximate realization of continuous mapping by neural network. *Neural Networks* **2,** 183–192.

Funahashi, S., and Kubota, K. (1994). Working memory and prefrontal cortex. *Neurosci. Res.* **21,** 1–11.

Goa, J.-H., Parsons, L. M., Bower, J. M., Xiong, J., Li, J., and Fox, P. T. (1996). The role of the cerebellum in sensory discrimination: A functional magnetic reasonance imaging study. *Science* **272,** 545–547.

Goldman-Rakic, P. S. (1992). Working memory and the mind. *Sci. Am.* **267,** 111–117.

Ito, M. (1970). Neurophysiological aspects of the cerebellar motor control system. *Int. J. Neurol.* **7,** 162–176.

Ito, M. (1984). "The Cerebellum and Neural Control." Raven Press, New York.

Ito, M. (1990). A new physiological concept on cerebellum. *Rev. Neurol. (Paris)* **146,** 564–569.

Ito, M. (1991). The cellular basis of cerebellar plasticity. *Curr. Opin. Neurobiol.* **1,** 616–620.

Ito, M. (1993). Movement and thought: Identical control mechanisms by the cerebellum. *Trends Neurosci.* **16,** 448–450.

Johnson-Laird, P. N. (1983). "Mental Model." Cambridge Univ. Press, Cambridge.

Kano, M., Rexhausen, U., Dreessen, J., and Konnerth, A. (1992). Synaptic excitation produces a long-lasting rebound potentiation of inhibitory synaptic signals in cerebellar Purkinje cells. *Nature* **356,** 601–604.

Kawato, M., Furukawa, K., and Suzuki, R. (1987). A hierarchical neural-network model for control and learning of voluntary movement. *Biol. Cybern.* **57,** 169–185.

Kawato, M., and Gomi, H. (1993). The cerebellum and VOR/OKR learning models. *Trends Neurosci.* **15,** 445–453.

Kimura, M. (1995). Role of basal ganglia in behavioral learning. *Neurosci. Res.* **22,** 353–358.

Lalonde, R. (1994). Cerebellar contribution to instrumental learning. *Neurosci. Biobehav. Rev.* **18,** 161–170.

Leiner, H. C., Leiner, A. L., and Dow, R. S. (1986). Does the cerebellum contribute to mental skills? *Behav. Neurosci.* **100,** 443–454.

Linden, D. J. (1994). Long-term synaptic depression in the mammalian brain. *Neuron* **12,** 457–472.

Middleton, F. A., and Strick, P. L. (1994). Anatomical evidence for cerebellar and basal ganglia involvement in higher cognitive function. *Science* **266,** 458–461.

Morishita, W., and Sastry, B. R. (1993). Long-term depression of IPSCs in rat deep cerebellar nuclei. *NeuroReport* **4,** 719–722.

Nagao, S., and Ito, M. (1991). Subdural application of hemoglobin to the cerebellum blocks vestibuloocular reflex adaptation. *NeuroReport* **2,** 193–196.

Nakazawa, K., Mikawa, S., Hashikawa, T., and Ito, M. (1995). Transient and persistent phosphorylation of AMPA-type glutamate receptor subunits in cerebellar Purkinje cells. *Neuron* **15,** 697–709.

Oscarsson, O. (1979). Functional units of the cerebellum-sagittal zones and microzones. *Trends Neurosci.* **2,** 143–145.

Petersen, S. E., Fox, P. T., Posner, M. I., Mintun, M., and Raichle, M. E. (1989). Positron emission tomographic studies of the processing of single-word. *J. Cogn. Neurosci.* **1,** 153–170.

Raymond, J. L., Lisberger, S. G., and Mauk, M. D. (1996). The cerebellum: A neuronal learning machine? *Science* **272,** 1126–1131.

Sasaki, K., Gemba, H., Nambu, A., and Matsuzaki, R. (1993). No-go activity in the frontal association cortex of human subjects. *Neurosci. Res.* **18,** 249–252.

Schmahmann, J. D. (1991). An emerging concept: The cerebellar contribution to higher function. *Arch. Neurol.* **48,** 1178–1187.

Shidara, M., Kawato, M., Gomi, H., and Kawato, K. (1995). Inverse-dynamics encoding of eye movements by Purkinje cells in the cerebellum. *Nature* **365,** 50–52.

Thach, W. T., Goodkin, H. P., and Keating, J. G. (1991). Cerebellum and the adaptive coordination of movement. *Annu. Rev. Neurosci.* **15,** 403–442.

Yanagihara, D., and Kondo, I. (1996). Nitric oxide plays a key role in adaptive control of locomotion in cats. *Proc. Natl. Acad. Sci. USA* **93,** 13292–13297.

CONTROL OF SENSORY DATA ACQUISITION

James M. Bower

California Institute of Technology, Division of Biology, Pasadena, California 91125

I. Introduction
II. Whiskers of the Rat as Viewed by the Cerebellum
   A. Physiological Organization of Cerebellar Cortical Circuitry
   B. Dual Role for Granule Cells in the Cerebellar Cortex
   C. Control of Tactile Data Acquisition: Putting It All Together
   D. Rat Mouths, Cat Paws, and Primate Fingers
   E. Tactile Sensory Data Acquisition in Humans
III. What is the Cerebellum Controlling?
   A. At What Level Is the Cerebellum Operating?
   B. Algorithmically, What Is Necessary?
   C. What about Cerebellar Outputs?
   D. What Are the Predicted Behavioral Effects of Cerebellar Involvement?
   E. How Does This Differ from Classical Motor Control?
IV. Implications for Cerebellar Function as a Whole
   A. The Cerebellum as a Support Device
   B. The Cerebellum Is Differentially Engaged
   C. Overall Cerebellar Activity May Be under Cerebral Cortical Control
   D. The Cerebellum Is Useful but Not Necessary
   E. Compensating for the Absence of the Cerebellum
V. Conclusion
   References

This chapter describes a new theory of cerebellar function which posits that the cerebellum is specifically involved in monitoring and adjusting the acquisition of most of the sensory data on which the rest of the nervous system depends. If correct, the cerebellum is not itself responsible for any particular behaviorally related function, whether "motor," "sensory," or "cognitive." Instead the cerebellum facilitates the efficiency with which other brain structures perform their own functions. In this way the cerebellum is seen as being useful but not necessary for many different kinds of brain functions. This chapter describes how this theory of cerebellar function has arisen from detailed study of the pattern of tactile afferent projections to the rat cerebellum as well as from an analysis of the neural circuitry that processes that information. It is proposed that the breadth of cerebellar involvement is reflected in the growing number of tasks which induce cerebellar activity, including cognitive tasks, even though the cerebellum is not itself directly involved in those tasks.

I. Introduction

It is the author's conviction that the debate concerning the function of the cerebellum will only finally be resolved when we understand the computations imbedded in its circuitry. Once we can look at the structure of a Purkinje cell and understand how this structure specifies a particular function, it will only be left to determine in what way that computation is useful to the rest of the brain. Of course, we are still a very long way from such an understanding. However, it is the author's view that, even today, careful study of the structure of the cerebellum is an essential ingredient in figuring out what this structure does. In the first place, as described in this chapter, careful consideration of the structure of cerebellar circuits can lead to new ideas concerning overall cerebellar function. In the second place, the closer the relationship between a theory and real anatomy and physiology, the easier it is to test experimental predictions. In order to understand our current thinking concerning the function of the cerebellum, this chapter reviews results concerning its anatomical and physiological organization. In the final sections, experiments intended to test this idea in humans as well as the implications of these results for overall cerebellar function are briefly described.

II. Whiskers of the Rat as Viewed by the Cerebellum

The idea that the cerebellum controls sensory data acquisition is based on detailed study of the lateral hemispheres of the cerebellum of the albino rat (Sasaki *et al.*, 1989; Bower and Woolston, 1983; Bower and Kassel, 1990; Gonzalez *et al.*, 1993; Jaeger and Bower, 1994; Paulin *et al.*, 1989a,b; Bower and Kassel, 1990; Morissette and Bower, 1996). It was demonstrated in the mid-1970s that the granule cell layer in these regions of the rat cerebellum contains a large representation of tactile body surfaces (reviewed in Welker, 1987). Somewhat surprisingly, however, the surfaces most represented are not from the fore or hindlimbs, as might be expected for the coordination of locomotion, but instead are the most sensitive perioral regions of the rat's face. In fact, Shambes and colleagues (1978b) found that only one hemispheric folium, the paramedian lobule, contained projections from the limbs, and these projections were almost exclusively from the forelimb. Even these projections were surrounded by representations of perioral surfaces (Fig. 1).

In addition to the surprising fact that the rat cerebellum contains an enormous tactile representation from perioral surfaces, Welker's laboratory

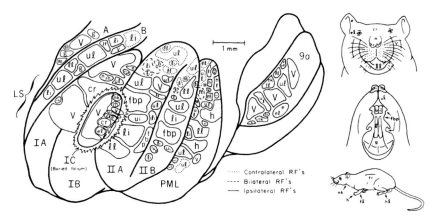

FIG. 1. Fractured somatotopy in the cerebellar cortex. The organization of tactile responsive regions of the cerebellum of the rat is summarized. The maps are based on multiunit recordings in the superficial granule cell layer of each folium shown. The region of the body surface projecting to each "patch" is indicated on the maps. The drawings of the rat face, perioral structures, and body on the right are similarly labeled. The type of boundary line surrounding each patch indicates ipsi-, bi-, and contralateral representations. The gaps between the patches are present for graphical clarity only. Modified from Bower and Woolston (1983).

also discovered that the topographic pattern of this projection was very different from that found in other known somatosensory structures (Shambes et al., 1978a). Instead of being a map of the topography of the body surface, tactile projections to the cerebellum have an unusual patchlike mosaic representation of different body parts (Welker, 1987). As shown in Fig. 1, adjacent cortical areas often receive projections from nonadjacent body parts, whereas within an area receiving projections from one body part (a "patch"), the body surface representation is somatotopically organized. Welker and colleagues referred to this projection pattern as "fractured somatotopy." Subsequent experiments demonstrated that projections from facial regions of the topographically organized maps in the somatosensory cortex (Bower et al., 1981) and the superior colliculus (Kassel, 1980) projected to the cerebellum in the same fractured pattern as the trigeminal nucleus. It has also been shown that the detailed patterns of these maps are very similar in different rats, and even among rats, mice, and guinea pigs (Bower and Kassel, 1990; Bower, 1997).

The consistencies in the pattern of the tactile maps suggest that their fractured organization has some important functional significance. Almost certainly, however, this significance is related to the physiological relationships between input and output within the cerebellar cortex itself. Accordingly, it was the logical next step to use our knowledge of the detailed pattern of tactile projections to explore the effects of the fractured maps on the response of overlying Purkinje cells (Bower and Woolston, 1983; Jaeger and Bower, 1994). As described in the next section, these results suggest that the cerebellar cortex is organized very differently than heretofore assumed.

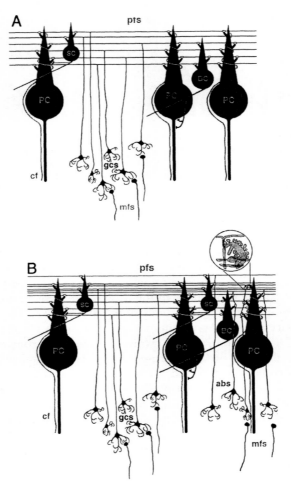

FIG. 2. A new view of cerebellar circuitry. This figure is a modification of the classic (A) graphical representation of cerebellar cortical circuitry (Eccles, 1973). It has been modified (B) to indicate the presense of synaptic contacts between the ascending branch of the granule cell axon and both Purkinje cells and the inhibitory interneurons of this cortex. Granule cell axons provide two different kinds of inputs to Purkinje cells; those associated with the ascending branch and the more distant contacts due to parallel fibers.

A. Physiological Organization of Cerebellar Cortical Circuitry

The afferent, mossy fiber input arriving in the granule cell layer is relayed via the granule cells to the cerebellar Purkinje cells (Palay and Chan-Palay, 1974). A predominant feature of this relay is the geometrical pattern of the granule cell axons in the molecular layer which all run parallel to each other (Fig. 2). Based on this geometry and the enormous number of parallel fiber synapses made on each Purkinje cell, it has been assumed by anatomists, physiologists, and theorists that any activation of the granule cell layer would result in a "beam" of activated Purkinje cells up and down the folium (Braitenberg and Atwood, 1958; Gabbiani et al., 1994; Karachot et al., 1994; Eccles, 1973). However, using our detailed knowledge of the fine pattern of tactile projections to the lateral hemispheres of the cerebellum, we have shown experimentally that the parallel fibers do not produce beams at all (Bower and Woolston, 1983). Instead, there is a strong vertical organization of Purkinje cell responses. Only those Purkinje cells immediately overlying an activated region of the granule cell layer demonstrate any response to tactile stimulation. No beams of Purkinje cell activity are ever produced.

While this might at first seem like a technical detail, the implications for cerebellar function are enormous. Almost all functional ideas concerning cerebellar circuitry even to this day assume that parallel fibers activate beams of Purkinje cells (cf. Eccles, 1973; Ito, 1996; Houk and Miller, 1995; Thach, 1996; Braitenberg, 1993; Chauvet, 1995; Strehler, 1990; Tyrrell and Willshaw, 1992). This includes the highly influential Marr–Albus cerebellar motor-learning hypothesis (Marr, 1969; Albus, 1971), which continues to have a substantial influence on cerebellar theorists (Killeen and Fetterman, 1993; Fujita, 1982; Kawato and Gomi, 1992a) and experimentalists (Ito, 1989; Ito, 1996; Thompson, 1988). The fact that focal activation of the granule cell layer does not produce beams required us to completely rethink the kind of computation that could be performed by the cerebellar cortex.

B. Dual Role for Granule Cells in the Cerebellar Cortex

Following the initial description of these results (Bower et al., 1980), Llinas (1984) was the first to propose that this vertical influence might be a result of activation of synapses associated with granule cell axons as they ascend into the molecular layer past the Purkinje cell dendrite. Motivated by this idea and previous results (Bower and Woolston, 1983), it was subsequently demonstrated that, as expected, the ascending branch synapses have a substantial physiological effect on overlying Purkinje cells (Jaeger and Bower, 1994). Further, electron microscopic level anatomical investigations in the author's laboratory have shown that these synapses exist in abundance, are ultramorphologically distinguishable from synapses on the parallel fiber axonal branch, and terminate on a different region of the Purkinje cell dendrite (Gundappa-Sulur and Bower, 1990).

Based on these anatomical and physiological studies, our view of cerebellar circuitry is quite different from that usually included in textbook descriptions. As shown in Fig. 2, we believe that descriptions of these circuits must include the synaptic contacts made by the ascending branch of the granule cell layer. Although it is beyond the scope of this chapter, research also suggests a much more prominent role for the inhibitory neurons within the cortex in cerebellar function. However, the most important point here is that we now believe that Purkinje cells receive two distinctly different types of granule cell axon influences: one from the ascending branch and one from the parallel fiber system. The ascending branch synapses appear to be responsible for most, if not all, of the classically recorded stimulus-locked Purkinje cell responses to afferent mossy fiber input. The parallel fibers, however, would appear to be more modulatory in nature. This proposal has been supported by modeling efforts which have demonstrated, using "realistic" numerical simulations of single Purkinje cells (De Schutter and Bower, 1994; Jaeger and Bower, 1997), that characteristic Purkinje cell responses to presumably ascending branch input are significantly modulated by the background spontaneous firing of parallel fibers and stellate cell synapses. Thus, in the model, it appears as if the background synaptic input of parallel and stellate cell synapses controls the physiological "state" of the Purkinje cell dendrite. Further, the variations in responses quite closely resemble experimentally observed trial-by-trial Purkinje cell response variability in response to peripheral tactile stimuli (Bower and Woolston, 1983). In summary, these results suggest that instead of driving Purkinje cell outputs directly, the parallel fiber system may instead modulate the responsive state of the Purkinje cell dendrite to the synaptic input received via the ascending branch of the granule cell axon.

C. CONTROL OF TACTILE DATA ACQUISITION: PUTTING IT ALL TOGETHER

Combining the pattern of tactile projections to the granule cell layers with our current interpretation of the physiological organization of cerebellar cortical circuitry, a fundamentally new view of the computation performed by this circuitry emerges. Through the modulatory influence of the parallel fibers, we would expect that the "dendritic state" of Purkinje cells would, at any one time, reflect sensory information arising from a large number of different tactile surfaces distributed over a considerable extent of cerebellar cortex. This information in turn is predicted to modulate the ascending branch-mediated response to each particular Purkinje cell to information originating in the specific region of the rat's face represented in the immediately underlying granule cell layer. In this way, the response of each Purkinje cell to its specific input should reflect the information arising from a diverse set of tactile surfaces. Generalizing, this analysis suggests that these regions of the cerebellar cortex as a whole may, in effect, place information arising from any particular tactile sensory surface in the larger sensory context provided by the other tactile sensory surfaces represented. Functionally the question then becomes, "for what purpose?"

The key to answering this question lies in an examination of the particular regions of the body surface represented in these maps. As already described, the cerebellar tactile representation in rats is largely limited to particular regions of the perioral surfaces of this animal. Figure 3, however, illustrates that not all tactile surfaces are represented. Instead, the represented regions correspond to those areas that are most in contact with peripheral stimuli during active tactile sensory exploration (Jacquin and Zeigler, 1983; Zeigler et al., 1984; Welker, 1964), suggesting, in turn, that this region of the cerebellum is involved in coordinating the use of these sensory surfaces during active sensory exploration. In keeping with this idea, Hartmann and Bower (1995) have shown that these regions of the rat's granule cell layer are, in fact, highly active during active tactile explora-

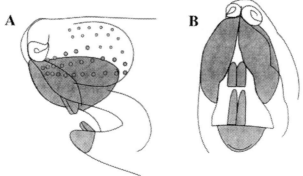

FIG. 3. The perioral regions of the rat consistently represented in folium crus IIA of the cerebellar hemispheres. Not all regions of the face are represented. Those regions that are represented are also those regions known to be in primary contact with objects during tactile sensory exploration.

tion and are not active at all during the stereotyped motor behavior involved in simple chewing.

D. RAT MOUTHS, CAT PAWS, AND PRIMATE FINGERS

Up until this point, this chapter has only considered original data from rats. However, our interpretation of the significance of the rat maps allows us to make predictions concerning the expected representation of tactile surfaces in other mammalian species. Specifically, animals that use their mouths for tactile sensory exploration should, like the rat, have predominant representations from perioral regions. As already described, this is the case in both the guinea pig and the mouse (Bower, 1997). At the same time, mammals like cats and primates that use their forelimbs more predominantly during sensory exploration should have tactile cerebellar representations dominated by these tactile sensory structures. In this regard, it has been known for some time that distal regions of the forelimb provide a large tactile representation in the lateral hemispheres of both cats (Snider and Stowell, 1944) and primates (Snider and Eldred, 1952). Welker's more recent detailed "micro-maps" of lateral granule cell regions confirms these results for cats (Kassel *et al.*, 1984) as well as for the primate *Galago crassicaudatus* (Welker *et al.*, 1988). These maps also demonstrate that tactile maps in both species are fractured. Further, these mapping studies failed to find representations from the hindlimb in either animal (Kassel *et al.*, 1984; Welker *et al.*, 1988). Although the lack of hindlimb representation is difficult to reconcile with a role for these regions of the cerebellum in the control of locomotion, the dominant forelimb representation is completely consistent with the hypothesis that the lateral regions of the cerebellar hemispheres are concerned with controlling the use of these surfaces in tactile sensory discrimination given that both cats and primates use their forelimbs for this purpose. Consistent with this interpretation, the kinds of tasks that monkeys have been asked to perform in cerebellar recording experiments are increasingly like sensory discrimination tasks, although they are still interpreted in the context of traditional movement control (Sinkjaer *et al.*, 1995; Robinson, 1995; Vankan *et al.*, 1994; Houk and Gibson, 1987; Dugas and Smith, 1992).

E. TACTILE SENSORY DATA ACQUISITION IN HUMANS

While the author believes that current experimental evidence is most consistent with a role of the lateral hemispheres of the cerebellum in the control of sensory acquisition using the motor system, the fact that the motor system is involved in tactile sensory data acquisition makes it difficult to distinguish unambiguously between the control of motor behavior using sensory information and the control of sensory information using motor behavior. This is especially true in experimental animals. However, for this reason, Gao et al. (1996) conducted a series of investigations using human subjects and functional magnetic resonance imaging (fMRI) to specifically test our tactile sensory data acquisition hypothesis. Although it has previously been reported with positron emission tomography (PET) studies that the cerebellum is somehow involved in the control of finger movements during tactile exploration (Seitz et al., 1991), the specific intention was to design finger-related tasks that, as far as possible, separated the simple control of finger movements from the use of the fingers as sensory devices. The specific intention was to test the hypothesis that lateral hemispheres were involved in the latter and not the former.

The measure chosen for the involvement of the lateral cerebellar hemispheres was the level of fMRI detectable activity in the lateral (dentate) nucleus. The dentate nucleus provides the sole output for the large lateral hemispheres of the primate cerebellum that are homologous to the region of tactile representation previously studied in rats (Larsell, 1972) and is assumed to reflect the activity of this entire cerebellar region. In order to distinguish between sensory and motor control, human subjects were asked to perform four specific tasks involving both passive and active sensory discrimination. The tasks were designed to test four specific predictions of the data acquisition hypothesis: (1) The dentate nucleus should show increased levels of activity following tactile sensory stimulation even when there are no accompanying overt finger movements; (2) finger movements not associated with tactile sensory discrimination should not induce substantial dentate activity; (3) the requirement to make a sensory discrimination with the fingers should induce an increase in dentate response, with or without accompanying finger movements; and (4) the dentate should be most strongly activated when there is the most opportunity to modulate the acquisition of the sensory data, i.e., when the human subjects could actively reposition tactile sensory surfaces through finger movements as part of the discrimination task. It is important to point out that these tasks were designed *before* the experiments themselves were conducted, specifically to test the hypothesis that is the subject of this chapter. It also should be noted that the predicted lack of dentate activity with fine finger movements alone is completely the opposite of what would be expected if existing theories of cerebellar motor control were correct.

The results of these experiments are discussed in Parsons and Fox (this volume). However, in brief, each of the predictions of the sensory data acquisition hypothesis were confirmed in these experiments. Activation of the dentate nuclues was most profound during sensory discrimination tasks, with or without movement. Somewhat to our surprise, there was absolutely no response above baseline levels with fine finger movements alone. Again, it is important to point out that these patterns of activity were predicted prior to the performance of the experiments themselves.

III. What Is the Cerebellum Controlling?

Based on this analysis of cerebellar circuitry and its afferent projections, we proposed that the lateral regions of the mammalian cerebellum are involved in the acquisition of tactile sensory information (Bower and Kassel, 1990). At the outset, it is obviously the case that the nervous system controls the form and nature of the tactile sensory data it receives. The most obvious form of control involves directing the fingers to the objects to be explored. This is the "reaching" studied in classic motor control. The author proposes, however, that the cerebellum operates at a much finer level of sensory acquisition control: specifically, in the fine positioning of multiple sensory receptor surfaces and/or the fine coordination of the position of single receptor surfaces based on data from multiple sensory modalities once contact is made with the surface being explored and once data are flowing into the nervous system. The author contends that such touching specifically requires active control of data being obtained from all the surfaces in contact with a particular object. Further, the use of any one surface almost certainly needs to be coordinated with the use of the other surfaces involved. In this regard, clear evidence from human psychophysical experiments shows that complex tactile descrimination tasks are performed based on sensory data obtained from multiple fingers (Evans and Craig, 1991; Klatzky *et al.*, 1993). In the context of rat tactile data acquisition specifically, Hartmann and Bower (1995), as well as previous studies (Jacquin and Zeigler, 1983), clearly demonstrate that rats maintain direct contact with objects at numerous perioral locations during sensory exploration. These are the locations represented in the cerebellum of the rat. Hartmann and Bower (1995) have shown that these cerebellar regions are specifically active during such exploratory activity.

A. At What Level Is the Cerebellum Operating?

Tactile receptors, like most neural receptors, operate at extreme levels of sensitivity (Cohen and Vierck, 1993; Edin *et al.*, 1995). For example, single rat whiskers can detect movements as small as a few tenths of a degree angular displacement (Gibson and Welker, 1983). This means that even slight changes in the position of sensory surfaces with respect to the objects they contact will produce substantial changes in the signals being sent to the nervous system (Cohen and Vierck, 1993). At the same time, the considerable variability found in the structure of real objects, especially at this level of resolution, means that the nervous system cannot rely on stereotyped tactile exploratory procedures to coordinate the multiple sensory surfaces involved. Instead, the author proposes that the cerebellum is responsible for monitoring incoming sensory data from these surfaces and adjusting their positions relative to each other and relative to the object being explored, in very close to real time. This should take place at a temporal and and spatial resolution appropriate to receptor sensitivities. In practice, we suspect this means in the range of tens to hundreds of milliseconds and involves movements on the scale of tens to a few hundred microns.

B. ALGORITHMICALLY, WHAT IS NECESSARY?

Algorithmically, a critical requirement at this intermediate level of sensory data acquisition control is that data arising from specific peripheral locations be interpreted in the context of data being acquired by other tactile surfaces involved in a particular episode of data acquisition. In this context, the fractured somatotopic pattern of tactile projections to the granule cell layer is seen to reflect the intermixing of tactile sensory data from the multiple sensory surfaces normally used by the rat for tactile sensory exploration. The representations in the cat and monkey can similarly be interpreted in this fashion. Within cerebellar cortical circuitry itself, the proposed dual role of the granule cell axon (Bower and Woolston, 1983) seems ideally suited to evaluate data arising from any one surface (via the ascending branch) in the context of data arising from the other surfaces involved (via the balance between the parallel fibers and the inhibitory input from the stellate cells).

C. WHAT ABOUT CEREBELLAR OUTPUTS?

With respect to the outputs of these regions of the cerebellar cortex, this hypothesis would predict that they would specifically influence motor centers which directly control the positions of tactile sensory surfaces. In the rat, for example, the hemispheres should have a more substantial influence on the facial motor nucleus which directly controls the location of the tactile receptors of the skin (Terashima *et al.,* 1993) than the trigeminal motor nucleus which controls opening and closing of the jaw (Westberg *et al.,* 1995). In the cat, Ekerot *et al.* (1995) compared tactile receptive fields to the movements invoked by cerebellar microstimulation in the same locations and demonstrated that evoked movements "act to withdraw the area of the skin corresponding to the . . . receptive field." This is completely in keeping with the idea that tactile regions of the cerebellum are concerned with controlling the contact being made by sensory surfaces.

In addition to the site and influence of cerebellar output, this hypothesis predicts that the cerebellar output "code" should reflect small position adjustment signals regardless of exact limb or finger location. In other words, cerebellar outputs should reflect instructions for relative, not absolute, changes in mouth, finger, or limb position. Several studies in awake behaving monkeys have suggested that this may be case (Vankan *et al.,* 1994; Houk and Gibson, 1987; Shidara *et al.,* 1993). Although these results are still interpreted in the context of traditional motor control, the author believes the "relative" and largely directionless cerebellar position signals that have been described are more compatible with the idea that the cerebellum is concerned with the kinds of small relative changes in the position of fingers and limbs that have a substantial effect on data being acquired by the sensory surfaces on these structure, but that would have little influence on larger scale movements.

D. WHAT ARE THE PREDICTED BEHAVIORAL EFFECTS OF CEREBELLAR INVOLVEMENT?

The principal behaviorally measurable consequence of this cerebellar involvement is expected to be the ongoing, potentially quite subtle, repositioning of tactile sensory surfaces to facilitate the acquisition of sensory data across all the sensory surfaces involved. During tactile exploration, this control is expected to be exerted through the modulation of ongoing "central" control signals to motorneurons. Internally, this cerebellar control should be evident as an enhanced coordination of sensory data obtained from different tactile surfaces. This coordination should be adaptive to the nature of the object being explored as well as to the expected use of data by other regions of the brain.

Under this hypothesis, the principal functional consequence of cerebellar computation is expected to be in the efficiency with which other somatosensory regions of the brain process tactile information. In this way, the cerebellum is seen as supporting the processing capabilities of the rest of the brain without contributing directly to those computations. This support should be particularly apparent during the performance of complex tactile discrimination tasks, as they are assumed to require better organized and coordinated data. However, cerebellar activity should be less apparent during the performance of simpler tasks or tasks that have been well learned previously. In other words, the cerebellum should be more active as tasks become more difficult and less active as tasks become rote. This prediction is consistent with several of the new imaging and psychophysical studies (Bower, 1995) and is the opposite of what would be expected from classical theories of cerebellar motor control (see later).

E. How Does This Differ from Classical Motor Control?

As already stated, for the somatosensory system, cerebellar control of sensory data acquisition is expected to be exerted through the use of muscles to finely control the relative positions of different sensory surfaces. Thus, in the case of tactile exploration (although not in all sensory systems, see later), it could be said that the cerebellum is still involved in some form of motor control. However, the type of control described here and its consequences are completely different than those previously proposed for these cerebellar regions (cf. Kawato and Gomi, 1992b, Horne and Butler, 1995; Thach *et al.*, 1992; Gilbert, 1974; Ito, 1993; Houk and Gibson, 1987). Specifically, in most previous theoretical and experimental work, the tactile responsiveness of the lateral cerebellar hemispheres has been interpreted as providing an *indirect* measure of the accuracy of motor performance (Kawato and Gomi, 1992b, Robinson, 1995; Lou and Bloedel, 1992; Horne and Butler, 1995), whereas cerebellar influences on the motor system were interpreted as a *direct* means of controlling the smoothness or accuracy of intended limb or finger movements (Kawato and Gomi, 1992b, Schwartz *et al.*, 1987; Lou and Bloedel, 1992; Houk and Miller, 1995; Horne and Butler, 1995). In contrast, under the data acquisition hypothesis, sensory data projecting to the cerebellar cortex are seen as providing a *direct* sample of the quality of sensory data being obtained by various sensory surfaces through exploratory movements. The influence of the cerebellar cortex on the motor system is seen as an *indirect* means of controlling data being obtained by these tactile sensory surfaces.

IV. Implications for Cerebellar Function as a Whole

Although it is beyond the scope of this chapter, this same interpretation can be applied to other "classical" cerebellar-involved systems including the control of eye movements, balance, and even the skeletal muscle system. Each are either dependent on sensory data (the skeletal muscle system) or specifically involved in stabalizing (vestibular) or acquiring (vision) sensory data. However, of particular relevence to this volume, this new hypothesis has a number of important implications for the interpretation of the imaging studies which many of the claims of cerebellar "cognitive" involvement are increasingly based.

A. THE CEREBELLUM AS A SUPPORT DEVICE

Perhaps the most important single implication of this hypothesis is that the cerebellum provides a support function for the rest of the nervous system, but does not itself directly participate in their computations. In this sense the current hypothesis is consistent with the general idea that the cerebellum is in some sense "metasystemic" with respect to the rest of the nervous system (MacKay and Murphy, 1979). It extends the idea considerably, however, by suggesting that the cerebellum has no direct responsibility for any behavior, including motor coordination or cognition, but instead facilitates the computational efficiency of a large number of other neural systems by supervising the acquisition of data on which these other systems and their functions depend.

While the idea that the cerebellar influence on behavior is indirect has been present in the literature for many years (cf. Holmes, 1917), cerebellar lesion, physiology, and imaging experiments continue to be interpreted as if the cerebellum played a specific role in whatever particular behavior is being studied. For example, if a lesion causes motor discoordination, the cerebellum is a motor control device (Bastian and Thach, 1995). If the cerebellum is activated during a task involving timing, it is involved in the control of timing, (Jueptner et al., 1995). If imaging and cerebellar damage are correlated with attention deficits, then the cerebellum is involved in coordinating shifts in attention (Courchesne et al., 1994). If cerebellar lesions produce deficits in classical conditioning, then the cerebellum is responsible for classical conditioning (Thompson, 1988).

In contrast to these straightforward relations between cause and effect, the data acquisition theory of cerebellar function suggests that the relationship between the goal-directed behavior of the animal, be that movement, attention, or perception, and the actual calculations performed in the cerebellum is at best indirect. A simple analogy might be a car radiator. The radiator does not contribute directly to the generation of the force necessary for the movement of the car down the road, although it appears to respond internally (flow of coolant, temperature) to the rate of that movement. Instead, the radiator maintains conditions under which the car's engine and other parts can perform their functions more efficiently. If the radiator malfunctions, or is removed, a wide range of mechanical consequences can be expected from disrupted car movement, to the malfunction of the air conditioning system and, eventually, the headlights. No one would claim, however, that these disrupted functions are directly reflected in the structure of the radiator or in its specific functional role in the car.

B. The Cerebellum Is Differentially Engaged

This hypothesis leads to the prediction that different sensory tasks will activate different cerebellar regions according to that region's afferent and efferent connections (see L. M. Parsons and P. T. Fox, this volume). It also suggests, however, that the extent of activation of any particular region should be dependent on the quality of sensory data needed by the rest of the brain. This is especially true for those regions of the cerebellum involved in active sensory acquisition (vermis and hemispheres). For example, relatively simple behaviors, requiring less accurate data, should not induce much cerebellar response. Similarly, on the assumption that learning involves some ability to generalize over sensory stimuli, behaviors that are well learned should not induce as much cerebellar responsiveness. However, during learning, or generally as task complexity increases, so should the level of cerebellar activity. As pointed out previously (Bower, 1995), the results of a number of cerebellar imaging studies already seem to support this prediction (cf. Jenkins et al., 1994; Raichle et al., 1994). There is also evidence in the fusimotor system that the sensitivity of muscle spindles is differentially regulated depending on the difficulty or unfamiliarity of the motor task (Hulliger, 1993). Given the influence of the cerebellum on pathways regulating the sensitivity of muscle spindle afferents, this theory predicts that much of this regulation will derive from the cerebellar influence.

C. Overall Cerebellar Activity May Be under Cerebral Cortical Control

If the degree of involvement of the cerebellar vermis and hemispheres is being regulated, it seems quite likely that the cerebral cortex plays a major role in that regulation (Morissette and Bower, 1996). After all, for many sensory systems, the cerebral cortex is the structure that makes the most sophisticated use of data. Although it is beyond the scope of this chapter, the proposed regional differences in the kinds of sensory data acquisition under cerebellar control allow interesting interpretations for regional differences in the patterns of projection from the cerebral cortex (Bower, 1997).

This proposed involvement of the cerebral cortex in the regulation of cerebellar activity has particularly important implications for the interpretation of data from imaging studies. For example, Gao et al. (1996) demonstrated that simply changing the significance of a tactile stimulus to the subject changed the level of dentate activation. If the cerebral cortex is regulating the level of cerebellar activation, it may be an error to assume that this means the cerebellum is involved directly in "cognitive functions," such as attention (Courchesne et al., 1994) or even imagined movements (Fiez et al., 1996; Ryding et al., 1993; Decety et al., 1990; Kim et al., 1994). Instead, increases in cerebellar activity may simply reflect an increased anticipated demand by some other region of the nervous system to monitor the quality of sensory data being obtained.

D. The Cerebellum Is Useful but Not Necessary

It is the view of the author that the role of the cerebellum proposed here has particularly important implications for the interpretation of cerebellar lesion data. Specifically, in most brain studies, an effect of a lesion on a particular behavior is taken as evidence that the lesioned structure has some direct responsibility for that behavior. It should be clear that if the current hypothesis is correct, this will not be the case with the cerebellum.

Although space does not allow even a cursory review of the enormous cerebellar lesion literature, several specific examples are worth considering. One particularly influential example of one-to-one lesion interpretations is a set of studies proposing a role for the cerebellum in associative learning (Thompson, 1988). Several authors have reported that lesions placed in specific cerebellar locations interfere with the ability of rabbits to associate conditioned stimuli with eye blinks. Although serious methodological concerns have been raised with respect to at least some of these experiments (cf. Bloedel and Bracha, 1995), the current concern is that none of these experiments have taken into account the possibility that the sensory data on which the classical conditioning is dependent have changed. If cerebellar lesions do disrupt the structure of sensory data, then it is not possible to draw conclusions about the location of any "memory trace" from these experiments. It should also be noted that cerebellar-related learning deficits due to the disruption of sensory inputs should be most noticeable in simple learning circuits or those that have to operate rapidly and therefore do not have the capacity to reorganize poorer data.

A role for the cerebellum in sensory data control also has important consequences for the large literature describing motor-related deficits following cerebellar lesions (Thach, 1996). If the cerebellum does coordinate the acquisition of proprioceptive information, then motor dysfunction following cerebellar lesions is more likely to be due to a disruption in the quality of data on which the motor system depends than on a direct contribution of the cerebellum to motor coordination. Some evidence suggests that this may be the case (Horak and Diener, 1994). Other studies of cerebellar-related "motor deficits" should carefully distinguish between deficits in sensory data and motor control. Parenthetically, the author points out that the fact that cerebellar removal impairs the ability of the visual system to adjust the gain of the vestibulo-occular reflex has never led to the suggestion that the cerebellum is itself directly involved in visual processing.

E. COMPENSATING FOR THE ABSENCE OF THE CEREBELLUM

Finally, the author suspects that one of the reasons for the diversity of theories of cerebellar function is that the consequences of cerebellar lesions are not as straightforward as for many other regions of the vertebrate brain. For example, although the removal of the visual cortex irreversibly and profoundly affects the ability of an animal to see, cerebellar lesions are often followed by a considerable recovery of general function, especially in young animals (cf. Schade and Ford, 1973; Berridge and Whishaw, 1992; Molinari and Petrosini, 1993). It appears as though the rest of the nervous system, including the motor system, compensates in some way for the absence of the cerebellum.

At least in the context of motor behavior, most would probably agree that the recovery of function following cerebellar lesions reflects adjustments in other brain regions (Molinari and Petrosini, 1993; Haggard et al., 1994). However, the uniqueness of the cerebellar cortical circuitry makes it unlikely that other structures actually replace cerebellar function. Instead, recovery must reflect the adjustment made in the absence of a functioning cerebellum. To return to the car radiator analogy, a driver realizing the car is overheating may attempt to prolong the length of time the engine will continue to run by slowing the car down. In this case, the driver is adjusting for the absence of the radiator, not replacing its function. Similarly, the current hypothesis specifically suggests that changes in other brain structures following cerebellar lesions reflect adaptations to the lack of well-controlled sensory data rather than replacement of the function normally provided by the cerebellum.

This interpretation of cerebellar lesion data leads to several specific predications. First, those behaviors requiring the highest quality sensory information should have more difficulty recovering from cerebellar lesions. For this reason, animals without a cerebellum may make behavioral modifications to avoid performing the task or, alternatively, may break a complex task down into a series of simpler ones that still complete the task. If the lesion in the cerebellum is restricted, we would expect animals to modify their behavior to use those sensory surfaces still under cerebellar control. Some evidence shows that reorganizations of this sort do take place in humans with cerebellar disease (Wessel et al., 1995; Haggard et al., 1994).

The second prediction that can be made is that, even with adjustments, the time it takes to do a wide variety of tasks should increase. Abundant evidence exists that cerebellar patients perform motor tasks more slowly (Wild and Dichgans, 1992; Hallett et al., 1991; Haggard et al., 1994; Muller and Dichgans, 1994; Wessel et al., 1994; Jahanshahi et al., 1993). However, this result is usually attributed to the disruption of generalized motor function. In the context of the current hypothesis, this slow down in function is predicted to result instead from the longer computing time necessary to process poorly controlled sensory data.

If this conjecture is correct, and if the cerebellum is providing sensory data control for a diverse set of sensory systems, then increased performance latencies should be observable in a wide range of sensory-related tasks (Keele and Ivry, 1990b). In an evoked potential study of cerebellar patients performing voluntary arm movements, Wessel et al. (1994) found that the delay in performance occurred specifically during the premotor phase of the task. These authors concluded that cerebellar patients "may try to compensate for their motor deficits by a longer cortical activation preceding voluntary movements." In the context of the current hypothesis, a more straightforward (causal) conclusion would be that with less well-controlled proprioceptive data, it takes longer for the motor cortex to organize the next movement. This speculation applies directly to another published evoked potential study of human cerebellar patients performing a semantic word recognition task (Tachibana et al., 1995). These authors specifically conclude that "these patient's impairment in cognitive information processing arose from the difficulty in pattern recognition of the stimuli (presented)." If the current hypothesis is correct, the careful examination of a wide range of goal-directed behaviors using evoked potentials should demonstrate a general slowing down in execution time attributable specifically to delays in the processing of sensory information.

For many discriminations and behaviors, successful completion of the task should occur regardless of the slow down due to the lack of good sensory data. However, for those tasks that specifically involve timing, any slow down in processing time would be expected to interfere with task completion itself. For this reason, this theory predicts that cerebellar dysfunction should have a particularly noticeable effect on discriminations of behavior specifically involving timing. Ample evidence already demonstrates that this is the case (Jueptner et al., 1995; Lundyekman et al., 1991; Keele and Ivry, 1990). However, once again, these data have been interpreted by some to suggest that the cerebellum itself contributes to timing functions (Keele and Ivry, 1990).

Finally, several imaging studies in humans have suggested a cerebellar involvement in autism (Courchesne et al., 1994; see also Filipek, 1995). Although there is no reasonable way to accommodate such an influence given classical theories of motor coordination, an association between autism and the cerebellum is completely natural for the current theory. It has been suggested many times that at least some forms of autism may be related to the inability of a child to deal with sensory data. Autistic children sometimes report that the world is simply too hard to predict or that sensory data are overwhelming or confusing. In the context of the hypothesis proposed in this chapter, cerebellar dysfunction could contribute to such confusion, especially if a dysfunctioning (as compared to a missing) cerebellum results in inconsistent or variable control over sensory data acquisition.

V. Conclusion

As described in the beginning of this chapter, the hypothesis that the cerebellum is involved in the control of sensory data acquisition specifically arose from a consideration of the physiological organization of cerebellar cortical circuitry and its afferent projections. Perhaps one of the most important long-standing questions regarding that circuitry is whether different regions of the cerebellum perform the same computation, as would seem to be implied by the uniformity of the circuit, or many different functions, as implied by regional variations in afferent and efferent projections. The diversity of lesion effects and PET and fMRI activations have led some authors to suggest that different regions of the cerebellum may, in fact, be performing distinct different functions (cf. Paulin, 1993; Bloedel, 1992).

It is the authors view, on principle, that structural uniformity requires computational uniformity. Furthermore, if the cerebellum supports but does not directly participate in the processing in many brain regions, then it is entirely possible that cerebellar circuitry everywhere performs the same computation, with variations only in the particular sensory surfaces involved and their afferent and efferent mappings. Further, it seems at least plausible that the kind of context-dependent evaluation of sensory data proposed for the tactile regions of the cerebellum may be generally applicable to the evaluation and control of data from most sensory systems. In any event, it is clear that the ultimate proof of any brain theory will lie in understanding how function is imbedded in structure. For this reason we will continue to use both modeling and experimental techniques to pursue the functional organization of cerebellar circuits.

References

Albus, J. S. (1971). A theory of cerebellar function. *Math. Biosci.* **10,** 25–61.
Bastian, A. J., and Thach, W. T. (1995). Cerebellar outflow lesions: A comparison of movement deficits resulting from lesions at the levels of the cerebellum and thalamus. *Ann. Neurol.* **38,** 881–892.
Berridge, K. C., and Whishaw, I. Q. (1992). Cortex, striatum and cerebellum: Control of serial order in a grooming sequence. *Exp. Brain Res.* **90,** 275–290.
Bloedel, J. R. (1992). Functional-heterogeneity with structural homogeneity: How does the cerebellum operate. *Behav. Brain Sci.* **15,** 666–678.
Bloedel, J. R., and Bracha, V. (1995). On the cerebellum, cutaneomuscular reflexes, movement control and the elusive engrams of memory. *Behav. Brain Res.* **68,** 1–44.
Bower, J. M. (1995). The cerebellum as a sensory acquisition controller. *Hum. Brain Mapp.* **2,** 255–256.
Bower, J. M. (1996). Perhaps it is time to completely rethink cerebellar function. Commentary on: Controversies in Neuroscience IV: Motor learning and synaptic plasticity in the cerebellum. *Behav. Brain Sci.* **19,** 438–439.
Bower, J. M. (1997). Is the cerebellum sensory for motor's sake, of motor for sensory's sake: The view from the whiskers of a rat? *In* "Progress in Brain Research" (C. I. de Zeeuw, P. Strata, and J. Voogd, eds.), pp. 483–516.
Bower, J. M., Beerman, D. H., Gibson, J. M., Shambes, G. M., and Welker, W. (1981). Principles of organization of a cerebro-cerebellar circuit: Micromapping the projections from cerebral (S1) to cerebellar (granule cell layer) tactile areas of rats. *Brain Behav. Evol.* **18,** 1–18.
Bower, J. M., and Kassel, J. (1990). Variability in tactile projection patterns to cerebellar folia crus IIa of the Norway rat. *J. Comp. Neurol.* **302,** 768–778.
Bower, J. M., and Woolston, D. C. (1983). Congruence of spatial organization of tactile projections to granule cell and Purkinje cell layers of cerebellar hemispheres of the albino rat: Vertical organization of cerebellar cortex. *J. Neurophysiol.* **49,** 745–766.
Bower, J. M., Woolston, D. C., and Gibson, J. M. (1980). Congruence of spatial patterns of receptive field projections to Purkinje cell and granule cell layers in the cerebellar hemispheres of the rat. *Soc. Neurosci. Abst.* **6,** 511.
Braitenberg, V. (1993). The cerebellar network: Attempt at a formalization of its structure. *Network* **4,** 11–17.

Braitenberg, V., and Atwood, R. P. (1958). Morphological observations on the cerebellar cortex. *J. Comp. Neurol.* **109,** 1–33.

Chauvet, G. A. (1995). On associative motor learning by the cerebellar cortex: From Purkinje unit to network with variational learning rules. *Math. Biosci.* **126,** 41–79.

Cohen, R. H., and Vierck, C. J. (1993). Population estimates for responses of cutaneous mechanoreceptors to a vertically indenting probe on the glabrous skin of monkeys. *Exp. Brain Res.* **94,** 105–119.

Courchesne, E., Townsend, J., Akshoomoff, N. A., Saitoh, O., Yeung-Courchesne, R., Lincoln, A. J., James, H. E., Haas, R. H., Schreibman, L., and Lau, L. (1994). Impairment in shifting attention in autistic and cerebellar patients. *Behav. Neurosci.* **108,** 848–865.

Decety, J., Sjoholm, H., Ryding, E., Stenberg, G., and Ingvar, D. H. (1990). The cerebellum participates in mental activity: Tomographic measurements of regional cerebral blood-flow. *Brain Res.* **535,** 313–317.

De Schutter, E., and Bower, J. M. (1994). Responses of cerebellar Purkinje cells are independent of the dendritic location of granule cell synaptic inputs. *Proc. Natl. Acad. Sci. USA* **91,** 4736–4740.

Dugas, C., and Smith, A. M. (1992). Responses of cerebellar Purkinje cells to slip of a hand-held object. *J. Neurophysiol.* **67,** 483–495.

Eccles, J. C. (1973). The cerebellum as a computer: Patterns in space and time. *J. Physiol. Lond.* **229,** 1–32.

Edin, B. B., Essick, G. K., Trulsson, M., and Olsson, K. A. (1995). Receptor encoding of moving tactile stimuli in humans. 1. Temporal pattern of discharge of individual low-threshold mechanoreceptors. *J. Neurosci.* **15,** 830–847.

Ekerot, C. F., Jorntell, H., and Garwicz, M. (1995). Functional relation between corticonuclear input and movements evoked on microstimulation in cerebellar nucleus interpositus anterior in the cat. *Exp. Brain Res.* **106,** 365–376.

Evans, P. M., and Craig, J. C. (1991). Tactile attention and the perception of moving tactile stimuli. *Percept. Psychophys.* **49,** 355–364.

Fiez, J. A., Raife, E. A., Balota, D. A., Schwarz, J. P., Raichle, M. E., and Petersen, S. E. (1996). A positron emission tomography study of the short-term maintenance of verbal information. *J. Neurosci.* **16,** 808–822.

Filipek, P. A. (1995). Quantitative magnetic-resonance-imaging in autism: The cerebellar vermis. *Curr. Opin. Neurol.* **8,** 134–138.

Fujita, M. (1982). Adaptive filter model of the cerebellum. *Biol. Cybern.* **45,** 195–206.

Gabbiani, F., Midtgaard, J., and Knîpfel, T. (1994). Synaptic integration in a model of cerebellar granule cells. *J. Neurophysiol.* **72,** 999–1009.

Gao, J. H., Parsons, L. M., Bower, J. M., Xiong, J., Li, J., and Fox, P. T. (1996). Cerebellum implicated in sensory acquisition and discrimination rather than motor control. *Science* **272,** 545–547.

Gibson, J. M., and Welker, W. I. (1983). Quantitative studies of stimulus coding in first-order vibrissae afferents of rats. I. Receptive field properties and threshold distributions. *Somatosens. Res.* **1,** 51–67.

Gilbert, P. F. C. (1974). A theory of memory that explains the function and structure of the cerebellum. *Brain Res.* **70,** 1–18.

Gonzalez, L., Shumway, C., Morissette, J., and Bower, J. M. (1993). Developmental plasticity in cerebellar tactile maps: Fractured maps retain a fractured organization. *J. Comp. Neurol.* **332,** 487–498.

Gundappa-Sulur, G., and Bower, J. M. (1990). Differences in ultramorphology and dendritic termination sites of synapses associated with the ascending and parallel fiber segments of granule cell axons in the cerebellar cortex of the albino rat. *Soc. Neurosci. Abst.* **16,** 896.

Haggard, P., Jenner, J., and Wing, A. (1994). Coordination of aimed movements in a case of unilateral cerebellar damage. *Neuropsychologia* **32**, 827–846.

Hallett, M., Berardelli, A., Matheson, J., Rothwell, J., and Marsden, C. D. (1991). Physiological analysis of simple rapid movements in patients with cerebellar deficits. *J. Neurol. Neurosurg. Psychiat.* **54**, 124–133.

Hartmann, M. J., and Bower, J. M. (1995). Cerebellar granule cell responses in the awake, freely-moving rat: The importance of tactile input. *Soc. Neurosci. Abstr.* **21**, 1910.

Holmes, G. (1917). The symptoms of acute cerebellar injuries due to gunshot wounds. *Brain* **40**, 461–535.

Holmes, G. (1939). The cerebellum of man. *Brain* **62**, 1–30.

Horak, F. B., and Diener, H. C. (1994). Cerebellar control of postural scaling and central set in stance. *J. Neurophysiol.* **72**, 479–493.

Horne, M. K., and Butler, E. G. (1995). The role of the cerebello-thalamo-cortical pathway in skilled movement. *Progr. Neurobiol.* **46**, 199–213.

Houk, J. C., and Gibson, A. R. (1987). Sensorimotor processing through the cerebellum. In "New Concepts in Cerebellar Neurobiology" (J. S. King, ed.), pp. 387–416. A. R. Liss, New York.

Houk, J. C., and Miller, L. E. (1995). Neural mechanisms for the generation of limb motor programs. *J. Physiol. Lond.* **487P**, P225–P226.

Hulliger, M. (1993). Fusimotor control of proprioceptive feedback during locomotion and balancing: Can simple lessons be learned for artificial control of gait. *Progr. Brain Res.* **97**, 173–180.

Ito, M. (1989). Long-term depression. *Annu. Rev. Neurosci.* **12**, 85–102.

Ito, M. (1993). Movement and thought: Identical control mechanisms by the cerebellum. *Trends Neurosci.* **16**, 448–450.

Ito, M. (1996). Cerebellar long-term depression. *Trends Neurosci.* **19**, 11–12.

Jacquin, M. F., and Zeigler, P. H. (1983). Trigeminal orosensation and ingestive behavior in the rat. *Behav. Neurosci.* **97**, 62–97.

Jaeger, D., and Bower, J. M. (1994). Prolonged responses in rat cerebellar Purkinje cells following activation of the granule cell layer: An intracellular in vitro and in vivo investigation. *Exp. Brain Res.* **100**, 200–214.

Jaeger, D., De Schutter, E., and Bower, J. M. (1997). The role of synaptic and voltage-gated currents in the control of Purkinje cell spiking: A modeling study. *J. Neurosci.* **17**, 91–106.

Jahanshahi, M., Brown, R. G., and Marsden, C. D. (1993). A comparative study of simple and choice-reaction time in parkinsons, huntingtons and cerebellar disease. *J. Neurol. Neurosurg. Psychiat.* **56**, 1169–1177.

Jenkins, I. H., Brooks, D. J., Nixon, P. D., Frackowiak, R. S. J., and Passingham, R. E. (1994). Motor sequence learning: A study with positron emission tomography. *J. Neurosci.* **14**, 3775–3790.

Jueptner, M., Rijntjes, M., Weiller, C., Faiss, J. H., Timmann, D., Mueller, S. P., and Diener, H. C. (1995). Localization of a cerebellar timing process using PET. *Neurology* **45**, 1540–1545.

Karachot, L., Kado, R. T., and Ito, M. (1994). Stimulus parameters for induction of long-term depression in in-vitro rat purkinje-cells. *Neurosci. Res.* **21**, 161–168.

Kassel, J. (1980). Superior colliculus projections to tactile areas of rat cerebellar hemispheres. *Brain Res.* **202**, 291–305.

Kassel, J., Shambes, G. M., and Welker, W. (1984). Fractured cutaneous projections to the granule cell layer of the posterior cerebellar hemisphere of the domestic cat. *J. Comp. Neurol.* **225**, 458–468.

Kawato, M., and Gomi, H. (1992a). The cerebellum and VOR/OKR learning models. *Trends Neurosci.* **15,** 445–453.

Kawato, M., and Gomi, H. (1992b). A computational model of four regions of the cerebellum based on feedback-error learning. *Biol. Cybernet.* **68,** 95–103.

Keele, S. W., and Ivry, R. (1990). Does the cerebellum provide a common computation for diverse tasks: A timing hypothesis. *Ann. N. Y. Acad. Sci.* **608,** 179–211.

Keifer, J., and Houk, J. C. (1994). Motor function of the cerebellorubrospinal system. *Physiol. Rev.* **74,** 509–542.

Killeen, P. R., and Fetterman, J. G. (1993). The behavioral-theory of timing: Transition analyses. *J. Exp. Anal. Behav.* **59,** 411–422.

Kim, S. G., Ugurbil, K., and Strick, P. L. (1994). Activation of a cerebellar output nucleus during cognitive processing. *Science* **265,** 949–951.

Klatzky, R. L., Loomis, J. M., Lederman, S. J., Wake, H., and Fujita, N. (1993). Haptic identification of objects and their depictions. *Percep. Psychophys.* **54,** 170–178.

Larsell, O. (1972). "The Comparative Anatomy and Histology of the Cerebellum from Monotremes through Apes." University of Minnesota Press, Minneapolis.

Llinas, R. (1984). Functional significance of the basic cerebellar circuit in motor coordination. *In* "Cerebellar Functions" (J. R. Bloedel ed.), pp. 171–185. Springer-Verlag, New York.

Lou, J.-S., and Bloedel, J. R. (1992). Responses of sagittally aligned Purkinje cells during perturbed locomotion: Relation of climbing fiber activation to simple spike modulation. *J. Neurophysiol.* **68,** 1820–1833.

Lundyekman, L., Ivry, R., Keele, S., and Woollacott, M. (1991). Timing and force control deficits in clumsy children. *J. Cogn. Neurosci.* **3,** 367–376.

MacKay, W. A., and Murphy, J. T. (1979). Cerebellar modulation of reflex gain. *Prog. Neurobiol. Oxford* **13,** 361–417.

Marr, D. (1969). A theory of cerebellar cortex. *J. Physiol. (Lond.)* **202,** 437–471.

Molinari, M., and Petrosini, L. (1993). Hemicerebellectomy and motor behavior in rats. 3. Kinematics of recovered spontaneous locomotion after lesions at different developmental stages. *Behav. Brain Res.* **54,** 43–55.

Morissette, J., and Bower, J. M. (1996). The contribution of somatosensory cortex to responses in the rat cerebellar granule cell layer following peripheral tactile stimulation. *Exp. Brain Res.* **109,** 240–250.

Muller, F., and Dichgans, J. (1994). Dyscoordination of pinch and lift forces during grasp in patients with cerebellar lesions. *Exp. Brain Res.* **101,** 485–492.

Palay, S. L., and Chan-Palay, V. (1974). "Cerebellar Cortex: Cytology and Organization." Springer, Berlin.

Paulin, M. G. (1993). The role of the cerebellum in motor control and perception. *Brain Behav. Evol.* **41,** 39–50.

Paulin, M. G., Nelson, M. E., and Bower, J. M. (1989a). Dynamics of compensatory eye movement control: An optimal estimation analysis of the vestibulo-ocular reflex. *Int. J. Neur. Syst.* **1,** 23–29.

Paulin, M. G., Nelson, M. E., and Bower, J. M. (1989b). Neural control of sensory acquisition: The vestibulo-ocular reflex. *In* "Advances in Neural Information Processing Systems" (D. Touretzky ed.), Vol. 1, pp. 410–418. Kaufmann, San Mateo, CA.

Raichle, M. E., Fiez, J. A., Videen, T. O., Macleod, A. M. K., Pardo, J. V., Fox, P. T., and Petersen, S. E. (1994). Practice-related changes in human brain functional-anatomy during nonmotor learning. *Cereb. Cortex* **4,** 8–26.

Robinson, F. R. (1995). Role of the cerebellum in movement control and adaptation. *Curr. Opin. Neurobiol.* **5,** 755–762.

Ryding, E., Decety, J., Sjoholm, H., Stenberg, G., and Ingvar, D. H. (1993). Motor imagery activates the cerebellum regionally: A SPECT RCBF study with tc-99m-HMPAO. *Cogn. Brain Res.* **1**, 94–99.

Sasaki, K., Bower, J. M., and Llinas, R. (1989). Multiple purkinje-cell recording in rodent cerebellar cortex. *Eur. J. Neurosci.* **1**, 572–586.

Schade, J. P., and Ford, D. H. (1973). "Basic Neurology." Elsevier, New York.

Schwartz, A. B., Ebner, T. J., and Bloedel, J. R. (1987). Responses of interposed and dentate neurons to perturbations of the locomotor cycle. *Exp. Brain Res.* **67**, 323–338.

Seitz, R. J., Roland, P. E., Bohm, C., Greitz, T., and Stoneelander, S. (1991). Somatosensory discrimination of shape: Tactile exploration and cerebral activation. *Eur. J. Neurosci.* **3**, 481–492.

Shambes, G. M., Beermann, D. H., and Welker, W. (1978a). Multiple tactile areas in cerebellar cortex: Another patchy cutaneous projection to granule cell columns in rats. *Brain Res.* **157**, 123–128.

Shambes, G. M., Gibson, J. M., and Welker, W. (1978b). Fractured somatotopy in granule cell tactile areas of rat cerebellar hemispheres revealed by micromapping. *Brain Behav. Evol.* **15**, 94–140.

Shidara, M., Kawano, K., Gomi, H., and Kawato, M. (1993). Inverse-dynamics model eye-movement control by Purkinje-cells in the cerebellum. *Nature* **365**, 50–52.

Sinkjaer, T., Miller, L., Andersen, T., and Houk, J. C. (1995). Synaptic linkages between red nucleus cells and limb muscles during a multijoint motor task. *Exp. Brain Res.* **102**, 546–550.

Snider, R. S., and Eldred, E. (1952). Cerebro-cerebellar relationships in the monkey. *J. Neurophysiol.* **15**, 27–40.

Snider, R. S., and Stowell, A. (1944). Receiving areas of the tactile, auditory, and visual systems in the cerebellum. *J. Neurophysiol.* **7**, 331–357.

Strehler, B. L. (1990). A new theory of cerebellar function: Movement control through phase-independent recognition of identities between time-based neural informational symbols. *Synapse* **5**, 1–32.

Tachibana, H., Aragane, K., and Sugita, M. (1995). Event-related potentials in patients with cerebellar degeneration: Electrophysiological evidence for cognitive impairment. *Cogn. Brain Res.* **2**, 173–180.

Terashima, T., Kishimoto, Y., and Ochiishi, T. (1993). Musculotopic organization of the facial nucleus of the reeler mutant mouse. *Brain Res.* **617**, 1–9.

Thach, W. T., Goodkin, H. P., and Keating, J. G. (1992). The cerebellum and the adaptive coordination of movement. *Annu. Rev. Neurosci.* **15**, 403–442.

Thach, W. T. (1996). Cerebellum, motor learning and thinking in man: PET studies. *Behav. Brain Sci.*, in press.

Thompson, R. F. (1988). The neural basis of basic associative learning of discrete behavioral responses. *Trends Neurosci.* **11**, 152–155.

Tyrrell, T., and Willshaw, D. (1992). Cerebellar cortex: Its stimulation and the relevance of Marr's theory. *Phil. Trans. R. Soc. Lond. Ser. B Biol. Sci.* **336**, 239–257.

Vankan, P. L. E., Horn, K. M., and Gibson, A. R. (1994). The importance of hand use to the discharge of interpositus neurons of the monkey. *J. Physiol. Lond.* **480(1)**, 171–190.

Welker, W. (1987). Spatial organization of somatosensory projections to granule cell cerebellar cortex: Functional and connectional implications of fractured somatotopy (summary of Wisconsin studies). In "New Concepts in Cerebellar Neurobiology" (J. S. King ed.), pp. 239–280. A. R. Liss, New York.

Welker, W., Blair, C., and Shambes, G. M. (1988). Somatosensory projections to cerebellar granule cell layer of giant bushbaby, Galago crassicaudatus. *Brain Behav. Evol.* **31**, 150–160.

Welker, W. I. (1964). Analysis of the sniffing behavior of the albino rat. *Behavior* **22**, 223–244.

Wessel, K., Verleger, R., Nazarenus, D., Vieregge, P., and Kompf, D. (1994). Movement-related cortical potentials preceding sequential and goal-directed finger and arm movements in patients with cerebellar atrophy. *Electroencephal. Clin. Neurophysiol.* **92,** 331–341.

Wessel, K., Zeffiro, T., Leu, J. S., Toro, C., and Hallett, M. (1995). Regional cerebral blood-flow during a self-paced sequential finger opposition task in patients with cerebellar degeneration. *Brain* **118,** 379–393.

Westberg, K. G., Sandstrom, G., and Olsson, K. A. (1995). Integration in trigeminal premotor interneurons in the cat. 3. Input characteristics and synaptic actions of neurons in subnucleus-gamma of the oral nucleus of the spinal trigeminal tract with a projection to the masseteric motoneuron subnucleus. *Exp. Brain Res.* **104,** 449–461.

Wild, B., and Dichgans, J. (1992). Why are cerebellar patients slow: A study of movement and EMG disorders of wrist flexions in humans with cerebellar lesions. *Eur. J. Neurosci.* 215–224.

Zeigler, P. H., Semba, K., and Jacquin, M. F. (1984). Trigeminal reflexes and ingestive behavior in the rat. *Behav. Neurosci.* **98,** 1023–1038.

NEURAL REPRESENTATIONS OF MOVING SYSTEMS

Michael G. Paulin

Department of Zoology and Center for Neuroscience, University of Otago, Dunedin, New Zealand

I. Introduction
II. What Is State Estimation?
III. Organization of the Cerebellum and the Motor System
IV. Oculomotor Vermis
V. The Vestibulo-Ocular Reflex
VI. Cerebellum and Motor Learning
VII. Bower's Hypothesis
VIII. Cerebellum and Cognition
IX. Implementation
X. Conclusions
References

The cerebellum is necessary for moving smoothly and accurately, but this does not imply that the cerebellum generates or modifies movement control signals. Cerebellar function can be explained by assuming that it is involved in constructing neural representations of moving systems, including the body, its parts, and objects in the environment. To draw a technological analogy, the cerebellum could be a neural analogue of a dynamical state estimator, or a part of one. This explanation is able to account not only for cerebellar involvement in motor control, motor learning, and certain kinds of reflex conditioning, but also cerebellar involvement in certain kinds of perceptual and cognitive tasks unrelated to the production of movements. Evidence for the hypothesis that the cerebellum is involved in a neural analogue of state estimation is (1) across phylogeny, cerebellar morphology reflects animals' use of particular sensory systems for analyzing their own movements and the movements of objects in the environment; (2) cerebellar "oculomotor" neurons are active in relation to movements of salient objects in the environment, regardless of whether the animal moves its eyes to look at them; (3) compensatory eye movements have dynamic characteristics indicating that the control signals are constructed from an underlying optimal head state representation; and (4) the motor symptoms of cerebellar dysfunction resemble the effects of faulty state estimation in artificial control systems. The state estimator hypothesis explains the participation of the cerebellum in controlling, perceiving, and imagining systems that move.

I. Introduction

Cerebellar neurons are active in relation to movements and cerebellar dysfunction leads to deficits in movements, but the cerebellum is not necessary for moving. Until recently, there have been no reports of obvious perceptual or cognitive deficits associated with cerebellar dysfunction except some which could be explained as consequences of movement deficits. Because of this the conventional understanding has been that the cerebellum fine-tunes or adjusts movement control signals in some way.

An alternative possibility is that the cerebellum is a neural analogue of a dynamical state estimator, or plays some essential role in state estimation. According to this hypothesis (Paulin, 1988, 1993a), the cerebellum is a sensory processing structure with a specific role in ensuring that central neural representations of moving systems, accurately reflect the behavior of those systems. The moving systems of interest include things in the environment as well as body parts. Accurate representations of these systems are needed to accurately compute control signals for movements, particularly those made in relation to other moving structures. Therefore the cerebellum is essential for accurate movement control.

Although state estimation underlies precise motor control, it would be a mistake to characterize state estimation as a motor control task. State estimators have a variety of applications beyond control. Evidence is presented here that the cerebellum has a common underlying role in sensorimotor, perceptual, and cognitive processes consistent with the state estimator hypothesis.

II. What Is State Estimation?

To compute a control signal that will change a system's output rapidly and efficiently from its current value to a specified target value it is necessary to know what the current value is and what the target is. Because of unpredictable disturbances and uncertainty about the system's dynamics it may not behave as intended. Therefore it is necessary to monitor the system and regulate control signals during the response to ensure that the movements are stable and accurate.

To effectively control and regulate the position of a moving system with one control input and one degree of freedom in its response, it is necessary to track not only the system's position but also its velocity and usually higher time derivatives of position as well. For systems with several inputs and outputs the situation becomes more complicated. The information needed in order to effectively control and regulate such a system is contained in a set of variables that may not be able to be measured directly or be computed from measurements in a simple way. They are called the system's state variables or, collectively, its state (Meditch, 1969; Mendel, 1987).

The rapid advances in control and guidance technology that occurred in the late 1950s and early 1960s are associated with the discovery of a mathematical relationship between the state variables of a dynamical system and measurements of its behavior (Kalman, 1960a,b). A device which directly solves Kalman's equations describing this relationship is called a Kalman filter. The Kalman filter has certain theoretical limitations and it is numerically difficult to implement, but is nonetheless widely used (Blackman, 1986; Mendel, 1987). There are a number of practical ways to approximate or generalize the Kalman filter, including neural network and fuzzy systems methods (Pacini and Kosko, 1992). Any device that does what a Kalman filter does—estimates state variables from data—is called a state estimator.

The motor symptoms of cerebellar dysfunction resemble the effects of faulty state estimation in artificial control systems (Paulin, 1993b). Optimal control signals for dynamical systems are functions of the current and intended state, $u = F(x_c, x_i)$. It is possible to move a system to a desired state without knowing its current state, but it is not possible to optimize the movement so that it is, for example, the fastest, smoothest, least stressful, or most energy-efficient movement toward the goal (or optimizes some tradeoff among criteria of this kind). If the goal state is defined by a moving target, then a state estimator is required to predict that state. Thus, if there is no concern about costs and benefits of reaching a goal in a particular way, and the goal can be specified without predicting the trajectory of a moving target, then state estimates are not required for computing control signals.

Conventional theories of cerebellar function may be summarized in this context as asserting that the cerebellum has something to do with computing the control function, $F()$, whereas the state estimator hypothesis asserts that the cerebellum is concerned with computing the operands of this function, x_c and x_i. In either case the result would be an incorrect control signal, $u = F(x_c, x_i)$, causing inaccuracy and instability in movements (Paulin, 1993b). Thus, in the light of modern control theory, evidence that cerebellar damage causes such deficits in movements is ambiguous about whether the cerebellum is a movement controller or a state estimator. The aim of this chapter is to present evidence and arguments suggesting that the ambiguity should be resolved in favor of the latter interpretation.

The key to understanding the theory that the cerebellum is (part of) a neural analogue of a dynamical state estimator is the observation that the Kalman filter is not merely a useful technological gadget, but a mathematical theorem about the relationship between measurements and states. No system that represents, predicts, or controls trajectories of multivariate stochastic dynamical systems, artificial or biological, is exempt from the implications of Kalman's theorem. A state estimator must incorporate knowledge of the observed system's dynamics and it must determine in real time which available signals contain information about the observed system's state. It couples these data accordingly to an internal representation in such a way as to ensure that its behavior matches that of the real system.

III. Organization of the Cerebellum and the Motor System

A classical rule about cerebellum is that its overall size and the relative sizes of its parts in different animals are correlated to the kinds and complexity of movements made by the animal (Larsell, 1970). It has been suggested that this rule is in fact an assumption used by early neuroanatomists to guide their interpretation of cerebellar morphology in the absence of modern physiological or tracer studies, and this rule is repeated in modern literature as if it were an empirical observation (Pearson, 1972). Evidence does not support the rule. It is a poor predictor of cerebellar morphology and it has contributed to a number of errors in earlier literature (Paulin, 1993a).

Whereas agility in particular motor behaviors is frequently associated with cerebellar enlargement, this is part of a wider pattern that can be seen by focusing on exceptions to the classical rule. A prominent exception to the rule is the association between cerebellum and electrosensory systems in certain fish. For example, *Gnathonemus petersii* is a bony fish about 10 cm long, whose brain is massive in comparison with other fish of the same size, and is mostly composed of cerebellum (Bell and Szabo, 1986). *Gnathonemus* has an electric sense, and the hypertrophied region of cerebellar cortex in this fish is linked with electrosensory structures in the midbrain, not with motor structures in the brain stem or spinal cord. This enlarged region is not the electrosensory lateral-line lobe, a structure that forms an extension of the cerebellum and has been called cerebellar like, but a more rostral structure that is unquestionably a part of the cerebellum. Neurons in this part of cerebellar cortex are active in relation to objects moving in the environment, while the fish itself is not moving. Weakly electric fish show deficits in electrolocation following cerebellar inactivation (Bombardieri and Feng, 1977).

This pattern of association between regions of cerebellum and sensory systems that track movements, whether or not those movements are made by the animal itself, is found throughout the vertebrates, including humans (Paulin, 1993a). One further example is given here, involving a comparison of the cerebellum in two kinds of mammals. These are chiropterans (bats) and cetaceans (whales and dolphins).

The medial lobe of the bat cerebellum, homologous to vermal lobules VI–VIII in other mammals, is relatively expanded in insect-eating or microchiropteran bats compared with fruit-eating or megachiropteran bats (Larsell, 1970; Henson, 1970). Lobule VIII is expanded in toothed whales and dolphins (odontocetes) when compared with baleen whales (balenopterans) (Larsell, 1970). The insect-eating bats are generally more agile than the fruit-eating bats, and toothed whales and dolphins are generally more agile than the filter-feeding baleen whales. While the shared agility might lead one to expect some shared characters in neural structures involved in movement control, the fact that the same regions of cerebellum are expanded in each case is incongruous with the classical rule about cerebellar expansion. These animals move in quite different ways. Microbats fly using only their forelimbs and use their hindlimbs only for grasping and not for manipulation. Whales and dolphins swim and steer using caudal trunk and tail muscles.

Insect-eating bats have more in common with toothed whales and dolphins than agility. Both groups use echolocation as their main sensory modality for tracking their own movements and the movements of other objects through space. Baleen whales and fruit-eating bats do not use echolocation. Neurons in the medial lobe of bat cerebellum are interested in acoustic parameters related to sound source trajectories (Sun *et al.*, 1990; Kamada and Jen, 1990), as they are in lobule VII of cat cerebellum (Wolfe, 1972). Although comparable neurophysiological data are not available for cetaceans, the shared derived expansion of caudal vermal lobules in the cerebellums of insectivorous bats and toothed cetaceans would appear to be related to the shared derived use of a certain kind of sensory system for locating things, including the animal itself, in space. This interpretation is strongly supported by the observation that certain weakly electric fish, distantly related to bats and cetaceans, have an anaologous cerebellar hypertrophy that is functionally related to an analogous sensory system. Other examples in various kinds of vertebrates have been discussed previously (Paulin, 1993a).

Within the conventional framework that cerebellar function is restricted to motor control, cerebellar hypertrophy in weakly electric fish is an anomaly. However, a pattern of association between cerebellar organization and motor system organization is to be expected if there is an underlying pattern of association between cerebellum and movement tracking systems because accurate movement control requires accurate information about the trajectory of the controlled system and its goal. From this point of view the gigantocerebellum of certain fish is not an exception to a rule but rather exemplifies the rule. The comparative evidence provides a choice between the idea that cerebellar function is some part of movement control, which carries some rather striking (and strikingly consistent) anomalies as baggage, and the idea that cerebellar function is some part of state estimation, having the corollary that cerebellum is important for movement control.

IV. Oculomotor Vermis

Lobules VIc and VII of the caudal vermis of the primate cerebellum are known as oculomotor vermis, and the underlying nucleus is known as the fastigial oculomotor region because neurons there are active in relation to saccades and smooth pursuit eye movements.

Oculomotor vermal neurons fire in bursts when the animal makes a saccade toward a visual target, but not when the animal makes a spontaneous saccade (Ohtsuka and Noda, 1995). The firing rate of Purkinje cells in oculomotor vermis varies sinusoidally when the eyes pursue a sinusoidally moving target, but most respond identically when the eyes fixate a point while the visual target moves near it (Suzuki *et al.*, 1981). The firing rate of fastigial oculomotor neurons varies sinusoidally during visual tracking of a sinusoidally moving target, but when the whole animal rotates during target movement such that the eyes do not need to move in order to fixate the target, these neurons respond in the same way. If the monkey glances away from the target during visual tracking, these neurons continue to behave as they do when the eyes are still smoothly tracking the target (Fuchs *et al.*, 1994). These observations suggest that this part of the cerebellum is interested in visual target movements, not eye movements.

Fuchs *et al.* (1994) reject the hypothesis that these are visual neurons because if the target briefly disappears during tracking, they continue to fire and the eyes continue to move as if the target were still present. Sometimes the neural firing patterns and the eye movements both diverge from the pattern seen when the target is present. Since in this situation the neuronal firing pattern matches the eye movements and not the sense data, Fuchs *et al.* (1994) conclude that these are oculomotor neurons. However, when a moving target suddenly vanishes, it is generally a good assumption that it is still out there but it has disappeared behind something or something has interfered with your sensors. This is in fact the problem that state estimators are designed to solve: Tracking a target using measurements which may be noisy, intermittent, and indirect. When data are poor, state estimators may predict badly, but they do not stop predicting (Meditch, 1969).

The observations of Fuch *et al.* (1994) on fastigial "oculomotor" neurons are consistent with those of Suzuki *et al.* (1981) and Ohtsuka and Noda (1995), and with the hypothesis that the cerebellum is involved in building central representations of moving systems. In the case of "oculomotor vermis" and the underlying section of the fastigial nucleus, the moving systems of interest include targets of the visual smooth pursuit system and perhaps also the eyes. This hypothesis has the corollary that these regions of cerebellum are important in the control of smooth pursuit eye movements and target-directed saccades. It is possible to interpret the evidence in terms of the hypothesis that this region of cerebellum generates or modulates control signals for smooth pursuit and target-directed saccades. However, this raises the question of why firing patterns of many neurons there are consistently related to the behavior of targets and only related to the behavior of the eyes when they move to look at targets. Futhermore, the outputs of the fastigial oculomotor region have been investigated and "Projections to structures related . . . to smooth pursuit eye movements were either few or practically nonexistent" (Noda *et al.*, 1990).

V. The Vestibulo-Ocular Reflex

The vestibulo-ocular reflex (VOR) counterrotates the eyes during head movements, holding the visual image steady on the retina. It is mediated by a neural circuit that includes cerebellar cortex and underlying vestibular nuclei (Ito, 1984). The spatial and dynamic characteristics of the VOR provide evidence that it is implemented using a neural analogue of a Kalman filter (Paulin *et al.*, 1989). Kalman filters predict state trajectories from noisy, biased, or conflicted data by making assumptions about the dynamics of the observed system and statistically estimating which state trajectory is most likely to have generated the data under these assumptions. A consequence of this strategy is that when the system does not behave according to those assumptions the state estimator makes errors of a predictable kind. These errors can be thought of as resulting from a preference for believing in a model based on prior experience rather than sense data when the sense data are unlikely according to the model. Such errors are seen in the behavior of the VOR, as discussed later.

In order to fixate a distant visual field during head rotation, eye counter-rotations should match head rotations as closely as possible. Under the simplifying assumption that VOR can be modeled as a single-input, single-output system it is commonly inferred that for optimal performance the central control circuits of the VOR should compensate for sensory and motor dynamics so that the overall reflex has unity gain and 180° phase shift over the bandwidth of head rotations. That is, the eyes should be instructed to rotate at the measured rotational velocity of the head, but in the opposite direction. This is not correct, as can be seen by considering a contrived example in which there is one unbiased sensor measuring angular velocity and assuming that the head always rotates with constant angular velocity. There are infinitely many possible trajectories for the head, and the sensor data can be averaged to estimate which trajectory it is on, i.e., to compute a central "belief" about how fast the head is rotating. If the assumption of constant rotation is correct, then no matter how small the measurement noise is, the belief will rapidly become more accurate than the observations. Simple calculations show that in this situation the influence of sensor data ought to decline linearly with time (Paulin, 1988). Thus the optimal VOR, giving the best compensatory eye velocity based on the available data under this particular assumption about how the head behaves, will have declining gain. At first a small shake of the head will cause a large eye movement but later the eyes will continue to move at constant velocity even if the head does not.

The tendency of optimal state estimators to fail badly if the underlying assumptions are wrong would seem to be a good argument against using them. Nevertheless they are effective in practice because real systems are constrained to move in particular ways and some of these constraints are consistent over time. The example in the previous paragraph involves an unrealistic constraint and so it is apparent why this estimator would not be able to track real head movements. However, real heads cannot move in arbitrary ways and by taking this into account it is possible not merely to track head movements in space more accurately than any sensor can measure them, but to accurately predict head trajectories into the near future. If VOR control signals are computed via a central head state estimate, rather than by directly transforming sense data as assumed in most VOR models (e.g., Ito, 1984), then the dynamics of the VOR should differ from unity gain with 180° phase lag. The discrepancy will depend on what information was available during the design process, i.e., evolution and learning.

Collewijn and Grootendorst (1979) placed rabbits on a turntable oscillating sinusoidally at 1/6 Hz in the light with a stationary visual surround. After 6 hr training at this frequency the rabbits' eyes oscillated at 1/6 Hz in the dark when the turntable moved, even if it oscillated at 1/12 Hz. The authors interpreted this by hypothesizing that there is a pattern generator in addition to a simple integrating filter underlying compensatory eye movements in rabbits. However, a more parsimonious possibility is that rabbit compensatory eye movement control signals are computed from an underlying optimal head state estimate, and training led the estimation circuitry to learn (what subsequently turned out to be) an invalid assumption about how the head moves when the rabbit is on the turntable. This situation is analogous to the constant-velocity situation described earlier. While the rabbit is in the testing arena the waveform of head movements is perfectly predictable. Once this has been determined, sense data about head movements in this context contain no information about those waveforms except for their timing or phase. The rabbits make fast repositioning eye movements, repeatedly attempting to phase lock their 1/6-Hz eye movements onto the 1/12-Hz stimulus (Collewijn and Grootendorst, 1979).

VORs of different animals may not behave in precisely the same way because the effect depends on how learning is constrained by the variant of VOR circuitry that has evolved in a particular species. The rabbit VOR appears to have excessive flexibility, being able to adapt to a kind of head movement dynamics that a rabbit is rather unlikely to encounter. Generally, a dynamical state estimator underlying head movements should reveal itself in subtle but systematic errors when head movement dynamics are subtley but systematically different from the natural dynamics to which the animal's VOR has been exposed during evolution and learning.

Paulin *et al.* (1989) calculated optimal head state estimator dynamics for a simple inverted pendulum model of primate head dynamics driven by flat-spectrum angular rotations, using a Kalman filter theory. We were able to reproduce the systematic deviations of VOR dynamics from unity gain, 180° phase shift that are observed when primate heads are forced to follow flat-spectrum rotation trajectories in the laboratory (Keller, 1978; Furman *et al.*, 1982). We interpreted this to mean that the deviations are not due to biological constraints leading to suboptimal performance in the laboratory situation, but are due to an underlying head state estimator which normally ensures optimal performance despite such constraints in natural situations (Paulin *et al.*, 1989).

A familiar example of incorrect compensatory eye movements occurs after prolonged head rotation in one direction. When the head stops the eyes continue as if the head were still moving, with rapid repositioning movements. This is called postrotatory nystagmus. In monkeys, if the head is tilted during postrotatory nystagmus then the plane of eye movement changes with respect to the head so that it remains in the same plane with respect to the earth. However, if the cerebellar nodulus is lesioned, the nystagmus plane remains fixed with respect to the head (Angelaki and Hess, 1995). As Angelaki and Hess (1995) put it, compensatory eye movements appear to be an "oculomotor footprint" of an underlying inertial representation of head movements. This representation exists in the vestibular nuclei and its accuracy depends on having an intact cerebellar nodulus. The view that emerges from this experiment is that the vestibulocerebellum is not a control system for the VOR, but is essential for maintaining accurate information about the dynamical state of the head, from which accurate compensatory eye movement control signals may be computed.

Pattern generation in rabbit VOR, excess high frequency gain in primate VOR, and postrotatory nystagmus are at first sight quite different phenomena, but they can be understood as particular manifestations of a common process underlying compensatory eye movement control in all species of vertebrates, namely dynamical head state prediction.

VI. Cerebellum and Motor Learning

The cerebellum must be adaptive in order to be able deal with novel dynamical systems, and clearly it is (Ito, 1984). However, it would be a mistake to say that cerebellar function is motor learning simply because it has a role in motor control and can learn. By analogy, the skin on my thumb has a callous where I hold my squash racket, so my skin is an adaptive component of a movement control system, but it is not an adaptive movement controller. Memory is not the function, it is a property that the structure must have in order to perform its function.

Thompson and colleagues (e.g., Thompson and Krupa, 1994) have investigated a well-defined example of cerebellar learning in relation to eyeblink reflex conditioning in rabbits. When a tone repeatedly precedes an airpuff directed onto a rabbit's cornea, the eyeblink reflex becomes conditioned to the tone and shields the eye from the airpuff. The mechanisms responsible for this reflex conditioning lie in the cerebellum or are at least critically dependent on its outputs (Thompson and Krupa, 1994; Krupa and Thompson, 1995).

This result can be interpreted in terms of the state estimator hypothesis by considering the natural role of the eyeblink reflex. The eyeblink reflex protects the cornea by shielding it from contact with (relatively) moving objects. An effective eyeblink reflex control system should incorporate a trajectory tracker able to predict when such contacts will occur. The problem presented to the rabbit in the tone-airpuff experiment can be solved by regarding it as the problem of predicting the trajectory of an object that can be heard but not seen approaching the eye. This is a rather unnatural trajectory prediction problem but solving it is no different in principle from what is required to predict head movements for VOR control or target movements for smooth pursuit. Since, by hypothesis, the rabbit has a trajectory predictor already linked to its eyeblink control circuitry, and normally engaged to solve problems of this kind, it is natural to further hypothesize that it is engaged in solving the tone-airpuff problem.

In general, state estimator malfunction will interfere with motor learning for two reasons. First, it must be difficult to learn to move accurately if a subsystem required for moving accurately is not functioning properly. Second, it must be difficult to learn how to match trajectories and states of limbs to goals if a subsystem necessary for accurately observing trajectories and states of limbs and targets is not functioning properly. That is, if the cerebellum is essential for state estimation, then cerebellar patients should neither move as precisely nor evaluate their movements as precisely as normal subjects. Because of this they should not be able to learn movements as rapidly or as well as normal subjects can.

VII. Bower's Hypothesis

The kind of evidence presented here and in earlier work (Paulin, 1993a) is difficult to reconcile with conventional ideals about the role of cerebellum in motor control. However, Bower has an unconventional idea that differs from the state estimator theory and encompasses evidence of this kind (Paulin et al., 1989; Bower and Kassel, 1990; Gao et al., 1996; Bower, this volume). The hypothesis is that the cerebellum generates motor control signals for a particular purpose, configuring and moving arrays of sense organs so that they acquire sense data in an optimal way. The state estimator hypothesis also implies that the cerebellum optimizes the quality of sensory information to the brain, but only a specific kind of information (about moving systems), for a specific purpose (representing and predicting their trajectories), and not necessarily involving any movements by the animal itself.

There is no doubt that vertebrates actively acquire sense data by moving, but cerebellar involvement in this can be explained in terms of conventional ideas about how the cerebellum is involved in movement control. Enhanced cerebellar activity when a subject moves to acquire sensory information as opposed to simply moving (Gao et al., 1996) can be explained in conventional terms by making the reasonable assumption that the sensory acquisition task requires more precision in movements.

The state estimator hypothesis can account for the result of Gao et al. (1996) on cerebellar involvement in active sensory acquisition in the same way because it agrees with the conventional view that the cerebellum is essential for accurate movement control. However, it can explain cerebellar involvement above and beyond that required to generate the movements because correctly interpreting data from moving sensors would seem to require information about their movements. For example, a cat must take into account its own head and pinnae movements in order to map sounds reaching its ear onto neural representations of the identity and location of the sound source.

Cerebellar involvement in the passive sensory discrimination task described by Gao et al. (1996) appears to be inconsistent with conventional ideas about the cerebellum because there is enhanced activity in the cerebellum when the subjects do not move. However, this appears to weigh against Bower's hypothesis for the same reason. The result can be explained using the state estimator hypothesis because the discrimination task involves analyzing data from moving objects, and therefore presumably involves computations with neural representatives of (aspects of) these objects. State estimation may be required in the machinery that keeps these representations in register with the moving reality in order to allow incoming sense data to be correctly incorporated into them. Thus, Bower's hypothesis has been fruitful in terms of providing new kinds of data about cerebellar function in situations that would not have been considered particularly interesting within other frameworks, but these data may be more elegantly explained by assuming that the reason the cerebellum is involved in active sensory acquisition is that it is involved in state estimation.

VIII. Cerebellum and Cognition

Cerebellar patients report that they cannot accurately control their movements, but generally do not report that they misperceive trajectories of moving objects or limbs. There are two ways to account for this. First, patients with general cerebellar dysfunction may have difficulty in verifying perceptions of trajectories because they lack neural machinery needed for correctly perceiving trajectories, while outcomes of goal-directed movements (e.g., finger-to-nose test) can be evaluated without being able to specify states during the movement. State estimator dysfunction does cause deficits in movement control and so there is no apparent need to look beyond this. Second, the belief that cerebellar dysfunction causes motor but not perceptual deficits may lead to perceptual deficits being downplayed or attributed to other causes. For example, an increased susceptibility to vertigo in some cerebellar patients could be called a cerebellar deficit in the perception of head movements and in the perception of movements of the environment and objects in it. However, the theory that cerebellar dysfunction does not entail perceptual deficits requires patients and physicians to deny that these perceptual deficits are cerebellar perceptual deficits.

Deficits in trajectory perception by cerebellar patients could be demonstrated experimentally by contriving situations in which the answer to some yes/no question depends on the correct perception of a trajectory. For example, a patient could be asked to listen to an acoustic target moving in space, then be asked to choose a graph of the movement pattern visually from a set.

State estimators are optimal noise-rejecting filters if the signal of interest can be modeled as the output of a known dynamical system. Thus the cerebellum could be involved in perceptual enhancement in certain circumstances. For example, a mechanic's "tuned ear" for detecting acoustic signatures of particular mechanical breakdowns in an engine could employ cerebellar circuitry. More likely, such tasks could involve the "cerebellar like" circuitry of the dorsal cochlear nucleus (Paulin, 1993a).

Cognitive tasks that most obviously may involve neural state estimation are those requiring trajectory analysis in space, such as predicting collisions. A related problem that could involve trajectory prediction is judging how long to leave a cup of coffee to cool before drinking it. More generally, the cerebellum may be employed in solving problems by spatial or geometric reasoning, where such reasoning requires the behavior of a dynamical system to be perceived visually or kinesthetically. Some things can be recognized by the way they move; state estimators are used in artificial target recognition systems, and the cerebellum could contribute to analogous neural computations.

Measurements made on or by moving systems will generally be parameterized by the movements, e.g., the doppler shift of a moving sound source or acoustic head shadow effects in a moving listener. Such effects need to be compensated for in order to provide an invariant perception of the source. Correctly interpreting sense data from moving sensors or sources presumably entails correctly coupling sensory information to a changing internal representation. Spatial patterns in the world may be presented to the nervous system as temporal patterns. For example, identifying an object by running your fingers over it would require accurate information about finger movements during the task, so that a bump detected by the fingertip could be incorporated at the right point on the representation.

It is not clear how to explain recently observed cerebellar involvement in cognitive tasks such as language (Leiner et al., 1991; Schmahmann, 1991) using the state estimator hypothesis (independent of associated motor behaviors or silent rehearsal of those behaviors), in large measure because the computational mechanisms underlying such tasks remain unclear. However, the hypothesis predicts that cognitive processes found to involve the cerebellum will be found to involve state estimation. Thus if the state estimator hypothesis is verified in other areas, it may provide clues that will help determine the neural computations underlying these cognitive tasks.

IX. Implementation

The suggestion that the cerebellum is a neural analogue of a (part of a) Kalman filter should not be taken to imply that the cerebellum physically resembles a Kalman filter. The claim is simply that the cerebellum is involved in providing the same information in the nervous system as state estimators provide in artifical systems. Two simple physical models are presented, illustrating the principles which may be involved.

Consider a simple pendulum, the movements of which are to be followed and predicted (Fig. 1). Another pendulum forms an "internal model" of the target. The model is coupled to the target through an ideal mechanical isolator (I) which prevents the model from reacting back on the target as it moves. Between the isolator and the model there is a resistive element. When this is locked (rigid) the isolator is powerful enough to force the model to mimic the target precisely. In that case it does not matter what the length (λ') or mass (m') parameters of the model are (if the optimality criterion is accuracy, but isolator power consumption is reduced by making the model pendulum short and light). This corresponds to a naive model of a sensory system in which sense data are simply mapped or projected onto an internal representation.

Now suppose there is some noise (ν) in the sense data. It seems intuitively reasonable that it might now pay to match the parameters of the model (λ', m') to those of the target (λ, m) so that it has a natural tendency to swing like the target does and to weaken the coupling (ω) so that the model is not forced to follow the noise. Kalman filtering generalizes this idea and takes it to the theoretical limit by asking what the coupling strengths and model parameters should be to make the model mimic or predict the target as accurately as possible. At the onset of tracking, for example, it

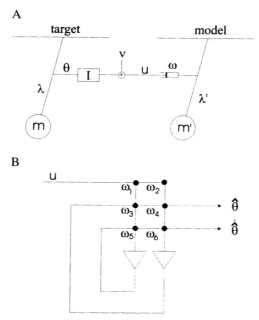

FIG. 1. (a) Mechanical model of optimal target tracking. With appropriate choices of parameters (ω, λ', m') the model pendulum mimics the target pendulum. I is a mechanical isolator that transmits forces from the target pendulum without allowing reaction forces from the model pendulum. The measurement u includes noise ν. The model pendulum is coupled to the target through a dashpot with resistance ω which weakens the effect of u on the model. (B) Electronic pendulum analogue for small swing angles. Setting the connection strengths ω_1–ω_6 allows the circuit to mimic any single input, second order linear system. In particular it could replace the mechanical pendulum in (A).

pays to lighten and shorten the model pendulum and have a high sensory coupling gain so that the model accelerates quickly. As the model trajectory converges onto the target trajectory, it pays to match model parameters more closely to target parameters and to weaken the sensory input gain so that the model is not affected by measurement noise.

If the target is unperturbed then it becomes theoretically possible to eliminate the sensory coupling altogether once the model has converged onto the target trajectory. The problem is that no matter how good the model is it will eventually diverge from the target's trajectory if it is not coupled to it. Because Kalman filters continuously test incoming data against expectations, it is straightforward to check for divergence and reset model and coupling parameters when this occurs. The rabbits in the experiments of Collewijn and Grootendorst (1979) described earlier make fast repositioning eye movements, repeatedly attempting to lock their 1/6-Hz eye movements onto the 1/12-Hz target, which may reflect a resetting strategy of this kind in neural head state estimation.

In general, Kalman filters converge until they are weakly coupled to the target. At this point they tend to discount sense data more than they should because their computations are based on the assumption that the internal model is correct. Because the internal model can never be perfect, the estimator will converge to give suboptimal performance and may diverge from the target. Therefore it is useful to check for divergence and reset the model parameters and coupling strengths when this occurs. The behavior of climbing fibers (firing associated with discrepancies between actual and expected movements; Ito, 1984) is consistent with their having a role in signaling such divergences.

The fully connected electronic integrator network in Fig. 1b can implement an arbitrary second order, single input linear dynamical transformation (it is a direct mapping of the general second order, single input linear system $\dot{x} = Ax + bu$ into hardware). For small movements, a pendulum is nearly linear and so the integrator network can be used to track the target pendulum instead of the mechanical model of Fig. 1a. Getting the integrator network to accurately track or predict the target trajectory is a matter of setting the connection strengths ($\omega_1, \omega_2, \omega_3, \omega_4, \omega_5, \omega_6$) in the network. The Kalman filter theory shows how to compute them.

These examples of trajectory tracking systems illustrate the point that systems with radically different architectures may nonetheless operate on the same principles. The vertebrate nervous system may use similar principles for trajectory tracking, and there is good evidence to support the hypothesis that the cerebellum has a key role in this, but this falls short of a theory of exactly what the cerebellum does and how it works.

Although the state estimator hypothesis does not prescribe an interpretation of cerebellar structure and function, interpretations consistent with the hypothesis are possible. In outline, one possibility is that firing patterns in cerebellar nuclei represent states of dynamical systems, i.e., the cerebellar nuclei are analogous to the integrator network in Fig. 1b. The cerebellar cortex could identify and evaluate information in sense data and in other parts of the nervous system about states of dynamical systems. On the basis of its evaluation the cerebellar cortex could adjust couplings among nuclear neurons and between nuclear neurons and their afferents so that the nuclei contain the most accurate possible (predictive) information about the states of the observed dynamical systems. This state information would be passed to other structures in the nervous system that require it to control, perceive, or imagine movements.

Neural models based on analogies with engineering control systems often assume that neural firing rates correspond to analogue or digital signal values in the engineering system, but it is important to realize that this is not necessarily the case. For example, suppose a certain group of neurons represents the state of a certain dynamical system. A conventional model might assume that the firing rate of a neuron in this group corresponds to the value of a state variable, but it could be that the firing rate represents the degree to which the state variable has a particular value. In that case, the group of neurons forms a patchwork covering the state space. Each member of the group fires to assert that the system is in its patch, and the spatial pattern of assertions over the group represents the state. There is also a question about whether there may be central representations of state or multiple representations used for different purposes. The answers to these questions have important implications for understanding how the cerebellum contributes to motor control and learning, among other things. For example, in a "distributed patchwork" model, adaptation to global perceptual modifications (e.g., prism spectacles which displace the visual field) could occur locally so that particular trajectories of particular limbs would adapt independently, depending on the particular implications of the perceptual change for the performance of the particular maneuver.

X. Conclusions

The idea that the cerebellum is directly involved in dynamical state estimation is both more specific and more general than the conventional idea that the cerebellum is involved in fine movement control. It is more specific because it explains the particular role of the cerebellum in movement control, and it is more general because it explains and predicts cerebellar involvement in other tasks.

State estimation is useful but not necessary for a variety of tasks. Only certain tasks, such as multivariate stochastic movement control, are critically dependent on dynamical state estimation and only these tasks should be critically dependent on the cerebellum. Differences may exist among normal subjects in whether they engage cerebellum for particular tasks, whereas cerebellar patients would be forced to adopt strategies not requiring the cerebellum. For example, one person might use their cerebellum in solving a mathematical problem by geometrical reasoning (visualizing objects moving in space), whereas another might solve the same problem using symbolic reasoning (applying rules). A cerebellar patient would be unable to use the cerebellar-based strategy, but may be better at solving the problem.

Even in movement control, cerebellar patients could adopt movement strategies that reduce dependence on state estimation, such as stereotypy in movements [allowing a memory-based strategy such as that proposed, ironically, as a model of cerebellar function by Marr (1969) and Albus (1971)] or decomposing movements with many degrees of freedom into sequences of movements with one or a few degrees of freedom, each of which can be stabilized without state feedback.

The state estimator hypothesis has the important implication that if the symptoms of cerebellar dysfunction are indeed consequences of the nervous system's inability to accurately compute and predict dynamical state information, then the symptoms may be alleviated by computing and providing this information electronically. For example, a miniature system analogous to multiple-target tracking systems already in use for air traffic control (Blackman, 1986) could provide predictive information about states of selected body parts and external systems via a visual display.

References

Albus, J. S. (1971). A theory of cerebellar function. *Math. Biosci.* **10**, 25–61.
Angelaki, D. E., and Hess, B. J. M. (1995). Inertial representation of angular motion in the vestibular system of rhesus monkeys. II. Otolith-controlled transformation that depends on an intact cerebellar nodulus. *J. Neurophys.* **73**(5), 1729–1751.
Bell, C. C., and Szabo, T. (1986). Electroreception in mormyrid fish: Central anatomy. *In* "Electroreception" (T. Bullock and W. Heiligenberg, eds.), pp. 375–422. Wiley, New York.
Blackman, S. S. (1986). "Multiple-Target Tracking with Radar Applications." Artech House, Inc., MA.
Bombardieri, R. A., and Feng, A. S. (1977). Deficit in object detection (electrolocation) following interruption of cerebellar function in the weakly electric fish *Apteronotus albifrons*. *Brain Res.* **130**, 343–347.
Bower, J. M., and Kassel, J. (1990). Variability in tactile projection patterns to cerebellar folia crus IIa of the norway rat. *J. Comp. Neurol.* **302**, 768–778.
Collewijn, H., and Grootendorst, A. F. (1979). Adaptation of optokinetic and vestibulo-ocular reflexes to modified visual input in the rabbit. *In* "Reflex Control of Posture and Movement" (R. Granit and O. Pompeiano, eds.), Elsevier, Amsterdam.
Fuchs, A. F., Robinson, F. R., and Straube, A. (1994). Participation of the caudal fastigial nucleus in smooth pursuit eye movements. I. Neuronal activity. *J. Neurophys.* **72**(6), 2714–2728.
Furman, J. M., O'Leary, D. P., and Wolfe, J. W. (1982). Dynamic range of the horizontal vestibulo-ocular reflex of the alert rhesus monkey. *Acta Otolaryngol.* **92**, 81–91.
Gao, J. H., Parsons, L. M., Bower, J. M., Xiong, J., Li, J., and Fox, P. T. (1996). Cerebellum implicated in sensory acquisition and discrimination rather than motor control. *Science* **272**, 545–547.
Henson, O. W. (1970). The central nervous system. *In* "The Biology of Bats" (W. A. Wimsatt, ed.), Vol. II, pp. 57–102. Academic Press, London.

Ito, M. (1984). "The Cerebellum and Neural Control." Raven Press, New York.
Kalman, R. E. (1960a). A new approach to linear prediction and filtering problems. *J. Basic Eng.* **82,** 35–45.
Kalman, R. E. (1960b). Contributions to the theory of optimal control. *Bol. Soc. Math. Mex.* **5,** 102–119.
Kamada, T., and Jen, P. H. S. (1990). Auditory response properties and directional selectivity of cerebellar neurons in the echolocating bat *Eptesicus fuscus, Brain Res.* **528,** 123–129.
Keller, E. L. (1978). Gain of the vestibulo-ocular reflex in monkey at high rotational frequencies. *Vis. Res.* **18,** 11–315.
Krupa, D. J., and Thompson, R. F. (1995). Inactivation of the superior cerebellar peduncle blocks expression but not acquisition of the rabbit's classically conditioned eye-blink reflex. *Proc. Natl. Acad. Sci. USA* **92**(11), 5097–5101.
Larsell, O. (1970). "The Comparative Anatomy and Histology of the Cerebellum from Monotremes through Apes" (J. Jansen, ed.). Univ. Minnesota Press, Minneapolis.
Leiner, H. C., Leiner, A. L., and Dow, R. S. (1991). The human cerebro-cerebellar system: Its computing, cognitive and language skills. *Behav. Brain Res.* **44,** 113–128.
Marr, D. (1969). A theory of cerebellar cortex. *J. Physiol.* **202,** 437–470.
Meditch, J. S. (1969). "Stochastic Optimal Estimation and Control." McGraw-Hill, New York.
Mendel, J. M. (1987). "Lessons in Digital Estimation Theory." Prentice-Hall, New York.
Noda, H., Sugita, S., and Ikeda, Y. (1990). Afferent and efferent connections of the oculomotor region of the fastigial nucleus in the macaque monkey. *J. Comp. Neurol.* **302,** 330–348.
Ohtsuka, K., and Noda, H. (1995). Discharge properties of Purkinje cells in the oculomotor vermis during visually guided saccades in the macaque monkey. *J. Neurophys.* **74**(5), 1828–1840.
Pacini, P. J., and Kosko, B. (1992). Comparison of fuzzy and kalman filter target-tracking control systems. *In* "Neural Networks and Fuzzy Systems" (B. Kosko, ed.), pp. 379–406. Prentice Hall, New Jersey.
Paulin, M. G. (1988). A kalman filter theory of the cerebellum. *In* "Dynamic Interactions in Neural Networks: Models and Data (M. Arbib and S. Amari, eds.), pp. 239–261. Springer-Verlag, Berlin/New York.
Paulin, M. G. (1993a). The role of the cerebellum in motor control and perception. *Brain Behav. Evolut.* **41,** 39–50.
Paulin, M. G. (1993b). A model of the role of the cerebellum in motor control. *Hum. Movement Sci.* **12,** 5–16.
Paulin, M. G., Nelson, M. E., and Bower, J. M. (1989). Dynamics of compensatory eye movement control: An optimal estimation analysis of the vestibulo-ocular reflex. *Intl. J. Neural Sys.* **1**(1), 23–29.
Pearson, R. (1972). "The Avian Brain." Academic Press, New York.
Schmahmann, J. D. (1991). An emerging concept: The cerebellar contribution to higher function. *Arch. Neurol.* **48,** 1178–1186.
Sun, D., Sun, X., and Jen, P. H. S. (1990). The influence of auditory cortex on acoustically evoked cerebellar responses in the CF-FM bat *Rhinolopus pearsonii chinensis. J. Comp. Physiol.* **166,** 477–488.
Suzuki, D. A., Noda, H., and Kase, M. (1981). Visual and pursuit eye movement-related activity in posterior vermis of monkey cerebellum. *J. Neurophys.* **46**(5), 1120–1139.
Thompson, R. F., and Krupa, D. J. (1994). The organization of memory traces in the brain. *Annu. Rev. Neurosci.* **17,** 519–549.
Wolfe, J. W. (1972). Responses of cerebellar auditory area to pure tone stimuli. *Exp. Neurol.* **36,** 295–309.

HOW FIBERS SUBSERVE COMPUTING CAPABILITIES: SIMILARITIES BETWEEN BRAINS AND MACHINES

Henrietta C. Leiner and Alan L. Leiner

Channing House, Palo Alto, California 94301

I. Introduction: "Hardware" and "Software" Capabilities
   A. Principles of Design
   B. Evolution of Cognitive Capabilities
II. "Hardware" in the Human Cerebellum
   A. Evolution of the Lateral Cerebellum
   B. Evolution of Connections to the Cerebral Prefrontal Cortex
III. "Software" Capabilities Inherent in Cerebro-Cerebellar Connections
   A. Communication Capabilities Inherent in Segregated Bundles of Fibers
   B. Communicating via Internal Languages
   C. Communications between the Cerebral Cortex and the Cerebellum
IV. Conclusions: Combined "Hardware" and "Software" Capabilities
   A. Complexity and Versatility
   B. Reconciliation of Diverse Theories
   References

Can the principles underlying the design of computers help to explain the cognitive capabilities of the human brain? This chapter shows that these principles can provide insight into the capabilities of the human cerebellum, the internal structure of which bears a remarkable resemblance to the design of a versatile computer. In computers, information processing is accomplished both by the hardware in the system (its circuitry) and by the software (the communication capabilities inherent in its circuitry), which in combination can produce a versatile information-processing system, capable of performing a wide variety of functions, including motor, sensory, cognitive, and linguistic ones. Such versatility of function is achieved by computer hardware in which many modules of similar circuits are organized into parallel processing networks; this structural organization is exemplified in the cerebellum by its longitudinal modules of similar circuits, which are arrayed in parallel zones throughout the structure. On the basis of this known cerebellar "hardware," it is possible to investigate the "software" capabilities inherent in the circuitry of the modules. Each module in the lateral cerebellum seems able to communicate with the cerebral cortex by sending out signals over a segregated bundle of nerve fibers, which is a powerful way of communicating information. We show why this bundling of fibers can enable the cerebellum to communicate with the cerebral cortex (including the prefrontal cortex) at a high level of discourse by using internal languages that are capable of conveying complex information about what to do and when to do it. We propose that such communication activity is reflected in the activation obtained on functional imaging of the cerebro-cerebellar system during the performance by humans of complex motor, sensory, cognitive, linguistic, and affective tasks. Further, we propose a new way of analyzing such cerebro-cerebellar activation, in order to ascertain whether the cerebellar circuitry can (like the circuitry in a versatile computer) perform a wide repertoire of computations on this wide range of information. It seems important to ascertain whether cerebellar circuitry is versatile in its computing capabilities because the demonstration of such versatile capabilities would enable theorists to resolve many of the current controversies about cognitive processing in the mammalian brain.

I. Introduction: "Hardware" and "Software" Capabilities

For decades it has been hotly debated whether the design of computers can provide insight into the cognitive capabilities of the human brain. A particularly good site in the brain for investigating this question is the cerebellum, whose neural organization bears a striking resemblance to the internal design of a versatile computer. The design of all such computers is based on certain fundamental principles that can explain, for example, how motor processing and cognitive processing both can be performed by the same circuitry. Although we do not equate the circuitry in the cerebellum with that in computers, we nevertheless are able to show how some principles of computer design can elucidate the capabilities of the human cerebro-cerebellar system.

A. Principles of Design

From theoretical principles and from practical experience, it is known how to construct versatile computing systems using basic components that possess only limited information-processing capabilities. Despite these limitations, the computing system can be endowed with powerful processing capabilities when the basic components are assembled into discrete modules and when large numbers of these modules are organized into parallel processing networks. Such a modular organization of computing "hardware" is exemplified in the mammalian cerebellum by its microzones of neurons, which are organized into longitudinal circuits that are arrayed in parallel zones throughout the structure (Ito, 1984). In the human cerebellum, the number of neurons is so large that it surpasses the number in all the rest of the nervous system combined (Andersen et al., 1992), and it surpasses by far the number of components in any existing computing machines. Theoretically, therefore, the human cerebellum should be endowed with extremely powerful capabilities as a result of possessing so large a number of components, organized into parallel longitudinal circuits.

In any species of mammal, the basic circuitry in each longitudinal module of its cerebellum is similar to the basic circuitry in every other module (Bloedel, 1992). What differs, however, is the incoming information that each module receives from other parts of the brain for processing and the output destinations in the brain to which the module can project the results of its processing (Dow, 1942). From these facts, the following two inferences can be drawn: (1) Because the basic circuitry in all the modules is similar, the basic sort of processing that can be performed on the incoming information would seem to be similar, no matter whether the incoming information represents motor, sensory, cognitive, linguistic, or any other kind of information. (2) Because the cerebellar output destinations are different in the brains of different species, the cerebellar processing can be put to different uses in their nervous systems.

These inferences about the mammalian nervous system are in accord with a fundamental principle of system design. This principle states that different capabilities can be designed into a processing system composed of the same modules merely by connecting the major parts of the system together in different ways. According to this principle, different capabilities would be built into the nervous system of different species when different connections evolved between the cerebellum and other parts of their brains. This could apply to cognitive capabilities no less than to motor and sensory capabilities; the way that cognitive capabilities could evolve is indicated below.

B. Evolution of Cognitive Capabilities

Although cerebellar connections in all vertebrates enable the cerebellum to contribute to vestibular and somesthetic processing, it is only in mammals that the cerebellar connections with the cerebral cortex become fully developed (Ito, 1984). In the mammals that evolved initially, which had small brains with little cerebral neocortex, the cerebellar connections were linked to sensorimotor areas of this cortex, which enabled cerebellar modules to participate in the processing of sensorimotor information; in the primates that evolved later, which had brains with larger cerebral "association" areas, the links to some newly evolved cerebral areas could enable the cerebellum to participate as well in the processing of some cognitive information, particularly in the cerebral prefrontal cortex (Leiner *et al.*, 1986).

When the human brain evolved, these cerebral prefrontal areas enlarged greatly in size, as did the lateral cerebellum (Passingham, 1975). Because the connections between these newly evolved areas could subserve cognitive functions, the neuroanatomical knowledge about these connections requires careful consideration. The following section therefore reviews the known neural connections between the lateral cerebellar modules and the prefrontal areas of the cerebral cortex. On the basis of this known "hardware," it becomes possible to assess the "software" capabilities of the modules for communicating information.

To assess these "software" capabilities, we examine how the output connections of cerebellar modules can transmit information to the cerebral cortex, and we note a significant fact: These output connections are composed of segregated bundles of nerve fibers, called fascicles. Such segregated bundles of fibers are also found in the design of computing machines, in which such bundling of transmission lines confers a communication advantage on the system. How this communication advantage could be conferred on the cerebro-cerebellar system by the bundling of the transmission lines is therefore explored. We show that such bundling can endow the output fibers of cerebellar modules with an increased capability for transmitting complex information to the cerebral cortex. Further, we indicate how such communications can attain a high level of complexity, comparable to the level of discourse in human language, which can convey complex sequential directions about what to do and when to do it.

We conclude that the combined "hardware" and "software" capabilities in the human cerebro-cerebellar system make it possible for the cerebellum to perform parallel processing on the information that is sent to it from the cerebral cortex (Fig. 1) and to send back to the cerebral cortex the results of this processing in the form of a wide variety of complex messages. Such messages can be transmitted repeatedly between the cerebellum and the cerebral cortex during the learning of complex skills. After repeated processing, these skills can be performed automatically by the system, and the automation can include both motor and cognitive skills, both of which underlie the evolution of language skills (J. A. Fiez and M. E. Raichle, this volume). How the human cerebellum can facilitate the evolution of such skills, by sending appropriate messages to appropriate places in the cerebral cortex at appropriate times, is discussed in detail in Sections II, III, and IV of this chapter.

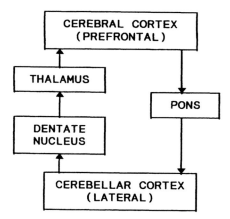

FIG. 1. A neural loop in the cerebro-cerebellar system. The lateral cortex of the cerebellum receives input from the cerebral cortex (via the pons) and sends output to the dentate nucleus, which sends its output (via the thalamus) to the cerebral cortex. Through such neural loops, the cortex of the cerebellum can receive information from the cerebral cortex for processing and can send the results of this processing back to the cerebral cortex via segregated bundles of nerve fibers. These bundles, which preserve the modularity of the cerebellum (see text), can transmit information from small aggregates of neurons in the dentate nucleus to small aggregates of neurons in the cerebral cortex, including the prefrontal cortex.

II. "Hardware" in the Human Cerebellum

A. Evolution of the Lateral Cerebellum

When the human brain evolved, several parts of the cerebellum enlarged greatly, including both the lateral part of the cerebellar cortex and the lateral nucleus to which this cortex sends its output: the dentate nucleus (Larsell and Jansen, 1972). The enlarged parts of the cerebellar cortex are indicated in Table I. They receive input from the cerebral cortex (Fig. 1) and, via the dentate nucleus and via the thalamus, send output to the cerebral cortex.

The enlargement of the dentate nucleus occurred primarily in its ventrolateral part, which is not homologous to the more primitive ventrolateral part that is found in animals such as the rabbit and cat (Ito, 1984). During the evolution of apes, the ventrolateral part of the dentate nucleus began to differentiate into a separate division of the nucleus; in humans this part became clearly differentiated, with unique characteristics (Dow, 1942, 1974, 1988). Its recent evolution and unique characteristics have aroused the interest of theorists in its functional capabilities (Leiner et al, 1993a,b; Ito, 1993; Kim et al., 1994; Gao et al., 1996). To quote from Dow's (1942) review:

TABLE I
Lobes and Lobules of the Cortex in the Human Cerebellum[a]

Medial cortex (vermis)		Lateral cortex (Hemispheres)	
Lobule I	Lingula	Lobule HI	Reduced in humans
Lobule II	Centralis	Lobule HII	Reduced in humans
Lobule III	Centralis	Lobule HIII	Reduced in humans
Lobule IV	Culmen	Lobule HIV	Quadrangular (anterior)
Lobule V	Culmen	Lobule HV	Quadrangular (anterior)
Lobule VI	Declive	Lobule HVI	Quadrangular (posterior)
Lobule VIIA	Folium	Lobule HVIIA	Semilunar
Lobule VIIB	Tuber	Lobule HVIIB	Gracile
Lobule VIII	Pyramis	Lobule HVIII	Biventer
Lobule IX	Uvula	Lobule HIX	Tonsila
Lobule X	Nodulus	Lobule HX	Flocculus

[a] Numbered according to Larsell: Anterior lobe includes lobules I through V; posterior lobe includes lobules VI through IX; and flocculonodular lobe consists of lobule X. The prefix H indicates a hemispheric lobule. Hemispheric lobules HVI, HVIIA, and HVIIB are the ones that grew to enormous size in the human brain. They differentiate late in embryonic life, as do the vermis lobules VI, VIIA, and VIIB, which indicates their late evolution in phylogeny (Larsell and Jansen, 1972).

There is embryological and histological evidence that the dentate nucleus of man consists of two parts: a dorso-medial older part, which is homologous to the so-called dentate nucleus of lower forms, and a very much expanded new part which comprises the bulk of the nucleus in man, the ventrolateral part. These two parts differ in regard to cell types found (Gans, 1924; Demolé, 1927), in regard to iron reaction (Gans, 1924), embryologically (Vogt and Astwazaturow, 1912; Brun, 1917; Demolé, 1927), myelogenetically (van Valkenburg, 1912), and under pathological conditions (Brouwer, 1913; Brun, 1917; Koster, 1926).

B. Evolution of Connections to the Cerebral Prefrontal Cortex

The neural projections from these enlarged parts of the human cerebellum are of particular interest theoretically, but they have not been extensively studied experimentally. More is known about the projections in the brains of nonhuman primates than in the human brain. Although experimental data on nonhuman primates are important, they do not suffice because the additional structures that evolved in the human brain, including the connections of these structures to each other, cannot be studied in experiments on the brains of nonhuman primates. Yet, precisely these additional connections in the human brain, particularly those linking the cerebellum to the language areas of the cerebral cortex, are of paramount importance to theorists who seek to explain the neural basis of human cognitive and linguistic capabilities.

One such cerebellar projection of particular interest is the projection to the cerebral triangular area of the ventrolateral frontal cortex, which is labeled area 45 in Fig. 2. This triangular area is considered to be a part of Broca's language area, which includes area 44 and area 45 on the left side of the brain. In nonhuman primates, a precursor of areas 44 and 45 seems to exist, in the arcuate sulcus of the rhesus monkey (Galaburda and Pandya, 1982; Petrides and Pandya, 1994; Petrides, 1994), which is connected to the pons and thereby can provide input to the cerebellum (Schmahmann and Pandya, 1995; J. D. Schmahmann and D. N. Pandya, this volume). Out-

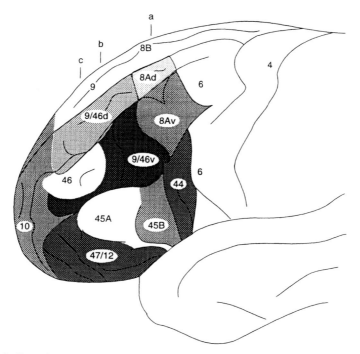

FIG. 2. Frontal areas of the human brain (numbered according to Petrides and Pandya, 1994). The enlargement of the frontal lobe in the human brain, along with the concomitant enlargement of the lateral cerebellum, made it possible for cerebellar projections to reach some additional frontal areas that evolved in the cerebral cortex. Besides reaching motor areas 4 and 6 and the frontal eye area 8, cerebellar projections can reach areas 44 and 45, thereby enabling the cerebellum to participate in human language functions.

put from the cerebellum can also be projected to this arcuate area by projections from the dentate nucleus via the thalamic nuclei, including nucleus X (Schell and Strick, 1984). These findings in nonhuman primates lend credence to the indications in the literature that neural connections have evolved in the human brain from the cerebellum to areas 44 and 45. This literature has been reviewed, and the available data summarized (Leiner *et al.*, 1986, 1987, 1989, 1991, 1993a,b, 1995). Although these data are fragmentary, they indicate not only that the dentate nucleus in some nonhuman primates is connected to cognitive areas of the cerebral cortex, such as prefrontal areas 46 and 9 (Middleton and Strick, 1994; F. A. Middleton and P. L. Strick, this volume), but also that the dentate nucleus in the human brain seems to be connected to the prefrontal area 45 (Van Buren and Borke, 1972), thereby enabling the cerebellum to participate in the evolution of human language skills.

An impediment to understanding how the cerebellum can contribute to human language capabilities is a lack of adequate neuroanatomical data about projections from the ventrolateral part of the human dentate nucleus to the thalamic nuclei that project to prefrontal linguistic areas (Fig. 1). Even without these data, however, it can be said that the conspicuous enlargement of such cerebellar and cerebral structures must confer important advantages on humans because, with the constraints that cranial size imposes on brain enlargement, natural selection would favor only those enlargements that subserve especially advantageous functions. This chapter does not include a discussion of anthropological factors that may explain why this enlarged design of the cerebro-cerebellar system came into existence. Rather, this chapter attempts to analyze what this system can do, based on what is now known about the principles underlying the design of versatile computing systems.

III. "Software" Capabilities Inherent in Cerebro-Cerebellar Connections

From theory and from practical experience, it is known how to design versatile computing systems using component modules that possess only a few basic properties. For example, at least some of the modules in such a system must be able to carry out these two essential processes: (1) They must perform transformations on the streams of information flowing into them, and (2) they must distribute the transformed streams to the right places in the system at the right times. In order to do this, the modules must be able to recognize and respond to three kinds of incoming messages, which can specify "which," "where," and "when." The "which" messages can tell the recipient module which kind of transformation it should perform on the inflowing stream of information; the "where" messages can tell it where in the system to send the transformed stream; and the "when" messages can tell it at what times and under what conditions it should transmit this output.

If a group of interconnected modules is able to recognize, respond to, and emit messages specifying these basic "which, where, and when" conditions, the modules become capable of functioning together as a fully versatile computer system. Such a system is able to perform the many functions that have been attributed to the cerebellum, e.g., timing, sequencing, predicting, planning, error detecting, adaptation, and learning. For this reason, we have tried to assess whether the cerebellum would be capable of communicating these "which, where, and when" messages to the cerebral cortex. To show that it seems able to do so, we now indicate how the output signals of the cerebellum could convey the requisite information.

A. Communication Capabilities Inherent in Segregated Bundles of Fibers

In order to communicate cerebellar information to the cerebral cortex, some modules in the cerebellum must be able to transmit neural signals that can activate or inhibit cerebral neurons. The essential job of the cerebellar module would be to select *which* particular cerebral neurons should or should not be activated in the course of performing the information-processing work. Through such selection, different patterns of activation could be transmitted from cerebellar modules to cerebral areas in the course of the work. To examine how this could be done, we focus first on the cerebellar transmission lines, which (as in computers) transmit signals over small bundles of fibers.

Each bundle, composed of several axons, can transmit at any one time not only a sequence of signals over each individual nerve fiber but also combinations of signals over the several fibers in the bundle (Fig. 3). As the remainder of this chapter will show, such combinations of signals provide the system with important functional advantages: each such combination can serve as a neural "symbol," comparable to the alphabetic symbols on this page, which can be combined into languages for communicating information. How such internal languages can be transmitted in the nervous system, and the advantages that they confer, are illustrated by the dentato-thalamic-cortical pathways (Fig. 1), which will therefore be described in greater detail.

The dentato-thalamic-cortical pathways provide channels through which the "symbols" can flow from specific cerebellar modules to specific places

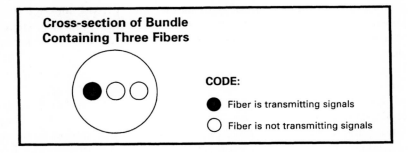

Fig. 3. A bundle of nerve fibers transmitting "symbols." From the small aggregate of neurons whose axons constitute a bundle of nerve fibers, information can be transmitted in a powerful way to the neurons at the receiving end of the bundle: the information can be transmitted not only as a sequence of signals over each individual nerve fiber but also as a combination of signals that can arrive together over the several fibers that compose the bundle. If the neurons at the receiving end can respond differently to each different combination that arrives at any moment, they can react distinctively to a wider variety of messages than can be transmitted over unbundled fibers (see text). For example: In the illustrated bundle of three fibers, eight different combinations of on/off signals can be transmitted, and each different combination can serve as a neural "symbol" ("A" through "H") in the particular neural language through which this bundle communicates information. In a bundle of only six

in the cerebral cortex. Each bundle of cerebellar output fibers apparently affords every module a segregated channel of communication, which does not overlap with the output of other cerebellar modules. This has been shown in experiments on nonhuman primates, where it was found that each narrow zone of the dentate nucleus terminates in a discrete thalamic target, called a "rod," which in turn projects to a discrete "column" in the cerebral cortex (Thach and Jones, 1979; Asanuma et al., 1983; Jones, 1985). Such a "rod-to-column" projection ensures that "symbols" from a small aggregate of neurons in the dentate nucleus can be transmitted to a small aggregate of neurons in the cerebral cortex. The neural bundles that serve as segregated channels of communication in such a pathway can confer an important functional advantage on the system in the following way.

The advantage of bundling the transmission fibers is that such bundling increases enormously the variety of messages that can be carried over a given number of fibers. Why this is so can be seen by considering the activity of a segregated bundle containing N fibers, each of which can transmit trains of pulses from the source of the bundle to the receiving end. At any particular moment, some of the neurons at the source of the bundle may be firing trains of signals into their output fibers while other neurons may be essentially silent. If each individual output fiber were able to convey from the source to the receiver only these two signals (on or off), and if the fibers were *not* bundled, the number of different signals arriving at the receiver at any time would be at most N. But if the fibers are bundled and the signals that reach the receiver can arrive as a combined unit, a *combination* of signals will be received at any time; each different combination of arriving signals can constitute a distinguishably different message (Fig. 3).

It is evident from combinatorial mathematics that the number of distinctive messages will be increased enormously by the bundling of fibers; it will be increased from the N signals, conveyed by the individual fibers, to the 2^N combinations of signals conveyed by the bundle of fibers. For example: a bundle of only 5 fibers would be able to convey $2^5 = 32$ different combinations of signals; a bundle of 10 fibers could convey 1024 combinations of signals; and a bundle of 20 fibers could convey over 1 million combinations of signals.

---

fibers, 64 different "symbols" could be communicated, which is greater than the number of distinguishable phonemes used in spoken languages. Even a small bundle of fibers therefore can provide a group of neurons with a powerful channel for communicating complex messages to another group of neurons.

Each of these different combinatorial messages possesses the potential for producing a significantly different effect on the recipient group of cortical neurons. As the number of such distinguishably different messages increases, therefore, the effects that they can elicit become more varied and complex. It is usually assumed that the information transmitted by a neuron is conveyed by the firing rate (or by the temporal pattern) of the signals coursing down its axon. We have now supplemented that view by suggesting that a stream of signals coursing down a segregated bundle of axons can provide a much more powerful way of communicating information.

B. Communicating via Internal Languages

In information-processing terms, each combination of signals transmitted by a bundle of fibers is called a "symbol" in order to distinguish it from the individual signals arriving over individual transmission lines. Internally, both in computing machines and in neural systems, these different "symbols" can serve as the "alphabet" of an internal language, just as externally, on this page, the different alphabetic symbols (A through Z) serve as the basic symbols of the written English language. Just as these 26 alphabetic symbols (A through Z) can be used for communicating messages in the English language, so too the analogous "alphabetic symbols" that are transmitted by a bundle of fibers can be used for communicating messages in the internal languages of nervous systems (Fig. 3). In the nervous system, a bundle of only six fibers can convey 64 "symbols," which is more than the number of distinguishable phonemes used in spoken languages.

In computing machines, sequences of the internal "alphabetic symbols" are assembled into "words," and the "words" are assembled into "sentences," "paragraphs," and "volumes," in ways that are analogous to the practices followed in the various spoken and written languages used by humans. In these various spoken and written languages, and in computing machines, all such collections of messages are constructed from basic alphabetic symbols that are joined together according to rules of syntax. In the languages used in computing machines, the larger collections of symbolic messages are used for communicating information between large assemblies of hardware, and the smaller collections of symbolic messages are used for small assemblies of hardware.

In the nervous system, too, a correlation can be drawn between hardware size and software size. At the microscopic scale, the combinatorial patterns called "symbols" can be regarded as messages being communicated between small aggregates of neurons, such as the messages communicated from a single cerebellar module (Fig. 3). At the macroscopic scale, the combinatorial patterns can be regarded as symbolic messages being communicated between gross anatomical parts of the system. In order to analyze neural communications at these different microscopic and macroscopic scales, a mathematical model embodying this hierarchy of language levels has been devised (A. L. Leiner, manuscript in preparation). The lowest level corresponds to the patterns of activity in a small group of neurons with an "alphabet" of minimal size, such as a single cerebellar module. The highest level corresponds to the patterns of activity in very large neural assemblies, such as volumes of brain anatomy that are spatially resolvable on functional imaging of the brain (i.e., clusters of pixels on functional magnetic resonance imaging or positron emission tomography).

The utility of this model is that it can provide further insight into the functioning of the internal communication system in the brain. In functional imaging of the brain, activations can be interpreted as manifestations of the symbolic messages being communicated internally at the macro level of the system. For purposes of analysis, these messages can be regarded as neural "words" and "sentences" in some high-level neural "language." Manifestations of these high-level messages can be obtained during the performance of various tasks by human subjects. By analyzing and interpreting such experimental data according to this model, testable predictions can be made concerning the messages that are to be expected during the performance of other related tasks. Testing these predictions would then shed light on the validity of applying these principles of computer design to the functioning of the human brain.

C. COMMUNICATIONS BETWEEN THE CEREBRAL CORTEX AND THE CEREBELLUM

Although the focus of this chapter has been on the cerebellum, our analysis of its communication capabilities may be relevant as well to the cerebral prefrontal cortex. Recent advances in understanding the structure and function of the primate prefrontal cortex (Goldman-Rakic, 1994, 1995) indicate that some similarities exist between it and the cerebellar cortex. Both cortices contain longitudinal "columns" of neurons that can be regarded as modules, and these modules are arrayed perpendicular to the surface layer of the cortex, which contains horizontal fibers that contact many modules. In each cortex, the constituent modules are thought to differ from each other in the domains of information that they receive for processing. Because some modules of the prefrontal cortex receive information from some modules of the cerebellum (e.g., prefrontal areas 9, 46, and 45; see Section II), our analysis of cerebellar communication capabilities may be relevant to these prefrontal portions of the neural loops in the cerebro-cerebellar system (Fig. 1).

In these neural loops, the cerebellar cortex and cerebral cortex are connected in such a way that the symbolic output from one cortex can provide symbolic input to the other cortex. That the cortices are able to communicate with each other by means of symbolic messages is indicated by the organization of their input–output connections. The projections of nerve fibers that descend from the cerebral cortex to the cerebellum appear to be organized into segregated bundles, in which axons descend from discrete areas of the cerebral cortex to discrete areas of the pons (Schmahmann and Pandya, 1992; J. D. Schmahmann and D. N. Pandya, this volume). Conversely, the ascending projections from the cerebellum also are organized into segregated bundles, in which axons from individual cerebellar modules ascend to discrete areas of the cerebral cortex via the thalamus (Jones, 1985; F. A. Middleton and P. L. Strick, this volume). Thus, the segregated organization of these connections makes it possible for the cerebral cortex and cerebellar cortex to communicate sequences of "symbols" to each other, thereby communicating via the internal languages of the system. These cortices can be viewed as carrying out linguistic "conversations" with each other in order to perform collaboratively the processing work that underlies the learning of complex skills.

IV. Conclusions: Combined "Hardware" and "Software" Capabilities

This chapter has attempted to fathom how the cerebellum functions at its highest level of organization, where it acts as a powerful processing partner of the human cerebral cortex. In trying to understand this cerebro-cerebellar system from an information-processing point of view, the requirements are not the same as the requirements for understanding it from a physiological point of view. Whereas the physiologist wishes to understand in detail how individual neurons function, another requirement in studying an information-processing system is to understand how the flow of information can be channeled through the participating modules of the system. This chapter, therefore, has intentionally bypassed the details of how the individual neurons function and concentrates instead on the activity of the participating modules in the system.

A. COMPLEXITY AND VERSATILITY

A key to understanding how the participating modules can function together as a coherent system lies in understanding how they can communicate with each other internally, using complex messages at the level of discourse manifest in human thought and speech. A start toward such understanding can be made by assessing how the system can process streams of neural "symbols" comparable to the alphabetic symbols of human language; how these basic neural symbols can be assembled into neural "words" that could convey complicated directions; and how these directions could specify a cognitive or linguistic action, as well as a motor or sensory action, which a recipient module could execute. For example, cerebellar messages could tell the cerebral cortex how to execute a procedural skill such as searching the memory stores to retrieve a suitable verb (Fiez, 1996).

Having indicated that the known output connections of the cerebellum seem capable of transmitting such complex messages to the cerebral cortex, we now face another important question: What computations does the cerebellum perform internally in order to send appropriate messages to the cerebral cortex? Several possibilities have been considered by theorists. One suggestion, for example, is that the processing performed in the cerebellar modules consists of carrying out a specific computation that is common to a wide variety of different tasks, such as the computation of precision timing (Keele and Ivry, 1990; R. Ivry, this volume). We have advocated a broader proposal, however, in which the cerebellar circuitry is more versatile and enables the module to perform a repertoire of computing operations (including timing). From this repertoire the module can select a sequence of symbolic responses that are appropriate to a specific situation. Such selection will depend both on the information that the module is currently receiving from outside sources and on the internal state of activity already existing within the module. Versatile computing machines are designed to make such selections, and they do it ubiquitously in today's world. In a situation calling for precision timing, for example, the selected sequence would consist mainly of "when" messages (see Section III). How a neural module could make appropriate selections is what we have tried to clarify in our analysis, by examining the "software" capabilities that are inherent in the neuroanatomical "hardware" of the system.

B. RECONCILIATION OF DIVERSE THEORIES

Because our view of human cerebellar capabilities is based both on cerebro-cerebellar evolution and on computing principles, it can offer a reasonable explanation of why the cerebellum has proved useful in the brains of so many other species, under so many different circumstances during the millions of years of vertebrate evolution. The reason, we suggest, is that the function of each cerebellar module is determined not merely by the circuitry within the module, which may be versatile in its computing capabilities, but also by the external input–output connections that evolved between the module and the other parts of the brain (see Section I,B and Section II,B). These external connections enable the modules to perform specific functions; the special information that may be needed for such functions can be transferred into and out of the modules via these external connections, in much the same way that data and special programs are transferred into and out of the modules of general-purpose computer systems.

This view of the cerebellum as a versatile computer can reconcile diverse theories about special-purpose vs general-purpose computing structures in the brain. Some anthropologists, for example, argue that the adaptive evolution of the brain would have produced special-purpose structures (Tooby and Cosmides, 1995), whereas some neuroscientists argue that the multimodal cortical areas seem to have general-purpose properties. In our view, the cerebellum exemplifies both: The circuitry within a cerebellar module, to the extent that it can perform a versatile repertoire of computations, can be regarded as providing some general-purpose capabilities while the input–output connections of the module can provide for special-purpose applications. In our view, too, the cerebellum exemplifies both the "selectionism" and the "instructionism" that some theorists regard as conflicting concepts (Sporns and Tononi, 1994). The cerebellum can exemplify both because the information conveyed to it can be regarded as instructions and data, while the transformation operations that are carried out within the module can be regarded as selections. Also, the cerebellum exemplifies both the parallel processing architecture that is characteristic of artificial neural networks and the symbol-processing capabilities that are characteristic of digital computers, which are both combined compatibly in this structure. We conclude, therefore, that further study of this impressive structure can help resolve these theoretical controversies about brain function in general as well as cerebellar function in particular (Barinaga, 1996).

To summarize: Connections that evolved in the mammalian brain and expanded extensively in higher primates have provided the cerebral cortex with extensive access to the computing capabilities of the cerebellum. These computing capabilities have increased greatly in the human brain, both because the "hardware" in the cerebro-cerebellar system has enlarged enormously and because a concomitant enlargement has occurred in the internal language capabilities of the system. Our assessment of these internal "software" capabilities, combined with data on "hardware" provided by neuroanatomy, has guided us during the past decade in making theoretical predictions, which have been confirmed experimentally, about the involvement of the human cerebellum in cognitive processing (Dow, 1995). These "software" capabilities, no less than the "hardware" of the brain, require further experimental investigation and theoretical analysis if the full capabilities of the human cerebellum are to be appreciated.

Acknowledgments

It is a pleasure to acknowledge our indebtedness to the late Robert S. Dow, the neurologist who collaborated with us for the past dozen years and who initiated the first research project to test our theoretical concepts of cerebellar function (Dow, 1988). We are also indebted to the many experimental neuroscientists who carried out subsequent investigations and sent us their data, which helped to confirm our predictions and strengthen our thesis.

References

Andersen, B. B., Korbo, L., and Pakkenberg, B. (1992). A quantitative study of the human cerebellum with unbiased stereological techniques, *J. Comp. Neurol.* **326,** 549–560.
Asanuma, C., Thach, W. T., and Jones, E. G. (1983). Anatomical evidence for segregated focal groupings of efferent cells and their terminal ramifications in the cerebellothalamic pathway of the monkey. *Brain Res. Rev.* **5,** 267–297.
Barinaga, M. (1996). The cerebellum: Movement coordinator or much more? *Science* **272,** 482–483.
Bloedel, J. R. (1992). Functional heterogeneity with structural homogeneity: How does the cerebellum operate? *Behav. Brain Sci.* **15,** 666–678.
Brouwer, B. (1913). Über hemiatrophia neocerebellaris. *Arch. Psychiat. Nervenkr.* **51,** 539–577.
Brun, R. (1917). Zur kenntnis der bildungsfehler des kleinhirns. *Schweiz. Arch. Neurol. Psychiat.* **1,** 61–123; **2,** 48–105; **3,** 13–88.
Demolé, V. (1927). Structure et connexion des noyaux dentelés du cervelet. *Schweiz. Arch. Neurol. Psychiat.* **20,** 271–294; and **21,** 73–110.
Dow, R. S. (1942). The evolution and anatomy of the cerebellum. *Biol. Rev. Cambridge Phil. Soc.* **17,** 179–220.
Dow, R. S. (1974). Some novel concepts of cerebellar physiology. *Mt. Sinai J. Med.* **41,** 103–119.
Dow, R. S. (1988). Contributions of electrophysiological studies to cerebellar physiology. *J. Clin. Neurophysiol.* **5,** 307–323.
Dow, R. S. (1995). Cerebellar cognition. *Neurology* **45,** 1785–1786.
Fiez, J. A. (1996). Cerebellar contributions to cognition. *Neuron* **16,** 13–15.
Galaburda, A. M., and Pandya, D.N. (1982). Role of architectonics and connections in the study of primate brain evolution. *In* "Primate Brain Evolution" (E. Armstrong and D. Falk, eds.), pp. 203–216, Plenum, New York.

Gans, A. (1924). Beitrag zur kenntnis des aufbaus des nucleus dentatus aus zwei teilen, namentlich auf grund von untersuchungen mit der eisenreaktion. *Z. Neurol. Psychiat.* **93,** 750–755.

Gao, J.-H., Parson, L. M., Bower, J. M., Xiong, J., Li, J. and Fox, P. T. (1996). Cerebellum implicated in sensory acquisition and discrimination rather than motor control. *Science* **272,** 545–547.

Goldman-Rakic, P. S. (1994). The issue of memory in the study of prefrontal function. *In* "Motor and Cognitive Functions of the Prefrontal Cortex" (A.-M. Thierry, J. Glowinski, P. S. Goldman-Rakic, and Y. Christen, eds.), pp. 112–121, Springer-Verlag, Berlin.

Goldman-Rakic, P. S. (1995). Architecture of the prefrontal cortex and the central executive. *In* "Structure and Functions of the Human Prefrontal Cortex" (J. Grafman, K. J. Holyoak, and F. Boller, eds.), Vol. 769, pp. 71–83. *New York Academy of Sciences,* New York.

Ito, M. (1984). "The Cerebellum and Neural Control." Raven Press, New York.

Ito, M. (1993). Movement and thought: Identical control mechanisms by the cerebellum. *Trends Neurosci.* **16,** 448–450.

Jones, E. G. (1985). The Thalamus, pp. 128–131. Plenum Press, New York.

Keele, S. W., and Ivry, R. (1990). Does the cerebellum provide a common computation for diverse tasks? *In* "The Development and Neural Basis of Higher Cognitive Function" (A. Diamond, ed.), Vol. 608, pp. 179–207, New York Academy of Sciences, New York.

Kim, S.-G., Ugurbil, K., and Strick, P.L. (1994). Activation of a cerebellar output nucleus during cognitive processing. *Science* **265,** 949–951.

Koster, S. (1926). Two cases of hypoplasia pontoneocerebellaris. *Acta Psychiat. Kbh.* **1,** 47–83.

Larsell, O., and Jansen, J. (1972). "The Comparative Anatomy and Histology of the Cerebellum," Vol. 3. University of Minnesota Press, Minneapolis.

Leiner, H. C., Leiner, A. L., and Dow, R. S. (1986). Does the cerebellum contribute to mental skills? *Behav. Neurosci.* **100,** 443–454.

Leiner, H. C., Leiner, A. L., and Dow, R. S. (1987). Cerebro-cerebellar learning loops in apes and humans. *Ital. J. Neurol. Sci.* **8,** 425–436.

Leiner, H. C., Leiner, A. L., and Dow, R. S. (1989). Reappraising the cerebellum: What does the hindbrain contribute to the forebrain? *Behav. Neurosci.* **103,** 998–1008.

Leiner, H. C., Leiner, A. L., and Dow, R. S. (1991). The human cerebro-cerebellar system: Its computing, cognitive and language skills. *Behav. Brain Res.* **44,** 113–128.

Leiner, H. C., Leiner, A. L., and Dow, R. S. (1993a). Cognitive and language functions of the human cerebellum, *Trends Neurosci.* **16,** 444–447.

Leiner, H. C., Leiner, A. L., and Dow, R. S. (1993b). The role of the cerebellum in the human brain. *Trends Neurosci.* **16,** 453–454.

Leiner, H. C., Leiner, A. L., and Dow, R. S. (1995). The underestimated cerebellum. *Hum. Brain Mapp* **2,** 244–254.

Middleton, F. A., and Strick, P. L. (1994). Anatomical evidence for cerebellar and basal ganglia involvement in higher cognitive function. *Science* **266,** 458–461.

Passingham, R. E. (1975). Changes in the size and organization of the brain in man and his ancestors. *Brain Behav. Evol.* **11,** 73–90.

Petrides, M. (1994). Frontal lobes and working memory: Evidence from investigations of the effects of cortical excisions in nonhuman primates, *In* "Handbook of Neurophysiology" (F. Boller, H. Spinnler, and J. A. Hendler, eds.) Vol. 9, pp. 59–82. Elsevier, Amsterdam.

Petrides, M., and Pandya, D. N. (1994). Comparative architectonic analysis of the human and the macaque frontal cortex. *In* "Handbook of Neuropsychology" (F. Boller, H. Spinnler, and J. A. Hendler, eds.), Vol. 9, pp. 17–58. Elsevier, Amsterdam.

Schell, G. R., and Strick, P. L. (1984). The origin of thalamic inputs to the arcuate premotor and supplementary motor areas. *J. Neurosci.* **4,** 539–560.

Schmahmann, J. D., and Pandya, D. N. (1992). Course of the fiber pathways to pons from parasensory association areas in the rhesus monkey. *J. Comp. Neurol.* **326,** 159–179.

Schmahmann, J. D., and Pandya, D. N. (1995). Prefrontal cortex projections to the basilar pons in rhesus monkey: Implications for cerebellar contribution to higher function. *Neurosci. Lett.* **199,** 175–178.

Sporns, O., and Tononi, G. (1994). "Selectionism and the Brain." Academic Press, San Diego.

Thach, W. T., and Jones, E. G. (1979). The cerebellar dentatothalamic connection: Terminal field, lamellae, rods and somatotopy. *Brain Res.* **169,** 168–172.

Tooby, J., and Cosmides, L. (1995). Mapping the evolved functional organization of mind and brain. *In* "The Cognitive Neurosciences" (M. S. Gazzaniga, ed.), pp. 1185–1210. MIT Press, Cambridge, MA.

Van Buren, J. M., and Borke, R. C. (1972). "Variations and Connections of the Human Thalamus," Vol. 1, p. 111. Springer, New York.

van Valkenburg, C. T. (1912). Bijdrage tot de kennis einer localisatie in de menschelijke kleine hersenen. *Ned. Tijdschr. Geneesk.* **1,** 6–24.

Vogt, H., and Astwazaturow, M. (1912). Über angeborene kleinhirnerkränkungen mit beiträgen zur entwicklungsgeschichte des kleinhirns. *Arch. Psychiat. Nervenkr.* **49,** 75–203.

CEREBELLAR TIMING SYSTEMS

Richard Ivry

Department of Psychology, University of California Berkeley, California 94720

I. A Modular Approach to Coordination
II. Cerebellar Contribution to Movement Timing
III. Perceptual Deficits in the Representation of Temporal Information
IV. Timing Requirements in Sensorimotor Learning
V. Characterizing the Cerebellar Timing System
VI. Interpreting Cerebellar Activation in Neuroimaging Studies: A Challenge for the Timing Hypothesis?
VII. Conclusions
References

Coordinated movement requires the normal operation of a number of different brain structures. Taking a modular perspective, it is argued that these structures provide unique computations that in concert produce coordinated behavior. The coordination problems of patients with cerebellar lesions can be understood as a problem in controlling and regulating the temporal patterns of movement. The timing capabilities of the cerebellum are not limited to the motor domain, but are utilized in perceptual tasks that require the precise representation of temporal information. Patients with cerebellar lesions are impaired in judging the duration of a short auditory stimulus or the velocity of a moving visual stimulus. The timing hypothesis also provides a computational account of the role of the cerebellum in certain types of learning. In particular, the cerebellum is essential for situations in which the animal must learn the temporal relationship between successive events such as in eyeblink conditioning. Modeling and behavioral studies suggest that the cerebellar timing system is best characterized as providing a near-infinite set of interval type timers rather than as a single clock with pacemaker or oscillatory properties. Thus, the cerebellum will be invoked whenever a task requires its timing function, but the exact neural elements that will be activated vary from task to task. The multiple-timer hypothesis suggests an alternative account of neuroimaging results implicating the cerebellum in higher cognitive processes. The activation may reflect the automatic preparation of multiple responses rather than be associated with processes such as semantic analysis, error detection, attention shifting, or response selection.

I. A Modular Approach to Coordination

The human brain can be described as an evolutionary device geared to make our interactions with the world more efficient. Although we have elaborate mechanisms for perceiving and learning about complex patterns, this information is only useful if we can respond to it in an appropriate manner. Action is the ultimate goal of cognition, and action systems are designed to allow us to achieve our goals in a coordinated and flexible manner.

Given this, it is not surprising that so many parts of the brain are implicated in motor control. A wide variety of neurological disorders can disrupt the production of coordinated behavior. In some of these disorders, such as apraxia, a loss of knowledge about the goal of behavior can be observed (Heilman *et al.,* 1981). However, in most movement disorders, the problem is a loss of coordination. The action may still be purposeful, but the control and execution of the action are disturbed. From this we can create a list of the neural systems involved in coordination and skilled movement. This list would include the motor cortex, the basal ganglia, various brain stem nuclei, and, of course, the cerebellum. But such a list would only provide a description of the functional domain of a neural structure. It would tell us little about how a particular structure contributes to the overall computations required to achieve coordinated behavior.

Research since the mid-1980s has focused on developing a psychological and neural model of the components of coordination (see Helmuth and Ivry, 1997). From this perspective, we would acknowledge that coordinated movement requires the normal operation of a number of different brain structures. However, the emphasis would be on identifying the specific contribution of these different structures. That is, we have worked from a starting assumption that there is a basic modularity to the organization of the motor system. Different neural structures contribute to movement by providing distinct computations, the sum of which will determine whether a particular action is coordinated or not. This modularity notion has been widely applied in the realm of perception. It has not been as well advanced in the motor domain. This is not to say that many, or any, researchers would argue that the basal ganglia and cerebellum perform the same function. Nonetheless, there has been a persistent tendency to describe the functional domain of motor structures in terms of tasks rather than computations. For example, the cerebellum may be described as essential for the production of well-learned movements whereas the cortex is essential for the acquisition of new movement patterns. A modular perspective, however, might emphasize that both structures are involved in both skilled and unskilled movements, but their relative contributions change in accord with varying computational demands.

Figure 1 provides an overview of some of the modules required for the performance of sequential movements. For each neural structure shown on the left side, a component operation is listed on the right side. For example, this overview characterizes the premotor cortex as playing a critical role in movement selection (Deiber *et al.*, 1991; Rizzolatti *et al.*, 1990). Thus, if one is to pick up a glass of water, the premotor areas will determine whether that gesture is made with the right or left hand. The specification of when that movement should occur and the fine tuning of the kinematics of the particular gesture, however, are assigned to other neural structures. For example, the basal ganglia may play a critical role in the switching from one action state to another (Hayes *et al.*, 1995; Robertson and Flowers, 1990).

In this conceptualization, the cerebellum is proposed to play a critical role in establishing the temporal patterns of muscular activation. This chapter reviews some of the evidence that supports the idea that the cerebellum plays a unique role in representing temporal information. A central theme to be emphasized is that this computational capability may be exploited in a variety of task domains. That is, the cerebellum can be viewed as an internal timing system that not only regulates the timing of muscular events, but is also used whenever a precise representation of temporal information is required. This computational demand may arise in percep-

Brain Region	Computations required for sequential movement
Frontal Lobes	• Goal Development
Parietal Cortex	• Spatial Representation and Planning
Premotor and Supplementary Motor Area	• Movement Selection
	Movement Specification and Initiation
Cerebellum	• Temporal patterns of activation
Basal Ganglia	• Switching between different patterns
Motor Cortex	• Movement Execution

FIG. 1. Overview of modules required for performance of sequential movements.

tion and learning, and as such, the cerebellum will be implicated in these nonmotor tasks. But this does not mean that the essential function of the cerebellum has changed. Rather, the domain of cerebellar function has become generalized because these other tasks utilize its timing capability (see Ivry, 1993).

II. Cerebellar Contribution to Movement Timing

While there has been much interest in nonmotor functions of the cerebellum, it is important not to lose sight of the lessons that have been garnered from a century of neurological observation. The foremost signs of cerebellar dysfunction involve a loss of coordination (Holmes, 1939). The springboard for the timing hypothesis stems from consideration of the unique movement problems that result from cerebellar lesions.

Figure 2 depicts the electromyographic (EMG) record associated with a series of movements produced by a patient with a unilateral cerebellar lesion (Hore *et al.*, 1991; see also Hallett *et al.*, 1975). As would be expected with this pathology, the patient's problem were restricted to the ipsilesional side, and thus the nomal records are from movements produced by the same

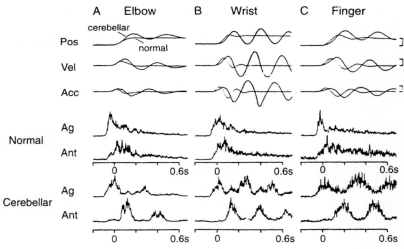

FIG. 2. Kinematic (top) and EMG (bottom) records from single-joint movements produced by a patient with the contralesional hand (Normal) or ipsilesional hand (Cerebellar). From Hore *et al.* (1991), with permission.

patient when using muscles on the contralesional side. On the unimpaired, normal side, the EMG record shows a biphasic pattern, with the antagonist muscle becoming active near the peak of the agonist activity. The antagonist provides the necessary braking force that will allow the movement to terminate at the target location. A different pattern emerges when we look at performance on the impaired side. Here the onset of the antagonist is delayed and fails to brake the movement. The patient ends up being hypermetric, i.e., overshooting the target. In addition, there is an intention tremor as the person hones in on the final goal of the movement, with the delayed antagonist activity producing a series of overshoots. Thus, the disruption of the temporal pattern of muscular events leads to both hypermetria and intention tremor. Other work has indicated that the problem is primarily in the timing of the muscle patterns. For example, these patients can scale the agonist burst when producing movements of different amplitudes (Hore *et al.*, 1991). Thus, while the timing hypothesis does not exclusively account for these results, it does meet the basic criterion of providing a consistent account of the coordination problems observed following cerebellar pathology.

In our research, we have looked for more direct evidence of a cerebellar involvement in timing. We began with a simple motor task in which patients with a variety of neurological disorders were tested on a timed tapping task (Ivry and Keele, 1989). Each trial began with a synchronization phase in which the subject tapped along with a series of computer-generated tones separated by 550 msec. After about 6 sec, the tones were terminated and the subject was instructed to continuing tapping, trying to maintain the target pace. Tapping continued until the subject had produced 30 unpaced intervals. Each subject completed at least 12 trials in this manner.

Overall, the patients tend to approximate the target interval in a consistent manner. The primary focus was on the standard deviation of the intertap intervals. Patients with either cerebellar or cortical lesions were more variable than control subjects on this task (Ivry and Keele, 1989). Although this result is not surprising given that all of the patients were selected because of their motor problems, this crude measure still proved sufficient to differentiate between the patient groups. In particular, patients with Parkinson's disease, a disorder of the basal ganglia, performed comparably to the control subjects.

There are many reasons why an individual may be inconsistent on the timed tapping task. Variability would, of course, be inflated if an internal timing system was damaged, creating noise in a process determining when each response should be initiated. However, a central timer might be intact, but its commands may be inconsistently executed due to problems in the motor implementation system. Wing and Kristofferson (1973) developed a formal model which partitions the total variability observed on this tapping task into two component parts. One component is associated with variability in central control processes including an internal timer. The second component is associated with variability arising from implementation processes. A description of this model and empirical justification for its primary assumptions can be found in Wing (1980). Ivry and Keele (1989; also Ivry *et al.*, 1988) describe neurological evidence in support of the model.

The Wing and Kristofferson model was used to analyze in detail the performance of patients with unilateral cerebellar lesions (Ivry *et al.*, 1988). This group was chosen because the patients could serve as their own control, i.e., their performance could be compared with the impaired, ipsilesional hand against that of their unimpaired, contralesional hand. In this analysis, the cerebellar group was separated into those with medial lesions and those with lateral lesions. Motivation came from consideration of the anatomy of the cerebellum (see Ghez, 1991). The output from the lateral regions is primarily ascending, ending up in the motor and premotor cortex. The output from the medial regions is primarily descending, ending up in brain stem nuclei or synapsing directly on spinal circuits.

The results showed a double dissociation (Fig. 3). When tapping with their ipsilesional hand, the increased variability in patients with lateral lesions was attributed to the central component (Ivry *et al.*, 1988). In contrast, the increased variability for the patients with medial lesions was attributed to the implementation component. This dissociation, coupled

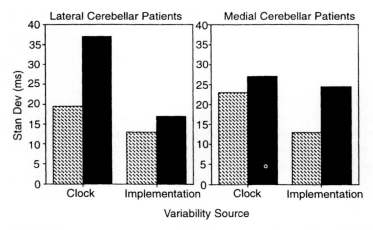

FIG. 3. Estimates of clock (central) and implementation variability on the repetitive tapping task. Hatched bars, unimpaired hand; solid bars, impaired hand. Modified from Ivry *et al.* (1988).

with the following perception results, led to the proposal that the lateral cerebellum plays a critical role in controlling the timing of these periodic movements. This does not exclude the possibility that the medial cerebellum is also involved in timing. Its contribution to coordination may also be time based, but in a manner that anticipates and corrects ongoing movements (e.g., efference copy) rather than one that initiates new motor commands (see Keele and Ivry, 1991).

III. Perceptual Deficits in the Representation of Temporal Information

The domain of the cerebellar timing system extends beyond motor control. One line of support for this hypothesis is that patients with cerebellar lesions are also impaired on perceptual tasks that require precise timing. This work was motivated by correlational studies showing that a common timing system was used in motor and perceptual timing tasks (Keele *et al.*, 1985; Ivry and Hazeltine, 1995). In our patient work, we employed a simple duration discrimination task (Ivry and Keele, 1989). On each trial, two pairs of two tones were presented. The first pair was separated by 400 msec; this provided a standard, reference interval. The second pair of tones formed an interval that was either shorter or longer than 400 msec. The subject made a two-alternative forced choice response. An adaptive psychophysical procedure was used to determine the difference threshold required for each subject to be accurate on approximately 72% of the trials (Pentland, 1980). For a control task, a similar stimulus configuration was used, but here the intensity of the second pair of tones was varied. The subject judged if the second pair was softer or louder than the first pair, and the same adaptive procedure was used to determine the difference threshold for loudness perception.

The results in this study provided a second double dissociation (Fig. 4). Only the patients with cerebellar lesions were impaired on the duration discrimination task (Ivry and Keele, 1989). Patients with Parkinson's disease were as accurate as controls and, more importantly, so were the cortical patients. In fact, the latter group was found to be impaired on the loudness discrimination task, perhaps because some of the lesions extended into the temporal lobe. While we were not particularly interested in the neural basis of loudness perception, the fact that the cortical group was selectively impaired on this task provides further weight to the claim that the cerebellar deficit on the duration discrimination task reflected a specific deficit in time perception.

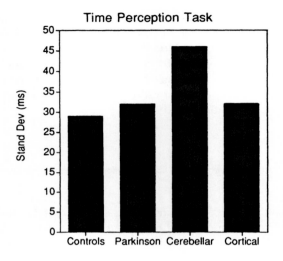
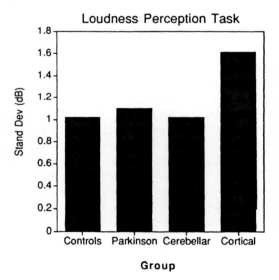

FIG. 4. Perceptual acuity on two psychophysical tasks, duration (top) and loudness (bottom), for four groups of age-matched subjects (Stand Dev, standard deviation). Modified from Ivry and Keele (1989).

Converging evidence for a role of the cerebellum in time perception comes from a positron emission tomography (PET) study (Jueptner et al., 1995). The experimental task in this study was essentially the same as that used in the patient study of duration discrimination. For their control task, the subjects simply listened to the stimuli and made alternating finger responses to control for motor output. Significant increases in blood flow were observed bilaterally, with the foci centered in the superior regions of the cerebellar hemispheres.

Patients with cerebellar lesions are also impaired in their ability to judge the velocity of a moving visual stimulus (Ivry and Diener, 1991). In these studies, the subjects viewed displays consisting of a series of dots that swept across the screen. As a dot reached the end of one side of the screen, a new dot appeared at the other end. This configuration was adopted to minimize tracking eye movements. The subjects were required to judge in which of two successive displays the dots moved fastest. A control task used a similar procedure, but here the location of the dots was adjusted in the vertical plane and the perceptual judgment was position based. The patients were significantly impaired only on the velocity task. This finding has been replicated in another laboratory (Nawrot and Rizzo, 1995), and similar velocity perception deficits have been found with somatosensory stimuli (Grill et al., 1994).

Ivry and Diener (1991) hypothesized that a faulty representation of the velocity of a moving stimulus, a time-based computation, may underlie some of the occulomotor problems observed in these patients (see Leigh and Zee, 1991). For example, in order to generate an appropriate saccade, it is necessary to have an accurate representation of the future position of a moving stimulus. By measuring eye movements, Ivry and Diener (1991) showed that the perceptual deficit was not an indirect consequence of a motor problem. Patients who were able to maintain fixation were as impaired as those who were unable to suppress intrusive eye movements.

IV. Timing Requirements in Sensorimotor Learning

Summarizing to this point, we have consistently observed impairments in patients with cerebellar lesions on tasks designed to require precise temporal processing. The timing hypothesis not only provides an account of the motor problems faced by these patients, but also leads to predicted perceptual deficits. A further source of evidence for this hypothesis comes from a very different paradigm: research showing that the cerebellum is involved in certain types of sensorimotor learning. This work also underscores the usefulness of thinking about brain structures in terms of their component operations rather than in terms of their task domains.

A large literature has been assembled since the mid-1980s demonstrating that the cerebellum plays a critical role in eyeblink conditioning (for reviews, see Thompson, 1990; Yeo, 1991, this volume; Woodruff-Pak, this volume). In the standard eyeblink conditioning paradigm, a tone is used as the conditioned stimulus and precedes an airpuff to the eye by a fixed interval such as 400 msec. Although the animal will make an unconditioned response to the airpuff, learning centers on the fact that, over time, the animal comes to make a conditioned response in anticipation of the airpuff.

A number of studies have demonstrated that animals with cerebellar lesions fail to learn the conditional response. Moreover, learned responses may be abolished following cerebellar lesions. The deficit does not appear to be a motor problem in that the same animals continue to produce the unconditioned response. Similar results have been reported in human literature. Patients with bilateral cerebellar lesions show a severe impairment in eyeblink conditioning (Daum et al., 1993; Topka et al., 1993). Patients with unilateral lesions are more severely disrupted on the side ipsilesional to the lesion (Woodruff-Pak et al., 1996).

Much of the research with this paradigm has focused on identifying the neural circuitry that is critical for this simple form of learning. Our interest in this phenomenon centers on the computational characteristics of eyeblink conditioning. In particular, one critical aspect is the need for a precise representation of the temporal interval between the conditioning stimulus and the unconditioned stimulus. The animal learns to make an anticipatory conditioned response. The fact that the response is anticipatory is what makes it adaptive: by blinking before the airpuff, the animal is able to attenuate the aversive consequences of the airpuff. Equally important, the conditioned eyeblink must be appropriately timed. The animal should not blink too soon or the blink may be finished before the airpuff is delivered. Thus, this paradigm demands that the animal learn not only to associate the two stimuli, but learn the precise temporal relationship between the tone and the airpuff. The evidence that they do just this is shown by the fact that the timing of the learned response is always just prior to the airpuff, regardless of the interstimulus interval (Kehoe et al., 1993; Wickens et al., 1969).

Thus, it could be argued that the cerebellum is not essential for eyeblink conditioning because of some general role in classical conditioning. Rather, the essential reason is because this type of learning requires precise timing and the cerebellum is uniquely suited for providing this type of computation. Classical conditioning of other responses that do not show the same temporal constraints do not involve the cerebellum (Lavond et al., 1984).

Two other points are relevant for the extension of the timing hypothesis to classical conditioning. First, it is of interest to note that at least four computational models of eyeblink conditioning have been proposed since the early 1990s (Bartha *et al.,* 1992; Buonamano and Mauk, 1994; Desmond and Moore, 1988; Grossberg and Schmajuk, 1989). A central feature of all four models is that they contain mechanisms which can provide an explicit representation of temporal information. This feature has not been part of neural models developed for other task domains.

Second, Perrett *et al.* (1993) have observed an important dissociation between the effects of lesions of the deep cerebellar nuclei and lesions of the cerebellar cortex on the conditioned response. Nuclear lesions abolish this learned response, presumably because these lesions destroy all of the output from the cerebellum. In contrast, cerebellar cortical lesions do not abolish the conditioned response. Rather, they disrupt the timing of the responses, with many of the eyeblinks occurring shortly after the onset of the tone. It is as if the delay imposed by the cortex to ensure that the eyeblink occurs at the right point in time is abolished. In this situation, the response is, of course, no longer adaptive.

V. Characterizing the Cerebellar Timing System

The timing hypothesis provides a general description of cerebellar function. This specifies a unique computational role of the cerebellum that is not limited to motor control, but also can account for perceptual and learning deficits associated with cerebellar lesions. An important question, of course, is how to best characterize the timing properties of the cerebellum. When we think of a timing system such as a clock, our first inclination is to think about oscillatory processes such as a pacemaker. Indeed, most models of internal timing systems center on a clock-counter system in which outputs from an endogenous oscillator are stored in a counter mechanism. The full models also include various memory and decision processes, as well as a gating process that can control whether the periodic outputs of the clock are stored (e.g., Gibbon and Church, 1990).

For the most part, these models have been developed to account for behaviors that span intervals considerably longer than those studied in motor and perceptual tasks, usually on the order of at least several seconds. A general assumption in the timing literature is that these same mechanisms would apply for millisecond timing. The basic idea of a pacemaker as a periodic process, however, may not provide the best description of the cerebellar timing system. Rather, the cerebellum may be viewed as providing a near-infinite set of hourglass or interval-type timers (see Ivry and Hazeltine, 1995). Each hourglass represents a particular interval. There may be some sort of organization to these units, a chronotopic map within the cerebellum. However, computational models such as that offered by Buonamano and Mauk (1994) capture the notion of multiple timers, but with a distributed representation that can be shaped as a function of the temporal demands of a particular task.

In our initial tapping studies, each patient served as his or her own control. The tapping performance was compared between trials in which the patients used their contralesional, unimpaired hand with trials in which they used their ipsilesional, impaired hand. Patients with lateral lesions were found to have higher clock variability on the impaired side. This suggested that there are at least two clocks: a damaged one on the lesioned side and an intact one on the normal side.

More recently, the notion of multiple timers has been explored in a variant of the repetitive tapping task. Franz et al. (1996b) examined what would happen when patients with unilateral hemispheric lesions tapped with both hands simultaneously. The results were quite surprising (Fig. 5). In the unilateral condition, our original findings were replicated. Variability was higher when tapping with the ipsilesional hand, and when the Wing–Kristofferson model was applied, the difference was attributed to the central component. However, this difference disappeared in the bimanual condition. Now the two hands were equally consistent. Most interesting, the bad hand became better. We were puzzled as to how to interpret the results. One possibility was that the patient could somehow rely on the good timer. Perhaps it provided a more salient signal and thus dominated performance.

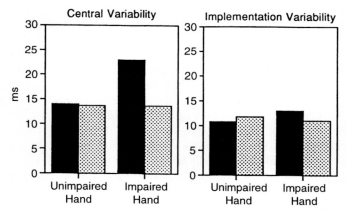

FIG. 5. Estimates of central and implementation variability in cerebellar patients during either unimanual (solid bars) or bimanual (hatched bars) repetitive tapping. Estimates are always of within-hand variability and are plotted for both unimpaired, contralesional hands and impaired, ipsilesional hands. Modified from Franz et al. (1996b).

However, this position was abandoned when the experiment was repeated in healthy subjects (Helmuth and Ivry, 1996). Thirty right-handed college students were asked to tap with their right hand, their left hand, or both hands. The results, in terms of total variability, are shown in Fig. 6. Two points stand out. First, subjects were slightly more consistent when tapping with their dominant hand. This effect was linked to higher implementation variability in the nondominant hand (see also Sergent et al., 1993). Second, and more important, timing variability was reduced when subjects tapped with both hands at the same time, i.e., each hand became more consistent when the two hands moved together. As with data from cerebellar patients, the Wing and Kristofferson (1973) model attributed this bimanual advantage to reduced variability in central control processes.

These results argue against the hypothesis that performance is determined by a single timer in bimanual tapping. If this were the case, we would not expect to see an improvement in the control subjects for both hands. Given these results, Helmuth and Ivry (1996) considered an alternative model to account for the bimanual advantage. This model centers on a simple, yet counterintuitive hypothesis. Specifically, Helmuth and Ivry (1996) postulate that there are two independent timers during bimanual movements: one associated with movements of the right hand and a second associated with movements of the left hand. The bimanual advantage emerges because of a central bottleneck that limits when central motor commands can be issued, a mechanism believed to underlie the ubiquitous

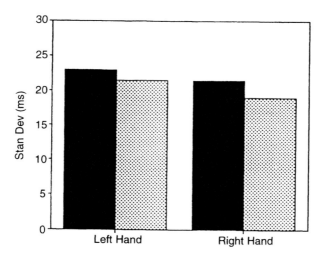

FIG. 6. Total variability in unimanual (solid bars) and bimanual (hatched bars) conditions in healthy subjects on the repetitive tapping task (Stand Dev, standard deviation). Modified from Helmuth and Ivry (1996).

temporal coupling observed in multi-effector actions. The authors propose that this bottleneck must integrate the two timing signals and issue a single command. The bimanual advantage results as a statistical consequence of this integration process (see Helmuth and Ivry, 1996). Thus, by this account, there is not a single timer in bimanual movements, but separate timers that are integrated by a constraint in terms of when central motor commands can be implemented.

As noted earlier, the timing of ipsilesional movements produced by cerebellar patients becomes less variable when accompanied by corresponding movements of the contralesional hand. Moreover, during bimanual movements, the patients show strong interlimb coupling. These two results suggest that the temporal integration process producing the bimanual advantage is not dependent on the cerebellum. Timing and temporal coupling appear to be associated with different neural systems (Franz *et al.*, 1996a).

In other work with normal subjects, a similar reduction in variability was found when subjects made simultaneous finger and foot movements regardless of whether the two limbs were on the same or different sides of the body (Helmuth and Ivry, manuscript in preparation). The generality of this effect is in accord with the notion that the cerebellum is best conceptualized as an array of a near-infinite set of timers. As long as the movements invoke nonoverlapping neural elements, the bimanual advantage will be obtained. In this sense, timing is assumed to reflect a general and unique computational capability of the cerebellum. This capability will be exploited whenever a task requires the timing function of the cerebellum, but the exact neural elements that will be activated will vary from task to task.

VI. Interpreting Cerebellar Activation in Neuroimaging Studies: A Challenge for the Timing Hypothesis?

The idea of multiple timers has potential implications beyond providing a characterization of the cerebellar timing system. It can also lead to a novel perspective on recent functional neuroimaging evidence that points to a role for the cerebellum in cognition. In these studies, the cerebellum is activated even when the experimental and control tasks are equated in terms of their motor requirements (e.g., Jenkins *et al.*, 1994; Kim *et al.*, 1994; Petersen *et al.*, 1988; Raichle *et al.*, 1994).

What do the metabolic events seen in PET and functional magnetic resonance imaging studies reflect? In the neuroimaging studies, there is a common denominator across those conditions that produce significant increases in metabolic activity in the cerebellum. This common denominator is that these conditions are invariably more difficult than the comparison conditions. That is, there seems to be a strong correlation between task difficulty and cerebellar activation. One operational definition of "task difficulty" would be to determine the possible set of responses. By this definition, difficult tasks are those associated with more response alternatives.

A review of the imaging literature indicates that experiments demonstrating a role of the cerebellum in cognition confound the experimental and control tasks in terms of the number of response alternatives. Consider the seminal language study of Petersen *et al.* (1988). In the two critical conditions, the stimuli were identical, the presentation of a single concrete noun. In the control, repeat condition, the subjects simply read the word. As such, there was only a single possible response. In the experimental, generate condition, the subjects had to name a verb that was a semantic associate of the stimulus. Here, we would expect there to be many possible responses, at least in the first trial. For example, if the target word was "apple," possible responses in the generate condition would be "eat," "peel," "throw," and "boot up."

The fact that the cerebellum was more active in the generate condition, even though both conditions required the subjects to articulate a single word, is frequently cited as demonstrating a cognitive role for the cerebellum.[1] However, an alternative, essentially motoric view would be that the cerebellar activation reflects the preparation of all of the possible responses. By this logic, the increased activation in the generate condition results from the fact that there are more potential responses and that the cerebellum does its part to prepare for each one.

Is this cognition? Would the fact that the cerebellum prepares all possible movements imply a cognitive role for this structure? On the other hand, we might imagine a motor theory of cognition in which the choice about which response to make requires the ability to plan that response. As such, the cerebellar contribution would seem to be cognitive. On the other hand, the cerebellum may be viewed as a system that simply goes about the business of preparing its contribution for candidate responses. The cerebellum could be entirely unrelated to the more cognitive aspects of the task.

---

[1] Fiez *et al.* (1992) have provided converging evidence based on a case report of a patient with a cerebellar lesion. This patient had great difficulty on a variety of semantic associate tasks, despite his superior performance on standard neuropsychological tests. However, Helmuth *et al.* (1997) failed to find similar deficits in a group study of cerebellar patients. The reason for this discrepancy remains unclear.

It may simply be a slave system in the sense that for any possible response generated in the cortex, the cerebellum helps prepare to make that response. Some responses will be selected, but the cerebellum need not be part of this more cognitive, decision-making process.

If this hypothesis is correct, the imaging results need not be at odds with the hypothesis that the primary function of the cerebellum involves timing. Perhaps the cerebellum faithfully goes about preparing the temporal patterning of the movements associated with all of the possible responses. The idea that each response requires its individual cerebellar activation is in accord with the hypothesis of Helmuth and Ivry (1996) regarding bimanual movements. In that work, the authors proposed that independent timers were invoked for each effector, even producing synchronous movements. In this reinterpretation of the imaging data, the authors propose that nonoverlapping motor plans are prepared for all candidate responses.

Raichle et al. (1994) have reported that, with practice, the cerebellar activation in the generate task diminishes. Given that subjects report the same semantic associate on successive trials, it would be expected that the number of potential responses also becomes reduced, eventually equal to that of the repeat condition (i.e., are possible response). Reductions in cerebellar activation with practice have been observed in other PET studies (Friston et al., 1992). These results are consistent with the idea that one aspect of skill automatization involves constraining the number of possible responses.

VII. Conclusions

Three main points emerge from this chapter. First, the cerebellum is part of a distributed system for motor control, and it is necessary to identify the component operations of the different structures involved in motor control. The timing hypothesis provides a specific functional role for the unique contribution of the cerebellum.

Second, this timing capability appears to extend beyond motor control into tasks focusing on perceptual processing or sensorimotor learning. As with motor function, these nonmotor tasks depend on the cerebellum when the task requires the precise representation of temporal information. This more cognitive view of the cerebellum still remains grounded in its prominent role in the motor system.

Third, within the cerebellum, time is represented in a distributed manner, with the exact elements required varying from task to task. The cerebellum performs its temporal computations whenever needed. This may include the programming of the temporal aspects for potential movements, regardless of whether that particular movement is produced.

To say that the cerebellum is involved in cognition does not propel our understanding of the system very far. We need to have specific ideas about what this contribution might be. This contribution may be timing, but timing in its many manifestations. This working hypothesis offers a concrete and testable idea about the role of the cerebellum in action, perception, and learning.

Acknowledgments

The author is grateful to Steve Keele and Scott Grafton for their comments on preliminary versions of this chapter. The preparation of this chapter was supported by NIH Grant NS30256.

References

Bartha, G. T., Thompson, R. F., and Gluck, M. A. (1992). Sensorimotor learning and the cerebellum. *In* "Visual Structures and Integrated Functions" (M. Arbib and J. Ewert, eds.). Spring-Verlag, Berlin.

Buonamano, D., and Mauk, M. (1994). Neural network model of the cerebellum: Temporal discrimination and the timing of motor responses. *Neur. Comput.* **6**, 38–55.

Daum, I., Schugens, M., Ackermann, H., Lutzenberger, W., Dichgans, J., and Birbaumer, N. (1993). Classical conditioning after cerebellar lesions in humans. *Behav. Neurosci.* **107**, 748–756.

Desmond, J., and Moore, J. (1988). Adaptive timing in neural networks: The conditioned response. *Biol. Cyber.* **58**, 405–415.

Deiber, M. P., Passingham, R. E., Colebatch, J. G., Friston, K. L. Nixon, P. D., and Frackowiak, R. S. J. (1991). Cortical areas and the selection of movement: A study with positron emission tomography. *Exp. Brain Res.* **84**, 393–402.

Franz, E., Eliassen, J., Ivry, R., and Gazzeniga, M. (1996a). Dissociation of spatial and temporal coupling in the bimanual movements of collosotomy patients. *Psychol. Sci.* **7**, 306–310.

Franz, E., Ivry, R., and Helmuth, L. (1996b). Reduced timing variability in patients with unilateral cerebellar lesions during bimanual movements. *J. Cogn. Neurosci.* **8**, 107–118.

Friston, K. J., Frith, C. D., Passingham, R. E., Liddle, P. F., and Frackowiak, R. S. J. (1992). Motor practice and neurophysiological adaptation in the cerebellum: A positron tomography study. *Proc. R. Soc. Lond. B* **248**, 223–228.

Ghez, C. (1991). The Cerebellum. *In* "Principles of Neural Science" (E. Kandel, J. Schwartz, and T. Jessell, eds.), pp. 626–646. Elsevier, New York.

Gibbon, J., and Church, R. (1990). Representation of time. *Cognition* **37**, 23–54.

Grill, S. E., Hallett, M., Marcus, C., and McShane, L. (1994). Disturbances of kinaesthesia in patients with cerebellar disorders. *Brain* **117**, 1433–1447.

Grossberg, S., and Schmajuk, N. A. (1989). Neural dynamics of adaptive timing and temporal discrimination during associative learning. *Neur. Networks* **2**, 79–102.

Hallett, M., Shahani, B., and Young, R. (1975). EMG analysis of patients with cerebellar lesions. *J. Neurol. Neurosurg. Psychiat.* **38**, 1163–1169.

Hayes, A., Davidson, M., Keele, S., and Rafal, R. (1995). Toward a functional analysis of the basal ganglia. *In* Cognitive Neuroscience Society 2nd annual meeting.

Heilman, K. M., Rothi, L. J., and Valenstein, E. (1981). Two forms of ideomotor apraxia. *Neurology* **32**, 342–346.

Helmuth, L., and Ivry, R. (1996). When two hands are better than one: Reduced timing variability during bimanual movements. *J. Exp. Psychol. Hum. Percept. Perform.* **22**, 278–293.

Helmuth, L., and Ivry, R. (1997). The neuropsychology of sequential behavior. *In* "The Neuropsychology of Movement" (R. Brown, ed.), in press.

Helmuth, L. L., Ivry, R. B., and Shimizu, N. (1997). Preserved performance by cerebellar patients on tests of word generation, discrimination learning, and attention. *Learn. Mem.*, in press.

Holmes, G. (1939). The cerebellum of man. *Brain* **62**, 1–30.

Hore, J., Wild, B., and Diener, H. (1991). Cerebellar dysmetria at the elbow, wrist, and fingers. *J. Neurophysiol.* **65**, 563–571.

Ivry, R. (1993). Cerebellar involvement in the explicit representation of temporal information. *In* "Temporal Information Processing in the Nervous System: Special Reference to Dyslexia and Dysphasia" (P. Tallal, A. M. Galaburda, R. R. Llinas, C. von Euler, eds.), Vol. 682, pp. 214–230. New York Academy of Sciences, New York.

Ivry, R., and Diener, H. C. (1991). Impaired velocity perception in patients with lesions of the cerebellum. *J. Cogn. Neurosci.* **3**, 355–366.

Ivry, R. B., and Hazeltine, R. E. (1995). The perception and production of temporal intervals across a range of durations: Evidence for a common timing mechanism. *J. Exp. Psychol. Hum. Percept. Peform.* **21**, 1–12.

Ivry, R. B., and Keele, S. W. (1989). Timing functions of the cerebellum. *J. Cogn. Neurosci.* **1**, 136–152.

Ivry, R. B., Keele, S. W., and Diener, H. C. (1988). Dissociation of the lateral and medial cerebellum in movement timing and movement execution. *Exp. Brain Res.* **73**, 167–180.

Jenkins, I. H., Brooks, D. J., Nixon, P. D., Frackowiak, R. S. J., and Passingham, R. E. (1994). Motor sequence learning: A study with positron emission. *J. Neurosci.* **14**(6), 3775–3790.

Jueptner, M., Rijntjes, M., Weiller, C., Faiss, J. H., Timmann, D., Mueller, S. P., and Diener, H. C. (1995). Localization of a cerebellar timining process using PET. *Neurology* **45**, 1540–1545.

Keele, S. W., and Ivry, R. (1991). Does the cerebellum provide a common computation for diverse tasks? *In* "The Development and Neural Bases of Higher Cognitive Functions" (A. Diamond, ed.), Vol. 608, pp. 179–211. New York Academy of Sciences, New York.

Keele, S. W., Pokorny, R., Corcos, D., and Ivry, R. (1985). Do perception and motor production share common timing mechanisms? *Acta Psychol.* **60**, 173–193.

Kehoe, E. J., Horne, P. S., and Horne, A. J. (1993). Discrimination learning using different CS-US intervals in classical conditioning of the rabbit's nictitating membrane response. *Psychobiology* **21**, 277–285.

Kim, S., Ugurbil, K., and Strick, P. (1994). Activation of a cerebellar output nucleus during cognitive processing. *Science* **265**, 949–951.

Lavond, D., Lincoln, D., McCormick, D., and Thompson, R. (1984). Effect of bilateral lesions of the dentate and interpositus nuclei in conditioning of heart rate and nictitating membrane/eyelid responses in the rabbit. *Brain Res.* **305**, 323–330.

Leigh, J., and Zee, D. (1991). "The Neurology of Eye Movements (2nd Ed.). Davis, Philadelphia.
Nawrot, M., and Rizzo, M. (1995). Motion perception deficits from midline cerebellar lesions in human. *Vision Res.* **35,** 723–731.
Pentland, A. (1980). Maximum likelihood estimation: The best PEST. *Percept. Psychophys.* **28,** 377–379.
Perrett, S., Ruiz, B., and Mauk, M. (1993). Cerebellar cortex lesions disrupt learning-dependent timing of conditioned eyelid responses. *J. Neurosci.* **13,** 1708–1718.
Petersen, S. E., Fox, P. T., Posner, M. I., Mintun, M., and Raichle, M. E. (1988). Positron emission tomographic studies of the cortical anatomy of single-word processing. *Nature* **331,** 585–589.
Raichle, M. E., Fiez, J. A., Videen, T. O., MacLeod, A. K., Pardo, J. V., Fox, P. T., and Petersen, S. E. (1994). Practice-related changes in human brain functional anatomy during nonmotor learning. *Cereb. Cortex.* **4,** 8–26.
Rizzolatti, G., Gentilucci, M., Camarda, R., Gallese, V., Luppino, G., Matelli, M., and Fogassi, L. (1990). Neurons related to reaching-grasping arm movements in the rostral part of area 6. *Exp. Brain Res.* **82,** 337–350.
Robertson, C., and Flowers, K. (1990). Motor set in Parkinson's disease. *J. Neurol. Neurosurg. Psychiat.* **53,** 583–592.
Sergent, V., Hellige, J., and Cherry, B. (1993). Effects of responding hand and concurrent verbal processing on time-keeping and motor-implementation processes. *Brain Cogn.* **23,** 243–262.
Thompson, R. (1990). Neural mechanisms of classical conditioning in mammals. *Phil. Trans. R. Soc. (Sect. B)* **329,** 161–170.
Topka, H., Valls-Sole, J., Massaquoi, S., and Hallett, M. (1993). Deficit in classical conditioning in patients with cerebellar degeneration. *Brain* **116,** 961–969.
Wickens, D. D., Nield, A. F., Tuber, D. S., and Wickens, C. (1969). Strength, latency, and form of conditioned skeletal and autonomic responses as functions of CS-US intervals. *J. Exp. Psychol.* **80,** 165–170.
Wing, A. (1980). The long and short of timing in response sequences. *In* "Tutorials in Motor Behavior" (G. Stelmach and J. Requin, eds.). North-Holland, New York.
Wing , A., and Kristofferson, A. (1973). Response delays and the timing of discrete motor responses. *Percept. Psychophys.* **14,** 5–12.
Woodruff-Pak, D. S., Papka, M., and Ivry, R. B. (1996). Cerebellar involvement in eyeblink classical conditioning in humans. *Neuropsychology*
Yeo, C. H. (1991). Cerebellum and classical conditioning of motor responses. *In* "Activity-Driven CNS Changes in Learning and Development" (J. R. Wolpaw, J. T. Schmidt, and T. M. Vaughan, eds.), Vol. 627, pp. 292–304. New York Academy of Sciences, New York.

ATTENTION COORDINATION AND ANTICIPATORY CONTROL

Natacha A. Akshoomoff,* Eric Courchesne,†,‡ and Jeanne Townsend†,‡

*Department of Psychology, Georgia State University, Atlanta, Georgia 30303;
†Department of Neurosciences, School of Medicine, University of California, San Diego,
California 92093; and ‡Autism and Brain Development Laboratory, Children's Hospital,
San Diego, California 92123

I. Attention and the Cerebellum
II. Shifting Attention and the Cerebellum
  A. Shifting Attention between Sensory Modalities
  B. Shifting Attention within the Visual Modality: Event-Related Potential and Behavioral Findings
  C. Shifting Attention within the Visual Modality: New Functional Magnetic Resonance Imaging Evidence
III. The Cerebellum and Attention Orienting
  A. Studies of Attention Orienting Using Response Speed
  B. Studies of Attention Orienting Using Response Accuracy
  C. The Issue of Eye Movements
IV. Conclusions
  References

The coordination of the direction of selective attention is an adaptive function that may be one of the many anticipatory tools under cerebellar control. This chapter presents neurobehavioral, neurophysiological, and neuroimaging data to support our hypothesis that the cerebellum plays a role in attentional functions. We discuss the idea that the cerebellum is a master computational system that anticipates and adjusts responsiveness in a variety of brain systems (e.g., sensory, attention, memory, language, affect) to efficiently achieve goals determined by cerebral and other subcortical systems.

I. Attention and the Cerebellum

The coordination of the direction of selective attention (the orienting, distributing, and shifting of attention) is an adaptive anticipatory function and may normally be one of the many anticipatory tools under cerebellar control (Courchesne *et al.*, 1994c). Selective attention to sensory information involves the enhancement of neural responsiveness to stimuli that are anticipated to have signal value for the task at hand, relative to other stimuli. To selectively adjust (orient or shift) the direction of attention requires the quick and accurate alteration of the pattern of enhanced neural responsiveness from the set of stimuli that had previously been anticipated to have signal value to a different set of stimuli that are now anticipated to have signal value.

The effective manipulation of attentional resources undoubtedly underlies much if not all of higher level cognitive function. The dynamic distribution and redistribution of attentional resources clearly require the integrated activity of many subcortical and cortical systems. While there may be overlap in function among these systems, there is also specificity, so that particular attentional operations have been associated with specific brain sites. Data from electrophysiological, cognitive, cerebral blood flow, developmental, animal, and brain lesion studies converge to suggest that in normal processing, selective attention operations require coordinated interactions among frontal, parietal, and brain stem–thalamic–reticular brain systems (e.g., Corbetta *et al.*, 1993; Crick, 1984; Harter and Aine, 1984; Hillyard and Picton, 1987; LaBerge, 1990; Mesulam, 1981; Posner and Petersen, 1990; Rafal and Posner, 1987; Wurtz *et al.*, 1992). Posner and Petersen (1990) have proposed functionally distinct but interrelated brain systems that are responsible for orienting, detecting, and alerting. The posterior attentional system, associated with visual orienting, includes posterior parietal cortex, the lateral pulvinar nucleus of the thalamus, and the superior colliculus. Posner (1988) suggests that the parietal cortex is specifically related to attentional disengagement, whereas midbrain areas are associated with the "moving" of attention, and the thalamus with the engagement of attention. Areas of lateral and medial frontal cortex are proposed to direct an anterior attentional system responsible for target detection, whereas a subsystem responsible for alertness depends on norepinephrine activation from the locus coeruleus and appears to be lateralized in the right hemisphere (Posner and Petersen, 1990).

Stimulation of cerebellar vermian lobules VI and VII in nonbehaving rats causes enhanced neural responsiveness in the superior colliculus to visual stimuli if the cerebellar stimulation occurs *in advance* of the visual stimulus (Crispino and Bullock, 1984). In a similar fashion, stimulation of cerebellar vermian lobules VI and VII in nonbehaving rats causes enhanced neural responsiveness in the hippocampus to somatosensory stimuli if the cerebellar stimulation occurs in advance of the somatosensory stimulus (Newman and Reza, 1979). Moreover, when background luminance is sufficient to reduce to noise levels the colliculus response to a flash stimulus, stimulation of cerebellar vermian lobules VI and VII causes the colliculus response to that flash to emerge above noise levels if the cerebellar stimulation occurs in advance of the visual stimulus (Crispino and Bullock, 1984). Also, these cerebellar influences on neural responsiveness to stimulation apparently apply to visual, auditory, and somatosensory systems at the brain stem, thalamic, and cerebral cortical levels (Crispino and Bullock, 1984).

These studies indicate that the cerebellum is in a position to enhance neural responsiveness in advance of stimulation. Enhanced responsiveness to important anticipated stimuli is a fundamental purpose of selective attention. That the cerebellum can respond selectively (as opposed to merely enhancing general responsiveness) is demonstrated by the numerous physiological studies of association learning in the cerebellum (e.g., McCormick and Thompson, 1984; Thompson, 1986; Yeo et al., 1985).

In addition to being in a physiological position to alter neural responsiveness in brain stem, thalamic, and cortical sensory pathways, the cerebellum also has connections with other systems thought to be even more directly involved in attention, including the reticular activating system, posterior parietal cortex, dorsolateral prefrontal cortex, superior colliculus, cingulate gyrus, and the pulvinar (Dow, 1988; Itoh and Mizuno, 1979; Middleton and Strick, 1994; Moruzzi and Magoun, 1949; Nieuwenhuys et al., 1988; Sasaki et al., 1972, 1979; Schmahmann, 1991; Schmahmann and Pandya, 1989, 1995; Steriade and Stoupel, 1960; Vilensky and Van Hoesen, 1981).

Despite anatomical and physiological data pointing toward an opportunity for the cerebellum to play a role in influencing attention, there have been very few animal studies of attention and the cerebellum. The first was a report of deficits in "sensory attention" in cats with lesions of vermian lobules VI and VII (Chambers and Sprague, 1955)—the same lobules first reported by magnetic resonance imaging (MRI) to be abnormal in infantile autism, a disorder involving severe deficits in sensory responsiveness and attention control (Courchesne et al., 1988, 1994b). There has been a dearth of direct experimentation until the study by Anderson (1994) which reported that the size of the cerebellar molecular layer in normal animals is positively correlated with attention to novelty.

It is clear that deciphering the functional roles of the cerebellum is now an active research area in the cognitive neurosciences (Barinaga, 1996). It is equally clear from neurobehavioral, neurodevelopmental, and neuroimaging evidence (the present volume), as well as older evidence and hypotheses (Moruzzi and Magoun, 1949; Snider, 1950; also see Reis et al., 1973; Watson, 1978; Newman and Reza, 1979; Bernston and Schumacher, 1980; Hamilton et al., 1983; Crispino and Bullock, 1984; Haines and Dietrichs, 1987), that old ideas of the cerebellum will have to step down in favor of more complex and challenging new ones that place the cerebellum in a position to influence perception, attention, cognition, and language.

Whereas in the past there were no published experiments directly testing the question of a role for the cerebellum in attention in humans, our laboratory has actively pursued this question through a series of neurobehavioral, neurophysiological, and neuroimaging studies, the results of which are summarized here.

II. Shifting Attention and the Cerebellum

Courchesne (1985) proposed that the cerebellum, particularly the neocerebellum, contributes to the dynamic, moment-to-moment control of shifts in the mental focus of attention, just as it coordinates motor operations in a precisely timed sequence. Of particular interest was the possibility that maldevelopment of the cerebellum in patients with autism significantly impairs their ability to adjust the mental focus of attention in following the rapidly changing verbal, gestural, postural, tactile, and facial cues that signal changes in social information (Courchesne, 1985, 1987, 1989; Courchesne et al., 1988; Courchesne et al., 1994a,d). If the cerebellum normally plays a role in coordinating shifts of attention, perhaps early damage to the cerebellum affects this ability in patients with autism and thus the subsequent development of joint social attention.

Recent identification of some of the neural abnormalities in patients with autism has allowed investigation of the association of these brain abnormalities with the behavioral deficits that characterize this disorder. The most consistently reported site of neuroanatomic abnormality in autism is the cerebellum. Sixteen MRI and autopsy studies involving more than 240 patients with autism have identified hypoplasia of the cerebellar vermis and hemispheres in the majority of these individuals (review: Courchesne et al., 1994e; Townsend et al., 1996a). In all autopsy cases reported to date, a significant reduction of Purkinje cells (40–60%) has been found in the cerebellar vermis and/or hemispheres of individuals with autism (Arin et al., 1991; Bauman, 1991; Bauman and Kemper, 1985, 1986, 1990; Ritvo et al., 1986; Williams et al., 1980). Because the cerebellar abnormality in autism is developmental, it is possible that maldevelopment of the cerebellum will result in additional structural abnormality of later developing brain systems such as cerebral cortex (Courchesne et al., 1993, 1995; Courchesne, 1995). However, to the extent that behavioral deficits and brain abnormalities in this developmentally disabled population are coincident with behavioral deficits resulting from acquired damage to the same structures, autism provides an interesting model within which to examine brain–behavior relationships. In the case of the cerebellum, autism provides a particularly rich source of information because the underlying cellular abnormality has been identified.

Our laboratory designed a paradigm that included a number of unique features to test these hypotheses. To verify that deficits were not simply the result of patients failing to recognize the cues signaling the command to shift attention to a different focus, each cue was presented within the principal focus of the patient's attention and required a response. To determine how much time was required for patients to fully and successfully execute a single shift of attention, the accuracy of detecting targets that were presented at varying intervals following cues to execute a shift of attention was recorded. To eliminate motor slowness as a possible factor, responses that occurred within 1400 msec of a target were allowed. To verify that patients had not mentally shifted their attention when they failed to respond to targets, the P3b of the event-related potential (ERP) was recorded. The presence of the P3b can be a sign of covert attention, namely the recognition of rare (infrequent) target information independent of overt motor action, and it is small or absent when a target stimulus is ignored or missed (for reviews, see Hillyard and Picton, 1987; Pritchard, 1981). A contrast focus-attention task was also designed to determine if the deficit was restricted to shifting attention.

A. SHIFTING ATTENTION BETWEEN SENSORY MODALITIES

The participants in two of our studies were 5 children with focal cerebellar damage, 1 young adult with an idiopathic cerebellar degenerative disorder, 8 adolescents diagnosed with early infantile autism, a control group of 10 normal children, and a control group of 8 normal adolescents (Akshoomoff and Courchesne, 1992; Courchesne *et al.*, 1994c). Each participated in two tasks that consisted of visual and auditory stimuli. In the shift-attention task (see Fig. 1), participants were instructed to respond to the first rare target stimulus in one sensory modality (e.g., visual), and then shift their attention to the other modality (e.g., auditory) and respond to the first rare target stimulus in that modality. The targets in the attended modality thus served to cue participants to disengage their attention from the current sensory modality and shift their attention as rapidly as possible to the other modality. In the focus-attention task, stimuli were presented in the same manner but participants were required to continuously maintain a focus of attention in one modality. In the "attend visual" portion of the task (see Fig. 1), participants responded to each of the rare visual target stimuli while ignoring all stimuli in the auditory modality. In the "attend auditory" portion, participants responded to each of the rare auditory

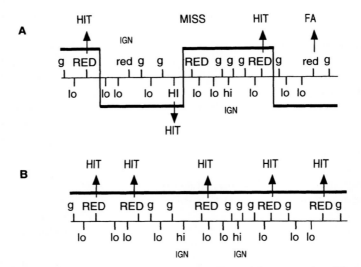

FIG. 1. Schematic drawings of the shift-attention (A) and focus-attention (B) paradigms used in our studies (Akshoomoff and Courchesne, 1992, 1994; Courchesne et al., 1994c). In this example of the auditory/visual shift-attention task, the participant presses a button (arrow) to the first rare visual target stimulus (a RED square). This serves as a cue to shift attention to the auditory stimuli, ignore ("IGN") the rare visual stimuli, and respond to the next rare auditory target (a HI tone). The visual target in the attended "channel" thus serves as a cue to shift attention as rapidly as possible to the other "channel." In the focus-attention task, participants maintained attention in one category of stimuli [e.g., visual stimuli (red and green squares) in the auditory/visual paradigm or form stimuli (circles and ellipses) in the form/color paradigm] and responded to the rare target stimuli ($p = 0.25$) while ignoring all stimuli in the other category [e.g., auditory stimuli (high and low tones) in the auditory/visual paradigm or color stimuli (cyan and dark blue squares) in the form/color paradigm]. HIT, correctly detected target; MISS, a failure to respond to a target; FA, an erroneous response to a rare stimulus that was to be ignored; IGN, a rare stimulus that was correctly ignored.

target stimuli while ignoring all stimuli in the visual modality. All stimuli were randomly ordered and 50 msec in duration. The interstimulus intervals varied randomly between 450 and 1500 msec (10 equal intervals).

In the shift-attention task, accuracy of performance when targets were close together in time (less than 2.5 sec) was compared with performance when targets were farther apart in time (2.5 to 30 sec). Within 2.5 sec of a cue to shift attention to the other modality, both patients with cerebellar damage and patients with autism were significantly worse than normal control participants in correctly detecting target information in the new focus (see Fig. 2). In contrast, all participants had similar performance in the focus-attention tasks which required continuous attention to only a

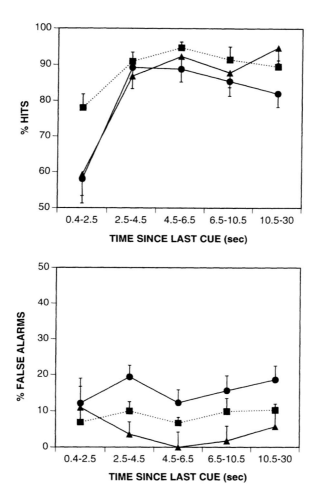

FIG. 2. Mean percentage of hits (top) and mean percentage of false alarms (bottom) for normal children (■), patients with autism (●), and patients with cerebellar damage (▲) in the auditory–visual shift-attention experiment. From Courchesne *et al.* (1994c), with permission from the American Psychological Association.

single principal focus of information for a long period of time. It is also important to note that the time-related deficit in the shift-attention task was not due to slowed response times among the patients. Furthermore, normal control participants were able to complete accurate attentional shifts in as little as 500 msec. This time course is consistent with that reported in previous visual shifting experiments in which controlled attentional shifts were accomplished in 300–400 msec (Reeves and Sperling, 1986; Weichselgartner and Sperling, 1987).

When patients with cerebellar damage and patients with autism missed target information following cues to shift attention, no P3b response was present (Akshoomoff and Courchesne, 1992; Courchesne et al., 1994c). When they did correctly detect such information, P3b responses were present. This is motor-free evidence that when these patients failed to respond to targets that followed a cue to shift attention, it was because they had failed to shift their attention rapidly enough to the new focus of attention (i.e., the other sensory modality) to detect that information.

We propose that the results of these studies suggest that the cerebellum plays an important role in shifting attention. We also propose that the shifting attention deficit we have described in patients with autism is linked to early cerebellar maldevelopment. It seems likely that this deficit is present early in life in these patients and contributes to their characteristic social and cognitive abnormalities. Cerebellar damage sustained after the normal development of affective and communicative reactions and knowledge would not necessarily lead to the regression of these abilities. This may explain why children with acquired cerebellar damage do not necessarily exhibit any autistic-like features but do exhibit a shifting attention deficit.

B. Shifting Attention within the Visual Modality: ERP and Behavioral Findings

In a subsequent study, we employed the same paradigm with 5 children with focal cerebellar damage and a control group of 10 normal children (Akshoomoff and Courchesne, 1994). In this experiment, the stimuli consisted of visual stimuli in two categories: color (cyan and dark blue squares) and form (white outline drawings of an ellipse and a circle). Large P3b responses to correctly detected rare stimuli in the shift-attention task and small responses to correctly ignored stimuli were found for the normal control children within 2.5 sec of a cue to shift attention. In contrast, the patients with cerebellar damage were significantly impaired in rapidly shifting their attention between color and form stimuli. When these patients successfully shifted their attention, their P3b responses to correctly detected targets and correctly ignored rare stimuli within 2.5 sec of a cue to shift attention were similar. These results suggest that their behavioral deficit reflects a deficit in the covert ability to selectively activate and deactivate attention rapidly. Greater attentional selectivity with additional time following a cue to shift attention was evidenced in these patients by improved hit rates and a differential P3b response. As in the auditory/visual shift-attention task, time-related false alarm rates were relatively low, suggesting that the patients' deficit was not in disengaging attention but rather in being able to rapidly move attention from the previously cued channel. Performance in the focus-attention tasks, which required continuous attention to only a single principal focus of information for a long period of time, was similar across both groups. The patients were also not significantly slower in responding to the targets that were correctly detected. The late positive ERP component associated with correct target detections in the shift-attention task in normal controls was diminished in patients with cerebellar damage, also indicative of a shifting attention deficit.

C. SHIFTING ATTENTION WITHIN THE VISUAL MODALITY: NEW FUNCTIONAL MAGNETIC RESONANCE IMAGING EVIDENCE

We now have evidence of neurofunctional activation of the cerebellum during a focused attention task (Allen et al., 1997). This task was similar to that employed in the Akshoomoff and Courchesne (1994) study, using color and shape visual stimuli. Using functional magnetic resonance imaging (fMRI), coronal multislice images of the cerebellum oriented perpendicular to the long axis of the brainstem were obtained during a motor task and a focus-attention task (see Fig. 3). Images were acquired while the task of interest was alternated with a contrasting task (passive visual stimulation during the attention task and rest during the motor task). Activity during the focus-attention task was most prominent in the left superior posterior cerebellum, whereas activity during the motor task was most prominent in the right anterior cerebellum (ipsilateral to movement). These results reflect a double dissociation between these two cerebellar regions with respect to their involvement in attention and motor operations.

Le and Hu (1996) used our shifting attention paradigm (Akshoomoff and Courchesne, 1994) in an fMRI study of the normal human cerebellum. Using the focus-attention task as the control for the shift-attention task (see earlier discussion), their design and analysis procedure eliminated motor contributions to the final reported activation effects. They found that shifting attention between visual stimuli significantly activated the right lateral cerebellar hemisphere in all subjects, and that in some subjects the ventral dentate nucleus was also activated. They concluded that the cerebellum in the normal human is involved in rapid shifting of attention as predicted by previous evidence (Akshoomoff and Courchesne, 1992, 1994; Courchesne et al., 1994c).

These findings are consistent with the hypothesis that the cerebellum plays an important role in tasks requiring anticipating rapid, continuous information. Future studies will use fMRI to determine what areas of the cerebellum are activated during focused and shifting attention tasks, and whether these areas are less activated in patients with cerebellar damage performing attention tasks compared with normal controls.

III. The Cerebellum and Attention Orienting

Data from our studies of the relationships of brain abnormalities and behavioral deficits in infantile autism have provided the basis for investigation of a new and unexpected role for the cerebellum in cognition: the rapid orientation and reorientation of attention.

Models for the role of frontal and parietal cortex in attentional systems have been derived from word processing or visual experimental tasks consisting of discrete trials in which attention is directed to a particular location, followed by a target at either the attended or an unattended location. Among the most widely used tasks to study normal and disordered visual attention are those patterned after a paradigm developed by Posner *et al.*

FIG. 3. Bar graphs show the median percentage activation for six right-handed, healthy, normal volunteers (three male, three female; mean age = 25.8 ± 2.1 years) in two volumes of interest (VOIs) during (A) a selective attention task without any motor response and (B) a motor task performed with the right hand. This demonstrates a statistically significant task by VOI interaction, reflecting a double dissociation between these two cerebellar regions with respect to their involvement in attention and motor operations. Functional activation maps showing the most common sites of activation (i.e., regions where the correlation between data and a model response function exceeded a threshold $r = 0.35$) across subjects overlaid on averaged coronal anatomical cerebellar images further demonstrate the differential neuroanatomical localization of attention and motor operations in the cerebellum. Yellow represents an overlap of three or more subjects, whereas blue represents any two subjects. (C) During the attention task, the most common site of activation was in the most posterior slice position in the left superior posterior cerebellum [the posterior portion of the quadrangular lobule (QuP) and the superior portion of the semilunar lobule (SeS); approximate Talairach coordinates of center of mass ($x = -37$, $y = -63$, $z = -22$)]. (D) During the motor task, the most common site was in the most anterior slice position ipsilateral to the moving right hand in the right anterior cerebellum [the anterior portion of the quadrangular lobule (QuA), the central lobule (C), and the anterior vermis (AVe); approximate Talairach coordinates of center of mass ($x = 7$, $y = -51$, $z = -12$)]. pf, primary fissure; hf, horizontal fissure; PVe, posterior vermis; SeI, inferior portion of the semilunar lobule; Gr, gracile lobule. As per radiological convention, the left cerebellum is on the reader's right and vice versa. Adapted from Allen *et al.* (1997).

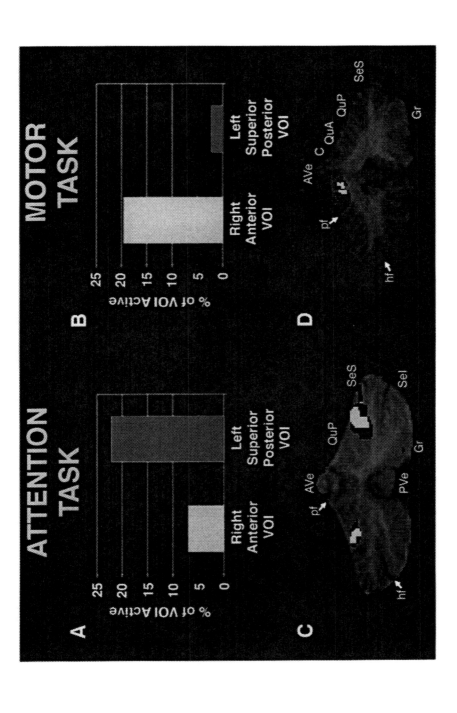

(1984, 1987). The basic implementation of this task uses a visual display of "boxes" to the right and left of a fixation point. The brightening of a box on one side serves as a cue, directing attention to the most probable location of a target stimulus. Results from these studies are robust and have consistently demonstrated that normal participants respond faster and more accurately to targets that appear at previously cued locations (Posner, 1988; Downing, 1988). The facilitated processing following an attention-directing cue is likely to be the result of enhanced sensory processing at the attended locus (e.g., Mangun and Hillyard, 1991).

The results that are most commonly reported from studies using spatial cueing tasks are analyses that have been conceptualized to reflect the engagement, disengagement, and movement of attention described in Posner's attention models (Posner et al., 1984, 1987; Posner and Petersen, 1990). For example, the well-known "validity effect" indexes the facilitation (or cost) associated with an attention-directing cue by comparing the speed of response to a target appearing in a validly cued location to the speed of response to a target appearing in an invalidly cued location. In a substantial body of neuropsychological literature, this effect has been used to demonstrate an association between parietal cortex and spatial attention operations. The "cost" in response time of an incorrect cue is typically small for normal participants, but is significantly greater for patients with damage to parietal cortex, an abnormality that is thought to reflect problems with the disengagement of attention from the cued location.

Validity effects are frequently the only measure examined or reported from spatial cueing tasks. However, these paradigms can be a rich source of additional information about visual attention operations. For example, the amount of time required to orient (i.e., shift and engage) attention can be indexed by the speed of response to targets at a validly cued location as a function of the cue-to-target delay. Normal participants are able to orient attention very rapidly and so respond to target information at a cued location almost as quickly after a 100 msec cue-to-target delay as they do with longer delays. Correspondingly, normal participants show maximal response facilitation from a peripheral cue within 100 msec and the effect of such a cue then diminishes with longer delays (Cheal and Lyon, 1991).

We have done a series of studies using both a traditional spatial orienting task in which attentional facilitation was indexed by speed of response and one in which the critical variable was accuracy of discrimination rather than a motor response. Participants for the studies reviewed here included patients diagnosed with autism, patients with focal brain damage, and normal control participants. The patients with autism are from a group in which we have reported damage to the cerebellar hemispheres bilaterally and to the cerebellar vermis (Courchesne et al., 1988, 1994d; Murakami et *al.*, 1989). All patients with autism had cerebellar vermian lobule VI–VII measures from MRI that were significantly different from those of age-matched normal control participants (see Figs. 4 and 5). Data from autopsy studies would suggest that the reduced size of these lobules as seen on MR reflects a reduced number of Purkinje cells. Data from four patients with acquired unilateral cerebellar damage (two left, two right) who participated in these studies are also reported (see Fig. 6).

A. STUDIES OF ATTENTION ORIENTING USING RESPONSE SPEED

These spatial attention studies were patterned after those developed by Posner *et al.* (1984). The basic visual display for these tasks was a central fixation point marked by a white cross (3 cm height and width) and flanked by 4-cm^2 green boxes on the left and right set at 5° visual angle from the central fixation point. An asterisk 2.5 cm in diameter presented in either box served as the target stimulus. A trial began with onset of the visual display, followed by an attention-directing cue (one of the peripheral boxes was brightened) or by a null cue condition in which there was no informa-

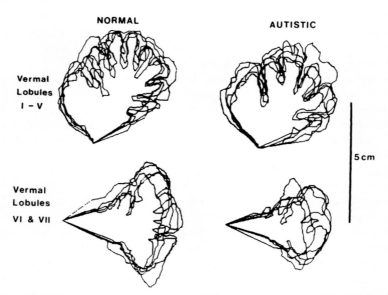

FIG. 4. Tracings show superimposed outlines of vermal lobules of five patients with autism and five control subjects. Lobules VI and VII of the patients are smaller than those of controls whereas lobules I–V are similar in size in both groups. From Courchesne *et al.* (1988), with permission.

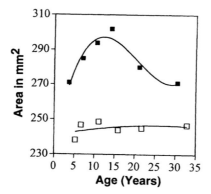

FIG. 5. MRI measures of cerebellar vermain lobules VI–VII in 200 subjects from Courchesne laboratory group studies. Mean area measurement of groups of 15–20 subjects are plotted at the center of age interval for those subjects (□, patients with autism; ■, normal control subjects). Vermian measures are smaller than those of normal subjects at age five and are relatively stable and smaller than normal subjects across the age ranges measured. From Haas *et al.* (1996), with permission.

tion about subsequent target location. Either 100 or 800 msec after the cue onset, the target stimulus was presented in either the right or the left box. The participant's task was to maintain fixation on the central cross and to press a button when the target was detected. In this task, the more quickly attention is directed to a cued location, the faster the response will be to a target at that location. Short cue-to-target delays provide little time to orient attention; longer delays provide more time to orient attention. An *index of attention orienting* is thus provided by the response time (RT) to a target at the cued location as a function of the delay between the onset of an attentional cue and delivery of the target stimulus (100 or 800 msec).

Patients with autism who have developmental damage to the cerebellum were significantly slower to detect a cued target presented shortly after an attentional cue than were control participants, but were as fast as control participants when given more time to orient their attention (Townsend *et al.*, 1996a). With more time to orient attention to a cued location (800 msec vs 100 msec) control participants were slightly faster (18%) whereas patients with autism improved by 42%.

It is this kind of result from patients with autism that prompted us to ask whether this deficit may also be found in patients with acquired cerebellar damage. As a first test, we studied four patients with acquired unilateral lesions of the posterior cerebellum (see Fig. 6). Like the patients with autism, those with acquired cerebellar damage showed significantly greater than normal improvement (36% to ipsilesional targets, 30% to contralesional targets) in RTs to target information at a precued location when they

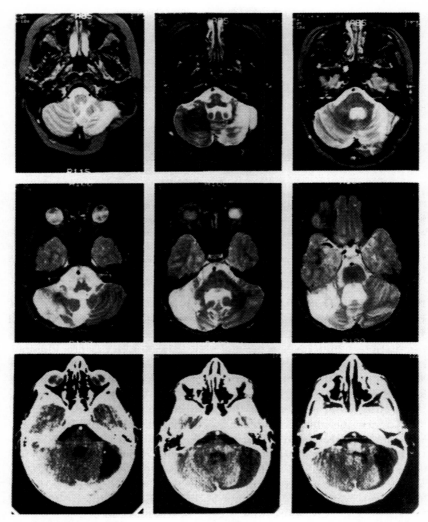

FIG. 6. (Top row) MRI scans of a 32-year-old female with a resected medulloblastoma of the left cerebellum. Lesion of the left lateral posterior cerebellar hemisphere probably extends to the dentate nucleus with the vermis and the left cerebellar tonsil intact. (Middle row) MRI scans of a 10-year-old male with a resected right cerebellar astrocytoma. Lesion of the right posterior lateral cerebellar hemisphere extends to the ventrolateral dentate nucleus. (Bottom row) MRI scans of a 12-year-old female with a resected left cerebellar astrocytoma. Lesion is of the left posterior lateral cerebellar hemisphere. Fourth patient (not shown): Eighteen year-old male with a resected right cerebellar astrocytoma. Lesion of right lateral cerebellar hemisphere extends to midline.

had more time (800 msec vs 100 msec) to orient their attention. Time-related response delays at the cued location have not been reported in the literature for other patient groups (e.g., those with parietal, frontal, midbrain, or thalamic lesions). Figure 7 displays these data along with comparison data for control participants and patients with cortical lesions.

Additional evidence that the patients with cerebellar damage were slow to orient attention to a cued location is that RT facilitation from attentional cueing (i.e., faster RT to target at the cued versus uncued location) was stronger in these patients after longer cue-to-target delays. In contrast, the effect of attentional precueing was maximal after 100 msec and decreased with longer delays among control participants (see Fig. 7). Patients with acquired unilateral cerebellar damage failed to use ipsilesional cuing infor-

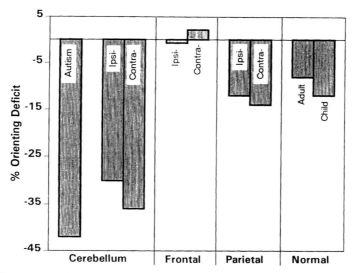

FIG. 7. Patients with cerebellar abnormality or damage are significantly slower to orient attention to a cued location than are age-matched normal control subjects or patients with cortical lesions. Orienting deficits are the percentage difference in speed of response to a target when the subject has only 100 msec to orient attention to the cued location compared to when there are 800 msec to orient attention to the cued location. Results for patients with cerebellar or cortical lesions are shown for responses to targets ipsilateral or contralateral to the lesion site. Patients with autism have bilateral cerebellar abnormality. Results are shown for patients with cerebellar damage [including patients with autism from Townsend et al. (1996) and four patients with unilateral cerebellar damage (see Fig. 6)]; pilot data are from five patients with unilateral frontal lobe damage; data for patients with parietal damage are from Posner et al. (1984); data from child control subjects are from Nichols et al. (1995); and adult control subjects are from Townsend et al. (1996). Note that the patients with frontal damage are slower to orient attention with longer cue-to-target delays, suggesting an inability to sustain attention over time.

mation with only 100-msec cue-to-target delays, i.e., their RTs to cued and uncued locations were not significantly different. By the 800-msec delay, however, these patients used cueing information in both visual fields.

B. Studies of Attention Orienting Using Response Accuracy

In the studies just discussed, patients with cerebellar damage had RTs that were as fast as those of age-matched normal control participants when given adequate time to orient their attention. This was a strong argument that the deficits we observed in this task in these patients were attributable to slowed attention orienting and not to slowed motor response processes. However, it is of some concern for the study of pathologic function (especially in the case of patients with brain lesions that may affect motor systems) that an impaired motor response may introduce artifact or mask attention-related effects. If, for example, the execution of a motor response is extremely slowed, differences in attentional processes may not be reflected in response time differences, i.e., it may be difficult to know whether long RTs are due to motor or processing or attentional impairment.

To control for these concerns, we developed a task in which performance depended on the speed of perceptual processing, not on the speed of motor response. The basic visual display was the same as in the previous study, but this new version required a target discrimination rather than a simple target detection. In this task, attention was cued (by brightening one of the boxes) to either the left or right of a fixation point, a target (a block letter "E") was presented at either the cued (80% probability) or the uncued location, and then masked in 50 msec. Participants were asked to move a joystick to indicate the direction (up, down, left, or right) in which the target pointed. Participants unable to orient attention to the location of the target before the target mask onset would perform at chance (25%).

Just as in the previous task, patients with autism who had developmental cerebellar abnormalities were slow to orient attention to a cued location (Townsend *et al.,* 1996b). Normal control participants performed well at the cued location within a very short time following the cue, and so were only slightly more accurate with more time to orient attention (from 89% accuracy at the 100-msec delay to 97% accuracy at the 800-msec delay). Patients with autism, however, performed poorly when they had little time to orient their attention to the target location, but were much improved with a longer delay between cue and target (from 58% accuracy at the 100-msec delay to 80% accuracy at the 800-msec delay). Additionally, this slowed orienting of attentional resources was reflected in the degree to which attention facilitated performance at a cued as compared to an uncued location. As in the previous study, normal control participants showed maximal performance facilitation within 100 msec following an attentional cue. Patients with autism showed no performance facilitation with only 100 msec to orient their attention, but significant performance facilitation when they had more time to orient to a cued location. Pilot testing of one patient with a large lesion of the left cerebellar hemisphere (see Fig. 8) showed similar but lateralized effects. This patient was quite slow to orient attention to the ipsilesional location. Note that this is the same effect that the four patients with unilateral cerebellar damage showed in the previously described study.

C. THE ISSUE OF EYE MOVEMENTS

Since the cerebellum may be involved in the control of saccadic eye movement (for review, see Noda, 1991), the results from these studies raise questions about the functional associations among spatial orienting of attention and eye movement. Typically, the head and eyes are moved with the focus of attention to maximize the processing of relevant information. Attention can be moved covertly, however, without the usual accompanying motor adjustments (Posner et al., 1980; for a review see Goldberg and Segraves, 1987). Although attention and gaze can be manipulated independently, it is a reasonable hypothesis that both overt and covert attention

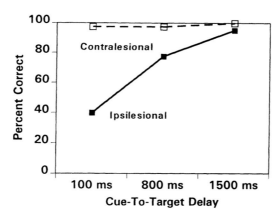

FIG. 8. Orienting deficit on spatial target discrimination task in a 10-year-old patient with a left unilateral cerebellar lesion (see Fig. 6). At the ipsilesional, but not the contralesional location, the patient is much less accurate at shorter than at longer cue-to-target delays, indicating slowed orienting of attention to ipsilesional information.

are supported by the same brain systems. Operations required to prepare for an overt or a covert orienting of attention (e.g., selection of a relevant location, encoding of that location, preparation of motor programs) would be identical until the final execution of a motor response. There is, in fact, some evidence that damage to systems controlling eye movement can produce concomitant damage to covert attentional movement as well. Patients with progressive supranuclear palsy have a progressive inability to make vertical eye movements that results from progressive degeneration of the superior colliculus. These patients are also unable to move attention covertly in the vertical dimension (Rafal et al., 1988).

Our data suggest that this sort of concomitant dysfunction of eye movement and covert attention may also occur with cerebellar damage. From animal models there is evidence for the involvement of the cerebellum in both overt and covert spatial attention. A study of spatial learning in mice with developmental Purkinje cell degeneration (*pcd*) concluded that the cerebellar abnormality in these animals results in significant spatial deficits that are independent of their motor dysfunction (Goodlett, et al., 1992). A study of orienting gaze shifts in cats used muscimol injections to inactivate the fastigial nucleus and found that subsequent saccadic movement away from the inactivated side was increasingly hypometric as the required movement amplitude increased (Goffart and Pelisson, 1994). Ipsiversive gaze shifts, however, were hypermetric with a constant error that was independent of the target eccentricity or the initial gaze position. The authors of this study concluded that the fastigial inactivation not only produced dysmetria, but also interfered with the process of encoding a target location.

IV. Conclusions

A recent attempt to account for a wide range of physioanatomical facts theorizes that the cerebellum is a master computational system that adjusts responsiveness in a variety of networks to attain a prescribed goal (Courchesne, 1995; Courchesne et al., 1994c). These networks include those thought to be involved in declarative memory, working memory, attention, arousal, affect, language, speech, homeostasis, and sensory modulation as well as motor control. This may require the cerebellum to implement a succession of precisely timed and selected changes in the pattern or level of neural activity in these diverse networks. We hypothesized that the cerebellum does this by encoding ("learning") temporally ordered sequences of multidimensional information about external and internal events (effector, sensory, affective, mental, autonomic), and, as similar sequences of external and internal events unfold, they elicit a readout of the full sequence in advance of the real-time events. This readout is sent to and alters, in advance, the state of each motor, sensory, autonomic, attentional, memory, or affective system which, according to the previous "learning" of this sequence, will soon be actively involved in the current real-time events. So, in contrast to conscious, longer time-scale anticipatory processes mediated by cerebral systems, output of the cerebellum provides moment-to-moment, unconscious, very short time-scale, anticipatory information.

The results from our neurobehavioral and neurophysiological studies showing deficits in shifting and orienting attention in patients with cerebellar damage, as well as new fMRI studies showing cerebellar activation during focused attention and shifting attention in normal adults, suggest that the cerebellum plays an important role in several aspects of selective attention. Cerebral cortical regions appear to be primarily responsible for generating the commands for enhancement and inhibition of different sources of information and sensory signals. Our data suggest that the cerebellum plays an important role in the execution of these commands in order to optimize the quality of sensory information for coordinating the direction of selective attention. We have demonstrated that this includes the shifting, distribution, and orienting of attention. We have also hypothesized that these findings may reflect one of the many anticipatory tools under cerebellar control. Indeed, modern neuroimaging (see Table I) and behavioral studies have shown that cerebellar activation is associated with a wide range of sensory and cognitive functions, in addition to motor ones.

TABLE I
HUMAN PET AND FUNCTIONAL MRI STUDIES THAT DEMONSTRATE CEREBELLAR ACTIVATION IS ASSOCIATED WITH A WIDE RANGE OF SENSORY AND COGNITIVE FUNCTIONS THAT ARE ALSO DEFICIENT IN PATIENTS WITH AUTISM

The cerebellum is active in	
Attention	Allen et al. (1997)
Working memory	Desmond et al. (1995); Klingberg et al. (1995)
Long-term memory	Andreasen et al. (1995); Grasby et al. (1993)
Paired-associated memory	Busatto et al. (1994)
Language generation	Raichle et al. (1994)
Learning novel skills	Raichle et al. (1994)
Complex problem solving	Kim et al. (1994)
Concept formation	Nagahama et al. (1995)
Sequencing	Rao et al. (1995)
Mental spatial exploration and imagery	Decety et al. (1994); Mellet et al. (1995)
Motor control	Allen et al. (1996); Cuenod et al. (1993)
Speech	Artiges et al. (1995)
Sensory discrimination	Gao et al. (1996)

References

Akshoomoff, N. A., and Courchesne, E. (1992). A new role for the cerebellum in cognitive operations. *Behav. Neurosci.* **106,** 731–738.

Akshoomoff, N. A., and Courchesne, E. (1994). ERP evidence for a shifting attention deficit in patients with damage to the cerebellum. *J. Cogn. Neurosci.* **6,** 388–399.

Allen, G., Buxton, R. B., Wong, E. C., and Courchesne, E. (1997). Attentional activation of the cerebellum independent of motor involvement. *Science* **275,** 1940–1943.

Anderson, B. (1994). The volume of the cerebellum molecular layer predicts attention to novelty in rats. *Brain Res.* **641,** 160–162.

Andreasen, N. C., O'Leary, D. S., Arndt, S., Cizadlo, T., Hurtig, R., Rezai, K., Watkins, G. L., Boles Ponto, L. L., and Hichwa, R. D. (1995). Short-term and long-term verbal memory: A positron emission tomography study. *Proc. Natl. Acad. Sci. USA* **92,** 5111–5115.

Arin, D. M., Bauman, M. L., and Kemper, T. L. (1991). The distribution of Purkinje cell loss in the cerebellum in autism. *Neurology* **41**(Suppl. 1), 307.

Artiges, E., Giraud, M. J., Mazoyer, B., Trichard, C., Mallet, L., de la Caffiniere, H., Verdys, M., Syrota, A. M., and Martinot, J. L. (1995). Word production: A brain activation study with H_2O^{15}. *Hum. Brain Map.* **3**(Suppl. 1), 227.

Barinaga, M. (1996). The cerebellum: Movement coordinator or much more? *Science* **272,** 482–483.

Bauman, M. L. (1991). Microscopic neuroanatomic abnormalities in autism. *Pediatrics* **87,** 791–796.

Bauman, M. L., and Kemper, T. L. (1985). Histoanatomic observations of the brain in early infantile autism. *Neurology* **35,** 866–874.

Bauman, M. L., and Kemper, T. L. (1986). Developmental cerebellar abnormalities: A consistent finding in early infantile autism. *Neurology* **36**(Suppl. 1), 190.

Bauman, M. L., and Kemper, T. L. (1990). Limbic and cerebellar abnormalities are also present in an autistic child of normal intelligence. *Neurology* **40**(Suppl. 1), 359.

Bernston, G. G., and Shumacher, K. M. (1980). Effects of cerebellar lesions on activity, social interactions, and other motivated behaviors in the rat. *J. Comp. Physiol. Psychol.* **94,** 706–717.

Busatto, G. F., Costa, D. C., Ell, P. J., Pilowsky, L. S., David, A. S., and Kerwin, R. W. (1994). Regional cerebral blood flow (rCBF) in schizophrenia during verbal memory activation: A 99m Tc-HMPAO single photon emission tomography (SPECT) study. *Psychol. Med.* **24,** 463–472.

Chambers, W. W., and Sprague, J. M. (1955). Functional localization in the cerebellum. II. Somatotopic organization in cortex and nuclei. *Arch. Neurol. Psychiat.* **74,** 653–680.

Cheal, M. L., and Lyon, D. R. (1991). Central and peripheral precuing of forced-choice discrimination. *Q. J. Exp. Psychol.* **43A,** 859–880.

Corbetta, M., Miezin, F. M., Dobmeyer, S., Shulman, G. L., and Petersen, S. E. (1993). A PET study of visuospatial attention. *J. Neurosci.* **13,** 1202–1226.

Courchesne, E. (1985). "The Missing Ingredients in Autism." Paper presented at the Conference on Brain and Behavioral Development, Biosocial Dimension, Elridge, MD.

Courchesne, E. (1987). A neurophysiologic view of autism. *In* "Neurobiological Issues in Autism" (E. Schopler and G. Mesibov, eds.). Plenum, New York.

Courchesne, E. (1989). Neuroanatomical systems involved in infantile autism: The implications of cerebellar abnormalities. *In* "Autism: New Perspectives on Diagnosis, Nature and Treatment" (G. Dawson, ed.), pp. 119–143. Guilford, New York.

Courchesne, E. (1995). Infantile autism. 2. A new neurodevelopmental model. *Int. Pediat.* **10,** 86–96.

Courchesne, E., Chisum, H., and Townsend, J. (1994a). Neural activity-dependent brain changes in development: Implications for psychopathology. *Dev. Psychopathol.* **6,** 697–722.
Courchesne, E., Press, G. A., and Yeung-Courchesne, R. (1993). Parietal lobe abnormalities detected on magnetic resonance images of patients with infantile autism. *Am. J. Roentgenol.* **160,** 387–393.
Courchesne, E., Saitoh, O., Yeung-Courchesne, R., Press, G., Haas, R., Lincoln, A., and Schreibman, L. (1994b). Abnormality of vermian lobules VI and VII in patients with infantile autism: Identification of hypoplastic and hyperplastic subgroups. *Am. J. Roentgenol.* **162,** 123–130.
Courchesne, E., Townsend, J., Akshoomoff, N. A., Saitoh, O., Yeung-Courchesne, R., Lincoln, A., James, H., Haas, R. H., Schreibman, L., and Lau, L. (1994c). Impairment in shifting attention in autistic and cerebellar patients. *Behav. Neurosci.* **108,** 848–865.
Courchesne, E., Townsend, J. P., Akshoomoff, N. A., Yeung-Courchesne, R., Press, G. A., Murakami, J. W., Lincoln, A. J., James, H. E., Saitoh, O., Egaas, B., Haas, R. H., and Schreibman, L. (1994d). A new finding: Impairment in shifting attention in autistic and cerebellar patients. *In* "Atypical Cognitive Deficits in Developmental Disorders: Implications for Brain Function" (S. H. Broman and J. Grafman, eds.), pp. 101–137. Lawrence Erlbaum, Hillsdale, NJ.
Courchesne, E., Townsend, J., and Chase, C. (1995). Neurodevelopmental principles guide research on developmental psychopathologies. *In* "A Manual of Developmental Psychopathologies" (D. Cicchetti and D. Cohen, eds.), Vol. 2, pp. 195–226. Wiley, New York.
Courchesne, E., Townsend, J., and Saitoh, O. (1994e). The brain in infantile autism: Posterior fossa structures are abnormal. *Neurology* **44,** 214–223.
Courchesne, E., Yeung-Courchesne, R., Press, G. T. A., Hesselink, J. R., and Jernigan, T. L. (1988). Hypoplasia of cerebellar lobules VI and VII in infantile autism. *N. Engl. J. Med.* **318,** 1349–1354.
Crick, F. (1984). Function of the thalamic reticular complex: The searchlight hypothesis. *Proc. Natl. Acad. Sci. USA* **81,** 4586–4590.
Crispino, L., and Bullock, T. H. (1984). Cerebellum mediates modality-specific modulation of sensory responses of the midbrain and forebrain in rat. *Proc. Natl. Acad. Sci. USA* **81,** 2917–2920.
Cuenod, C. A., Zeffiro, T., Pannier, L., Pose, S., Bonnerot, V., Jezzard, P., Turner, R., Frank, J. A., and LeBihan, D. (1993). "Functional Imaging of the Human Cerebellum during Finger Movement with a Conventional 1 5 T MRI Scanner." Abstract from the 12th Annual Meeting of the Society of Magnetic Resonance in Medicine Workshop.
Decety, J., Perani, D., Jeannerod, M., Bettinardi, V., Tadary, B., Woods, R., Mazziotta, J. C., and Fazio, F. (1994). Mapping motor representations with positron emission tomography. *Nature* **371,** 600–602.
Desmond, J. E., Babrieli, J. D. E., Ginier, B. L., Demb, J. B., Wagner, A. D., Enzman, D. R., and Glover, G. H. (1995). A functional MRI (fMRI) study of cerebellum during motor and working memory tasks. *Soc. Neurosci. Abstracts* **21,** 1210.
Dow, R. S. (1988). Contributions of electrophysiological studies to cerebellar physiology. *J. Clin. Neurophysiol.* **5,** 307–323.
Downing, C. J. (1988). Expectancy and visual-spatial attention: Effects on perceptual quality. *J. Exp. Psychol. Hum. Percept. Perform.* **14,** 188–202.
Gao, J.-H., Parsons, L. M., Bower, J., Xiong, J., Li, J., and Fox, P. T. (1996). Cerebellum implicated in sensory acquisition and discrimination rather than motor control. *Science* **272,** 545–547.
Goffart, L., and Pelisson, D. (1994). Cerebellar contribution to the spatial encoding of orienting gaze shifts in the head-free cat. *J. Neurophysiol.* **72,** 2547–2550.

Goldberg, M. E., and Segraves, M. A. (1987). Visuospatial and motor attention in the monkey. *Neuropsychologia* **25**, 107–118.

Goodlett, C. R., Hamre, K. M., and West, J. R. (1992). Dissociation of spatial navigation and visual guidance performance in Purkinje cell degeneration (pcd) mutant mice. *Behav. Brain Res.* **47**, 129–141.

Grasby, P. M., Frith, C. D., Friston, K. J., Bench, C. F., Frackowiak, R. S. J., and Dolan, R. J. (1993). Functional mapping of brain areas implicated in auditory-verbal memory function. *Brain* **116**, 1–20.

Haas, R. H., Townsend, J., Courchesne, E., Lincoln, A. J., Schreibman, L., and Yeung-Courchesne, R. (1996). Neurologic abnormalities in infantile autism. *J. Child Neurol.* **11**, 84–92.

Haines, D. E., and Dietrichs, E. (1987). On the organization of interconnections between the cerebellum and hypothalamus. *In* "New Concepts in Cerebellar Neurobiology" (J. S. King, ed.), pp. 113–149. R. Liss, New York.

Hamilton, N. G., Frick, R. B., Takahashi, T., and Hopping, M. W. (1983). Psychiatric symptoms and cerebellar pathology. *Am. J. Psychiat.* **140**, 1322–1326.

Harter, M. R., and Aine, C. J. (1984). Brain mechanisms of visual selective attention. *In* "Varieties of Attention" (R. Parasuraman and D. R. Davies, eds.), pp. 293–321. Academic Press, New York.

Hillyard, S. A., and Picton, T. W. (1987). Electrophysiology of cognition. *In* "Handbook of Physiology" (F. Plum, ed.), Vol. 5, pp. 519–584. American Physiological Society, Bethesda, MD.

Itoh, K., and Mizuno, N. (1979). A cerebello-pulvinar projection in the cat as visualized by the use of antero-grade transport of horseradish peroxidase. *Brain Res.* **106**, 131–134.

Kim, S.-G., Ugurbil, K., and Strick, P. L. (1994). Activation of a cerebellar output nucleus during cognitive processing. *Science* **265**, 949–951.

Kitano, K., Ishida, Y., Ishikawa, T., and Murayama, S. (1976). Responses of extralemniscal thalamic neurones to stimulation of the fastigial nucleus and influences of the cerebral cortex in the cat. *Brain Res.* **106**, 172–175.

Klingberg, T., Roland, P. E., and Kawashima, R. (1995). The neural correlates of the central executive function during working memory: A PET study. *Hum. Brain Mapp.* **3**(Suppl. 1), 414.

LaBerge, D. (1990). Thalamic and cortical mechanisms of attention suggested by recent positron emission tomographic experiments. *J. Cogn. Neurosci.* **2**, 358–372.

Le, T. H., and Hu, X. (1996). Involvement of the cerebellum in intramodality attention shifting. *NeuroImage* **3**, s246.

Mangun, G. R., and Hillyard, S. A. (1991). Modulations of sensory-evoked brain potentials indicate changes in perceptual processing during visual-spatial priming. *J. Exp. Psychol. Hum. Percep. Perform.* **17**, 1057–1074.

McCormick, D. A., and Thompson, R. F. (1984). Cerebellum: Essential involvement in the classically conditioned eyelid response. *Science* **223**, 296–299.

Mellet, E., Crivello, F., Tzourio, N., Joliot, M., Petit, L., Laurier, L., Denis, M., and Mazoyer, B. (1995). Construction of mental images based on verbal description: Functional neuroanatomy with PET. *Hum. Brain Mapp.* **3**(Suppl. 1), 273.

Mesulam, M. M. (1981). A cortical network for directed attention and unilateral neglect. *Ann. Neurol.* **10**, 309–315.

Middleton, F. A., and Strick, P. L. (1994). Anatomical evidence for cerebellar and basal ganglia involvement in higher cognitive function. *Science* **266**, 458–461.

Moruzzi, G., and Magoun, H. W. (1949). Brainstem reticular formation and activation of the EEG. *Electroencephalogr. Clin. Neurophysiol.* **1**, 455–473.

Murakami, J. W., Courchesne, E., Press, G. A., Yeung-Courchesne, R., and Hesselink, J. R. (1989). Reduced cerebellar hemisphere size and its relationship to vermal hypoplasia in autism. *Arch. Neurol.* **46,** 689–694.

Nagahama, Y., Fukuyama, H., Yamamuchi, Y., Matsuzaki, S., Ouchi, Y., Kimura, J., Yonekura, Y., and Shibasaki, H. (1995). Functional localization and lateralization of the activated cortex during the Wisconsin card sorting test. *Hum. Brain Mapp.* **3**(Suppl. 1), 196.

Newman, P. P, and Reza, H. (1979). Functional relationships between the hippocampus and the cerebellum: An electrophysiological study of the cat. *J. Physiol. (London)*, **287,** 405–426.

Nieuwenhuys, R., Voogd, J., and van Huijzen, C. (1988). "The Human Central Nervous System: A Synopsis and Atlas." Springer-Verlag, Berlin.

Nichols, S. L., Townsend, J., and Wulfeck, B. (1995). "The Development of Covert Visual Attention in Normal School-Age Children." Project in Cognitive and Neural Development, Technical Report CND-9501, Center for Research in Language, University of California, San Diego, La Jolla, CA.

Noda, H. (1991). Cerebellar control of saccadic eye movements: Its neural mechanisms and pathways. *Jap. J. Physiol.* **41,** 351–368.

Posner, M. I. (1988). Structures and functions of selective attention. *In* "Master Lectures in Clinical Neuropsychology" (T. Boll and B. Bryant, eds.), pp. 173–202. American Psychological Association, Washington, D.C..

Posner, M. I., and Petersen, S. E. (1990). The attention system of the human brain. *Annu. Rev. Neurosci.* **13,** 25–42.

Posner, M. I. Snyder, C. R., and Davidson, B. J. (1980). Attention and the detection of signals. *J. Exp. Psychol. Gen.* **21,** 160–174.

Posner, M. I., Walker, J. A., Freidrich, F. A., and Rafal, R. D. (1987). How do the parietal lobes direct covert attention? *Neuropsychologia* **25,** 135–145.

Posner, M. I., Walker, J. A., Freidrich, F. A., and Rafal, R. D. (1984). Effects of parietal injury on covert orienting of attention. *J. Neurosci.* **4,** 1863–1874.

Pritchard, W. S. (1981). Psychophysiology of P300. *Psychol. Bull.* **89,** 506–540.

Rafal, R. D., and Posner, M. I. (1987). Deficits in human visual spatial attention following thalamic lesions. *Proc. Natl. Acad. Sci USA* **84,** 7349–7353.

Rafal, R. D., Posner, M. I., Freidman, J. H., Inhoff, A. W., and Bernstein, E. (1988). Orienting of visual attention in progressive supranuclear palsy. *Brain* **111,** 267–280.

Raichle, M. E., Fiez, J. A., Videen, T. O., MacLeod, A. K., Pardo, J. V., Fox, P. T., and Petersen, S. E. (1994). Practice-related changes in human brain functional anatomy during nonmotor learning. *Cereb. Cortex* **4,** 8–26.

Rao, S. L., Harrington, D. L., Haaland, K. Y., Bobholz, J. A., Binder, J. R., Hameke, T. A., Frost, J. A., Myklebust, B. M., Jacobson, R. D., Bandettini, P. A., and Hyde, J.S. (1995). Functional MRI correlates of cognitive-motor learning. *Hum. Brain Mapp.* **3**(Suppl. 1), 412.

Reeves, A., and Sperling, G (1986). Attention gating in short-term visual memory. *Psychol. Rev.* **93,** 180–206.

Reis, D. J., Doba, N., and Nathan, M. A. (1973). Predatory attack, grooming, and consummatory behaviors evoked by electrical stimulation of cat cerebellar nuclei. *Science* 845–847.

Ritvo, E. R., Freeman, B. J., Scheibel, A. B., Duong, T., Robinson, H., Guthrie, D., and Ritvo, A. (1986). Lower Purkinje cell counts in the cerebella of four autistic subjects: Initial findings of the UCLA-NSAC autopsy research report. *Am. J. Psychiat.* **143,** 862–866.

Sasaki, K., Matsuda, Y., Kawaguchi, S., and Mizuno, N. (1972). On the cerebello-thalamo-cerebral pathway for the parietal cortex. *Exp. Brain Res.* **16,** 89–103.

Sasaki, K., Jinnai, K., Gemba, H., Hashimoto, S., and Mizuno, N. (1979). Projection of the cerebellar dentate nucleus onto the frontal association cortex in monkeys. *Exp. Brain Res.* **37,** 193–198.

Schmahmann, J. D. (1991). An emerging concept: The cerebellar contribution to higher function. *Arch. Neurol.* **48,** 1178–1187.

Schmahmann, J., and Pandya, D. N. (1989). Anatomical investigation of projections to the basis pontis from the posterior parietal association cortices in rhesus monkey. *J. Comp. Neurol.* **289,** 53–73.

Schmahmann, J., and Pandya, D. N. (1995). Prefrontal cortex projections to the basilar pons in rhesus monkey: Implications for the cerebellar contribution to higher function. *J. Comp. Neurol.* **289,** 53–73.

Snider, R. S. (1950). Recent contributions to the anatomy and physiology of the cerebellum. *Arch. Neurol. Psychiat.* **64,** 196–219.

Steriade, M., and Stoupel, N. (1960). Contribution a l'etude des relations entre l'aire auditive du cerebelet et l'ecore cerebrale chez le chat. *Electroencephalogr. Clin. Neurophysiol.* **12,** 119–136.

Thompson, R. F. (1986). The neurobiology of learning and memory. *Science* **233,** 941–947.

Townsend, J., Courchesne, E., and Egaas, B. (1996a). Slowed orienting of covert visual-spatial attention in autism: Specific deficits associated with cerebellar and parietal abnormality. *Dev. Psychopathol.* **8,** 563–584.

Townsend, J., Singer-Harris, N. S., and Courchesne, E. (1996b). Visual attention abnormalities in autism: Delayed orienting to location. *J. Int. Neuropsychol. Soc.* **2,** 541–550.

Vilensky, J. A., and Van Hoesen, G. W. (1981). Corticopontine projections from the cingulate cortex in the rhesus monkey. *Brain Res.* **205,** 391–395.

Watson, P. J. (1978). Nonmotor functions of the cerebellum. *Psychol. Bull.* **85,** 944–967.

Weichselgartner, E., and Sperling, G. (1987). Dynamics of automatic and controlled visual attention. *Science* **238,** 778–780.

Williams, R. S., Hauser, S. L., Purpura, D. P., DeLong, R., and Swisher, C. N. (1980). Autism and mental retardation: Neuropathological studies performed in four retarded persons with autistic behavior. *Arch. Neurol.* **37,** 749–753.

Wurtz, R. H., Goldberg, M. E., and Robinson, D. L. (1992). Behavioral modulation of visual responses in the monkey: Stimulus selection for attention and movement. In "Frontiers in Cognitive Neuroscience" (S. M. Kosslyn and R. A. Andersen, eds.), pp. 346–365. MIT Press, Cambridge, MA.

Yeo, C., Hardiman, M., and Glickstein, M. (1985). Classical conditioning of the nictitating membrane response of the rabbit. II. Lesions of the cerebellar cortex. *Exp. Brain Res.* **60,** 99–113.

CONTEXT–RESPONSE LINKAGE[1]

W. Thomas Thach

Department of Anatomy and Neurobiology, Washington University School of Medicine,
St. Louis, Missouri 63110

I. Motor Learning and the Cerebellum
II. Cognitive Functions of the Cerebellum
   A. Does the Cerebellum Only Modulate What Others Have Begun or Can It Run the Whole Show?
   B. What Does the Cerebellum Contribute to Mental Movement?
   C. Context Triggering and Combination
   D. Mental Rehearsal of Motor Performance
III. Conclusions
    References

Brindley (1969) proposed that we initially generate movements "consciously," under higher cerebral control. As the movement is practiced, the cerebellum learns to link within itself the context in which the movement is made to the lower level movement generators. Marr, (1969) Albus (1971), and Ito (1972) proposed that the linkage is established by a special input from the inferior olive, which plays upon an input–output element within the cerebellum during the period of the learning. When the linkage is complete, the occurrence of the context (represented by a certain input to the cerebellum) will trigger (through the cerebellum) the appropriate motor response. The "learned" movement is distinguished from the "unlearned" conscious movement by being automatic, rapid, and stereotyped. Another important variable can be added to the idea of the context–response linkage: novel combinations of downstream elements. With regard to the motor system, this could explain how varied combinations of muscles may become active in precise time–amplitude specifications so as to produce coordinated movements appropriate to specific contexts. This chapter further extends this idea to the premotor parts of the brain and their role in cognition. These areas receive influences from the cerebellum and are active both in planning movements that are to be executed and in thinking about movements that are not to be executed. Evidence shows that the cerebellar output extends even to what has been characterized as the ultimate frontal planning area, the "prefrontal" cortex, area 46. The cerebellum thus may be involved in context–response linkage and in response combination even at these higher levels. The implication

---

[1] Adapted and reprinted with the permission of Cambridge University Press. © 1996 Cambridge University Press.

would be that, through practice, an experiential context would automatically evoke a certain mental action plan. The plan would be in the realm of thought and could—but need not—lead to execution. The specific cerebellar contribution would be one of the context linkage and the shaping of the response through trial and error learning. The prefrontal and premotor areas could still plan without the help of the cerebellum, but not so automatically, rapidly, stereotypically, so precisely linked to context, or so free of error. Nor would their activities improve optimally with mental practice.

I. Motor Learning and the Cerebellum

Several theories and a number of lines of evidence have pointed to a crucial role of the cerebellum in the adaptation and learning of movement. Brindley (1969) first suggested that the acquisition of skilled movements, such as playing the piano, begins as a conscious act mostly under the control of the cerebral cortex, without help from the cerebellum (Fig. 1), but the cerebellum itself can also initiate the performance, and it immediately begins to acquire control of the task. It recognizes the contexts in which each "piece" of consciously initiated movement occurs. After repeated tries it links that context within itself to the movement generators so that the occurrence of the context automatically triggers the movement. Thus, with time and practice, the cerebellum largely controls the process, with little or no help from the cerebrum. The cerebrum and the conscious mind are free to do and think about other things. Control of the task has been shifted from a conscious cerebral cortical process to a subconscious one mostly under the control of the cerebellum.

Marr (1969), Albus (1971), and Ito (1972) independently modeled the process using the cerebellar circuit design and function as sketched by Ramon y Cajal and updated by Eccles *et al.* (1967) and Llinas (1981). Gilbert (1974) added that synapses were not simply turned "on" (Marr, 1969) or "off" (Albus, 1971) but were adjusted to give the continuum of Purkinje cell firing frequencies that are actually observed in awake behaving animals (cf. Thach *et al.*, 1992). The circuit models were based on the great differences between the two main cerebellar input systems. The highly convergent mossy fiber–parallel fiber–Purkinje cell system brought information from most parts of the nervous system, and the information was represented as modulations in high-frequency firing of 0–500/sec. The relatively one-to-one climbing fiber Purkinje cell system arose exclusively from neurons of the inferior olive; the synaptic contact was very powerful, but the firing rates were so low (around 1/sec) as to raise questions about the information content.

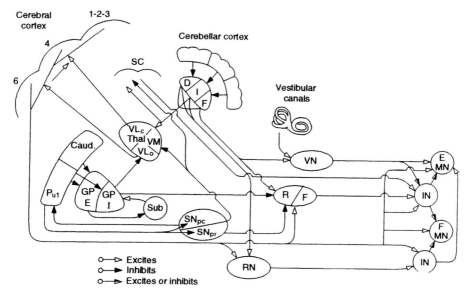

FIG. 1. Descending pathways to the spinal cord and their origins in brain stem, cerebellum, and cerebrum. Caud., caudate; Put., putamen; GPe, globus pallidus, external segment; GPi, globus pallidus, internal segment; Thal., thalamus; VLc caudal ventrolateral nucleus; VLo, oral ventrolateral nucleus; VM, ventromedial nucleus; Sub, subthalamic nucleus; D, dentate nucleus; I, interposed nucleus; F, fastigial nucleus; SNpc, substantia nigra pars compacta; SNpr, substantia nigra pars reticulata; RN, red nucleus; RF, reticular formation (e.g., reticular nucleus of the pontine tegmentum); VN, vestibular nuclei; IN, interneuron; E MN, extensor motoneuron; F MN, flexor motoneuron (from Thach and Montgomery, 1990).

Mugnaini (1983) showed that the parallel fiber is 6 to 10 times longer than had been supposed and contacts Purkinje cells along a beam spanning one-third to one-half the width of the cerebellar cortex. Animal studies have shown at least one more or less complete body map in each of nucleus fastigius, interpositus, and dentatus (Asanuma *et al.* 1983; Fig. 2). Further, each nucleus appears to control a different aspect or mode of movement for the entire body it maps (Thach *et al.*, 1992): dentate, synergist muscles in visually guided movements (e.g., pinch and reach); interpositus, agonist–antagonist synergy and stretch reflexes at a single joint (Frysinger *et al.*, 1984; Schieber and Thach, 1985a,b; Smith and Bourbonnais, 1981; Wetts *et al.*, 1985); and fastigius, synergists in upright stance and locomotion (Arshavsky *et al.*, 1980; Thach *et al.*, 1992). Thach *et al.* (1992) and Goodkin *et al.* (1993) reported that cerebellar lesions impair compound movements more than simple and suggested that a cardinal role of the cerebellum is to combine (through the learning mechanisms above) the elements of

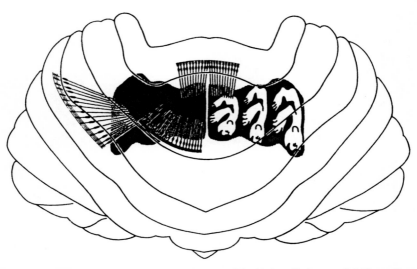

FIG. 2. Diagram showing linkage into beams of Purkinje cells by parallel fibers. Beams project down onto the somatotopically organized nuclei. Purkinje cell beams thus link body parts together within each nucleus and link adjacent nuclei together. Such linkage could be the mechanism of the cerebellar role in movement coordination (from Thach *et al.*, 1992).

movement using the parallel fiber contacts on the long beam of Purkinje cells.

The proposed mechanisms of this cerebellar function are that the conditions and "context" in which a movement is to be learned and performed are represented in the modulate discharge of the mossy fibers (Fig. 3), which not only monitor sensory information, but also much of the ongoing activity of most of the nervous system. This afferent information is transmitted to the parallel fiber, which branches to contact thousands of Purkinje cells, which in turn project to the somatotopic motor representations within each of the deep nuclei. The parallel fiber then is the critical middle layer between sensory and other input and motor output, representing both the context in which the movement is made and being a chief instrument for organizing the motor response.

The climbing fiber detects and corrects errors in performance, and changes the strength of parallel fiber–Purkinje cell synapses, thereby creating novel context–response linkages and response combinations. When a new movement needs to be learned or an old one adapted, the climbing fiber, which normally fires irregularly at a rate of around 1 Hz and in no particular relation to movement, suddenly [driven by error between intended movement and actual movement (Albus, 1971)] begins to fire

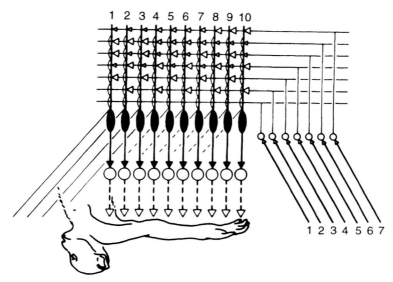

FIG. 3. Diagram of intrinsic cerebellar circuitry. Mossy fibers ascend from the lower right to the cerebellar cortex carrying input information. They contact and excite granule cells, which give rise to parallel fibers. The parallel fibers branch and contact many Purkinje cells along the "beam" in either direction. The strength of the synaptic contact is diagrammed as being variable. Purkinje cells receive many parallel fibers, carrying many different kinds of information. This parallel fiber input creates the "context" which each Purkinje cell is capable of recognizing. Climbing fibers ascend from the lower left to excite Purkinje cells, one on one. Climbing fiber activity paired in time with mossy fiber–parallel fiber activity weakens the strength of that particular parallel fiber synapse. Climbing fiber activity may thus create and change the mossy fiber–parallel fiber activity context that drives the Purkinje cell (from Thach *et al.*, 1992).

(once) immediately after the error occurs, reliably time after time. The effect of this low frequency but synaptically very powerful climbing fiber firing is to reduce (Albus, 1971) the strength of the synapse on the Purkinje cell of those parallel fibers that are active at the time (and helping to cause the inappropriate movement). What is left after practice, and repeated firings, are those parallel fibers the actions of which cause the correct movement, gradually improving performance. Once the behavior has changed to correct behavior, the error that drove the change is eliminated, the climbing fiber returns to its random background firing, and the remaining potent parallel fibers are left to drive the system in the particular movement context.

Nearly a million parallel fibers contact the human Purkinje cell. Over mossy fibers, vestibular, somatosensory, visual, and auditory sensory information arrives in medial and intermediate zones, and cerebral cortical (presumably "cognitive") information in the lateral zone cerebellar cortex. There, the information is conveyed to parallel fibers, but these fibers are so long (Mugnaini, 1983) that any one Purkinje cell could conceivably receive the mossy fiber information from all three zones via parallel fibers. The input—and the context that any one Purkinje cell "sees"—could include virtually all the sensory modalities, all the feedback from movement of each of the moving body parts, *and* the feedforward of the plans for the control of each body part, as well as the environmental conditions as the subject perceives them and contemplates what to do about them. Thus the context triggering conforms to a fan-in and fan-out model of organization.

Learned context triggering also suggests the possibility of generation of sequences of behavior automatically across time. It has been said that "reflex chains" as Sherrington (1906) envisaged were improbable because of the time involved (see Keele, 1981). Thus, in the performance of a sequence of musical notes A,B,C,D,E,F, and G, E cannot be triggered by feedback from the movement D because it would take too long to run over real neural pathways. If E could be triggered by the intent to play and contexts surrounding B or C, then each of the notes could be triggered by the preconditions for some of the earlier elements in the sequence, and the process may be made to play out with realistic speed. The phrase—or indeed and entire piece—might thus be generated automatically by preceding contexts. Evidence shows that behavioral sequences do show such linkages and are generated in "chunks" of linked elements, whose linkages are less dissociable than are the boundaries of the chunk (see Keele, 1981). In this scheme, the triggering information would come from several different levels in the hierarchical scheme; this is consistent with the observations from psychological experiments that different chunks are controlled from different levels.

II. Cognitive Functions of the Cerebellum

The "motor association" areas of the cerebral cortex are presumably concerned with motor planning. These include the supplementary motor area (medial area 6), the premotor area (lateral area 6), the frontal eye fields (area 8) and the accessory frontal eye fields of Schlag, and the motor speech areas in humans (areas 44 and 45). Anatomically, they receive from posterior areas associated with perception and awareness (see later), and project to middle level motor pattern generators such as the motor cortex, area 4. They receive multimodal sensory inputs and send to different movement generators. These parts are active in animal recording and human positron emission tomography studies during the movement and their ablation impairs movement.

Evidence shows that these areas may be active in anticipating or rehearsing a movement without actually performing it [the studies of anticipatory signals in monkey supplementary motor area (SMA) and premotor cortex (PMC) (Tanji and Evarts, 1976; Tanji, 1985), of mimicry signals in monkey PMC (Di Pelligrino *et al.*, 1992), and of mental motor rehearsal signals in human SMA (Roland, 1987)]. The two roles in purely mental imaging of movement and in movement planning would appear to go together.

Various studies have added projections from the cerebellum to include virtually all levels of the motor system: spinal motor and interneurons (cf. Asanuma *et al.*, 1983a,b,c), the superior colliculi (May *et al.*, 1993), and (via thalamus) the cerebral cortical "motor association" areas (Lynch *et al.*, 1992; Middleton and Strick, 1994; Schell and Strick, 1983; Yamamoto *et al.*, 1992). As such, the cerebellum is upstream from movement pattern generators at all levels, but since each of these movement generators has other prominent excitatory inputs and since cerebellar ablation impairs but never abolishes movement, cerebellar functions have traditionally been characterized as regulatory and modulatory instead of executive.

A. Does the Cerebellum Only Modulate What Others Have Begun or Can It Run the Whole Show?

These anatomic connections place the cerebellum in a position to excite any or all of the major motor generators, at every level from motor neuron to motor association cortex. Why has it thus been delegated to the role of the regulator and modulator rather than executor? The historic arguments include: (1) cerebellar lesions impair but never abolish movement, and often cause little or no observed motor defect; (2) until recently cerebellar electrical stimulation has not been reliable in causing movement; and (3) cerebellar target motor structures each have an additional major excitatory input, which has been credited with the major driving effect. However, newer knowledge questions all three premises. First, focal cerebellar lesions do indeed abolish particular categories of movement without affecting others. The effect of the lesions depends on its precise localization and our knowledge of the particular region and its control functions. The effects may be disabling (cf. Thach *et al.*, 1992). Second, electrical stimulation of the output nuclei does not reliably cause movement (cf. Thach *et al.*, 1993). Stimulation of the cortex may not cause movement because of mixed excitatory and inhibitory effects. Third, cerebellar excitatory effects on a target structure are often stronger than those of the "second" excitatory input. This is true for the red nucleus. Also, the dentate nucleus fires before and apparently helps initiate output from the motor cortex. These facts shed new light on the cerebellar control of movement and on the so-called cerebellar motor learning theories, suggesting that the cerebellum may operate at the highest level of direction and coordination.

B. What Does the Cerebellum Contribute to Mental Movement?

Leiner *et al.* (1993) and H. C. Leiner and A. L. Leiner (this volume) have argued for a cerebellar influence on frontal lobe mentation. Much of their careful reasoning depends on there being anatomical projections from the cerebellum via the thalamus to the cerebral frontal association cortex. Whereas evidence for these connections was at the time rather scant, evidence for the connections has steadily increased. Anatomical connections from cerebellum to far frontal association areas were proposed in monkey by Sasaki *et al.* (1976) using electroanatomical methods. In humans, a phylogenetically new and unique posterolateral part of the dentate was proposed by Hassler (1950) and Leiner *et al.* (1991) to project via the thalamus to the far frontal cortex. Leiner *et al.* (1993) inferred that the cerebellum could contribute to whatever was processed in these areas. Although they did not specify the type of mental operation performed in these areas, they did specify regions that are now identified as being higher order motor. Even though they did not say exactly what the cerebellum might provide to these areas, because of the uniformity of the cerebellar architecture and the likelihood that the algorithms were similar or identical to those used for motor control, Leiner *et al.* (1993) used the terms "coordination" and "skill."

Let us now return to the fact that the cerebellum is active during mental movement (Decety *et al.*, 1990; Ryding *et al.*, 1993), which in itself could have been a spurious correlation; the cerebellum could have been active in relation to some unobserved synergic muscle activity that was associated with, but was not necessary for, the performance of the task. It is now clear that ablation of the cerebellum does indeed impair mental task performance. So somehow the cerebellum is involved, and the question is "how?"

C. Context Triggering and Combination

For motor learning, the author has pursued the idea that the cerebellum may link a behavioral context to a motor response. The response may be a combination of many downstream neural elements firing together. Both the context–response linkage and the response composition would be achieved through trial and error learning. After practice, the occurrence of the context would trigger the occurrence of the response. This would explain how combinations of muscles may become active all at once, especially in skilled movements outside the capabilities of the motor pattern generators so as to produce coordinated behavior appropriate to a specific context.

This idea has been extended to include the premotor parts of the brain, to which the cerebellum is now known to project, and which are known to play a role in mental movement and cognition. These areas are active in the planning of movements that are then executed, and they plan movements that are not to be executed. They "think" movements. Evidence shows that the cerebellar output extends even to what has been characterized as the ultimate frontal planning area, the "prefrontal" cortex, area 46. The cerebellum may be involved in combining these cellular elements, so that, through practice, an experiential context can automatically evoke an action plan. The plan would be in the realm of thought. The plan either could or need not lead to execution. Again, the specific cerebellar contribution would be one of linkage of the context to a specific response, and the combination of the response from simpler elements, accomplished through repeated practice. The prefrontal and premotor areas could still plan without the help of the cerebellum, but not so rapidly, automatically, or so precisely liked to context. At some level of task complexity, cerebellar damage would reveal itself in behavioral errors. This would have nothing to do with other cognitive activities—visual, auditory, attention, and so on—and areas of the brain to which the cerebellum does not project.

D. MENTAL REHEARSAL OF MOTOR PERFORMANCE

It has been well documented that mental rehearsal improves motor performance (see Jeannerod, 1993). This is common knowledge among musicians, athletes, chess players, actors, and lecturers. Repetitively playing through the performance in one's mind can remarkably improve the next actual performance. Is there any way in which the cerebellum may play a role in motor learning in which only mental movement is practiced?

A hitherto curious and unexplained oddity in the structure of the primate motor system is the evolutionary change in the connectivity of the red nucleus. As mentioned previously, the magnocellular (phylogenetically older) red nucleus gives rise to a prominent rubrospinal tract in carnivores and receives from the motor cortex and the cerebellar interposed nucleus.

It is supposed to play a role similar to that of the corticospinal pathway in providing voluntary control of small distal muscle groups. In contrast, the parvocellular (phylogenetically newer) red nucleus receives from the premotor cortex (area 6) and the cerebellar dentate nucleus. Its output is thought to go not to the spinal cord but exclusively to the principal portion of the inferior olive, which in turn projects back to the lateral hemisphere of the cerebellum. This has seemed odd that a system capable of firing at high sustained frequencies should funnel into and dead end in the conspicuously low frequency inferior olive. The suggestion had been made that this might fit into the general cerebellar role of motor learning (Kennedy, 1990), but a tighter rationale has not been provided. Mental motor rehearsal could well be the specific framework for its role in learning motor performances that are only practiced internally and not overtly expressed. The context would be brain states corresponding to prior elements of the performance, the combined responses could be of premotor neuron assemblies in area 6 (and 8, 9, 44, 45, 46?), and the linkage would be established through repetitive practice. It is not clear what mental performance errors might consist of or how they might be detected.

III. Conclusions

The cerebellar contribution to mental (and motor) performance would become critical (1) when the task involves imagined movement that is rapidly and automatically triggered by context, (2) when the task contains a number of linked neural (or body part) components, and (3) when these properties are being adapted or newly acquired. Paradoxically, the context–response coupling and the response combinations would be "unconscious" aspects of thought. They would have been at a level of awareness only during the learning phases. Ironically, this would be something that we have learned and no longer "know" anything about because it has been given over to the cerebellum for the implementation of automatic motor control, actual or imagined.

The cognitive functions of the cerebellum may thus hinge entirely on its involvement in imagined movement. Our thesis is that imagined movement is similar/identical to the early initiatory phases or actual movement. The cerebellum plays the same specific role in "coordinating" these imagined movements that it does in actual movements. The author supports the Brindley–Marr–Albus–Ito–Gilbert theories and suggests that these cerebellar pathways are used to (1) build through trial and error learning behavioral context–response linkages and (2) build up appropriate responses from simpler constitutive elements.

As cerebellar ablation does not abolish actual movements, only their fine control and coordination, neither does it abolish imagined movements, only their automaticity, speed of response, stereotypy, and ability to improve them with practice. It is presumably due to the some inexactness in the use of central representation of movement that, after cerebellar damage, errors are made in the movement-associated mental activities that are required for some "cognitive" task performance.

References

Albus, J. S. (1971). A theory of cerebellar function. *Math. Biosci.* **10,** 25–61.
Arshavsky, Y. I., Orlovsky, G. N., Pavlova, G. A., and Perret, C. (1980). Activity of neurons of cerebellar nuclei during fictitious scratch reflex in the cat. II. The interpositus and lateral nuclei. *Brain Res.* **200,** 249–258.
Asanuma, C., Thach, W. T., and Jones, E. G. (1983a). Anatomical evidence for segregated focal groupings of efferent cells and their terminal ramifications in the cerebellothalamic pathway of the monkey. *Brain Res. Rev.* **5,** 267–299.
Asanuma, C., Thach, W. T., and Jones, E. G. (1983b). Distribution of cerebellar terminations and their relation to other afferent terminations in the ventral lateral thalamic region of the monkey. *Brain Res. Rev.* **5,** 237–265.
Asanuma, C., Thach, W. T., and Jones, E. G. (1983c). Brainstem and spinal projections of the deep cerebellar nuclei in the monkey, with observations on the brainstem projections of the dorsal column nuclei. *Brain Res. Rev.* **5,** 299–322.
Brindley, G. S. (1969). The use made by the cerebellum of the information that it receives from the sense organs. *Int. Brain. Res. Org. Bull.* **3,** 80.
Decety, J., Sjoholm, H., Ryding, E., Stenberg, G., and Ingvar, D. H. (1990). The cerebellum participates in mental activity: Tomographic measurements of regional cerebral blood flow. *Brain Res.* **535,** 313–317.
Di Pellegrino, G., Fadiga, L., Fogassi, L., Gallese, V., and Rizzolatti, G. (1992). Understanding motor events: A neurophysiological study. *Exp. Brain Res.* **91,** 176–80.
Eccles, J. C., Ito, M., and Szentagothai, J. (1967). *The Cerebellum as a Neuronal Machine.* Springer-Verlag, New York.
Frysinger, R. C., Bourbonnais, D., Kalaska, J. F., and Smith, A. M. (1984). Cerebellar cortical activity during antagonist cocontraction and reciprocal inhibition of forearm muscles. *J. Neurophysiol.* **51,** 32–49, 1984.
Gilbert, P. F. C. (1974). A theory of memory that explains the function and structure of the cerebellum. *Brain Res.* **70,** 1–18.
Goodkin, H. P., Keating, J. G., Martin, T. A., and Thach, W. T. (1993). Preserved simple and impaired compound movement after infarction in the territory of the superior cerebellar artery. *Can. J. Neurol. Sci.* **20**(Suppl. 3), S93–S104.
Hassler, R. (1950). Uber kleinhirnprojektionen zum mittlehirn und thalamus beim menschen. *Deut. Zeitschr. Nervenheilk.* **163,** 629–671.
Ito, M. (1972). Neural design of the cerebellar control system. *Brain Res.* **40,** 80–82.

Jeannerod, M. (1993). The representing brain: Neural correlates of motor intention and imagery. *Behav. Brain Sci.*

Keele, S. W. (1981). Behavioral analysis of movement. In "Handbook of Physiology" (J. M. Brookhart V. B. Mountcastle and V. B. Brooks, eds.), Vol. III, pp. 1391–1414. American Physiological Society, Bethesda, MD.

Kennedy, P. R. (1990). Corticospinal, rubrospinal, and rubroolivary projections: A unifying hypothesis. *Trends Neurosci.* **13**, 474–479.

Leiner, H. C., Leiner, A. L., and Dow, R. S. (1991). The human cerebrocerebellar system: Its computing, cognitive, and language skills. *Behav. Brain Res.* **44**, 113–128.

Leiner, H. C., Leiner, A. L., and Dow, R. S. (1993). Cognitive and language functions of the human cerebellum. *Trends Neurosci.* **16**, 444–454.

Llinas, R. (1981). Electrophysiology of cerebellar networks. In "Handbook of Physiology" (V. B. Brooks, ed.), Vol. II, pp. 831–876. American Physiological Society, Bethesda, MD.

Lynch, J. C., Hoover, J. E., and Strick, P. L. (1992). The primate frontal eye field is the target of neural signals from the substantia nigra, superior colliculus, and dentate nucleus. *Soc. Neurosci. Abstr.* **18**, 855.

Marr, D. (1969). A theory of cerebellar cortex. *J. Physiol.* **202**, 437–470.

May, P. J., Hall, W. C., Porter, J. D., and Sakai, S. T. (1993). The comparative anatomy of nigral and cerebellar control over tectally initiated orienting movements. In "Role of the Cerebellum and Basal Ganglia in Voluntary Movement" (N. Mano I. Hamada and M. R. DeLong, eds.), pp. 121–131. Excerpta Medica, Amsterdam.

Middleton, F. A., and Strick, P. L. (1994). Anatomical evidence for cerebellar and basal ganglia involvement in higher cognitive function. *Science* **226**, 448–451.

Mugnaini, E. (1983). The length of cerebellar parallel fibers in chicken and rhesus monkey. *J. Comp. Neurol.* **220**, 7–15.

Roland, P. E. (1987). Metabolic mapping of sensorimotor integration in the human brain, In "Motor Areas of the Cerebral Cortex" (G. Bock M. O'Connor and J. Marsh, eds.), pp. 251–268. Wiley, Chichester, UK.

Ryding, E., Decety, J., Sjoholm, H., Stenberg, G,. and Ingvar, D. H. (1993). Motor imagery activates the cerebellum regionally: A SPECT rCBRF study with 99m Tc-HMPAO. *Cogn. Brain Res.* **1**, 94–99.

Sasaki, K. S., Kawaguchi, S., Oka., H., Saki, M., and Mizuno, N. (1976). Electrophysiological studies on the cerebellocerebral projections in monkeys. *Exp. Brain Res.* **24**, 495–507.

Schell, G. R. and Strick, P. L. (1983). The origin of thalamic inputs to the arcuate premotor and supplementary motor areas. *J. Neurosci.* **4**, 539–560.

Schieber, M. H. and Thach, W. T. (1985a). Trained slow tracking. I. Muscular production of wrist movement. *J. Neurophysiol.* **55**, 1213–1227.

Schieber, M. H., and Thach, W. T. (1985b). Trained slow tracking. II. Bidirectional discharge patterns of cerebellar nuclear, motor cortex, and spindle afferent neurons. *J. Neurophysiol.* **55**, 1228–1270.

Sherrington, C. S. (1906). "The Integrative Action of the Nervous System." Constable, London.

Smith, A. M. and Bourbonnais, D. (1981). Neuronal activity in cerebellar cortex related to control of prehensile force. *J. Neurophysiol.* **45**, 286–303.

Tanji, J., (1985). Comparison of neural activities in the monkey supplementary and precentral motor areas. *TINS* **18**, 137.

Tanji, J. and Evarts, E.V. (1976). Anticipatory activity of motor cortex in relation to direction of an intended movement. *J. Neurophysiol.* **39**, 1062–1068.

Thach, W. T. (1996). On the specific role of the cerebellum in motor learning and cognition: Clues from PET activation and lesion studies in man. *Behav. Brain Sci.* **19**, 411–431.

Thach, W. T., Goodkin, H. P., and Keating, J. G. (1992). The cerebellum and the adaptive coordination of movement. *Annu. Rev. Neurosci.* **15,** 403–442.

Thach, W. T., and Montgomery, E. B. (1990). Motor system. *In* Neurobiology of Disease (A. L. Pearlman and R. C. Collins, eds.), p. 170. Oxford University Press, New York.

Thach, W. T., Perry, J. G., Kane, S. A., and Goodkin, H. P. (1993). Cerebellar nuclei: Rapid alternating movement, motor somatotopy, and a mechanism for the control of muscle synergy. *Rev Neurol.* **149,** 607–628.

Wetts, R., Kalaska, J. F., and Smith, A. M. (1985). Cerebellar nuclear cell activity during antagonist co-contraction and reciprocal inhibition of forearm muscles. *J. Neurophysiol.* **54,** 231–244.

Yamamoto T., Yoshida K., Yoshikawa H., Kishimoto Y., and Oka H. (1992). The medial dorsal nucleus is one of the thalamic relays of the cerebellocerebral responses to the frontal association cortex in the monkey: Horseradish peroxidase and fluorescent dye double staining study. *Brain Res.* **579,** 315–320.

DUALITY OF CEREBELLAR MOTOR AND COGNITIVE FUNCTIONS

James R. Bloedel and Vlastislav Bracha

Division of Neurobiology, Barrow Neurological Institute, Phoenix, Arizona 85013

I. Cerebellum and Cognition: A Historical Perspective
II. Motor Function and Cognition: A Dichotomy Worth Saving?
III. Task Dependency and Context Dependency: Determinants of Cerebellar Involvement in Regulating Behaviors
IV. Cerebellar Functions: Implications from Distributed Circuits
V. Conclusions
References

This chapter develops a specific perspective regarding the interrelationship of the cerebellum, motor behaviors, and cognitive processes. The advent of the proposals regarding the cerebellum and cognition has challenged many investigators to examine this issue aggressively and to address the extent to which current concepts, definitions, and experimental approaches are adequate for deriving new insights into the interfaces between the domains of motor execution and adaptive modifications in behavior. This chapter contends that the dichotomy often made between motor processes and cognitive processes is inconsistent with the organization of behaviors in general and that, when a broader, more integrative view is adopted, a role of the cerebellum in "cognitive" processes is not only expected but also necessary given this structure's contribution to motor coordination and behavioral adaptations.

I. Cerebellum and Cognition: A Historical Perspective

Postulates pertaining to the role of the cerebellum in cognition fall very naturally into the historical evolution of theories oriented toward identifying *the* (or at least *a*) predominant function(s) of this structure. Similar to most of the fundamental postulates that have been offered, the suggestion that the cerebellum is involved in cognitive processes is based in part on observations made following cerebellar lesions in patients. In the authors' view, the development of this concept follows predictably the historical evolution of ideas derived from examining the paradigm-specific behaviors affected by ablations of certain cerebellar regions. For the purpose of this discussion, it is most convenient to divide the behaviors evaluated in these types of experiments into five classes.

The first class of behaviors includes a variety of motor behaviors with a strong focus on those related to the coordinated performance of voluntary, goal-oriented movements as well as posture and orientation of the head and body in space (Holmes, 1939; Brooks and Thach, 1981; Goldberger and Growdon, 1973; Dow and Moruzzi, 1958). The second class consists of abnormalities in several types of cutaneomuscular and proprioceptive reflexes as well as postural support responses dependent on the integration of sensory inputs for their proper execution (Rademaker, 1980; Chambers and Sprague, 1951, 1955; Goldberger and Growdon, 1973; Amassian and Ross, 1972; Amassian et al., 1972; Bloedel and Bracha, 1995). Adaptive modifications of vestibulo-ocular and postural reflexes represent the third class (Robinson, 1976; Lisberger et al., 1994; Horak and Diener, 1994). As will be emphasized later, experiments supporting this class were among the first to stimulate discussions regarding the possibility that the cerebellum may contribute to adaptive modifications of specific behaviors rather than serving only to coordinate the performance of ongoing movements. The fourth class consists of the classically conditioned withdrawal reflexes. Studies examining the effects of ablating cerebellar regions on these behaviors revealed that this structure actually was involved in associative processes (McCormick et al., 1981; Kolb et al., 1994; Supple and Kapp, 1993). Furthermore, initial discussions pertaining to the role of the cerebellum in eyeblink conditioning were the first to propose this structure as being involved only in processes related to the establishment and storage of memory traces to the exclusion of an involvement in the control and execution of the movement itself. The fifth class of behaviors includes what have been referred to as "cognitive processes" (Thach, 1996; Akshoomoff and Courchesne, 1992; Fietz et al., 1992; Leiner et al., 1986, 1993; Schmahmann, 1991; Watson, 1978; Botez et al., 1989). This designation implies that these behaviors are not only separable from the motor behaviors in the first class but also that they are distinct from those characterizing either reflex adaptation or classical conditioning.

The abnormalities produced by ablating different cerebellar regions on the first class of behaviors are now classical. These include ablation syndromes in which deficits in posture, volitional movements, and balance were noted and studied. These observations led to theories focused on the role of the cerebellum in the coordination of goal-directed and spontaneous movements as well as in postural regulation and vestibular function. Across all of these experiments (Luciani, 1915; Dow and Moruzzi, 1958; Holmes, 1939; Chambers and Sprague, 1951, 1955; Sherrington, 1906) the paradigms employed resulted in the accumulation of data related specifically to these types of behaviors. Although deficits of postural reactions requiring the processing of sensory information were noted in some laboratories, the emphasis was directed toward building proposals on the role of the cerebellum in the execution of ongoing motor behaviors (for a review see Dow and Moruzzi, 1958).

Postulates of cerebellar function based on the second class of behavioral tasks, the regulation of reflex systems, were unique in their attempt to incorporate the known neuroanatomy and, to some extent, the evolving knowledge of cerebellar physiology. The concepts that evolved also reflected the landmark contributions of Brodal (1972), Eccles *et al.* (1967), and many others regarding the interrelationship between cerebellar systems and other parts of the brain. In addition to demonstrating that the cerebellum is a component of several loops relating this structure to brain stem nuclei and the thalamocortical system, these studies provided insights into the circuits by which the cerebellum could influence segmental reflexes. In a major review on this subject, MacKay and Murphy (1979) offered a "metasystem" hypothesis suggesting a specific role for this structure in regulating reflex gain. This notion attempted to combine information regarding the known involvement of the cerebellum in reflex control, theories of control systems, and the known organization of the pathways by which reflexes could be modified. It also was rooted in observations indicating that the cerebellum was not essential for the performance of most behaviors but rather played its role in providing the integration necessary for optimizing the performance of a variety of motor tasks.

A third generation of cerebellar theories, derived primarily from the third class of behaviors, was propelled by two different types of contributions. The first was a series of discoveries demonstrating that cerebellar lesions in mammals result in the loss of the capacity to compensate for deficits induced by lesions in the vestibular system (Magnus, 1924; Llinas *et al.*, 1975; Carpenter *et al.*, 1959). Subsequent studies showed that the cerebellum was required for adaptive changes in the gain of the vestibuloocular reflex (VOR) (Robinson, 1976) and for modifications required to overcome the shift in a visual image produced by wearing prisms (Baizer and Glickstein, 1974). The second contribution consisted of theoretical analyses (Albus, 1971; Marr, 1969) proposing, based mostly on the anatomical architectonics of the cerebellar cortex and its major inputs, that circuits contained in the cerebellar cortex could serve as a substrate for learning and information storage.

The synthesis of these two contributions resulted in the initial proposal that the function of the cerebellum encompassed more than a role in the online regulation of motor performance. Rather, its function was expanded to include a contribution to the adaptive modifications of these processes. Consequently, a family of new theories of cerebellar function was developed (Ito, 1984; Lisberger, 1988, 1994; Peterson and Houk, 1991; Peterson et al., 1991). To accommodate the learning process and to describe the mechanisms of the participation of the cerebellum in the change of existing behaviors, these postulates introduced new terms like motor error, error detection, "teaching" inputs, and modifiable synapses. This view did not exclude a role for the cerebellum in regulating movements: it expanded the existing concepts of cerebellar function to include a direct role in adaptive processes with the evolution of broad terms, including motor learning.

The introduction of the fourth class of behaviors challenged the traditional concepts of cerebellar function by demonstrating that cerebellar lesions dramatically affected the classically conditioned eyeblink reflex. Several laboratories (e.g., Lavond et al., 1984; Yeo et al., 1985) demonstrated that lesions of small parts of the cerebellar interposed nuclei in the rabbit completely blocked the expression of previously learned, classically conditioned eyeblinks as well as the capacity to acquire these responses in naive rabbits. The unique and timely features of these observations were twofold. First, in contrast to most previously reported cerebellar deficits, the performance of the conditioned eyeblink reflex was completely incapacitated; it could not be generated despite the fact that spontaneous eyeblinks as well as unconditioned reflex eyeblinks still could be elicited. The second important feature of this work was that it generated the necessity for a field immersed previously in the concepts and terminology of motor control to interact with the concepts and terminology from the field of physiological psychology. Descriptions of the involvement of the cerebellum in the control of this reflex introduced new terms like memory trace, memory engram, association, reinforcement, conditioned response pathway, unconditioned stimulus pathway, and effector systems. The original postulate that the cerebellum serves as a storage site for memory traces used for the production of the learned conditioned response (Thompson, 1986) confronted the more traditional notions rooted in motor control with a new class of processes responsible for the experience-dependent extraction of temporal associations for the purpose of establishing new, previously nonexistent, automatic responses. In this context the cerebellum is considered to be important for the establishment, retrieval, and use of associations between stimuli to generate new responses in a context-dependent and adaptive manner.

The findings related to the role of the cerebellum in the classically conditioned eyeblink response have not yet been incorporated into cerebellar theories that also address the role of this structure in regulating behaviors included in the first two classes of behaviors introduced earlier. Nevertheless, the concepts pertaining to learning, plasticity, and memory storage stimulated by these studies contributed substantially to the discovery that cerebellar lesions in patients could produce changes in the performance of cognitive tasks (for reviews see Thach, 1996; Daum and Ackermann, 1995; Schmahmann, 1991). This discovery suggested that there was yet one more class of behaviors in which the cerebellum participates, the so-called cognitive tasks. This development brought the evolution of cerebellar theories to a new crossroad. To many this appeared to be a radical departure that was inconsistent with the traditional views of cerebellar function. Consequently, a question must be raised: is this view such a departure that it must be considered incorrect from the outset based on first principles; or is it a view that is not only tenable, but also instructive, suggesting the need to broaden our concepts of cerebellar function, revisit what has become a dichotomy between motor and cognitive functions, and consider the heterogeneous contributions the cerebellum can make to nervous system function?

II. Motor Function and Cognition: A Dichotomy Worth Saving?

To a significant degree, the concepts pertaining to the involvement of the cerebellum in cognition are derived from the evaluation of patients with cerebellar pathology using paradigms that previously had not been employed in addressing the function of this structure. For example, patients with cerebellar pathology were reported deficient in the tests of noun–verb associations and other rule-based word generation tasks (Fietz *et al.*, 1992); in judging the relative duration of two short, sequentially presented stimuli (Ivry *et al.*, 1988), in improving reaction times in a task requiring detection of spatiotemporal patterns between triggering stimuli (Pascual-Leone *et al.*, 1993); or in tasks requiring a rapid shift of attention between sensory modalities (Akshoomoff and Courchesne, 1992).

Deficits demonstrated using paradigms of this type are frequently conceptualized as characterizing "cognitive" cerebellar functions rather than the more traditional "motor" functions attributed to this structure. This approach raises a fundamental question: do the cerebellar deficits in the "motor" and "cognitive" tasks represent two distinct types of cerebellar functions? The following paragraphs develop an argument supporting the suggestion that the answer to this question is "no" because the "motor–cognitive" distinction is artificial and follows from inadequacies of current terminology.

Consider a hypothetical human subject with a well-localized, large unilateral cerebellar lesion who is evaluated using a battery of tasks: (1) finger-to-nose test; (2) classical conditioning of the eyeblink response; (3) adaptation to prisms producing a lateral displacement of a visual image; and (4) solving the Tower of Hanoi task. In the first task the patient would exhibit ataxia of voluntary movements and frequently miss the tip of the nose. Most observers would conclude, along with Holmes (1939), that the patient exhibits a "motor" deficit. In the classical conditioning of the eyeblink response the patient would fail to develop conditioned responses (Lye et al., 1988; Topka et al., 1993). Some would conclude that the patient displays a deficiency in a process related to "motor" learning (Thach, 1996). Others would suggest that there is an impairment of a mental skill related to associative motor learning (Bracke-Tolkmitt et al., 1989). Still others would conclude that the failure to exhibit the conditioned responses reflects a cognitive defect (Ivry and Baldo, 1992). In the task employing prisms, the cerebellar patient would not be able to adapt and consequently would not redirect the throw of the darts from the apparent target position to the real one (Thach et al., 1992; Bloedel et al., 1996). Some would argue that the failure to adapt to the lateral displacing prisms represents a "motor" learning deficit (Weiner et al., 1983; Thach et al., 1992). However, we prefer to argue that the cerebellar deficit in this paradigm is not "motor" but "cognitive" because the patient fails to acquire and use specific procedural knowledge, as has been observed in other cerebellar-dependent tasks (Pascual-Leone et al., 1993). In the fourth task the patient would require consistently more steps than a normal subject to assemble the pieces of the puzzle correctly (Grafman et al., 1992). Taking into consideration the nature of the task, we would conclude that the patient suffers from a deficit in cognitive planning (Grafman et al., 1992). However, it could be argued that this abnormality could be considered a "motor" deficit because the task requires planning of motor sequences (Thach, 1996). In summary, based on the patient's performance in these tests, it could be concluded that there were deficits in both motor and cognitive tasks.

In all four of the experimental tasks the patient was verbally instructed to act either passively or actively in order to achieve a specific goal or state (touch the nose with a finger, sit and endure an unpleasant stimulation of the cornea and periocular region, throw a dart at a target, manipulate and rearrange pieces of a puzzle into a specific spatial relationship). In responding to each experimental condition, the patient was required to generate an overt response, an observable behavior, and in each instance the patient's behavior was abnormal when compared with that of normal subjects. The critical part of the experiment remains: inferring, on the basis of the behavioral (movement) abnormalities exhibited by the patient, which hypothetical functions of the cerebellum were affected by the lesion.

The inferences drawn from this experiment would be highly dependent on the prevailing behavioral concepts and the notions of nervous system function on which they are based. If the accepted basis of behavior is a system of stimulus–response interactions with a strong emphasis on the role of reflexes (e.g., Hull, 1943; Watson, 1914), a classification of the cerebellar lesion-induced behavioral abnormalities based on the pattern of reflex organization would be expected. We propose that this tradition led to the primary subdivision of observed deficits into sensory and motor abnormalities, as these terms also designate the principal components of the reflex arc. This tendency was apparent in classical studies performed at the beginning of this century (Luciani, 1915; Holmes, 1917). If, however, the patient is considered to have the capacity to utilize complex knowledge (internal representations of self, representations of the external environment, and representations of the history of interaction with the environment) to solve the experimental tasks, the observed behavioral abnormalities can be ascribed to deficiencies in the handling of this complex information and in relating it to the processing of conceptual knowledge or, stated differently, deficiencies in cognitive functions (Tolman, 1949). Based on a rigid application of these two approaches, an emphasis would have to be placed on inferring whether a given task is more motor or more cognitive, and on that basis a decision would have to be made regarding which function is affected by the lesion. In the finger-to-nose task, most traditionalists would interpret the abnormality as motor, although the existence of appropriate internal and external representations of execution space cannot be denied (for relation of these representations to cerebellar processing see Bloedel, 1992). The inverse is true for the Tower of Hanoi puzzle. Thus, the ultimate interpretation depends on the assignment of one of the prevailing schemes of behavioral organization and nervous system function on a selective if not exclusive basis. To us this approach is at best confusing and more importantly fails to interpret the observations in the most meaningful context.

It must be asked whether the traditional emphasis on concepts pertaining to motor function results in an excessively restrictive attempt to categorize deficits produced by cerebellar lesions solely as dysfunctions in the motor domain. The answer to this question is twofold. Such a tendency does exist. The allocation of the cerebellum to the "motor" system and the designation of the cerebellar-dependent learning paradigms as "motor" learning paradigms are but a few examples. Alternatively, there is a progressive trend toward recognizing that certain effects of cerebellar ablation do not fit this categorization. Particularly important are the cerebellar-dependent learning paradigms in which the lesion results in the incapacity to produce the required behavior, although related behaviors can be performed reasonably well. For example, human subjects with cerebellar lesions are incapable of acquiring the classically conditioned eyeblink on the side ipsilateral to the lesion (Lye et al., 1988; D. S. Woodruff-Pak, this volume), but they can blink to the corneal airpuff and they can produce both spontaneous and voluntary eyeblinks. Consequently, the observed deficits in this classical conditioning paradigm are more related to the associative or cognitive domain of the behavior. Similar arguments can be made relative to the adaptation of the VOR and adaptation to the laterally displacing prisms. To emphasize a point that to some may be obvious, these selected examples illustrating the cognitive domain of specific behaviors indicate that processes related to this domain are not restricted to tasks performed at the conscious level.

Any dichotomy that is applied to the distinction between motor and cognitive functions is probably outdated and excessively aligned to the distinctions based on the traditional roots of the field. More problematic, adherence to these traditions leads to an inadequate categorization of functions related to the acquisition and performance of new or modified motor behaviors. In our view, on occasion it is necessary to modify the concepts of motor behavior and motor learning used as a basis for interpreting data by infusing them with experimental psychological notions. In some cases only an effort to derive a more explicit basis for defining a function regulated by the cerebellum may be necessary. Whatever the scheme, it must acknowledge that cognitive and motor processes are integral constituents of the mechanisms governing animal behavior. There is not movement without cognition, and there is not cognition without movement. In fact, substantial data support the view that, if the cerebellum is important for executing a specific behavior, it also participates in any long- or short-term modification of the characteristics of the behavior (Bloedel et al., 1996). Viewed from this perspective, any necessity for a dichotomy between "motor" and "cognitive" notions of cerebellar function vanishes because both deal with the involvement of the cerebellum in the causes (the objective of behavior, the state and processes in the external and internal environment, the history of interaction with the environment) as well as the expression of behaviors in cerebellar-dependent tasks.

III. Task Dependency and Context Dependency: Determinants of Cerebellar Involvement in Regulating Behaviors

The preceding discussion provides a strong rationale for the premise that segregating the functions in which the cerebellum is involved into cognition-related and motor-related processes is artificial at best and basically is inconsistent with the integrative features of the behaviors regulated by this structure. One of the clearest facts regarding the integrative role of the cerebellum is that it is involved in the regulation of a large number of different behaviors, i.e., the cerebellum displays substantial functional heterogeneity (Dow and Moruzzi, 1958; Watson, 1978; Schmahmann, 1991; Bloedel, 1992). It is important to emphasize that this heterogeneity of function also applies to the role of the cerebellum in the specific processes involved in the acquisition, retention, and recall of specific motor tasks. Intriguingly, the fact that it is involved in any one of these processes does not define automatically the functional nature of its contribution to a specific behavior. An important principle emerges from data pertinent to this issue, namely that the precise action of the cerebellum is task dependent (see also Thach, 1996).

One of the clearest examples of this task dependency is the contrast between the role of this structure in the modification of the vestibulo-ocular and eyeblink reflexes and its role in the acquisition of certain operantly conditioned, voluntary movements. Although the cerebellum is critical for the acquisition of both the conditioned nictitating membrane reflex (NMR) (Lincoln *et al.*, 1982) and the adaptation of the VOR (Robinson, 1976), lesions of this structure do not dramatically impair the acquisition of certain complex, operantly conditioned tasks (Shimansky *et al.*, 1994). The importance of the interactions of the cerebellum with other central structures in determining this type of dependency is apparent from the demonstration that, although intact rabbits cannot acquire the conditioned NMR without the cerebellum (Lincoln *et al.*, 1982), decerebrate rabbits can (Kelly *et al.*, 1990), a finding that also has been reported in the turtle (Keifer, 1993). In all of these behaviors the cerebellum is involved in both cognitive and motor aspects of their expression. However, its specific function in each of these behaviors may be quite different, as suggested by the fact that for some behaviors this structure is essential for task acquisition and for others it is not.

Not only is the role of the cerebellum in regulating different behaviors task dependent, but its role in a specific behavior can be context dependent. For example, effects of the same cerebellar lesion on the eyeblink reflex depend entirely on the specific context in which it is evoked. Inactivation of the cerebellar interposed nuclei suppresses the performance of this reflex when it is classically conditioned (Thompson and Krupa, 1994a). However, the same procedure reduces but does not suppress the amplitude of the unconditioned eyeblink reflex (Bracha *et al.*, 1994) and only minimally affects the performance of spontaneous, naturally occurring eyeblinks.

Related observations have also been made in studies of the involvement of the cerebellum in withdrawal reflex systems. Similar to the NMR, the amplitude of the classically conditioned forelimb withdrawal response can be reduced substantially following the inactivation of the interposed nuclear region in cats without producing comparable effects on voluntary flexion movements (Winters *et al.*, 1995; Bloedel and Bracha, 1995). In this group of behaviors, it also is clear that the action of the cerebellum and its efferent pathways can be changed by modifying the paradigm. In studies examining the mechanisms underlying the alpha conditioning of a forelimb withdrawal reflex response, the phasic activation of efferent projections from the cerebellum actually can serve as the conditioned stimulus (Rispal-Padel and Meftah, 1992). In this experiment, a small response evoked by the conditioned stimulus was enhanced when this stimulus was paired with an unconditioned stimulus applied to the forelimb. Furthermore, unlike the conditioned eyeblink reflex acquired using the delay paradigm, the alpha-conditioned limb withdrawal reflex requires the thalamocortical structures and may involve plastic changes within these regions (Meftah and Rispal-Padel, 1994). The importance of the cerebellum to the regulation of cutaneomuscular reflexes is not limited to its role in conditioned responses. The properties of unconditioned postural responses dependent on these reflexes such as the contact placing response, are changed dramatically following cerebellar lesions (for a review see Bloedel and Bracha, 1995).

Our laboratory has been very interested in a specific aspect of context dependency: the dependency on a specific feature of the learning process (i.e., acquisition, storage, retention, consolidation, expression) or on a specific component of the behavior being studied. One of the paradigms used in addressing this type of dependency requires that cats acquire and subsequently execute a complex forelimb movement in which they reach for a vertical manipulandum and move it through a template consisting of two to three consecutive straight grooves (Milak *et al.*, 1995). Unlike the classically conditioned eyeblink reflex, this behavior could be expressed despite the inactivation of the interposed and dentate nuclei. These experiments also determined whether a sequence of movements required to execute the template task could be acquired under the same condition, namely while the interposed and dentate nuclei were inactivated (Shimansky *et al.*, 1994). Despite the animals' ataxia, they were able to learn to perform this behavior.

Two lines of experiments provided insights into the nature of the involvement of the cerebellum in the acquisition process related to the learning of this behavior. First, experiments were performed in which multiple single unit responses were recorded from each of the cerebellar nuclei while cats learned to perform this template task, and changes in the event-related modulation were connected to the time course of the acquisition process (Milak *et al.*, 1995). Interestingly, the modulation of neurons in each nucleus increased dramatically at the time the movement was first performed reasonably well, i.e., approximately at the time the task first was executed over successive trials. This trend was found for approximately 85% of the cells studied. Furthermore, it occurred for responses associated with all components of the movement, including those such as paw lift-off that preceded the performance of the template, the component of the task requiring the learning of the specified motor sequence.

Second, studies were undertaken to determine if there were behavioral differences between the movements performed by cats that learned with the cerebellum intact and those that learned with the ipsilateral dentate and interposed nuclei inactivated (Shimansky *et al.*, 1994). Assessments of the movements' kinematics in these two groups of animals revealed an interesting difference. Although the movements learned with the cerebellum intact were performed in a very stereotypic fashion, those acquired during nuclear inactivation were executed with much greater variability over successive trials. This difference could not be accounted for only by the dysmetria afflicting the cats whose nuclei were inactivated.

Together, these two sets of findings suggest that, although the cerebellum is not essential for acquisition to occur, it may contribute substantially to the selection and subsequent learning of a specific motor pattern with which the desired movement can be executed efficiently. This is a process-specific role for the cerebellum in this behavior: its role is one that is specific to the acquisition process and is important, not for determining whether or not a task can be learned at all, but rather for the organization of the motor pattern selected for executing the behavior.

Additional experiments demonstrated that the specific role of the cerebellum is dependent not only on the process involved and the behavior being learned, but also on the nature of the condition requiring modification. Using the same basic paradigm, Shimansky *et al.* (1995) perturbed the reaching component of the task. The perturbation consisted of activating an elastic band as the reach was initiated in a manner that resulted in the application of a progressively greater force at a 45° angle to the direction of the movement. Both normal cats and those with the dentate and interposed nuclei inactivated with muscimol were capable of adapting to these perturbations when applied in each successive trial. Additional tests, however, revealed that the two groups of animals implemented different compensatory strategies. Normal cats acquired a strategy that was triggered by the on-line processing of proprioceptive information evoked by the perturbation and consequently were able to respond successfully even when the perturbations were applied in random trials. In contrast, the strategy acquired by cats with the dentate and interposed nuclei inactivated was initiated at the very beginning of the reaching movement and therefore likely depended on the appropriate motor set. These animals responded successfully only when the occurrence of the perturbation was predictable, a condition that could be produced by applying the perturbation in several successive trials. Thus, in the acquisition of this type of behavior, the availability of the cerebellum does not determine whether an adaptive process occurs at all. Rather, the availability of this structure determines the specific type of adaptation that will occur as a task is practiced.

This final example emphasizes the interrelationship among the task being modified, the context in which it is being performed, and the underlying process in determining the role of the cerebellum in regulating the behavior. Clearly the cerebellum is functionally heterogeneous, and this heterogeneity relates substantially to its role in the modification of behaviors, a class of functions which is related to cognitive processes.

IV. Cerebellar Functions: Implications from Distributed Circuits

The foregoing sections illustrated the historical evolution of concepts based on deficits observed in cerebellar-dependent behavioral tasks following lesions in this structure, the interrelationship between "motor" and "cognitive" processes inherent in the behaviors organized to perform these tasks, and the task dependency governing the participation of the cerebellum in specific behaviors. This section addresses issues related to the challenge of deriving concepts regarding *how* the cerebellum contributes to the control of a variety of cerebellar-dependent behaviors. Even though it is not yet possible to characterize specific mechanisms in detail, particularly with regard to the functions addressed in this essay, it is feasible to consider the concepts that govern how these mechanisms must be derived.

The term "cerebellar function" is often used in two common but not well-differentiated contexts which we will designate *operational cerebellar functions* and *systemic cerebellar functions*. Operational cerebellar functions encompass the *mechanisms* by which the cerebellum contributes to the organization of animal behavior. Examples of operational cerebellar functions from the literature are the proposed role of the cerebellum as a necessary and sufficient site of plastic changes subserving classical conditioning of the eyeblink response (Thompson, 1991) and the notion that the cerebellum is a part of the cortico-rubro-cerebellar recurrent network generating elemental movement commands (Houk *et al.*, 1996). These operational cerebellar functions should not be confused with systemic cerebellar functions, which reflect the *global tasks* or *behaviors* of the organism in which this structure participates. Examples of such systemic functions are motor coordination and cognition.

One of the primary approaches to investigating the mechanisms underlying the involvement of the cerebellum in systemic functions, i.e., in the derivation of operational cerebellar functions, is based on an assumption that there are localized substrates and perhaps singular brain structures subserving functions defined on the systemic level. This approach is used frequently in experimental psychological studies. Conceptually, first a specific systemic process is defined, then a model of the task examining this process is identified, and finally the brain is investigated using lesioning and other neurophysiological methods to identify neural substrates subserving this function.

Perhaps the best known example of this strategy is a series of investigations examining substrates of learning and memory using the model of the classically conditioned eyeblink response in the rabbit (for reviews see Thompson and Krupa, 1994b; Bloedel and Bracha, 1995; Bracha and Bloedel, 1996; R. F. Thompson *et al.*, this volume). This paradigm appeared to be ideal for attempts to localize traces of memory because the circuits relevant to eyeblink conditioning are relatively restricted. After the discovery that rabbits with lesions of the intermediate cerebellum are not capable of producing previously learned conditioned responses and that they are not capable of acquiring new conditioned eyeblinks, it was proposed that the intermediate cerebellum is the necessary and sufficient substrate for this form of learning (Thompson, 1986). If true, this specification would represent a unique situation in which a function defined on a systemic level (in terms of cognitive psychology, associative memory; in behaviorist terms, classical conditioning; or in motor control terms, motor learning) would be the sole responsibility of a single brain structure. The contribution of other related structures to this function would be confined to their role in serving simple, hard-wired relay circuits.

Most investigators, including us, agree with the argument that the cerebellum is a required structure for eyeblink conditioning in otherwise normal animals. Nevertheless, the notion that plasticity in the cerebellum is sufficient for eyeblink conditioning remains a highly controversial issue. A solution to this controversy is elusive because of two principles specifically related to deriving operational cerebellar functions imbedded in and hence related to the more global systemic functions. The first principle is that the cerebellar-dependent systemic functions as well as the related operational functions are not products of processes local to the cerebellum but rather reflect emergent or collective properties of neural networks that include but are not confined to this structure. The second principle is that the same networks with many parallel components are involved in the expression of the animals' behavior as well as in other functions inferable from this behavior (e.g., learning, motor, and cognitive planning). These characteristics make the functional isolation of a particular central nervous system structure extremely challenging experimentally.

This problem becomes apparent from an examination of the circuits involving the intermediate cerebellum in the control of eyeblink reflexes (Fig. 1). Even if one assumes that plasticity important for eyeblink reflex conditioning occurs in the cerebellum, singular methods are not available to determine directly whether comparably important plasticity occurs at any of the cerebellar input sources (e.g., pontine nuclei or trigeminal nuclei) because of the difficulty in manipulating them without affecting the operation of the cerebellum itself. Similar logic applies to evaluating the role of structures which receive efferents from the cerebellum, particularly given the reciprocal patterns of cerebellar connectivity existing for some important brain stem target nuclei. Furthermore, from the current literature it is apparent that several nodes of the network depicted in Fig. 1 receive inputs required for learning (i.e., information about the conditioned and unconditioned stimuli) and also demonstrate a capacity for plastic changes in related paradigms (for reviews see Bracha and Bloedel, 1996; Bloedel and Bracha, 1995).

These characteristics of this distributed circuitry together with the multiple features of the behavior with which the cerebellum is involved pose difficult problems for the experimentalist. First, because the intermediate cerebellum is involved in both the acquisition and the execution of the conditioned eyeblinks, it is difficult to determine whether any of the manipulations of the cerebellum-related network differentially affect the capacity to execute the learned behavior or the capacity to acquire it. Second, given the organization of this network and the known aspects of its physiology, it is difficult, if not impossible, to determine whether changes in behavior during local manipulations of specific parts of the system are a consequence of disrupting a process localized to the site of the manipulation or whether they result from a malfunction at some other site within the circuitry or of the network as a whole (for a review of these issues see Bracha and Bloedel, 1996; and Bloedel and Bracha, 1995).

Based on this reasoning, we proposed previously an alternative model of the participation of the cerebellum in the classical conditioning of the eyeblink response in the rabbit: the multiple pathway model (Fig. 1, adopted from Bracha and Bloedel, 1996). Two crucial postulates serve as the basis for this model: (1) that plasticity which is causal for the classical conditioning is distributed and (2) that the behavioral capacity to acquire and express the learned eyeblink response cannot be explained by a process local to any of the participating structures but only by the collective opera-

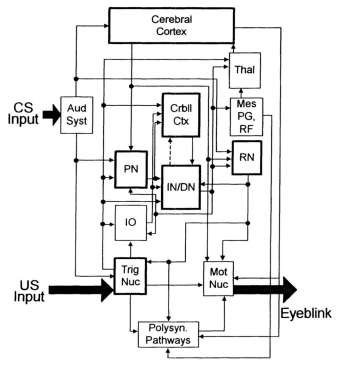

FIG. 1. Schematic of the cerebellar-related circuits involved in the control of the classically conditioned eyeblink in the rabbit. (Bold boxes) Putative sites of plasticity related to NMR conditioning. All of these putative sites receive inputs activated by both the conditioned and the unconditioned stimulus, and learning-related plasticity was reflected by observations at most of them. This "multiple pathway model" presumes that the learning of the conditioned response is subserved by modifications distributed across multiple nodes of the network and that the acquisition and retention of the conditioned responses are a result of a collective operation of the whole network. Aud Syst, auditory system; CS, conditioned stimulus; Crbll Ctx, cerebellar cortex; IN/DN, cerebellar interposed and dentate nuclei; IO, inferior olive; Mes PG, RF, mesencephalic periaqueductal gray and reticular formation; Mot Nuc, motor nuclei mediating the eyeblink/nictitating membrane response; Polysyn. Pathways, medullary polysynaptic pathways mediating the unconditioned eyeblink; PN, pontine nuclei; RN, red nucleus; Thal, thalamus; Trig Nucl, trigeminal sensory nuclei; US, unconditioned stimulus. Adapted from Bracha and Bloedel (1996).

tion of the whole network. Based on this view, the cerebellum serves as part of a higher order network responsible for systemic functions. Interestingly, this notion was inherent in the classical works of several early investigators such as Holmes (1939), Rademaker (1931, 1980), and Popov (1929). Similar assumptions are also intrinsic to the original suggestions for the cerebellar participation in cognitive tasks (Leiner et al., 1986, 1991; for reviews and discussion, see Bracke-Tolkmitt et al., 1989; Schmahmann, 1991; Watson, 1978). Given the likelihood (if not the fact) that these concepts derived from the distributed nature of the circuitry apply to virtually all behaviors and systemic functions, the same limitations inherent in determining operational functions related to the mechanisms of the participation of the cerebellum in eyeblink conditioning also apply to determining the mechanisms underlying other systemic functions such as cognition.

Importantly, these features of this system underlie one of the critical points of this essay, namely that network components participating in the so-called cognitive functions simultaneously subserve the organization of the behavior and its execution as well. This postulated imbedding of different operational functions within the same network can be illustrated from a series of interesting observations related to the dentate nucleus of the cerebellum. Throughout phylogeny there is a clear relationship between the complexity of this nucleus and the capacity of individual species to perform complex arm, hand, and finger movements (Holmes, 1939; Massion, 1973). The same evolutionary trends as well as considerable morphological data have led to the postulate that the neocerebellum may also play a role in cognitive processes (Schmahmann and Pandya, 1991, 1993, 1995; Schmahmann, 1991; Leiner et al., 1986, 1991, 1993; Thach, 1996). Studies by Middleton and Strick (1994) have provided support for this view through the demonstration of a specific projection from the dentate nucleus to the prefrontal cortex. Furthermore, this cerebellar nucleus is activated in functional imaging studies during the attempts of human subjects to solve a complex puzzle (Kim et al., 1994).

Nevertheless, substantial evidence remains in the literature that the dentate nucleus is also involved in the control of ongoing movements. Considerable evidence shows that permanent ablations of this region produce deficits in ongoing movements of the extremities (for a review see Gilman et al., 1981). Even more dramatic findings result from the temporary inactivation of this structure (Milak et al., 1992, 1993; Brooks et al., 1973; Brooks, 1984). In addition, the activity of dentate neurons can be modulated with respect to specific features of overtrained complex arm and wrist movements (Thach, 1978). Importantly, recording studies also support the conclusions based on the imaging study reviewed earlier. Modulation of dentate neurons was also found to be related to higher order features of a flexion–extension movement of the wrist (Strick, 1983). Changes in activity were found to be related to intended movement direction rather than to actual movement direction (Strick, 1983; see also Thach, 1978). Taken together, these findings support a joint role of the dentate nucleus in cognitive as well as motor processes.

Thus, the organization of these circuits substantiates the inferences made earlier regarding the integral nature of the multiple processes regulated by the cerebellum and the improbability of segregating the role of this structure into cognitive and noncognitive functions. Even though fundamental features of certain behaviors, particularly reflex behaviors, can be organized within identifiable brain stem and spinal pathways, the nuclei which mediate them usually have multiple interactions with nuclei outside these fundamental circuits. Which of these interactions are cognitive? Which are noncognitive? It is impossible to support such a separation on the basis of either the circuit organization or the characteristics of the behaviors.

V. Conclusions

This chapter emphasized the issue of the cerebellum and cognition from several perspectives: the historical, the behavioral, the organizational, and the functional. At the very least it is clear that this issue is not a simple one. At its simplest, this issue has been reduced to questions regarding the presence or absence of plastic changes in the cerebellum. In fact, we are certain it will be surprising to some that we not only acknowledge but support the view that the cerebellum plays a role in (so-called) cognitive processes. Based on the arguments presented earlier, merely the inseparable nature of the motor and cognitive domains compels the acceptance of this notion, particularly since an "involvement in" is all that is required for the verity of this notion (Bloedel, 1993).

If one accepts the view that the function of any central structure including the cerebellum must be considered from a behavioral context, at least as far as functional classifications are concerned, several inferences become apparent. Relating the function of a structure to one process such as acquisition or execution can be done only in the context of the specific task with which the process is associated. For example, it is apparent that the cerebellum plays a major role in retention only in relation to certain behaviors. Second, the function of the cerebellum is always broader than, and therefore cannot be confined to, a single process. Third, classical functional concepts such as "motor" function, "sensory" function, and "cognitive" function are no longer appropriate descriptors of the contributions of the cerebellum to the well-being and behavior of an organism. Each of these are incorporated indivisibly into the integration performed by the cerebellum in relation to each type of behavior: VOR, cutaneomuscular reflexes, volitional goal-directed movements, and postural control. It becomes clear even from this list that these behaviors themselves become integrated when a particular task is performed: eye–head control, posture, reaching, associated autonomic changes, and modification in reflex systems, to name a few. Thus, it is the synthesis of behaviors in the context of achieving a specific behavioral objective that ultimately defines "function." Clearly, cognitive processes, which can be conscious or subconscious, play a role in virtually all of these behaviors. As emphasized earlier, for any behavior in which it is involved, the cerebellum participates both in regulating its execution and in modifying its characteristics to optimize its performance in a specific context.

Acknowledgment

Research from our laboratory was supported by NIH Grants R01 NS21958 and P01 NS30013.

References

Akshoomoff, N. A., and Courchesne, E. (1992). A new role for the cerebellum in cognitive operations. *Behav. Neurosci.* **106,** 731–738.
Albus, J. S. (1971). A theory of cerebellar function *Math. Biosci.* **10,** 25–61.
Amassian, V. E., and Ross, R. (1972). Cerebellothalamocortical interrelations in contact placing and other movements in cats. In "Corticothalamic Projections and Sensorimotor Activities" (T. Frigyesi, E. Rinvik, and M. D. Yahr, eds.), pp. 395–444. Raven Press, New York.
Amassian, V. E., Weiner, H., and Rosenblum, M. (1972). Neural systems subserving the tactile placing reaction: A model for the study of higher level control of movement. *Brain Research.* **40,** 171–178.
Baizer, J. S., and Glickstein, M. (1974). Role of the cerebellum in prism adaptation. *J. Physiol. (Lond.)* **234,** 34P–35P.
Bloedel, J. R. (1992). Functional heterogeneity with structural homogeneity: How does the cerebellum operate? *Behav. Brain Sci.* **15,** 666–678.
Bloedel, J. R. (1993). "Involvement in" versus "storage of." *Trends Neurosci.* **16,** 451–452.
Bloedel, J. R., and Bracha, V. (1995). On the cerebellum, cutaneomuscular reflexes, movement control and the elusive engrams of memory. *Behav. Brain Res.* **68,** 1–44.
Bloedel, J. R., Ebner, T. J., and Wise, S. P. (eds.) (1996). "Acquisition of Motor Behavior in Vertebrates," MIT Press, Boston.
Botez, M. I. (1989). Role of the cerebellum in complex human behavior. *Neurol. Sci.* **10,** 291–300.
Bracha, V., and Bloedel, J. R. (1996). The multiple pathway model of circuits subserving the classical conditioning of withdrawal reflexes. In "Acquisition of Motor Behavior in Vertebrates" (J. R. Bloedel, T. J. Ebner, and S. P. Wise, eds.). MIT Press, Boston.

Bracha, A. V., Webster, M. L., Winters, N. K., Irwin, K. B., and Bloedel, J. R. (1994). Effects of muscimol inactivation of the cerebellar nucleus interpositus on the performance of the nictitating membrane response in the rabbit. *Exp. Brain Res.* **100**, 453–468.
Bracke-Tolkmitt, R., Linden, A., Canavan, A. G. M., Rockstroh, B., Scholz, E., Wessel, K., and Diener, H.-C. (1989). The cerebellum contributes to mental skills. *Behav. Neurosci.* **103**, 442–446.
Brodal, A. (1972). Cerebellar pathways: Anatomical data and some functional implications. *Acta Neurol. Scand. Suppl.* **51**, 153–196.
Brooks, V. B. (1984). Cerebellar functions in motor control. *Hum. Neurobiol.* **2**, 251–260.
Brooks, V. B., Kozlovskaya, I. B., Atkin, A., Horvath, F. E., and Uno, M. (1973). Effects of cooling dentate nucleus on tracking-task performance in monkeys. *J. Neurophysiol.* **36**, 974–995.
Brooks, V. B., and Thach, W. T. (1981). Cerebellar control of posture and movement. *In* "Handbook of Physiology" (J. M. Brookhart, and V. B. Mountcastle, eds.), Vol. II, pp. 877–946. American Physiology Society, Bethesda, MD.
Carpenter, M. B., Fabrega, H., and Glinsmann, W. (1959). Physiological deficits occurring with lesions of labyrinth and fastigial nuclei. *J. Neurophysiol.* **22**, 222–234.
Chambers, W. W. and Sprague, J. M. (1951). Differential effects of cerebellar anterior lobe cortex and fastigial nuclei on postural tonus in the cat. *Science* **114**, 324–325.
Chambers, W. W., and Sprague, J. M. (1955). Functional localization in the cerebellum. II. Somatotopic organization in cortex and nuclei. *Arch. Neurol. Psychiatr.* **74**, 653–680.
Daum, I., and Ackermann, H. (1995). Cerebellar contributions to cognition. *Behav. Brain Res.* **67**, 201–210.
Dow, R. S., and Moruzzi, G. (1958). "The Physiology and Pathology of the Cerebellum." University of Minnesota Press, Minneapolis.
Eccles, J. C., Ito, M., and Szentagothai, J. (1967). "The Cerebellum as a Neuronal Machine." Springer-Verlag, Berlin.
Fietz, J. A., Petersen, S. E., Cheney, M. K., and Raichle, M. E. (1992). Impaired nonmotor learning and error detection associated with cerebellar damage. *Brain* **115**, 155–178.
Gilman, S., Bloedel, J. R., and Lechtenberg, R. (1981). "Disorders of the Cerebellum." Davis Co., Philadelphia.
Goldberger, M. E., and Growdon, J. H. (1973). Pattern of recovery following cerebellar deep nuclear lesions in monkeys. *Exp. Neurol.* **39**, 307–322.
Grafman, J., Litvan, I., Massaquoi, S., Stewart, M., Sirigu, A., and Hallett, M. (1992). Cognitive planning deficit in patients with cerebrallar atrophy. *Neurology* **42**, 1493–1496.
Holmes, G. (1917). The symptoms of acute cerebellar injuries due to gunshot injuries. *Brain* **40**, 461–535.
Holmes, G. (1939). The cerebellum of man. *Brain* **62**, 1–30.
Horak, F. B., and Diener, H. C. (1994). Cerebellar control of postural scaling and central set in stance. *J. Neurophysiol.* **72**, 479–493.
Houk, J. C., Buckingham, J. T., and Barto, A. G. (1996). Models of the cerebellum and motor learning. *Behav. Brain Sci.* **19**(3), 368–383.
Hull, C. L. (1943). "Principles of Behavior," Appleton, New York.
Ito, M. (1984). "The Cerebellum and Neural Control," Raven Press, New York.
Ivry, R. B., and Baldo, J. V. (1992). Is the cerebellum involved in learning and cognition? *Curr. Opin. Neurobiol.* **2**, 212–216.
Ivry, R. B., Keele, S. W., and Diener, H. C. (1988). Dissociation of the lateral and medial cerebellum in movement timing and movement execution. *Exp. Brain Res.* **73**, 167–180.
Keifer, J. (1993). The cerebellum and red nucleus are not required for classical conditioning of an in vitro model of the eye-blink reflex. *Soc. Neurosci. Abstr.* **19**, 1001.

Kelly, T. M., Zuo, C.-C., and Bloedel, J. R. (1990). Classical conditioning of the eyeblink reflex in the decerebrate-decerebellate rabbit. *Behav. Brain Res.* **38,** 7–18.

Kim, S.-G., Urgubil, K., and Strick, P. L. (1994). Activation of a cerebellar output nucleus during cognitive processing. *Science* **265,** 949–951.

Kolb, F. P., Irwin, K. B., Winters, N. K., Bloedel, J. R., and Bracha, V. (1994). Involvement of the cat cerebellar interposed nucleus in the control of conditioned and unconditioned withdrawal reflexes. *Soc. Neurosci. Abstr.* **20,** 1746.

Lavond, D. G., Lincoln, J. S., McCormick, D. A., and Thompson, R. F. (1984). Effect of bilateral lesions of the dentate and interpositus cerebellar nuclei on conditioning heart rate and nictitating membrane/eyelid responses in the rabbit. *Brain Res.* **305,** 323–330.

Leiner, H. C., Leiner, A. L., and Dow, R. S. (1986). Does the cerebellum contribute to mental skills. *Behav. Neurosci.* **100,** 443–454.

Leiner, H. C., Leiner, A. L., and Dow, R. S. (1991). The human cerebro-cerebellar system: Its computing, cognitive, and language skills. *Behav. Brain Res.* **44,** 113–128.

Leiner, H. C., Leiner, A. L., and Dow, R. S. (1993). Cognitive and language functions of the human cerebellum. *Trends Neurosci.* **16,** 444–447.

Lincoln, J. S., McCormick, D. A., and Thompson, R. F. (1982). Ipsilateral cerebellar lesions prevent learning of the classically conditioned nictitating membrane eyelid response. *Brain Res.* **242,** 190–193.

Lisberger, S. G. (1988). The neutral basis for motor learning in the vestibulo-ocular reflex in monkeys. *Trends Neurosci.* **11,** 147–152.

Lisberger, S. G. (1994). Neural basis for motor learning in the vestibuloocular reflex of primates. III. Computational and behavioral analysis of the sites of learning. *J. Neurophysiol.* **72,** 974–998.

Lisberger, S. G., Pavelko, T. A., and Broussard, D. M. (1994). Neural basis for motor learning in the vestibuloocular reflex of primates. I. Changes in the responses of brain stem neurons. *J. Neurophysiol.* **72,** 928–953.

Llinas, R., Walton, K., Hillman, D. E., and Sotelo, C. (1975). Inferior olive: Its role in motor learning. *Science* **190,** 1230–1231.

Luciani, L. (1915). The hind-brain. *In* "Human Physiology" (G. M. Holmes, eds.) pp. 419–485. MacMillan, London.

Lye, R. H., O'Boyle, D. J., Ramsden, R. T., and Schady, W. (1988). Effects of a unilateral cerebellar lesion on the acquisition of eye-blink conditioning in man. *J. Physiol. (Lond.)* **403,** 58.

MacKay, W. A., and Murphy, J. T. (1979). Cerebellar modulation of reflex gain. *Prog. Neurobiol.* **13,** 361–417.

Magnus (1924). "Korperstellung," Springer, Berlin.

Marr, D. (1969). A theory of cerebellar cortex. *J. Physiol. (Lond.)* **202,** 437–470.

Massion, J. (1973). Intervention des voies cerebello-corticales et corticocerebelleuses dans l'organisation et al regulation du mouvement. *J. Physiol. (Paris)* **67,** 117A–170A.

McCormick, D. A., Lavond, D. G., Clark, G. A., Kettner, R. E., Rising, C. E., and Thompson, R. F. (1981). The engram found? Role of the cerebellum in classical conditioning of nictitating and eyelid response. *Bull. Psychon.* **18,** 103–105.

Meftah, E. M., and Rispal-Padel, L. (1994). Synaptic plasticity in the thalamocortical pathway as one of the neurobiological correlates of forelimb flexion conditioning: Electrophysiological investigation in the cat. *J. Neurophysiol.* **72,** 2631–2647.

Middleton, F. A., and Strick, P. L. (1994). Anatomical evidence for cerebellar and basal ganglia involvement in higher cognitive function. *Science* **266,** 458–461.

Milak, M. S., Bracha, V., and Bloedel, J. R. (1993). Effects of temporary inactivation of the specific cerebellar nuclei on the organization of EMG activity during a complex forelimb movement. *Neurosci. Abstr.* **19,** 979.

Milak, M. S., Bracha, V., and Bloedel, J. R. (1995). Relationship of simultaneously-recorded cerebellar nuclear neuron discharge to the acquisition of a complex, operantly conditioned forelimb movement in cats. *Exp. Brain Res.* **105**, 325–330.

Milak, M. S., Bracha, V., Kolb, F., McAlduff, J. D., and Bloedel, J. R. (1992). Selective effects of muscimol injections into cerebellar nuclei in cats performing both a locomotor and a reaching task. *Neurosci. Abstr.* **18**, 1550.

Pascual-Leone, A., Grafman, J., Clark, K., Stewart, M., Massaquoi, S., Lou, J.-S., and Hallett, M. (1993). Procedural learning in Parkinson's disease and cerebellar degeneration. *Ann. Neurol.* **34**, 594–602.

Peterson, B. W., Baker, J. F., and Houk, J. C. (1991). A model of adaptive control of vestibuloocular reflex based on properties of cross-axis adaptation. *Ann. N.Y. Acad. Sci.* **627**, 319–337.

Peterson, B. W. and Houk, J. C. (1991). A model of cerebellar-brainstem interaction in the adaptive control of the vestibuloocular reflex. *Acta Otolaryngol. (Stockholm)* **111**(Suppl. 481), 428–432.

Popov, N. F. (1929). Notes for the study of cerebellar function. In "Vishaya N'ervnaya D'eyatel'-nost': Sbornik Trudov Instituta" (D. S. Fursikov, M. O. Gurevich, and A. N. Zalmanzon, eds.) pp. 93–139. Komunisticheska Akademiya, Institut Vyshey Nervnoy D'eyatel'nosti, Moscow.

Rademaker, G. G. J. (1931). "Das Stehen: Statische Reaktionen, Gleichgewichtreaktionen and Muskeltonus unter besonderer Berücksichtigung ihres Verhaltens bei kleinhirnlosen Tieren," Julius Springer, Berlin.

Rademaker, G. G. J. (1980). "The Physiology of Standing (Das Stehen): Postural Reactions and Equilibrium with Special Reference to the Behavior of Decerebellate Animals," University of Minnesota Press, Minneapolis.

Rispal-Padel, L., and Meftah, E.-M. (1992). Changes in motor responses induced by cerebellar stimulation during classical forelimb flexion conditioning in cat. *J. Neurophysiol.* **68**, 908–926.

Robinson, D. A. (1976). Adaptive gain control of vestibuloocular reflex by the cerebellum. *J. Neurophysiol.* **39**, 954–969.

Schmahmann, J. D. (1991). An emerging concept: The cerebellar contribution to higher function. *Arch. Neurol.* **48**, 1178–1187.

Schmahmann, J. D. and Pandya, D. N. (1991). Projections to the basis pontis from the superior temporal sulcus and superior temporal region in the rhesus monkey. *J. Comp. Neurol.* **308**, 224–248.

Schmahmann, J. D., and Pandya, D .N. (1993). Prelunate, occipitotemporal, and parahippocampal projections to the basis pontis in the monkey. *J. Comp. Neurol.* **337**, 94–112.

Schmahmann, J. D. and Pandya, D. N. (1995). Prefrontal cortex projections to the basilar pons in the rhesus monkey: Implications for the cerebellar contribution to higher function. *Neurosci. Lett.* **199**, 175–178.

Sherrington, C. S. (1906). "The Integrative Action of the Nervous System," Constable, London.

Shimansky, Y., Wang, J.-J., Bloedel, J. R., and Bracha, V. (1994). Effects of inactivating the deep cerebellar nuclei on the learning of a complex forelimb movement. *Soc. Neurosci. Abstr.* **20**, 21.

Shimansky, Yu., Wang, J.-J., Bracha, V., and Bloedel, J. R. (1995). Cerebellar inactivation abolishes the capability of cats to compensate for unexpected but not expected perturbations of a reach movement. *Neurosci. Abstr.* **21**, 914.

Strick, P. L. (1983). The influence of motor preparation on the response of cerebellar neurons to limb displacements. *J. Neurosci.* **3**, 2007–2020.

Supple, W. F., Jr., and Kapp, B. S. (1993). The anterior cerebellar vermis: Essential involvement in classically conditioned bradycardia in the rabbit. *J. Neurosci.* **13**, 3705–3711.

Thach, W. T. (1978). Correlation of neural discharge with pattern and force of muscular activity, joint position, and direction of intended next movement in motor cortex and cerebellum. *J. Neurophysiol.* **41,** 654–676.

Thach, W. T. (1996). On the specific role of the cerebellum in motor learning and cognition: Clues from PET activation and lesion studies in man. *Behav. Brain Sci.* **19**(3), 411–431.

Thach, W. T., Goodkin, H. P., and Keating, J. G. (1992). The cerebellum and the adaptive coordination of movement. *Annu. Rev. Neurosci.* **15,** 403–442.

Thompson, R. F. (1986). The neurobiology of learning and memory. *Science* **223,** 941–947.

Thompson, R. F. (1991). Are memory traces localized or distributed? *Neuropsychologia* **29,** 571–582.

Thompson, R. F., and Krupa, D. J. (1994a). Organization of memory traces in the mammalian brain. *Annu. Rev. Neurosci.* **17,** 519–549.

Thompson, R .F., and Krupa, D. J. (1994b). Organization of memory traces in the mammalian brain. *Annu. Rev. Neurosci.* **17,** 519–549.

Tolman, E. C. (1949). There is more than one kind of learning. *Psychol. Rev.* **27,** 217–233.

Topka, H., Valls-Solé, J., Massaquoi, S. G., and Hallett, M. (1993). Deficit in classical conditioning in patients with cerebellar degeneration. *Brain* **116,** 961–969.

Watson, J. B. (1914). "Behavior: An Introduction to Comparative Psychology," Holt, New York.

Watson, P. J. (1978). Nonmotor functions of the cerebellum. *Psychiatr. Bull.* **85,** 944–967.

Weiner, M. J., Hallett, M., and Funkenstein, H. H. (1983). Adaptation to lateral displacement of vision in patients with lesions of the central nervous system. *Neurology* **33,** 766–772.

Winters, N. K., Irwin, K. B., Kolb, F. P., Bloedel, J. R., and Bracha, V. (1995). Involvement of the cerebellar interposed nucleus in reflexive and voluntary forelimb movements. *Neurosci. Abstr.* **21,** 915.

Yeo, C. H., Hardiman, M. J., and Glickstein, M. (1985). Classical conditioning of the nictitating membrane response of the rabbit. I. Lesions of the cerebellar nuclei. *Exp. Brain Res.* **60,** 87–98.

SECTION VII
FUTURE DIRECTIONS

THERAPEUTIC AND RESEARCH IMPLICATIONS

Jeremy D. Schmahmann, M.D.

Department of Neurology, Massachusetts General Hospital and Harvard Medical School, Boston, Massachusetts 02114

I. Introduction
II. Therapeutic Implications
   A. The Need to Know
   B. Rehabilitation Efforts
   C. Potential Future Therapeutic Modalities
III. Research Implications
IV. Conclusions
   References

Investigations into the relationship between the cerebellum and nonmotor processing have produced a substantial body of evidence which seems to require a revision of accepted notions about the functional role of the cerebellum. This chapter presents a perspective on the contemporary and possible future therapeutic and research implications of these findings. These include the need for patients and their families to know of the behavioral consequences of cerebellar disease processes; potential approaches for improvement through rehabilitation therapies; and future treatment strategies, such as electrical stimulation of the cerebellum and psychosurgical approaches applied to the cerebellum. In addition, some areas of basic science investigation that could prove informative in understanding this relationship are addressed. It will be important to obtain a more complete characterization of the anatomy, physiology, and functional topography of the cerebellum in humans and in animal models, and a greater understanding of the clinical consequences of cerebellar lesions.

I. Introduction

The realization that the role of the cerebellum in nervous system function is not limited to the coordination of voluntary movement has ushered in an exciting new area of cognitive neuroscience investigation. There are presently more questions than answers, but there is now a clearer notion of where future research efforts may reasonably be directed. There is also some hint of the clinical significance of the findings to date and of the potential for therapeutic intervention. This chapter presents a view of therapeutic and research implications derived from our current state of knowledge in this field.

II. Therapeutic Implications

A. The Need to Know

Dual clinical imperatives motivate the clinician in the management of neurologically impaired individuals. The central directive is to provide care in all cases, and cure when possible. A second consideration of great importance to patients and their families is the need to know the diagnosis, understand the disease process, and participate actively in their own medical treatment. This facilitates a physician–patient partnership that is an essential ingredient in both acute and long-term management. The knowledge that seemingly bizarre behavior, inappropriate social interactions, disinhibited personality style, and limited intellectual flexibility and abilities may be explained by the cerebellar insult itself (Bracke-Tolkmitt *et al.*, 1989; Wallesch and Horn, 1990; Pollack, 1995; see chapter by J. D. Schmahmann and J. C. Sherman), can provide the patient and family with an explanation and understanding that was previously unavailable. It is too early to be definitive in presenting this association, but there is sufficient evidence to inform patients and families of the discussion of the probable role of the cerebellum in the modulation of behavior. This is also valuable in providing reassurance regarding the reversibility of the syndrome following acute lesions (Botez-Marquard *et al.*, 1994; see chapter by J. D. Schmahmann and J. C. Sherman) although the full extent and time course of the cognitive recovery have yet to be adequately studied.

We do not yet understand the range of manifestations and rate of progression of the cognitive and emotional disturbances that may accompany degenerative cerebellar disease. Early indicators show that there is an intellectual and emotional decline with time (Grafman *et al.*, 1992; Appollonio *et al.*, 1993; Kish *et al.*, 1994; Botez-Marquard and Botez, 1995; see chapter by J. D. Schmahmann and J. C. Sherman) but the clinical relevance and impact on occupational performance and personal life situations are essentially unknown. It will be important to establish these facts in order to share them with patients and their families.

B. REHABILITATION EFFORTS

Cognitive rehabilitation has been helpful in patients who have suffered closed head injury (Levin, 1992) and in those with aphasia (Wertz et al., 1981; Pring, 1986). It is conceivable that patients with a behavioral syndrome from cerebellar lesions may also benefit from such approaches, but one first has to recognize that this syndrome exists before being able to treat it.

The awareness of a cerebellar role in sensory (Gao et al., 1996), autonomic (Martner, 1975; see chapter by D. E. Haines et al.), emotional (Cooper et al., 1978; Heath et al., 1979; see chapter by J. D. Schmahmann and J. C. Sherman), and intellectual processing (Bracke-Tolkmitt et al., 1989; Grafman et al., 1992; Appollonia et al., 1993; see chapters by T. Botez-Marquard and M. I. Botez and by J. D. Schmahmann and J. C. Sherman) may facilitate new approaches to the neurologically disabled patient that capitalize on cerebrocerebellar communication. Could physical/occupational/cognitive/vestibular therapy approaches exploit the cerebellar component of the cerebrocerebellar circuit to help compensate for functions lost by cerebral hemispheric damage? That is, could vestibular physical therapy (Shephard et al., 1993) improve dexterity in a partially paralyzed extremity, e.g., by promoting cross-modal integration within cerebellum, and thus benefiting the motor system? Could enhanced sensory stimulation (posterior column or muscle spindle input) promote recovery from motor incapacity by cerebellar as well as by cerebral mechanisms? These may be fanciful notions, but the potential for utilizing compensatory mechanisms across different modalities is at least an interesting possibility, derived in large part from the evolving understanding of the convergence within the cerebellum of afferents from multiple domains of neurologic function and the tightly linked, highly organized, cerebro–cerebellar interactions (Hampson et al., 1952; Henneman et al., 1952; Sasaki et al., 1975; Brodal, 1978, 1979; Haines and Dietrichs, 1984; Glickstein et al., 1985; Schmahmann, 1991, 1996; Leiner et al., 1993; Middleton and Strick, 1994; see chapter by J. D. Schmahmann and D. N. Pandya).

C. POTENTIAL FUTURE THERAPEUTIC MODALITIES

1. Cerebellar Stimulation

The improvement in mood and aggression induced in some patients by cerebellar cortical stimulation (Cooper et al., 1978) was believed to be related to cerebellar connections with the limbic and autonomic systems (Snider and Maiti, 1975; Martner, 1975). The techniques and results of cerebellar cortical stimulation were not always consistent, possibly reflecting the fact that placement of the stimulating electrodes generally did not respect established anatomic boundaries or the topographic (zonal) pattern of cerebellar cortical afferent and efferent systems (Haines, 1981). Furthermore, the pathophysiology of the improvement, the extent and duration of the clinical recovery, and the nature of the side effects or complications were also not fully established.

Fastigial nucleus (FN) stimulation has been pursued experimentally in the study of vasomotor influences of the cerebellum (Doba and Reis, 1972; Martner, 1975; McKee *et al.*, 1976; Chida *et al.*, 1986; see chapter by D. J. Reis and E. V. Golanov), and recent findings indicating a neuroprotective effect of FN stimulation in the setting of cerebrovascular ischemia are intriguing (Reis *et al.*, 1991; see chapter by D. J. Reis and E. V. Golanov). Stroke is a major cause of morbidity and mortality, and efforts to prevent or limit neuronal injury and death from ischemic infarction are receiving intense scrutiny. Could there be a role for some modified and clinically applicable version of FN stimulation in this patient population? Are there lessons to be learned about mechanisms of neuronal protection from these studies that could be more broadly applied to the protection of ischemic neural tissue without having to resort to the use of a brain stimulator?

There is a contemporary precedent for the therapeutic use of electrical stimulation of the nervous system. The transcutaneous electrical nerve stimulator (TENS), based on the gate theory of Melzack and Wall (1965), has substantially helped some individuals with intractable pain syndromes (Katz *et al.*, 1989); and electroconvulsive therapy is still valuable in the treatment of catatonia and profound depression not responsive to medication (Greenblatt, 1977; Frankel, 1984; Casey, 1994). As we learn more about the dynamic interactions between the cerebellum and other neural systems that subserve cognition, emotion, and autonomic function, there appears to be sufficient reason, based on both early and contemporary work in this field, to warrant a scientific reevaluation of the indications and techniques of therapeutic cerebellar stimulation and of more sophisticated future adaptations thereof.

2. *Cerebellar Psychosurgery*

Psychosurgery has reemerged as a valid and effective treatment of selected diseases. In vogue in the first half of this century as prefrontal leukotomy for schizophrenia (Landis, 1949), this modality lost favor until recent advances in neuroscience facilitated its careful and judicious use. Knowledge of the functional neuroanatomy of obsessive compulsive disorder (Breiter *et al.*, 1996; Jenike *et al.*, 1996) has led to the successful use of therapeutic cingulotomy in those severely afflicted individuals who do not respond to medications (Spangler *et al.*, 1996). Additionally, stereotactic pallidotomy guided by magnetic resonance imaging and performed with the assistance of physiologic recording has proven to be successful for the treatment of disabling tremor and bradykinesia in Parkinson's disease (Laitinen, 1995; Baron *et al.*, 1996). Furthermore, temporal lobectomy, which when first introduced resulted in the amnesia typified by patient H.M. (Scoville and Milner, 1957; Milner *et al.*, 1968), is now routinely employed in epilepsy units utilizing contemporary diagnostic techniques (Fried, 1993; Spencer, 1996).

The amelioration of aggression in monkeys by vermis and archicerebellar lesions but not by neocerebellar lesions was an extraordinary finding (Peters and Monjan, 1971; Berman *et al.*, 1974). Is there a relationship between the vermis/archicerebellum/fastigial nucleus complex and emotional dyscontrol? This is surely a difficult area of study because aggression and violence are so interwoven with psychological, social, and biological factors. Nevertheless, there appears to be sufficient scientific rationale at least to address the question in the context of hypothesis-driven research protocols, and these observations in nonhuman primate need to be repeated. It may also be reasonable to open the debate regarding the use of cerebellar psychosurgery for the management of unremitting affective disorders, including emotional dyscontrol, in humans.

3. *Cerebellar Transplantation*

The study of transplanted cerebellar tissue has been ongoing for some years (Sotelo and Alvarado-Mallart, 1987), and recent findings suggest some hope for anatomic and functional success with this approach (Triarhou *et al.*, 1996). As the understanding of the neural circuitry of the cerebrocerebellar system evolves, and the appreciation of the scope of cerebellar function becomes evident, the transplantation of cerebellar tissue in neurodegenerative disorders may find a role in clinical practice for its potential motor as well as its nonmotor consequences. Furthermore, as neuroprotective treatment options (Schultz *et al.*, 1996), and eventually gene therapy (Hahania *et al.*, 1995), become available for cerebellar neurodegenerative disorders, it is quite likely that the cognitive and affective improvements resulting from these interventions will need to be monitored as closely as the course of the cerebellar motor phenomena.

III. Research Implications

It is apparent that a number of issues related to therapeutic intervention discussed earlier require thorough clinical and basic science investigation.

In addition, there are many intriguing questions in each discipline of the neurosciences that have yet to be explored in understanding the relationship between the cerebellum and higher order function.

Considerable strides have been made in understanding the complexities of cerebrocerebellar organization in the nonhuman primate (Allen and Tsukuhara, 1974; Brodal, 1979; Voogd and Bigaré, 1980; Haines *et al.*, 1982; Glickstein *et al.*, 1985; Dore *et al.*, 1990; Schmahmann, 1994, see chapters by F. A. Middleton and P. L. Strick, by J. D. Schmahmann and D. N. Pandya, by D. E. Haines *et al.*, and by J. D. Schmahmann). More detailed knowledge of cerebellar interactions with the neuraxis will be gained by further studies of the projections from pons to the cerebellum, the cerebellar cortical to nuclear projection, and the cerebellar nuclear projection to thalamus. There is currently only limited information available, for example, regarding the anatomic relationship between discrete architectonic regions of the cerebral hemispheres (Pandya and Yeterian, 1985) and individual cerebellar lobules, folia, and nuclei (Schmahmann *et al.*, 1996).

In order to establish the functional relevance of the various anatomic nodes within the distributed cerebrocerebellar circuitry, experimental investigations using physiologic recordings and behavioral studies need to be undertaken. Current technical achievements permit multiple simultaneous electrode recordings in different neural structures of awake behaving animals (e.g., Gardiner and Kitai, 1992). Cerebellar neuronal discharges have been extensively studied in relation to motor activity (e.g., van Kan *et al.*, 1993). It would be of great interest to determine whether stimuli that challenge attention, motivation, learning, visual–spatial analysis, working memory, and other higher order functions also result in activation of the cerebrocerebellar circuitry and, if so, in what temporal sequence. Furthermore, measures of cognitive function in the nonhuman primate (Mountcastle *et al.*, 1975; Mishkin, 1982; Petrides 1987; Desimone and Ungerleider, 1989) could be employed in lesion studies of the cerebellum or its connections in order to perform lesion–behavior correlation analyses.

Studies of large groups of patients with discrete and well-characterized cerebellar pathology (such as stroke) will be essential in further exploring lesion-deficit correlations in humans. More searching tasks derived from experimental psychology will be valuable in the study of the mechanisms that subserve the cerebellar influence upon nonmotor function. This approach will be important in patients as well as in normal volunteers undergoing functional neuroimaging. What is the functional topographic map of the human cerebellum? Is there regional specialization in the cerebellum as predicted from studies to date (Schmahmann, 1996), and how focused are the various higher order functional properties within each folium or lobule? Do motor and sensory cerebellar areas (primary, secondary, and perhaps others) overlap or interdigitate with cognitive, affective, and autonomic areas or are they separate and distinct from each other? What are the mechanisms of cognitive recovery following acute cerebellar lesions? Are there compensatory cerebellar mechanisms, or is recovery dependent on cerebrocerebellar interactions as is the case in the sensory-motor realm (Growdon et al., 1967; Mackel, 1987)?

Acceptance of the notion that the cerebellum was important only in motor control precluded the consideration of its role in diseases with principally behavioral manifestations. This essentially excluded all of psychiatry. It is now apparent that the pathology of early infantile autism includes consistently abnormal morphological features in the cerebellum (Bauman and Kemper, 1985; Courchesne et al., 1988); schizophrenia has been associated with cerebellar vermis abnormalities (Heath et al., 1979; Weinberger et al., 1980); and mood and emotional dyscontrol are improved by cerebellar cortical stimulation (Cooper et al., 1978). In addition, sham rage and predatory attack can be induced by stimulation of the fastigial nucleus in cats (Zanchetti and Zoccolini, 1954; Reis et al., 1973), and aggression was ameliorated in monkeys by destructive lesions of the vermis and flocculonodular lobes (Peters and Monjan, 1971; Berman et al., 1978). The possibility of a cerebellar role in psychiatric disease having been more frankly stated, it may now be reasonable to use the available anatomic and functional neuroimaging techniques, as well as morphologic and immunohistochemical pathologic study, to challenge old assumptions and provide new insights into these psychiatric diseases. In this vein, it has been shown that methylphenidate, the drug of choice in the treatment of attention deficit hyperactivity disorder (ADHD), significantly increases brain metabolism, most consistently in the cerebellum, as well as in frontal and temporal lobes (Volkow et al., 1997). Children with ADHD demonstrate impairments of attention, as well as of memory and learning (Barkley et al., 1992), and the possibility of a cerebellar role in the manifestations or pathogenesis of this condition is intriguing.

IV. Conclusions

There are a multitude of avenues to investigate in the evolving understanding of the cerebellum and its role in the normally functioning nervous system. The rapidly increasing sophistication of contemporary investigative tools and concepts has facilitated the exploration of old ideas and novel discoveries concerning the cerebellum. The study of the relationship between the cerebellum and cognition opens a new and exciting chapter in contemporary cognitive neuroscience.

Acknowledgment

The author is grateful to Drs. Margaret L. Bauman, Aaron J. Berman, and Duane E. Haines for their helpful comments regarding this manuscript.

References

Allen, G. I., and Tsukuhara, N. (1974). Cerebrocerebellar communication systems. *Physiol. Rev.* **54,** 957–1008.
Appollonio, I. M., Grafman, J., Schwartz, V., Massaquoi, S., and Hallett, M. (1993). Memory in patients with cerebellar degeneration. *Neurology* **43,** 1536–1544.
Barkley, R. A., Grodzinsky, G., and Du Paul, G. J. (1992). Frontal lobe functions in attention deficit disorder with and without hyperactivity: A review and research report. *J. Abnorm. Child Psychol.* **20,** 163–188.
Baron, M. S., Vitek, J. L., Bakay, R. A., Green, J., Kaneoke, Y., Hashimoto, T., Turner, R. S., Woodard, J. L., Cole, S. A., McDonald, W. M., and DeLong, M. R. (1996). Treatment of advanced Parkinson's disease by posterior Gpi pallidotomy: 1-year results of a pilot study. *Ann. Neurol.* **40,** 355–366.
Bauman, M., and Kemper, T. L. (1985). Histoanatomic observations of the brain in early infantile autism. *Neurology* **35,** 866–874.
Berman, A. J., Berman, D., and Prescott, J. W. (1974). The effects of cerebellar lesions on emotional behavior in the rhesus monkey. In "The Cerebellum, Epilepsy and Behavior" (I. S. Cooper, M. Riklan, and R. S. Snider, eds.), pp. 277–284, Plenum Press, New York.
Botez-Marquard, T., and Botez, M. I. (1995). Reaction time and intelligence in patients with olivopontocerebellar atrophy. *Neuropsychiat. Neuropsychol. Behav. Neurol.* **8,** 168–175.
Botez-Marquard, T., Léveillé J., and Botez, M. I. (1994). Neuropsychological functioning in unilateral cerebellar damage. *Can. J. Neurol. Sci.* **21,** 353–357.
Bracke-Tolkmitt, R., Linden, A., Canavan, A. G. M., Rockstroh, B., Scholz, E., Wessel, K., and Diener, H.-C. (1989). The cerebellum contributes to mental skills. *Behav. Neurosci.* **103,** 442–446.
Breiter, H. C., Rauch, S. L., Kwong, K. K., Baker, J. R., Weiskoff, R. M., Kennedy, D. N., Kendrick, A. D., Davis, T. L., Jiang, A., Cohen, M. S., Stern, C. E., Belliveau, J. W., Baer, L., O'Sullivan, R. L., Savage, C. R., and Rosen, B. R. (1996). Functional magnetic resonance imaging of symptom provocation in obsessive-compulsive disorder. *Arch. Gen. Psychiat.* **53,** 595–606.
Brodal, P. (1978). The corticopontine projection in the rhesus monkey: Origin and principles of organization. *Brain* **101,** 251–283.
Brodal, P. (1979). The pontocerebellar projection in the rhesus monkey: An experimental study with retrograde axonal transport of horseradish peroxidase. *Neuroscience* **4,** 193–208.
Casey, D. A. (1994). Depression in the elderly. *South. Med. J.* **87,** 559–563.

Chida, K., Iadecola, C., Underwood, M. D., and Reis, D. J. (1986). A novel vasodepressor response elicited from the rat cerebellar fastigial nucleus: The fastigial depressor response. *Brain Res.* **370,** 378–382.

Cooper, I. S., Riklan, M., Amin, I., and Cullinan, T. (1978). A long term follow-up study of cerebellar stimulation for the control of epilepsy. *In* "Cerebellar Stimulation in Man" (I. S. Cooper, ed.), pp. 19–38, Raven Press, New York.

Courchesne, E., Yeung-Courchesne, R., Press, G. A., Hesselink, J. R., and Jernigan, T. L. (1988). Hypoplasia of cerebellar vermal lobules VI and VII in autism. *N. Engl. J. Med.* **318,** 1349–1354. Elsevier, New York.

Desimone, R., and Ungerleider, L. G. (1989). Neural mechanisms of visual processing in monkeys. *In* "Handbook of Neuropsychology" (F. Boller and J. Grafman eds.), Vol. **2,** pp. 267–299. Elsevier, New York.

Doba, N., and Reis, D. J. (1972). Changes in regional blood flow and cardiodynamics evoked by electrical stimulation of the fastigial nucleus in the cat and their similarity to orthostatic reflexes. *J. Physiol. (Lond.)* **227,** 729–747.

Dore, L., Jacobson, C. D., and Hawkes, R. (1990). Organization and postnatal development of Zebrin II antigenic compartmentation in the cerebellar vermis of the grey opossum, *Monodelphis domestica. J. Comp. Neurol.* **291,** 431–449.

Frankel, F. H. (1984). The use of electroconvulsive therapy in suicidal patients. *Am. J. Psychother.* **38,** 384–391.

Fried, I. (1993). Anatomic temporal lobe resections for temporal lobe epilepsy. *Neurosurg. Clin. N. Am.* **4,** 233–242.

Gao, J.-H., Parsons, L. M., Bower, J. M., Xiong, J., Li, J., Brannon, S., and Fox, P. T. (1996). Cerebellar dentate-nucleus activated by sensory and perceptual discrimination, imagined hand movement, and mental rotation of objects. *Science* **272,** 545–547.

Gardiner, T. W., and Kitai, S. T. (1992). Single-unit activity in the globus pallidus and neostriatum of the rat during performance of a trained head movement. *Exp. Brain Res.* **88,** 517–530.

Glickstein, M., May, J. G., and Mercier, B. E. (1985). Corticopontine projection in the macaque: The distribution of labelled cortical cells after large injections of horseradish peroxidase in the pontine nuclei. *J. Comp. Neurol.* **235,** 343–359.

Grafman, J., Litvan, I., Massaquoi, S., Stewart, M., Sirigu, A., and Hallett, M. (1992). Cognitive planning deficit in patients with cerebellar atrophy. *Neurology* **42,** 1493–1496.

Greenblatt, M. (1977). Efficacy of ECT in affective and schizophrenic illness. *Am. J. Psychiat.* **134,** 1001–1005.

Growdon, J. H., Chambers, W. W., and Liu, C. N. (1967). An experimental study of cerebellar dyskinesia in the rhesus monkey. *Brain* **90,** 603–632.

Hahania, E. G., Kavanagh, J., Hortobagyi, G., Giles, R. E., Champlin, R., and Deisseroth, A. B. (1995). Recent advances in the application of gene therapy to human disease. *Am. J. Med.* **99,** 537–552.

Haines, D. E. (1981). Zones in the cerebellar cortex: Their organization and potential relevance to cerebellar stimulation. *J. Neurosurg.* **55,** 254–264.

Haines, D. E., and Dietrichs, E. (1984). An HRP study of hypothalamo-cerebellar and cerebellohypothalamic connections in squirrel monkey (*Saimiri sciureus*). *J. Comp. Neurol.* **229,** 559–575.

Haines, D. E., Patrick, G. W., and Satrulee, P. (1982). Organization of cerebellar corticonuclear fiber systems. *In* "The Cerebellum: New Vistas" (S. L. Palay, and V. Chan-Palay eds.), pp. 320–371. Springer-Verlag, Berlin.

Hampson, J. L., Harrison, C. R. and Woolsey, C. N. (1952). Cerebro-cerebellar projections and somatotopic localization of motor function in the cerebellum. *Res. Publ. Assn. Nerv. Ment. Dis.* **30,** 299–316.

Heath, R. G., Franklin, D. E., and Shraberg, D. (1979). Gross pathology of the cerebellum in patients diagnosed and treated as functional psychiatric disorders. *J. Nerv. Ment. Dis.* **167,** 585–592.

Henneman, E., Cooke, P. M., and Snider, R. S. (1952). Cerebellar projections to the cerebral cortex. *Res. Publ. Assn. Nerv. Ment. Dis.* **30,** 17–333.

Jenike, M. A., Breiter, H. C., Baer, L., Kennedy, D. N., Savage, C. R., Olivares, M. J., O'Sullivan, R. L., Shera, D. M., Rauch, S. L., Keuthen, N., Rosen, B. R., Caviness, V. S., and Filipek, P. A. (1996). Cerebral structural abnormalities in obsessive-compulsive disorder: A quantitative morphometric magnetic resonance imaging study. *Arch. Gen. Psychiat.* **53,** 625–632.

Katz, J., France, C., and Melzack, R. (1989). An association between phantom limb sensations and stump skin conductance during transcutaneous electrical nerve stimulation (TENS) applied to the contralateral leg: A case study. *Pain* **36,** 367–377.

Kish, S. J., El-Awar, M., Stuss, D., Nobrega, J., Currier, R., Aita, J. F., Schut, L., Zoghbi, H. Y., and Freedman, M. (1994). Neuropsychological test performance in patients with dominantly inherited spinocerebellar ataxia: Relationship to ataxia severity. *Neurology* **44,** 1738–1746.

Laitinen, L. V. (1995). Pallidotomy for Parkinson's disease. *Neurosurg. Clin. N. Am.* **6,** 105–112.

Landis, C. (1949). Psychology. *In* "Selective Partial Ablation of the Frontal Cortex" (F. A. Mettler ed.), pp. 492–296. Paul B. Hoeber, New York.

Leiner, H. C., Leiner, A. L., and Dow, R. S. (1993). Cognitive and language functions of the human cerebellum. *Trends Neurosci.* **16,** 444–454.

Levin, H. S. (1992). Head injury and its rehabilitation. *Curr. Opin. Neurol. Neurosurg.* **5,** 673–676.

Mackel, R. (1987). The role of the monkey sensory cortex in the recovery from cerebellar injury. *Exp. Brain Res.* **66,** 638–652.

McKee, J. C., Denn, M. J., and Stone, H. L. (1976). Neurogenic cerebral vasodilatation from electrical stimulation of the cerebellum in the monkey. *Stroke* **7,** 179–186.

Melzack, R., and Wall, P. D. (1965). Pain mechanisms: A new theory. *Science* **150,** 971–979.

Martner, J. (1975). Cerebellar influences on autonomic mechanisms. *Acta. Physiol. Scand.* (Suppl.) **425,** 1–42.

Middleton, F. A., and Strick, P. L. (1994). Anatomical evidence for cerebellar and basal ganglia involvement in higher cognitive function. *Science* **266,** 458–451.

Milner, B., Corkin, S., and Teuber, H. L. (1968). Further analysis of the hippocampal amnesic syndrome: 14 years follow-up study of H.M. *Neuropsychologia* **6,** 215–234.

Mishkin, M. (1982). A memory system in the monkey. *Trans. R. Soc. Lond. B* **298,** 85–95.

Mountcastle, V. B., Lynch, J. C., and Georgopoulos, A. (1975). Posterior parietal association cortex of the monkey: Command functions for operations within extrapersonal space. *J. Neurophysiol.* **38,** 871–908.

Pandya, D. N., and Yeterian, E. H. (1985). Architecture and connections of cortical association areas. *In* "Cerebral Cortex" (A. Peters and E. G. Jones, eds.), Vol. **4,** pp. 3–61, Plenum Press, New York.

Peters, M., and Monjan, A. A. (1971). Behavior after cerebellar lesions in cats and monkeys. *Physiol. Behav.* **6,** 205–206.

Petrides, M. (1987). Conditional learning and the primate frontal cortex. *In* "The Frontal Lobes Revisited" (E. Perecman, ed.), pp. 91–108, IRBN Press, New York.

Pollack, I. F. (1995). Mutism and pseudobulbar symptoms after resection of posterior fossa tumors in children: Incidence and pathophysiology. *Neurosurgery* **37,** 885–893.

Pring, T. R. (1986). Evaluating the effects of speech therapy for aphasics: Developing the single case methodology. *Br. J. Disord. Commun.* **21,** 103–115.

Reis, D. J., Doba, N., and Nathan, M. A. (1973). Predatory attack, grooming and consummatory behaviors evoked by electrical stimulation of cat cerebellar nuclei. *Science* **182,** 845–847.

Sasaki, K., Oka, H., Matsuda, Y., Shimono, T., and Mizuno, N. (1975). Electrophysiological studies of the projections from the parietal association area to the cerebellar cortex. *Exp. Brain Res.* **23,** 91–102.

Schmahmann, J. D. (1991). An emerging concept: The cerebellar contribution to higher function. *Arch. Neurol.* **48,** 1178–1187.

Schmahmann, J. D. (1994). The cerebellum in autism: Clinical and anatomic perspectives. *In* "The Neurobiology of Autism" (M. L. Bauman, and T. L. Kemper, eds.), pp. 195–226, Johns Hopkins University Press, Baltimore, MD.

Schmahmann, J. D. (1996). From movement to thought: Anatomic substrates of the cerebellar contribution to cognitive processing. *Hum. Brain Mapp.* **4,** 174–198.

Schmahmann, J. D., Doyon, J., Holmes, C., Makris, N., Petrides, M., Kennedy, D. N., and Evans, A. C. (1996). An MRI atlas of the human cerebellum in Talairach space. *NeuroImage* **3,** S122.

Schultz, J. B., Matthews, R. T., Henshaw, D. R., and Beal, M. F. (1996). Neuroprotective strategies for treatment of lesions produced by mitochondrial toxins: Implications for neurodegenerative diseases. *Neuroscience* **71,** 1043–1048.

Scoville, W. B., and Milner, B. (1957). Loss of recent memory after bilateral hippocampal lesions. *J. Neurol. Neurosurg. Psychiat.* **20,** 11–21.

Shephard, N. T., Telian, S. A., Smith-Wheelock, M., and Raj, A. (1993). Vestibular and balance rehabilitation therapy. *Ann. Otol. Rhinol. Laryngol.* **102,** 198–205.

Snider, R. S., and Maiti, A. (1976). Cerebellar contributions to the *Papez* circuit. *J. Neurosci. Res.* **2,** 133–146.

Sotelo, C., and Alvarado-Mallart, R. M. (1987). Reconstruction of the defective cerebellar circuitry in adult Purkinje cell degeneration mutant mice by Purkinje cell replacement through transplantation of solid embryonic implants. *Neuroscience* **20,** 1–22.

Spangler, W. J., Cosgrove, G. R., Ballantine, H. T., Jr., Casem, E. H., Rauch, S. L., Nierenberg, A., and Price, B. H. (1996). Magnetic resonance image-guided stereotactic cingulotomy for intractable psychiatric disease. *Neurosurgery* **38,** 1071–1076.

Spencer, S. S. (1996). Long-term outcome after epilepsy surgery. *Epilepsia* **37,** 807–813.

Triarhou, L. C., Zhang, W., and Lee, W. H. (1996). Amelioration of the behavioral phenotype in genetically ataxic mice through bilateral intracerebellar grafting of fetal Purkinje cells. *Cell Transplant.* **5,** 269–277.

Van Kan, P. L., Houk, J. C., and Gibson, A. R. (1993). Output organization of intermediate cerebellum of the monkey. *J. Neurophysiol.* **69,** 57–73.

Volkow, N. D., Wang, G.-J., Fowler, J. S., Logan, J., Angrist, B., Hitzemann, R., Lieberman J., and Pappas, N. (1997). Effects of methylphenidate on regional brain glucose metabolism in humans: Relationship to Dopamine D_2 receptors. *Am. J. Psychiat.* **154,** 50–55.

Voogd, J., and Bigaré, F. (1980). Topographical distribution of olivary and corticonuclear fibers in the cerebellum: A review. *In* "The Inferior Olivary Nucleus" (E. Courville, C. de Montigny, and Y. Lamarre, eds.), pp. 207–234. Raven Press, New York.

Wallesch, C.-W., and Horn, A. (1990). Long-term effects of cerebellar pathology on cognitive functions. *Brain Cogn.* **14,** 19–25.

Weinberger, D. R., Kleinman, J. E., Luchins, D. J., Bigelow, L., and Wyatt, R. (1980). Cerebellar pathology in schizophrenia: A controlled postmortem study. *Am. J. Psychiat.* **137,** 359–361.

Welker, W. (1987). Spatial organization of somatosensory projections to granule cell cerebellar cortex: Functional and connectional implications of fractured somatotopy (summary of Wisconsin studies). *In* "New Concepts in Cerebellar Neurobiology" (J. S. King, ed.), pp. 239–280. Liss, New York.

Wertz, R. T., Collins, M. J., Weiss, D., Kurtzke, J. F., Friden, T., Brookshire, R. H., Pierce, J., Holzapple, P., Hubbard, D. J., Porch, B. E., West, J. A., Davis, L., Matovich, V., Morley, G. K., and Resurrection, E. (1981). Veterans Administration cooperative study on aphasia: A comparison of individual and group treatment. *J. Speech Hear. Res.* **24,** 580–594.

Zanchetti, A., and Zoccolini, A. (1954). Autonomic hypothalamic outbursts elicited by cerebellar stimulation. *J. Neurophysiol.* **17,** 475–483.

INDEX

A

ADHD, *see* Attention deficit hyperactivity disorder
Aging
  eyeblink classical conditioning effects, 352–353
  Purkinje cell alterations, 350–352
Agrammatism, cerebellum dysfunction as cause, 332–335, 403, 464–465
Alzheimer's disease, eyeblink classical conditioning in patients, 359–360
γ-Aminobutyric acid (GABA), hypothalamocerebellar neurotransmission, 91
Anarthria, *see* Mutism
Arterial pressure, fastigial nucleus regulation
  electrical stimulation studies, 122
  fastigial pressure response, 122–124, 140
  rostral ventrolateral quadrant control, 123–125, 140
Articulation, *see also* Dysarthia; Mutism; Verbal fluency
  activation studies, 235, 241–243, 246–247
  covert versus overt articulation, 239–240
  dysarthia, 235
  nonverb word tasks, 243–245
  verbal learning, 245–249
  verb generation tasks, 241–245
  working memory activation studies, 236–238
Aspartate, hypothalamocerebellar neurotransmission, 90
Ataxia
  cerebellar lesions as cause, 10–11, 96
  Friedreich's ataxia, *see* Friedreich's ataxia
Attention
  cerebellar atrophy patients, 313
  cerebellum role
    history of animal studies, 381, 576–577
  orienting attention
    autism deficits, 585–587, 590–591
    cerebellar lesion effects, 587, 589–590
    Posner paradigm, 584–585
    response accuracy studies, 590–591
    response speed studies, 586–587, 589–590
  shifting attention
    autism deficits, 578–580, 582
    behavioral findings, 582–583
    functional magnetic resonance imaging, 583–584
    P3b event-related potential, 579–583
  coordination of brain resources, 576–577
  eye movement dysfunction association with deficits, 591–592
Attention deficit hyperactivity disorder (ADHD), cerebellum in pathogenesis, 643
Autism
  cerebellar abnormalities, 360, 507, 578
  cerebellum pathways affected, 381–382
  eyeblink classical conditioning in patients, 360–361
  histoanatomic abnormalities, 371–373, 375, 378–382
  intelligence quotient testing, 370–371
  neuroimaging and cerebellum role, 16, 368–371, 382–383, 593
  orienting attention deficits, 585–587, 590–591
  shifting attention deficits, 578–580, 582
  symptoms, 367
Autonomic nervous system
  clinical evidence of cerebellar influence on visceral function
    case studies, 96–100
    magnetic resonance imaging of lesions, 96–98

connections with pons, 42, 44
fastigial nucleus regulation, *see* Fastigial nucleus
history of cerebellum interaction studies, 5, 7, 13

B

Basal ganglia
　cognition role, 14, 227
　stabilization role, 477
Bower hypothesis, cerebellum and cognition, 525–526

C

CCA, *see* Cerebellar cortical atrophy
Central nervous system
　basic duties, 476–477
　cerebellum support role, 502–503, 508
　computer analogy, *see* Computer
　hierarchical organization, 477
Centralis lateralis nucleus, cortical connections, 50–51
Cerebellar cognitive affective syndrome
　bedside mental state testing, 436–437
　electroencephalography, 435, 439–440
　elementary neurologic examination, 435–436
　features, 438
　neuroimaging, 434, 439
　neuropsychological testing, 437, 439
　posterior inferior cerebellar artery stroke and effects, 435–437
　study design, 434–435
　superior cerebellar artery stroke and effects, 435–437
Cerebellar cortical atrophy (CCA)
　clinical features, 443
　cognition deficits, 443–445, 449–450
　diagnosis, 443–445
　heredity, 443
Cerebellar microcomplex
　evolution, 476–477
　fiber inputs, 475, 483–484
　long-term depression and adaption, 484–485
　plasticity, 484
　reflex arc connection, 477–479
　structure, 483–484

thought roles, 481–483
voluntary movement roles, 480–481
Cerebellar mutant mice
　Morris water maze testing
　　anatomic correlates, 204–206
　　neuropathology, 198–201
　　performance, 201–204, 219, 442–443
　　radial maze performance, 206–207
　　spatial alternation test performance, 207–208
　types, 198–201, 442–443
Cerebellar psychosurgery, therapy in cerebellar disorders, 640–641
Cerebellar stimulation, therapy in cerebellar disorders, 639–640
Cerebellar transplantation, therapy in cerebellar disorders, 641
Cerebellin, hypothalamocerebellar neurotransmission, 91–92
Cerebral glucose metabolism (rCGU)
　areas activated in learning, 153
　fastigial nucleus regulation
　　chemical stimulation studies, 129, 142
　　electrical stimulation studies, 127–129
　　rostral ventrolateral reticular nucleus role, 130–133
Cerebrocerebellar system
　climbing and mossy fiber system interaction in learning, 52–53
　communication capabilities in cerebrocerebellar connections
　　fiber bundles, 543, 545–546
　　internal languages, 546–547
　　prefrontal cortex loops, 547–548
　computer analogy, 538
　evolution, 540–542
　feedback limb
　　circuit, 32, 49, 54
　　corticonuclear projection, 49
　　thalamic inputs, 50–52
　feedforward limb
　　circuit, 32, 54
　　corticopontine projections
　　　autonomic connections, 42, 44
　　　fiber pathways to pons, 45
　　　occipitotemporal projections, 40
　　　origin, 33
　　　parahippocampal projections, 40
　　　paralimbic connections, 42, 44
　　　parastriate projections, 40

parietopontine connections, 35, 37–38
prefrontopontine connections, 41–42
temporopontine connections, 38–39
pontocerebellar projections, 45–47
imaging of dentate activation, 76–77
neuron recording in awake trained primates, 71–76
output mapping with herpes simplex virus, 64–65, 67–71
Classical conditioning, *see also* Conditioned response pathway; Conditioned stimulus pathway; Eyeblink response; Unconditioned stimulus pathway
error-correcting algorithm, 173–174
eyeblink classical conditioning
aging effects, 350–353
Alzheimer's disease patients, 359–360
autism patients, 360–361
cerebellar lesion effects, 344–350, 616
human circuitry comparison to rabbit, 342–343, 346–350
Huntington's disease patients, 358–359
interference during dual-task conditions, 355–356
interpositus role, 353
medial temporal lobe lesion effects, 356–358
methodology, 460–461
motor-cognitive functioning, 618–621
multiple pathway model, 626–629
neuropsychological predictors, 354–355
positron emission tomography of cerebellum, 353–354, 361
Purkinje cell role, 352–353, 361
systemic functioning of cerebellum, 625–626
timing requirements, 563–565
history of studies, 343
long-term depression in memory storage, 176–178
state estimator hypothesis, 525–525
Climbing fiber
cerebellar microcomplex input, 475, 483–484
mossy fiber interaction in learning, 52–53

Cognition
basal ganglia role, 14
cerebellum, *see also* Cerebellar cognitive affective syndrome
anatomy, *see* Cerebrocerebellar system
history of studies, 257–258, 456–457, 613–617
imaging, *see* Functional magnetic resonance imaging; Positron emission tomography
role overview, 14–17, 54–55, 255–256, 266–267, 287–291, 318–319, 433–434, 485–486, 526–528, 608–609
climbing and mossy fiber system interaction in learning, 52–53
evolution of capabilities, 537–538
motor function similarity, 617–620, 629–630
neuroimaging, 16–17, 265, 568–570
rehabilitation therapy in cerebellar disorders, 639
Computed tomography (CT)
cerebellar cognitive affective syndrome, 434, 439
posterior fossa syndrome, 419–420
unilateral cerebellar damage patients, 398–399
Computer, cerebellar analogy
cerebrocerebellar system, 538
design principles, 536–537, 542
hardware
cerebral prefrontal cortex connections, 540–542
lateral cerebellum, 539–540
reconciliation of diverse theories, 549–550
software communication capabilities in cerebrocerebellar connections
fiber bundles, 543, 545–546
internal languages, 546–547
prefrontal cortex loops, 547–548
versatility, 548–549
Conditioned response pathway (CR)
decerebration effects, 158–159
lesions and abolition
cortical lesions, 157–158
interpositus, 153, 157–158, 171
performance hypothesis, 171–172
red nucleus, 154

reversal by receptor antagonists, 154
superior cerebellar peduncle, 154
reversible inactivation studies of memory trace, 160–161, 163, 169
Conditioned stimulus pathway (CS)
circuits, 154–156
pontine nuclei electrical stimulation studies, 155
Purkinje neuron response, 156–157
reversible inactivation studies, 163
Coordination, cerebellum role
loss in neurological disorders, 556, 558
mental movement, 606
modular view, 556–558
Corticopontine pathway
autonomic connections, 42, 44
fiber pathways to pons, 45
nonmotor function role, history of study, 8–9
occipitotemporal projections, 40
origin, 33
parahippocampal projections, 40
paralimbic connections, 42, 44
parastriate projections, 40
parietopontine connections, 35, 37–38
prefrontopontine connections, 41–42
temporopontine connections, 38–39
CR, *see* Conditioned response pathway
CS, *see* Conditioned stimulus pathway
CT, *see* Computed tomography
Culmen, emotional effects of lesions, 114, 116

D

DAO, *see* Dorsal accessory olive
Dementia
cerebellar dysfunction, 458–459
olivopontocerebellar atrophy, 445–447, 458
Dentate nucleus, *see also* Neodentate nucleus
anatomy, 9, 50
evolution, 539–540
functional magnetic resonance imaging activation, 76–77
cognition, 262–264
tactile response, 497–498
motor control, 605–606, 628

neuron recording in awake trained primates
cerebellar output channels, 73–76
instruction related neurons, 74–76
remembered sequence task, 71–72
tracking task, 73
Dorsal accessory olive (DAO)
classical conditioning role, 174
lesions and unconditioned stimulus pathway abolition, 154
Down's syndrome, vermal hypoplasia and speech, 326
Dysarthia, cerebellar damage as cause, 235, 249, 326
Dysgraphia, cerebellum dysfunction as cause, 335–336

E

Echolocation, cerebellum role, 519
Electrolocation, cerebellum role in fish, 518–519
Emotion, cerebellum role
electrode stimulation effects, 12–13, 112–113
history of studies, 11–13
infant isolation and aggression, 111, 113–114, 116–117
taming effect of lesions, 111–112, 114, 116–117, 641, 643
Epilepsy, bilateral cerebellar damage patients
intelligence quotient assessment, 390
movement time testing, 394, 396
phenytoin treatment in damage, 389–390
reaction time testing, 394, 396
ERP, *see* Event-related potential
Event-related potential (ERP), P3b response in attention shifting, 579–583
Evolution
cerebellar microcomplex, 476–477
cerebral prefrontal cortex connections, 540–542
cognition capabilities, 537–538
lateral cerebellum, 539–540
red nucleus connectivity, 607–608
volitional movement complexity and cerebellar structure, 518–519

Executive function
  bilateral cerebral damage patients and impairment, 405
  cerebellar atrophy patients
    attention testing, 313
    memory retrieval dysfunction, 311-313
    temporal order processing, 313-314
    time estimation
      bisection tasks, 315-317
      time production task, 314-315
    tower-type planning task performance, 309-311
  cerebellar microcomplex role, 481-483
  functional neuroimaging, 317
Eyeblink response, *see also* Classical conditioning
  aging effects, 350-353
  Alzheimer's disease patients, 359-360
  autism patients, 360-361
  cerebellar involvement, 156, 179, 181, 361
  cerebellar lesion effects, 344-350, 565, 616
  human circuitry comparison to rabbit, 342-343, 346-350
  Huntington's disease patients, 358-359
  interference during dual-task conditions, 355-356
  interpositus role, 353
  medial temporal lobe lesion effects, 356-358
  methodology, 460-461
  motor-cognitive functioning, 618-621
  multiple pathway model, 626-629
  neuropsychological predictors, 354-355
  positron emission tomography of cerebellum, 353-354, 361
  Purkinje cell role, 352-353, 361
  systemic functioning of cerebellum, 625-626
  timing requirements, 563-565

F

FA, *see* Friedreich's ataxia
Fastigial nucleus (FN)
  arterial pressure regulation
    electrical stimulation studies, 122
    fastigial pressure response, 122-124, 140
    rostral ventrolateral quadrant control, 123-125, 140
  cerebral glucose metabolism regulation
    chemical stimulation studies, 129, 142
    electrical stimulation studies, 127-129
    rostral ventrolateral reticular nucleus role, 130-133
  electrode stimulation effects, 13
  heart rate effects, 122
  lesion effects
    emotions, 113
    spatial learning, 205
  neuroprotection in focal cerebral ischemia
    biological significance, 142
    electrical stimulation and infarction reduction, 134, 142
    mechanisms
      cerebral blood flow independence, 135
      inflammation reactivity impairment, 138-139
      neuronal excitability reduction, 135, 137-138
  regional cerebral blood flow regulation
    chemical stimulation studies, 129
    electrical stimulation studies, 125-129
    measurement techniques, 125
    rostral ventrolateral reticular nucleus role, 130-133
  stimulation therapy in cerebellar disorders, 640
Floculondular lobe, emotional effects of lesions, 114, 116
FN, *see* Fastigial nucleus
Focal cerebral ischemia
  animal models, 133-134
  fastigial nucleus neuroprotection
    biological significance, 142
    electrical stimulation and infarction reduction, 134, 142
    mechanisms
      cerebral blood flow independence, 135
      inflammation reactivity impairment, 138-139
      neuronal excitability reduction, 135, 137-138

Friedreich's ataxia (FA)
　bilateral cerebellar damage patients
　　damage correlation with cognition
　　　deficits, 391–393
　　movement time testing, 396–398
　　neuropsychological testing, 391
　　reaction time testing, 396–398
　clinical features, 447–448
　cognition deficits, 448–450
　pathology, 447–449
　single photon emission computed
　　tomography, 401
Frontal lobe
　cognitive functions, 463, 466
　dysfunction in olivopontocerebellar
　　atrophy, 446
Frontal operculum
　lesions and anarthria, 238–239
　phonological storage, 239
　working memory activation studies,
　　236–238
Functional magnetic resonance imaging,
　cerebellar imaging
　attention shifting studies, 583–584
　dentate nucleus, 76–77, 262–264
　dissociating somatomotor functions from
　　cognition, 259–260
　double dissociation of cerebellar
　　functions and motor processing,
　　262–265
　field of view and cerebellum imaging,
　　233–234, 256
　history of study, 257–258
　skill learning, 285–286, 305
　tactile response, 497–498
　working memory activation studies,
　　236–238

G

GABA, *see* γ-Aminobutyric acid
GFAP, *see* Glial fibrillary acidic protein
Glial fibrillary acidic protein (GFAP),
　learning role, 178
Granule cell, modulation of Purkinje cell
　response, 492–494

H

Hearing, cerebellum role, 10
Heart rate, fastigial nucleus effects, 122

Herpes simplex virus type 1 (HSV1),
　transneuronal transport and linked
　neuron labeling
　cerebellar output mapping, 64–65,
　　67–71
　principle, 63
　strains, 63–65
Hippocampus, Morris water maze testing of
　spatial learning with lesions
　afferents, 194–195, 197
　efferents, 195–198
　overview, 180, 191, 193–194
Histamine, hypothalamocerebellar
　neurotransmission, 90–91
Hole board test, *see* Spatial learning
HSV1, *see* Herpes simplex virus type 1
Huntington's disease
　eyeblink classical conditioning in
　　patients, 358–359
　neuropathology, 358
Hypothalamocerebellar system
　cerebellar projections to hypothalamus,
　　92–94
　collateral branches to other targets, 94
　cortical projections from hypothalamus
　　labeling, 85, 87–89
　　responses to hypothalamic
　　　stimulation, 89
　　topography, 88
　feedback circuits, 101–102
　feedforward circuits, 101–102
　indirect connection mediation
　　basilar pontine nuclei, 94–95
　　lateral reticular nucleus, 95
　labeling of neurons, 84–85, 93–94
　neurotransmitters, 90–92
　nuclear projections from hypothalamus,
　　89–90
　species similarity, 84–85

I

IGF-I, *see* Insulin-like growth factor I
Inferior olive cell
　histology in autism, 372, 375,
　　378–370
　lesion effects on spatial learning,
　　198
　mouse mutants, 199

Inferior parietal lobule (IPL)
  functions, 35
  projections, 37–38
Insulin-like growth factor I (IGF-I), learning role, 178–179
Intelligence quotient (IQ)
  autism, 370–371
  cerebellar dysfunction effects, 462–463
  epilepsy, 390
  matching in control subjects, 465–466
Interpositus nucleus
  classical conditioning role, 174, 353
  lesions and conditioned response pathway abolition, 153, 157–158, 171
  memory storage role, 173
  reversible inactivation studies of memory trace, 161, 163, 166, 169, 173
IPL, see Inferior parietal lobule
IQ, see Intelligence quotient

K

Kalman filter, see State estimator hypothesis

L

Language, see Articulation
Lateral hypothalamic area (LHAr), projections to cerebellum, 85, 89–90, 93–94
Learning, see Motor learning; Skill learning; Spatial learning; Supervised learning
LHAr, see Lateral hypothalamic area
Lidocaine, neuronal inactivation studies, 161, 163, 166, 169
Lobusus simplex, emotional effects of lesions, 114, 116
Long-term depression (LTD)
  cerebellar microcomplex adaption, 484–485
  knockout mice studies
    glial fibrillary acidic protein, 178
    protein kinase Cρ, 177–178
  memory storage role, 177–178
  plasticity mechanism, 176–177, 484
  temporal properties, 177
LTD, see Long-term depression

M

M1, see Primary motor cortex
Magnetic resonance imaging (MRI), see also Functional magnetic resonance imaging
  autism cerebellum, 368–371
  cerebellar cognitive affective syndrome, 434, 439
  cerebellar lesions, 96–98
  posterior fossa syndrome, 419–420
Maze test, see Spatial learning
Medial dorsal thalamic nucleus, cerebellar input, 51
Medial temporal lobe, lesion effects on eyeblink classical conditioning, 356–358
Memory trace
  localization in brain by reversible inactivation, 159–161, 163, 166–169, 171
  validity of studies, 504
Mental dysfunction, history of cerebellum studies, 5–7, 15–17
Mental retardation, cerebellar malformations, 457–458
Morris water maze (MWM), see Spatial learning
Mossy fiber
  cerebellar microcomplex input, 475, 483–484
  climbing fiber interaction in learning, 52–53
Motor learning
  adaptation learning
    comparison to skill learning, 299
    definition, 298, 459
    elbow movement task, 301
    prism glasses in assessment, 300, 460
    vestibulo-ocular reflex gain, 299, 459–460, 505
  components, 298, 308
  eye blink conditioning, 300
  habituation, 460
  models, 459, 600–602
  serial reaction time test and sequence learning, 305–308, 464
  skill learning, see Skill learning
Movement, see Coordination; Movement time; Volitional movement

Movement time (MT)
  epileptics, 394, 396
  Friedreich's ataxia patients, 396–398
  olivopontocerebellar atrophy patients, 396–398
  rationale and methodology of testing, 394
  unilateral cerebellar damage patient testing, 398, 400, 403
MRI, *see* Magnetic resonance imaging
MT, *see* Movement time
Muscinol, neuronal inactivation studies, 160–161, 166–168
Mutism
  frontal operculum lesions as cause, 238–239
  posterior fossa syndrome, *see* Posterior fossa syndrome
MWM, *see* Morris water maze

N

Navigation
  cerebellum role, 204–206, 226–227, 466
  maze testing, *see* Spatial learning
Neodentate nucleus, discovery, 9, 50
Nitric oxide synthase (NOS), fastigial nucleus effects on induction, 138–139
Nonverbal communication, cerebellum role, 54
NOS, *see* Nitric oxide synthase

O

Oculomotor vermis
  neuron firing in tracking, 520–521
  structure, 520
Olivopontocerebellar atrophy (OPCA)
  bilateral cerebellar damage patients
    damage correlation with cognition deficits, 391–393
    movement time testing, 396–398
    neuropsychological testing, 391, 406
    reaction time testing, 396–398
  classification, 390–391
  clinical features, 445
  cognition deficits, 445–447, 450, 465
  dementia association, 445–447, 458
  diagnosis, 443–445
  frontal lobe dysfunction, 446

heredity, 445
single photon emission computed tomography, 401
OPCA, *see* Olivopontocerebellar atrophy

P

Parallel fiber, Purkinje cell interactions in learning, 600–604
Paramedian lobule, emotional effects of lesions, 114, 116
Parietal cortex, attention role, 576
Peripheral sensory afferents, discovery in cerebellum, 9–10
PET, *see* Positron emission tomography
PICA, *see* Posterior inferior cerebellar artery
PMv, *see* Ventral premotor area
Positron emission tomography (PET), cerebellar activation studies
  brain regions in time perception, 562–563
  complementation with lesion data, 467
  dissociating somatomotor functions from cognition, 258–262
  double dissociation of cerebellar functions and motor processing, 260–265
  eyeblink conditioning, 353–354
  field of view and cerebellum imaging, 233–234, 256
  grasping tasks, 264–265
  history of study, 258
  mental rotation of abstract objects, 260–261
  sensitivity, 290–291
  skill learning
    motor maze test, 278–279
    repeated sequence tests, 279–285, 304
    rotor pursuit test, 277–278
    serial reaction time test, 307–308
    trajectorial movement test, 279
  verb generation tasks, 241
Posterior fossa syndrome
  delay of onset, 424
  imaging analysis, 419–421, 425–426
  mutism
    anatomic basis, 423–426
    incidence, 412–413
    neuropsychiatric dysfunction association, 417–418, 421–423, 427

INDEX 657

resolution, 412, 419, 421, 426, 429
tumor resection, approaches and
  mutism, 413, 416, 422-423,
  428-429
neurobehavioral deficits
  anatomic basis, 427-428
  resolution, 412, 419, 421, 426, 429
  types, 413-415
neuropsychological testing, 420-421
Posterior inferior cerebellar artery (PICA),
  see Cerebellar cognitive affective
  syndrome
Posture, cerebellum role, 614-615
Prefrontal cortex, cerebellar outputs, 70-71
Primary motor cortex (M1)
  cerebellar outputs, 69-70, 78
  functional imaging in skill learning,
    277-278
  ventrolateral thalamus projection, 62
Prism testing
  adaptation learning assessment, 300, 460
  motor-cognitive functioning, 620
Procedural memory, see Skill learning;
  Spatial learning
Protein kinase Cγ, learning role, 177-178
Purkinje cell
  alterations in aging, 350-352
  autism histology, 372-373, 379-380
  conditioned stimulus pathway, 156-157
  eyeblink classical conditioning role,
    352-353, 361
  granule cell modulation, 492-494
  parallel fiber interactions in learning,
    600-603
  vertical organization of tactile response,
    492-493
Pyramis, emotional effects of lesions, 114,
  116

R

rCBF, see Regional cerebral blood flow
rCGU, see Cerebral glucose metabolism
Reaction time (RT), see also Serial reaction
  time test
  epileptics, 394, 396
  Friedreich's ataxia patients, 396-398
  olivopontocerebellar atrophy patients,
    396-398, 406

rationale and methodology of testing,
  394
unilateral cerebellar damage patient
  testing, 398, 400, 402-403
Red nucleus
  evolution of connectivity, 607-608
  lesions and conditioned response
    pathway abolition, 154
  motor control, 605-606
  reversible inactivation studies of memory
    trace, 161, 172
Regional cerebral blood flow (rCBF), see
  also Positron emission tomography
  cerebral glucose metabolism regulation
    chemical stimulation studies, 129, 142
    electrical stimulation studies,
      127-129
    rostral ventrolateral reticular nucleus
      role, 130-133
  fastigial nucleus regulation
    chemical stimulation studies, 129
    electrical stimulation studies, 125-129
    measurement techniques, 125
    rostral ventrolateral reticular nucleus
      role, 130-133
Reticular formation, history of cerebellum
  modulation studies, 8
Rostral ventrolateral reticular nucleus, see
  Fastigial nucleus
RT, see Reaction time

S

SCA, see Superior cerebellar artery
Schizophrenia, cerebellum role, 15
Serial reaction time test (SRTT)
  methodology, 305-306
  motor learning assessment, 306-308, 464
  positron emission tomography, 307-308
Sham rage, history of cerebellum
  modulation studies, 7-8
Single photon emission computed
  tomography (SPECT)
  cerebellum imaging in skill learning,
    276-277
  findings in unilateral cerebellar damage
    patients, 398-402
  Friedreich's ataxia patients, 401
  olivopontocerebellar atrophy patients,
    401

Skill learning
  cerebellum imaging
    functional magnetic resonance imaging, 285–286, 290, 304–305
    positron emission tomography, 277–285, 290
    single photon emission computed tomography, 276–277, 304
    tasks, 275, 503
  cerebellum role, 286–287, 302–303, 460, 622–624
  cortical takeover, 291
  definition, 274, 299, 302
  mirror vision tasks, 302
  steps in process, 275–276, 288
  trajectory generation tasks, 302–303, 526–527
SMA, see Supplementary motor area
Spatial learning
  declarative memory, 226
  hole board test of cerebellectomized rats, 208
  Morris water maze test
    cerebellar mutant mice
      anatomic correlates, 204–206
      neuropathology, 198–201
      performance, 201–204, 219, 442–443
    cortical ablation studies, 218–219
    hemicerebellectomized rats
      cues, 221, 225
      searching behavior, 220–221, 223–224
      spatial map building, 224–227
    hippocampal lesion studies
      afferents, 194–195, 197
      efferents, 195–198
      overview, 180, 191, 193–194
    methodology, 192–193, 217–218
  procedural memory, 226
  radial maze performance of cerebellar mutant mice, 206–207
  spatial alternation test performance of cerebellar mutant mice, 207–208
SPECT, see Single photon emission computed tomography
SPL, see Superior parietal lobule
SRTT, see Serial reaction time test
State estimator hypothesis
  classical conditioning, 524–525
  comparison with Bower hypothesis, 525–526
  oculomotor vermis, 520–521
  state estimation
    cerebellar functions, 517–518, 531–532
    definition, 516–517
    Kalman filter, 517, 521, 528–530
    necessity, 531–532
  target tracking, mechanical model, 528–530
  trajectory perception, 526–527
  vestibulo-ocular reflex, 521–524
Stroke, see Cerebellar cognitive affective syndrome; Focal cerebral ischemia
STS, see Superior temporal sulcus
Superior cerebellar artery (SCA), see Cerebellar cognitive affective syndrome
Superior cerebellar peduncle, reversible inactivation studies of memory trace, 169, 171
Superior parietal lobule (SPL)
  functions, 35
  projections, 37–38
Superior temporal sulcus (STS)
  functions, 38–40
  projections, 39
Supervised learning
  cerebellum role, 175–176
  plasticity effects, 175–176
Supplementary motor area (SMA)
  covert versus overt articulation role, 239–240
  functional imaging in skill learning, 277–279, 281–282
  working memory activation studies, 236–238

T

Tactile response
  behavioral effects of cerebellar involvement, 500–501
  cat, 496
  cerebellar outputs, 499–500
  cerebral cortex control of cerebellar activity, 503–504
  data acquisition control by cerebellum
    algorithmic necessities, 499
    comparison to classic motor control, 501

functional magnetic resonance
 imaging, 497
humans, 496–498
rats, 494–496, 498
primate, 496
rat, 490–496
regions of cerebellum, 490–494
sensitivity, 498–499
TENS, *see* Transcutaneous electrical nerve stimulator
Therapy, cerebellar disorders
cerebellar lesion studies, 642–643
cerebellar psychosurgery, 640–641
cerebellar stimulation, 639–640
cerebellar transplantation, 641
cognitive rehabilitation, 639
prognosis, 638
Thought
cerebellar microcomplex role, 481–483
modeling, 483
Time, *see also* Reaction time
cerebellar lesions and perceptual deficits, 561–563
estimation in cerebellar atrophy patients
bisection tasks, 315–317
time production task, 314–315
neuroimaging of activated brain regions in perception, 562–563
temporal order processing in cerebellum, 313–314, 319, 461–462, 570–571, 592–593
Timing, *see also* Serial reaction time test
electromyography studies, 558–559
modeling of cerebellar system, 565–568
requirements in sensorimotor learning, 563–565
timed interval tapping task, 350, 559–561, 566–568
volitional movement, timing role of cerebellum, 180–181, 297, 334, 381, 558–561
Tower testing
cerebellar atrophy patient performance, 309–311
motor-cognitive functioning, 618–619
Tracking
deficits in cerebellar lesion patients, 563
mechanical model, 528–530
neuron recording in awake trained primates, 73

oculomotor vermis neuron firing, 520–521
trajectorial movement tasks, 279, 302–303, 526–527
Transcutaneous electrical nerve stimulator (TENS), therapy in cerebellar disorders, 640

U

Unconditioned stimulus pathway (US)
conditioned response pathway relationship, 171–172
dorsal accessory olive lesions and abolition, 154
reversible inactivation studies, 160–161
US, *see* Unconditioned stimulus pathway
Uvula, emotional effects of lesions, 114, 116

V

Ventral premotor area (PMv), cerebellar outputs, 69–70, 78
Ventromedial nuclei (VMNu), projections to cerebellum, 85, 89–90
Verbal fluency
deficits in cerebellar patients, 327–330, 463
definition, 327
letter fluency, 330–331
phonemic clustering, 330–332
semantic fluency, 330–331
Vermal hypoplasia
autism, 368–370, 382–383, 578
Down's syndrome, 326
mental retardation association, 457–458
Vestibulo-ocular reflex (VOR)
gain in adaptation learning, 299, 459–460, 505, 615, 621
prostrotary nystagmus, 523–524
rabbit flexibility, 523
state estimation, 521–524
Visceral function, *see* Autonomic nervous system
Vision, *see* Spatial learning
VMNu, *see* Ventromedial nuclei

Volitional movement, *see also* Coordination; Timing
  cerebellar microcomplex role, 480–481
  cerebral cortex motor association areas, 604–605
  cognition function similarity in cerebellum, 617–620, 629–630
  compensating for absence of cerebellum, 505–507
  complexity and cerebellar structure, 518–519
  context triggering, 606–608, 621
  history of cerebellum studies, 4, 17
  mental rehearsal and performance, 607–608
  multijointed movement and cerebellum role in memory, 179
VOR, *see* Vestibulo-ocular reflex

W

Whisker, rat
  behavioral effects of cerebellar involvement, 500–501
  sensitivity, 498–499
  tactile responsive regions of cerebellum, 490–495

CONTENTS OF RECENT VOLUMES

Volume 31

Animal Models of Parkinsonism Using Selective Neurotoxins: Clinical and Basic Implications
  Michael J. Zigmond and Edward M. Stricker

Regulation of Choline Acetyltransferase
  Paul M. Salvaterra and James E. Vaughn

Neurobiology of Zinc and Zinc-Containing Neurons
  Christopher J. Frederickson

Dopamine Receptor Subtypes and Arousal
  Ennio Ongini and Vincenzo G. Longo

Regulation of Brain Atrial Natriuretic Peptide and Angiotensin Receptors: Quantitative Autoradiographic Studies
  Juan M. Saavedra, Eero Castrén, Jorge S. Gutkind, and Adil J. Nazarali

Schizophrenia, Affective Psychoses, and Other Disorders Treated with Neuroleptic Drugs: The Enigma of Tardive Dyskinesia, Its Neurobiological Determinants, and the Conflict of Paradigms
  John L. Waddington

Nerve Blood Flow and Oxygen Delivery in Normal, Diabetic, and Ischemic Neuropathy
  Phillip A. Low, Terrence D. Lagerlund, and Philip G. McManis

INDEX

Volume 32

On the Contribution of Mathematical Models to the Understanding of Neurotransmitter Release
  H. Parnas, I. Parnas, and L. A. Segel

Single-Channel Studies of Glutamate Receptors
  M. S. P. Sansom and P. N. R. Usherwood

Coinjection of *Xenopus* Oocytes with cDNA Produced and Native mRNAs: A Molecular Biological Approach to the Tissue-Specific Processing of Human Cholinesterases
  Shlomo Seidman and Hermona Soreq

Potential Neurotrophic Factors in the Mammalian Central Nervous System: Functional Significance in the Developing and Aging Brain
  Dalia M. Araujo, Jean-Guy Chabot, and Rémi Quirion

Myasthenia Gravis: Prototype of the Antireceptor Autoimmune Diseases
  Simone Schönbeck, Susanne Chrestel, and Reinhard Hohlfeld

Presynaptic Effects of Toxins
  Alan L. Harvey

Mechanisms of Chemosensory Transduction in Taste Cells
  Myles H. Akabas

Quinoxalinediones as Excitatory Amino Acid Antagonists in the Vertebrate Central Nervous System
  Stephen N. Davies and Graham L. Collingridge

Acquired Immune Deficiency Syndrome and the Developing Nervous System
  Douglas E. Brenneman, Susan K. McCune, and Illana Gozes

INDEX

Volume 33

Olfaction
S. G. Shirley

Neuropharmacologic and Behavioral Actions of Clonidine: Interactions with Central Neurotransmitters
Jerry J. Buccafusco

Development of the Leech Nervous System
Gunther S. Stent, William B. Kristan, Jr., Steven A. Torrence, Kathleen A. French, and David A. Weisblat

$GABA_A$ Receptors Control the Excitability of Neuronal Populations
Armin Stelzer

Cellular and Molecular Physiology of Alcohol Actions in the Nervous System
Forrest F. Weight

INDEX

Volume 34

Neurotransmitters as Neurotrophic Factors: A New Set of Functions
Joan P. Schwartz

Heterogeneity and Regulation of Nicotinic Acetylcholine Receptors
Ronald J. Lukas and Merouane Bencherif

Activity-Dependent Development of the Vertebrate Nervous System
R. Douglas Fields and Phillip G. Nelson

A Role for Glial Cells in Activity-Dependent Central Nervous Plasticity? Review and Hypothesis
Christian M. Müller

Acetylcholine at Motor Nerves: Storage, Release, and Presynaptic Modulation by Autoreceptors and Adrenoceptors
Ignaz Wessler

INDEX

Volume 35

Biochemical Correlates of Long-Term Potentiation in Hippocampal Synapses
Satoru Otani and Yehezkel Ben-Ari

Molecular Aspects of Photoreceptor Adaptation in Vertebrate Retina
Satoru Kawamura

The Neurobiology and Genetics of Infantile Autism
Linda J. Lotspeich and Roland D. Ciaranello

Humoral Regulation of Sleep
Levente Kapás, Ferenc Obál, Jr., and James M. Krueger

Striatal Dopamine in Reward and Attention: A System for Understanding the Symptomatology of Acute Schizophrenia and Mania
Robert Miller

Acetylcholine Transport, Storage, and Release
Stanley M. Parsons, Chris Prior, and Ian G. Marshall

Molecular Neurobiology of Dopaminergic Receptors
David R. Sibley, Frederick J. Monsma, Jr., and Yong Shen

INDEX

Volume 36

Ca^{2+}, N-Methyl-D-aspartate Receptors, and AIDS-Related Neuronal Injury
Stuart A. Lipton

Processing of Alzheimer Aβ-Amyloid Precursor Protein: Cell Biology, Regulation, and Role in Alzheimer Disease
Sam Gandy and Paul Greengard

Molecular Neurobiology of the $GABA_A$ Receptor
Susan M. J. Dunn, Alan N. Bateson, and Ian L. Martin

The Pharmacology and Function of Central GABA$_B$ Receptors
David D. Mott and Darrell V. Lewis

The Role of the Amygdala in Emotional Learning
Michael Davis

Excitotoxicity and Neurological Disorders: Involvement of Membrane Phospholipids
Akhlaq A. Farooqui and Lloyd A. Horrocks

Injury-Related Behavior and Neuronal Plasticity: An Evolutionary Perspective on Sensitization, Hyperalgesia, and Analgesia
Edgar T. Walters

INDEX

Volume 37

Section I: Selectionist Ideas and Neurobiology

Selectionist and Instructionist Ideas in Neuroscience
Olaf Sporns

Population Thinking and Neuronal Selection: Metaphors or Concepts?
Ernst Mayr

Selection and the Origin of Information
Manfred Eigen

Section II: Development and Neuronal Populations

Morphoregulatory Molecules and Selectional Dynamics during Development
Kathryn L. Crossin

Exploration and Selection in the Early Acquisition of Skill
Esther Thelen and Daniela Corbetta

Population Activity in the Control of Movement
Apostolos P. Georgopoulos

Section III: Functional Segregation and Integration in the Brain

Reentry and the Problem of Cortical Integration
Giulio Tononi

Coherence as an Organizing Principle of Cortical Functions
Wolf Singer

Temporal Mechanisms in Perception
Ernst Pöppel

Section IV: Memory and Models

Selection versus Instruction: Use of Computer Models to Compare Brain Theories
George N. Reeke, Jr.

Memory and Forgetting: Long-Term and Gradual Changes in Memory Storage
Larry R. Squire

Implicit Knowledge: New Perspectives on Unconscious Processes
Daniel L. Schacter

Section V: Psychophysics, Psychoanalysis, and Neuropsychology

Phantom Limbs, Neglect Syndromes, Repressed Memories, and Freudian Psychology
V. S. Ramachandran

Neural Darwinism and a Conceptual Crisis in Psychoanalysis
Arnold H. Modell

A New Vision of the Mind
Oliver Sacks

INDEX

Volume 38

Regulation of GABA$_A$ Receptor Function and Gene Expression in the Central Nervous System
A. Leslie Morrow

Genetics and the Organization of the Basal Ganglia
Robert Hitzemann, Yeang Olan, Stephen Kanes, Katherine Dains, and Barbara Hitzemann

Structure and Pharmacology of Vertebrate GABA$_A$ Receptor Subtypes
Paul J. Whiting, Ruth M. McKernan, and Keith A. Wafford

Neurotransmitter Transporters: Molecular Biology, Function, and Regulation
Beth Borowsky and Beth J. Hoffman

Presynaptic Excitability
Meyer B. Jackson

Monoamine Neurotransmitters in Invertebrates and Vertebrates: An Examination of the Diverse Enzymatic Pathways Utilized to Synthesize and Inactivate Biogenic Amines
B. D. Sloley and A. V. Juorio

Neurotransmitter Systems in Schizophrenia
Gavin P. Reynolds

Physiology of Bergmann Glial Cells
Thomas Müller and Helmut Kettenmann

INDEX

Volume 39

Modulation of Amino Acid-Gated Ion Channels by Protein Phosphorylation
Stephen J. Moss and Trevor G. Smart

Use-Dependent Regulation of $GABA_A$ Receptors
Eugene M. Barnes, Jr.

Synaptic Transmission and Modulation in the Neostriatum
David M. Lovinger and Elizabeth Tyler

The Cytoskeleton and Neurotransmitter Receptors
Valerie J. Whatley and R. Adron Harris

Endogenous Opioid Regulation of Hippocampal Function
Michele L. Simmons and Charles Chavkin

Molecular Neurobiology of the Cannabinoid Receptor
Mary E. Abood and Billy R. Martin

Genetic Models in the Study of Anesthetic Drug Action
Victoria J. Simpson and Thomas E. Johnson

Neurochemical Bases of Locomotion and Ethanol Stimulant Effects
Tamara J. Phillips and Elaine H. Shen

Effects of Ethanol on Ion Channels
Fulton T. Crews, A. Leslie Morrow, Hugh Criswell, and George Breese

INDEX

Volume 40

Mechanisms of Nerve Cell Death: Apoptosis or Necrosis after Cerebral Ischemia
R. M. E. Chalmers-Redman, A. D. Fraser, W. Y. H. Ju, J. Wadia, N. A. Tatton, and W. G. Tatton

Changes in Ionic Fluxes during Cerebral Ischemia
Tibor Kristian and Bo K. Siesjo

Techniques for Examining Neuroprotective Drugs *in Vivo*
A. Richard Green and Alan J. Cross

Techniques for Examining Neuroprotective Drugs *in Vitro*
Mark P. Goldberg, Uta Strasser, and Laura L. Dugan

Calcium Antagonists: Their Role in Neuroprotection
A. Jacqueline Hunter

Sodium and Potassium Channel Modulators: Their Role in Neuroprotection
Tihomir P. Obrenovich

NMDA Antagonists: Their Role in Neuroprotection
Danial L. Small

Development of the NMDA Ion-Channel Blocker, Aptiganel Hydrochloride, as a Neuroprotective Agent for Acute CNS Injury
Robert N. McBurney

The Pharmacology of AMPA Antagonists and Their Role in Neuroprotection
Rammy Gill and David Lodge

GABA and Neuroprotection
Patrick D. Lyden

Adenosine and Neuroprotection
Bertil B. Fredholm

Interleukins and Cerebral Ischemia
Nancy J. Rothwell, Sarah A. Loddick, and Paul Stroemer

Nitrone-Based Free Radical Traps as Neuroprotective Agents in Cerebral Ischemia and Other Pathologies
Kenneth Hensley, John M. Carney, Charles A. Stewart, Tahera Tabatabaie, Quentin Pye, and Robert A. Floyd

Neurotoxic and Neuroprotective Roles of Nitric Oxide in Cerebral Ischemia
Turgay Dalkara and Michael A. Moskowitz

A Review of Earlier Clinical Studies on Neuroprotective Agents and Current Approaches
Nils-Gunnar Wahlgren

Lightning Source UK Ltd.
Milton Keynes UK
UKOW040717160911

178749UK00005B/97/A